# Storm and Cloud Dynamics

## 2nd Edition

This is Volume 99 in the
INTERNATIONAL GEOPHYSICS SERIES
A series of monographs and textbooks
Edited by RENATA DMOWSKA, DENNIS HARTMANN and H. THOMAS ROSSBY
A complete list of books in this series appears at the end of this volume.

# Storm and Cloud Dynamics

## The Dynamics of Clouds and Precipitating Mesoscale Systems

**2nd Edition**

William R. Cotton
Department of Atmospheric Science
Colorado State University
Fort Collins, Colorado 80523

George H. Bryan
National Center for Atmospheric Research
P.O. Box 3000
Boulder, Colorado 80307

Susan C. van den Heever
Department of Atmospheric Science
Colorado State University
Fort Collins, Colorado 80523

ELSEVIER

AMSTERDAM • BOSTON • HEIDELBERG • LONDON
NEW YORK • OXFORD • PARIS • SAN DIEGO
SAN FRANCISCO • SINGAPORE • SYDNEY • TOKYO

Academic Press is an Imprint of Elsevier

Academic Press is an imprint of Elsevier
30 Corporate Drive, Suite 400, Burlington, MA 01803, USA
525 B Street, Suite 1900, San Diego, CA 92101-4495, USA
84 Theobald's Road, London WC1X 8RR, UK

Second edition 2011



**Library of Congress Cataloging-in-Publication Data**
A catalog record for this book is available from the Library of Congress

**British Library Cataloguing in Publication Data**
A catalogue record for this book is available from the British Library

ISBN: 978-0-12-0885428

For information on all Academic Press publications
visit our website at books.elsevier.com

Printed and bound in USA

11 12 10 9 8 7 6 5 4 3 2 1

# Contents

## 8. Cumulonimbus Clouds and Severe Convective Storms



## 9. Mesoscale Convective Systems

## 10. The Mesoscale Structure of Extratropical Cyclones and Middle and High Clouds



## 11. The Influence of Mountains on Airflow, Clouds, and Precipitation

# Preface

The focus of this book is on the dynamics of clouds and of precipitating mesoscale meteorological systems. Mesoscale meteorology is concerned with weather systems that have spatial and temporal scales between the domains of macro-and micrometeorology. Generally, macrometeorology is concerned with weather systems having spatial scales greater than 1000 km and temporal scales on the order of several days or longer. Micrometeorology is the science dealing with atmospheric dynamics having spatial scales of tens to hundreds of meters and time scales on the order of minutes. Mesoscale meteorology can therefore be thought of as the science dealing with any weather system lying between these two extreme temporal and spatial scales. Orlanski (1975) subdivided the classification of mesoscale systems into three scales: meso-$\alpha$, meso-$\beta$, and meso-$\gamma$ (see Fig. 1). He suggested that the term meso-$\alpha$ should be applied to weather systems such as frontal systems and hurricanes having horizontal scales of 200-2000 km and temporal scales of 1 day to 1 week. The term meso-$\beta$ should be applied to such systems as the nocturnal low-level jet, squall lines, inertial waves, cloud clusters, and mountain and lake/coastal circulations. These systems have horizontal scales on the order of 20-200 km and temporal scales on the order of several hours to 1 day. Finally, he suggested that the meso-$\gamma$ regime should include thunderstorms, internal gravity waves, clear air turbulence, and urban effects with horizontal scales of 2-20 km and temporal scales on the order of one-half hour to several hours. In this book, we shall generally adhere to this terminology.

Clouds and precipitation mesoscale systems represent some of the most important and scientifically exciting weather systems in the world. These are the systems that produce torrential rains, severe winds including downbursts and tornadoes, hail, thunder and lightning, and major snow storms. Forecasting such storms represents a major challenge since they are too small to be adequately resolved by conventional observing networks and numerical prediction models.

Less well known is the role that clouds and storms play in atmospheric chemistry on regional and global scales. Not only do they sweep large quantities of pollutants residing in the boundary layer into the middle and upper troposphere and even the lower stratosphere, they also mix the pollutants with cloud particles that form in the rising air parcels, thus serving as wet chemical reactors. It is now generally recognized, for example, that the annual budget of acid precipitation at a particular location is dominated by a few major storm events that strip the boundary layer of large quantities of sulfates and oxides of nitrogen, where they become incorporated in precipitation elements, undergo chemical reactions, and then settle to the ground in raindrops or snowflakes as

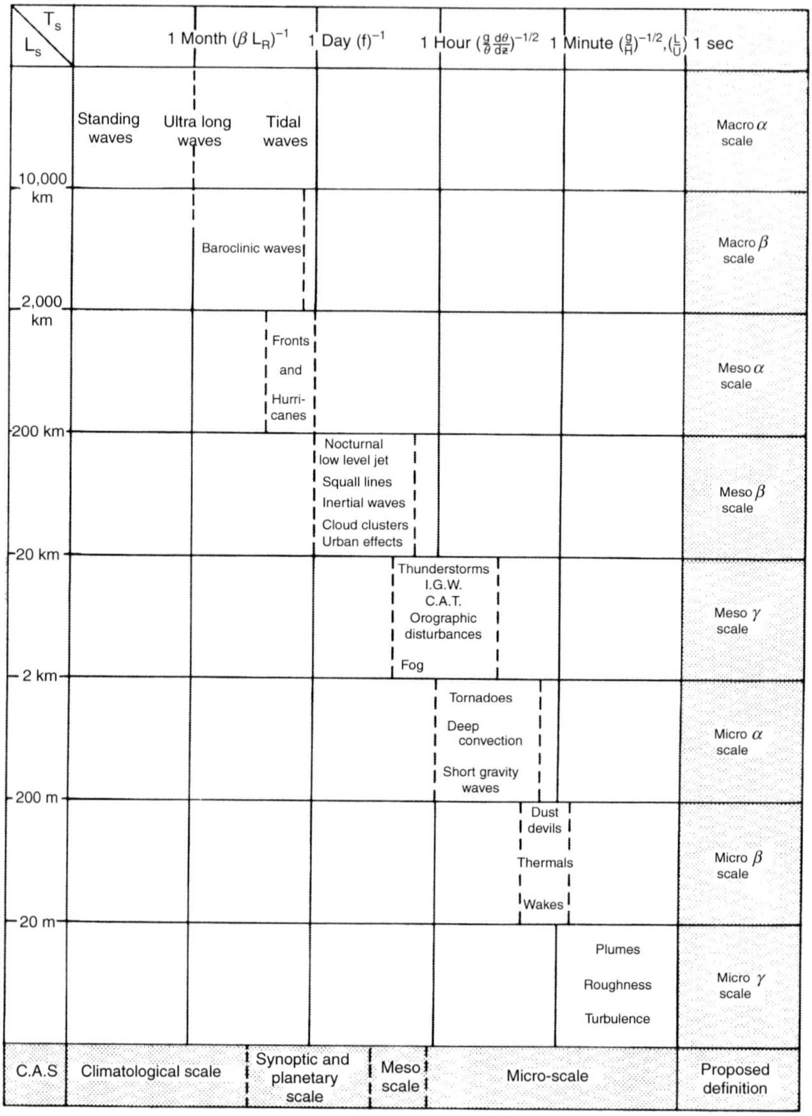


**FIGURE 1 Scale definitions and different processes with characteristic time and horizontal scales.** *(Adapted from Orlanski (1975))*

highly concentrated acidic precipitation elements. Even the lightning produced in convective storms contributes significantly to global production of nitrates, $NO_x$, and other chemical species. Here again, one major mesoscale convective system can produce as much as 25% of the annual number of cloud-to-ground lightning strokes at a given location.

Clouds and precipitating mesoscale systems play an important role in the earth's general circulation, climate, and climatic variability. The latent heats released in clouds serve as "engines" which drive the global atmospheric circulations. Moreover, they are dominant components in the earth's hydrologic cycle, transporting water vertically and horizontally, changing the phase of water from vapor to liquid and solid phases, removing water from the atmosphere through precipitation, and shielding the earth's surface from direct solar radiation, thus altering surface evaporation and evapotranspiration rates. Clouds are also a major factor in determining the earth's radiation budget by reflecting incoming solar radiation and absorbing upwelling terrestrial radiation. Variations in the coverage of clouds, the heights of clouds, and even the number and sizes of individual cloud particles all have large effects on the radiation budget of the earth.

Thus, clouds and precipitating mesoscale systems are not only fascinating subjects of study for their own sake, but the knowledge gained from studying them is essential to improving short-and long-range forecasting and gaining a quantitative understanding of atmospheric chemistry, the earth's general circulation, and climatic variability.

This, the second edition of *Storm and Cloud Dynamics* represents our attempt to update the original text with the results of over 20 years of research. Much has been learned about clouds and cloud systems during this period. You will note that the authorship on this edition has changed with Dr. Anthes being unable to join in this effort owing to his commitments as director of the University Corporation for Atmospheric Research (UCAR) and as President of the American Meteorological Society (AMS)as we prepared this edition. Dr. George Bryan and Dr. Susan van den Heever have joined me (Cotton) in preparing this revised text.

The book comprises three parts. Part I is a survey of the general theory that is appropriate to describing clouds and precipitating mesoscale systems. The theory may be described as small-scale dynamics and is, for the most part, concerned with nonhydrostatic dynamics. In addition to a review of the general theory of cloud dynamics, a review of the various physical processes relevant to the modeling of clouds and mesoscale systems is given. These processes include precipitation processes, radiative transfer, and turbulent transport. They are described not so much from the point of view of a detailed examination of the physics of the process as from the perspective of what essential elements of the processes most strongly influence the dynamics of clouds and cloud systems and the modeling of clouds and mesoscale systems.

In Part II, we describe the physics and dynamics of various cloud systems, ranging from the least dynamic, fogs and stratocumulus clouds, to the most dynamic, severe convective clouds. A discussion follows of the physics and dynamics of the dominant precipitating mesoscale systems, including squall lines, cloud clusters, mesoscale convective systems,the formation of tropical cyclones, and the subsynoptic structure of extratropical cyclones including the physics and dynamics of middle and high clouds. We then discuss the physics and dynamics of stable orographic clouds. In Part III we discuss the role of clouds on global climate. In our discussion of each cloud type, we attempt to describe the current state of knowledge as derived from the marriage of theoretical modeling and observational analysis. The detailed examination of the theory in Part I is not essential to the interpretation of Part II. However, we recommend that the reader become familiar with the notations in Part I in order to follow more readily the discussions in Part II.

We appreciate the essential help of Brenda Thompson in processing the manuscript, requesting figure authorizations, editing, and keeping figures and the many drafts of this text organized.

Original research results reported in the text were supported by numerous grants provide by the National Science Foundation (NSF) over the years to William Cotton and Susan van den Heever. George Bryan acknowledges the support of the National Center for Atmospheric Research, which is sponsored by the National Science Foundation.

## REFERENCES

Orlanski, I. (1975). A rational subdivision of scale for atmospheric processes. Bull. Am. Meteorol. Soc. 56, 527–530.

Chapter 1

# Clouds

*Clouds are pictures in the sky*
*They stir the soul, they please the eye*
*They bless the thirsty earth with rain,*
*which nurtures life from cell to brain—*
*But no! They're demons, dark and dire,*
*hurling hail, wind, flood, and fire*
*Killing, scarring, cruel masters*
*Of destruction and disasters*
*Clouds have such diversity—*
*Now blessed, now cursed,*
*the best, the worst*
*But where would life without them be?*

Vollie Cotton

## 1.1. INTRODUCTION

Since the late 1940s, when the experiments by Langmuir (1948) and Schaefer (1948) suggested that seeding of certain types of clouds could release additional precipitation, there has been intensive investigation into the physics of clouds. The major focus of these studies has been on the microphysical processes involved in cloud formation and the production of precipitation. As the studies have unraveled much about the detailed microphysics of clouds, it has become increasingly apparent that these processes are affected greatly by macroscale dynamics and thermodynamics of the cloud systems. We have also learned to appreciate that the microphysical processes can alter the macroscale dynamic and thermodynamic structure of clouds. Thus, while the focus of this book is on the dynamics of clouds, we cannot neglect cloud microphysical phenomena. The title of this book implies a perspective from which we view the cloud or cloud system as a whole. From this perspective, cloud microphysical processes can be seen as a swarm or ensemble of particles that contribute collectively, and in an integrated way, to the macroscale dynamics and thermodynamics of the cloud.

We take a similar perspective of small-scale air motions in clouds. Again, we will not use our highest power magnifying lens to view the smallest scale motions or turbulent eddies in clouds. We will instead examine the collective behavior or statistical contributions of the smallest cloud eddies (i.e. those less than a few hundred meters or so) to the energetics of clouds and to transport processes in clouds.

Following the same analogy, we view the meso-$\beta$-scale and meso-$\alpha$-scale with a wide-angle lens, thus encompassing their contributions to the energetics and transport processes of a particular cloud as well as to neighboring clouds or cloud systems. The meso-$\beta$-scale and meso-$\alpha$-scale can, for the most part, be considered the environment of the cloud scale that is generally the meso-$\gamma$-scale.

## 1.2. THE CLASSIFICATION OF CLOUDS

Cloud types are generally defined according to the phases of water present and the temperature of cloud top [AMS Glossary]. Clouds are referred to as warm clouds, or as liquid phase clouds if all portions of a cloud have temperatures greater than 0 °C. For clouds extending above the 0 °C level, precipitation formation can be either by ice phase or droplet coalescence processes. Clouds consisting entirely of ice crystals are called ice-crystal clouds. Analogously, a cloud composed entirely of liquid water drops is called a water cloud and a mixed-phase cloud contains both water drops (supercooled at temperatures below 0 °C) and ice crystals, without regard to their actual spatial distributions (coexisting or not) within the cloud. Convective clouds extending into air colder than about −10 °C are generally mixed clouds. Supercooled droplets may coexist with ice particles until temperatures are cold enough to support homogeneous freezing or below about −40 °C.

Since this text emphasizes the dynamics of clouds, it would seem appropriate that we adopt a classification of clouds that is based on the "dynamic" characteristics of clouds rather than on the physical appearance of clouds from the perspective of a ground observer. In fact, several scientists (Scorer, 1963; Howell, 1951; Scorer and Wexler, 1967) have attempted to design such a classification scheme based on cloud motions. However, since we wish to label the various cloud forms for later discussion, we shall generally adhere to the classifications given in the "International Cloud Atlas" (World Meteorological Society, 1956). This classification is based on ten main groups called *genera*, and most of the genera are subdivided into *species*. Each subdivision is based on the shape of the clouds or their internal structure. The species is sometimes further divided into *varieties*, which define special characteristics of the clouds related to their transparency and the arrangements of the macroscopic cloud elements.

The definitions of the ten genera are as follows:

*Cirrus*—Detached clouds in the form of white, delicate filaments or white or mostly white patches or narrow bands. These clouds have a fibrous (hairlike) appearance, or a silky sheen, or both.

*Cirrocumulus*—Thin, white patches, sheets, or layers of cloud without shading, composed of very small elements in the form of grains or ripples, merged or separate, and more or less regularly arranged; most of the elements have an apparent width of less than 1°.

*Cirrostratus*—Transparent, whitish cloud veil of fibrous or smooth appearance, totally or partially covering the sky, and generally producing halo phenomena.

*Altocumulus*—White or grey, or both white and grey, patches, sheets, or layers of cloud, generally with shading, composed of laminae, rounded masses, or rolls, which are sometimes partially fibrous or diffuse and which may or may not be merged; most of the regularly arranged small elements usually have an apparent width of 1°-5°.

*Altostratus*—Greyish or bluish cloud sheet or layer of striated, fibrous, or uniform appearance, totally or partially covering the sky, and having parts thin enough to reveal the sun at least dimly, as through ground glass. Altostratus does not produce halos.

*Nimbostratus*—Grey cloud layer, often dark, the appearance of which is rendered diffuse by more or less continuously falling rain or snow, which in most cases reaches the ground. It is thick enough to completely obscure the sun. Low, ragged clouds frequently occur below the nimbostratus layer.

*Stratocumulus*—Grey or whitish, or both grey and whitish, patches, sheets, or layers of cloud which almost always have dark parts, composed of crenellations, rounded masses, or rolls, which are nonfibrous (except when virga-inclined trails of precipitation-are present) and which may or may not be merged; most of the regularly arranged small elements have an apparent width of more than 5°.

*Stratus*—Generally grey clouds with a fairly uniform base, which may produce drizzle, ice prisms, or snow grains. If the sun is visible through the cloud, its outline is clearly discernible. Stratus clouds do not produce halo phenomena except, possibly, at very low temperatures. Sometimes stratus clouds appear in the form of ragged patches.

*Cumulus*—Detached clouds, generally dense and with sharp outlines developing vertically in the form of rising mounds, domes, or towers, of which the bulging upper part often resembles a cauliflower. The sunlit parts of these clouds are mostly brilliant white; their base is relatively dark and nearly horizontal. Sometimes cumulus clouds are ragged.

*Cumulonimbus*—Heavy, dense clouds, with a considerable vertical extent, in the form of a mountain or huge tower. At least part of their upper portion is usually smooth, fibrous, or striated and is nearly always flattened; this part often spreads out in the shape of an anvil or vast plume. Under the base of these clouds, which is generally very dark, there are frequently low ragged clouds and precipitation, sometimes in the form of virga.

In general we will not have to refer to the definitions of the clouds species or varieties used in the "International Cloud Atlas." The exceptions mainly concern cumulus clouds, which we refer to as follows:

*Cumulus humilis*—Cumulus clouds of only a slight vertical extent; they generally appear flattened.
*Cumulus mediocris*—Cumulus clouds of moderate vertical extent, the tops of which show fairly small protuberances.
*Cumulus congestus*—Cumulus clouds which exhibit markedly vertical development and are often of great vertical extent; their bulging upper part frequently resembles a cauliflower. We also may have occasion to refer to the following supplementary features and accessories of clouds:
*Mamma*—Hanging protuberances, like udders, on the under surface of a cloud.
*Virga*—Vertical or inclined trails of precipitation (fall streaks) falling from the base but reaching the earth's surface.
*Pileus*—An accessory cloud of small horizontal extent, in the form of a cap or hood above the top or attached to the upper part of a cumuliform cloud which often penetrates it.

Fog is not treated as a separate cloud genus in the "International Cloud Atlas." Instead it is defined in terms of its microstructure, visibility, and proximity to the earth's surface as follows:

*Fog*—Composed of very small water droplets (sometimes ice crystals) in suspension in the atmosphere; it reduces the visibility at the earth's surface generally to less than 1000 m. The vertical extent of fog ranges between a few meters and several hundred meters.

We include the discussion of fog in the chapter on stratocumulus clouds, since we shall see there is not always a clear distinction between the formative mechanisms of a marine stratocumulus cloud whose base is elevated from the surface, and a fog which reaches the surface.

Another cloud form discussed in this text that is not treated in the "International Cloud Atlas" as a separate cloud genus is the *orographic* cloud. According to the "Glossary of Meteorology" (Huschke, 1959), an orographic cloud is a cloud whose form and extent is determined by the disturbing effects of orography upon the passing flow of air. Since orography can also initiate convective clouds, we shall often refer to a stable orographic cloud as the cloud form typically encountered in the wintertime during periods when the atmosphere is stably stratified. The *cap* cloud is the least complicated form of the orographic cloud and refers to a nearly stationary cloud that hovers over an isolated peak. The *crest* cloud is like the cap cloud with the exception that it hovers over a mountain ridge. The *chinook arch* or *foehn* wall cloud refers to a bank or wall of clouds associated with a chinook or foehn wind storm. Finally, the lenticular cloud, or *lenticularis*, is a lens-shaped cloud that forms over, or to the lee of, orographic barriers as a result of mountain waves. As the name implies, lenticular clouds generally have a smooth shape with sharp outlines, sometimes vertically-stacked with clear air separating each lenslike element.

## 1.3. CLOUD TIME SCALES, VERTICAL VELOCITIES, AND LIQUID-WATER CONTENTS

In this section we examine certain cloud characteristics that have a major controlling influence upon whether or not precipitation processes are important and whether diabatic processes such as condensational heating and radiative transfer dominate the cloud energetics. Because these physical processes affect the dynamics of the cloud, it is important to recognize under what conditions and in which cloud types these processes are important.

Saturation vapor pressure decreases when the temperature decreases (the Clausius-Clapeyron relation). Cloud formation occurs when the saturation vapor pressure becomes smaller than the actual partial pressure of the water vapor in the air. The greater the difference, the stronger the forcing for liquid droplets or ice particles to grow by vapor deposition. Clouds generally form when a buoyant parcel of air is lifted (convective ascent) and cooled by adiabatic expansion. As a parcel of air inside a cloud ascends, temperature decreases following a moist adiabatic lapse rate which is slightly less (0.65 C per 100 m) than in clear air adiabatic ascent (1 C per 100 m), because of the latent heat released by condensation. The rate of condensation depends on the temperature and pressure of the cloud. At 900 m and 20 °C, for example, it is approximately 2 g $kg^{-1}$ per km of ascent. Assuming no mixing, the mixing ratio of condensed water at any level above cloud base can be derived as the difference between the water vapor mixing ratio at cloud base and the saturation water vapor mixing ratio at that level. This is referred to as the *adiabatic water-mixing ratio.* Owing to the mixing processes, the actual condensed water-mixing ratio is generally lower than the adiabatic value. At the cloud base, the condensation of the available water vapor is not instantaneous and the actual water vapor partial pressure is higher than the saturation vapor pressure leading to supersaturation. Supersaturation plays a critical role near cloud base for the activation of cloud condensation nuclei (CCN) and ice nuclei (IN) that initiate cloud droplets or ice crystals. The amount of condensed water content, either liquid or ice, is a key parameter for precipitation formation. Precipitation is most likely to form in the regions of largest condensed mixing ratio, i.e. in the least diluted cloud cells.

The macroscopic parameters of clouds that characterize precipitation and diabatic process are (1) cloud time scales, (2) cloud vertical velocities, (3) cloud liquid-water contents, (4) cloud temperature, and (5) cloud turbulence. Time scales are important because precipitation processes are time dependent. Therefore, if the cloud lifetime is too short for the time it takes to form precipitation, the cloud will not precipitate even though other properties, such as liquid-water content, are sufficient to support precipitation. Two time scales are critical. One is the cloud lifetime, which we shall label $T_C$. The other, called the parcel lifetime, represents the time it takes a parcel to enter the cloud and exit its top or sides. We shall label this time scale as $T_P$.

Cloud vertical velocities are important because the updrafts control the time scale $T_P$ and determine the cloud's ability to suspend precipitation particles. The magnitude of vertical velocity also provides an estimate of the wet (saturated) adiabatic cooling rate. For example, in the middle troposphere the wet adiabatic lapse rate $\gamma_m$ is approximately 0.5 °C/100 m. Thus the wet adiabatic cooling rate $CR_\gamma$ is

$$CR_\gamma \simeq (0.5\ ^\circ C/100\ m) \times W, \tag{1.1}$$

where $W$ is the cloud vertical velocity in meters per second.

Both the potential for precipitation formation and the cooling rates of clouds depend on the liquid-water content (LWC) in a cloud, for two reasons. First, the LWC determines the ultimate potential for a cloud to produce precipitation. Generally speaking, unless a cloud generates a liquid-water content in excess of 0.5 g $m^{-3}$, it is unlikely to precipitate. Of course, other factors, such as aerosol concentrations (see Chapter 4) and whether the cloud is supercooled, also control the critical LWC for initiating precipitation. Second, the LWC is important because it determines the rates of shortwave or longwave radiational heating and cooling.

Cloud temperature also represents an important parameter in precipitation potential. The cloud-base temperature indicates the liquid-water producing potential of the cloud. For example, a cloud with a base temperature of +20 °C has a cloud-base saturation mixing ratio of ~15 g $kg^{-1}$, while a cloud with a base temperature of +4 °C has a cloud-base saturation mixing ratio of only 5 g $kg^{-1}$. Thus, if these two clouds have equal depths, the one with the warmer cloud base has a much greater potential for producing rainfall. Cloud-top temperature is important for similar reasons, because the greater the difference between the cloud-base temperature and cloud-top temperature, the greater the potential for rainfall. Furthermore, if the cloud-top temperature is below 0 °C, then ice is possible, which greatly affects precipitation and radiation processes.

Turbulence, the last consideration in our discussion of macroscopic cloud parameters, is important because it mixes properties of the cloud and interacts closely with the other parameters. When we speak of "characteristic" time scales, vertical velocities, liquid-water contents, and temperatures, the level of turbulence determines how representative these "characteristic" scales really are. For example, in some convective clouds the average updraft velocity may be 1 m $s^{-1}$, while the standard deviation of the vertical velocity may be as large as 3 m $s^{-1}$. The level of turbulence also affects the precipitation processes, due to the formation of higher peak supersaturations and to increased interactions among cloud particles of different types and sizes. Turbulence is also likely to affect the radiative properties of a cloud. In a turbulent cloud the cloud top is likely to be very lumpy, and large fluctuations in liquid water will exist. As a consequence, the cloud-top radiative emittance and absorptance will differ significantly from that found in a more homogeneous cloud.

Let us next consider how the characteristics just introduced differ in several cloud forms that we will study in this book.

### 1.3.1. Fog

Fog may be considered the least dynamic of clouds. Fogs typically have lifetimes ($T_c$) of 2 to 6 h. The mean vertical velocity in fog is usually quite small. If we assume a mean updraft of 0.01 m s$^{-1}$ for a 100-m-deep fog, the time scale for a parcel entering the cloud base and exiting the cloud top would be

$$T_P = 100 \text{ m}/0.01 \text{ m s}^{-1} = 10^4 \text{ s}, \tag{1.2}$$

that is, $T_P$ is on the order of 3 h. This represents the time scale in which cloud microphysical processes must operate in order to generate precipitation.

However, the liquid-water content in fog typically ranges from 0.05 to 0.2 g m$^{-3}$. Thus, precipitation is unlikely in all but the deepest, wettest, and most maritime fogs, even though the mean vertical velocity might indicate a potential for precipitation. If we use our estimate of $W$, of 0.01 m s$^{-1}$, we determine that the cooling rate due to wet adiabatic cooling is of the order of

$$\text{CR}_\gamma = (0.5\ ^\circ\text{C}/100 \text{ m})0.01 \text{ m s}^{-1} = 5 \times 10^{-5}\ ^\circ\text{C s}^{-1}, \tag{1.3}$$

which is approximately 0.2 °C h$^{-1}$. By comparison, the rate of cooling by longwave radiation flux divergence at the top of the fog can easily range from 1 to 4 °C h$^{-1}$. Thus, we see that fogs can be dominated by radiative cooling.

The absolute magnitude of turbulence in fogs is usually small, although there have been reports of vertical velocity fluctuations in some valley fogs as large as 1 m s. However, if we consider turbulence in terms of fluctuations from the mean motions, it appears that because both horizontal and vertical mean velocities are typically small in fogs, a fog is dominated by turbulence. Thus, turbulence affects transport and nearly all physical processes in fogs, even though its absolute magnitude is generally small.

### 1.3.2. Stratus and Stratocumulus Clouds

Stratus clouds and stratocumulus clouds do not differ markedly from fogs in terms of time scales, liquid-water contents, or turbulence levels. The lifetimes of stratus and stratocumulus clouds are longer, ranging from 6 to 12 h. As in fog, the time scale for a parcel to enter a stratus having a mean vertical velocity of 0.1 m s$^{-1}$ and rising through a depth of, say, 1000 m may be 3 h. Typical liquid-water contents in stratus clouds range from 0.05 to 0.25 g m$^{-3}$, with some maxima of over 0.6 g m$^{-3}$ reported. This combination of time scales and liquid-water contents results in precipitation in the deepest, wettest stratus and stratocumulus clouds in the form of drizzle.

Again, assuming vertical velocities of 0.1 m s$^{-1}$, the wet adiabatic cooling rates are of the order of 2 °C h$^{-1}$. Thus, radiation and wet adiabatic cooling are approximately equal contributors to the destabilization of stratus and stratocumulus clouds.

The turbulence level in stratus clouds is low in absolute magnitude, just as it is in fog. However, since mean vertical velocities are also small, turbulence is a significant contributor to vertical transport processes, energetics, and the physics of stratus clouds.

### 1.3.3. Cumulus (Humilis and Mediocris) Clouds

Cumulus clouds whose vertical extent may be 1500 m have a lifetime ($T_c$) of 10-30 min, which is shorter than that for the preceding two types of clouds. If we consider an average vertical velocity of 3 m s$^{-1}$, the time scale for a parcel to enter the cloud base and exit the cloud top is of the order of

$$T_P = 1500\ \text{m}/3\ \text{m s}^{-1} = 500\ \text{s} \simeq 10\ \text{min}. \tag{1.4}$$

The liquid-water content of small cumuli rarely exceeds 1.0 g m$^{-3}$ and is typically approximately 0.3 g m$^{-3}$. Thus, for such short time scales and low liquid-water contents, precipitation is unlikely in all but the most maritime or cleanest airmass, wettest cumuli.

Comparing wet adiabatic cooling rates to cloud-top radiation cooling, we estimate

$$CR_\gamma \simeq (0.5\ ^\circ\text{C}/100\ \text{m}) \times 3\ \text{m s}^{-1} = 1.5 \times 10^{-2}\ ^\circ\text{C s}^{-1} \simeq 50\ ^\circ\text{C h}^{-1}, \tag{1.5}$$

which is considerably greater than the cloud-top radiation cooling rates for clouds of such liquid-water contents ($CR_{IR} \sim 4$ °C h$^{-1}$). Thus, wet adiabatic cooling dominates radiative effects in such clouds.

The turbulence levels in small cumuli is relatively moderate, with root-mean-square (RMS) velocities ranging from 1 to 3 m s$^{-1}$. Thus, turbulence plays an important role in such clouds.

### 1.3.4. Cumulus Congestus Clouds

The lifetime of cumulus congestus clouds exceeds that of cumuli, from 20 to 45 min. However, the transit time $T_P$ for a parcel entering the cloud base, rising at 10 m s$^{-1}$, and exiting the cloud top is similar to that of small cumuli, since

$$T_P = 5000\ \text{m}/10\ \text{m s}^{-1} = 500\ \text{s} \simeq 10\ \text{min}. \tag{1.6}$$

That is, higher updraft velocities in cumulus congestus clouds offset their greater depth in determining $T_P$. Because of the small $T_P$, precipitation would be unlikely if it were not for the higher liquid-water content of cumulus congestus clouds, which ranges from 0.5 to 2.5 g $m^{-3}$. Because the turbulence level in such clouds is often quite strong, it is possible for air parcels to spend a considerably longer residence time in the cloud than would be implied by $T_P$.

As in the smaller cumuli, radiative effects are secondary to wet adiabatic processes in the energetics of cumulus congestus clouds.

### 1.3.5. Cumulonimbus Clouds

Cumulonimbi are the longest living convective clouds. They have lifetimes from 45 min to several hours. However, the time scale for a parcel of air to enter the cloud base and commence forming precipitation before exiting the top remains relatively short. Let us take, as an example, a cumulonimbus cloud that is 12,000 m deep and has an average updraft velocity of 30 m $s^{-1}$. The Lagrangian time scale is only

$$T_P = 12{,}000\ \text{m}/30\ \text{m s}^{-1} = 400\ \text{s}, \tag{1.7}$$

which is actually less than that in the smaller cumuli. Because of the enormous cooling of air parcels rising through the great depths of the cloud, typical liquid-water contents in cumulonimbi range from 1.5 to 4.5 g $m^3$ and often greater. These high liquid-water contents compensate, to some extent, for the short time scale. The short time scale sometimes limits the formation of precipitation, which accounts for the weak echo regions (WERs) that are often observed by radars. It should be noted that such intense updrafts as those present in WERs are not characteristic of the entire convective storm. Because turbulence levels can be so intense, there is considerable opportunity for air parcels to experience much longer lifetimes than are encountered rising in the main updraft.

With the exception of the anvil outflow region of cumulonimbi, wet adiabatic processes dominate over radiative cooling. However, radiative cooling may contribute significantly to the destabilization and maintenance of the weak updraft regions of cumulonimbus anvils.

### 1.3.6. Stable Orographic Clouds

Let us consider now a wintertime stable orographic cloud that is above a 1400-m-high mountain with a half-width of 18 km (Fig. 1.1). For this type of cloud, the cloud lifetime could be many hours or even days. However, the time scale for precipitation processes to operate if the winds are about 15 m $s^{-1}$ is only

$$T_P = 18{,}000\ \text{m}/15\ \text{m s}^{-1} = 1200\ \text{s} = 20\ \text{min}. \tag{1.8}$$

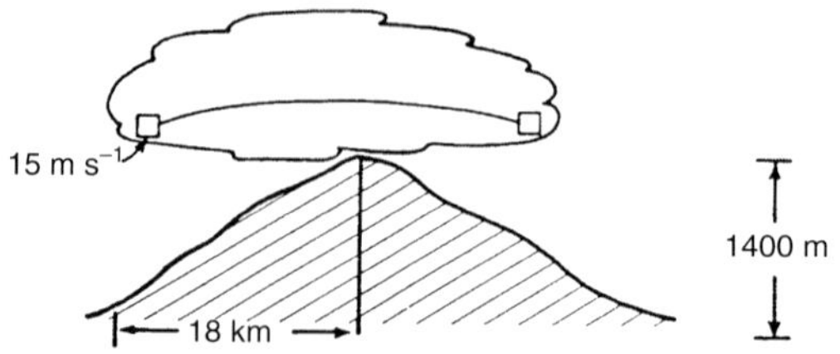


FIGURE 1.1 **Schematic diagram of a stable orographic cloud.**

Thus, the time scale $T_p$ is longer than that for cumuli but considerably shorter than that for stratus clouds. The liquid-water contents of wintertime stable orographic clouds do not differ substantially from those of stratocumuli; they are typically less than 0.2 g $m^{-3}$. It is only in highly efficient maritime clouds or colder and efficient ice-phase-dominated clouds that precipitate occurs.

If we consider typical updraft speeds near the mountain barrier to be about 1 m $s^{-1}$, we have an estimate of wet adiabatic cooling rates of

$$CR_\gamma = 18\,^\circ\mathrm{C\,h^{-1}}, \tag{1.9}$$

which is greater by an order of magnitude than radiative cooling rates. Thus, near the barrier crest, wet adiabatic processes remain dominant. At distances removed from the barrier crest, however, where a blanket cloud may reside, or in weaker wind situations, one can anticipate that radiative processes become more significant in such clouds.

There has been very little characterization of the levels of turbulence in wintertime stable orographic clouds. At cloud levels near the barrier crest, surface-generated turbulence could be quite significant. At higher cloud levels, however, turbulence levels can be expected to be relatively weaker under the typically stable conditions.

We can see from these simple comparisons and contrasts that clouds form in a broad range of conditions that control the ultimate destiny of the cloud. Depending on the vertical velocity, liquid-water content, and cloud time scale, precipitation processes may or may not affect significantly the dynamics of the cloud. Similarly, radiation processes may or may not be an important destabilizing influence on the cloud. It should be remembered that these are only rough estimates and that one can expect considerable variability within a given cloud category. To account for such variability we must construct sophisticated models of each of the cloud types. In the following chapters, we present the foundation for constructing such models of the dynamics and physics of various cloud systems.

## REFERENCES

Howell, W. E. (1951). The classification of cloud forms. Clouds, fogs and aircraft icing. In "Compendium of Meteorology" (T. F. Malone, Ed.), pp. 1162–1166. Am. Meteorol. Soc., Boston, Massachusetts.

Huschke, R. E., (Ed.) (1959). In "Glossary of Meteorology" Am. Meteorol. Soc., Boston, Massachusetts.

Langmuir, I. (1948). The growth of particles in smokes and clouds and the production of snow from supercooled clouds. Proc. Am. Philos. Soc. 92, 167.

Schaefer, V. J. (1948). The production of clouds containing supercooled water droplets or ice crystals under laboratory conditions. Bull. Am. Meteorol. Soc. 29, 175.

Scorer, R. S. (1963). Cloud nomenclature. Q. J. R. Meteorol. Soc. 89, 248–253.

Scorer, R. S., and Wexler, H. (1967). "Cloud Studies in Colour." Pergamon, Oxford.

World Meteorological Society (1956). "International Cloud Atlas, Vol. 1", Geneva.

# Part I

# Fundamental Concepts and Parameterizations

Chapter 2

# Fundamental Equations Governing Cloud Processes

*If your work is to withstand*
*tests of time and season*
*Your foundation must be planned*
*with patience, insight, reason.*

Vollie Cotton

## 2.1. INTRODUCTION

In this chapter we develop a set of equations that govern kinematic and thermodynamic processes in clouds and cloud systems. Our primary focus is on the application of these equations. Some equations are derived from first principles, but in most cases references are provided instead for interested readers. The following equations could be used to formulate a numerical model of a cloud. To make this possible, numerical techniques, boundary conditions, and initial conditions are necessary. Some integration strategies are discussed briefly, but it is not our intent to discuss numerical modeling in detail.

To begin, we note that the fundamental equations governing dry air are generally accepted in fluid dynamics communities. By contrast, the governing equations for moist atmospheric flows remain a source of some controversy and active research. The fundamental issue is that water in the atmosphere exists in three phases: gas, liquid, and solid. In the gas phase, water vapor can be treated in the same manner as dry air; the continuum hypothesis can be made, and water vapor becomes simply another component of air. However, when water is in liquid or solid phase in the atmosphere, the continuum hypothesis is no longer valid: liquid water drops and solid water particles (i.e. snow/ice) are not in physical contact with each other. Liquid and solid water particles in the atmosphere must therefore be treated with special mathematical techniques, which will be treated separately in Chapter 4. Furthermore, water is continuously changing phases in clouds, and the adjustments of energy and momentum during phase changes is what makes moist flows difficult to model.

## 2.2. STATE PROPERTIES

The properties of cloudy air can be quantified by a set of state variables. These include the air's mass, temperature, pressure, etc. Consider here a volume of cloudy air. The total mass ($m$) is the sum of the masses of dry air ($m_a$), water vapor ($m_v$), liquid water ($m_l$), and snow/ice ($m_i$). The following equations use primarily the density, which is the mass per unit volume: $\rho \equiv m/V$.

### 2.2.1. State Properties: Gas

Consider a volume of air consisting of dry air with density $\rho_a$ and water vapor with density $\rho_v$. The gaseous components of the air obey the ideal gas law and Dalton's law of partial pressure, and therefore the equation of state can be expressed as

$$p = p_d + e = \rho_a R_a T + \rho_v R_v T \tag{2.1}$$

where $p$ is the total air pressure, $p_d$ is the partial pressure of dry air, $e$ is the partial pressure of water vapor, $R_a$ is the gas constant for dry air, $R_v$ is the gas constant for water vapor, and $T$ is the temperature. It is assumed that the temperature of dry air and water vapor are always the same.

Equation (2.1) is often rewritten as

$$p = \rho_a R_a T \left(1 + \frac{R_v}{R_a}\frac{\rho_v}{\rho_a}\right). \tag{2.2}$$

Using the definition of mixing ratio, $r_v \equiv \rho_v/\rho_a$, and using $\epsilon \equiv R_a/R_v \approx 0.622$, then (2.2) can be further rewritten in the common form

$$p = \rho_a R_a T(1 + r_v/\epsilon). \tag{2.3}$$

Another form uses the density of moist air, $\rho_m \equiv \rho_a + \rho_v$, and the specific humidity, $q_v \equiv \rho_v/\rho_m$, so that

$$p = \rho_m R_a T \left[1 + \left(\frac{1}{\epsilon} - 1\right) q_v\right]. \tag{2.4}$$

No approximations have been made in any of these forms of the equation of state. When applying these equations, care must be taken to ensure that the proper density ($\rho_a$ or $\rho_m$), the proper measure of water vapor ($r_v$ or $q_v$), and the proper coefficient [$1/\epsilon \approx 1.61$ or $(1/\epsilon - 1) \approx 0.61$] are used.

### 2.2.2. State Properties: Liquid and Ice

As discussed earlier, the continuum hypothesis does not apply to liquid and ice particles (i.e. condensate) in the atmosphere. Consequently, condensate does not exert any pressure on a volume of cloudy air, and thus the state properties of condensate do not appear in the equation of state above. Nevertheless, the state properties of condensate are very important. The density of liquid water in cloudy air is defined here as the mass of liquid water per unit volume of cloudy air: $\rho_l \equiv m_l/V$. (This is not to be confused with the density of an individual liquid water drop, which is about 1000 kg m$^{-3}$.) Similarly, the density of ice in cloudy air is given by $\rho_i \equiv m_i/V$.

Mixing ratios are defined as follows: $r_l \equiv \rho_l/\rho_a$ and $r_i \equiv \rho_i/\rho_a$. Specific humidities are defined as follows: $q_l \equiv \rho_l/\rho_m$ and $r_i \equiv \rho_i/\rho_m$. Note that a specific humidity is related to a mixing ratio as follows: $q_\chi = r_\chi/(1+r_\chi)$, where $\chi$ is either $v$ for vapor, $l$ for liquid, or $i$ for ice/snow.

The temperatures of liquid and ice are given here as $T_l$ and $T_i$, respectively. It is important to note that the temperature of condensate may differ from the temperature of air, especially when particle sizes are large (i.e. when diameters are greater than a few millimeters).

### 2.2.3. Other Useful Properties of State

Meteorologists will sometimes define indirect properties of state. For example, the *virtual temperature* is given by

$$T_v \equiv T\frac{(1+r_v/\epsilon)}{1+r_v}. \tag{2.5}$$

Note that the equation of state can be written without approximation as $p = \rho_m R_a T_v$, from which it becomes apparent that virtual temperature is the temperature dry air would need to be in order to have the same density as moist air at the same pressure. Similarly, the *density temperature* is

$$T_\rho \equiv T\frac{(1+r_v/\epsilon)}{1+r_t} \tag{2.6}$$

where $r_t \equiv r_v + r_l + r_i$ is the mixing ratio of total water. The equation of state can be written without approximation as $p = \rho R_a T_\rho$, where $\rho \equiv \rho_a + \rho_v + \rho_l + \rho_i = \rho_a(1+r_t)$ is the total density. Density temperature is the temperature dry air would need to be in order to have the same density as cloudy air at the same pressure. Note that $T_v \geq T$ and that moist air is always less dense than dry air at the same pressure and temperature.

It is sometimes useful to use a nondimensional pressure defined as

$$\pi \equiv \left(\frac{p}{p_{00}}\right)^{\frac{R_a}{c_{pa}}} \tag{2.7}$$

where $p_{00}$ is a constant reference pressure (usually 1000 mb) and $c_{pa}$ is the specific heat of dry air at constant pressure. The variable $\pi$ is sometimes referred

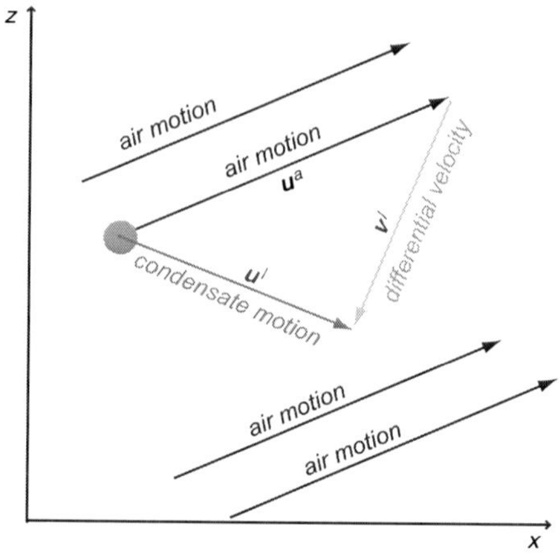


**FIGURE 2.1** **Illustration of the velocity notation used in the text.** Air is moving up and to the right with velocity $\mathbf{u}^a$. Condensate (a liquid-water drop) is moving down and to the right with velocity $\mathbf{u}^l$. The differential velocity of the condensate, relative to the air, is down and to the left; see vector $\mathbf{v}^l$.

to as the Exner function. A very useful measure of temperature is the potential temperature

$$\theta \equiv \frac{T}{\pi} \tag{2.8}$$

which is the temperature that air would have if brought isentropically to pressure $p_{00}$. Meteorologists will sometimes use the virtual potential temperature, $\theta_v \equiv T_v/\pi$, and the density potential temperature, $\theta_\rho \equiv T_\rho/\pi$. Using (2.7) and (2.8), the equation of state can be expressed without approximation as

$$\pi = \left(\frac{\rho_a R_a \theta}{p_{00}} (1 + r_v/\epsilon)\right)^{\frac{R_a}{c_{va}}} \tag{2.9}$$

where $c_{va}$ is the specific heat of dry air at constant volume.

## 2.3. VELOCITY

Consider the volume of cloudy air shown in Fig. 2.1. The dry air and water vapor move with the same velocity, $\mathbf{u}^a$, which in this case is up and to the right. (Herein, bold variables are vectors.) The liquid water drops move with velocity $\mathbf{u}^l$ (which may or may not be different from $\mathbf{u}^a$), and in this case is down and to the right. The differential velocity of the liquid water drops compared to that of the air is $\mathbf{v}^l \equiv \mathbf{u}^l - \mathbf{u}^a$. Note that the vertical component of $\mathbf{v}^l$ is always negative (i.e. condensate always falls toward the ground relative to air motion). Often, the *horizontal* velocity of liquid is about the same as the air; in this case, the differential velocity is pointed straight downwards, and is then called the *terminal fall velocity*, $v_{tl}$. The same vector notation applies to ice, but with $l$ above replaced by $i$.

Throughout this chapter, basic Cartesian tensor notation is used for convenient, abbreviated notation. The indices $i = 1, 2, 3$ correspond to Cartesian vector coordinates $x$, $y$, and $z$, respectively. The following simple rules apply:

- Repeated indices are summed (e.g. $a_{ii} = a_{11}+a_{22}+a_{33}$ in three-dimensional space).
- Single indices in a term are called free indices and refer to the order of a tensor. The maximum value a free index can attain in a three-dimensional system is 3.
- Only tensors of the same order can be added.
- Multiplication of tensors can be performed as for scalars. They are commutative with respect to addition and multiplication.
- The delta function parameter is defined to simplify writing certain terms; thus

$$\delta_{ij} = \begin{cases} 1 & \text{for } i = j. \\ 0 & \text{for } i \neq j. \end{cases}$$

Similarly the permutation symbol $\varepsilon_{ijk}$ is defined to express vector cross products. Thus,

$$\varepsilon_{ijk} = \begin{cases} 0 & \text{if } i = j, \text{ or } j = k, \text{ or } i = k. \\ 1 & \text{if } i, j, k \text{ are an even permutation of } 1, 2, 3 \\ -1 & \text{if } i, j, k \text{ are an odd permutation of } 1, 2, 3. \end{cases}$$

Hence,

$$\varepsilon_{1,2,3} = 1, \qquad \varepsilon_{1,3,2} = -1, \qquad \varepsilon_{3,1,2} = 1, \qquad \varepsilon_{3,2,1} = -1,$$
$$\varepsilon_{3,3,1} = \varepsilon_{2,1,2} = 0.$$

Cartesian tensor notation yields the following equivalents with Cartesian vector notation:

$$\begin{array}{ll} \text{Divergence:} & \nabla \cdot \mathbf{V} \rightarrow \partial u_j/\partial x_j. \\ \text{Advection:} & -\mathbf{V} \cdot \nabla\phi \rightarrow -u_j(\partial\phi/\partial x_j). \\ \text{Coriolis:} & -\mathbf{f} \times \mathbf{V} \rightarrow -\varepsilon_{ijk} u_j f_k. \end{array}$$

Hereafter, the tensor notation for velocity is $u_\beta^\alpha$ where $\alpha$ refers to the constituent type (i.e. $a$ for dry air, $l$ for liquid water, $i$ for ice/snow) and $\beta$ is the tensor notation index. For example, $\mathbf{u}^a$ is equivalent to $u_i^a$, and $\mathbf{v}^l$ is equivalent to $v_i^l$.

## 2.4. CONSERVATION OF MASS

A general equation for mass conservation has the form

$$\frac{\partial \chi}{\partial t} + \frac{\partial}{\partial x_j}\left(\chi u_j^\chi\right) = \dot{\chi} \tag{2.10}$$

where $\chi$ is a flow property per unit volume (e.g. density), $u_j^\chi$ is the velocity following $\chi$, and $\dot{\chi}$ is the rate of change of $\chi$ (from sources and sinks, such as phase changes of water). Because there is no source/sink of dry air, the conservation equation for dry air density is simply given by

$$\frac{\partial \rho_a}{\partial t} + \frac{\partial}{\partial x_j}\left(\rho_a u_j^a\right) = 0. \tag{2.11}$$

Water vapor mass has several possible sources/sinks. For example, water vapor may condense to form liquid water, or rain drops may evaporate, etc. The conservation equation for water vapor density is thus given by

$$\frac{\partial \rho_v}{\partial t} + \frac{\partial}{\partial x_j}\left(\rho_v u_j^a\right) = -\dot{\rho}_{lv} - \dot{\rho}_{iv} \tag{2.12}$$

where vapor is assumed to move at the same velocity as dry air, and where $\dot{\rho}_{lv}$ and $\dot{\rho}_{iv}$ are the rates of change of vapor density to/from liquid and ice, respectively. Similarly, the conservation equations for liquid and ice are

$$\frac{\partial \rho_l}{\partial t} + \frac{\partial}{\partial x_j}\left(\rho_l u_j^l\right) = +\dot{\rho}_{lv} - \dot{\rho}_{il} \tag{2.13}$$

and

$$\frac{\partial \rho_i}{\partial t} + \frac{\partial}{\partial x_j}\left(\rho_i u_j^i\right) = +\dot{\rho}_{iv} + \dot{\rho}_{il} \tag{2.14}$$

where $\dot{\rho}_{il}$ is the rate of change of density between liquid and ice. Adding (2.11)–(2.14) and using the vector identities from the previous section yields an equation for conservation of total density,

$$\frac{\partial \rho}{\partial t} + \frac{\partial}{\partial x_j}\left(\rho u_j^a\right) = -\frac{\partial}{\partial x_j}\left(\rho_l v_j^l\right) - \frac{\partial}{\partial x_j}\left(\rho_i v_j^i\right), \tag{2.15}$$

where $\rho \equiv \rho_a + \rho_v + \rho_l + \rho_i$.

By using (2.11) and the definition of a mixing ratio, the equations for water can be easily expressed in terms of mixing ratios. For example, the governing equation for $r_l$ is

$$\frac{\mathrm{D}r_l}{\mathrm{D}t} = +\dot{r}_{lv} - \dot{r}_{il} - \frac{1}{\rho_a}\frac{\partial}{\partial x_j}\left(\rho_a r_l v_j^l\right) \tag{2.16}$$

where the material derivative is

$$\frac{\mathrm{D}}{\mathrm{D}t} \equiv \frac{\partial}{\partial t} + u_j^a \frac{\partial}{\partial x_j} \tag{2.17}$$

and $\dot{r}$ represents the rate of change of the mixing ratio. If a terminal velocity is assumed (see previous section), and if $v_{tl}$ is assumed to be positive downward (which is the standard convention in the atmospheric sciences), then (2.16) can be expressed as

$$\frac{\mathrm{D}r_l}{\mathrm{D}t} = +\dot{r}_{lv} - \dot{r}_{il} + \frac{1}{\rho_a}\frac{\partial}{\partial z}\left(\rho_a r_l v_{tl}\right) \tag{2.18}$$

which is a common form used in numerical models.

## 2.5. CONSERVATION OF MOMENTUM

Most current debate and research about the governing equations for moist flows concerns the momentum equations. The reason is that changes in momentum during a phase change remain poorly understood. Consider, for example, a volume of motionless cloudy air consisting of some small liquid-water drops (Fig. 2.2a). The total momentum of the cloudy air is zero: that is, $M \equiv \rho_a w^a + \rho_l w^l = 0$, where in this example we will consider only vertical velocities. If the water drops collect together to make a smaller number of larger drops, they may begin to fall relative to the air (Fig. 2.2b). Clearly, the momentum of the drops will be nonzero: that is, $\rho_l w^l < 0$. But what about the air? If the air is assumed to retain zero velocity (as is commonly assumed) then the total momentum of the air would be negative: that is, $M = \rho_a w^a + \rho_l w^l < 0$. The problem with this assumption is that $M$ was previously zero, and so $M$ has not been conserved.

Instead, if the velocity of air becomes positive (dashed lines in Fig. 2.2c) then the total momentum can be conserved. This may not seem intuitive. However, consider the micro-scale flow around the liquid water drops: air must move upward to occupy the space previously occupied by the falling water drops, because a vacuum is not permitted. Such micro-scale motions are not considered further here, but it is important to note that these effects can influence the macro-scale motions of interest to meteorologists. Simple calculations using reasonable values for condensate mass and fall velocity show that $w^a$ of order

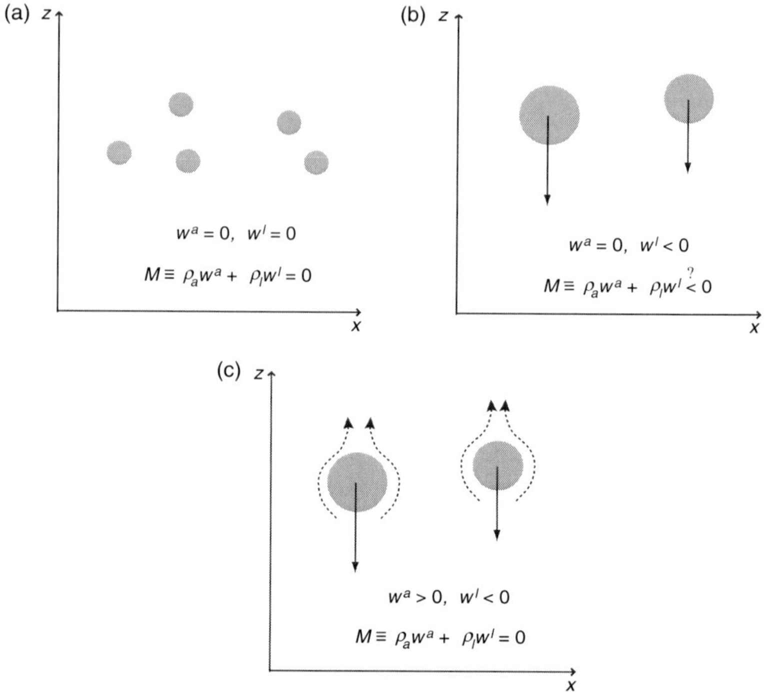


**FIGURE 2.2 Illustration of momentum conservation as condensate begins to fall.** (a) Air motion and condensate motion are both zero; total momentum ($M$) is zero. (b) Condensate then collects into larger drops that begin to fall; if air velocity remains zero, then $M$ is is less than zero, and thus total momentum is not conserved. (c) Air velocity (dashed lines) becomes positive; $M$ is zero, and thus total momentum is conserved.

1-10 cm s$^{-1}$ may be incurred by this process; this is comparable to vertical motions associated with synoptic-scale weather systems.

A conservation equation for total momentum, $M_i \equiv \rho_a u_i^a + \rho_v u_i^v + \rho_l u_i^l + \rho_i u_i^i = \rho u_i^a + \rho_l v_i^l + \rho_i v_i^i$, is given by:

$$\frac{\partial M_i}{\partial t} + \frac{\partial \left(M_i u_j^a\right)}{\partial x_j} + \frac{\partial p}{\partial x_i} - \varepsilon_{ijk} M_j f_k + \delta_{i3}\rho g - \frac{\partial \tau_{ij}}{\partial x_j} = -\frac{\partial \left(\rho_l u_i^l v_j^l\right)}{\partial x_j} - \frac{\partial \left(\rho_i u_i^i v_j^i\right)}{\partial x_j} \tag{2.19}$$

where $f_k$ is the Coriolis vector, $g$ is acceleration due to gravity, and $\tau_{ij}$ is the viscous stress tensor. This equation has slightly different forms in the literature (e.g. Ooyama, 2001; Bannon, 2002; Satoh, 2003) depending on how the momentum of condensate is treated.

Solving for the velocity of air, a governing equation for $u^a$ can be written as follows:

$$\frac{\mathrm{D}u_i^a}{\mathrm{D}t} + \frac{1}{\rho}\frac{\partial p}{\partial x_i} - \varepsilon_{ijk}u_j^a f_k + \delta_{i3}g - \frac{1}{\rho}\frac{\partial \tau_{ij}}{\partial x_j} = W_i, \tag{2.20}$$

where $W_i$ represents changes in momentum due to moisture effects (including that illustrated in Fig. 2.2). Most meteorologists omit $W_i$ from (2.20), assuming implicitly that air velocity is not significantly affected by phase changes or by differential fallout of condensate. Some simple calculations and numerical simulations by Satoh (2003) and Hausman et al. (2006) seem to support this assumption. The neglect of these effects may be acceptable for most short-term simulations, but it may be more problematic for long-term simulations. Further studies are needed to fully understand these effects.

For the remainder of this chapter, condensate-air momentum transfer is not discussed further. More information is available in specialty texts such as Crowe et al. (1998). The momentum equation thus reduces to

$$\frac{\mathrm{D}u_i^a}{\mathrm{D}t} + \frac{1}{\rho}\frac{\partial p}{\partial x_i} - \varepsilon_{ijk}u_j^a f_k + \delta_{i3}g - \frac{1}{\rho}\frac{\partial \tau_{ij}}{\partial x_j} = 0. \tag{2.21}$$

The viscous stress tensor is

$$\tau_{ij} = \mu\left[\left(\frac{\partial u_i}{\partial x_j} + \frac{\partial u_j}{\partial x_i}\right) - \frac{2}{3}\delta_{ij}\frac{\partial u_k}{\partial x_k}\right] \tag{2.22}$$

where $\mu$ is the dynamic viscosity. The kinematic viscosity $\nu$ is related by $\mu = \rho\nu$. The viscous term in (2.21) is only significant at small scales in the atmosphere (of order 1 mm). Thus, it is neglected in many numerical models of the atmosphere. When the viscous term is considered important, the small-scale flow is usually considered to be incompressible [i.e. $\partial u_i^a/\partial x_i = 0$] and $\nu$ is typically considered constant. Further, local variations in density are assumed to be negligible, and thus $\rho$ may be replaced with a constant value in the viscous term. Under these assumptions, the final term on the left-hand side of (2.21) may be replaced by

$$-\nu\frac{\partial^2 u_i}{\partial x_j^2} \tag{2.23}$$

which is the form of Fickian diffusion.

Sometimes it becomes convenient to write the pressure gradient term of (2.21) in terms of $\pi$. Making use of the ideal gas law, then (2.21) can be rewritten without approximation as

$$\frac{\mathrm{D}u_i^a}{\mathrm{D}t} + c_{pa}\theta_\rho\frac{\partial \pi}{\partial x_i} - \varepsilon_{ijk}u_j^a f_k + \delta_{i3}g - \frac{1}{\rho}\frac{\partial \tau_{ij}}{\partial x_j} = 0. \tag{2.24}$$

## 2.6. CONSERVATION OF INTERNAL ENERGY

The thermodynamics of moist air is a complex subject. There are several textbooks devoted entirely to atmospheric thermodynamics. Furthermore, thermodynamic equations in atmospheric research can vary greatly from one study to the next, depending on the level of complexity (i.e. the degree of accuracy) that is needed.

The treatment here will begin with some basic rules. The specific heats of dry air and water vapor are related to gas constants as follows:

$$c_{pa} - c_{va} = R_a, \qquad c_{pv} - c_{vv} = R_v, \tag{2.25}$$

where $c_{pv}$ and $c_{vv}$ are the specific heats of water vapor at constant pressure and volume, respectively. Although the specific heats and gas constants vary slightly under normal tropospheric conditions, we shall assume them to be constant herein. Typical values are listed in Table 2.1. The specific heats of condensate—$c_l$ for liquid and $c_i$ for ice—are proportional to temperature: they vary by 10-20% under normal tropospheric conditions. As discussed below, they are sometimes assumed to be approximately constant.

The change in internal energy of air during a phase change of water, assuming constant pressure, is called the enthalpy. More commonly referred to as "latent heat," it will be denoted herein by $L_{ab}$ for a-to-b phase changes. That is, $L_{lv}$ is a liquid-to-vapor phase change (or vice versa), and is commonly called the latent heat of vaporization. Similarly, $L_{iv}$ is the latent heat of sublimation, and $L_{il}$ is the latent heat of freezing/melting. Conservation of internal energy requires that

$$L_{iv} = L_{il} + L_{lv}. \tag{2.26}$$

The latent heats are functions of temperature only, and are related to specific heats by Kirchoff's relations:

$$\begin{aligned} \left.\frac{\mathrm{d}L_{iv}}{\mathrm{d}T}\right|_p &= c_{pv} - c_i(T), \qquad \left.\frac{\mathrm{d}L_{il}}{\mathrm{d}T}\right|_p = c_l(T) - c_i(T). \\ \left.\frac{\mathrm{d}L_{lv}}{\mathrm{d}T}\right|_p &= c_{pv} - c_l(T). \end{aligned} \tag{2.27}$$

If $c_l$ and $c_i$ are assumed to be constant (which is common in numerical models), then (2.27) may be easily integrated with respect to temperature and the latent heats become linear functions of temperature. For example, $L_{lv}$ would be given by: $L_{lv}(T) = L_{lv}(T_0) + (c_{pv} - c_l)(T - T_0)$ where $T_0$ is a reference temperature (typically 273.15 K). Reference values of the latent heats at $T_0$ are listed in Table 2.1.

If the temperature dependence of $c_l$ and $c_i$ is retained, then further information is required to determine numerical values of the latent heats. Most

**TABLE 2.1** Values of some thermodynamic variables. For variables that are strong functions of temperature, values are listed at $T = T_0$ only

| Symbol | Description | Value |
|---|---|---|
| $c_i$ | Specific heat of ice | $2106 \text{ J kg}^{-1} \text{ K}^{-1}$ (at $T = T_0$) |
| $c_l$ | Specific heat of liquid water | $4218 \text{ J kg}^{-1} \text{ K}^{-1}$ (at $T = T_0$) |
| $c_{pa}$ | Specific heat of dry air at constant pressure | $1005 \text{ J kg}^{-1} \text{ K}^{-1}$ |
| $c_{pv}$ | Specific heat of water vapor at constant pressure | $1850 \text{ J kg}^{-1} \text{ K}^{-1}$ |
| $c_{va}$ | Specific heat of dry air at constant volume | $718 \text{ J kg}^{-1} \text{ K}^{-1}$ |
| $c_{vv}$ | Specific heat of water vapor at constant volume | $1390 \text{ J kg}^{-1} \text{ K}^{-1}$ |
| $e_{si}$ | Saturation vapor pressure with respect to ice | 610.6 Pa (at $T = T_0$) |
| $e_{sl}$ | Saturation vapor pressure with respect to water | 610.7 Pa (at $T = T_0$) |
| $L_{il}$ | Latent heat of freezing | $0.334 \times 10^6 \text{ J kg}^{-1}$ (at $T = T_0$) |
| $L_{iv}$ | Latent heat of sublimation | $2.501 \times 10^6 \text{ J kg}^{-1}$ (at $T = T_0$) |
| $L_{lv}$ | Latent heat of vaporization | $2.835 \times 10^6 \text{ J kg}^{-1}$ (at $T = T_0$) |
| $R_a$ | Gas constant for dry air | $287.05 \text{ J kg}^{-1} \text{ K}^{-1}$ |
| $R_v$ | Gas constant for water vapor | $461.51 \text{ J kg}^{-1} \text{ K}^{-1}$ |
| $T_0$ | Reference temperature | 273.15 K |
| $\epsilon$ | $\epsilon \equiv R_a/R_v$ | 0.622 |

**Source:** Iribarne and Godson (1981)

convenient for this purpose are the Clausius-Clapeyron equations:

$$\frac{\mathrm{d} \ln e_{sl}}{\mathrm{d}T} = \frac{L_{lv}}{R_v T^2}, \qquad \frac{\mathrm{d} \ln e_{si}}{\mathrm{d}T} = \frac{L_{iv}}{R_v T^2}, \tag{2.28}$$

where $e_{sl}(T)$ and $e_{si}(T)$ are, respectively, the saturation vapor pressures with respect to liquid and ice (that is, the pressure that water vapor has at a state of saturated equilibrium with respect to a plane surface of liquid or ice). Values of $e_{sl}$, $e_{si}$ have been determined experimentally (e.g. Wexler, 1976, 1977), which can be used to determine $L_{lv}$ and $L_{iv}$ using (2.28). Consistent values for $c_l$ and $c_i$ can then be determined using (2.27). [See, for example, Ooyama (1990).]

A governing equation for the total internal energy of cloudy air ($E$) can be expressed as

$$\frac{\partial E}{\partial t} + \frac{\partial}{\partial x_j}\left(E u_j^a + \rho_l e_l v_j^l + \rho_i e_i v_j^i\right) + p\frac{\partial u_j^a}{\partial x_j} - \rho_a Q_r = \tau_{ij}\frac{\partial u_i^a}{\partial x_j} + Q_{v_t} + \frac{\partial}{\partial x_j}\left(\kappa\frac{\partial T}{\partial x_j}\right) \quad (2.29)$$

where $E \equiv \rho_a e_a + \rho_v e_v + \rho_l e_l + \rho_i e_i$ is the sum of the internal energies of dry air, water vapor, liquid water, and ice, and $Q_r$ represents radiative tendencies. Terms on the right-hand side of (2.29) are usually neglected in numerical weather prediction models for various reasons, which are discussed below.

The first term on the right-hand side of (2.29) is the dissipative heating term, i.e. the increase in internal energy associated with the dissipation of kinetic energy. This term tends to be small at low wind speeds ($< 40$ m s$^{-1}$) and is generally considered to be negligible over short time scales ($< 1$ day). Hence, it has typically been neglected in numerical cloud models. However, this effect is significant near the surface in tropical cyclones, and can increase tropical cyclone intensity by 10-20% (e.g. Bister and Emanuel, 1998). Furthermore, this term should be included in long-term (climate) simulations so that total energy can be conserved.

The second term on the right-hand side of (2.29) represents the tendencies to internal energy from hydrometeor/air interaction processes as condensate falls at terminal velocity. For example, dissipative heating occurs around condensate due to micro-scale frictional drag effects. The magnitude of this term has been argued to be significant in some cases (e.g. Pauluis et al., 2000; Bannon, 2002), although this claim has been disputed (Rennó, 2001). Nevertheless, a proper numerical investigation has yet to be conducted. The effect is typically excluded from numerical models because of uncertainty about how exactly to formulate this term.

The final term on the right-hand side of (2.29) is the heat conduction term, where $\kappa$ is the thermal conductivity. This effect is only important at small scales (of order 1 mm) and hence is usually neglected in numerical weather prediction models.

To form an equation for air temperature, (2.29) is modified by rearranging the first two terms on the left-hand side. Utilizing the mass-conservation equations, (2.11)–(2.14), and the definition of $E$, we have, without approximation,

$$\frac{\mathrm{D}e_a}{\mathrm{D}t} + r_v\frac{\mathrm{D}e_v}{\mathrm{D}t} + r_l\frac{\mathrm{D}e_l}{\mathrm{D}t} + r_i\frac{\mathrm{D}e_i}{\mathrm{D}t} + r_l v_j^l\frac{\partial e_l}{\partial x_j} + r_i v_j^i\frac{\partial e_i}{\partial x_j} + (e_l - e_v)\dot{r}_{lv} + (e_i - e_v)\dot{r}_{iv} + (e_i - e_l)\dot{r}_{il} = -\frac{p}{\rho_a}\frac{\partial u_j^a}{\partial x_j} + Q_r + \dot{Q}, \quad (2.30)$$

where $\dot{Q}$ represents the entire right-hand side of (2.29) divided by $\rho_a$. By definition, the specific heats at constant volume are related to the changes in internal energy as follows:

$$\frac{\mathrm{d}e_a}{\mathrm{d}T} = c_{va}, \qquad \frac{\mathrm{d}e_v}{\mathrm{d}T} = c_{vv}, \qquad \frac{\mathrm{d}e_l}{\mathrm{d}T_l} = c_l(T_l), \qquad \frac{\mathrm{d}e_i}{\mathrm{d}T_i} = c_i(T_i). \tag{2.31}$$

Thus, (2.30) can be rewritten as an equation for air temperature,

$$(c_{va} + c_{vv}r_v + c_l r_l + c_i r_i)\frac{\mathrm{D}T}{\mathrm{D}t} = -\frac{p}{\rho_a}\frac{\partial u_j^a}{\partial x_j} + Q_r + \dot{Q} + (L_{lv} - R_v T)\dot{r}_{lv} + (L_{iv} - R_v T)\dot{r}_{iv} + L_{il}\dot{r}_{il} - r_l v_j^l \frac{\partial e_l}{\partial x_j} - r_i v_j^i \frac{\partial e_i}{\partial x_j}, \tag{2.32}$$

where it has been assumed that the air temperature ($T$) and the hydrometeor temperatures ($T_l$ and $T_i$) are always the same, and where we have used the relations

$$e_v - e_l = L_{lv} - R_v T, \qquad e_v - e_i = L_{iv} - R_v T, \qquad e_l - e_i = L_{iv}, \tag{2.33}$$

which follow from (2.25), (2.27), and (2.31). The assumption that $T = T_l = T_i$ is made for simplicity, but also because numerical model timesteps are typically greater than the timescale of hydrometeor-air temperature diffusion. Bannon (2002) has noted that this assumption fails for large condensate (diameter of order 1 cm).

The final two terms on the right-hand side of (2.32) represent the transport of internal energy by condensate that falls relative to air. They can be rewritten approximately as

$$\begin{aligned} W_T &\equiv -r_l v_j^l \frac{\partial e_l}{\partial x_j} - r_i v_j^i \frac{\partial e_i}{\partial x_j} \\ &\cong +r_l v_{tl}\frac{\partial (c_l T)}{\partial z} + r_i v_{ti}\frac{\partial (c_i T)}{\partial z} \end{aligned} \tag{2.34}$$

where $v_{tl}$ and $v_{ti}$ are terminal fall velocities (which are positive downward by convention), and where we again assume that hydrometeor temperature instantly equilibrates with air temperature. Note that these terms are negligible if the terminal fall velocities are negligible (i.e. if the condensate is not falling relative to air). Furthermore, although it may not be obvious, these terms are also negligible if terminal fall velocities are very large (of order 10 m s$^{-1}$); this is because hydrometeor content (i.e. $r_l$) is necessarily small for very large fall velocities (see the pseudoadiabatic approximation, discussed in Section 2.10.2). For intermediate values of $v_{tl}$ (of order 3 m s$^{-1}$), typical values in (2.34) reveal that this effect can cause *cooling* of several K h$^{-1}$. Hence, these terms should generally not be neglected in numerical simulations, although their neglect is very common (probably because of the added expense of calculating this term, and/or because the term is thought to be negligible).

A simplified thermodynamic equation shall prove useful later for developing moist thermodynamical relations. The equation is derived by setting $Q_r = \dot{Q} = W_T = 0$. Using (2.32), the mass continuity equations, and the equation of state, then we find

$$c_{pm}\frac{\mathrm{D}\ln T}{\mathrm{D}t} - R_m\frac{\mathrm{D}\ln p}{\mathrm{D}t} + \frac{L_{lv}}{T}\frac{\mathrm{D}r_v}{\mathrm{D}t} - \frac{L_{il}}{T}\frac{\mathrm{D}r_i}{\mathrm{D}t} = 0 \tag{2.35}$$

where $c_{pm} \equiv c_{pa}+c_{pv}r_v+c_l r_l+c_i r_i$ is the specific heat of cloudy air at constant pressure, and $R_m \equiv R_a + R_v r_v$. Note from (2.25) that $c_{pm} = c_{vm} + R_m$ where $c_{vm} \equiv c_{va} + c_{vv}r_v + c_l r_l + c_i r_i$ is the specific heat of cloudy air at constant volume.

## 2.7. EQUATION FOR PRESSURE

Some numerical models of deep moist convection use a prognostic equation for pressure. It is derived by using the equation of state, the mass continuity equation, and the internal energy equation. The governing equation is

$$\begin{aligned}\frac{\mathrm{D}p}{\mathrm{D}t} &= -\frac{c_{pm}}{c_{vm}}p\frac{\partial u_j^a}{\partial x_j} + \frac{p}{c_{vm}T}\left(Q_r + \dot{Q} + W_T\right) + \frac{p}{c_{vm}T}\left(L_{lv}\dot{r}_{lv}\right.\\ &\left.\quad + L_{iv}\dot{r}_{iv} + L_{il}\dot{r}_{il}\right) - \frac{R_v c_{pm}}{R_m c_{vm}}p\left(L_{lv}\dot{r}_{lv} + L_{iv}\dot{r}_{iv}\right).\end{aligned} \tag{2.36}$$

A governing equation for nondimensional pressure ($\pi$) is given simply by

$$\frac{\mathrm{D}\pi}{\mathrm{D}t} = \frac{R_a}{c_{pa}}\frac{\pi}{p}\frac{\mathrm{D}p}{\mathrm{D}t}. \tag{2.37}$$

The use of a pressure equation in nonhydrostatic numerical models has been very common in the atmospheric sciences, primarily because pressure is needed to calculate pressure gradient terms in the velocity equations. Recently, the need to enforce mass conservation has led to the development of efficient techniques that can use the compressible mass continuity equation, (2.11), instead of a pressure equation (e.g. Klemp et al., 2007).

## 2.8. EQUATION FOR POTENTIAL TEMPERATURE

Atmospheric numerical models commonly use $\theta$ as a predictive variable, rather than $T$. Using (2.32), the definition of $\theta$, and the equation of state, then the governing equation for $\theta$ is found to be

$$\frac{\mathrm{D}\theta}{\mathrm{D}t} = -\theta\left(\frac{R_m}{c_{vm}} - \frac{R_a c_{pm}}{c_{pa}c_{vm}}\right)\frac{\partial u_j^a}{\partial x_j} + \frac{c_{va}}{c_{vm}c_{pa}\pi}\left(Q_r + \dot{Q} + W_T\right)$$

$$
\begin{aligned}
&+ \frac{c_{va}}{c_{vm} c_{pa} \pi} \left( L_{lv} \dot{r}_{lv} + L_{iv} \dot{r}_{iv} + L_{il} \dot{r}_{il} \right) \\
&- \theta \frac{R_v}{c_{vm}} \left( 1 - \frac{R_a c_{pm}}{R_m c_{pa}} \right) \left( \dot{r}_{lv} + \dot{r}_{iv} \right)
\end{aligned} \tag{2.38}
$$

(Bryan and Fritsch, 2002).

## 2.9. APPROXIMATE EQUATIONS

Rigorously derived conservation equations for mass, momentum, and energy are rarely used in atmospheric sciences. For various reasons, approximate equations sets are used instead. For example, when computing power was less advanced, approximate equations were developed to reduce computational expense (e.g. Wilhelmson, 1977; Tripoli and Cotton, 1981). In some applications, terms are unimportant and are therefore neglected (e.g. dissipative heating in short-term simulations). Sometimes an effect is neglected because it is poorly understood (e.g. viscous heating around falling water drops). In other cases, approximate equations are excellent proxies for the complete equations, and may sometimes eliminate effects that are not directly important in atmospheric flows (e.g. acoustic waves). The following sections present some commonly used approximate equations.

### 2.9.1. Approximate Equations for $\theta$ and $p$

The most common approximation in atmospheric numerical models is to neglect all terms with $v_i^l$ and $v_i^i$ (i.e. the differential fall velocities) except when appearing in mass continuity equations. The neglect of hydrometeor sedimentation terms from the velocity equation (i.e. the $W_i$ term) seems to be a good assumption (e.g. Satoh, 2003; Hausman et al., 2006). Thus, recent debate (e.g. Ooyama, 2001; Bannon, 2002) about the exact form of the momentum equation for a precipitating atmosphere may be moot. On the other hand, the hydrometeor sedimentation terms in the internal energy equation ($W_T$) are not always negligible. For example, Hausman et al. (2006) reported a change in hurricane structure and intensity when a term analogous to $W_T$ was included in their model. The neglect of $\dot{Q}$ is also very common in cloud models.

Because (2.38) is so complex, it is usually simplified further. By setting $c_{pv} = c_{vv} = c_l = c_i = R_v = 0$ in (2.38), the most common approximate equation for $\theta$ is obtained:

$$
\frac{D\theta}{Dt} = \frac{1}{c_{pa} \pi} \left( L_{lv} \dot{r}_{lv} + L_{iv} \dot{r}_{iv} + L_{il} \dot{r}_{il} \right) + \frac{1}{c_{pa} \pi} Q_r . \tag{2.39}
$$

To conserve mass, the pressure equation consistent with (2.39) is

$$
\frac{Dp}{Dt} = - \frac{c_{pa}}{c_{va}} p \frac{\partial u_i^a}{\partial x_i} + \frac{p}{c_{va} T} \left( L_{lv} \dot{r}_{lv} + L_{iv} \dot{r}_{iv} + L_{il} \dot{r}_{il} \right) + \frac{p}{c_{va} T} Q_r . \tag{2.40}
$$

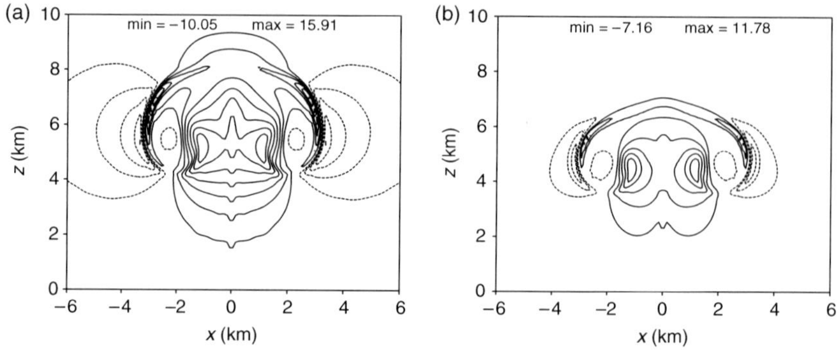


**FIGURE 2.3 Vertical velocity of air ($w^a$) from simulations of a rising warm bubble in a saturated environment [following Bryan and Fritsch (2002)] using (a) Eq. (2.38) and (b) Eq. (2.39).** Contour interval is 2 m s$^{-1}$. Negative contours are dashed. The zero contour is excluded.

The primary assumption inherent in (2.39)–(2.40) is that the heat content of water is negligible. Numerical integrations with a simple one-dimensional model show this assumption to be appropriate for shallow clouds (Bryan and Fritsch, 2004). However, for deep convection, integrations with (2.39) have a pronounced cool bias compared to integrations with (2.38), which leads to weaker vertical velocity in updrafts, as shown in Fig. 2.3. This artificial cooling in turn leads to less rainfall (Bryan and Fritsch, 2002) and weaker tropical cyclone intensity (Bryan and Rotunno, 2009).

If conservation of mass is not crucial in a numerical simulation, then (2.40) can be simplified further. Specifically, terms associated with advection, phase changes of water, and radiation can be reasonably neglected (Klemp and Wilhelmson, 1978). Noting also that the speed of sound ($c_s$) is given by ${c_s}^2 = \left(c_{pa}/c_{va}\right) R_a T$, the following approximate equation can be derived

$$\frac{\partial p}{\partial t} = -{c_s}^2 \rho_a \frac{\partial u_i^a}{\partial x_i}. \tag{2.41}$$

For short-term simulations of a few hours, results using (2.41) can be amazingly similar to results using (2.36). Klemp and Wilhelmson (1978) suggested that *gradients* in $p$ can be accurately simulated using approximate equations such as (2.41), and that the inclusion of advection and diabatic terms acts merely to shift the mean pressure inside a model's domain by a small amount for short-term simulations.

Under typical tropospheric conditions, $c_s \approx 330$ m s$^{-1}$. Some researchers have found that $c_s$ can be artificially lowered in (2.41), thereby reducing the propagation speed of simulated acoustic waves. This allows a model user to increase the timestep, thereby reducing computational cost. In practice, values of $c_s$ in (2.41) should be greater than $\sim$ 100 m s$^{-1}$, as it seems to be important for acoustic waves to propagate significantly faster than the atmospheric flow of interest (e.g. thunderstorm updrafts, the jet stream, etc.).

Numerical models cannot conserve mass if they use a pressure equation that neglects diabatic terms. Some studies have demonstrated a significant error in some quantitative model output (such as precipitation) for numerical models that do not conserve mass (e.g. Qiu et al., 1993; Lackmann and Yablonksy, 2004). The effect is most pronounced for strongly precipitating weather systems such as tropical cyclones and mesoscale convective systems. However, other studies have found mass conservation to have a very small effect in nonhydrostatic models (e.g. Bryan and Fritsch, 2002; Bryan and Rotunno, 2009). The different conclusions in these studies may be attributable to hydrostatic versus nonhydrostatic processes; that is, the larger spatial and temporal scales of hydrostatic phenomena may make them more susceptible to the errors inherent in approximate equation sets.

## 2.9.2. Anelastic Equations

To circumvent the need for special numerical techniques that account for acoustic waves, it is instead possible to use an approximate equation set that filters acoustic modes. The anelastic equations were developed for this purpose. Originally presented by Ogura and Phillips (1962), the equation set has since been refined by Dutton and Fichtl (1969), Lipps and Hemler (1982), and Bannon (1996). The derivation begins by assuming a fixed, horizontally homogeneous reference state (or "base state") that is representative of the mean environment around clouds. This reference state is denoted here by subscript "0" and perturbations from this reference state are denoted with superscript primes; that is, $\alpha(x, y, z, t) = \alpha_0(z) + \alpha'(x, y, z, t)$, where $\alpha$ represents a state variable (e.g. density, pressure, temperature, etc.). The reference state is assumed to obey the equation of state, $p_0 = \rho_0 R_a T_0$, and it is in hydrostatic balance, $\mathrm{d}p_0/\mathrm{d}z = -\rho_0 g$. Perturbations from the reference state are assumed to be negligibly small compared to the reference state at any height; that is, $\alpha'/\alpha_0 \ll 1$. Consequently, perturbations can be reasonably neglected in the velocity and mass-continuity equations. One important exception is that perturbations should not be neglected when multiplied by gravity, as the buoyancy force is considered to be an important term in the vertical momentum equation that should not be approximated. Under these assumptions and conditions, the mass continuity equation becomes

$$\frac{\partial}{\partial x_i}\left(\rho_0 u_i^a\right) = 0, \tag{2.42}$$

the velocity equation can be written as

$$\frac{\mathrm{D}u_i^a}{\mathrm{D}t} = -\frac{\partial \phi}{\partial x_i} + \delta_{i3} g\left(\frac{\theta_v{}'}{\theta_0} - r_l - r_i\right) + \varepsilon_{ijk} u_j^a f_k + \frac{1}{\rho_0}\frac{\partial \tau_{ij}}{\partial x_j} \tag{2.43}$$

where $\phi \equiv p'/\rho_0$ is the anelastic pressure perturbation, and here $\tau_{ij} = \rho_0 \nu (\partial u_i/\partial x_j + \partial u_j/\partial x_i)$. [See Bannon (1996) for details.] The internal energy equation is typically expressed in terms of potential temperature,

$$\frac{\mathrm{D}\theta}{\mathrm{D}t} = \frac{1}{c_{pa}\pi}\left(L_{lv}\dot{r}_{lv} + L_{iv}\dot{r}_{iv} + L_{il}\dot{r}_{il} + Q_r\right), \tag{2.44}$$

and the equation of state is given by

$$\frac{p'}{p_0} = \frac{\rho'}{\rho_0} + \frac{T'}{T_0} + \frac{r_v}{\epsilon} \tag{2.45}$$

where the base-state is assumed dry ($r_{v0} = 0$). Furthermore, the constraint

$$\frac{\theta_v{}'}{\theta_0} = \frac{p'}{\rho_0 g H_\rho} - \frac{\rho'}{\rho_0} - r_v \tag{2.46}$$

is needed to guarantee conservation of mass and energy (Dutton and Fichtl, 1969; Bannon, 2002), where $H_\rho \equiv (-\mathrm{d}\ln\rho_0/\mathrm{d}z)^{-1} \approx 8$ km is the density scale height. In this system, anelastic pressure ($\phi$) is a *diagnostic* variable that needs to be determined consistently with the governing equations (2.42)–(2.43). Re-writing the velocity equation as $\partial u_i^a/\partial t = F_i$ (where $F_i$ includes all terms on the right-hand side of (2.43) plus advection terms), then using (2.42), and then rearranging yields the elliptic equation

$$\frac{\partial}{\partial x_i}\left(\rho_0 \frac{\partial \phi}{\partial x_i}\right) = \frac{\partial}{\partial x_i}\left(\rho_0 F_i\right). \tag{2.47}$$

A solution for $\phi$ can be obtained from (2.47) provided that the right-hand side of (2.47) is known and given suitable boundary conditions. As is typical with elliptic equations, $\phi$ can only be determined from (2.47) to within an unknown constant. This is usually not a problem, as only gradients of $\phi$ are needed in (2.43). However, in cloudy conditions temperature is needed for microphysical calculations. Bannon et al. (2006) showed that mass conservation can be used as a constraint to determine $\phi$ uniquely, and thus $T$ can be obtained using (2.45) and the definition of $\phi$.

### 2.9.3. Incompressible Equations

If vertical displacements are of order 1 km or less, then the incompressible equations can be used. They are similar to the anelastic equations except the reference density is assumed to be constant, and thus the mass continuity equation reduces to

$$\partial u_i^a/\partial x_i = 0. \tag{2.48}$$

See Bannon (1996) for further details. The incompressible equations may be used to study shallow clouds, and hence are used in some large eddy simulation (LES) models.

### 2.9.4. Pseudo-incompressible Equations

As an alternative to the anelastic equations, Durran (1989) suggested replacing the anelastic mass continuity equation (2.42) with

$$\frac{\partial}{\partial x_i}\left(\rho_0\theta_0 u_i^a\right) = \frac{H}{c_{pa}\pi_0}, \tag{2.49}$$

where $H$ represents the heating rate (e.g. from phase changes of water, radiation, etc.). Unlike the anelastic equations, in Durran's system the pressure-gradient term in the velocity equations remains unmodified, i.e. they have the form $(1/\rho)\partial p/\partial x_i$ if pressure is used or $(c_{pa}\theta_\rho)\partial\pi/\partial x_i$ if nondimensional pressure is used. The primary advantage of this system is that baroclinic effects are represented accurately (because the pressure-gradient term is unmodified), and hence this system performs better than the anelastic system when $\theta'/\theta_0$ is relatively large (of order 0.1).

## 2.10. DERIVED THERMODYNAMIC VARIABLES

By applying some simple constraints it is possible to derive variables that are constant under moist adiabatic conditions. The following assumptions must be made: viscous effects and turbulent mixing are neglected; radiative effects are neglected; saturated equilibrium is imposed (that is, supersaturation is not permitted); and the ice phase is typically neglected. This may seem like a restrictive list of assumptions. Nevertheless, experience shows the following variables to be quite useful.

Additionally, an important assumption must be made about the fate of condensate. One of two extreme assumptions can be made: (1) reversible thermodynamics, in which all condensate moves with the air at all times; or (2) pseudoadiabatic thermodynamics, in which all condensate is immediately removed from air by fallout upon formation. These two cases are treated separately in the next two sections.

### 2.10.1. Reversible Thermodynamics

The reversible thermodynamics assumption is most applicable to shallow clouds and to cumulus congestus before precipitation has developed. Total water is conserved following air motion, and thus $\mathrm{D}r_t/\mathrm{D}t = 0$, where $r_t = r_v + r_l$ is the total water mixing ratio. There are several exact solutions for reversible thermodynamics (see, e.g. Emanuel, 1994). The most fundamental of these variables are enthalpy (typically represented by $k$, although sometimes by $h$) and entropy (typically represented by $s$). They can be derived easily from a small set of fundamental thermodynamics principles (see, e.g. Bohren and Albrecht, 1998).

Neglecting the ice phase, moist enthalpy ($k$) is derived from (2.35) by assuming that pressure is held constant. We find that

$$k = \left(c_{pa} + c_l r_t\right) T + L_{lv} r_v \tag{2.50}$$

is a conserved variable. Enthalpy cannot be used in models of deep convection because of the assumption of constant pressure. Enthalpy is most often used in studies of air-sea interaction (e.g. water vapor flux from the ocean to the atmosphere).

More commonly used is total moist entropy

$$s = \left(c_{pa} + c_l r_t\right) \ln T - R_a \ln p_d + \frac{L_{lv} r_v}{T} - R_v r_v \ln(\mathcal{H}) \tag{2.51}$$

where $p_d$ is the partial pressure of dry air and $\mathcal{H} \equiv e/e_{sl}$ is relative humidity. Technically, $s$ is known only to an arbitrary constant, because it is derived via integration of a differential equation for moist thermodynamical processes. In practical applications this constant is essentially zero, which implicitly assumes that the integration is performed from a very low temperature and pressure. Few assumptions are required to derive $s$. Therefore, the use of entropy as a predictive variable for numerical models has been advocated passionately by some researchers (e.g. Hauf and Höller, 1987; Ooyama, 1990). It is more commonly used as a diagnostic tool for numerical model output, because it is conserved along trajectories (under the assumptions listed above).

The most commonly used moist thermodynamic variable in meteorology is the equivalent potential temperature

$$\theta_e = T \left(\frac{p_{00}}{p_d}\right)^{R_a/(c_{pa}+c_l r_t)} (\mathcal{H})^{-R_v r_v/(c_{pa}+c_l r_t)} \exp\left[\frac{L_{lv} r_v}{\left(c_{pa} + c_l r_t\right) T}\right]. \tag{2.52}$$

This variable is actually a different measure of entropy, as it can be shown that $\left(c_{pa} + c_l r_l\right) \ln \theta_e = s + R_a \ln p_{00}$. Conceptually, $\theta_e$ is the potential temperature an air parcel would have if all the water vapor were condensed by lifting the parcel to zero pressure. $\theta_e$ is often used to infer the source region of air parcels in clouds.

It is possible to derive formulations for $s$ and $\theta_e$ that are appropriate for ice microphysical processes. In this case, $r_t = r_v + r_i$, and in (2.51)–(2.52) $L_{lv}$ is replaced by $L_{iv}$, $c_l$ is replaced by $c_i$, and $H$ is replaced by $H_i \equiv e/e_{si}$.

It is possible to use $\theta_e$ as a predictive variable in a numerical model. In this case, the thermodynamic equation would simply have the form $D\theta_e/Dt = 0$. However, this is rarely done, for two primary reasons: (1) temperature needs to be retrieved from (2.52), and the exponential terms in (2.52) are prohibitively expensive; and (2) it is difficult to include ice microphysical processes accurately, and so the solutions are typically less accurate for deep convection. Consequently, $\theta_e$ is primarily used as a diagnostic tool in meteorological research.

Another useful thermodynamic variable is the liquid-water potential temperature

$$\theta_l = T\left(\frac{p_{00}}{p}\right)^{\chi}\left(1-\frac{r_l}{\epsilon + r_t}\right)^{\chi}\left(1-\frac{r_l}{r_t}\right)^{-\gamma}\exp\left[\frac{L_{lv}r_l}{\left(c_{pa}+c_{pv}r_t\right)T}\right] \tag{2.53}$$

where here

$$\chi \equiv \frac{R_a + R_v r_t}{c_{pa}+c_{pv}r_t}, \qquad \gamma \equiv \frac{R_v r_t}{c_{pa}+c_{pv}r_t}. \tag{2.54}$$

Conceptually, $\theta_l$ is the potential temperature an air parcel would have if all liquid water were evaporated. In the absence of liquid water (i.e. if $r_l = 0$), $\theta_l$ is equivalent to $\theta_m$ (moist-air potential temperature), a variable that is conserved in subsaturated conditions.

It is possible to derive formulations of $\theta_l$ that are appropriate for ice microphysical processes. In this case, $r_t = r_v + r_i$, $L_{lv}$ is replaced by $L_{iv}$, and $r_l$ is replaced by $r_i$ in (2.53).

It is important to note that (2.53) is an exact solution. Because of its complex (and therefore expensive) form, it is common to use approximate formulations of $\theta_l$. One example, presented by Betts (1973), is

$$\theta_l \approx \theta\left(1-\frac{L_{lv}(T_0)r_l}{c_{pa}T}\right) \tag{2.55}$$

where $L_{il}(T_0)$ is a constant value of $L_{il}$ valid at $T_0 = 273.15$ K. This formulation of $\theta_l$ becomes equivalent to $\theta$ (dry-air potential temperature) in the absence of liquid water. It is reasonably accurate for small liquid water contents (e.g. shallow clouds) and hence is most often used in large eddy simulation (LES) models.

Several attempts have been made to incorporate both liquid and ice in a formulation of $\theta_l$. Usually referred to as ice-liquid water potential temperature ($\theta_{il}$), they are necessarily approximate formulations. One formulation, derived by Tripoli and Cotton (1981), is used in several cloud models. A review of their formulation and several others was conducted by Bryan and Fritsch (2004).

They found the following formulation to be most accurate:

$$\theta_{il} \approx T\left(\frac{p_{00}}{p}\right)^{\chi}\left(1-\frac{r_l+r_i}{\epsilon+r_t}\right)^{\chi}\left(1-\frac{r_l+r_i}{r_t}\right)^{-\gamma} \times \exp\left[\frac{-L_{lv}r_l - L_{iv}r_i}{\left(c_{pa}+c_{pv}r_t\right)T}\right] \quad (2.56)$$

where, here, $r_t = r_v + r_l + r_i$.

Some numerical models use $\theta_l$ or $\theta_{il}$ as a predictive variable, where in this case the thermodynamic equation has the form $D\theta_l/Dt = 0$. Irreversible processes such as condensate fallout and turbulent mixing must also be considered, but this is usually straightforward (see, e.g. Tripoli and Cotton, 1981). For shallow clouds, it is sufficient to use $\theta_l$, but simulations of deep convection require the use of $\theta_{il}$. These variables are more attractive as a predictive variable than $\theta_e$ because they reduce to sub-saturated variables (e.g. $\theta$) in the absence of condensate. Because clouds typically occupy a small portion of a model's domain, these models only need to activate special procedures to retrieve $T$ from $\theta_l$ on a small fraction of grid points, and hence models that use $\theta_l$ are more efficient than models that use $\theta_e$. However, the necessarily approximate formulation of $\theta_{il}$ has made its use in numerical models less common recently.

It is possible to derive several other variables that are conserved under vertical displacements. One common variable is the ice-liquid water static energy

$$h_{il} = \left(c_{pa}+c_{pv}r_t\right)T - L_{lv}r_l - L_{iv}r_i + (1+r_t)gz \quad (2.57)$$

where, here, $r_t = r_v + r_l + r_i$. One important caveat about $h_{il}$ is that it has been derived by assuming hydrostatic vertical displacements. Thus, its application as a predictive variable in a nonhydrostatic numerical model seems to be questionable. Nevertheless, $h_{il}$ is a useful diagnostic tool in the same manner as $\theta_e$.

A major drawback to the reversible thermodynamic assumption invoked throughout this section is that it does not apply straightforwardly to deep precipitating convection. This is not necessarily a problem for numerical models of convection because irreversible processes such as hydrometeor sedimentation (i.e. fallout) can be incorporated every timestep by simple procedures. However, for diagnostic purposes such as estimating the source region of air parcels, the use of reversible formulations can introduce large errors. A different constraint that is suitable to deep convection is presented next.

### 2.10.2. Pseudoadiabatic Thermodynamics

The opposite extreme assumption about condensate is referred to as pseudoadiabatic thermodynamics. In this case, condensate is immediately removed from air, presumably by fallout. Thus, the liquid or ice water content is always zero ($r_l = r_i = 0$).

Under the pseudoadiabatic assumption, rigorously derived formulas for thermodynamic variables typically have integral terms associated with the entropy of water vapor. For example, moist entropy under the pseudoadiabatic assumption is

$$s^p = c_{pa} \ln T + c_l \int_{T_o}^{T} r_v \, \mathrm{d} \ln(T) - R_a \ln p_d + \frac{L_{lv} r_v}{T} - R_v r_v \ln(\mathcal{H}) \quad (2.58)$$

where $T_o$ here represents a reference temperature for integration, which is usually chosen to be a very low value. The integral terms in exact pseudoadiabatic variables [e.g. second term on the right-hand side of (2.58)] arise because total water is not conserved following a parcel; that is, $r_t = r_v$ everywhere.

To avoid expensive calculation of integral terms, several researchers have derived approximate formulations of quasi-conserved variables. The most accurate of these is the pseudoadiabatic equivalent potential temperature presented by Bolton (1980)

$$\theta_e{}^p = T \left( \frac{p_{00}}{p} \right)^{0.2854(1-0.28 r_v)} \exp \left[ \left( \frac{3376}{T_L} - 2.54 \right) r_v \left( 1 + 0.81 r_v \right) \right] \quad (2.59)$$

where $T_L$ is the absolute temperature at the lifting condensation level. By comparison against exact numerical integration of the governing equations for pseudoadiabatic thermodynamics, Bolton (1980) found (2.59) to be accurate to within $\sim$0.3 K. Because of its accuracy and its relatively simple formulation, (2.59) is very useful for study with observations and numerical model output. By contrast, this formulation has been difficult to apply in some theoretical studies because (2.59) was derived by fitting empirical functions to numerical output. Thus, associated thermodynamic variables such as moist entropy are not available.

An approximate but accurate set of formulas for pseudoadiabatic thermodynamics was presented by Bryan (2008). They are derived by neglecting part of the entropy of water vapor and then compensating for this error by using a constant (but relatively large) value for $L_{lv}$. The final formulas are

$$s^p = c_{pa} \ln T - R_a \ln p_d + \frac{L_0 r_v}{T} - R_v r_v \ln(\mathcal{H}) \quad (2.60)$$

and

$$\theta_e^p = T \left( \frac{p_{00}}{p_d} \right)^{R_a/c_{pa}} (\mathcal{H})^{-R_v r_v / c_{pa}} \exp \left[ \frac{L_0 r_v}{c_{pa} T} \right] \quad (2.61)$$

where $L_0 \cong 2.555 \times 10^6$ J kg$^{-1}$ is an empirically determined constant. These variables are consistent because $c_{pa} \ln \theta_e{}^p = s^p + R_a \ln p_{00}$. Note that the pseudoadiabatic formulas, (2.60) and (2.61), have the same form as their reversible counterparts, (2.51) and (2.52), except $L_{lv}$ is replaced by a constant $L_0$ and all terms with $c_l$ are neglected. By comparison against numerical calculations, Bryan (2008) found that (2.61) produces errors slightly larger than, but comparable to, those from (2.59). The derivation of (2.60)–(2.61) from first principles, and the corresponding first law of thermodynamics, makes them attractive for use in theoretical studies of deep precipitating convection.

### 2.10.3. Other Derived Variables

Other thermodynamic variables that are frequently used in diagnostic studies of cloud processes are the wet-bulb temperature $T_w$ and the wet-bulb potential temperature $\theta_w$. The wet-bulb temperature results when air is cooled isobarically by the evaporation of liquid water. Thus, for an isobaric process in which the ice phase is absent, and ignoring radiative processes and molecular dissipation, we can utilize the moist enthalpy, (2.50), and find that

$$(c_{pa} + c_l r_t)(T - T_w) = L_{lv}(T_w) r_s(T_w) - L_{lv}(T) r_v, \tag{2.62}$$

where $r_s$ is the saturation mixing ratio at $T_w$, and $r_t$ is the mixing ratio of total water (in this case, the sum of $r_v$ and the total amount of condensate needed to saturate the air).

Determination of $T_w$ from (2.62) requires iteration, and thus can be expensive. Consequently, $T_w$ is often evaluated by use of a thermodynamic diagram. Consider the skew $T$-log $p$ diagram illustrated in Fig. 2.4, where the air at, say, 800 mb has a temperature of $+10°$C and a mixing ratio of $\sim$6.9 g kg$^{-1}$. If the air is first lifted to the lifting condensation level (LCL) and we follow the pseudoadiabat to its original pressure (800 mb), the temperature arrived at is called the *wet-bulb* temperature. If we continue following the pseudoadiabat, which passes through $T_w$ until it intersects the 1000-mb level, the temperature we arrive at is $\theta_w$.

Notice that both $\theta_w$ and $\theta_e^p$ are conserved during moist and dry adiabatic processes. In practice, therefore, use of these two variables is interchangeable, and the investigators' selection criteria are generally based upon personal preference.

## 2.11. BUOYANCY

Consider the pressure-gradient and gravitational terms in the vertical velocity equation:

$$\frac{\partial w^a}{\partial t} = -\frac{1}{\rho}\frac{\partial p}{\partial z} - g. \tag{2.63}$$

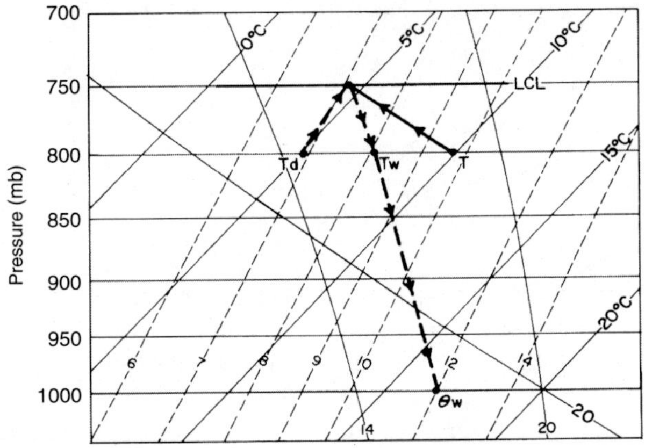


**FIGURE 2.4 The wet-bulb potential temperature is often determined by use of a thermodynamic diagram.** Consider this skew $T$-log $p$ diagram, where the air at, say, 800 mb has a temperature of +10 °C and a mixing ratio of 6.9 g kg$^{-1}$. If the air is first lifted to the lifting condensation level (LCL) and we follow the pseudoadiabat to its original pressure (800 mb), then the temperature obtained is called the *wet-bulb* temperature. If we continue following the pseudoadiabat that passes through $T_w$ until it intersects the 1000-mb level, the temperature we obtain is $\theta_w$.

For practical applications, it is convenient to rewrite pressure in terms of a horizontally homogeneous reference profile that is in hydrostatic balance. Using the same notation as in Section 2.9.2, then (2.63) can be rewritten without approximation as

$$\frac{\partial w^a}{\partial t} = -\frac{1}{\rho}\frac{\partial p'}{\partial z} - g\frac{\rho'}{\rho}. \tag{2.64}$$

The final term in (2.64) is the buoyancy term ($B$). If the reference profile is characteristic of the environment around a cloud, and if the pressure gradient term is negligible, then a parcel will accelerate upward if it is less dense than its environment. By linearizing the equation of state and neglecting pressure perturbations, $B$ can be approximated as

$$B = -g\frac{\rho'}{\rho} \approx g\left(\frac{T'}{T_0} + \frac{r_v{}'}{\epsilon}\right). \tag{2.65}$$

Thus, a parcel can accelerate upwards by buoyancy if it is warmer and/or moister than its environment.

It is usually more convenient to work with the vertical velocity equation when the pressure gradient term is written in terms of $\pi$. Without approximation, (2.63) can be re-written as

$$\frac{\partial w^a}{\partial t} = -c_{pa}\theta_\rho\frac{\partial \pi'}{\partial z} + g\frac{\theta_\rho{}'}{\theta_{\rho 0}} \tag{2.66}$$

where $\theta_\rho$ is density potential temperature (see Section 2.2.3). In this case, buoyancy is given by

$$\begin{aligned} B &= g\frac{\theta_\rho{}'}{\theta_{\rho 0}} \\ &\approx g\left(\frac{\theta'}{\theta_0} + 0.61 r_v{}' - r_l - r_i\right), \end{aligned} \tag{2.67}$$

where the approximate form [second line of (2.67)] is derived using the binomial approximation and then neglecting products of perturbations.

In actual clouds, the pressure gradient term is usually not negligible. In fact, pressure gradient accelerations typically oppose buoyancy. To demonstrate how this is so, we take the divergence of the momentum equation and use (2.42) (the anelastic mass continuity equation) to eliminate the time-tendency terms; considering only pressure-gradient and buoyancy terms, we find:

$$\frac{\partial}{\partial x_i}\left(c_{pa}\rho_0\theta_{\rho 0}\frac{\partial \pi'}{\partial x_i}\right) = \frac{\partial}{\partial z}\left(\rho_0 B\right). \tag{2.68}$$

Given a spatial distribution of $B$ and appropriate boundary conditions, (2.68) can be used to solve for $\pi'$. Typically, $\pi'$ thus determined is referred to as the buoyancy pressure perturbation ($\pi_b{}'$), i.e. the portion of the pressure field that is attributable to the distribution of buoyancy. Some examples are shown in Fig. 2.5. Note that $\pi_b{}'$ is positive where $B$ decreases with height and $\pi_b{}'$ is negative where $B$ increases with height. As a consequence, the vertical pressure gradient term [first term on the right side of (2.66)] opposes the acceleration due to buoyancy. The total acceleration attributable to buoyancy can thus be expressed as

$$F_B = -c_{pa}\theta_\rho\frac{\partial \pi_b{}'}{\partial z} + B. \tag{2.69}$$

Examples of $F_B$ are shown in Fig. 2.6. Note that the magnitude of $F_B$ is always less than the magnitude of $B$.

The *distribution* of buoyancy is crucial for determining vertical accelerations. Consider Fig. 2.5b as compared to Fig. 2.5c: $\pi_b{}'$ is larger in magnitude as the buoyancy field becomes wider (all else remaining equal). As shown by comparing Fig. 2.6b and Fig. 2.6c, the total buoyant acceleration thus decreases. This relationship between $F_B$ and the distribution of $B$ is ultimatelyattributable to the mass conservation equation, which was used to derive (2.68). There is simply more mass to move around when $B$ distributions are wider. Note also from Fig. 2.6 that $F_B$ can be negative in some places even when $B > 0$ everywhere (or, conversely, $F_B$ can be positive even when $B < 0$ everywhere); this behavior also follows from mass-conservation, as air must be displaced to allow buoyant parcels to move.

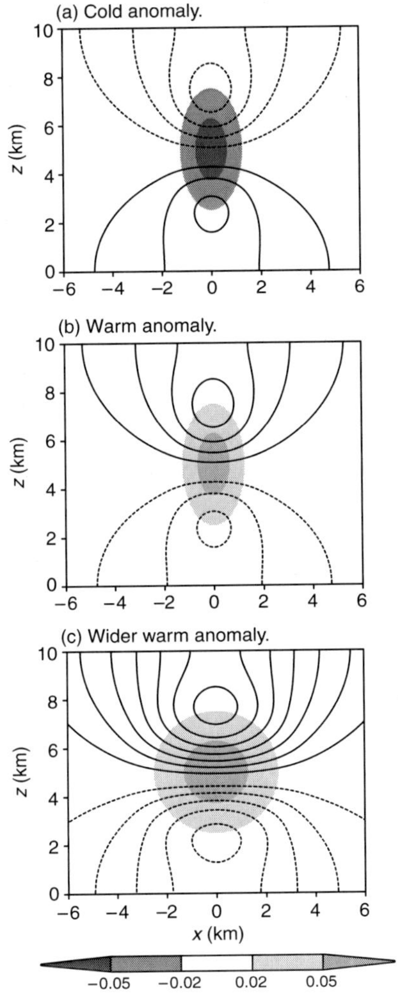


**FIGURE 2.5 Examples of $\pi_b{}'$ (contours) retrieved from specified distributions of $B$ (shading, see legend at bottom).** The contour interval is $3 \times 10^{-5}$ (nondimensional). The zero contour is excluded.

Interesting solutions occur at two extreme limits of buoyancy width. For the case of infinitely thin buoyancy, $\pi_b{}' \rightarrow 0$. This case, for which pressure perturbations are negligible, is usually referred to as "parcel theory." In the opposite extreme, for infinitely wide buoyancy the vertical pressure gradient cancels exactly buoyancy. This is the hydrostatic limit, for which the total vertical acceleration of a parcel is negligible (compared to terms in the hydrostatic reference state).

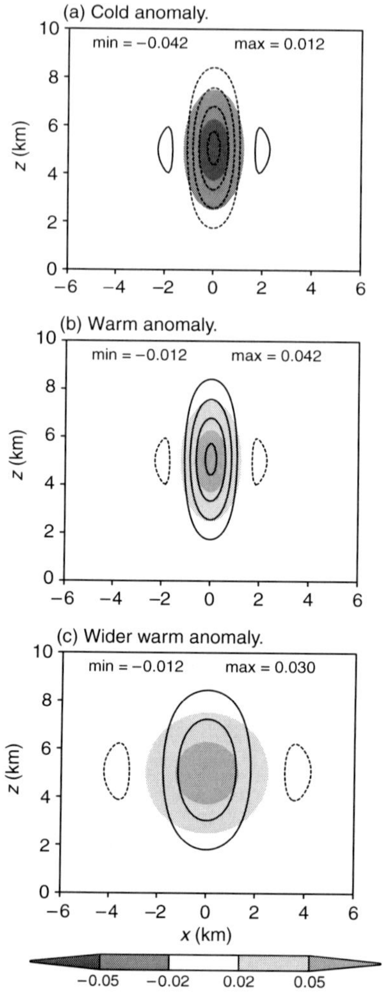


**FIGURE 2.6 Examples of total buoyant acceleration ($F_B$) for specified distributions of $B$ (shading, see legend at bottom).** The contour interval is 0.01 m s$^{-2}$. The zero contour is excluded. Maximum and minimum values of $F_B$ are provided at the top of each panel.

## 2.12. LAPSE RATES AND STABILITY

Using equations developed in preceding sections, we now consider lapse rates and stability in cloudy environments. Consider an infinitely small parcel of air that does not interact with its environment. In the following thought experiments, the parcel is moved upward or downward through an environment having a constant temperature lapse rate,

$$\gamma \equiv -\partial T/\partial z. \tag{2.70}$$

As the parcel moves vertically, it experiences adiabatic expansion or compression, thus cooling or warming accordingly. The rate of cooling of the rising parcel can be evaluated using (2.35) by assuming that radiative heating and molecular dissipation are negligible. If air motions are strictly vertical, then (2.35) can be expressed as

$$c_{pm}\frac{\mathrm{d}\ln T}{\mathrm{d}z} - R_m\frac{\mathrm{d}\ln p}{\mathrm{d}z} + \frac{L_{lv}}{T}\mathrm{d}r_v = 0. \tag{2.71}$$

[Notice that the ice phase is neglected in this section, for simplicity.] It is further assumed that pressure changes experienced by the parcel are hydrostatic ($\mathrm{d}p/\mathrm{d}z = -\rho_m g$). Using also the equation of state ($p = \rho_a R_m T$), then (2.71) can be solved for the lapse rate of a vertically displaced parcel.

Consider first a completely dry case, for which $c_{pm} = c_{pa}$, $R_m = R_a$, $\rho_m = \rho_a$, and $r_v = 0$ everywhere. In this case, we find the dry-adiabatic lapse rate :

$$-\mathrm{d}T/\mathrm{d}z \equiv \gamma_d = \frac{g}{c_{pa}} = 9.8\ \mathrm{K\,km^{-1}}. \tag{2.72}$$

Next, consider a moist but subsaturated case, for which $\mathrm{d}r_v = r_l = 0$. Repeating the steps above yields

$$-\mathrm{d}T/\mathrm{d}z \equiv \gamma_m = \frac{(1+r_v)}{\left(1+r_v c_{pv}/c_{pa}\right)}\frac{g}{c_{pa}}. \tag{2.73}$$

Notice that the lapse rate in a moist (but subsaturated) environment is always less than the dry adiabatic lapse rate ($\gamma_m \leq \gamma_d$), and the difference becomes greater as $r_v$ increases. This distinction is typically ignored in practice because the difference is usually small. In very moist tropical environments, $\gamma_m$ may be as small as 9.5 K km$^{-1}$.

For a moist environment with phase changes (i.e. in saturated conditions), the derivation of the appropriate lapse rate ($\gamma_w$) is more complicated. As with the derived thermodynamic variables in Section 2.10, a decision must be made about condensate. Assuming reversible thermodynamics, then the total water is constant (i.e. $\mathrm{d}r_t/\mathrm{d}z = 0$). If saturated equilibrium is also assumed, then $\mathrm{d}r_v = -\mathrm{d}r_{sl}$, where $r_{sl} = \epsilon e_{sl}/(p - e_{sl})$ is the saturation mixing ratio. Using also the Clausius-Clapeyron equation, a formulation for the saturated adiabatic lapse rate can be derived:

$$-\mathrm{d}T/\mathrm{d}z \equiv \gamma_w = \gamma_d\,(1+r_t)\,\frac{1+\frac{L_{lv}r_{sl}}{R_a T}}{\frac{c_{pm}}{c_{pa}} + \frac{L_{lv}{}^2(\epsilon+r_{sl})r_{sl}}{c_{pa}R_a T^2}} \tag{2.74}$$

(e.g. Durran and Klemp, 1982). In moist tropical environments, $\gamma_w$ is of order 5 K km$^{-1}$.

The static stability of an environment can now be considered by examining the buoyant acceleration experienced by vertically displaced parcels. Starting from (2.66) and expanding about a reference environment that is in hydrostatic balance, the gravitational term can be expressed as

$$\frac{\mathrm{d}w}{\mathrm{d}t} = g\left(\frac{{\theta_\rho}^p - {\theta_\rho}^e}{{\theta_\rho}^e}\right) \tag{2.75}$$

where $\theta_\rho$ is the density potential temperature (see Section 2.2.3), superscript $e$ denotes conditions in the environment, and superscript $p$ denotes conditions along a parcel. Using Taylor series to determine values of $\theta_\rho$ at a displaced location $\delta z$, the following relations can be derived:

$$\frac{\mathrm{d}w}{\mathrm{d}t} = -N^2 \delta z \tag{2.76}$$

where

$$N^2 = g\left(\frac{\mathrm{d}\ln {\theta_\rho}^e}{\mathrm{d}z} - \frac{\mathrm{d}\ln {\theta_\rho}^p}{\mathrm{d}z}\right), \tag{2.77}$$

and $N$ is the Brunt-Väisälä frequency.

Consider first a completely dry case, for which $\theta_\rho = \theta$. Using (2.71) to determine conditions along the parcel path, we find $\mathrm{d}\theta^p/\mathrm{d}z = 0$, and thus

$$N^2 = \frac{g}{\theta}\frac{\mathrm{d}\theta}{\mathrm{d}z} \tag{2.78}$$

where we have excluded the superscript $e$ because this equation (and all subsequent equations in this section) refer to the environment. From (2.76) and (2.78), we find that displaced parcels will accelerate in the direction of displacement—and hence the environment is absolutely unstable—if $\mathrm{d}\theta/\mathrm{d}z < 0$.

For a moist but subsaturated case, $\theta_\rho = \theta_m \equiv T\,(p_{00}/p)^{R_m/c_{pm}}$ and $\mathrm{d}{\theta_m}^p/\mathrm{d}z = 0$. Thus,

$$N^2 = \frac{g}{\theta_m}\frac{\mathrm{d}\theta_m}{\mathrm{d}z} \tag{2.79}$$

and the environment is absolutely unstable if $\mathrm{d}\theta_m/\mathrm{d}z < 0$.

For a moist and saturated case, the development of $N^2$ is more complex. Details are available in Durran and Klemp (1982), who found that

$$ {N_m}^2 = \frac{g}{T}\left(\frac{\mathrm{d}T}{\mathrm{d}z} + \gamma_w\right)\left(1 + \frac{L_{lv} r_{sl}}{R_a T}\right) - \frac{g}{1+r_t}\frac{\mathrm{d}r_t}{\mathrm{d}z} \tag{2.80}$$

where ${N_m}^2$ is the traditional notation for $N^2$ in a moist saturated environment, and $\gamma_w$ is given by (2.74). A moist saturated environment is absolutely unstable if ${N_m}^2 < 0$.

It is important to recognize that the final term in (2.80) is not negligible. Consequently, the stability of a moist saturated environment cannot be determined solely by the saturated-adiabatic lapse rate $\gamma_w$: that is, considering the first term on the left-hand side of (2.80), the value of $\gamma_w$ compared with the value of $\mathrm{d}T/\mathrm{d}z$ is insufficient to determine the sign of ${N_m}^2$. This dilemma was addressed by Richiardone and Giusti (2001) who found that if the *condensate mixing ratio* is negligible (which is a good assumption for shallow clouds) then the following lapse rate can be used to assess stability:

$$\gamma_w^* = \gamma_d \frac{1 + 1.18\frac{L_{lv} r_{sl}}{R_a T}}{1 + 0.93\frac{L_{lv}^2 \epsilon r_{sl}}{c_{pa} R_a T^2}}. \tag{2.81}$$

In warm, moist conditions, the magnitude of $\gamma_w$ can be 0.6 °C lower than the magnitude of $\gamma_w^*$, meaning that the use of $\gamma_w$ could incorrectly identify a layer of air as being unstable.

A summary of methods to determine static stability is provided in Table 2.2. Note that stability can be determined using either a stratification condition (based on $N^2$) or by a comparison of lapse rates.

## 2.13. CAPE

Deep convective clouds are most likely if a parcel can become positively buoyant with respect to its environment. A useful measure of the total possible buoyant energy in an environment is the convective available potential energy (CAPE), calculated as follows:

$$\mathrm{CAPE} = \int_{\mathrm{LFC}}^{\mathrm{EL}} B\,\mathrm{d}z \tag{2.82}$$

where LFC is the level of free convection (i.e. where $B$ first becomes positive) and EL is the equilibrium level (i.e. the level above the LFC where $B$ becomes zero); see example in Fig. 2.7. Here, $B \equiv g[(\theta^p - \theta^e)/\theta^e + 0.61({r_v}^p - {r_v}^e) - {r_l}^p - {r_i}^p]$ where the superscript $p$ represents conditions experienced by an ascending parcel and the superscript $e$ represents the environment. The parcel is assumed to follow a moist-adiabatic lapse rate during saturated ascent, and pressure perturbations are neglected. There are several different ways to calculate the moist-adiabatic lapse rate. For example, the parcel could obey reversible thermodynamics or pseudoadiabatic thermodynamics (see Section 2.10). Regarding ice processes, they can be neglected (which is common) or they could be included via a model of mixed-phase microphysics.

The most common method for calculating CAPE is to use pseudoadiabatic thermodynamics without ice processes (CAPE-1). (Recall that pseudoadiabatic

**TABLE 2.2** Criteria for determination of static stability

| Stability state | Stratification | Lapse rate |
|---|---|---|
| **Dry air:** | | |
| Absolutely unstable | $d\theta/dz < 0$ | $\gamma > \gamma_d$ |
| Neutral | $d\theta/dz = 0$ | $\gamma = \gamma_d$ |
| Absolutely stable | $d\theta/dz > 0$ | $\gamma < \gamma_d$ |
| **Moist, subsaturated air:** | | |
| Absolutely unstable | $d\theta_m/dz < 0$ | $\gamma > \gamma_m$ |
| Neutral | $d\theta_m/dz = 0$ | $\gamma = \gamma_m$ |
| Absolutely stable | $d\theta_m/dz > 0$ | $\gamma < \gamma_m$ |
| **Moist, saturated air:** | | |
| Absolutely unstable | ${N_m}^2 < 0$ | $\gamma \gtrsim \gamma_w^*$ |
| Neutral | ${N_m}^2 = 0$ | $\gamma \cong \gamma_w^*$ |
| Absolutely stable | ${N_m}^2 > 0$ | $\gamma \lesssim \gamma_w^*$ |

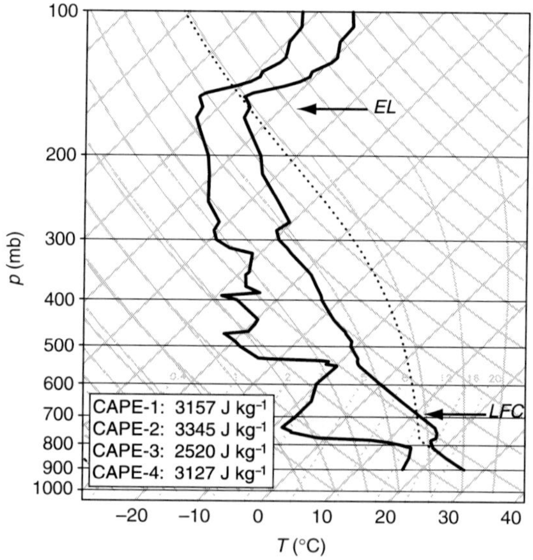


**FIGURE 2.7 A sounding from Dodge City, Kansas, plotted on a skew $T$-log $p$ diagram.** Black-dashed line illustrates parcel temperature under pseudoadiabatic liquid-only thermodynamics. Also shown are the level of free convection (LFC) and the equilibrium level (EL) for this case.

**TABLE 2.3** Summary of some methods to calculate CAPE

| Name | Method to calculate moist adiabat |
|---|---|
| CAPE-1 | Pseudoadiabatic thermodynamics, without ice |
| CAPE-2 | Pseudoadiabatic thermodynamics, with ice |
| CAPE-3 | Reversible thermodynamics, without ice |
| CAPE-4 | Reversible thermodynamics, with ice |

thermodynamics assumes instantaneous removal of condensate by fallout; see Section 2.10.2.) The temperature profile along this adiabat is shown as a dashed line in Fig. 2.7. Three other methods to calculate the moist adiabat are listed in Table 2.3 and are considered below. To illustrate the differences, CAPE was calculated using 732 soundings from Key West, Florida, in 2007 in the following analysis.

Ice processes can be included in a simple way; here, we use the method from Bryan and Fritsch (2004). CAPE calculated using pseudoadiabatic thermodynamics and ice processes is referred to here as CAPE-2. Comparison of CAPE-1 and CAPE-2 from the Key West soundings (Fig. 2.8a) shows that including the latent heat of fusion results in a positive increase in buoyancy that increases CAPE by a small amount. On average, for these soundings, the increase is 9% (considering only cases in which CAPE-1 is greater than 1000 J $kg^{-1}$). This small difference seems to justify the common neglect of ice processes for calculation of CAPE.

Returning to liquid-only processes, we consider now the effects of switching from pseudoadiabatic thermodynamics to reversible thermodynamics. (Recall that all condensate moves with the air under the reversible thermodynamics assumption; see Section 2.10.1.) CAPE calculated using reversible thermodynamics without ice processes is referred to here as CAPE-3. For the Key West soundings, CAPE-3 is significantly different from CAPE-1 (Fig. 2.8b), especially for low-CAPE environments. Note that CAPE-3 is always lower that CAPE-1; this is attributable to the negative contribution to buoyancy from condensate loading (e.g. Xu and Emanuel, 1989; Mapes, 2001). In cases where CAPE-3 is negligible but CAPE-1 is of order 1000 J $kg^{-1}$, water loading roughly negates the positive contribution to buoyancy from latent heat release.

CAPE calculated using reversible thermodynamics *and* ice processes is referred to here as CAPE-4. For the Key West soundings, CAPE-4 is very similar to CAPE-1 (Fig. 2.8); this is because the inclusion of the latent heat of fusion for CAPE-4 (a positive contribution to buoyancy) roughly compensates for the drag of the lifted condensate (a negative contribution to buoyancy) (e.g. Williams and Rennó, 1993). Ice processes are more important under reversible thermodynamics (cf., Fig. 2.8b and c) than under pseudoadiabatic

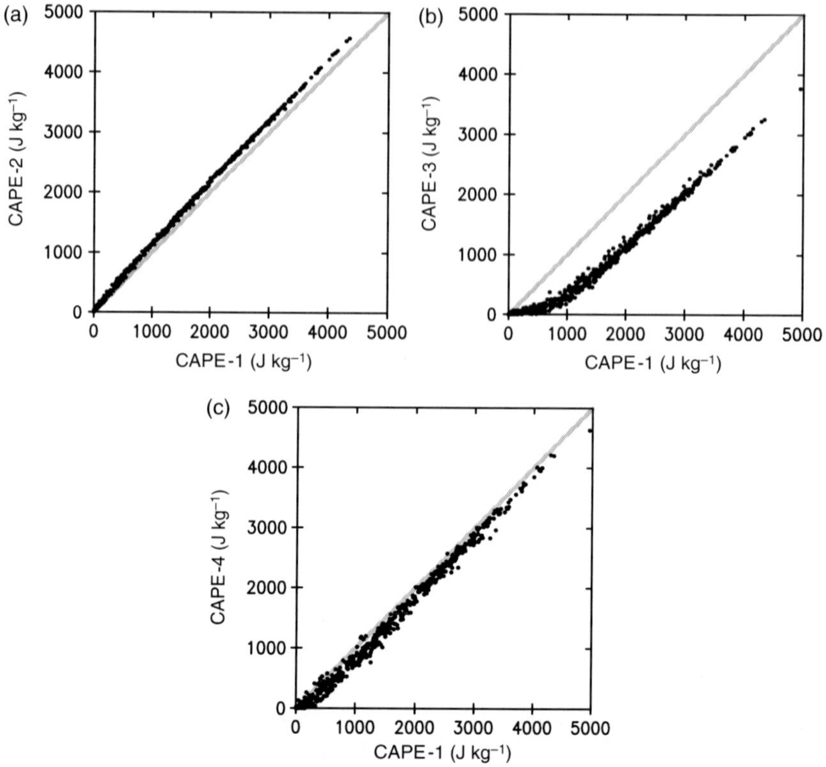


**FIGURE 2.8 Scatterplot comparing CAPE-1 (pseudoadiabatic liquid-only) against: (a) CAPE-2 (pseudoadiabatic with ice processes), (b) CAPE-3 (reversible liquid-only), and (c) CAPE-4 (reversible with ice processes).** This analysis uses 732 soundings from Key West, Florida, from 2007.

thermodynamics (Fig. 2.8a) because condensate is lofted above the freezing level with reversible thermodynamics; the freezing of this condensate (which is not possible under the pseudoadiabatic assumption) is a significant positive contribution to buoyancy.

These concepts are summarized using the Dodge City sounding (Fig. 2.7); the vertical profiles of buoyancy for this case are shown in Fig. 2.9. The most common method (CAPE-1) uses pseudoadiabatic thermodynamics without ice processes; buoyancy for this case is shown by the solid black line in Fig. 2.9. Adding ice processes (CAPE-2: black-dashed line in Fig. 2.9) results in a small increase in buoyancy above the freezing level and a 6% increase in CAPE. Neglecting ice process but using reversible thermodynamics (CAPE-3: grey-dashed line in Fig. 2.9) leads to significantly lower buoyancy throughout the column, and a 20% reduction in CAPE. The temperature along the CAPE-3 adiabat is actually *warmer* than the temperature along the CAPE-1 adiabat (not shown), but the buoyancy is lower because of condensate loading.

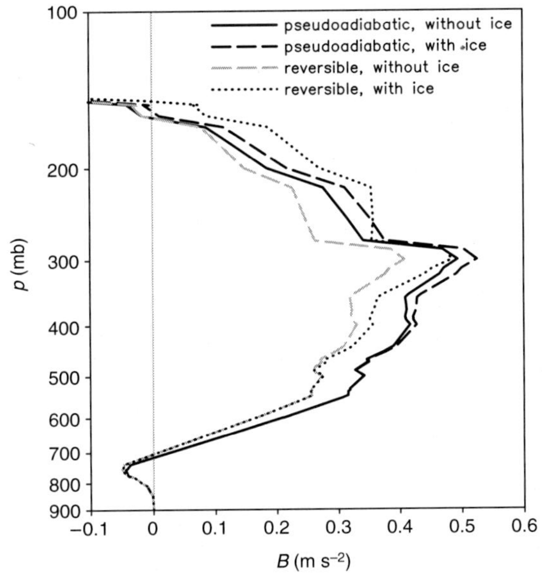


**FIGURE 2.9 Buoyancy profiles used for calculation of CAPE for the sounding shown in Fig. 2.7.**

Finally, adding ice processes with reversible thermodynamics (CAPE-4: dotted line in Fig. 2.9) leads to significantly larger buoyancy in upper levels when the lofted condensate is frozen (a warming effect); the effects of ice processes compensate for the condensate loading, and CAPE-4 is nearly the same as CAPE-1.

Considering only the effects of buoyancy, the maximum possible vertical velocity in clouds would be

$$w_{\max} = (2 \times \mathrm{CAPE})^{1/2} . \tag{2.83}$$

For the Dodge City sounding (Fig. 2.7), this would give $w_{\max}$ between 70-80 m $s^{-1}$. Thunderstorm updrafts rarely reach this upper limit because of vertical pressure gradients (see Section 2.11) and because of mixing/entrainment with environmental air.

These calculations reinforce the importance of thermodynamic assumptions in meteorology. Researchers and forecasters need to choose their assumptions carefully, hopefully choosing a set of assumptions that closely represents the phenomenon that is being studied. For actual convective clouds, other processes become important, the most notable being mixing/entrainment with the environment, and radiative effects. These processes are difficult to incorporate analytically, although they can have a significant impact on the profiles of buoyancy in clouds (e.g. Zipser, 2003).

## 2.14. SUMMARY

The system of equations that we have developed thus far is nonlinear and time dependent. In addition, the system applies to very small parcels of air (of order 1 mm) rather than to properties averaged over several thousand meters, which is a typical grid size in a numerical model of convection. Thus, further modification of the equations must be accomplished before some form of integrated solution can be obtained. Specifically, it is necessary to filter (or, average) the equation set (see next chapter).

To integrate these equations numerically, additional techniques are needed, including: a specification for vertical coordinate (e.g. height above surface, pressure, potential temperature, or a hybrid combinations thereof); a time integration method (e.g. explicit, implicit, split-explicit, etc.); and numerical methods to calculate spatial derivatives (e.g. finite differences, Galerkin techniques, etc.). Further details can be found in specialty texts on these topics (e.g. Durran, 1999; Pielke, 2002).

## REFERENCES

Bannon, P. R. (1996). On the anelastic approximation for a compressible atmosphere. J. Atmos. Sci. 54, 3618–3628.

Bannon, P. R. (2002). Theoretical foundations for models of moist convection. J. Atmos. Sci. 59, 1967–1982.

Bannon, P. R., Chagnon, J. M., and James, R. P. (2006). Mass conservation and the anelastic approximation. Mon. Wea. Rev. 134, 2989–3005.

Betts, A. K. (1973). Non-precipitating cumulus convection and its parameterization. Quart. J. Roy. Meteor. Soc. 99, 178–196.

Bister, M., and Emanuel, K. A. (1998). Dissipative heating and hurricane intensity. Meteor. Atmos. Phys. 65, 233–240.

Bohren, C. F., and Albrecht, B. A. (1998). "Atmospheric Thermodynamics." Oxford, 402 pp.

Bolton, D. (1980). The computation of equivalent potential temperature. Mon. Wea. Rev. 108, 1046–1053.

Bryan, G. H. (2008). On the computation of pseudoadiabatic entropy and equivalent potential temperature. Mon. Wea. Rev. 136, 5239–5245.

Bryan, G. H., and Fritsch, J. M. (2002). A benchmark simulation for moist nonhydrostatic numerical models. Mon. Wea. Rev. 130, 2917–2928.

Bryan, G. H., and Fritsch, J. M. (2004). A reevaluation of ice-liquid water potential temperature. Mon. Wea. Rev. 132, 2421–2431.

Bryan, G. H., and Rotunno, R. (2009). The maximum intensity of tropical cyclones in axisymmetric numerical model simulations. Mon. Wea. Rev. 137, 1770–1789.

Crowe, C., Sommerfeld, M., and Tsuji, Y. (1998). "Multiphase Flows with Droplets and Particles." CRC Press, 471 pp.

Durran, D. R. (1989). Improving the anelastic approximation. J. Atmos. Sci. 46, 1453–1461.

Durran, D. R. (1999). "Numerical Methods for Wave Equations in Geophysical Fluid Dynamics." Springer, 465 pp.

Durran, D. R., and Klemp, J. B. (1982). On the effects of moisture on the Brunt-Väisälä frequency. J. Atmos. Sci. 39, 2152–2158.

Dutton, J. A., and Fichtl, G. H. (1969). Approximate equations of motion for gases and liquids. J. Atmos. Sci. 47, 241–254.

Emanuel, K. A. (1994). "Atmospheric Convection." Oxford, 580 pp.

Hauf, T., and Höller, H. (1987). Entropy and potential temperature. J. Atmos. Sci. 44, 2887–2901.

Hausman, S. A., Ooyama, K. V., and Schubert, W. H. (2006). Potential vorticity structure of simulated hurricanes. J. Atmos. Sci. 63, 87–108.

Iribarne, J. V., and Godson, W. L. (1981). "Atmospheric Thermodynamics." 2nd ed., D. Reidel, 259 pp.

Klemp, J. B., Skamarock, W. C., and Dudhia, J. (2007). Conservative split-explicit time integration methods for the compressible nonhydrostatic equations. Mon. Wea. Rev. 135, 2897–2913.

Klemp, J. B., and Wilhelmson, R. B. (1978). The simulation of three-dimensional convective storm dynamics. J. Atmos. Sci. 35, 1070–1096.

Lackmann, G. M., and Yablonksy, R. M. (2004). The importance of the precipitation mass sink in tropical cyclones and other heavily precipitating systems. J. Atmos. Sci. 61, 1674–1692.

Lipps, F. B., and Hemler, R. S. (1982). A scale analysis of deep moist convection and some related numerical calculations. J. Atmos. Sci. 39, 2192–2210.

Mapes, B. E. (2001). Water's two height scales: The moist adiabat and the radiative troposphere. Quart. J. Roy. Meteor. Soc. 127, 2353–2366.

Ogura, Y., and Phillips, N. A. (1962). Scale analysis of deep and shallow convection in the atmosphere. J. Atmos. Sci. 19, 173–179.

Ooyama, K. V. (1990). A thermodynamic foundation for modeling the moist atmosphere. J. Atmos. Sci. 47, 2580–2593.

Ooyama, K. V. (2001). A dynamic and thermodynamic foundation for modeling the moist atmosphere with parameterized microphysics. J. Atmos. Sci. 58, 2073–2102.

Pauluis, O., Balaji, V., and Held, I. M. (2000). Frictional dissipation in a precipitating atmosphere. J. Atmos. Sci. 57, 989–994.

Pielke Sr., R. A. (2002). "Mesoscale Meteorological Modeling." 2nd ed., Academic Press, 676 pp.

Qiu, C.-J., Bao, J.-W., and Xu, Q. (1993). Is the mass sink due to precipitation negligible?. Mon. Wea. Rev. 121, 853–857.

Rennó, N. O. (2001). Comments on Frictional dissipation in a precipitating atmosphere. J. Atmos. Sci. 58, 1173–1177.

Richiardone, R., and Giusti, F. (2001). On the stability criterion in a saturated atmosphere. J. Atmos. Sci. 58, 2013–2017.
Satoh, M. (2003). Conservative scheme for a compressible nonhydrostatic model with moist processes. Mon. Wea. Rev. 131, 1033–1050.
Tripoli, G. J., and Cotton, W. R. (1981). The use of ice-liquid water potential temperature as a thermodynamic variable in deep atmospheric models. Mon. Wea. Rev. 109, 1094–1102.
Wexler, A. (1976). Vapor pressure formulation for water in range 0 to 100°C. A revision. J. Res. Natl. Bur. Stand. 80A, 775–785.
Wexler, A. (1977). Vapor pressure formulation for ice. J. Res. Natl. Bur. Stand. 81A, 5–20.
Wilhelmson, R. B. (1977). On the thermodynamic equation for deep convection. Mon. Wea. Rev. 105, 545–549.
Williams, E., and Rennó, N. (1993). An analysis of the conditional instability of the tropical atmosphere. Mon. Wea. Rev. 121, 21–36.
Xu, K.-M., and Emanuel, K. A. (1989). Is the tropical atmosphere conditionally unstable?. Mon. Wea. Rev. 117, 1471–1479.
Zipser, E. J. (2003). Some views on "hot towers" after 50 years of tropical field programs and two years of TRMM data. In "Cloud Systems, Hurricanes, and the Tropical Rainfall Measuring Mission (TRMM)." In Meteor. Monogr., vol. 51. pp. 49–58. Amer. Meteor. Soc.

Chapter 3

# Turbulence

## 3.1. INTRODUCTION

The nonlinear equations that are presented in Chapter 2 can permit atmospheric motion and transport over a broad range of spatial scales, ranging from the largest eddies on the globe (of order 5000 km) to the smallest eddies associated with molecular dissipation (of order 1 mm or less). The smallest-scale motions in a flow are usually considered to be random and chaotic. We call these small-scale structures *turbulence*.

The extraordinarily large range of scales in the atmosphere is consistent with a large Reynolds number:

$$\mathrm{Re} = \frac{UL}{\nu} \tag{3.1}$$

where $U$ is a characteristic velocity scale, $L$ is a characteristic length scale, and $\nu$ is kinematic viscosity. In terms of physical processes, Re represents the ratio of the magnitude of inertial forces to the magnitude of viscous forces. Consider a cumulonimbus cloud, such as that shown in Fig. 3.1. The largest scale (referred to as the *large eddy scale*) is roughly the cloud size, which is of order 10 km. The velocity scale varies as $\sqrt{2 \times \mathrm{CAPE}}$ and is of order 40 m s$^{-1}$. Because $\nu$ is of order $1 \times 10^{-6}$ m$^2$ s$^{-1}$, it follows that Re is of order $10^{11}$. A flow becomes turbulent when Re exceeds a certain value, referred to as the critical Reynolds number ($\mathrm{Re}_{\mathrm{crit}}$). Although $\mathrm{Re}_{\mathrm{crit}}$ varies by fluid and by certain types of flows, it is generally of order of $10^3$. Clearly, $\mathrm{Re} \gg \mathrm{Re}_{\mathrm{crit}}$ for the cumulonimbus cloud shown in Fig. 3.1. Many atmospheric motions, including many types of clouds, are similarly turbulent.

The many scales of motion in the cumulonimbus cloud give it a cauliflower-like appearance. Consider the scale of the *smallest* eddies (i.e. coherent motions) in this flow, which is denoted by $\eta$. These are the scales that are associated with viscous dissipation. Kolmogorov (1941) found that this scale (often called the Kolmogorov microscale) scales with $\nu$ and the rate of dissipation of kinetic

**FIGURE 3.1** **Photograph of a cumulonimbus cloud showing turbulent eddies of many scales.**

energy per unit mass ($\epsilon$) as follows:

$$\eta \approx \left(\frac{\nu^3}{\epsilon}\right)^{1/4}. \tag{3.2}$$

Note that $\nu$ is a property of the fluid (i.e. air), and $\epsilon$ is a property of a flow (i.e. a cumulus cloud, or a stratus cloud, etc.). Studies have found that $\epsilon \sim U^3/L$. Solving, then, for the ratio of the large-eddy scale to the dissipative scale yields

$$\frac{L}{\eta} \approx \left(\frac{UL}{\nu}\right)^{3/4} = \mathrm{Re}^{3/4}. \tag{3.3}$$

From the scaling provided above, we find that $\eta \approx 0.1$ mm in the cumulonimbus. Therefore, to simulate all scales in the flow shown in Fig. 3.1, a numerical model with grid spacing of order 0.1 mm would be required (e.g. Bryan et al., 2003). The direct simulation of all scales in a turbulent flow, using the unmodified governing equations from Chapter 2, is called *direct numerical simulation* (DNS) . The number of grid points ($N$) needed for DNS of turbulent flow is proportional to Re,

$$N^3 \geq 4\mathrm{Re}^{9/4}, \tag{3.4}$$

(Pope, 2000). Because Re $> 10^9$ for many types of clouds (e.g. cumulus, stratocumulus, etc.), DNS of atmospheric flow is clearly not feasible with current or foreseeable computing resources, as the largest simulations (in terms of grid points) today use of order $10^{10}$ grid points. Moreover, current observational systems are incapable of resolving or defining all scales of motion in the atmosphere.

If Re is low (of order $10^3$-$10^4$) then DNS is possible. For example, DNS was used in the studies by Moeng and Rotunno (1990), Coleman (1999), and Fritts et al. (2009). Nevertheless, usage of DNS in atmospheric science is fairly rare.

Because of the existence (and importance) of small-scale flows in the atmosphere, and because of our inability to resolve and/or observe them, meteorologists are forced to distinguish between those eddies that are in a sense "resolvable", either by our observation systems, or by some form of finite-difference representation of the governing equations (here we include finite-element and truncated spectral representations of the equations). The remaining eddies that are not fully resolved, either observationally or computationally, are "unresolved."

Furthermore, it is important to realize that we must be content with describing, or predicting, only the statistical properties of turbulent flows. Otherwise a complete, unique description or prediction of their characteristics brings us back into the perspective of "resolvable" eddies.

For the development of concepts in this chapter we will make use of the incompressible Boussinesq equations

$$\frac{\partial u_i}{\partial t} = -u_j \frac{\partial u_i}{\partial x_j} - \frac{1}{\rho_0}\frac{\partial p'}{\partial x_i} + \delta_{i3} g \frac{\theta'_\rho}{\Theta_0} + \varepsilon_{ijk} u_j f_k + \nu \frac{\partial^2 u_i}{\partial x_j^2}, \tag{3.5}$$

$$\frac{\partial u_i}{\partial x_i} = 0, \tag{3.6}$$

where $\rho_0$ is a constant reference density, $\theta_\rho$ is density potential temperature [which includes the effects of water vapor and condensate on buoyancy, because $\theta_\rho \approx \theta(1 + 0.61 r_v - r_l - r_i)$], $\Theta_0(z)$ is a one-dimensional reference profile of $\theta_\rho$ that is in hydrostatic balance with a reference pressure profile (following Section 2.9.2), and primes indicate departures from the hydrostatic reference profile. Extension to the anelastic equations is usually considered to be straightforward, but the incompressible equations are used here because of their relative simplicity. We will also make use of equations for a thermodynamic scalar ($s$) and a mixing ratio of water ($r$)

$$\frac{\partial s}{\partial t} = -u_i \frac{\partial s}{\partial x_i} + p[s] + \gamma_s \frac{\partial^2 s}{\partial x_i^2}, \tag{3.7}$$

$$\frac{\partial r}{\partial t} = -u_i \frac{\partial r}{\partial x_i} + p[r] + \gamma_r \frac{\partial^2 r}{\partial x_i^2}. \tag{3.8}$$

Here, $s$ could be either total moist entropy, potential temperature, ice-liquid water potential temperature, etc., and $r$ is a mixing ratio of water such as water vapor, cloud water, snow, etc. The terms $p[s]$ and $p[r]$ represent sources and sinks to $s$ and $r$, respectively, due to phase changes, hydrometeor sedimentation, or (in the case of $s$) radiation. The terms with $\gamma$ represent molecular diffusion. Note that liquid water drops and/or frozen ice particles do not diffuse molecularly (because the liquid/ice particles are not necessarily in contact with one another, as discussed in Chapter 2) and so $\gamma_r = 0$ for these variables.

## 3.2. REYNOLDS AVERAGING

Perhaps the most common approach to deal with turbulence is called Reynolds averaging. Conceptually, this method separates a variable into a mean component (indicated by an overbar) and a rapidly fluctuating component (indicated herein by double prime superscript). The goal of Reynolds averaging is to study the mean component while indirectly accounting for the effects of the turbulent fluctuations. That is, the effects of the fluctuating component are *parameterized.* Because the mean component is smooth, it can be studied with coarse resolution, or it can even be collapsed onto one dimension (as is the case of a turbulent boundary layer simulated in a mesoscale model, for example).

We consider first the velocity equation. It proves convenient to write the advection term—the first term on the right-hand side of (3.5)—in flux-form. Making use of (3.6) we have

$$\frac{\partial u_i}{\partial t} = -\frac{\partial}{\partial x_j}(u_j u_i) - \frac{1}{\rho_0}\frac{\partial p'}{\partial x_i} + \delta_{i3} g \frac{\theta'_\rho}{\Theta_0} + \varepsilon_{ijk} u_j f_k + \nu \frac{\partial^2 u_i}{\partial x_j^2}, \tag{3.9}$$

We now decompose each variable into a mean and a turbulent fluctuating component:

$$\begin{aligned} u_i &= \overline{u}_i + u''_i, \\ p' &= p - p_0(z) = \overline{p'} + p'', \\ \theta'_\rho &= \theta_\rho - \Theta_0(z) = \overline{\theta'_\rho} + \theta''_\rho. \end{aligned} \tag{3.10}$$

The averaging operator shall not be specified herein, but the choice is important. It could be a temporal average, or an ensemble average (i.e. an average over many different "snapshots" or "realizations" of the same type of flow). In general, a spatial averaging operator *can not* be used for Reynolds averaging, which will be discussed further in Section 3.3; however, spatial averaging procedures in which the filtered component is de-correlated from the unfiltered component—such as a wave-cutoff filter in spectral space—could, in principle, be used for Reynolds averaging. Whatever filter is used, the Reynolds averaging procedure is to assume that for variables $a$ and $b$

$$\begin{aligned} &\overline{\overline{a}} = \overline{a}, && \overline{a''} = 0, \\ &\overline{a + b} = \overline{a} + \overline{b}, && \overline{\overline{a}b} = \overline{a}\overline{b}, \\ &\overline{a''\overline{b}} = 0 = \overline{\overline{a}b''}, && \overline{\frac{\partial a}{\partial t}} = \frac{\partial \overline{a}}{\partial t}. \end{aligned} \tag{3.11}$$

In the case of an ensemble averaging operator, all turbulence in the mean field is removed as long as the number of realizations, $N$, is large enough. Thus, mean and fluctuating variables are uncorrelated, and the rules (3.11) are exactly satisfied. Typically one assumes that the rules (3.11) are satisfied even for a temporal or spatial averaging operation. The magnitude of error that is introduced by this assumption, however, is not generally known.

The first step in the Reynolds averaging procedure is to substitute (3.10) into (3.9):

$$\frac{\partial}{\partial t}(\overline{u}_i + u_i'') = -\frac{\partial}{\partial x_j}\left(\overline{u}_j\overline{u}_i + \overline{u}_j u_i'' + u_j''\overline{u}_i + u_j'' u_i''\right) - \frac{1}{\rho_0}\frac{\partial}{\partial x_i}(\overline{p'} + p'') + \delta_{i3} g\left(\frac{\theta_\rho'}{\Theta_0} + \frac{\theta_\rho''}{\Theta_0}\right) + \varepsilon_{ijk}\left(\overline{u}_j + u_j''\right) f_k + \nu\frac{\partial^2 \overline{u}_i}{\partial x_j^2} + \nu\frac{\partial^2 u_i''}{\partial x_j^2}. \quad (3.12)$$

The second step is to average every term

$$\overline{\frac{\partial}{\partial t}(\overline{u}_i + u_i'')} = -\overline{\frac{\partial}{\partial x_j}\left(\overline{u}_j\overline{u}_i + \overline{u}_j u_i'' + u_j''\overline{u}_i + u_j'' u_i''\right)} - \frac{1}{\rho_0}\overline{\frac{\partial}{\partial x_i}(\overline{p'} + p'')} + \delta_{i3} g\overline{\left(\frac{\theta_\rho'}{\Theta_0} + \frac{\theta_\rho''}{\Theta_0}\right)} + \varepsilon_{ijk}\overline{\left(\overline{u}_j + u_j''\right) f_k} + \nu\overline{\frac{\partial^2 \overline{u}_i}{\partial x_j^2}} + \nu\overline{\frac{\partial^2 u_i''}{\partial x_j^2}}. \quad (3.13)$$

The final step is to use the Reynolds averaging rules, (3.11), to simplify the equation. We find

$$\frac{\partial \overline{u}_i}{\partial t} = -\frac{\partial}{\partial x_j}\left(\overline{u}_j\overline{u}_i\right) - \frac{\partial}{\partial x_j}\left(\overline{u_j'' u_i''}\right) - \frac{1}{\rho_0}\frac{\partial \overline{p'}}{\partial x_i} + \delta_{i3} g\frac{\overline{\theta_\rho'}}{\Theta_0} + \varepsilon_{ijk}\overline{u}_j f_k + \nu\frac{\partial^2 \overline{u}_i}{\partial x_j^2}. \quad (3.14)$$

For large Reynolds number flow the effects of molecular diffusion on the mean dependent variables is negligible, and so the final term in (3.14) can be neglected.

Notice that (3.14) is similar in form to (3.9) except every variable is replaced by an averaged variable and there is one additional term. This new term—the second term on the right-hand side—is called the Reynolds stress term: it represents the effects of turbulent velocity fluctuations on the mean flow $\overline{u}_i$.

Using the same procedure for the equations governing mass-continuity, entropy, and water mass leads to

$$\frac{\partial \overline{u}_i}{\partial x_i} = 0, \quad (3.15)$$

$$\frac{\partial \overline{s}}{\partial t} = -\frac{\partial}{\partial x_i}(\overline{u}_i \overline{s}) - \frac{\partial}{\partial x_i}\left(\overline{u_i'' s''}\right) + \overline{p[s]}, \tag{3.16}$$

$$\frac{\partial \overline{r}}{\partial t} = -\frac{\partial}{\partial x_i}(\overline{u}_i \overline{r}) - \frac{\partial}{\partial x_i}\left(\overline{u_i'' r''}\right) + \overline{p[r]}, \tag{3.17}$$

and so on for any scalar variable. The terms $\overline{p[s]}$ and $\overline{p[r]}$ represent grid-averaged sources and sinks to $\overline{s}$ and $\overline{r}$. Equations (3.14)–(3.17) are not, however, closed. Some means of defining the correlations among velocity fluctuations and among velocity and scalar fluctuations must be found.

### 3.2.1. First-Order Closure

The simplest approach to closing the Reynolds-averaged equations is to assume that the Reynolds stresses and eddy transport terms can be expressed in terms of gradients of the mean variables. In atmospheric sciences this is usually referred to as a first-order closure model. (In some other disciplines it is referred to as a zero-equation model or an algebraic model.)

Boussinesq (1877) introduced the concept of eddy viscosity, in which the Reynolds stress is considered analogous to the viscous stress but where the kinematic viscosity is replaced by an *eddy viscosity*. Assuming that the deviatoric Reynolds stress tensor is proportional to the mean rate of the strain tensor then

$$-\overline{u_i'' u_j''} + \frac{2}{3}\delta_{ij}\overline{e} = 2K_m D_{ij}, \tag{3.18}$$

where $K_m$ is the eddy viscosity (or eddy exchange coefficient) for momentum, $D_{ij}$ is the mean rate of strain tensor

$$D_{ij} = \frac{1}{2}\left(\frac{\partial \overline{u}_i}{\partial x_j} + \frac{\partial \overline{u}_j}{\partial x_i}\right), \tag{3.19}$$

and $\overline{e} = \overline{u_k'' u_k''}/2$ is the turbulence kinetic energy (i.e. the kinetic energy associated with the fluctuating motions). The second term on the left-hand side of (3.18) is often neglected in atmospheric science studies but it is required for physical consistency (Hinze, 1975); specifically, it ensures that both the left-hand side and the right-hand side are zero when $i$ is contracted on $j$ (wherein the right-hand side becomes zero because of the incompressible mass-continuity equation). In practice, the term with $\overline{e}$ is combined with the pressure-gradient term and thus the velocity equation can be written

$$\frac{\partial \overline{u}_i}{\partial t} = -\frac{\partial}{\partial x_j}(\overline{u}_j \overline{u}_i) + \frac{\partial}{\partial x_j}(2K_m D_{ij}) - \frac{\partial}{\partial x_i}\left(\frac{\overline{p'}}{\rho_0} + \frac{2}{3}\overline{e}\right)$$

$$+\delta_{i3}g\frac{\overline{\theta'_\rho}}{\Theta_0}+\varepsilon_{ijk}\overline{u}_j f_k. \tag{3.20}$$

It is customary to define $\phi \equiv \overline{p'}/\rho_0 + (2/3)\overline{e}$ and then to form an elliptic equation for $\phi$ utilizing (3.15) and (3.20), which is standard procedure for incompressible numerical models.

Following a similar line of reasoning, it is assumed that the turbulent transport of any scalar property $s$ can be expressed in terms of a scalar eddy diffusivity ($K_s$) as

$$-\overline{u_i''s''}=K_s\frac{\partial\overline{s}}{\partial x_i}. \tag{3.21}$$

Now, the only unknown variables are $K_m$ and $K_s$. There are several models for determining these variables. Early cloud and mesoscale models (in the 1970s) used values of $K_m$ and $K_s$ which were invariant in both time and space. Because the eddy viscosity must depend on the flow variables, and is not a property of the fluid itself (as is the case with molecular viscosity), such an assumption is unjustified.

One of the first physical models was introduced by Prandtl (1925). It has the form

$$K_m=l^2\left[\left(\frac{\partial u}{\partial z}\right)^2+\left(\frac{\partial v}{\partial z}\right)^2\right]^{1/2} \tag{3.22}$$

where $l$ is a characteristic length scale of the turbulent eddies. This model is often used near surfaces (i.e. in boundary layers). The value for $l$ is highly dependent on the flow.

More generally, an energy budget for the turbulence kinetic energy can be formed following Lilly (1962). More details are provided in the next section. Assuming steady homogeneous turbulence, one finds that

$$K_mS^2-K_sN^2=\epsilon \tag{3.23}$$

where $\epsilon$ is the rate of dissipation of kinetic energy at molecular scales, $S^2$ is the flow deformation

$$S^2=\frac{\partial\overline{u}_i}{\partial x_j}\left(\frac{\partial\overline{u}_i}{\partial x_j}+\frac{\partial\overline{u}_j}{\partial x_i}\right)=2D_{ij}D_{ij}, \tag{3.24}$$

and $N^2$ is the squared Brunt-Väisälä frequency (or $N_m^2$ in saturated conditions; see Section 2.12). On dimensional grounds $\epsilon\sim K_m^3/l^4$ and thus

$$K_m=l^2S\,(1-\mathrm{Ri}/\mathrm{Pr})^{1/2} \tag{3.25}$$

where $\mathrm{Pr}=K_m/K_s$ is the Prandtl number (which is ~1/3) and $\mathrm{Ri}=N^2/S^2$ is the Richardson number. Note from (3.25) that turbulence occurs (i.e. $K_m > 0$) only if $\mathrm{Ri} < \mathrm{Pr}$. (If $\mathrm{Ri} > \mathrm{Pr}$ in a numerical simulation, then $K_m$ is set to zero.)

It is difficult to choose appropriate values for $l$ with this model, and the final choice is often accomplished by trial-and-error. Nevertheless, $l$ should in principle be set according to the size of the largest turbulent eddies in a flow. For an ordinary cumulus cloud $l$ would probably be of order 100 m, and in cumulonimbus clouds $l$ would probably be of order 1000 m.

One disadvantage of the eddy viscosity closure approach in a cloud system is that eddy exchange coefficients $K_s$ for all mean prognostic variables (including water vapor, cloud water, cloud ice, etc.) must be specified or predicted. Generally, the specification of any $K_s$ for water conservation equations is done with little basis from observation or theory, but is typically based only on the Prandtl number,

$$K_s = K_m/\mathrm{Pr}, \tag{3.26}$$

where Pr is a specified constant. There is little evidence to support or refute such an assumption when applied to deep moist convection.

### 3.2.2. 1.5-order Closure

One major shortcoming of the first-order closure model, (3.25), is that it predicts zero turbulence intensity ($K_m = 0$) in the absence of mean-flow deformation ($S = 0$) for neutral stability. To overcome this shortcoming, Kolmogorov (1942) and Prandtl (1945) suggested that $K_m$ could be determined directly from turbulence kinetic energy, $\overline{e} = \overline{u_i''u_i''}/2$, using a relation of the form

$$K_m = c_m l \overline{e}^{1/2}, \tag{3.27}$$

where $c_m$ is a constant. A governing equation can be developed for $\overline{e}$, as will be shown below. This model thus introduces one additional predictive equation. In atmospheric sciences, this approach is typically referred to as a 1.5-order closure model. (In some disciplines it is referred to as a "one-equation model".)

The first step in the derivation is to form a governing equation for $u_i''$ by subtracting Eq. (3.14) from (3.12):

$$\begin{aligned}\frac{\partial}{\partial t}(u_i'') = &-\frac{\partial}{\partial x_j}(\overline{u}_j u_i'') - \frac{\partial}{\partial x_j}(u_j''\overline{u}_i) - \frac{\partial}{\partial x_j}(u_j''u_i'') + \frac{\partial}{\partial x_j}\overline{(u_j''u_i'')} \\ &- \frac{1}{\rho_0}\frac{\partial p''}{\partial x_i} + \delta_{i3} g\frac{\theta_\rho''}{\Theta_0} + \varepsilon_{ijk} u_j'' f_k + \nu\frac{\partial^2 u_i''}{\partial x_j^2}.\end{aligned} \tag{3.28}$$

Then, we multiply (3.28) by $u_i''$ and rearrange. We make use of the fact that both the mean flow and the turbulent fluctuations behave incompressibly, i.e.

$$\frac{\partial \overline{u}_i}{\partial x_i} = \frac{\partial u_i''}{\partial x_i} = 0. \tag{3.29}$$

We find

$$\begin{aligned} \frac{\partial}{\partial t}\left(\frac{u_i''u_i''}{2}\right) &= -\overline{u}_j \frac{\partial}{\partial x_j}\left(\frac{u_i''u_i''}{2}\right) - u_i''u_j''\frac{\partial \overline{u}_i}{\partial x_j} - \frac{\partial}{\partial x_j}\left(u_j''\frac{u_i''u_i''}{2}\right) + u_i''\frac{\partial}{\partial x_j}\left(\overline{u_j''u_i''}\right) \\ &\quad - \frac{1}{\rho_0}\frac{\partial}{\partial x_i}(u_i''p'') + \frac{g}{\Theta_0}\theta_\rho''w'' + \underbrace{\cancel{\nu\frac{\partial^2}{\partial x_j^2}\left(\frac{u_i''u_i''}{2}\right)}}_{\to 0} - \nu\frac{\partial u_i''}{\partial x_j}\frac{\partial u_i''}{\partial x_j}. \end{aligned} \tag{3.30}$$

Notice that the Coriolis terms sum to zero. The molecular diffusion term is negligible for high Reynolds number flows. Equation (3.30) is then averaged and, using (3.11) and the definition of $\overline{e}$, we find:

$$\begin{aligned} \frac{\partial \overline{e}}{\partial t} &= \overset{\text{(a)}}{-\overline{u}_j\frac{\partial \overline{e}}{\partial x_j}} \overset{\text{(b)}}{-\overline{u_i''u_j''}\frac{\partial \overline{u}_i}{\partial x_j}} \overset{\text{(c)}}{-\frac{\partial}{\partial x_i}\overline{(u_i''e)}} \overset{\text{(d)}}{-\frac{1}{\rho_0}\frac{\partial}{\partial x_i}\overline{(u_i''p'')}} \\ &\quad \overset{\text{(e)}}{+\frac{g}{\Theta_0}\overline{\theta_\rho''w''}} \overset{\text{(f)}}{-\nu\overline{\frac{\partial u_i''}{\partial x_j}\frac{\partial u_i''}{\partial x_j}}}. \end{aligned} \tag{3.31}$$

Term (a) is the transport of $\overline{e}$ by the mean flow. Term (b) is the mean-gradient production term, which represents the transfer of kinetic energy from the mean flow to the subgrid turbulence. Terms (c) and (d) both represent the redistribution of turbulence: term (c) by eddy transport and term (d) through the interaction of velocity fluctuations and pressure fluctuations. Term (e) is the buoyancy term; it represents the source/sink of turbulence intensity by local stratification. Notice that $\theta_\rho''/\Theta_0 \approx \theta''/\theta_0 + 0.61r_v'' - r_l'' - r_i''$, and so term (e) includes the correlation of vertical velocity fluctuations with fluctuations in temperature, water vapor, and condensate. Term (f), which is negative-definite, is the dissipation of kinetic energy by viscous forces ($\epsilon$).

Terms (b)-(f) contain unknown variables that must be modeled (parameterized). The eddy viscosity model, (3.18), can be used for term (b) and the down-gradient diffusion model for scalars, (3.21), can be used for term (e), although the buoyancy term is usually generalized as

$$\frac{g}{\Theta_0}\overline{\theta_\rho''w''} = -K_s N^2 \tag{3.32}$$

where $N^2$ is replaced by $N_m^2$ in saturated conditions, so that the effects of moisture on stability can be incorporated properly. Terms (c) and (d) can be modeled diffusively,

$$-\overline{u_i''e} - \frac{\overline{u_i''p''}}{\rho_0} = 2K_m \frac{\partial \overline{e}}{\partial x_i}. \tag{3.33}$$

To finish closing (3.31), a model must be provided for $\epsilon$. Closures for $\epsilon$ take many forms. The most common method follows from dimensional arguments:

$$\epsilon \sim K_m^3 / l^4. \tag{3.34}$$

Using this relation and (3.27) it follows that

$$\epsilon = c_\epsilon \overline{e}^{3/2} / l, \tag{3.35}$$

where $c_\epsilon$ is a constant. It is also possible to formulate the dissipation rate based on a time scale (rather than a length scale), such as

$$\epsilon = \mu \overline{e} / \tau_t \tag{3.36}$$

(e.g. Chen and Cotton, 1983) where $\mu$ is a constant and $\tau_t$ is a timescale for turbulence energy dissipation.

Notice that the first-order closure model, (3.25), can be readily derived from these relations by setting the left-hand side of (3.31) to zero (i.e. assuming steady flow) and by neglecting terms (a), (c), and (d) (i.e. assuming statistically homogeneous turbulence).

We also note that $l$ (or $\tau_t$) still needs to be specified in this model as well as the constants $c_m$ and $c_\epsilon$. Thus, the 1.5-order closure has more degrees of freedom than a first-order model. However, the 1.5-order model does not assume steadiness and, furthermore, turbulent intensity can be transported across a numerical model grid. Both effects can be important in mesoscale model simulations.

Some model developers have used (3.34) to form an additional predictive equation for $\epsilon$. Usually referred to as an "$e$-$\epsilon$" model (or, in some disciplines, as a "two-equation model"), the primary advantage is that the length scale $l$ can be eliminated from the two equations, leaving only a handful of constants to be specified. Further details can be found in the review article by Detering and Etling (1985).

### 3.2.3. Higher Order Closure

A fundamental weakness of eddy exchange theory is its underlying assumption that turbulence always acts to diffuse a property down the mean gradient of the property. There are many examples of counter gradient transport in planetary boundary layer studies and planetary-scale eddy transport analyses. While there exist few, if any, documented cases of countergradient turbulent transport associated with deep convection, most researchers would not be surprised to find not only that it exists, but that it prevails at times. As a consequence, a number of investigators are seeking alternate means of closing the averaged equations. The approach which we shall discuss next is called *higher order closure theory*.

Rather than make a simple first-order approximation to the correlations $\overline{u_i''u_j''}$ or $\overline{u_i''s''}$ such as the down-gradient approximations [Eqs. (3.18) and (3.21)], higher order closure theory instead involves the formulation of prognostic equations for these covariances. Triple-correlation terms such as $\overline{u_i''u_j''u_k''}$ or $\overline{u_i''u_j''s''}$ appear in these models (as shown below); if a simple diagnostic model is used to evaluate these terms, the model is referred to as a *second-order closure model*. If a predictive equation is formed for the triple-correlation terms and a diagnostic equation is formed on the resulting quadruple-correlation terms, the model is referred to as a *third-order closure model*, and so forth.

To illustrate the procedure we multiply Eq. (3.28) by $u_j''$ giving

$$
\begin{aligned}
u_j''\frac{\partial u_i''}{\partial t} &= -u_j''\frac{\partial}{\partial x_k}(\overline{u}_k u_i'') - u_j''\frac{\partial}{\partial x_k}(u_k''\overline{u}_i) - u_j''\frac{\partial}{\partial x_k}(u_k''u_i'') + u_j''\frac{\partial}{\partial x_k}(\overline{u_k''u_i''}) \\
&\quad - \frac{u_j''}{\rho_0}\frac{\partial p''}{\partial x_i} + \delta_{i3}\frac{g}{\Theta_0}\theta_\rho''u_j'' + \varepsilon_{imk}f_k u_m''u_j'' + \nu u_j''\frac{\partial^2 u_i''}{\partial x_k^2},
\end{aligned}
\tag{3.37}
$$

(where we now use $k$ as the free index in the flux-divergence terms, and we have substituted $m$ for $j$ in the Coriolis term, without changing the results). The next step is to write a governing equation similar to (3.37) but interchanging the indices $i$ and $j$:

$$
\begin{aligned}
u_i''\frac{\partial u_j''}{\partial t} &= -u_i''\frac{\partial}{\partial x_k}(\overline{u}_k u_j'') - u_i''\frac{\partial}{\partial x_k}(u_k''\overline{u}_j) - u_i''\frac{\partial}{\partial x_k}(u_k''u_j'') + u_i''\frac{\partial}{\partial x_k}(\overline{u_k''u_j''}) \\
&\quad - \frac{u_i''}{\rho_0}\frac{\partial p''}{\partial x_j} + \delta_{j3}\frac{g}{\Theta_0}\theta_\rho''u_i'' + \varepsilon_{jmk}f_k u_m''u_i'' + \nu u_i''\frac{\partial^2 u_j''}{\partial x_k^2}.
\end{aligned}
\tag{3.38}
$$

We then add Eq. (3.37) and Eq. (3.38) to obtain

$$
\begin{aligned}
\frac{\partial}{\partial t}(u_i''u_j'') &= u_i''\frac{\partial u_j''}{\partial t} + u_j''\frac{\partial u_i''}{\partial t} \\
&= -u_i''\frac{\partial}{\partial x_k}(\overline{u}_k u_j'') - u_i''\frac{\partial}{\partial x_k}(u_k''\overline{u}_j) - u_i''\frac{\partial}{\partial x_k}(u_k''u_j'') + u_i''\frac{\partial}{\partial x_k}(\overline{u_k''u_j''}) \\
&\quad - u_j''\frac{\partial}{\partial x_k}(\overline{u}_k u_i'') - u_j''\frac{\partial}{\partial x_k}(u_k''\overline{u}_i) - u_j''\frac{\partial}{\partial x_k}(u_k''u_i'') + u_j''\frac{\partial}{\partial x_k}(\overline{u_k''u_i''}) \\
&\quad - \frac{u_i''}{\rho_0}\frac{\partial p''}{\partial x_j} - \frac{u_j''}{\rho_0}\frac{\partial p''}{\partial x_i} + \delta_{i3}\frac{g}{\Theta_0}\theta_\rho''u_j'' + \delta_{j3}\frac{g}{\Theta_0}\theta_\rho''u_i''
\end{aligned}
$$

$$+\varepsilon_{jmk} f_k u''_m u''_i + \varepsilon_{imk} f_k u''_m u''_j + \nu u''_i \frac{\partial^2 u''_j}{\partial x_k^2} + \nu u''_j \frac{\partial^2 u''_i}{\partial x_k^2}. \tag{3.39}$$

We rearrange the pressure-velocity term as follows

$$-u''_i \frac{\partial p''}{\partial x_j} - u''_j \frac{\partial p''}{\partial x_i} = -\left[\frac{\partial}{\partial x_j}(u''_i p'') + \frac{\partial}{\partial x_i}(u''_j p'')\right] + p''\left(\frac{\partial u''_i}{\partial x_j} + \frac{\partial u''_j}{\partial x_i}\right), \tag{3.40}$$

and we rearrange the viscous terms as follows

$$\nu\left(u''_i \frac{\partial^2 u''_j}{\partial x_k^2} + u''_j \frac{\partial^2 u''_i}{\partial x_k^2}\right) = \nu\left[\frac{\partial^2 (u''_i u''_j)}{\partial x_k^2} - 2\left(\frac{\partial u''_i}{\partial x_k}\right)\left(\frac{\partial u''_j}{\partial x_k}\right)\right]. \tag{3.41}$$

We can neglect the first term on the right-hand side of Eq. (3.41), the molecular diffusion term. After some rearranging using the mass-continuity relations, (3.29), the equation is then averaged to yield:

$$\begin{aligned}
\frac{\mathrm{d}}{\mathrm{d}t}\left(\overline{u''_i u''_j}\right) &= \frac{\partial}{\partial t}\left(\overline{u''_i u''_j}\right) + \overline{u}_k \frac{\partial}{\partial x_k}\left(\overline{u''_i u''_j}\right) \\
&= \overset{\text{(a)}}{-\overline{u''_j u''_k}\frac{\partial \overline{u}_i}{\partial x_k}} \overset{\text{(b)}}{- \overline{u''_i u''_k}\frac{\partial \overline{u}_j}{\partial x_k}} \\
&\quad \overset{\text{(c)}}{+\delta_{i3}\frac{g}{\Theta_0}\overline{u''_j \theta''_\rho}} \overset{\text{(d)}}{+\delta_{j3}\frac{g}{\Theta_0}\overline{u''_i \theta''_\rho}} \overset{\text{(e)}}{-\frac{\partial}{\partial x_k}\left(\overline{u''_i u''_j u''_k}\right)} \\
&\quad \overset{\text{(f)}}{-\left(\frac{\partial}{\partial x_j}\overline{u''_i p''} + \frac{\partial}{\partial x_i}\overline{u''_j p''}\right)} \overset{\text{(g)}}{+\overline{p''\left(\frac{\partial u''_i}{\partial x_j} + \frac{\partial u''_j}{\partial x_i}\right)}} \\
&\quad \overset{\text{(h)}}{+\left(\varepsilon_{jmk} f_k \overline{u''_m u''_i} + \varepsilon_{imk} f_k \overline{u''_m u''_j}\right)} \\
&\quad \overset{\text{(i)}}{-2\nu\overline{\left(\frac{\partial u''_i}{\partial x_k}\right)\left(\frac{\partial u''_j}{\partial x_k}\right)}}.
\end{aligned} \tag{3.42}$$

Equation (3.42) is called the *Reynolds stress equation*. Thus instead of making an assumption regarding the behavior of the Reynolds stress $\overline{u''_i u''_j}$, we have instead formed a prognostic equation for it which can be integrated in a numerical model. We note that the subgrid turbulence kinetic energy equation, (3.31), can be retrieved from (3.42) by simply contracting the indices $i$ and $j$ and then dividing by 2. The meaning of the various terms contributing to the time variation of the Reynolds stress is given below.

The left-hand side of Eq. (3.42) represents the substantive derivative of $\overline{u_i'' u_j''}$ or the rate of change following a trajectory of mean velocity $\overline{u}_k$.

Terms (a) and (b) on the right-hand side of Eq. (3.42) are the *mean-gradient production* terms. They represent the production of new velocity correlations by the interaction of the turbulence with the mean velocity field, as well as the modification of existing correlations by the variation of the mean field.

Terms (c) and (d), the *buoyancy terms*, represent the production or reduction of the velocity correlation due to interactions of velocity fluctuations with fluctuations in temperature and/or water content. In a cloud system, buoyancy production of turbulence can arise from the mixing of saturated and subsaturated air parcels. The resultant mixed parcels contain evaporating and/or condensing volumes of air which, in turn, can lead to further buoyant production of turbulence by the latent heats exchanged during the phase changes.

Term (e) represents the turbulent transport of the covariance $\overline{u_i'' u_j''}$ by the turbulent fluctuations, and is often called the *velocity diffusion term*. It is the action of this term which causes the movement of air parcels from a turbulent region into a quiescent environment. This overshooting of turbulent air parcels causes the entrainment of quiescent air with differing properties into the turbulent field.

Term (f) is called the *pressure diffusion* term. The term is a redistribution term that often acts to destroy the existing stress.

Term (g) is called the *tendency toward isotropy* term. Its name arises from the fact that, in the total turbulent energy equation, this term vanishes because the divergence is zero in incompressible flow. Thus the term is thought to behave in such a way that the turbulent energy is rearranged among the various components.

Term (h) represents the production or reduction of $\overline{u_i'' u_j''}$ due to the earth's rotation.

Term (i) is the *molecular dissipation term* and represents the conversion of the kinetic energies $\overline{u''^2}$, $\overline{v''^2}$, and $\overline{w''^2}$ into internal energy by the action of molecular viscosity.

It should be noted that since $\theta_\rho''/\Theta_0 \approx \theta''/\theta_0 + 0.61 r_v'' - r_l'' - r_i''$, the buoyancy production terms (c) and (d) introduce correlations among all the thermodynamic components of the system. Thus in a cloud system, fluctuations in total condensate water $r_w''$, and phase changes causing anomalies in $\theta''$, $r_v''$, $r_l''$ and $r_i''$, can make substantial contributions to buoyancy production of Reynolds stress. To model such processes requires the formulation of equations for scalar fluxes such as $\overline{u_i'' r_w''}$, $\overline{u_i'' \theta_{il}}$, and $\overline{u_i'' r_v''}$, which is detailed below.

One of the problems of second-order turbulence theory is that Eq. (3.42) is not closed as it stands, even when equations forecasting correlations among velocity fluctuations and thermodynamic variable fluctuations are formulated. The triple-correlation production term (e) either requires the formulation of a prognostic equation for $(\partial/\partial t)\overline{u_i''u_j''u_k''}$ or the modeling of this term in analogy to the Reynolds stress terms [i.e. (3.18)]. In the former case, this introduces the formulation of an infinite set of moment equations, while, in the latter case, the artificiality of eddy diffusion is again introduced, but this time on the third-order terms. Closures must also be devised for terms (f), (g), and (i).

Many different closure models, of varying complexity and effectiveness, have been devised for terms (e), (f), (g), and (i). In a recent review article, Mironov (2009) referred to the literature on this topic as "extremely voluminous." Thus, we refer readers to specialty texts or review articles on the topic [e.g. Pope (2000), Mironov (2009), and references therein]. Debate about these closures, and further research, will likely continue for many years, especially as new observational studies become available to evaluate these closures.

In a complete cloud system, closure models must be formulated for mean prognostic equations for heat and water substance. Moreover, closure of the Reynolds stress equation, (3.42), requires models for covariances with velocity fluctuations in order to model the buoyancy terms.

Following a procedure similar to that used in deriving Eq. (3.42), one can form tendencies on scalar covariances and fluxes on scalar quantities. Thus, for a cloud system composed of only cloud water (no rain) and a thermodynamic energy equation using $\theta_{il}$, a second-order turbulence model can be expressed as follows. For the scalar covariances, $\overline{\theta_{il}''^2}$, $\overline{r_t''^2}$, $\overline{\theta_{il}''r_t''}$, etc., the tendencies written in terms of free variables $a''$ and $b''$ are

$$\frac{\partial}{\partial t}(\overline{a''b''}) = -\overline{u}_i\frac{\partial}{\partial x_i}(\overline{a''b''}) - \overline{u_i''a''}\frac{\partial \overline{b}}{\partial x_i} - \overline{u_i''b''}\frac{\partial \overline{a}}{\partial x_i} - \frac{\partial}{\partial x_i}(\overline{u_i''a''b''}) - (\gamma_a + \gamma_b)\overline{\frac{\partial a}{\partial x_i}\frac{\partial b}{\partial x_i}}, \tag{3.43}$$

where the last term on the right-hand side represents the rate of molecular dissipation of the scalar correlation $\overline{a''b''}$. Similarly, the tendencies for the fluxes of the scalar quantities $\overline{u_i'''\theta_{il}''}$, $\overline{u_i''r_t''}$, etc., have the form

$$\frac{\partial}{\partial t}(\overline{u_i''a''}) = -\overline{u}_j\frac{\partial}{\partial x_j}(\overline{u_i''a''}) - \overline{a''u_j''}\frac{\partial \overline{u}_i}{\partial x_j} - \overline{u_i''u_j''}\frac{\partial \overline{a}}{\partial x_j} - \frac{\partial}{\partial x_j}(\overline{u_j''u_i''a''}) - \frac{1}{\rho_0}\frac{\partial}{\partial x_i}(\overline{p''a''}) + \frac{1}{\rho_0}\overline{p''\frac{\partial a''}{\partial x_i}} + \delta_{i3}\frac{g}{\Theta_0}\left(\overline{\theta_\rho''a''}\right) + \varepsilon_{ijk}f_k\overline{u_j''a''} - (\gamma_a + \nu)\overline{\frac{\partial u_i''}{\partial x_j}\frac{\partial a''}{\partial x_j}}. \tag{3.44}$$

Notice that subgrid sources and sinks (e.g. $p[s] - \overline{p[s]}$ and $p[r] - \overline{p[r]}$) are typically neglected, for simplicity, although subgrid condensation is considered later in Section 3.4.

As the physics of the cloud system is expanded to include rainwater and ice phase, the number of equations for scalar correlations and fluxes increases enormously. Note also that Equations (3.43) and (3.44) have additional unclosed terms (e.g. new triple-moment terms, pressure-correlation terms, and dissipation terms). Models must be developed for these unclosed terms before (3.43) and (3.44) can be integrated in a numerical model.

It should also be noted that physical processes such as precipitation and radiation processes become intimately coupled in Eqs. (3.43) and (3.44). For example, the variance and transport of water substance quantities will vary, depending on the models of cloud physical processes. Furthermore, as will be discussed in Chapter 4, the rates of production of precipitation will change, depending on the turbulent structure of the cloud systems. Thus turbulence affects the precipitation structure of a cloud system, and precipitation affects its turbulent structure.

A number of approximations to the complete system of equations can be made. For example, assuming locally steady and homogeneous turbulence, the time-tendency and flux-divergence terms can be neglected in the Reynolds stress equation, (3.42). Consequently, there will be a balance between the mean-gradient production, the buoyancy terms, the Coriolis terms, and dissipation. A diagnostic equation can then be formed for the Reynolds stresses with a few more approximations. The same technique can be applied to (3.43) and (3.44). Furthermore, some planetary boundary layer models truncate these equations further by considering only vertical fluxes, which substantially reduces the number of equations and calculations (and, thus, reduces the cost of the turbulence model).

It is obvious that these approximations may be overly restrictive when applied to various cloud systems. However, when dealing with the complete set of Eqs. (3.42), (3.43), and (3.44), one can experiment with various combinations of prognostic equations and diagnostic equations derived from equilibrium approximations. Mellor and Yamada (1974) have discussed the impact of forming various levels of prognostic/diagnostic equations in modeling the evolution of the planetary boundary layer. Similar analyses must be done for combinations of prognostic and diagnostic equations for models of stratocumuli, trade-wind cumuli, or other cloud systems based on the higher ordered equation set.

## 3.3. LARGE EDDY SIMULATION

In the previous section all of the turbulent detail was removed by an averaging operator. The predicted fields are thus smooth and the effects of turbulent eddies must be accounted for by a parameterization of turbulence.

An alternative approach is to resolve *some* of the turbulence, specifically only the largest turbulent features that contain most of the turbulent kinetic energy of a flow and that carry most of the scalar flux. Consider, for example, the cumulonimbus cloud shown in Fig. 3.1. If there were ~100 grid points across the cloud in each direction (which would be computationally tractable), then at least the largest of the turbulent eddies could be simulated. This method of simulating turbulent flows is called *large eddy simulation* (LES).

In LES, we apply a *spatial filter* to all variables $a$, decomposing them into a filtered (i.e. resolved) part and a sub-filter-scale (i.e. subgrid-scale) part.

$$a = \overline{a} + a''. \tag{3.45}$$

The most common filters in atmospheric sciences have been the Gaussian filter, the top-hat (or "box") filter, and the wave-cutoff filter in spectral space. There have been a few direct comparisons of the choice of filter (e.g. Moeng and Wyngaard, 1988, 1989), although the final choice seems to be tied intimately to a model's numerical techniques (e.g. whether the model is based on finite-difference techniques, spectral methods, etc.). Nevertheless, the details of the filter are not important at this point in the development, as long as the filtering operator commutes with differentiation, e.g.

$$\overline{\frac{\partial a}{\partial x}} = \frac{\partial \overline{a}}{\partial x}. \tag{3.46}$$

It is important to recognize that no assumptions have been made about the filtering operator at this point, and thus no rules can be imposed about the sub-filter-scale fluctuations. For example, filtering of a sub-filter-scale flow does not necessarily yield a zero result, and multiple applications of the filter may not yield the same result: that is,

$$\overline{a''} \neq 0, \qquad \overline{\overline{a}} \neq \overline{a}. \tag{3.47}$$

This is in sharp contrast to the Reynolds averaging procedure discussed earlier in this chapter.

We begin with the equations for mass continuity and velocity. Filtering of (3.6) gives simply

$$\frac{\partial \overline{u}_i}{\partial x_i} = 0. \tag{3.48}$$

By subtracting (3.48) from (3.6) we find

$$\frac{\partial u_i''}{\partial x_i} = 0. \tag{3.49}$$

For the velocity equation we begin with (3.9) which we simply filter:

$$\frac{\partial \overline{u}_i}{\partial t} + \frac{\partial \left(\overline{u_i u_j}\right)}{\partial x_j} - \varepsilon_{ijk}\overline{u}_j f_k + \frac{1}{\rho_0}\frac{\partial \overline{p'}}{\partial x_i} - \delta_{i3} g \frac{\overline{\theta'_\rho}}{\Theta_0} = \nu \frac{\partial^2 \overline{u}_i}{\partial x_j^2}. \tag{3.50}$$

Recall that viscous dissipation occurs only at the smallest scales in a flow (i.e. at scales of order $\eta$). The filtering operator removes these scales, and hence the term on the right-hand side of (3.50) can be neglected. The second term in (3.50) is particularly interesting; using (3.45) it can be expanded:

$$\overline{u_i u_j} = \overline{\overline{u}_i \overline{u}_j} + \overline{\overline{u}_i u''_j} + \overline{u''_i \overline{u}_j} + \overline{u''_i u''_j}. \tag{3.51}$$

The first term on the right-hand side of (3.51) can be inconvenient to calculate, as it requires filtering of two variables and then filtering of their product. Therefore, (3.51) is often rewritten as

$$\begin{aligned}\overline{u_i u_j} &= \overline{u}_i \overline{u}_j - \overline{u}_i \overline{u}_j + \overline{\overline{u}_i \overline{u}_j} + \overline{\overline{u}_i u''_j} + \overline{u''_i \overline{u}_j} + \overline{u''_i u''_j} \\ &= \overline{u}_i \overline{u}_j + \tau_{ij} + \frac{2}{3}\delta_{ij} e\end{aligned} \tag{3.52}$$

where $e = (\overline{u_k u_k} - \overline{u}_k \overline{u}_k)/2$ is the sub-filter-scale turbulence kinetic energy, and the turbulent stress tensor for LES, $\tau_{ij}$, is defined as follows:

$$\tau_{ij} \equiv \left[\overline{\overline{u}_i \overline{u}_j} - \overline{u}_i \overline{u}_j\right] + \left[\overline{\overline{u}_i u''_j} + \overline{u''_i \overline{u}_j}\right] + \overline{u''_i u''_j} - \frac{2}{3}\delta_{ij} e. \tag{3.53}$$

The first bracketed term in (3.53) is the Leonard stress [after (Leonard, 1974)] , the second bracketed term is the cross-term stress, and the third term in (3.53) is the familiar Reynolds stress. The term with $e$ is included in the definition of $\tau_{ij}$ to ensure that it is a deviatoric tensor (i.e. to ensure that $\tau_{ii} = 0$). The final form of the filtered velocity equation can thus be written as

$$\frac{\partial \overline{u}_i}{\partial t} + \frac{\partial \left(\overline{u}_i \overline{u}_j\right)}{\partial x_j} - \varepsilon_{ijk}\overline{u}_j f_k + \frac{\partial \phi}{\partial x_i} - \delta_{ij} g \frac{\overline{\theta'_\rho}}{\Theta_0} = -\frac{\partial \tau_{ij}}{\partial x_j}, \tag{3.54}$$

in which $\phi = \overline{p'}/\rho_0 + (2/3)e$ is the modified pressure variable. Notice that (3.54) is very similar in form to the original velocity equation, (3.9), except that all the variables have been replaced by filtered variables, the viscous term has been replaced by a turbulent stress divergence term, and pressure has been replaced by a modified pressure.

Similarly, equations are developed for the thermodynamic and moisture equations:

$$\frac{\partial \overline{s}}{\partial t} + \frac{\partial \left(\overline{u}_i \overline{s}\right)}{\partial x_i} = -\frac{\partial \tau_{si}}{\partial x_i} + \overline{p[s]}, \tag{3.55}$$

$$\frac{\partial \bar{r}}{\partial t} + \frac{\partial \left(\bar{u}_i \bar{r}\right)}{\partial x_i} = -\frac{\partial \tau_{ri}}{\partial x_i} + \overline{p[r]}. \qquad (3.56)$$

To integrate these equations, the unknown terms $\tau_{ij}$, $\tau_{si}$, $\tau_{ri}$, etc., must be parameterized. There are many models for doing this. Some of the models presented below are very similar in form (and in derivation) to the subgrid models presented in the previous section for Reynolds averaging. However, it is important to understand that the derivations are different in concept (i.e. *all* turbulence is removed by averaging in Reynolds averaging, and only small-scale turbulent eddies are removed by the filter in LES). Furthermore the constants, length-scales, and other parameters in LES subgrid models are quite different, sometimes by several orders of magnitude.

The standard method is analogous to the first-order closure for Reynolds averaging where the sub-filter-scale stress is assumed to be proportional to the rate of strain of the resolved flow, i.e.

$$\tau_{ij} = -K_m \left( \frac{\partial \bar{u}_i}{\partial x_j} + \frac{\partial \bar{u}_j}{\partial x_i} \right), \qquad (3.57)$$

where $K_m$ is the eddy viscosity for momentum, and scalar fluxes are modeled by downgradient diffusion

$$\tau_{si} = -K_s \frac{\partial \bar{s}}{\partial x_i}, \qquad (3.58)$$

where $K_s$ is the scalar eddy diffusivity.

A common model for determining $K_m$ and $K_s$ originated with Lilly (1962) and Smagorinsky (1963). By assuming the subgrid turbulent flow is statistically steady and homogeneous, and assuming the production of subgrid energy is balanced by dissipation, the following model is obtained:

$$K_m = (C_s l)^2 S \left( 1 - \frac{\mathrm{Ri}}{\mathrm{Pr}} \right)^{1/2} \qquad (3.59)$$

where $C_s$ is called the Smagorinsky constant and

$$S^2 = \frac{\partial \bar{u}_i}{\partial x_j} \left( \frac{\partial \bar{u}_i}{\partial x_j} + \frac{\partial \bar{u}_j}{\partial x_i} \right), \qquad \mathrm{Ri} = \frac{N^2}{S^2}, \qquad \mathrm{Pr} = K_m / K_s, \qquad (3.60)$$

where $S$ is the deformation, $N^2$ is the squared Brunt-Väisälä frequency (although $N_m^2$ is used in saturated conditions), Ri is the Richardson number, and Pr is the Prandtl number. Note that $K_m$ is set to zero if the term in parentheses in (3.59) is less than zero (i.e. if Ri > Pr). This model is called the Smagorinsky model. Technically, the length scale, $l$, should be set to the filter scale (i.e. the scale where the filter begins to remove turbulent fluctuations). In practice, $l$ is typically set to the grid spacing or, if the horizontal and vertical grid spacings are different, to $l = (\Delta x \Delta y \Delta z)^{1/3}$. This leaves only two variables that need to be set in (3.59): $C_s$ and Pr. For homogeneous, isotropic subgrid turbulence, Lilly (1967) showed that $C_s = (1/\pi)[2/(3\alpha)]^{3/4}$, where $\alpha$ is the Kolmogorov constant. For $\alpha = 1.5$, $C_s = 0.17$. There is little guidance as to how to set Pr, and cloud modelers typically use $\mathrm{Pr} \approx 1/3$.

Another model, introduced by Lilly (1967) and implemented by Deardorff (1980), is to relate the variables $K_m$ and $K_s$ to the subgrid turbulence kinetic energy,

$$K_m = c_m l \overline{e}^{1/2}, \tag{3.61}$$

$$K_s = c_h l \overline{e}^{1/2} \tag{3.62}$$

where $l$ is the same length-scale that is used in the Smagorinsky model. A predictive equation for $e$ can be determined using (3.9), (3.54), and the definition of $e$. The exact result depends on the filter being used but is usually of the form

$$\frac{\partial \overline{e}}{\partial t} = -\frac{\partial\left(\overline{u}_i \overline{e}\right)}{\partial x_i} - \tau_{ij}\frac{\partial \overline{u}_i}{\partial x_j} + \frac{g}{\Theta_0}\overline{w''\theta''_\rho} - \frac{\partial(\overline{u''_i e} + \overline{u''_i \phi})}{\partial x_i} - \epsilon \tag{3.63}$$

where $\epsilon$ is the dissipation rate. Three unclosed terms appear in (3.63). The term attributable to buoyancy is proportional to the squared Brunt-Väisälä frequency,

$$\frac{g}{\Theta_0}\overline{w''\theta''_\rho} = -K_s N^2, \tag{3.64}$$

the triple-moment terms are modeled assuming downgradient diffusion,

$$\overline{u''_i e} + \overline{u''_i \phi} = -2K_m \frac{\partial \overline{e}}{\partial x_i}, \tag{3.65}$$

and dissipation is given, on dimensional grounds, as

$$\epsilon = c_\epsilon \overline{e}^{3/2}/l. \tag{3.66}$$

Assuming homogeneous and isotropic subgrid turbulence, Lilly (1967) showed that $c_m = (1/\pi)[2/(3\alpha)]^{3/2} = 0.094$ and that $c_\epsilon = \pi^2 c_m = 0.93$. The variable $c_h$ is more uncertain, and values between roughly 0.1 and 0.3 have been used. The variables $c_m$ and $c_h$ can be tuned by small amounts in numerical models to produce desirable results, which becomes necessary if numerical algorithms and/or spatial filters are changed (e.g. Moeng and Wyngaard, 1988).

The $\overline{e}$-based subgrid model requires the integration of an additional predictive equation, (3.63), which makes it more expensive to implement than the Smagorinsky model. However, the ability to transport $\overline{e}$ across the grid (by advective and diffusive transport) has been shown to be advantageous for some situations, such as cloud edges and entrainment zones (e.g. Stevens et al., 1999).

It is possible to develop higher-order models for LES, analogous to the higher-order models discussed for Reynolds-averaging in Section 3.2.3. A few researchers, such as Deardorff (1974) and Schmidt and Schumann (1989), have implemented higher-order models of various complexity for LES. However, these schemes are costly, so LES modelers have traditionally used either the Smagorinsky or $\overline{e}$-based models and have used the additional computer resources to increase resolution.

Earliest large-eddy simulations used constant values for the parameters $l$, $C_s$, $c_h$, and $c_\epsilon$. However, it was found that such a configuration was overly diffusive in areas of strong static stability, presumably because the development of turbulence is suppressed under stable stratification. Hence, most models now use stability-dependent values, e.g.

$$l = \min\left[0.82\left(\frac{\overline{e}}{N^2}\right)^{\frac{1}{2}}, \Delta\right], \tag{3.67}$$

with

$$c_\epsilon = 0.19 + 0.74\frac{l}{\Delta}, \qquad c_h = 0.1 + 0.2\frac{l}{\Delta} \tag{3.68}$$

where $\Delta = (\Delta x \Delta y \Delta z)^{1/3}$. The lower limit of $c_\epsilon$ (= 0.19) is set to ensure that subgrid turbulence intensity goes to zero for $\mathrm{Ri} > 0.25$. Note also that $l$ is typically set to $\Delta$ if $N^2 < 0$. More details are available in Deardorff (1980) and Stevens et al. (1999). Stability-dependence can also be built into the Smagorinsky model (e.g. Mason and Brown, 1999; Stevens et al., 1999), although there are many different models for doing so.

It is very important to recognize that LES has certain computational requirements. For example, LES *must* be conducted using a three-dimensional numerical model. This is because turbulent processes are inherently three-dimensional: the energy cascade from large scales to small scales occurs primarily through vortex stretching. The consequences of using a two-dimensional model to simulate turbulent flows was evaluated by Moeng et al. (2004), who found that simulated velocity fields had the greatest errors. Furthermore, LES requires that the grid spacing ($\Delta$) be sufficiently smaller than the characteristic scale of the turbulent flow ($L$) in order for the energy cascade to be reproduced. Bryan et al. (2003) and Wyngaard (2004) showed that this criterion could be interpreted in terms of the Reynolds number for a simulated flow:

$$\mathrm{R}_t \sim \left(\frac{L}{\Delta}\right)^{4/3}. \tag{3.69}$$

For the simulated flow to be turbulent (i.e. for $R_t \gg 1$) the grid spacing must be much less than the characteristic scale of the flow (i.e. $\Delta \ll L$). If this criterion is not met then simulated convection tends to be too intense (*if* it develops), which generally leads to excessive precipitation in simulations of deep convection [(Bryan and Rotunno, 2005) see also discussion at the end of Chapter 12].

The subgrid models discussed thus far have all assumed that kinetic-energy is transferred from resolved scales to unresolved scales where it is ultimately dissipated by unresolved turbulence. That is, there is a one-way (down-scale) transfer of energy. However, in turbulent flows there can occasionally be transfer of kinetic-energy from small-scales to large-scales that cannot be reproduced by the above models. Called "backscatter," the effect appears to be most significant in stable flows with large shear. Because LES, by definition, does not have any information about small-scale flows, the effect must be included stochastically in numerical model simulations. Efforts to include *stochastic backscatter* have been reported by Brown et al. (1994) and Weinbrecht and Mason (2008), among others. These studies have found that backscatter can improve statistics from LES, and may also be important for relatively coarse-resolution simulations (i.e. simulations in which $L/\Delta$ is only of order 10).

Although LES modeling with the above subgrid-scale closures has been shown to work very well at simulating many turbulent atmospheric flows, there are some notable deficiencies. The most difficult problem to overcome has been the behavior of LES models near the surface and within strong stable layers (e.g. the inversions at the top of stratocumulus layers). The problem arises, in part, because turbulence becomes small (i.e. unresolvable) and anisotropic near these boundaries. Thus, although the underlying framework of LES may be adequate, it has been shown that subgrid-scale models do not perform well (e.g. Sullivan et al., 1994). To remedy this problem, a number of different solutions have been proposed, including dynamic eddy viscosity models that *predict* coefficients in the above relations (e.g. Germano et al., 1991; Zhou et al., 2001).

A further problem with traditional LES is that the eddy diffusivity models, (3.57) and (3.58), do not always accurately model small-scale turbulence, as demonstrated by a comparison of DNS and LES simulations by Clark et al. (1979). Consequently, some modelers have developed entirely different models for the subgrid stress tensor. For example, Sullivan et al. (1994) developed a "two-part" eddy viscosity model of the form

$$\tau_{ij} = -2\nu_t \gamma S_{ij} - 2\nu_T \langle S_{ij} \rangle, \tag{3.70}$$

where $\nu_t$ is the fluctuating eddy viscosity which uses the standard model (3.61) (with, here, $K_m = \nu_t$), $\nu_T$ is a mean-field eddy viscosity which is calculated based on the mean shear, and is adjusted dynamically during a simulation to match similarity theory in the surface layer, angled brackets denote horizontal averaging, and $\gamma$ is an isotropy factor that essentially reduces $\nu_t$ near the surface to account for the anisotropy of turbulent eddies. Sullivan et al. (1994) found that such a model was necessary to produce expected results near the surface for stable boundary layers.

Kosović (1997) developed a nonlinear subgrid model in which small-scale isotropy is not imposed and where the subgrid stresses are related to resolved velocity gradients. The model has the form:

$$\begin{aligned}\tau_{ij} = &-C_e l 2\bar{e}^{1/2} S_{ij} - C_1 l^2 \left( S_{ik} S_{kj} - \frac{\delta_{ij}}{3} S_{mn} S_{nm} \right) \\ &- C_2 l^2 \left( S_{ik} \Omega_{kj} - \Omega_{ik} S_{kj} \right)\end{aligned} \tag{3.71}$$

where $C_e$, $C_1$, and $C_2$ are constants, and $\Omega_{ij}$ is the rotation tensor. This model allows for non-isotropic subgrid eddies, and also permits energy transfer from small scales to large scales (i.e. backscatter). The constants $C_1$ and $C_2$ were set by Kosović (1997) to produce a specified amount of backscatter, and tests demonstrated improved solutions for a shear-driven boundary layer.

Another potential shortcoming of the subgrid models presented above is the lack of explicit treatment of cloud microphysical processes. This issue is addressed in the next section.

## 3.4. SUBGRID CONDENSATION AND EVAPORATION

Traditionally, various cloud models have been formulated such that the amount of cloud water is diagnosed by an "all or nothing" scheme. Using this scheme, condensation occurs only when the mixing ratio of the air, averaged over a grid volume, reaches the saturation mixing ratio determined from the average temperature of the grid volume. In a turbulent cloudy environment, however, one can expect to find local regions of condensate bordering subsaturated regions, all within a given averaging volume. An obvious example is the case where the averaging volume is large enough to encompass a field of fair-weather cumuli. Thus, if one were to make a horizontal traverse across the domain, one would expect to encounter local saturated cloudy regions followed by subsaturated regions. The "all or nothing" scheme might diagnose $\overline{r}_t - r_s(\overline{T})$ (where $r_s$ is the saturation mixing ratio) as being subsaturated and therefore free of clouds. In actual fact there would exist an average cloud liquid-water content $\overline{r_c}$ in the domain which would contribute to the thermodynamics and dynamics of the cloud field even though $\overline{r}_t - r_s(\overline{T}) < 0$.

The application of such an "all or nothing" procedure in cloud or mesoscale models can lead to problems, since this treatment of condensation delays the release of latent heat during the early stages of cloud formation until a grid point is driven to saturation. Selected grid points then release latent heat explosively while neighboring ones may not. The result is that latent heat is released on the smallest horizontal scales resolved by the model, with a resultant generation of computational noise.

Several authors have derived expressions for average cloud water and fractional cloud coverage for a fluctuating water and thermal field [e.g. Sommeria and Deardorff (1977); Mellor (1977); Manton and Cotton (1977); Oliver et al. (1978); Bougeault (1981b); Randall (1987); Lewellen and Yoh (1993); Tompkins (2002); Larson et al. (2002)]. Let us consider first the approach taken by Manton and Cotton (1977) (MC). For simplicity, we assume the system is nonprecipitating and that water substance is distributed into $r_t = r_v + r_c$. MC assumed that the zero-mean random variable $r_t'' - r_s''$ is normally distributed with variance

$$\sigma_c^2 = \overline{(r_t'' - r_s'')^2}. \tag{3.72}$$

The corresponding average liquid-water content is

$$\overline{r_c} = (\overline{r_t} - \overline{r_s})h + (2\pi)^{1/2} \exp\left\{-(\overline{r_t} - \overline{r_s})^2/2\sigma_c^2\right\}, \tag{3.73}$$

where

$$h = \frac{1}{2}\left\{a + \mathrm{erf}(\overline{r_t} - \overline{r_s})/2^{1/2}\sigma_c\right\}. \tag{3.74}$$

The function $h$ behaves as a Heaviside unit step function as $\sigma_c$ approaches zero; i.e. $h = H(\overline{r_t} - \overline{r_s})$ as $\sigma_c \to 0$. Equation (3.73) allows the formation of cloud water contributing to a nonzero $\overline{r_c}$ even though $\overline{r_t} - \overline{r_s} < 0$. To define $\sigma_c^2$, we must determine $\overline{r_t''^2}$, $\overline{r_s''^2}$, and $\overline{r_t'' r_s''}$, since $\sigma_c^2 = \overline{r_t''^2} + \overline{r_s''^2} - 2\overline{r_t'' r_s''}$. The usefulness of this approach hinges on the validity of the assumption that $r_t'' - r_s''$ is normally distributed.

Banta (1979) examined this hypothesis by computing the distribution of $r_t'' - r_s''$ from aircraft data collected in the subcloud layer. Cloud-layer data were not used at that time because of the difficulty of simultaneously measuring $r_t = r_v + r_c$ and $r_s(T)$ in cloudy air and clean air with instrument responses of comparable sensitivities. Figures 3.2 and 3.3 illustrate that for 1000-m averaging legs near the surface and well into the interior of the atmospheric boundary layer (ABL), all $r_t'' - r_s''$ are well represented by a normal distribution. While Banta's analysis lends confidence to MC's hypothesis, it does not confirm that $r_t'' - r_s''$ is normally distributed in the cloud layer or in more complex cloud regimes.

Banta also tested the validity of Sommeria and Deardorff's (SD) hypothesis that $\theta_l$ and $r_t$ have joint normal probability distributions within a given grid volume, where $\theta_l = \theta[1 - L_{lv} r_l/(c_{pa}T)]$ is a simplified form of $\theta_{il}$ that can

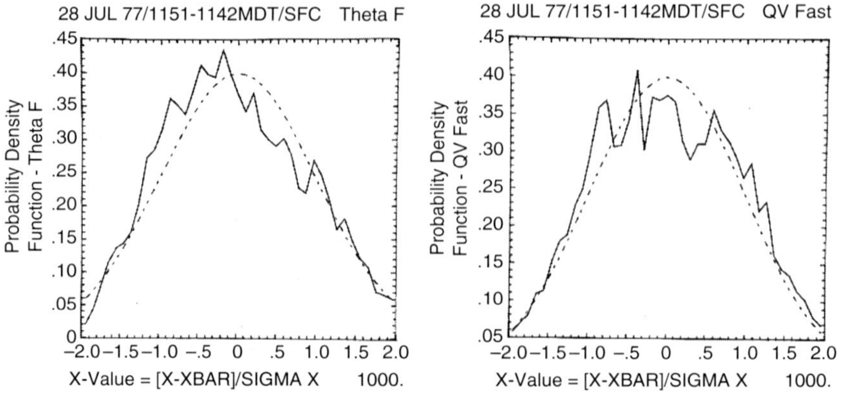


**FIGURE 3.2 Estimated probability density functions for $\theta$ (left) and $r_v$ (right), superimposed on a dashed curve of the standard normal density.** Data are from 28 July for a flight leg over South Park, Colorado, at a height of 50 m off the surface, starting at 1152 MDT. Averaging interval was 1000 m. *(From Banta (1979))*

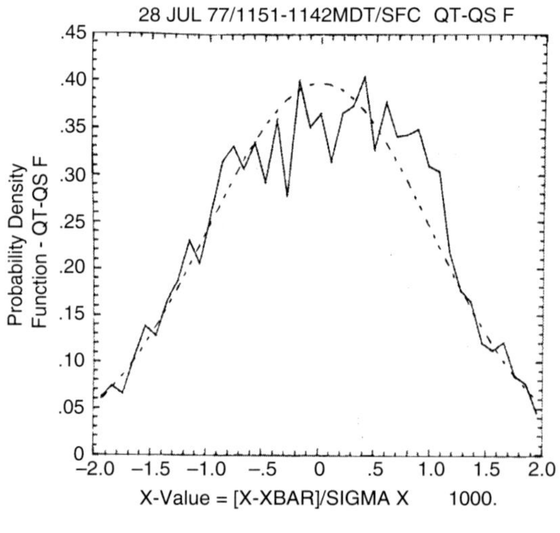


**FIGURE 3.3 Estimated probability density function for $r_t - r_s$ superimposed on standard normal curve (dashed line).** Data are from the same flight leg as in Fig. 3.2. *(From Banta (1979))*

be used to study shallow, liquid-only clouds. Under this assumption SD showed that

$$\overline{r_c} = \frac{1}{1+\beta_1 \overline{r_s}} \left\{ R(\overline{r_t} - \overline{r_s}) + \frac{\sigma_c}{\sqrt{2\pi}} \exp\left[ -\frac{(\overline{r_t} - \overline{r_s})^2}{2\sigma_c^2} \right] \right\}, \tag{3.75}$$

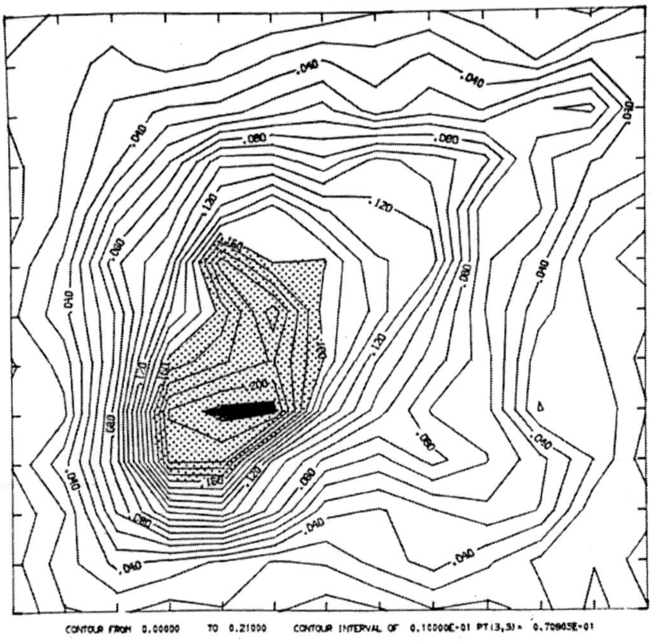


**FIGURE 3.4 Estimated joint probability density function (PDF) for $\theta$ (abscissa) and $r_v$ (ordinate).** The scale on both axes runs from −2 standard deviations to +2 standard deviations ($x$ and $y$ go from −2 to +2). The shaded region indicates where PDF values exceed 0.16, which represents more than 72 observations per point. The darkened area indicates PDF values exceeding 0.21. The correlation coefficient for this data set is +0.155. *(From Banta (1979))*

where $\beta_1 = 0.622(L_{lv}/R_\mathrm{a}T_l)(L_{lv}/c_{pa}T_l)$, $T_l = T - L_{lv}r_l/c_{pa}$ is the liquid-water temperature,

$$\sigma_\mathrm{c} = (\sigma_{r_\mathrm{t}}^2 + \alpha_1\sigma_{\theta_l}^2 - 2\phi\alpha_1\sigma_{r_\mathrm{t}}\sigma_{\theta_l})^{1/2} = \left(\overline{r_\mathrm{t}''^2} + \overline{r_\mathrm{s}''^2} - 2\overline{r_\mathrm{t}''r_\mathrm{s}''}\right)^{1/2},$$

$$\phi = \overline{r_\mathrm{t}''\theta_l''}/(\sigma_{r_\mathrm{t}}\sigma_{\theta_l}) \quad \text{and} \quad \alpha_1 = (\overline{p}/p_0)^{0.286}\left(\frac{\partial r_\mathrm{s}}{\partial T}\right)_{T=T_l}.$$

The cloud fraction $R$ is

$$R = \frac{1}{2\sqrt{2\pi}\,\sigma_{\theta_l}}\int_{-\infty}^{\infty}\exp\left[-\frac{(\theta_l - \overline{\theta_l})^2}{2\sigma_{\theta_l}^2}\right] \times \left\{1 + \mathrm{erf}\left[\frac{\overline{r_\mathrm{t}} - \overline{r_\mathrm{s}} - \alpha_1(\theta_l - \overline{\theta_l})}{\sqrt{2}\sigma_\mathrm{T}}\right]\right\}\mathrm{d}\theta_l. \tag{3.76}$$

Figure 3.4 illustrates Banta's evaluation of the joint distribution of $\theta_l$ and $r_\mathrm{t}$ in the subcloud layer. Departures from bivariant normality are quite pronounced.

This does not refute SD's hypothesis, but it does introduce some doubt that the hypothesis can be substantiated in general. In both cases, MC's and SD's formulations are quite complex.

Using the assumption that $\theta_l$ and $r_t$ can be represented by a bivariate Gaussian distribution, Bougeault (1981a) showed that the liquid-water content $r_c$ depends only on a linear combination of $\theta_l''$ and $r_t''$. Thus,

$$\begin{aligned} r_c &= a\Delta\overline{r_t} + 2s \quad \text{if } s > -a\Delta\overline{r_t}/2 \\ r_c &= 0 \qquad \text{otherwise} \end{aligned} \tag{3.77}$$

where $s = a(r_t'' - \alpha_1\theta_l'')/2$ and $a = (1 + L_{lv}^2\overline{r_s}/R_v c_{pa} T_l^2)^{-1}$. Recognizing the absence of field observational data to evaluate thermodynamic and water substance parameters controlling cloud coverage and average liquid-water content, Bougeault (1981b) instead analyzed the statistics of data produced by Sommeria's (1976) three-dimensional (3-D) model of the trade-wind cumulus layer.

Bougeault (1981a) (hereafter referred to as B81) used the 3-D model data set obtained in a trade-wind cumuli simulation reported by Sommeria and LeMone (1978). Cloud cover in the 3-D simulations did not exceed 10%. To develop and test the cloud fraction model, B81 averaged the 3-D model-generated data over a given horizontal level and over a time interval sufficiently long to smooth out most of the variability associated with the largest eddies (or clouds) simulated by the model. B81 evaluated three proposed models of the distribution of temperature and moisture fluctuations in a cloud layer. If we let $t$ represent the normalized variable $s/\sigma_s$ where

$$\sigma_s = (a/2)(\overline{r_t''^2} + \alpha_1^2\overline{\theta_l''^2} - 2\alpha_1\overline{\theta_l'' r_t''})^{1/2}, \tag{3.78}$$

$G(t)$ is the probability density and $Q_1$ is a measure of the departure of the mean state from saturation ($Q_1 = a\Delta\overline{r_t}/2\sigma_s$), then Eq. (3.77) can be written as

$$\begin{aligned} r_c/2\sigma_s &= Q_1 + t \quad \text{if } t > -Q_1, \\ &= 0 \qquad \text{otherwise.} \end{aligned} \tag{3.79}$$

The cloud fraction $R$ and correlations between fluctuating liquid-water content and other variables can then be expressed in terms of $G(t)$ as follows

$$R = \int_{-Q_1}^{\infty} G(t)\,\mathrm{d}t, \tag{3.80}$$

$$\overline{r_c}/2\sigma_s = \int_{-Q_1}^{\infty} (Q_1 + t)G(t)\,\mathrm{d}t, \tag{3.81}$$

$$\overline{s r_c''}/2\sigma_s^2 = \int_{-Q_1}^{\infty} t(Q_1 + t)G(t)\,\mathrm{d}t, \tag{3.82}$$

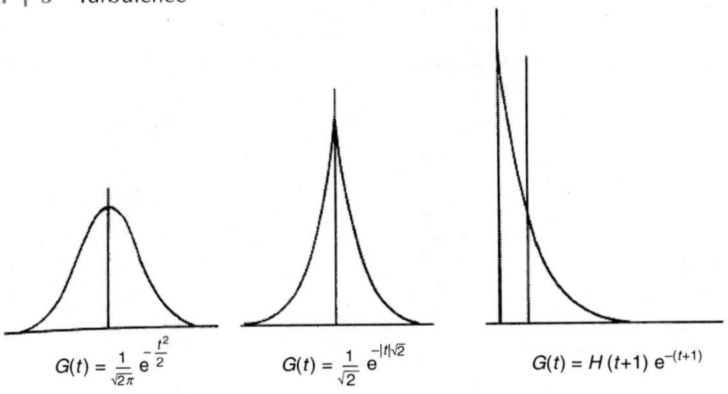


**FIGURE 3.5 Schematic representation of three proposed models of distribution.** *(From Bougeault (1981a))*

$$\overline{s^2 r_c''}/2\sigma_s^3 = \int_{-Q_1}^{\infty} t^2(Q_1 + t)G(t)\,\mathrm{d}t - \overline{r_c}/2\sigma_s. \tag{3.83}$$

B81 postulated three models to be tested against the 3-D data. They are the Gaussian model

$$G_1(t) = \frac{1}{\sqrt{2\pi}} \exp(-t^2/2), \tag{3.84}$$

the exponential model

$$G_2(t) = \frac{1}{\sqrt{2}} \exp(-|t|\sqrt{2}), \tag{3.85}$$

and a positively skewed distribution of the form

$$G_3(t) = H(t+1)\exp[-(t+1)]. \tag{3.86}$$

Each of these distributions is illustrated in Fig. 3.5. The computed distribution functions using the 3-D model data are illustrated in Fig. 3.6 along with the distribution functions for the three models at three different levels in the cloud layer. At the lowest level (500 m), the Gaussian model fits remarkably well. Higher in the cloud (1000 m), the computed distribution exhibits a pronounced tail. Thus, the skewed distribution function, Eq. (3.86), fits best. At 1250 m the distribution is not only skewed but a secondary mode in the distribution is evident. Bougeault concluded that a positively skewed distribution was more appropriate for a trade-wind cumulus layer. He suggested that the distribution of fluctuations of conservative variables should be expected to be skewed in a cumulus layer, since strong updrafts in clouds are compensated by slow descending motions outside clouds.

Bougeault's analyses were based on numerical model output. There have been disappointingly few studies of variability within observed clouds. Perhaps

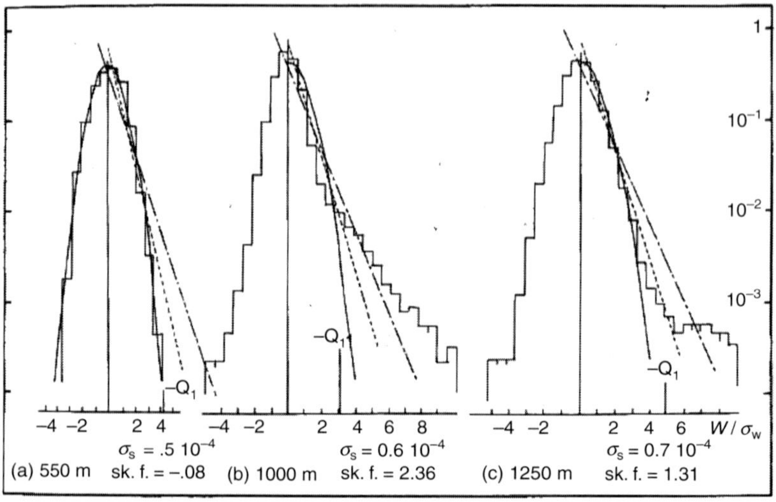


**FIGURE 3.6 Histograms of *t* from 3-D data at three levels inside the cloud layer. The three theoretical models have also been plotted: (a) 550 m, (b) 1000 m, and (c) 1250 m.** *(From Bougeault (1981a))*

the most comprehensive analysis was conducted by Larson et al. (2001, 2002) who used direct aircraft observations of cumulus and stratocumulus clouds. They used the data to evaluate several types of analytic probability density function (PDF) models.

Larson et al. (2001) found that observed distributions of scalar variability were best modeled by one-dimensional analytic PDFs that have three parameters: e.g. mean, variance, and skewness. The three-parameter PDFs worked well over a remarkably broad range of scales (from 2 km to 50 km) with no change in adjustable coefficients, which is advantageous for use in numerical weather prediction and regional-climate models. They also found that a single-Gaussian scheme based on two parameters (mean and standard deviation), such as that used by SD, worked surprisingly well, with a notable exception being for cumulus clouds with large grid boxes (greater than ~25 km).

To fully close the high-order closure model presented in Section 3.2.3, joint probability density functions between velocity and scalars are needed. Most important are the joint PDFs between vertical velocity and conserved scalars, which is needed for the buoyancy-flux terms that are important in clouds. Larson et al. (2002) analyzed joint PDFs from stratocumulus clouds (e.g. Fig. 3.7) and cumulus clouds (e.g. Fig. 3.8). For the stratocumulus cloud shown in Fig. 3.7, the joint PDFs have a roughly Gaussian shape: that is, there is generally one maximum, and the distributions are relatively unskewed. In contrast, the field of cumulus clouds shown in Fig. 3.8 is clearly not Gaussian: there are long, highly skewed tails to the distributions in certain parts of the parameter space. Further

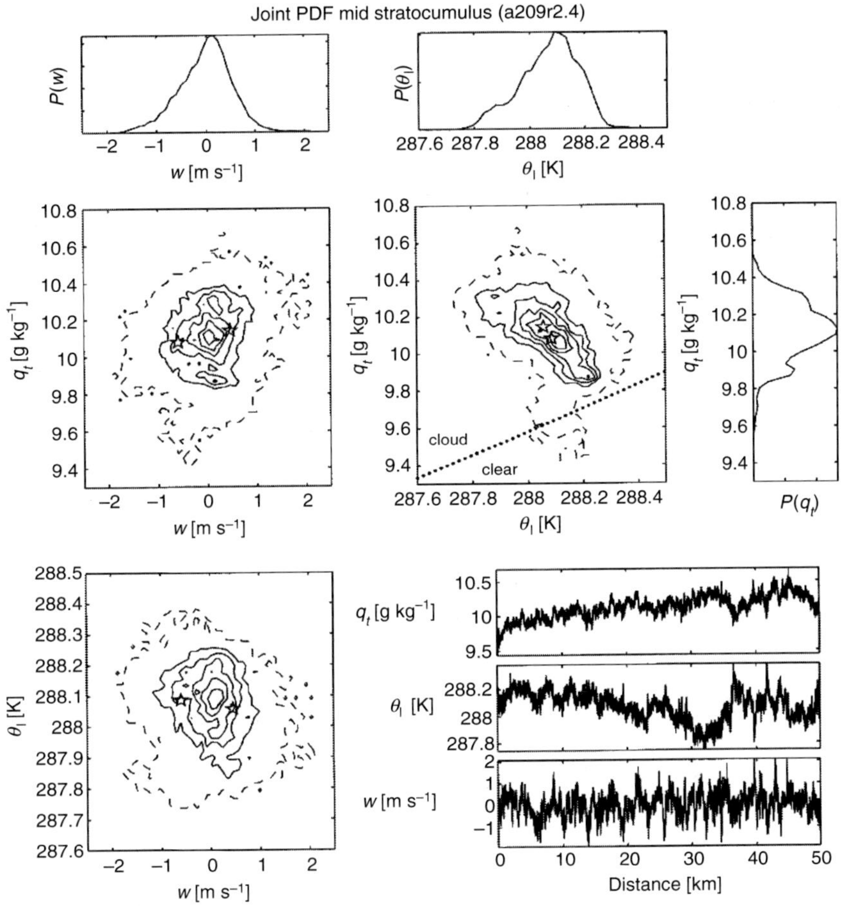


**FIGURE 3.7 One- and two-dimensional PDFs from a flight through a stratocumulus cloud.** Shown in the lower-right corner are the flight-level data. The dotted line in the $q_t - \theta_l$ panel corresponds to saturation (i.e. points above are saturated, points below are sub-saturated). *(From Larson et al. (2002))*

statistical analysis by Larson et al. (2002) found that double-Gaussian analytic PDFs could model these clouds acceptably well (e.g. Fig. 3.9).

The utility of analytic PDFs for cloud variability was demonstrated by Golaz et al. (2002a,b) who designed and tested a boundary layer model based on high-order closure equations. They found good agreement compared to LES results for a variety of cases, despite the fact that no case-specific tuning was made in their PDF-based model.

Progress is clearly being made in modeling of clouds within grid volumes. It is obvious, however, that further advances in modeling cloud fractional coverage and mean cloud properties using higher order closure models requires additional

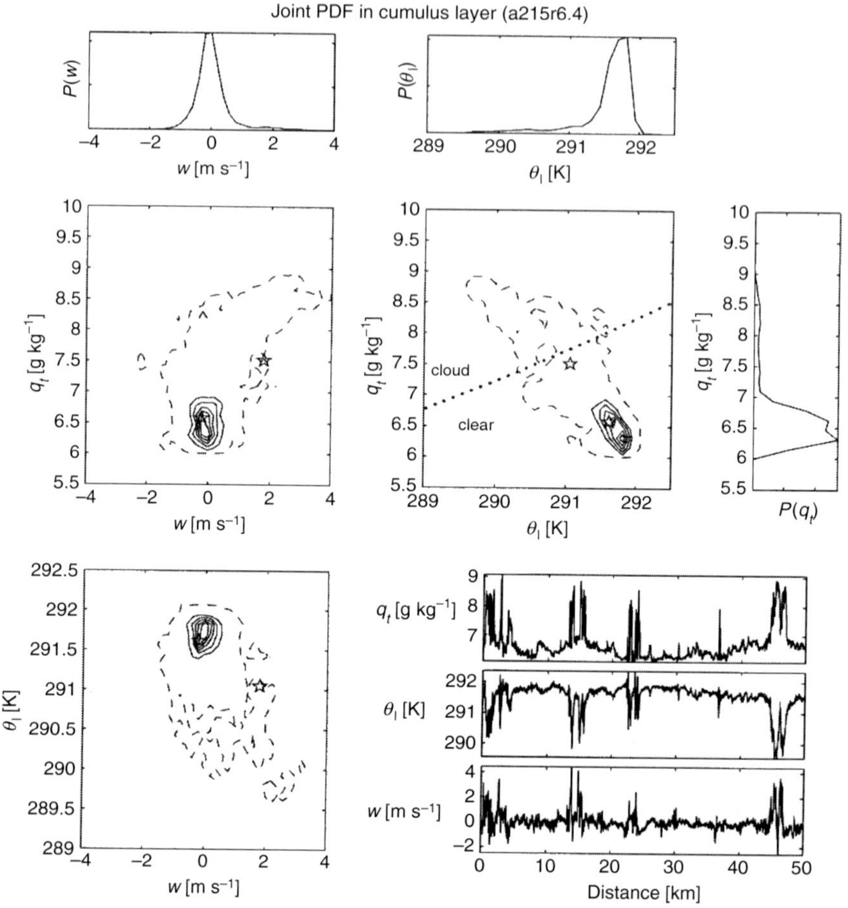


**FIGURE 3.8 As in Fig. 3.7 but from a flight through a field of cumulus clouds.** *(From Larson et al. (2002))*

direct observations of the fluctuating thermal and water substance quantities at various levels in a cloud field. Analyses of deep clouds (e.g. cumulus congestus and cumulonimbus) are also badly needed for parameterization of deep clouds in climate models.

## 3.5. IMPLICATIONS OF FILTERING TO THE INTERPRETATION OF MODEL-PREDICTED DATA

In this chapter we have examined the concepts of averaging (or filtering) the governing equations to study turbulent processes. One consequence of averaging the governing equations is that inherent uncertainties are revealed in model forecasts. In the case of Reynolds-averaged numerical models,

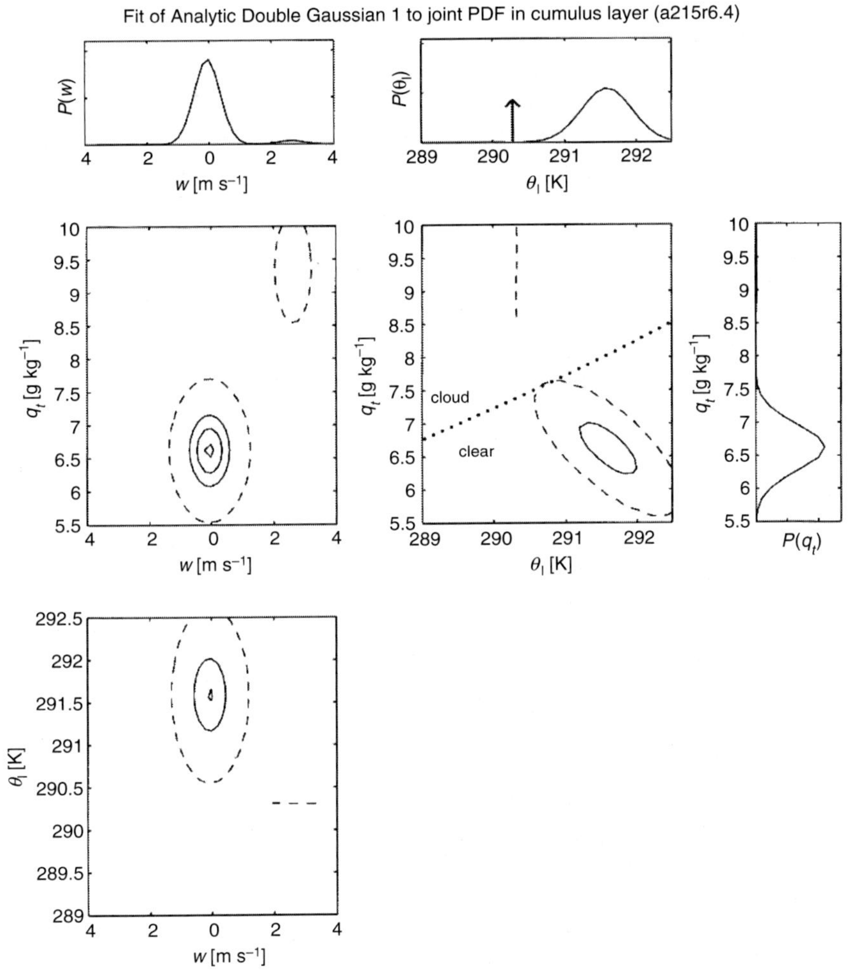


**FIGURE 3.9 A double-Gaussian analytic fit to the data shown in Fig. 3.8.** The arrow in the top-right panel denotes a delta function. *(From Larson et al. (2002))*

uncertainties are present not only due to the initial and boundary conditions supplied to the model, but also due to variability across the model grid volume. The variances are a measure of our inherent uncertainty in defining the existence criteria for a given cloud or cloud distribution. Until we can develop operational models with sufficient resolution to be able to simulate explicitly the finest details of cloud motions and physical processes, we must recognize that such inherent uncertainty exists in any forecast of a cloud system.

In the next chapter we examine cloud microphysical processes and the parameterization of those processes. We shall see that the concepts of averaging also affect our perspective of how to model these processes.

## REFERENCES

Banta, R. M., (1979). Subgrid condensation in a cumulus cloud model. In "6th Conf. Probab. Stat. Atmos. Sci. Banff, Alta.", pp. 197–202. Preprints.

Bougeault, P. (1981a). Modeling the trade-wind cumulus boundary layer. Part I: Testing the ensemble cloud relations against numerical data. J. Atmos. Sci. 38, 2414–2428.

Bougeault, P. (1981b). Modeling the trade-wind cumulus boundary layer. Part II: A high-order one-dimensional model. J. Atmos. Sci. 38, 2429–2439.

Boussinesq, J. (1877). Essai sûr la theorie des eaux courantes. Mem. Acad. Sci. 23, 24–46.

Brown, A. R., Derbyshire, S. H., and Mason, P. J. (1994). Large-eddy simulation of stable atmospheric boundary layers with a revised stochastic subgrid model. Q. J. R. Meteor. Soc. 120, 1485–1512.

Bryan, G. H., (2005). Statistical convergence in simulated moist absolutely unstable layers. In "11th Conf. on Mesoscale Processes", Amer. Meteor. Soc., 1M.6. Preprints.

Bryan, G. H., Wyngaard, J. C., and Fritsch, J. M. (2003). Resolution requirements for the simulation of deep moist convection. Mon. Wea. Rev. 131, 2394–2416.

Chen, C., and Cotton, W. R. (1983). A one-dimensional simulation of the stratocumulus-capped mixed layer. Bound.-Layer Meteor. 25, 289–321.

Clark, R. A., Ferziger, J. H., and Reynolds, W. C. (1979). Evaluation of subgrid-scale models using an accurately simulated turbulent flow. J. Fluid Mech. 91, 1–16.

Coleman, G. N. (1999). Similarity statistics from a direct numerical simulation of the neutrally stratified planetary boundary layer. J. Atmos. Sci. 56, 891–900.

Deardorff, J. W. (1974). Three-dimensional numerical study of the height and mean structure of a heated planetary boundary layer. Bound.-Layer Meteor. 7, 81–106.

Deardorff, J. W. (1980). Stratocumulus-capped mixed layer derived from a three-dimensional model. Bound.-Layer Meteor. 18, 495–527.

Detering, H. W., and Etling, D. (1985). Application of the E-$\epsilon$ turbulence model to the atmospheric boundary layer. Bound.-Layer Meteor. 33, 113–133.

Fritts, D. C., Wang, L., Werne, J., Lund, T., and Wan, K. (2009). Gravity wave instability dynamics at high Reynolds numbers. Part I: Wave field evolution at large amplitudes and high frequencies. J. Atmos. Sci. 66, 1126–1148.

Germano, M., Piomelli, U., Moin, P., and Cabot, W. H. (1991). A dynamic subgrid-scale eddy viscosity model. Phys. Fluids 3, 1760–1765.

Golaz, J.-C., Larson, V. E., and Cotton, W. R. (2002a). A PDF-based model for boundary layer clouds. Part I: Method and model description. J. Atmos. Sci. 59, 3540–3551.

Golaz, J.-C., Larson, V. E., and Cotton, W. R. (2002b). A PDF-based model for boundary layer clouds. Part II: Model results. J. Atmos. Sci. 59, 3552–3571.

Hinze, J. O. (1975). "Turbulence." McGraw-Hill, 790 pp.
Kolmogorov, A. N. (1941). The local structure of turbulence in incompressible viscous fluid for very large Reynolds numbers. Dokl. ANSSSR 30, 301–305.
Kolmogorov, A. N. (1942). The equations of turbulent motion in an incompressible fluid. Izv. Acad. Sci. USSR 6, 56–58.
Kosović, B. (1997). Subgrid-scale modelling for the large-eddy simulation of high-reynolds-number boundary layers. J. Fluid Mech. 336, 151–182.
Larson, V. E., Golaz, J.-C., and Cotton, W. R. (2002). Small-scale and mesoscale variability in cloudy boundary layers: Joint probability density functions. J. Atmos. Sci. 59, 3519–3539.
Larson, V. E., Wood, R., Field, P. R., Golaz, J.-C., Vonder Haar, T. H., and Cotton, W. R. (2001). Small-scale and mesoscale variability of scalars in cloudy boundary layers: One-dimensional probability density functions. J. Atmos. Sci. 58, 1978–1994.
Leonard, A. (1974). Energy cascade in large-eddy simulations of turbulent fluid flows. Adv. Geophys. 18A, 237–248.
Lewellen, W. S., and Yoh, S. (1993). Binormal model of ensemble partial cloudiness. J. Atmos. Sci. 50, 1228–1237.
Lilly, D. K. (1962). On the numerical simulation of buoyant convection. Tellus 14, 148–172.
Lilly, D. K., (1967). The representation of small-scale turbulence in numerical simulation experiments. In "IBM Scientific Symp. on Environmental Sciences, Yorktown Heights, NY, IBM DP Division." pp. 195–210. Preprints.
Manton, M. J., and Cotton, W. R., (1977). Formulation of approximate equations for modeling moist deep convection on the mesoscale. ATS Pap. 266, Colorado State University, 42 pp.
Mason, P. J., and Brown, A. R. (1999). On subgrid models and filter operations in large eddy simulations. J. Atmos. Sci. 56, 2101–2114.
Mellor, G. L. (1977). The Gaussian cloud model relations. J. Atmos. Sci. 34, 356–358; Corrigendum, 1483–1484.
Mellor, G. L., and Yamada, T. (1974). A hierarchy of turbulence closure models for planetary boundary layers. J. Atmos. Sci. 31, 1971–1806; Corrigendum, 34, 1482.
Mironov, D. V. (2009). Turbulence in the lower troposphere: Second-order closure and mass-flux modelling frameworks. Lect. Notes Phys. 756, 161–221.
Moeng, C.-H., McWilliams, J. C., Rotunno, R., Sullivan, P. P., and Weil, J. (2004). Investigating 2D modeling of atmospheric convection in the PBL. J. Atmos. Sci. 61, 889–903.
Moeng, C.-H., and Rotunno, R. (1990). Vertical-velocity skewness in the buoyancy-driven boundary layer. J. Atmos. Sci. 47, 1149–1162.
Moeng, C.-H., and Wyngaard, J. C. (1988). Spectral analysis of large-eddy simulations of the convective boundary layer. J. Atmos. Sci. 45, 3573–3587.

Moeng, C.-H., and Wyngaard, J. C. (1989). Evaluation of turbulent transport and dissipation closures in second-order modeling. J. Atmos. Sci. 46, 2311–2330.
Oliver, D. A., Lewellen, W. S., and Williamson, G. G. (1978). The interaction between turbulent and radiative transport in the development of fog and low-level stratus. J. Atmos. Sci. 35, 310–316.
Pope, S. B. (2000). "Turbulent Flows." Cambridge University Press, 771 pp.
Prandtl, L. (1925). Bericht über die Entstehung der Turbulenz. Z. Agnew. Math. Mech. 5, 136–139.
Prandtl, L. (1945). Über ein neues Formelsystem für die ausgebildete Turbulenz. Nachr. Akad. Wiss. Göttingen, Math.-Phys. K1, 6–19.
Randall, D. A. (1987). Turbulent fluxes of liquid water and buoyancy in partly cloudy layers. J. Atmos. Sci. 44, 850–858.
Schmidt, H., and Schumann, U. (1989). Coherent structure of the convective boundary layer derived from large-eddy simulations. J. Fluid Mech. 200, 511–562.
Smagorinsky, J. (1963). General circulation experiments with the primitive equations. I. The basic experiment. Mon. Wea. Rev. 91, 99–164.
Sommeria, G. (1976). Three-dimensional simulation of turbulent processes in an undisturbed trade-wind boundary layer. J. Atmos. Sci. 33, 216–241.
Sommeria, G., and Deardorff, J. W. (1977). Subgrid-scale condensation in models of nonprecipitating clouds. J. Atmos. Sci. 34, 344–355.
Sommeria, G., and LeMone, M. A. (1978). Direct testing of a three-dimensional model of the planetary boundary layer against experimental data. J. Atmos. Sci. 35, 25–39.
Stevens, B., Moeng, C.-H., and Sillivan, P. P. (1999). Large-eddy simulations of radiatively driven convection: Sensitivities to the representation of small scales. J. Atmos. Sci. 56, 3963–3984.
Sullivan, P. P., McWilliams, J. C., and Moeng, C.-H. (1994). A subgrid-scale model for large-eddy simulations of boundary layers. Bound.-Layer Meteor. 71, 247–276.
Tompkins, A. M. (2002). A prognostic parameterization for the subgrid-scale variability of water vapor and clouds in large-scale models and its use to diagnose cloud cover. J. Atmos. Sci. 59, 1917–1942.
Weinbrecht, S., and Mason, P. J. (2008). Stochastic backscatter for cloud-resolving models. Part I: Implementation and testing in a dry convective boundary layer. J. Atmos. Sci. 49, 1826–1847.
Wyngaard, J. C. (2004). Toward numerical modeling in the terra incognita. J. Atmos. Sci. 61, 1816–1826.
Zhou, Y., Brasseur, J. G., and Juneja, A. (2001). A resolvable subfilter-scale model specific to large-eddy simulation of under-resolved turbulence. Phys. Fluids 13, 2602–2610.

Chapter 4

# The Parameterization or Modeling of Microphysical Processes in Clouds

## 4.1. INTRODUCTION

As we have seen, the formulation of a model of a cloud or field of clouds requires a number of value judgments or compromises. The need for compromise, however, becomes most obvious when one is faced with the task of formulating models of the microstructure of clouds. If a modeler with access to the most advanced levels of computer power is developing a 3-D cloud model, the modeler is likely to come to the conclusion that a sophisticated, explicit prediction of the evolution of cloud microstructure is either impossible or impractical.

The alternative is to develop a simple parameterization of cloud microphysical processes. As a general approach, detailed theoretical models and/or experimental data are used to formulate parameterizations of the physics. Ideally, the parameterizations should capture the essence of the known microphysics in simple formulations of the processes. The problem is that, in some cases, the physics is not sufficiently well enough known, or is too complex, to fully capture its essence in simple formulations.

In this chapter we review briefly the concepts and general theory of the microphysics of clouds. We then summarize approaches to parameterizing cloud microphysical processes. We conclude by discussing the interaction of cloud microphysical processes and the dynamics of clouds.

## 4.2. GENERAL THEORY OF THE MICROPHYSICS OF CLOUDS

By "warm" clouds we refer to clouds in which the ice phase does not play a significant role in either the thermodynamics or precipitation processes. In general, the term refers to clouds whose tops are no colder than 0 °C. However, the physical processes that are prevalent at temperatures warmer than 0 °C can also operate quite effectively at colder temperatures or in supercooled clouds. Thus, we will not limit our discussion to clouds that are entirely warmer than 0 °C.

Noting the differences in droplet concentration in cumuli formed in maritime and continental air masses, Squires (1956, 1958) introduced the concept of *colloidal stability* of warm clouds. He pointed out that similar clouds forming in a maritime air mass are more likely to produce rain than clouds forming in a continental air mass. Thus, maritime clouds are less colloidally stable than are their continental counterparts. The relationship between the cloud droplet concentration and the cloud nucleus population was demonstrated by Twomey and Squires (1959) and Twomey and Warner (1967). Hence, in a nucleus-rich continental air mass, a given liquid-water content must be distributed over numerous small droplets having small collection kernels or collection cross sections (i.e. low terminal velocities, collection efficiencies, and cross-sectional areas). Thus, in general, the collision and coalescence process is inhibited in nucleus-rich continental air masses. These concepts form the basis of many of the parameterizations of warm-cloud processes that we shall discuss shortly. The fundamental premise of many of the parameterizations is that the cloud droplet concentration or activated cloud condensation nuclei (CCN) concentration, at cloud base, determine whether or not a cloud will precipitate.

Johnson (1980) pointed out that the cloud-base temperature also influences the activation of cloud droplets. Other things being the same (i.e. aerosol distribution, cloud-base updraft velocity), clouds with colder cloud bases will activate more cloud droplets than those having warmer bases. This is a consequence of the nonlinear variation of saturation vapor pressure with temperature, which results in *higher peak supersaturations in cold based clouds* than in warm-based clouds that are otherwise the same. It is the direction of this effect that is most interesting, because the temperature effect will accentuate the tendency for colloidal stability in continental clouds if those clouds also have cold bases (a common occurrence in mid-latitude, continental regions). The aerosol distribution and updraft velocity at cloud base, however, remain as the most important factors controlling the concentration of activated droplets and, hence, the colloidal stability of a cloud.

Cloud-base temperature also figures in a cloud's colloidal stability because a cloud with a warm cloud-base temperature has a larger saturation mixing ratio at cloud base. Other things being the same (i.e. cloud depth, activated CCN concentration, etc.), a warm-based cloud will have a greater potential for producing a significant amount of condensed liquid-water and, therefore, a greater chance of generating a few big droplets.

For some time it was thought that collision and coalescence could not proceed until the radius of droplets exceeded 19 μm (Hocking, 1959). Later calculations of collision efficiency by Klett and Davis (1973), Hocking and Jonas (1970), and Jonas and Goldsmith (1972) suggest that droplets smaller than 19 μm in radius do exhibit finite collection efficiencies. Due to their small fall velocities and cross-sectional areas, however, the rate of collision among droplets of such a small size is very low. Thus, it is still thought that a few larger droplets $r > 20$ μm must form in a cloud in order to initiate significant growth rates through the relatively random collisions among small, comparably sized droplets. This initial phase of collision and coalescence has been modeled as a stochastic process (Telford, 1955; Gillespie, 1972) or what is now referred to as quasistochastic process (Berry, 1967).

Since condensation theory for a smooth, unmixed updraft predicts a narrowing of the droplet spectrum with time (Howell, 1949; Mordy, 1959; Neiburger and Chien, 1960; Fitzgerald, 1974), the search continues to explain the formation of larger droplets that can sustain vigorous collision and coalescence growth of precipitation.

There are four hypothesized mechanisms to explain the formation of droplets large enough to be considered precipitation embryos:

- Turbulence influences on condensation growth.
- Role of giant cloud condensation nuclei.
- Turbulence influence on droplet collision and coalescence.
- Radiative cooling of droplets to form precipitation embryos.

### 4.2.1. Turbulence Influences on Condensation Growth

Simple adiabatic parcel models assume a uniform updraft and develop a constant droplet concentration a few meters above cloud base. Convective clouds are, in fact, made up of a series of updrafts of diverse intensities. The least vigorous updraft at cloud base also produces the lowest droplet concentration during activation of the available CCN. Turbulence further contributes to entrainment of dry environmental air in clouds, hence reducing LWC below adiabatic values and diluting the droplet concentration below its initial value after CCN activation. Convective cells with a significantly reduced concentration might generate bigger droplets than adiabatic cores, if they are further experiencing a convective ascent (Baker et al., 1980; Telford and Chai, 1980; Telford et al., 1984). Turbulence, however, also contributes to the continuous mixing of the convective cells, hence smoothing out their differences in terms of concentration and droplet growth.

Airborne measurements performed in cumulus clouds with a high resolution droplet spectrometer Fast-FSSP reveal the occurrence of very narrow droplet spectra in cloud cores (Brenguier and Chaumat, 2001). However, droplet spectra are often much broader than the narrow adiabatic reference, with droplet sizes extending from zero to the maximum predicted by the adiabatic model. This reflects the impact of mixing between convective cells that have experienced various levels of dilution with the entrained air. There are also airborne observations in stratocumulus clouds showing negative correlations between droplet concentration and droplet sizes (see for example Fig. 7 in (Pawlowska and Brenguier, 2000). This corroborates the hypothesis that fluctuations of the updraft intensity at cloud base or the ascent of diluted cloud cells might contribute to the formation of droplets bigger than the adiabatic prediction based on the mean droplet concentration. In deeper clouds, airborne cloud traverses at a given level all look rather similar in term of droplet concentration and sizes, suggesting that concentration/size correlations progressively dissipate due to continuous mixing.

Numerical simulations (Vaillancourt et al., 2002; Shaw et al., 1998) suggest that turbulence may also generate concentration fluctuations at the microscale by inertia in the regions of high vorticity, hence leading to superadiabatic droplet growth in the microcells with the lowest concentrations. This hypothesis has not be supported by in situ measurements of the droplet spatial distribution at the microscale (Chaumat and Brenguier, 2001).

In summary, the observational evidence that turbulence contributes to the formation of precipitation embryos by enhanced condensation growth is lacking. In fact, stochastic processes induced by turbulence are not likely to enhance droplet growth because condensation is a cumulative process. To experience superadiabatic growth, droplets need to remain isolated in regions of higher supersaturation for a significant part of their ascent. The odds of this are low as mixing continuously redistributes droplets in the cloud. Turbulence is more likely to affect a discontinuous process like collision, since once droplets coalesce, they cannot be separated.

### 4.2.2. Role of Giant Cloud Condensation Nuclei (GCCN)

Observations reported by Woodcock (1953), Nelson and Gokhale (1968), Hindman (1975), Johnson (1976, 1982), and Hobbs et al. (1980) have shown the presence of potentially significant concentrations of aerosol particles of sizes as large as 100 μm. Their concentrations are about $10^{-3}$ $cm^{-3}$ (Woodcock, 1953), that is, about one in $10^5$ or $10^6$ CCN are giant particles. Nevertheless, these particles can have a significant effect on the development of precipitation by serving as coalescence embryos (Johnson, 1982; Feingold et al., 1999; Yin et al., 2000). The droplet that forms is large enough for coalescence to start immediately even before the droplet reaches its critical size based on the Köhler equation. This can occur if the nuclei are completely soluble (e.g. sea-salt particles) or are mixed particles with a soluble coating (e.g. mineral dust with a coating of sulfate, (Levin et al., 1996) or are very large and wettable. The presence of GCCN on precipitation formation has been investigated in a number of cloud resolving models, which show that their contribution to rain formation may be appreciable in polluted clouds but has little influence in clouds forming in clean air masses (i.e. Feingold et al., 1999; Khain et al., 2000).

### 4.2.3. Turbulence Influence on Droplet Collision and Coalescence

Turbulence can influence the collision and coalescence process in three ways:

- By enhancing collision efficiencies.
- By enhancing the collection kernels.

- By producing inhomogeneities in droplet concentration.

Collision efficiencies are generally calculated in laminar or stagnant flow. In turbulent flow droplets will be accelerating and thereby be able to cross streamlines more efficiently than in laminar flow resulting in enhanced collision efficiencies. Large droplets, having more inertia, will be affected more by turbulence than smaller drops. Calculations by Koziol and Leighton (1996) suggest that this effect is small for droplets smaller than 20 μm diameter. However, turbulence can also cause fluctuations on vertical fall speeds and horizontal motions, such that the collection kernel is enhanced (Pinsky and Khain, 1997; Khain and Pinsky, 1997), relative to that defined in laminar flow. Because the collection rate is proportional to the square of droplet concentrations, inhomogeneities in droplet concentrations due to turbulence can produce enhanced regions of collection where the droplet concentrations are locally enhanced in, say, regions of low vorticity (Pinsky and Khain, 1997).

The problem is that there is little known about the details of turbulence in real clouds on the scales of a few centimeters and less. We know that inhomogeneities exist on those scales (Baker, 1992), but when it comes to either laboratory simulations or theoretical calculations, our ability to simulate high Reynolds number turbulence and its effects on droplet collection is still very rudimentary.

### 4.2.4. Radiative Cooling of Droplets to Form Precipitation Embryos

Consider a population of droplets that resides near a cloud top for a sufficiently long time. Those droplets will emit radiation to space quite effectively if the atmosphere above is relatively dry and cloud free. As a result, the droplets will be cooler than they would be without considering radiative effects. This means that the saturation vapor pressure at the surface of the droplet will be lowered and the droplets will grow faster.

But radiation cooling is proportional to the cross sectional area of a droplet so that its effect is much greater on larger droplets than small ones (Roach, 1976; Barkstrom, 1978; Guzzi and Rizzi, 1980; Austin et al., 1995). In fact, Harrington et al. (2000) have shown that in a marine stratocumulus environment, when droplets are competing for a limited supply of water vapor, the larger droplets grow so rapidly by radiative enhancement that droplets smaller than 10 μm in radius evaporate producing a bimodal size spectrum. This process is only effective in clouds where droplets reside near the cloud top for time scales of 12 minutes or longer such as fogs, stratus, and stratocumulus. Cumulus clouds with vigorous overturning expose droplets to space for too short a time.

Whatever the nature of the process of formation of embryonic droplets, the portions of a given cloud most favorable for the initiation of precipitation in warm clouds are the regions of highest liquid-water content. Twomey (1976) showed that if locally enhanced regions of LWC comprise only 1% of the cloud volume, and exist for periods of a few minutes, such regions can produce significant concentrations of large drops averaged over the entire volume of the cloud. Thus, the presence of protected updrafts having nearly wet adiabatic liquid-water contents (Heymsfield et al., 1978) can have significant bearing upon the initiation of precipitation in warm clouds. Of course, the ultimate amount of rainfall from a given cloud is controlled by the overall time-space character of its updrafts and its liquid-water content.

Langmuir (1948) suggested that once raindrops grow to a critical size of approximately 6 mm in diameter, they will break up due to hydrodynamic instability. He hypothesized that each breakup fragment will act as a new precipitation embryo which can grow to breakup size and create more raindrop embryos. He referred to this process as the chain reaction theory of warm-rain formation. Other observations (Blanchard, 1948; Magarvey and Geldhart, 1962; Cotton and Gokhale, 1967; Brazier-Smith et al., 1972; McTaggert-Cowan and List, 1975) have suggested that collisions among droplets of the order of 2–3 mm in diameter and smaller can initiate breakup.

Computations of the evolution of raindrop spectra reported by Brazier-Smith et al. (1973), Young (1975), and Gillespie and List (1976) have indicated the greater importance of collision-induced breakup over spontaneous breakup to the evolution of raindrop spectra. Srivastava (1978) calculated that for rainwater contents $M > 1\ \mathrm{g\ m^{-3}}$, collision breakup results in raindrop size distributions which are approximately constant in slope and have an intercept of the distribution function which is proportional to $M$. Using a numerical cloud model, Farley and Chen (1975) have concluded that a necessary condition for the development of a Langmuir chain reaction requires that a cloud must develop sustained updrafts in excess of $10\ \mathrm{m\ s^{-1}}$

## 4.3. APPROACHES TO MODELING CLOUD MICROPHYSICS IN CLOUD MODELS

### 4.3.1. Bin-resolving Microphysics

The approach to modeling microphysics in clouds ranges from explicit bin-resolving approaches in which the evolution of droplet and ice particle spectra are explicitly resolved, to bulk approaches in which the size distribution of hydrometeors is determined by a specified basis function in which one or more moments of the basis function is predicted.

The most simple class of model in which droplet spectra are explicitly resolved is the Lagrangian (or moving mass-grid) method. In this method particles at discrete sizes and concentrations follow the growth by condensation on a moving mass grid. This approach eliminates numerical diffusion and allows for a smooth transition from aerosol to haze to cloud droplets without artificial distinctions between these classes. The cloud supersaturation is calculated based on source (cooling, which is related to updraft velocity) and sink (condensation) terms, enabling accurate determination of the number of activated droplets.

These models typically focus on the initial growth phases from haze to droplet and include detailed representation of aerosol sizes and compositions (Mordy, 1959; Fitzgerald, 1974; Facchini et al., 1999; Feingold and Kreidenweis, 2000; Feingold and Chuang, 2002; Lohmann, 2004). They frequently also consider the effect of trace gases on activation (Kulmala et al., 1993). Aqueous production of sulfate has also been represented in studies that examine the effects of cloud processing on the aerosol size distribution (Hegg et al., 1991; Bower and Choularton, 1993; Feingold and Kreidenweis, 2000). The Lagrangian method is used almost exclusively in kinematic cloud parcel models, where a parcel of air is moved either adiabatically or according to some known trajectory through a cloud. It is not easily adapted to the study of growth by processes such as collision-coalescence, and it is not suitable for general application in Eulerian dynamical models.

Another class of models that explicitly resolve the evolution of hydrometeor size spectra is the so called *bin-resolving* technique. Assuming that the droplet spectrum is continuous on the scale of the averaging domain of a cloud model, one can formulate integral-differential equations that describe the evolution of droplet spectra. For example, we can formulate a prognostic equation for the variation of the spectral density $f(x)$ of cloud droplets of mass $x$ to $x \pm \delta x/2$ at a given geometric position and at a given instant. An integral differential equation describing the evolution of the droplet spectrum takes the form

$$\begin{aligned}\frac{\partial f(x)}{\partial t} &= N(x) - \frac{\partial\,[\dot{x} f(x)]}{\partial x} + G(x)|_{\text{gain}} + G(x)|_{\text{loss}} \\ &\quad + B(x)|_{\text{gain}} + B(x)|_{\text{loss}} + \tau(x), \end{aligned} \tag{4.1}$$

where $N$ represents nucleation, $G$ represents collection, $B$ represents breakup, and $\tau$ represents the sum of both mean and turbulent transport processes.

The first term on the right-hand side of Eq. (4.1) is the production of droplets of mass $x$ by the nucleation of such droplets on activated CCN. This term appears in Eq. (4.1) only if the droplet spectrum $f(x)$ is truncated at some small droplet mass.

The second term on the right-hand side of Eq. (4.1) is the divergence of $f(x)$ due to continuous vapor mass deposition on droplets growing at the rate $\dot{x}$, where $\dot{x}$ is a function of the droplet mass, its solubility in water, and the local cloud supersaturation as well as other factors (Byers, 1965; Mason, 1971; Pruppacher and Klett, 1978). If one chooses to extend the droplet spectrum to include soluble particles of mass $x_s$, the nucleation term ($N$) would be included explicitly in the second term. However, since $\dot{x}$ is a function of the solubility of a droplet, one should use a two-dimensional density function $f(x, x_s)$, as was done by Clark (1973).

The third and fourth terms on the right-hand side of Eq. (4.1) represent, respectively, the gain and loss integrals of $f(x)$ due to the collision and coalescence of cloud droplets. The gain and loss terms were formulated by

Berry (1967) as follows

$$G(x)|_{\text{gain}} = \frac{1}{2}\int_0^x K(x_c, x')f(x_c)f(x')\mathrm{d}x' \tag{4.2}$$

where $x_c = x - x'$ and

$$G(x)_{\text{loss}} = -\int_0^\infty K(x, x')f(x)f(x')\mathrm{d}x'. \tag{4.3}$$

The term $K(x, x')$ is the collision cross section, or collection kernel, often taken to be

$$K(x, x') = \pi(R_x + R_{x'})^2[v(x) - v(x')]E(x, x'), \tag{4.4}$$

where $E(x, x')$ is the collection efficiency, $R_x$ and $R_{x'}$ are the radii of droplets of mass $x$ and $x'$, and $v(x)$ and $v(x')$ are the average terminal velocities of droplets of mass $x$ and $x'$.

Integration of the droplet spectra evolution using Eqs (4.2)–(4.4) has been referred to by Gillespie (1975) as the *quasistochastic* model because it predicts a unique spectrum at a given time and point in space, whereas the pure stochastic model predicts fluctuations in the droplet spectrum at a given time and point in space.

The fifth and sixth terms on the right-hand side of Eq. (4.1) represent, respectively, the gain and loss of spectral density $f(x)$ due to the breakup of droplets.

Thus, the breakup of larger droplets whose fragments are of mass $x \pm \delta x/2$ is formulated as

$$B(x)|_{\text{gain}} = \int_x^\infty p(x')f(x')g(x|x')\mathrm{d}x', \tag{4.5}$$

where $p(x')$ is the probability per unit/time that a droplet of mass $x'$ to $x' \pm \delta x/2$ will break up due to internal hydrodynamic instability. The function $g(x|x')$ represents the number of fragments of mass $x \pm \delta x/2$ formed by the breakup of a droplet of mass $x' \pm \delta x/2$.

The loss of $f(x)$ due to the breakup of droplets of mass $x \pm \delta x/2$ is formulated as

$$B(x)|_{\text{loss}} = -p(x)f(x). \tag{4.6}$$

It was mentioned earlier that breakup due to collision of raindrops appears to be the dominant contributor to drop breakup, at least if the rainwater contents exceed 1 g m$^{-3}$. Brazier-Smith et al. (1973) formulated collision breakup by combining the collision and breakup terms in Eq. (4.1) to formulate a general stochastic interaction equation of the form

$$\begin{aligned} G(x)|_{\text{gain}} &+ G(x)|_{\text{loss}} + B(x)|_{\text{gain}} + B(x)|_{\text{loss}} \\ &= \frac{1}{2}\int_0^\infty \int_0^\infty K(x|x_\alpha, x_\beta) f(x_\beta) f(x_\alpha) \mathrm{d}x_\alpha \mathrm{d}x_\beta \\ &- f(x) \int_0^\infty \left| \int_0^\infty K(x_\alpha|x, x_\beta) x_\alpha \mathrm{d}x_\alpha \right| \frac{f(x_\beta)}{x + x_\beta} \mathrm{d}x_\beta, \end{aligned} \quad (4.7)$$

where $K(x|x_\alpha, x_\beta)$ may be considered a generalized interaction kernel related to the probability $[h(x|x_\alpha, x_\beta)]$ of forming a droplet of mass $x$ because of the interaction of droplets of mass $x_\alpha$ and $x_\beta$. In this instance, the interaction could represent a pure coalescence problem in which case Eq. (4.7) degenerates to Eqs (4.2) and (4.3). Alternatively, it could represent a collision event which promotes breakup.

A common approach to integrating Eq. (4.1) is to discretize $f(x)$ into 40 to 70 elements and then integrate the equations by finite-difference methods (Twomey, 1964, 1966; Bartlett, 1966, 1970; Warshaw, 1967; Berry, 1967; Kovetz and Olund, 1969; Bleck, 1970; Chien and Neiburger, 1972). Great care must be taken in representing collision-coalescence to avoid numerical diffusion in the mass-transfer equations and rapid (spurious) acceleration of growth to precipitation-sized particles.

Multi-moment representations of cloud processes have been developed (Tzivion et al., 1987; Hounslow et al., 1988; Chen and Lamb, 1994b) in which two or more moments in each individual drop bin are predicted. This approach significantly reduces numerical diffusion and has the added benefit of conserving more than one moment in each bin of the size distribution.

Another approach to bin-resolved microphysical modeling is the hybrid approach (Cooper et al., 1997; Jacobson, 1999; Pinsky and Khain, 2002) where the advantages of moving grids for condensational growth are combined with fixed grid approaches for collection processes. In spite of their large computational demands, bin-resolving microphysical models have been applied to large eddy simulation models (Stevens et al., 1998), mixed-phase three dimensional models of thunderstorms (Fridlind et al., 2004) and even mesoscale models (Lynn et al., 2005). However,for complicated cloud or mesoscale models, forecast models, and climate models, there still remains a strong desire to develop simplified techniques for predicting the evolution of the droplet spectrum to form rain along with its sedimentation through the cloud.

### 4.3.2. Bulk Warm-cloud Physics

The concept of bulk microphysics was first introduced by Kessler (1969). He assumed that the raindrop size-spectrum can be represented by a simple exponential function (with fixed pre-exponent) in which a single-moment of the hydrometeor spectra is predicted. Self-collection among cloud droplets is

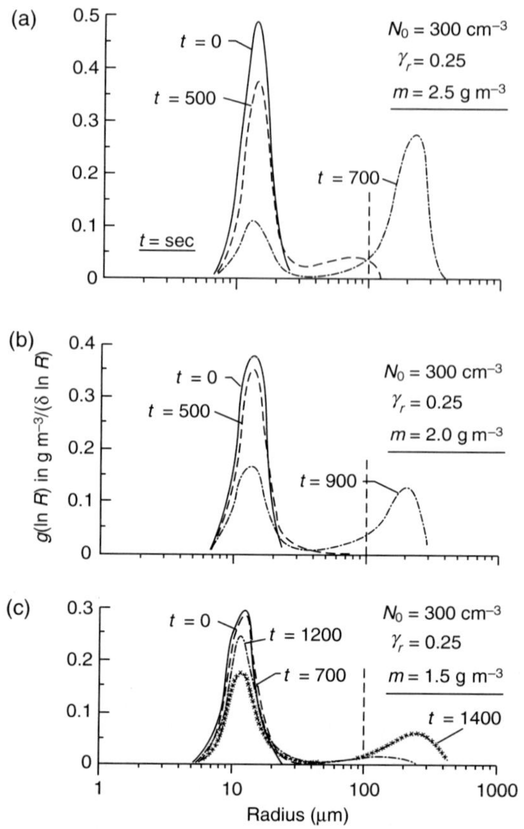


**FIGURE 4.1 Computed variation in the distribution of water mass density for an initial concentration of 100 cm$^{-3}$, a radius dispersion of 0.25, and LWCs of (a) 2.0, (b) 1.5, and (c) 1.0 g m$^{-3}$.** *(From Cotton (1972a))*

parameterized using an autoconversion formulation, and large hydrometeors such as raindrops are assumed to collect smaller drops by continuous accretion. Moreover, all classes of large hydrometeors are assumed to fall with a constant fall speed, usually a water content-weighted fall speed. To illustrate this approach consider Fig. 4.1 in which condensed liquid-water is partitioned into two domains. Note that the area under the curve is the liquid-water content.

Generally, we distinguish between cloud droplets, assumed to have sufficiently small terminal velocities to generally move with air parcels, and raindrops, which have significant settling speeds $v(R)$, where $R$ is the drop radius. Kessler's (1969) identification of an arbitrary threshold, separating cloud droplets falling at small terminal velocities from higher terminal velocity raindrops, is illustrated as a vertical dashed line in Fig. 4.3. To the left of the vertical dashed line, water vapor is condensed on low-terminal-velocity cloud droplets, where droplet collection forms raindrops which are on the right-hand side of the line. Berry and Reinhardt (1974) demonstrated that a natural break between cloud and raindrops occurs at a radius of 50 μm.

### 4.3.3. Conversion Parameterizations

Kessler (1969) first developed a simple parameterization of the rate of autoconversion of condensed liquid water from cloud droplets, having liquid-water content $m$ (mass/volume), to raindrops, having water content $M$. He assumed that the rate of conversion was a linear function of $m$ for liquid-water contents greater than 1.0 g m$^{-3}$.

Transformed into mixing ratio quantities, Kessler's autoconversion formula can be written

$$CN_{cr} = -\rho_o^{-1}(\mathrm{d}m/\mathrm{d}t) = K_1(r_c - a), \tag{4.8}$$

where $K_1 > 0$ if $r_c > a$ and $K_1 = 0$ if $r_c \leq a$. In our notation we have adopted the subscript $cr$, which indicates that raindrops having mixing ratios $r_r$ are supplied by cloud droplets having mixing ratios $r_c$.

Over the years there has been a variety of autoconversion formulations that have either been ad hoc [e.g. Kessler (1969), Manton and Cotton (1977), Cotton et al. (1986)], or derived from detailed bin-microphysics simulations [e.g. Berry (1967), Cotton (1972a,b), Berry and Reinhardt (1974), Beheng (1994), Khairoutdinov and Kogan (2000)]. Liu and Daum (2004) showed that many of the Kessler type of autoconversion formulations are special cases of a more general formulation. Note that for a given liquid-water content these formulations differ by several orders of magnitude.

Application of the conversion rate formulas to deep convective clouds reveals that these formulas serve mainly as triggers for warm-rain initiation. Once precipitation-size droplets form, the rain formation process is dominated by raindrops accreting smaller cloud droplets. As such, the absolute magnitude of the conversion rate may not be very important.

Only in stratocumuli and shorter lived, low-liquid-water cumuli will the magnitude of the conversion rate be essential to determining if such a cloud will produce precipitation.

### 4.3.4. Parameterization of Accretion

Once embryonic precipitation particles are formed, Kessler (1969) hypothesized that the water content converted to rain is distributed in an exponential distribution function formulated by Marshall and Palmer (1948) as

$$N(D) = N_0 \mathrm{e}^{-\lambda D}, \tag{4.9}$$

where $N(D)$ represents the number of raindrops per unit/volume of diameter $D \pm \delta D/2$. Figure 4.2 illustrates the Marshall-Palmer distribution function.

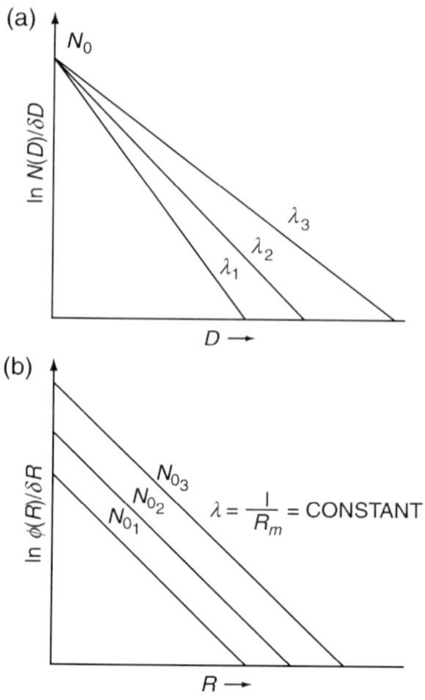


**FIGURE 4.2 Schematic illustration of the exponential drop-size distribution function of Marshall-Palmer drop-size distribution. (a) Kessler's model in which $N_0$ is assumed constant and $\lambda$ varies with rainwater content; (b) Manton-Cotton's model in which the slope $\lambda = 1/R_m$ is a constant and $N_0 = N_R/R_m$ varies with rainwater content.**

Kessler assumed that the rate of mass growth of raindrops is primarily by accretion of cloud droplets of the form

$$(\mathrm{d}/\mathrm{d}t)[x(D)] = (\pi D^2/4)\overline{E} V_D \rho_0 r_c, \tag{4.10}$$

where $x(D)$ is the mass of a raindrop of diameter $D$, $V_D$ is the terminal velocity of a raindrop of diameter $D$, and $\overline{E}$ represents an average collection efficiency between raindrops and cloud droplets.

The rate of change of rainwater mixing ratio by accretion or collection of cloud droplets is then

$$CL_{cr} = \frac{1}{\rho_0} \int_0^\infty \frac{\mathrm{d}}{\mathrm{d}t}[x(D)] N(D) \mathrm{d}D, \tag{4.11}$$

or

$$CL_{cr} = \frac{1}{\rho_0} \int_0^\infty \frac{\pi D^2}{4} \overline{E} V_{D\rho_0 r_c} N(D) \mathrm{d}D. \tag{4.12}$$

Kessler further assumed that $VD$ is given by

$$VD = 130.0D^{1/2}\ \mathrm{m\,s^{-1}}, \tag{4.13}$$

from Spilhaus (1948). Then, after substituting Eq. (4.19) into Eq. (4.18) and integrating, we find that

$$CL_{cr} = \frac{130.0\rho}{4} N_0 \overline{E} \rho_0 r_c \frac{\Gamma(3.5)}{\lambda^{3.5}}, \tag{4.14}$$

Under the assumptionthat $N_0$ is a constant, the total rainwater mixing ratio $r_r$ may be obtained as

$$r_r = \frac{1}{\rho_0}\int_0^\infty X(D)N(D)\mathrm{d}(D). \tag{4.15}$$

After substituting Eq. (4.15) and recognizing that

$$X(D) = \pi D^3 \rho_1/6, \tag{4.16}$$

Eq. (4.21) becomes, after integrating,

$$r_r = \pi \rho_1 N_0 \Gamma(4) \rho_0 6\lambda^4. \tag{4.17}$$

Solving Eq. (4.23) for $\lambda$ and substituting into Eq. (4.20) gives us

$$CL_{cr} = \left\{ \frac{130.0\pi^{0.125}}{4} \left[\frac{6}{\Gamma(4)}\right]^{0.875} \Gamma(3.5) \right\} \times N_0^{0.125} \rho_0^{1.875} \overline{E} r_c r_r^{0.875}. \tag{4.18}$$

Equation (4.24) shows that raindrops accreting cloud droplets produce a rainwater mixing ratio at a rate proportional to $r_c r_r^{0.875}$.

Over the years there has been a number of refinements and modifications to the simple Kessler formula. They include assuming that the slope of the exponential function is a constant (Manton and Cotton (1977), see also Tripoli and Cotton (1980)) to the use of gamma or log-normal basis functions Clark (1976); Clark and Hall (1983); Nickerson et al. (1986); Ferrier (1994); Meyers et al. (1997); Reisner et al. (1998); Seifert et al. (2006); Milbrandt and Yau (2005a,b), and then explicitly predicting the evolution of two or more moments. The advantage of the multimoment schemes is that they predict number concentration, mass mixing ratio (and sometimes higher order moments) and therefore are able to derive the broad features of the drop size distribution. In so doing they improve the representation of growth processes and precipitation formation.

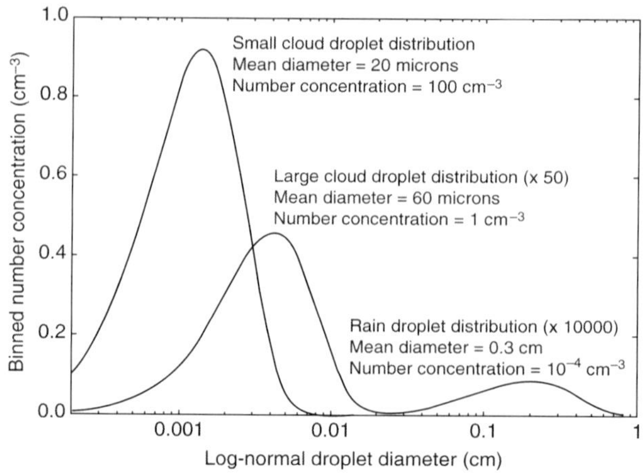


**FIGURE 4.3 Droplet spectra for the rain and dual cloud droplet hydrometeor categories.** For comparison to the small cloud droplet mode, the number concentration for the given diameter is exaggerated for the rain and large cloud droplet modes. *(From Saleeby and Cotton (2004))*

Departing from the Kessler bulk parameterization philosophy, the bulk scheme in RAMS (Regional Atmospheric Modeling System, Cotton et al. 2003) essentially emulates a full-bin microphysics model. The evolution of this modeling approach can be found in Clark and Hall (1983), Verlinde et al. (1990), Walko et al. (1995), Meyers et al. (1997), Feingold et al. (1998), and Saleeby and Cotton (2004). Instead of using continuous accretion approximations, which has been common in cloud parameterizations, Feingold et al. (1998) showed that full stochastic collection solutions for self-collection among cloud droplets and for rain (drizzle) drop collection of cloud droplets can be obtained for realistic collection kernels by making use of look-up tables. Saleeby and Cotton (2004) refined this approach by introducing two cloud modes, one for newly nucleated droplets on CCN having diameters less than 40 μm, and a second larger droplet mode from 40 to 80 μm in which giant CCN (GCCN) are nucleated. The activation of CCN in RAMS is parameterized using a look-up table derived from an ensemble of a Lagrangian parcel model calculations that considers ambient cloud conditions for the activation of cloud droplets from aerosol.The use of two cloud droplet modes provides a quantitative representation of the collection process and permits representing both CCN and GCCN as contributors to the collection process. Figure 4.3 illustrates schematically the prescribed basis functions. This approach has been extended to all hydrometeor class interactions by collection, including the growth of graupel and hail by riming (Saleeby et al., 2007). The philosophy of bin representation of collection has been extended to calculations of drop sedimentation (Feingold et al., 1998). Bin sedimentation is simulated by dividing the basis function into discrete bins and then building look-up tables to calculate how much mass and number in a given grid cell fall into each cell beneath a given level in a given time step. This permits the representation of size-sorting of hydrometeors which is not done in standard bulk schemes.

## 4.4. FUNDAMENTAL PRINCIPLES OF ICE-PHASE MICROPHYSICS

The representation of ice-phase microphysical processes in a cloud model is greatly complicated by the variety of forms of the ice phase, as well as by the numerous physical processes that determine the crystal forms. Moreover, in contrast to the physics of warm clouds, our understanding of ice-phase physics is far less complete. This means that in many cases the formulation of simple parameterized models of the ice phase cannot be done using information derived from detailed theoretical/numerical models or from observations.

The physical processes that should be considered in formulating a model or parameterization of the ice phase are as follows:

(1) Primary and secondary nucleation of ice crystals.
(2) Vapor deposition growth of ice crystals.
(3) Riming growth of ice crystals.
(4) Graupel or hail particle initiation from heavily rimed crystals.
(5) Graupel or hail particle initiation by the freezing of supercooled raindrops.
(6) Graupel or hail particle riming and vapor deposition growth.
(7) Graupel or hail particle-particle collision with supercooled raindrops.
(8) Shedding of water drops from hailstones growing by wet growth or from partially wetting ice particles.
(9) The initiation of aggregates of ice crystals by collision among ice crystals.
(10) Aggregate collection of ice crystals.
(11) Aggregate riming of cloud droplets.
(12) Melting of all forms of ice particles.

### 4.4.1. Nucleation of Ice Crystals

Ice particles can form either homogeneously or heterogeneously on some form of ice nuclei (IN). Homogeneous nucleation can take place either directly from the vapor or by freezing of cloud droplets. However, homogeneous nucleation of ice crystals from the vapor, or the chance formation of an embryo of ice-like structure of critical size, requires very high supersaturations with respect to ice and such low temperatures that it does not take place in the troposphere. On the other hand, homogeneous freezing of supercooled droplets by the chance formation of a cluster of ice-like embryos can occur in the atmosphere.

For homogeneous freezing to occur, enough ice-like water molecules must come together within the droplet to form an embryo of ice large enough to survive and grow. Because the numbers and sizes of the ice embryos that form by chance increase with decreasing temperature, below a certain temperature (which depends on the volume of water considered) freezing by homogeneous nucleation becomes a virtual certainty. Homogeneous nucleation occurs in about one second at about −41°C for droplets about 1 μm in diameter, and at about −35 °C for drops 100 μm in diameter. An analogous freezing process occurs for unactivated droplets or haze particles at temperatures below −40°C, a process for ice formation in cirrus clouds (DeMott, 2002). Hence, in the atmosphere, homogeneous nucleation by freezing generally occurs only in high clouds or high latitudes.

The presence of some form of nucleus or mote is required at temperatures warmer than −35 °C, in a process called *heterogeneous nucleation*. The principle ice nucleation mechanisms are (1) vapor-deposition nucleation, (2) condensation-freezing, (3) immersion-freezing nucleation, and (4) contact-freezing nucleation.

*Vapor deposition nucleation* refers to the direct transfer of water vapor to a nucleus that results in the formation of an ice crystal. *Condensation-freezing nucleation* refers to the condensation of water vapor on an internally-mixed nucleus to form an embryonic droplet, followed by freezing. This is viewed as a two-step process, so the name "condensation freezing" is often used. In practice it is not easy to distinguish between these two modes of nucleation.

*Immersion freezing* refers to the nucleation of a cloud droplet or raindrop on an ice nucleus which is immersed within the drop. Two theories of immersion freezing have emerged. One theory views the freezing process as stochastic, such that at a given degree of supercooling, not all drops of a population of drops will freeze at the same time Bigg (1953a,b, 1955); Carte (1956); Dufuor and Defay (1963). The second theory, called the singular theory, holds that at a given degree of supercooling, the probability that a drop of a given size will freeze depends solely on the likelihood that the drop contains an active freezing nucleus. This process is independent of time. Laboratory experiments reported by Vali and Stansbury (1966) have indicated that the freezing process is time dependent, though the amount of time dependence is small. Both theories predict that the probability of freezing increases exponentially with the degree of supercooling and with the volume (or size) of the drop. Thus, for small cloud droplets at small degrees of supercooling, this mechanism of nucleation is not very effective.

*Contact nucleation* refers to the nucleation of a supercooled drop by a nucleus that makes contact with the surface of the drop in the supercooled state. Observations have shown that dry particles, such as clays, sand, CuS, and organic compounds which make contact with a supercooled drop, are much more effective as contact-freezing nuclei than when immersed within the drop (Rau, 1950; Fletcher, 1962; Levkov, 1971; Gokhale and Spengler, 1972; Pitter and Pruppacher, 1973; Fukuta, 1972a,b). In the atmosphere, contact nucleation can produce considerable time dependence in the nucleation process, since it depends both on the probability that an aerosol particle makes contact with a supercooled drop and on the probability that the aerosol particle acts as an active freezing nucleus. Young (1974a) modeled contact nucleation by naturally and artificially generated aerosols. He considered cloud droplet scavenging of nuclei panicles by *Brownian diiffusion* and by the combined effects of *thermophoresis* and *diffusiophoresis*.

Brownian diffusion refers to the chance encounter between a supercooled cloud droplet and an active nucleus (aerosol particle) due to the random motion of both species as a consequence of the thermal bombardment with gas molecules. The rate of collision is a function of the kinetic energy of the air molecules (or temperature) and the mobilities of both species. Thermophoresis refers to a net transport of particles in a thermogradient from warm toward colder regions. Diffusiophoresis refers to the net transport of aerosol particles in the direction of a vapor flux. In the case of a droplet growing by vapor deposition, diffusiophoresis is directed toward the droplet, whereas thermophoresis acts away from the droplet. The reverse is true of an evaporating droplet. Young (1974a) has noted that since thermophoresis dominates over diffusiophoresis, contact nucleation by nuclei in the size range $0.15 < 1.0$ μm would be suppressed in a growing cumulus cloud. He cites observations reported by Mee and Takeuchi (1968) and Koenig (1962) which indicate that ice is most prevalent in downdrafts and at the edges of clouds. This gives evidence, he suggests, that phoretic-contact nucleation is an effective mechanism in natural clouds.

Historically, ice nuclei concentrations have been assumed as predictors of ice particle concentrations. Laboratory studies in the 1940's and 1950's using expansion chambers, mixing cloud chambers or aerosol collected on filters, indicated that the variability in ice nuclei concentration was strongly a function of temperature. Fletcher (1962) derived an empirical relationship relating the increase of concentration of ice nuclei with decreasing temperature:

$$N_{\mathrm{IN}} = A\ \exp(\beta T_s), \tag{4.19}$$

where $N$ is the concentration of active ice nuclei per liter of air, $T_s$ is the degree of supercooling, $\beta$ varies from about 0.3 to 0.8 and $A$ is about $10^{-5}$ liter$^{-1}$. For $\beta = 0.6$, Eq. (4.19) predicts that the concentration of ice nuclei increases by about a factor of 10 for every 4 °C decrease in temperature. In urban air, the total concentration of aerosol is on the order of 108 liter$^{-1}$ and only about one particle in 108 acts as an ice nucleus at $-20$ °C.

The activity of a particle as a condensation-freezing or a deposition nucleus depends not only on the temperature but also on the supersaturation of the ambient air. Thus, at a given temperature, Gagin (1972), Huffman (1973), and Huffman and Vali (1973) found that the concentration of IN varies with the supersaturation with respect to ice. The effect of supersaturation on measurements of ice nucleus concentrations is shown in Fig. 4.4, where it can be seen that at a constant temperature the greater the supersaturation with respect to ice the more particles serve as ice nuclei. The empirical equation to

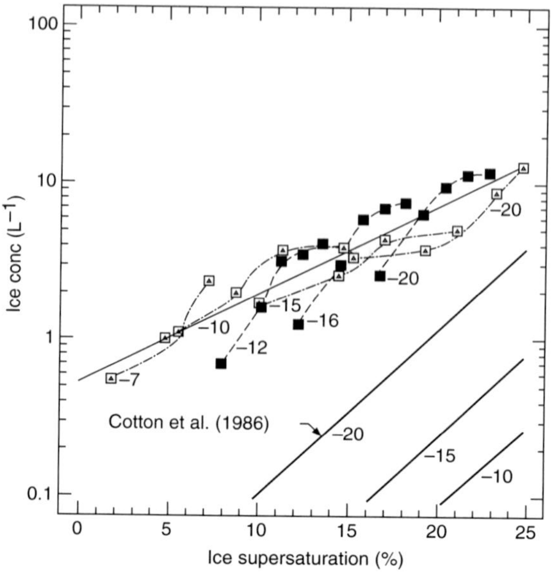


**FIGURE 4.4 Continuous-flow diffusion-chamber ice-nucleus concentration measurements versus ice supersaturation from (open square) Rogers (1982), and from (filled square) Al-Naimi and Saunders (1985).** Constant-temperature measurement series are indicated and the regression given in Eq. (4.20) is shown. Also presented are constant temperature values predicted by the deposition-condensation-freezing nucleation model formulated by Cotton et al. (1986). *(From Meyers et al. (1992))*

the best-fit line to these measurements is

$$N_i = \exp\{a + b[100(S_i - 1)]\}, \tag{4.20}$$

where $N_i$ is the concentration of ice nuclei per liter, and $S_i - 1$ is the supersaturation with respect to ice, $a = 0.639$ and $b = 0.1296$ (Meyers et al., 1992). These measurements were obtained using a continuous flow diffusion chamber (CFDC), whose limited data exhibits roughly a factor of ten higher concentrations of IN at warmer temperatures than that found with older devices such as the filter-processing systems. Recognizing the need to allow vertical and horizontal variations in IN concentrations in mesoscale model simulations, Cotton et al. (2003) modified Eq. (4.20) to include the prognostic variable $N_{\rm IN}$:

$$Ni = N_{\rm IN} \exp[12.96(Si - 1)], \tag{4.21}$$

where $T < -5\,°\mathrm{C}$. The variable $N_{\rm IN}$ can be deduced from the continuous flow diffusion chamber data and used as a forecast variable in regional simulations (i.e. van den Heever et al., 2006; van den Heever and Cotton, 2007).

As noted above, there is considerable evidence that freezing of drops by contact nucleation is much more efficient than immersion freezing of drops.

 

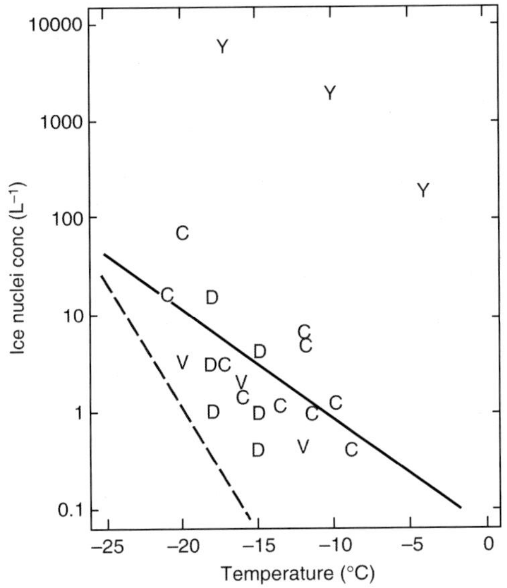


**FIGURE 4.5 Summary of measurements of contact-freezing ice-nuclei concentrations made by various authors [Cooper (C), Deshler (D) and Vali (V)].** For comparison, the estimates of Young (1974a) (Y) are shown. The regression line is an exponential fit to measurements. Fletcher (1962) ice-nucleus curve is also shown for reference (—). *(From Meyers et al. (1992))*

Unfortunately, there are no field-deployable devices for measuring contact nuclei concentrations. All that is available are a few exploratory laboratory experiments for estimating contact nuclei IN concentrations or the effectiveness of various aerosols to serve as contact nuclei. Using data from vertical wind tunnel experiments reported by Blanchard (1957), Young (1974a) deduced that contact IN concentrations $N_{ic}$ can be described by

$$N_{ic} = N_{a0}(270.16T_c)^{1.3} \tag{4.22}$$

where $T_c$ is the cloud droplet temperature, and $N_{a0} = 2.0 \times 10^2$ liter$^{-1}$ at sea level. Meyers et al. (1992) fitted data measured by three different prototype devices simulating contact nucleation by Vali (1974, 1976), Cooper (1980), and Deshler (1982), that are plotted in Fig. 4.5.

Note that those measurements suggest that contact nuclei concentrations are much less than Young's estimates. Meyers et al. (1992) approximated the data with an equation of the form:

$$N_{ic} = \exp[a = b(273.15T_c)], \tag{4.23}$$

where $a = -2.80$ and $b = 0.262$ when $N_{ic}$ has units of liter$^{-1}$. Remember that the data used to derive Eq. (4.23) are based on a limited number of air samples.

We expect that the concentrations of contact nuclei will vary temporarily and spatially as much as condensation-freezing nuclei do.

Equations (4.19), (4.21) and (4.23) only represent initial ice particle formation on IN and do not necessarily represent actual ice particle concentrations because other processes such as ice multiplication (see Section 4.4.2), sedimentation, breakup, and advection can greatly influence the concentrations of ice particles. Note that Eq. (4.21) allows for both horizontal and vertical variations in IN. Because the aerosol contributing to IN are large and large aerosols generally decrease with height (Georgi and Kleinjung, 1968; DeMott et al., 2003), we expect that IN concentrations generally decrease with height as well.

One should not therefore be surprised that many observations of ice particle concentrations do not show a good correlation with temperature (Gultepe et al., 2001; Field et al., 2005). Gultepe et al. (2001), for example, compiled data from the glaciated regions of stratus clouds for a number of field campaigns and found that ice crystals smaller than 1000 μm diameter do not show a good correlation with temperature and that the concentrations of these smaller ice particles varied up to three orders of magnitude, for a given temperature. On the other hand, measurements of ice particle concentrations in wave clouds where only initial ice particles are likely, Cooper and Vali (1981) showed ice particle concentrations increasing with decreasing temperature.

In summary, we see that there remain many unanswered questions regarding the concentrations of ice nuclei, their composition, their activation relative to different environmental factors, and their relationship to ice crystal concentrations. Work has been hampered by severe difficulties in the precise measurement of IN. Moreover, current field deployable devices for measuring IN do not take into account activation of IN by contact freezing. Fletcher's or Meyers empirical curves are still used and often misused in numerical models, ranging from cloud-resolving models to GCMs. These curves, at best, only represent the concentrations of ice particles initially formed in clouds and probably rarely represent ice particle concentrations in most clouds.

### 4.4.2. Ice Multiplication

As we have seen, it has become increasingly evident that concentrations of ice crystals in real clouds are not always represented by the concentrations of IN measured or expected to be activated in such environments. In particular, it has been found that at temperatures warmer than −10 °C, the concentration of ice crystals can exceed the concentration of IN activated at cloud top temperature by as much as three or four orders of magnitude (Braham, 1964; Koenig, 1963; Mossop et al., 1970; Mossop and Ono, 1969; Mossop et al., 1967, 1968, 1972; Magono and Lee, 1973; Auer and Marwitz, 1969; Hobbs, 1969, 1974). The effect is greatest in clouds with broad drop-size distributions (Koenig, 1963; Mossop et al., 1968, 1972; Hobbs, 1974).

Some explanations or hypotheses that have been proposed to account for the high ice particle concentrations observed in some clouds are as follows:

- Ice multiplication by fracturing of fragile ice crystals, which may breakup during collision with each other. (Vardiman, 1978).
- Fragmentation of large drops during freezing. (Mason and Maybank, 1960).
- Secondary ice particle formation during ice particle riming. ((Hallett and Mossop, 1974; Mossop and Hallett, 1974).
- Enhanced ice nucleation in the presence of spuriously high supersaturations. (Hobbs and Rangno, 1985).
- Secondary ice particle generation during evaporation of ice particles (Oraltay and Hallett, 1989; Dong et al., 1994).

Of these processes, the one that has been given the most attention and quantified in models is secondary ice particle formation by the rime-splinter process. Laboratory studies by Hallett and Mossop (1974) and Mossop and Hallett (1974), confirmed by Goldsmith et al. (1976) have indicated that copious quantities of splinters are produced during ice particle riming under very selective conditions. These conditions are:

- Temperature in the range of $-3$ °C to $-8$ °C.
- A substantial concentration of large cloud droplets ($D > 24$ μm)
- Large droplets coexisting with small cloud droplets ($D < 12.3$ μm).

An optimum average splinter production rate of 1 secondary ice particle for 250 large droplet collisions occurred at a temperature of $-5$ °C.

This process is consistent with observations of the greatest departure from IN estimates of ice crystals when clouds contain graupel particles and frozen raindrops, and is consistent with field observations (Hobbs and Cooper, 1987).

There is much indirect or inferential evidence that evaporation enhances ice crystal concentrations. This evidence is perhaps more intriguing than it is compelling. Some field studies have related unusually high ice nuclei numbers, or unusual increases in ice crystal numbers, to circumstances in which clouds were evaporating. Cooper (1995), for example, found a 100-fold increase in ice crystal concentrations in the evaporation region of orographic layer clouds. The largest ice enhancements in the Cooper study were observed in clouds with temperatures approaching the onset temperature for homogeneous freezing. Smaller enhancements were found in warmer clouds and no enhancements were found warmer than about $-20$ °C. Further evidence of the possible role of evaporation nucleation has been presented by Field et al. (2001) and Cotton and Field (2002). They show observational evidence and supporting parcel modeling calculations from wave cloud studies that suggest ice had to form close to the downstream edge of the wave cloud. Ice production coincident with the start of the liquid cloud, or earlier, would have suppressed the observed liquid cloud. Some of the observations of rapid ice crystal concentration enhancement versus expected IN concentrations in cumulus cloud studies of Hobbs and Rangno (1985, 1990) and Rangno and Hobbs (1994) were also observed to originate in close proximity to regions of cloud evaporation. Stith et al. (1994) followed the development of ice in a cumulus turret near its top at −18 °C. During the updraft stages, low ice concentrations were observed in the turret (similar to what would be expected from primary ice nucleation), but during the downdraft stages, the ice concentrations increased by an order of magnitude. This observation cannot be explained by rime splintering.

In summary, it is unlikely that all primary and secondary ice-forming processes have been quantitatively identified. Other mechanisms may sometimes operate, but their exact nature remains a mystery. Moreover, our ability to measure small ice crystals, in particular, has significant errors and need improvement. Consequently, there are large uncertainties associated with our ability to simulate the affect of aerosols on the initiation of ice, and subsequent impacts on precipitation. This remains as one of the critical problems in cloud physics.

### 4.4.3. Ice-Particle Growth by Vapor Deposition

Once ice crystals are nucleated by some mechanism of primary or secondary nucleation, and if the environment is supersaturated with respect to ice, the crystals can then grow by vapor deposition. Because the saturation vapor pressure with respect to ice is less than the saturation vapor pressure with respect to water, a cloud which is saturated with respect to water will be supersaturated with respect to ice. Figure 4.6 shows the variation in supersaturation with respect to ice as a function of temperature for a water-saturated cloud. Note that ice crystals in a water-saturated cloud can experience supersaturation in excess of 10% This leads to a process commonly known as the *Bergeron Findeisen mechanism*.

If a cloud is at water saturation at −10 °C, for example, the supersaturation with respect to ice will be 10%. As the ice crystals grow by vapor deposition, they deplete the vapor content, thereby driving the environment below water saturation. The cloud droplets will then evaporate, which helps sustain a vapor pressure difference between ice and water. By this process ice crystals are then said to grow at the expense of cloud droplets. This process has been described by Wegener (1911), Bergeron (1935), and Findeisen (1938).

It should be noted that because the Bergeron-Findeisen process causes the evaporation of cloud droplets, it can also affect the contact nucleation process. Young (1974a,b) and Cotton et al. (1986) have pointed out that if the Bergeron-Findeisen process results in the partial evaporation of larger cloud droplets, phoretic scavenging of potential contact nuclei will be enhanced. Thus the Bergeron-Findeisen process and the contact nucleation process can form a positive-feedback loop. Vapor deposition growth of ice Crystals lowers the saturation ratio below water saturation, causing droplet evaporation. This

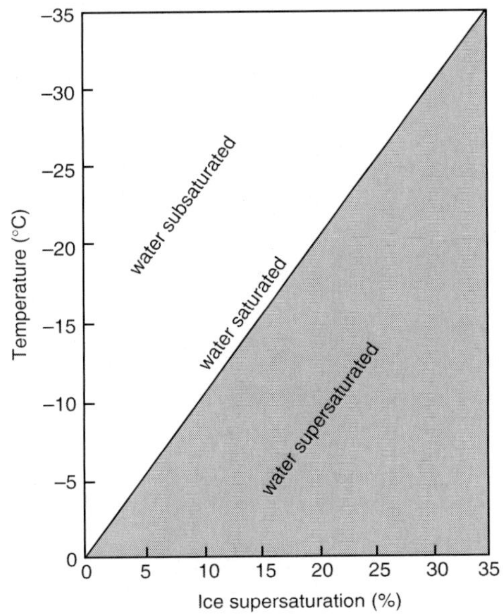


**FIGURE 4.6 Supersaturation with respect to ice as a function of temperature for a water-saturated cloud.** The shaded area represents a water-supersaturated cloud.

evaporation favors the phoretic-contact nucleation of the supercooled droplets that have not fully evaporated, and that in turn, grow as ice crystals, further lowering the saturation ratio below water saturation, and so on. This process is favored whenever the droplet spectrum is broad enough to allow the partial evaporation of the largest droplets.

It should be noted also that the rate of depletion of supercooled liquid water by ice crystals growing at the expense of cloud droplets is dependent on the concentration and size of droplets. In a study of the affects of aerosol pollution on orographic clouds, Saleeby and Cotton (2005) found that in a polluted cloud where cloud droplets were numerous and small, ice crystals depleted liquid water much more rapidly than when the cloud was clean and droplet concentrations were less and droplets were larger. This is simply because, for a given liquid-water content, more numerous small droplets expose a larger net surface area to the depleted vapor content of the air than a cloud containing few, larger droplets.

Prediction of the vapor deposition growth of ice crystals is complicated by the fact that ice crystals exhibit differing habits or shapes depending upon the temperature and supersaturation (with respect to ice) of the environment. For example, the results of the laboratory experiments shown in Fig. 4.7 illustrate that the ice crystal habit can change from plates to needles or prisms to plates over less than 1 °C [for the definition of the various ice crystal habits, see Pruppacher and Klett (1978) and Mason (1971)]. This figure also shows that

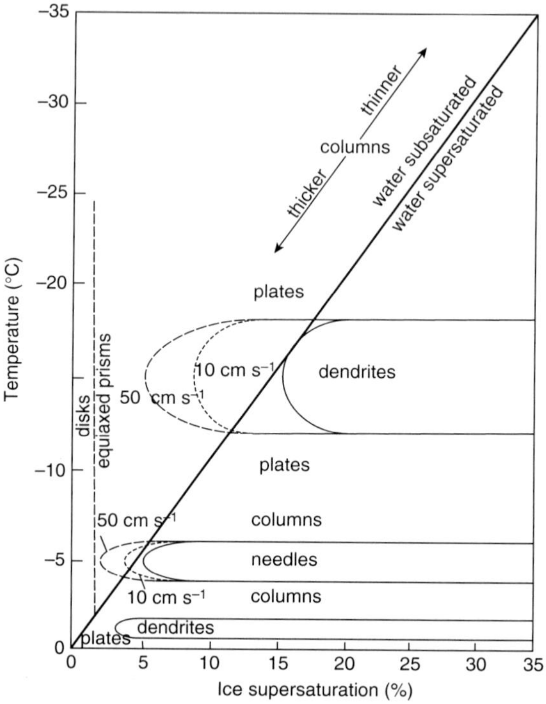


**FIGURE 4.7 The shape of a crystal is related to environmental conditions in a complicated manner: temperature, supersaturation of the atmosphere with water vapor, and speed of falling all have an effect.** Most crystals grow under natural conditions not far removed from the diagonal line representing water saturation. The dotted lines to the left of "dendrites" and "needles" show how the speed of falling extends the zones in which those elongated forms grow. *(After Keller and Hallett (1982); figure from Hallett (1984))*

the ambient temperature and supersaturation are not the only properties that determine the crystal habit. The fall velocity of the crystals and the associated ventilation can also extend the regimes of needle and dendritic forms of crystals into regions of subsaturation with respect to water. The habit of growth of an ice crystal can have a pronounced influence on the rates of vapor-deposition growth. This is especially true for dendritic and needle growth habits, which greatly accelerate the deposition growth rate over that of an equivalent spherical particle.

Thus, the vapor-deposition (evaporation) equation, Eq. (4.24), for spherical particles must be modified to include the role of ice crystal habit. The conventional approach is to assume that the diffusion of heat and water vapor in the vicinity of a complex-shaped ice crystal behaves in a manner analogous to the rate of electrical charge dissipation from an electrically charged capacitor of similar shape (Jeffreys, 1916). Under this assumption, the rate of mass vapor

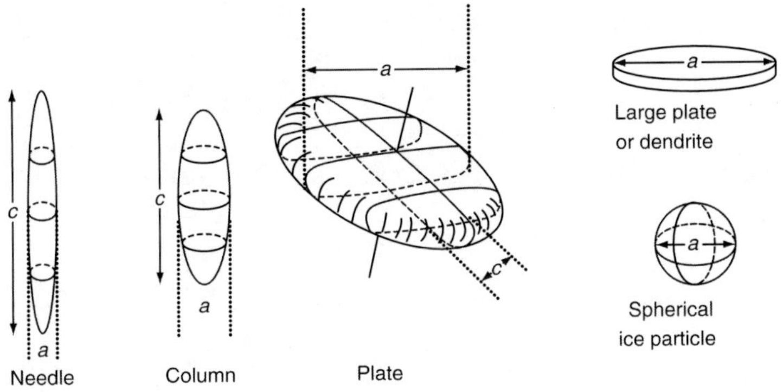


**FIGURE 4.8** **Illustration of approximations to crystal shapes by spheroids of revolution.**

deposition (sublimation) on an ice crystal can be formulated as

$$\left.\frac{\mathrm{d}x_i}{\mathrm{d}t}\right]_{VD} = 4\pi C G_i(T, P)(S_i - 1) f(Re) - \frac{M_w L_s L_f G(T, P)}{K_i R_a T^2} \left.\frac{\mathrm{d}x_i}{\mathrm{d}t}\right]_{RM}, \tag{4.24}$$

where $x_i$ is the crystal mass, $C$ is the "capacitance" an ice crystal, $S_i$ is the saturation ratio with respect to ice, $f(Re)$ represents a ventilation function of an ice crystal, and $G(T, P)$ is a thermodynamic function similar to that for water drops, but modified to include the saturation vapor pressure and the latent heat between vapor and ice. Both the ventilation function and the capacitance vary depending on the particular habit of the ice crystal. The second term on the right-hand side of. Eq. (4.24) represents the contribution to the crystal heat balance by the latent heat released during riming. Cotton (1970) found that for clouds with liquid-water contents greater than 0.3 g $\mathrm{m}^{-3}$, ice crystal riming $(\mathrm{d}x_i/\mathrm{d}t)_{RM}$ can warm the crystal sufficiently to drive the surface of the crystal below ice saturation, causing the crystal to sublimate mass. At this point, however, crystal riming growth dominates the growth of the crystal.

The crystal capacitance is generally computed from theoretical electrostatic capacitance models for simplified shapes such as spheres, disks, and prolate or oblate spheres of revolution. Thus, if we consider $a$ to be the length of the basal plane, and $c$ to be the length of the prism plane of an ice crystal as shown in Fig. 4.8, the capacitance can be approximated as follows:
for needles,

$$C = c/\ln(4c^2/a^2); \tag{4.25}$$

for prismatic columns,

$$C = ce/[\ln|(1+e)/(1-e)|] \quad \text{where } e = \sqrt{1 - a^2/c^2} \tag{4.26}$$

for hexagonal plates,

$$C = ae/2 \sin^{-1} e, \quad \text{where } e = \sqrt{1 - c^2/a^2} \tag{4.27}$$

for thin hexagonal plates or dendrites, the disk approximation

$$C = a/\pi \tag{4.28}$$

may be used, while for spheres,

$$C = a/2. \tag{4.29}$$

In order to evaluate the capacitance of each crystal, the bulk geometry of the crystals must be diagnosed or predicted. This is generally done by using laboratory and/or field data to empirically determine aspect ratios $c/a$ of the crystals or crystal mass versus major dimension (Koenig, 1971; Cotton, 1972a; Young, 1974a; Scott and Hobbs, 1977; Cotton et al., 1982; Jayaweera, 1971). Alternately, one can follow a more fundamental approach as proposed by Chen and Lamb (1994b). They made use of direct measurements of the individual growth rates of the prism and basal faces of ice crystals to determine the individual condensation coefficients of the prism and basal faces. They included the effects of ventilation of the crystals which enhances the vapor fluxes especially on the prominent edges of the ice crystals. This allows the crystals to grow to several hundred microns whereupon ventilation forces the crystals to grow primarily along their long axis.

### 4.4.4. Riming Growth of Ice Particles

Once ice crystals become large enough, they can settle through a population of supercooled cloud droplets, colliding and coalescing with them. When they impinge upon the ice surface, the droplets immediately freeze, since ice is an "ideal" nucleator. A deposit of frozen droplets called "rime" accumulates on the surface of the ice crystal. The riming growth process is a collision-coalescence process, analogous to the collision-coalescence growth of liquid cloud droplets. The rate of growth of a single ice crystal of mass $x_i$ can be thus described as

$$\left.\frac{\mathrm{d}x_i}{\mathrm{d}t}\right]_{RM} = \int_0^{\infty} A_i'(V_i - V_c)E(x_i/x)xf(x)\mathrm{d}x, \tag{4.30}$$

where $A_i'$ represents the geometric cross-sectional area that an ice crystal sweeps out relative to a cloud droplet of mass $x$, $V_i$ is the terminal velocity of an ice crystal of mass $x_i$, $V_c$ is the terminal velocity of a cloud droplet of mass $x$, $E(x_i/x)$ is the collection efficiency between the ice crystal and the cloud droplet $x$, and $f(x)$ is the spectral density of cloud droplets of mass $x \pm \delta x$.

Because ice crystals typically fall with their major dimension in the horizontal plane, the geometric cross section can be approximated as

$$A_i' = (a + r)(c + r)$$

for needles and columns of dimensions $a$ and $c$ falling through cloud droplets of radius $r$, and

$$A_1' = \pi(a/2 + r)^2$$

for plates, dendrites, and spherical ice panicles. By selecting a suitable spectral density function $f(x_i)$ to keep track of the mass and geometry of ice particles, one can compute the evolution of the ice spectrum by solving quasistochastic integral-differential growth equations similar to Eq. (4.1) used for cloud droplets. In this manner, Young (1974b) and Scott and Hobbs (1977) have simulated the evolution of the ice-particle spectrum by storing information about particle mass and geometry in a series of continuous "bins."

If the ice particles are considerably larger than the cloud droplets, then Eq. (4.30) may be simplified by ignoring the cloud droplet radius in evaluating $A_i'$, by assuming $V_i \gg V_c$, and that an average $\overline{E}$ can be used between all cloud droplets and ice particles of mass $x_i$. Under these assumptions Eq. (4.30) becomes

$$\left.\frac{\mathrm{d}x_i}{\mathrm{d}t}\right]_{RM} = A_i' V_i \overline{E(i/c)} \rho_0 r_c, \tag{4.31}$$

where $r_c$ is the cloud droplet mixing ratio.

It should be noted that the ice crystal vapor-deposition growth habits affect $A_i'$, $V_i$, and $E$. Details on how to evaluate the terminal velocity and collection efficiency of ice crystals collecting cloud droplets can be found in texts by Pruppacher and Klett (1978) and Mason (1971) or in papers like Saleeby et al. (2007). As an ice crystal grows by riming, the geometry of the ice crystal is modified, which, in turn, alters $A_i'$, $V_i$, and $E$.

In convective clouds having high liquid-water contents (generally greater than 1.0 g m$^{-3}$) the riming growth of ice crystals can proceed to the point that the rime deposit nearly obscures the original crystal habit. Often such a particle tumbles and rimes to become nearly symmetric or conical in shape. We refer to such heavily rimed particles as *graupel*. The density of small graupel particles may be as low as 0.13 g m$^{-3}$ (Magono, 1953; Nakaya and Terada, 1935). However, as the graupel particle grows larger and falls faster, the density of the rime deposit may become as high as 0.9 g m$^{-3}$. It is often thought that graupel forms from those ice crystals which have grown for a relatively long time by vapor deposition and riming. These "large" crystals are candidates for forming graupel. Reinking (1975) has observed that the embryos of graupel particles are predominantly a select few of relatively smaller crystals that collect droplets at comparably rapid rates. Perhaps these select few crystals randomly collect one or more larger than average cloud droplets. The large cloud droplets could upset the hydrodynamic stability of the crystals, causing them to tumble and form graupel particles. In contrast, crystals which have undergone longer growth times by vapor deposition would have large aspect ratios and be less susceptible to hydrodynamic instability by the chance collision with a droplet slightly larger in size than the average. Furthermore laboratory experiments performed by Fukuta et al. (1982, 1984) suggest that in the temperature range $-6\ ^\circ$C to $-10\ ^\circ$C, where more isometrically shaped, columnar forms of vapor-grown crystals prevail, ice particles switch over more readily to the graupel mode of riming growth. It is interesting that the more rapid switchover to the graupel mode of growth compensates to some degree for the otherwise suppressed precipitation growth by vapor deposition in that temperature range.

There is also some indication (Holroyd, 1964) that aggregates of ice crystals can serve as embryos for graupel particles. We will discuss aggregation more fully in the next section. Graupel can also originate as frozen large cloud droplets, drizzle drops, or raindrops. The drops may freeze by contact or immersion freezing or by collecting a small ice crystal. These large frozen drops have a high density (perhaps greater than 0.9 g cm$^{-3}$ and therefore rapidly fall through the population of cloud droplets, collecting them to become high-density graupel particles.

If the cloud contains a high liquid-water content and vigorous updrafts as in cumulonimbi, such graupel particles can serve as embryos for hailstone growth. The rate of mass growth of a hailstone may be estimated from the accretion equation, Eq. (4.31), where $A_i' = \pi R_h^2$ and $R_h$ is the radius of the hailstone.

However, as the hailstone grows by collecting cloud droplets, the latent heat of freezing is liberated. This latent heat can warm the hailstone to such a degree that not all the accreted water can freeze. If this occurs, some of the unfrozen water can be shed from the hailstone, thereby limiting its growth (Ludlam, 1951). If the hailstone sheds all unfrozen water, it is said to be growing in the "wet regime," and its mass accumulation can be estimated from the thermodynamic budget of the hailstone (Pruppacher and Klett, 1978; Mason, 1971). If all the accreted water can freeze, then Eq. (4.30) represents the hailstone growth rate and the hailstone is said to be growing in the "dry regime." There remains, however, some uncertainty as to what fraction of the unfrozen water is actually "shed" from a hailstone growing in the wet regime. Macklin (1961) carried out a set of laboratory experiments simulating the growth of hail during the wet regime. He found that much of the unfrozen water was not shed in the wake of the accreting object, but instead was incorporated into the ice structure as a spongy or mushy ice deposit containing a substantial amount of liquid water. This process is apparently enhanced by the collection of ice crystals which, Macklin postulated, can form an interwoven mesh of dendritic crystal structures which trap liquid water. Macklin noted that only at temperatures of $-1\ ^\circ$C to $-2\ ^\circ$C was any of the unfrozen accreted liquid shed in the wake of the simulated hailstones. At colder temperatures, all the unfrozen water was retained in the "spongy" ice deposit.

### 4.4.5. Aggregation Growth of Ice Particles

Snowflakes, a common form of precipitation in the wintertime in mid-latitudes, are made up of clusters or aggregates of "pristine" ice crystals. These aggregates have formed by the collision and coalescence among ice crystals. The collection process can be described by a quasistochastic collection model similar to Eqs (4.1)–(4.4). The problem, however, is complicated by the complex geometries and orientations of the falling pristine crystals, which affect the formulation of the collection kernel. An additional complication arises from the fact that once they have made physical contact, ice crystals do not always coalesce. In the case of cloud droplets, there is considerable evidence that the coalescence efficiency is reasonably high, approaching unity. In this case, the major problem in evaluating the collection kernel, Eq. (4.4), is in estimating the hydrodynamic collision efficiency between cloud droplets. This is a rather straightforward procedure but by no means a simple problem (Pruppacher and Klett, 1978). In the case of ice crystal aggregation, in addition to the challenging problem of estimating the hydrodynamic efficiency among ice crystals, we must also estimate their probability of sticking. Laboratory experiments and inferences from field studies suggest that the coalescence efficiency among ice crystals is higher at warm temperatures (Hallgren and Hosler, 1960; Hosler and Hallgren, 1960) and is a function of the habit of the ice crystals, with delicately branched dendrites, having the highest efficiencies (Rogers, 1974). However, Latham and Saunders (1971) found no such temperature dependence in their laboratory experiments. Rauber (1985) has also suggested, from field observational evidence, that a mixture of plane dendrites and spatial dendrites, which have different fall-velocity spectra, favor the formation of aggregates. The exact magnitude of coalescence efficiencies among ice crystals and their variation with crystal habit and temperature remains unknown at the present time.

Another complicating aspect of snowflake aggregation theory is that ice crystals and aggregates of ice crystals exhibit both horizontal velocity fluctuations and fluctuations in terminal velocity. These velocity fluctuations are caused by the complex geometry of the crystals, which can produce aerodynamic lifting forces similar to that of an aircraft wing. Further causes of velocity fluctuations include changes in aerodynamic drag forces due to variations in orientation of the ice crystals and to the shedding of turbulent eddies in the wake of the particle. All of these factors influence the rate of collision among ice crystals. For example, when we discussed the coalescence among cloud droplets, we defined a collection kernel, Eq. (4.4), which was a function of the area swept out by the "parent" droplet and the differences in mean terminal velocities. However, in the case of ice crystals, the collection kernel is influenced by horizontal and vertical velocity fluctuations. Sasyo (1971) examined this effect by laboratory and field experiments as well as numerical calculations. He applied the kinetic theory of gases to this problem and formulated a collection kernel as

$$V(r_i, r_j) = 2\sigma\sqrt{\pi}(r_i + r_j)^2, \tag{4.32}$$

where $r_i$ and $r_j$ represent the effective radii of ice crystals or aggregates, and $\sigma$ is the standard deviation of horizontal and vertical velocity fluctuations. Sasyo estimated a standard deviation of 5 cm $s^{-1}$ from field experiments. Based on these results he suggested that collisions of this type would be important during the first stages of aggregate formation when ice crystals are of similar size and shape and before a broad spectrum of crystals has formed.

Passarelli and Srivastava (1979) hereafter referred to as PS have also noted that due to the particular structure of snowflakes, snowflakes of a given mass can have a spectrum of sizes, fall speeds, and shapes. In contrast to the standard collection kernel, Eq. (4.4), where particles of the same mass have zero probability of colliding, such a spectrum leads to a finite probability of collision of particles of the same mass. PS considered two models of aggregation. In the first model snowflakes of a given mass are considered to be spherical and have a unique diameter but a spectrum of fall speeds. This model bears some resemblance to Sasyo's model except that horizontal motions of snowflakes are not considered. In the second model snowflakes are also assumed to be spherical, but snowflakes of a given mass are assumed to have a spectrum of bulk densities, which results in a spectrum of diameters and fall speeds. Numerical experiments with the first model suggested that its contribution to snowflake aggregation was only 10% of that due to the standard kernel. However, experiments with the second model, which includes a spectrum of particle densities, indicated that the magnitude of the modified kernel is always greater than that of the standard kernel. Depending on the spectrum width, the modified kernel results in substantially more rapid aggregation. This work suggests that new approaches to modeling aggregation may be needed. For example, if aggregate and pristine ice crystals are only described in a model as a function of their mass, it may be necessary to develop probability density functions (PDF's) of their collection kernels for each mass bin.

It is clear from this discussion that there remains a great deal to be learned about snowflake aggregation processes. In addition to the uncertainties mentioned above, it should also be noted that aggregation commences from the collision among pristine crystals whose concentration cannot be consistently predicted even within several orders of magnitude of observed values. This is especially true at warmer temperatures, where aggregation seems to be the most prevalent. As can be seen from the collection equations [Eqs (4.2) and (4.3)], the aggregation rate is proportional to the product of the concentration of ice crystals. Thus a threefold order of magnitude uncertainty in ice crystal concentration results in roughly a sixfold order of magnitude uncertainty in our estimates of the initial rate of aggregation. However, there is considerable motivation to continue to improve models of the aggregation process, since aggregates of dendrites are the most common type of snowfall in mid-latitudes.

### 4.4.6. Melting of Ice Particles

The melting of ice particles (and associated cooling due to the latent heat of fusion), contributes substantially to the formation of thunderstorms (Knupp, 1985) and mesoscale downdrafts in tropical squall lines (Houze, 1977) and in wintertime orographic cloud systems (Marwitz, 1983).

The melting of an ice particle is basically a thermodynamic process. Consider a graupel particle of mass $X_g$ which has fallen through the 0 °C isotherm into warmer temperatures. Suppose further that the layer that the graupel has fallen into contains cloud droplets at the ambient temperature $T$. Assuming a steady state and that the graupel maintains a surface temperature $T_f$ of 0 °C, the rate of latent heat release due to melting must be balanced by the rate of heat transfer through the layer of water on the graupel surface. Thus,

$$L_{li}\left[\frac{\mathrm{d}X_g}{\mathrm{d}t}\right]_{\text{melt}} = -2\pi D_g K_T f(R_e)(T - T_f) - 2\pi L_{lv} D_g D_v f(R_e)(\rho_{vsfc}) - \left[\frac{\mathrm{d}X_g}{\mathrm{d}t}\right]_{RM} c_w(T - T_f), \tag{4.33}$$

where $D_g$ is the graupel diameter, $K_T$ and $D_v$ are the diffusivities for heat and water vapor, respectively, and $c_w$ is the thermal conductivity of liquid water. The first term on the right-hand side of Eq. (4.33) represents the diffusion of heat to the surface of the melting graupel particle at temperature $T_f$. The second term on the right-hand side of Eq. (4.33) represents the diffusion of water vapor and the corresponding transfer of latent heat from the graupel surface. A ventilation term $f(R_e)$ is included in both diffusion terms. The third term on the right-hand side of Eq. (4.33) represents the transfer of sensible heat to the graupel as it accretes cloud droplets at the rate $(\mathrm{d}X_g/\mathrm{d}t)_{RM}$. Mason (1956) considered the effects of the first two terms in his study of the melting process. Wisner et al. (1972) and Cotton et al. (1982) have included the last term in their cloud models. The transfer of sensible heat during collection can greatly accelerate the melting process, especially in clouds with low cloud bases and corresponding deep layers of cloud warmer than 0 °C.

## 4.5. SUMMARY OF CLOUD MICROPHYSICAL PROCESSES

In preceding sections, we have seen that the evolution of precipitation in clouds can take on a variety of forms and involve numerous physical processes. The evolution of ice-phase precipitation processes is greatly dependent upon the

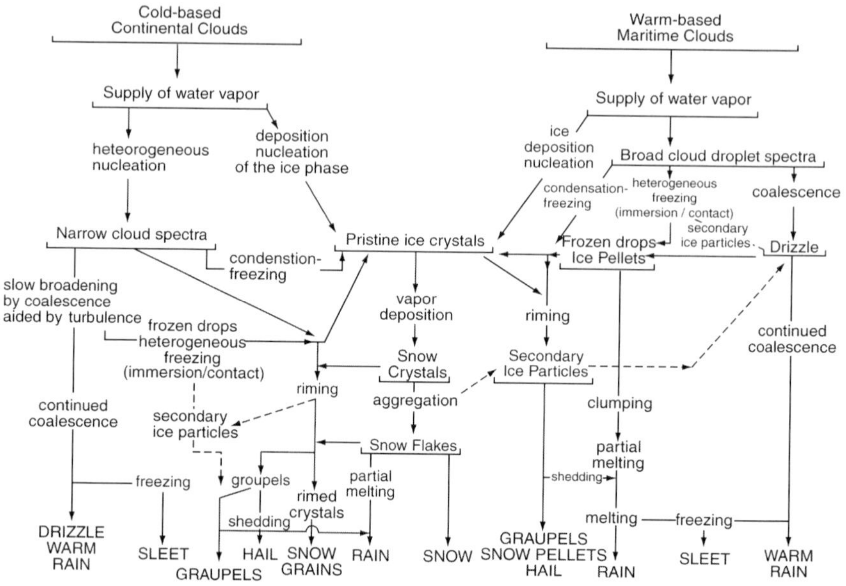


**FIGURE 4.9 Flow diagram describing microphysical processes, including paths for precipitation formation.** *(Adapted from Braham (1968))*

prior or concurrent evolution of the liquid-phase. These processes, in turn, are dependent upon the characteristics of the air mass (i.e. aerosols), the liquid-water production of the cloud, the vertical motion of air within the cloud, the turbulent structure, and the time scales of the cloud. We characterize the liquid water production of the cloud by its base temperature. Clouds having warm bases will have higher cloud base mixing ratios and thus will produce more condensate than a cold-based cloud over a given depth of cloud. Illustrated in Fig. 4.9 are the different precipitation paths that may occur depending upon whether the cloud is a cold-based continental cloud versus a warm-based maritime cloud. We use the term maritime cloud to represent a very clean air mass and continental to represent one with much higher CCN concentrations. A polluted cloud would have still higher CCN concentrations. The figure does not note the speed by which these regimes can produce precipitation. We have seen that a cloud with a vigorous warm rain process or, what we refer to in the figure as a warm-based, maritime cloud will produce precipitation much faster than a cold-based, continental cloud. The rapidity of glaciation of a warm-based maritime cloud is much faster than the cold-based maritime cloud since the presence of drizzle drops and supercooled raindrops once frozen can rapidly transform a cloud from an all-water cloud to an ice-dominated cloud. This should not be interpreted to mean that the largest precipitation elements such as hail would occur in a warm-based maritime cloud. In fact, just the opposite can take place, as a vigorous precipitation process lower in the cloud can deplete supercooled water amounts higher up and lower the trajectory of precipitation elements leading to smaller sized hailstones.

This diagram is used to illustrate two distinctly different regimes. In fact, there is a continuum of cloud types between these two extreme states. The heaviest precipitating clouds, and generally the most efficient, are those that are warm-based and form in clean, maritime air masses. By contrast, clouds developing in a polluted air mass should be less efficient in producing precipitation than similar clouds with the same cloud base temperatures. This simple reasoning is valid only for single clouds and storms. As we will see, in some cases a suppression or retardation of precipitation in a primary convective cell could lead to a transformation of a cloud into a longer-lived storm through the interaction of cold-pools and in some cases lead to greater amounts of precipitation.

## 4.6. MODELING AND PARAMETERIZATION OF ICE-PHASE MICROPHYSICS

Modeling the microphysics of ice-phase or mixed-phase clouds generally mirrors the approaches used to model liquid-only clouds. The ice-phase, however, is more complicated owing to the departures of ice particles from being spheres and to the variations in particle densities. The most advanced models which keep track of ice particle habits such as Chen and Lamb (1994a) are so computationally demanding that they are generally limited to the simple Lagrangian parcel type of dynamical frameworks. Single-moment bin-resolving models such as Khain and Sednev (1995) and multi-moment bin models (Reisin et al., 1996a,b) have been extended to include the ice-phase but with simple parameterizations of ice particle shapes like assuming a single crystal habit, and making use of mass-diameter empirical relationships such as that used in bulk models. These models have been applied to two and three dimensional cloud models (Reisin et al., 1996a,b; Harrington et al., 2000; Khain et al., 2004; Lynn et al., 2005) even though they are very computationally demanding. In fact, a full bin scheme has been implemented into the Earth Simulator General Circulation Model (GCM)!

Most cloud and storm models, mesoscale models and especially GCMs follow a bulk microphysics approach that is an extension of the Kessler (1969) warm cloud microphysics philosophy. Like the warm cloud schemes, ice clouds are assumed to be ice saturated and produce "cloud ice," the size distribution of all ice hydrometeors is assumed to follow an exponential size-distribution, collection is modeled as a continuos accretion process, and all hydrometeor classes settle as a mass-weighted mean fall velocity (Cotton, 1972a,b; Lin et al., 1983; Rutledge and Hobbs, 1984). Power-law relationships are generally used to relate ice particle mass to the dimensions of the ice particles (Locatelli and Hobbs, 1974; Lin et al., 1983; Rutledge and Hobbs, 1984; Mitchell et al., 1990; Cotton et al., 2003). We must recognize that there are many differences between the liquid and ice-phase that make the extension of the Kessler approach to the ice phase less desirable. For one thing, while supersaturations with respect to water are believed to be less than 1%, that is not the case for the ice phase where supersaturations with respect to ice can exceed 10%. Thus while assuming water clouds remain saturated allowing the diagnosis of cloud water may not lead to large errors, extension to the ice-phase can lead to much greater errors in diagnosing the mixing ratios of vapor grown ice crystals and their thermodynamic consequences. Thus Cotton et al. (1986, 2003) replace the concept of "cloud ice" with predicted pristine ice which retains substantial supersaturations. Likewise, while modeling collection as a continuous accretion process may be valid for graupel particles and hailstones collecting cloud droplets, it is not a good approximation when hailstones collect graupel particles or raindrops, or aggregates collect pristine ice crystals as the differences in terminal velocity between the collector and collectee can be quite small, and the sizes of both species is appreciable. A somewhat ad hoc approximation was proposed by Wisner et al. (1972) in which the differences in terminal velocities in the collection equation is approximated by the differences in mass-weighted fall speeds. As shown by Verlinde et al. (1990) this can lead to large errors in computed collection rates.

Due to the storage requirements and computing time required to compute detailed microphysics, explicit prediction of the habits of ice crystals and their size distribution have been limited to use in simple one-dimensional, steady-state parcel models (Cotton, 1972b; Young, 1974b, 1975; Chen and Lamb, 1994b), at the most in one-dimensional, time-dependent models (Scott and Hobbs, 1977). It is generally assumed that the ice particles are either spherical or have a single crystal habit such as a hexagonal plate. Cotton et al. (2003) allow different crystal habits depending on temperature but these habits are not allowed to advect or diffuse in the cloud or undergo transformations to mixed habits like capped-columns.

Important to any ice parameterization is the number of ice categories. Because each ice category affects particle fall speeds and collection rates, a larger number of ice categories better represents a cloud composed of, say, vapor-grown ice crystals, partially-rimed ice crystals, aggregates, graupel particles of varying densities, frozen raindrops, and hailstones of varying densities (McCumber et al., 1991). Of course each category of ice added increases the computational cost. Even the use of bin-resolving microphysics models does not eliminate the need to select the appropriate number of ice categories. Ferrier et al. (1995) for example, found that using separate graupel and hail categories greatly improved their storm characteristics in comparison to observations. An example of a large number of ice categories is that of Straka and Mansell (2005) who decomposed ice into ten categories consisting of ice crystals (three habits), aggregates/snow, graupel (three densities), frozen drops, and hail (small and large).

It has become increasingly popular to expand the number of prognostic parameters or moments describing the basis functions that define the size-distribution of hydrometeors (Nickerson et al., 1986; Ferrier et al., 1995; Meyers et al., 1997; Cotton et al., 2003; Reisner et al., 1998; Morrison et al., 2005a,b). Clark (1974) and more recently Milbrandt and Yau (2005a,b) have implemented triple-moment prognostic systems. Milbrandt and Yau applied their triple-moment model to simulating hailstorms in which the sixth-moment or radar reflectivity was the prognostic variable. In addition, while the earlier ice parameterizations were "hard-wired" to the exponential basis function (Cotton et al., 1986; Lin et al., 1983; Rutledge and Hobbs, 1984) more flexible and more general basis functions such as gamma and log-normal basis functions are now being used (Ziegler, 1985; Ferrier, 1994; Walko et al., 1995; Clark, 1976; Feingold and Levin, 1986; Feingold et al., 1998).

Finally, as mentioned earlier, in the RAMS model Saleeby et al. (2007) extended the bin-emulating approach to the ice-phase in which realistic (at least the best available in the literature) collection kernels were used for graupel particle and hailstone collection of cloud droplets. All these improvements in the bulk parameterization schemes makes them more flexible and closer in character to the full bin-resolving models.

We now focus briefly on hail parameterization. When we consider parameterization of most hydrometeor species the emphasis is on estimating the average water contents and perhaps concentrations. But for hail the emphasis shifts to the tail of the distribution, or largest, most damaging particles. Thus conventional one- or two-moment bulk parameterizations are less suitable. Certainly the most appropriate approach is to use a full quasi-stochastic bin-resolving model. At the time of this writing no one has implemented a full bin-resolving microphysics model for use in two and three dimensional storm models that includes hail thermodynamics (see below). There have been a couple of bin-resolving models of hail growth that have been applied to simple Lagrangian parcel models (Young, 1978; Danielsen et al., 1972). In the past several researchers have implemented a hybrid approach to modeling hail (Farley and Orville, 1986; Johnson et al., 1993). In this approach all non-hail species are modeled following the Kessler-type philosophy. Hail, however, is treated using finite bins with a continuous accretion approximation for hail collecting cloud droplets. Another strategy is to use a bulk microphysics approach but to divide the hail spectrum into a small-hail and large-hail modes (Straka and Mansell, 2005). Then there is the bulk approach of Milbrandt and Yau (2005a,b) in which a bulk approach is used but three moments of the hail basis function are predicted. The relative advantages of the hybrid bulk/bin approach, versus the two-mode bulk approach, versus three-moment bulk approaches have not been quantitatively determined.

One final comment on hail modeling is the thermodynamics of hail growth. Hail collects supercooled drops at such a high rate that the latent heating of freezing can warm the hailstone to near 0 °C. At this point the hailstone is said to undergo "wet growth" such that only a fraction of the collected supercooled water freezes. Some of the unfrozen water can shed, although the amount of water shed can be small (List, 1959, 1960; Macklin, 1961) with some of it being incorporated into what is called a "spongy hail" mesh (List, 1965). Nonetheless determination of whether a hailstone is in wet or dry growth regimes is important for the shedding calculations and furthermore the radar reflectivity of hailstones is dependent on whether the hailstone surface is ice or water coated. A common "textbook" approach to estimating hailstone temperatures (i.e. Pruppacher and Klett), is to calculate the rate of collection of supercooled water by continuous accretion and then calculate the heat budget of the hailstone by the latent heat of freezing of water and removal of latent and sensible heat by diffusion, and conduction. Normally this is done following Schumann (1938) and Ludlam (1958) in which the heat storage of the hailstone is neglected, which permits forming a balance equation between heat released and heat absorbed by the hailstone. However, this is a very poor assumption for massive hydrometeors such as hailstones, and probably even graupel particles. To overcome this deficiency, Walko et al. (1995) developed a heat budget equation for each ice category including hail that retained heat storage. This was done by defining and diagnosing a reference category energy from which the fraction of ice vs water can be diagnosed. Such a procedure could be adapted to a multi-category hail model such as Straka and Mansell's (2005) model or even bin-resolving models.

It has become increasingly evident that aerosols can have an important influence on cloud microphysics and especially on precipitation (Levin and Cotton, 2007). Most of the studies simulating aerosol/cloud-system interactions have followed a bin-resolving microphysics approach where not only the hydrometeor spectra are described in discrete bins but so is the aerosol spectrum (Feingold et al., 1999; Yin et al., 2000; Reisin et al., 1996a,b; Khain et al., 1999). The more sophisticated of these models treat CCN as a size-resolved prognostic species and track soluble material inside drops (Flossman et al., 1985; Chen and Lamb, 1994b; Feingold et al., 1996). This permits not only examining aerosol effects on clouds but also cloud effects on aerosol.

As far as bulk microphysics models are concerned aerosol effects on clouds have been mainly examined by varying droplet concentrations which then impact on autoconversion processes. This approach can be useful when a single class of cloud is represented, such as a thunderstorm or an orographic cloud. Once an ensemble of clouds is represented, or within a model domain layer clouds and deep convective clouds simultaneously occur, this approach ignores the role of cloud vertical velocities in determining the concentration of cloud droplets nucleated. Thus Saleeby and Cotton (2005) were motivated to develop a parameterization of CCN activation that takes into account not only the variability of CCN concentrations but also vertical velocity and temperature. First, RAMS had to be changed from a pure diagnostic model of cloud liquid water using a saturation adjustment approach to predicting the mixing ratio of cloud liquid water (Walko et al., 2000). Then an ensemble of Lagrangian parcel model calculations were performed with the Feingold and Heymsfield (1992) parcel model. Assuming that the aerosol could be chemically characterized as ammonium sulfate particles and that their size-distribution could be described by a log-normal basis function with a constant breadth parameter, an ensemble of realizations was performed, varying temperature, vertical velocity, CCN concentration, and median radius of the CCN distribution. The results of those calculations were put into a look-up table that is accessed by RAMS in cloud and storms simulations. Furthermore a second cloud droplet mode was defined wherein GCCN particles were nucleated assuming they are sodium chloride particles. Like the more advanced bin-resolving microphysics models, the mass of soluble material was kept track of so that evaporating drops can recharge the atmosphere with CCN, albeit with an altered size. In summary, the RAMS bulk microphysics model has three prognostic aerosol species, the concentrations of CCN, GCCN, and IN using Eq. (4.21).

## 4.7. IMPACT OF CLOUD MICROPHYSICAL PROCESSES ON CLOUD DYNAMICS

What level of complication in the formulation of cloud microphysical processes is needed to simulate a particular cloud system? This question arises naturally, and the answer is influenced, in part, by the scientific objectives. If the goal is to simulate the average precipitation over a mesoscale region, it may not be necessary to simulate microphysics in great detail. If, however, it is important to distinguish among the various forms of precipitation—such as freezing rain versus graupel, or aggregates versus pristine or lightly rimed ice crystals—then greater sophistication in the formulation of cloud microphysics is desirable. The answer to the above question is also driven by the impact of cloud microphysical processes on the dynamics of the cloud system. If cloud microphysical processes impact strongly on the dynamics of a cloud system, then an incorrect simulation of a cloud microphysical process could lead, for example, to the simulation of short-lived multicellular thunderstorms in cases where severe, steady supercell storms are actually observed, or weak storm downdrafts where severe downbursts are observed. Frequently, numerical models of clouds experience a bifurcation in dynamic behavior, depending on the cloud microphysical structure. That is, local changes in cloud microphysical processes can eventually lead to a cloud system having completely different dynamical characteristics. Let us, therefore, summarize some of the interactions between cloud microphysical processes and cloud dynamics. More detailed examinations of such interactions will be made in Part II of this volume when we study the dynamics of various cloud systems.

### 4.7.1. Water Loading

In Chapter 2 we noted that the vertical equation of motion for a cloud system contains an additional term in the buoyancy term which accounts for the weight of suspended condensate having combined liquid and ice mixing ratios $r_d$ and $r_i$, respectively. This term arises from the fact that if cloud droplets and precipitation particles are falling on average at nearly their terminal velocities, then the sum of the drag forces on a parcel of air due to settling hydrometeors is equal to the weight of the condensate. As a rule of thumb, roughly 3 g $kg^{-1}$ of condensate is equivalent to 1 K of negative thermal buoyancy. One consequence of a precipitation process, therefore, is that it unloads the cloud updraft at higher levels of its condensate and redistributes it to low levels. The additional condensate at lower levels may turn a thermally buoyant updraft into a downdraft.

### 4.7.2. Redistribution of Condensed Water into Subsaturated Regions

A greater consequence of a precipitation process is the redistribution of condensation and evaporation processes associated with the condensate redistribution. Numerical experiments by Liu and Orville (1969) and Murray and Koenig (1972) suggest that the thermodynamic consequences of the precipitation process are far more important than the water loading effects.

Precipitation may settle into unsaturated air beneath the cloud base where evaporation commences. The evaporatively chilled air, in turn, can stimulate and intensify downdrafts in the subcloud layer, which can lead to the decay of the cloud or contribute to the propagation of the storm as air is lifted along the gust front.

As will be discussed more fully in Chapter 8, The strength of downdrafts and cold pools beneath clouds, in turn, is related to the dryness of the downdraft air as a result of entrainment of dry air in the lower levels of the cloud, the precipitation rate and to the size-distribution of raindrops. The dependence on drop size is related to the fact that the surface to volume ratio is greater for smaller drops than for larger drops. Thus for a given rainwater content, a precipitating downdraft composed of smaller drops will evaporate more total water when exposed to subsaturated air than one composed of a few larger drops, leading to colder downdraft air (Hookings, 1965; Knupp, 1985).

### 4.7.3. Cloud Supersaturation and Cloud Droplet Evaporation

As mentioned earlier it is often assumed that a cloud does not become supersaturated with respect to water. This is in accordance with theoretical studies (Howell, 1949; Squires, 1952; Mordy, 1959; Neiburger and Chien, 1960) and observations (Warner, 1969) in non-precipitating clouds, that peak supersaturations are generally less than 1% and more typical supersaturations are between 0.1% and 0.2%. Numerical experiments in a precipitating system suggest that there is a much greater tendency for the supersaturation to exceed nominal values (Clark, 1973; Young, 1974c). This is a result of the fact that the magnitude of supersaturation is a result of two competing processes: (1) supersaturation production by adiabatic cooling and (2) supersaturation reduction by condensation on cloud droplets. In a non-precipitating cloud, numerous small cloud droplets are able to readily deplete supersaturation as condensation occurs on them. Because precipitation particles form at the expense of cloud droplets, the net surface area over which condensation takes place is reduced substantially when water is converted from numerous cloud droplets to fewer, large precipitation elements. As a result, supersaturations may exceed 5%, causing a delay in latent heat release compared to situations in which the cloud's humidity remains close to 100%.

Perhaps more important is that the concept that clouds do not become highly supersaturated, motivated a saturation adjustment approach in which cloud water contents are determined by the difference between total water mixing ratios and saturation mixing ratios. The procedure eliminates the need to know anything about the details of cloud droplet concentrations and sizes. However, detailed bin-resolving microphysics models of Xue and Feingold (2006) and Jiang and Feingold (2006) show that increasing concentrations of CCN and droplets, produced smaller droplets and suppressed drizzle and led to enhanced evaporation of droplets by entrainment. Because, for a given LWC, smaller droplets evaporate more readily than larger droplets, owing to the greater net surface area exposed to subsaturated air, entrainment induced evaporative cooling was enhanced when CCN droplet concentrations were high. This led to greater entrainment rates in clouds, lower cloud liquid-water contents and reduced cloud depth. A model that uses the saturation adjustment approach to determine cloud liquid-water contents would not be able to simulate this process.

### 4.7.4. Latent Heat Released during Freezing and Sublimation

At levels in the atmosphere colder than 0 °C, the potential exists for ice crystals to be nucleated and grow by vapor deposition and collection of cloud drops. Also, supercooled cloud droplets and raindrops may freeze. The freezing of cloud droplets and raindrops results in the release of the latent heat of fusion.

The amount of heat liberated is proportional to the amount of supercooled liquid water frozen. Furthermore, as ice crystals grow by vapor deposition, they release the latent heat of sublimation. If the cloud is water saturated and contains a substantial amount of supercooled cloud droplets, however, the full latent heat of sublimation is not absorbed by the cloudy air. This is because ice crystals grow by vapor deposition at the expense of cloud droplets, which must evaporate as the saturation vapor pressure is lowered locally below water saturation. As a result, the evaporating droplets absorb the latent heat of condensation. The net result of ice crystals growing by vapor deposition and of cloud droplets evaporating is that the cloud experiences heating only in proportion to the latent heat of fusion (i.e. $L_f = L_s - L_c$, where $L_f$ is latent heat of fusion, $L_s$ is latent heat of sublimation, and $L_c$ is latent heat of condensation) and the water mass deposited on the ice crystals. In a glaciated cloud where liquid- water droplets are absent, a cloud will experience the full latent heat of sublimation as vapor is deposited on ice crystals.

What this means, of course, is that any cloud will experience an additional source of buoyancy as freezing and ice vapor deposition takes place. The latent heating can be rather smoothly released if ice crystals grow by vapor deposition in an air mass that is cooling adiabatically by large-scale lifting. In contrast, it can take place as a burst of energy release in convective towers if large quantities of supercooled water suddenly freeze.

It is important to recognize that ice-phase-related latent heating begins to become important at levels in the atmosphere where the latent heat of condensation is greatly reduced. This is due to the fact that at colder temperatures the vertical variation of saturation vapor pressure with respect to water diminishes substantially as a parcel of air is cooled by adiabatic expansion. Ice-phase latent heating is also important in disturbed environments such as tropical cyclones (Lord et al., 1984) and mesoscale convective systems (Chen and Cotton, 1988), where the environmental sounding is nearly wet adiabatic. In such an environment the cloud realizes little buoyancy gain from the latent heat of condensation, while ice-phase latent heating can contribute to substantial convective instability. We will discuss the importance of ice-phase latent heating more fully in Chapters 7–9.

### 4.7.5. Cooling by Melting

Melting is distinctly different from the freezing process, in which freezing is distributed through a considerable vertical depth. By contrast, cooling by melting is quite localized. In the case of the more stratiform precipitation, melting can result in a well-defined isothermal layer. The stability of this isothermal layer can inhibit the downward penetration of upper tropospheric winds to lower tropospheric levels.

As will be discussed more fully in Chapter 8, cooling by melting can be an appreciable contributor to thunderstorm downdraft strengths (Knupp, 1985). Furthermore, as shown by van den Heever and Cotton (2004) the size of melting hydrometeors is important to the cooling rate of storm downdrafts. This is because melting is a result of heat exchanges at the surface of the ice particle. Thus, like evaporation of raindrops, a precipitating downdraft composed of a large number of small hailstones will melt faster than one with the same amount of condensed mass but composed of a few hailstones. We will discuss the consequences of this on the dynamics of thunderstorms in Chapter 8.

### 4.7.6. Radiative Heating/Cooling

We shall see in the next chapter that cloud radiative heating and cooling rates are strongly modulated by the microphysical structure of a cloud. The reflection of incident solar radiation at the tops of clouds is dependent upon the concentration and size of cloud droplets or the concentration, size, and habit of ice crystals. The absorption and heating by solar radiation, in turn, is a function primarily of the integrated liquid-water path and secondarily of the size of the droplets. Likewise, the absorption of terrestrial radiation is related to the integrated liquid-water path. As a consequence, both the rates of solar and terrestrial radiative cooling are primarily related to the vertical distribution of condensate in a cloud.

Because precipitation processes alter the vertical distribution of condensate, they also strongly impact upon longwave and shortwave radiative heating/cooling rates in clouds. The transformation of a cloud from the liquid phase to the ice phase can impact the radiative properties of a cloud. Furthermore, the particular habit of ice crystal growth can affect the rate of absorption of solar radiation which, in turn, can alter the thermodynamic stability of the cloud system. We will examine these processes more quantitatively in the next chapter as well as in Chapter 6 and Chapter 10.

### 4.7.7. Electrical Effects

The influence of cloud electrification processes on the dynamics of clouds is discussed more fully in Chapter 8. Here we will simply enumerate some of the hypothesized ways in which cloud electrification can affect cloud dynamics:

(1) Localized heating arising from lightning discharges.
(2) Levitation of cloud particles, which alters the terminal velocity of particles, causing a redistribution of condensate.
(3) Enhancement of droplet and ice-particle coalescence, thus enhancing precipitation formation in a redistribution of condensate.

## REFERENCES

Al-Naimi, R., and Saunders, C. P. R. (1985). Measurements of natural deposition and condensation-freezing ice nuclei with a continuous flow chamber. Atmos. Environ. 19, 1871–1882.

Auer, A. H., and Marwitz, John D. (1969). Comments on the collection and analysis of freshly fallen hailstones. J. Appl. Meteorol. 8, 303–304.

Austin, P., Wang, Y., Pincus, R., and Kujala, V. (1995). Precipitation in stratocumulus clouds: Observational and modeling results. J. Atmos. Sci. 52, 2329–2352.

Baker, B. A. (1992). Turbulent entrainment and mixing in clouds: A new observational approach. J. Atmos. Sci. 49, 387–404.

Baker, M. B., Corbin, R. G., and Latham, J. (1980). The influence of entrainment on the evolution of cloud droplet spectra: I. A model of inhomogeneous mixing. Q. J. R. Meteorol. Soc. 106, 581–598.

Barkstrom, B. R. (1978). Some effects of 8-12μm radiant energy transfer on the mass and heat budgets of cloud droplets. J. Atmos. Sci. 35, 665–673.

Bartlett, J. T. (1966). The growth of cloud droplets by coalescence. Q. J. R. Meteorol. Soc. 92, 93–104.

Bartlett, J. T. (1970). The effect of revised collision efficiencies on the growth of cloud droplets by coalescence. Q. J. R. Meteorol. Soc. 96, 730–738.

Beheng, K. D. (1994). A parameterization of warm cloud microphysical conversion processes. Atmos. Res. 33, 193–206.

Bergeron, T. (1935). On the physics of cloud and precipitation. In "Proc. Assem. Int. Union Geodesy Geophys., 5th, Lisbon." pp. 156–178.

Berry, E. X. (1967). Cloud droplet growth by collection. J. Atmos. Sci. 24, 688–701.

Berry, E. X., and Reinhardt, R. L. (1974). An analysis of cloud drop growth by collection: Part I. Double distributions. J. Atmos. Sci. 31, 1814–1824.

Bigg, E. K. (1953a). The supercooling of water. Proc. Phys. Soc. London, Sect. B 66.

Bigg, E. K. (1953b). The formation of atmospheric ice crystals by the freezing of droplets. Q. J. R. Meteorol. Soc. 79, 510–519.

Bigg, E. K. (1955). Ice-crystal counts and the freezing of water drops. Q. J. R. Meteorol. Soc. 81, 478–479.

Blanchard, D. C., (1948). Observations of the behavior of water drops at terminal velocity in air. Occas. Rep., pp. 100–110. Proj. Cirrus, General Electric Res. Labs., Schenectady, NY.

Blanchard, D. C. (1957). The supercooling, freezing and melting of giant waterdrops at terminal veclodity in air. In "Artificial Stimulation of Rain" (H. Weickmann and W. Smith, Eds.), pp. 233–249. Pergamon Press.

Bleck, R. (1970). A fast, approximate method for integrating the stochastic coalescence equation. J. Geophys. Res. 75, 5165–5171.

Bower, K. N., and Choularton, T. W. (1993). Cloud processing of the cloud condensation nucleus spectrum and its climatological consequences. Q. J. R. Meteorol. Soc. 119, 655–679.

Braham, R. R. (1964). What is the role of ice in summer rain showers? J. Atmos. Sci. 21, 640–645.

Braham, R. R. (1968). Meteorological bases for precipitation development. Bull. Am. Meteorol. Soc. 49, 343–353.

Brazier-Smith, P. R., Jennings, S. G., and Latham, J. (1972). The Interaction of falling water drops: Coalescence. Proc. R. Soc. London, Ser. A 326, 393–408.

Brazier-Smith, P. R., Jennings, S. G., and Latham, J. (1973). Raindrop interactions and rainfall rates within clouds. Q. J. R. Meteorol. Soc. 99, 260–272.

Brenguier, J. L., and Chaumat, L. (2001). Droplet spectra broadening in cumulus clouds. Part I: Broadening in adiabatic cores. J. Atmos. Sci. 58, 628–641.

Byers, H. R. (1965). "Elements of Cloud Physics." University of Chicago Press, Chicago, Illinois.

Carte, A. E. (1956). The freezing of water droplets. Proc. Phys. Soc. London 69, 1028–1037.

Chaumat, L., and Brenguier, J. L. (2001). Droplet spectra broadening in cumulus clouds. Part II: Micro-scale droplet concentration heterogeneities. J. Atmos. Sci. 58, 642–654.

Chen, S., and Cotton, W. R. (1988). The simulation of a mesoscale convective system and its sensitivity to physical parameterizations. J. Atmos. Sci. 45, 3897–3910.

Chen, J.-P., and Lamb, D. (1994a). The theoretical basis for the parameterization of ice crystal habits: Growth by vapor deposition. J. Atmos. Sci. 51, 1206–1221.

Chen, J.-P., and Lamb, D. (1994b). Simulation of cloud microphysical and chemical processes using a multicomponent framework. Part I: Description of the microphysical model. J. Atmos. Sci. 51, 2613–2630.

Chien, E. H., and Neiburger, M. (1972). A numerical simulation of the gravitational coagulation process for cloud droplets. J. Atmos. Sci. 29, 718–727.

Clark, T. L. (1973). Numerical modeling of the dynamics and microphysics of warm cumulus convection. J. Atmos. Sci. 30, 857–878.

Clark, T. L. (1974). A study in cloud phase parameterization using the gamma distribution. J. Atmos. Sci. 31, 142–155.

Clark, T. L. (1976). Use of log–normal distributions for numerical calculations of condensation and collection. J. Atmos. Sci. 33, 810–821.

Clark, T. L., and Hall, W. D. (1983). A cloud physical parameterization method using movable basis functions: Stochastic coalescence parcel calculations. J. Atmos. Sci. 40, 1709–1728.

Cooper, W. A. (1980). A method of detecting contact ice nuclei using filter samples. In "Eighth International Conf. on Cloud Physics. Clermont-Ferrand, France." pp. 665–668, Preprints.

Cooper, W. A. (1995). Ice formation in wave clouds: Observed enhancement during evaporation. In "Proc. Conf. on Cloud Physics." pp. 147–152. Amer. Met. Soc, Dallas.

Cooper, W. A., and Vali, G. (1981). The origin of ice in mountain cap clouds. J. Atmos. Sci. 38, 1244–1259.

Cooper, W. A., Bruintjes, R. T., and Mather, G. K. (1997). Calculations pertaining to hygroscopic seeding with flares. J. Appl. Meteorol. 36, 1449–1469.

Cotton, R. J., and Field, P. R. (2002). Ice nucleation characteristics of an isolated wave cloud. Q. J. R. Meteorol. Soc. 128, 2417–2437.

Cotton, W. R., (1970). A numerical simulation of precipitation development in supercooled cumuli. Ph.D. Thesis, Pennsylvania State University.

Cotton, W. R. (1972a). Numerical simulation of precipitation development in supercooled cumuli, Part I. Mon. Weather Rev. 100 (11), 757–763.

Cotton, W. R. (1972b). Numerical simulation of precipitation development in supercooled cumuli, Part II. Mon. Weather Rev. 100 (11), 764–784.

Cotton, W. R., and Gokhale, N. R. (1967). Collision, coalescence, and breakup of large water drops in a vertical wind tunnel. J. Geophys. Res. 72 (16), 4041–4049.

Cotton, W. R., Stephens, M. A., Nehrkorn, T., and Tripoli, G. J. (1982). The Colorado State University three-dimensional cloud/mesoscale model 1982. Part II. An ice-phase parameterization. J. Rech. Atmos. 16, 295–320.

Cotton, W. R., Tripoli, G. J., Rauber, R. M., and Mulvihill, E. A. (1986). Numerical simulation of the effects ofvarying ice crystal nucleation rates and aggregation processes on orographic snowfall. J. Clim. Appl. Meteorol. 25, 1658–1680.

Cotton, W. R., Pielke Sr., R. A., Walko, R. L., Liston, G. E., Tremback, C. J., Jiang, H., McAnelly, R. L., Harrington, J. Y., Nicholls, M. E., Carrió, G. G., and McFadden, J. P. (2003). RAMS 2001: Current status and future directions. Meteor. Atmos. Phys. 82, 5–29.

Danielsen, E. F., Bleck, R., and Morris, D. A. (1972). Hail growth by stochastic collection in a cumulus model. J. Atmos. Sci. 29, 135–155.

DeMott, P. J. (2002). Laboratory studies of cirrus cloud processes. In “Cirrus” (D. K. Lynch, K. Sassen, D. O. C. Starr and G. Stephens, Eds.), Oxford University Press, New York, Chp. 5.

DeMott, P. J., Cziczo, D. J., Prenni, A. J., Murphy, D. M., Kreidenweis, S. M., Thomson, D. S., Borys, R., and Rogers, D. C. (2003). Measurements of the concentration and composition of nuclei for cirrus formation. Proc. Natl. Acad. Sci. 100 (25), 14655–14660.

Deshler, T., (1982). Contact ice nucleation by submicron atmospheric aerosols. Ph.D. Dissertation, Dept. of Physics and Astronomy, University of Wyoming, 107 pp.

Dong, Y. Y., Oraltay, R. G., and Hallett, J. (1994). Ice particle generation during evaporation. Atmos. Res. 32, 45–53.

Dufuor, R., and Defay, L. (1963). “Thermodynamic of Clouds.” Academic Press, New York.

Facchini, M. C., Mircea, M., Fuzzi, S., and Charlson, R. J. (1999). Cloud albedo enhancement by surface-active organic solutes in growing droplets. Nature 401, 257–259.

Farley, R. D., and Chen, C. S. (1975). A detailed microphysical simulation of hydroscopic seeding on the warm rain process. J. Appl. Meteorol. 14, 718–733.

Farley, R. D., and Orville, H. D. (1986). Numerical modeling of hailstorms and hailstone growth. Part I: Preliminary model verification and sensitivity tests. J. Clim. Appl. Meteorol. 25, 2014–2035.

Feingold, G., and Heymsfield, A. J. (1992). Parameterizations of condensational growth of droplets for use in general circulation models. J. Atmos. Sci. 49, 2325–2342.

Feingold, G., and Levin, Z. (1986). The longitudinal fit to raindrop spectra from frontal convective clouds in Israel. J. Clim. Appl. Meteorol. 25, 1346–1363.

Feingold, G., and Chuang, P. Y. (2002). Analysis of the influence of film-forming compounds on droplet growth: Implications for cloud microphysical processes and climate. J. Atmos. Sci. 59, 2006–2018.

Feingold, G., and Kreidenweis, S. M. (2000). Does heterogeneous processing of aerosol increase the number of cloud droplets? J. Geophys. Res. 105 (24), 351–324, 361.

Feingold, G., Kreidenweis, S. M., Stevens, B., and Cotton, W. R. (1996). Numerical simulation of stratocumulus processing of cloud condensation nuclei through collision-coalescence. J. Geophys. Res 101 (21), 391–321, 402.

Feingold, G., Walko, R. L., Stevens, B., and Cotton, W. R. (1998). Simulations of marine stratocumulus using a new microphysical parameterization scheme. Atmos. Res. 4748, 505–528.

Feingold, G., Cotton, W. R., Kreidenweis, S. M., and Davis, J. T. (1999). The impact of giant cloud condensation nuclei on drizzle formation in stratocumulus: Implications for cloud radiative properties. J. Atmos. Sci. 56, 4100–4117.

Ferrier, B. S. (1994). A double-moment multiple-phase four-class bulk ice scheme. Part I: Description. J. Atmos. Sci. 51, 249–280.

Ferrier, B. S., Tao, W.-K., and Simpson, J. (1995). A double-moment multiple-phase four-class bulk ice scheme. Part II: Simulations of convective storms in different large-scale environments and comparisons with other bulk parameterizations. J. Atmos. Sci. 52, 1001–1033.

Field, P. R., Cotton, R. J., Noone, K., Glantz, P., Kaye, P. H., Hirst, E., Greenaway, R. S., Jost, C., Gabriel, R., Reiner, T., Andreae, M., Saunders, C. P. R., Archer, A., and Choularton, T. (2001). Ice nucleation in orographic wave clouds: Measurements made during INTACC. Q. J. R. Meteorol. Soc. 127 (575A), 1493–1512.

Field, P. R., Hogan, R. J., Brown, P. R. A., Illingworth, A., Choularton, T. W., and Cotton, R. J. (2005). Parameterization of ice-particle size distributions for mid-latitude stratiform cloud. Q. J. R. Meteorol., Soc. 131, 1997–2017.

Findeisen, W. (1938). Die Kolloidmeteorologischen Vorgange der Niederschlagsbildung. Meteorol. Z. 55, 121–133.

Fitzgerald, J. W. (1974). Effect of aerosol composition on cloud droplet size distribution: A numerical study. J. Atmos. Sci. 31, 1358–1367.

Fletcher, N. H. (1962). "The Physics Of Rainclouds." Cambridge University Press, London.

Flossman, A. I., Hall, W. D., and Pruppacher, H. R. (1985). A theoretical study of the wet removal of atmospheric pollutants. Part I: The redistribution of aerosol particlescaptured through nucleation and impaction scavenging by cloud droplets. J. Atmos. Sci. 42, 582–606.

Fridlind, A. M., Ackerman, A. S., Jensen, E. J., Heymsfield, A. J., Poellot, M. R., Stevens, D. E., Wang, D., Miloshevich, L. M., Baumgardner, D., Lawson, R. P., Wilson, J. C., Flagan, R. C., Seinfeld, J. H., Jonsson, H. H., VanReken, T. M., Varutbangkul, V., and Rissman, T. A. (2004). Evidence for the predominance of mid-tropospheric aerosols as subtropical anvil cloud nuclei. Science 304, 718–722.

Fukuta, N., (1972a). Growth theory of a population of droplets and the supersaturation in clouds. In "Abstr. Vol., Int. Cloud Phys. Conf., ICCP, IAMAP, IUGG, R. Meteorol. Soc., WMO, London." pp. 147–148.

Fukuta, N. (1972b). Advances in organic ice nuclei generator technology. J. Rech. Atmos. Dessens Mem. (1–3), 155–164.

Fukuta, N., Kowa, M., and Gong, N.-H. (1982). Determination of ice crystal growth parameters in a new supercooled cloud tunnel. In "Conf. Cloud Phys. Chicago, Ill." pp. 325–328. Am. Meteorol. Soc., Boston, Massachusetts. Preprint.

Fukuta, N., Gong, H.-H., Wang, A.-S., (1984). A microphysical origin of graupel and hail. In "Proc. Int. Conf. Cloud Phys., 9th, Acad. Sci. USSR, Sov. Geophys. Comm., Acad. Sci. Eston. SSR, Inst. Astrophys. Atmos. Phys., USSR State Comm. Hydrometeorol. Control Nat. Environ., Cent. Aerol. Obs., Tallinn, USSR." pp. 257–260.

Gagin, A. (1972). Effect of supersaturation on the ice crystal production by natural aerosols. J. Rech. Atmos. 6, 175–185.

Georgi, H. W., and Kleinjung, E (1968). Relations between the chemical composition of atmospheric aerosol particles and the concentration of natural ice nuclei. J. Rech. Atmos. 3, 145–156.

Gillespie, D. T. (1975). Three models for the coalescence growth of cloud drops. J. Atmos. Sci. 32, 600–607.

Gillespie, J. R. (1972). The stochastic coalescence model for cloud droplet growth. J. Atmos. Sci. 29, 1496–1510.

Gillespie, J. R., and List, R. (1976). Evolution of raindrop size distribution in steady state rainshafts. In "Proc. Int. Cloud Phys. Conf. Boulder, Colo." pp. 472–477. Am. Meteorol. Soc, Boston, Massachusetts.

Gokhale, N. R., and Spengler, J. D. (1972). Freezing of freely suspended supercooled water drops by contact nucleation. J. Appl. Meteorol. 11, 157–160.

Goldsmith, P., Goster, J., and Hume, C. (1976). The ice phase in clouds. In "Int. Conf. Cloud Phys. Boulder, Colo." pp. 163–167. Am. Meteorol. Soc, Boston, Massachusetts. Preprint.

Gultepe, I., Isaac, G. A., and Cober, S. G. (2001). Ice crystal number concentration versus temperature for climate studies. Internat. J. Climatology 21, 1281–1302.

Guzzi, R., and Rizzi, R. (1980). The effect of radiative exchange on the growth of a population of droplets. Contrib. Atmos. Phys. 53, 351–365.

Hallett, J. (1984). How snow crystals grow. Am. Sci. 72, 582–589.

Hallett, J., and Mossop, S. C. (1974). Production of secondary ice particles during the riming process. Nature (London) 249, 26–28.

Hallgren, R. E., and Hosler, C. L. (1960). Preliminary results on the aggregation of ice crystals. Geophys. Monogr., Am. Geophys. Union No. 5, 257–263.

Harrington, J. Y., Feingold, G., and Cotton, W. R. (2000). Radiative impacts on the growth of a population of drops within simulated summertime arctic stratus. J. Atmos. Sci. 57, 766–785.

Hegg, D. A., Yuen, P.-F., and Larson, T. V. (1991). Modeling the effects of heterogeneous cloud chemistry on the marine particle size distribution. J. Geophys. Res. 97, 12,92712,933.

Heymsfield, A. J., Johnson, D. N., and Dye, J. E. (1978). Observations of moist adiabatic ascent in northeast Colorado cumulus congestus clouds. J. Atmos. Sci. 35, 1689–1703.

Hindman II, E. E., (1975). The nature of aerosol particles from a paper mill and their effects on clouds and precipitation. Ph.D. Thesis, University of Washington.

Hobbs, P. V. (1969). Ice multiplication in clouds. J. Atmos. Sci. 26, 315–318.

Hobbs, P. V. (1974). High concentrations of ice particles in a layer cloud. Nature (London) 251, 694–696.

Hobbs, R. I., and Cooper, W. A. (1987). Field evidence supporting qauntitative predictions of secondary ice production rates. J. Atmos. Sci. 44, 1071–1082.

Hobbs, P. V., and Rangno, A. L. (1985). Ice particle concentrations in clouds. J. Atmos. Sci. 42, 2523–2549.

Hobbs, P. V., and Rangno, A. L. (1990). Rapid development of high ice particle concentrations in small polar maritime cumuliform clouds. J. Atmos. Sci. 47, 2710–2722.

Hobbs, P. V., Matejka, T. J., Herzegh, P. H., Locatelli, J. D., and Houze, R. A. (1980). The mesoscale and microscale structure and organization of clouds and precipitation in midlatitude cycles. I: A case of a cold front. J. Atmos. Sci. 37, 568596.

Hocking, L. M. (1959). The collision efficiency of small drops. Q. J. R. Meteorol. Soc. 85, 44–50.

Hocking, L. M., and Jonas, P. R. (1970). The collision efficiency of small drops. Q. J. R. Meteorol. Soc. 96, 722–729.

Holroyd, E. W. (1964). A suggested origin of conical graupel. J. Appl. Meteorol. 3, 633–636.

Hookings, G. A. (1965). Precipitation-maintained downdrafts. J. Appl. Met. 4, 190–195.

Hosler, C. L., and Hallgren, R. E. (1960). The aggregation of small ice crystals. Discuss. Faraday Soc. 30, 200–208.

Hounslow, M. J., Ryall, R. L., and Marshall, V. R. (1988). A discretized population balance for nucleation, growth and aggregation. AIChE J. 34, 1821–1832.

Houze Jr., R. A. (1977). Structure and dynamics of a tropical squall-line system. Mon. Weather Rev. 15, 1540–1567.

Howell, W. E. (1949). The growth of cloud drops in uniformly cooled air. J. Meteorol. 54, 134–149.

Huffman, P. J. (1973). Supersaturation spectra of AgI and natural ice nuclei. J. Appl. Meteorol. 12, 1080–1082.

Huffman, P. J., and Vali, G. (1973). The effect of vapor depletion on ice nucleus measurements with membrane filters. J. Appl. Meteorol. 12, 1018–1024.

Jacobson, M. Z. (1999). Studying the effects of calcium and magnesium on size-distributed nitrate and ammonium with EQUISOLV II. Atmos. Environ. 33, 3635–3649.

Jayaweera, K. (1971). Calculations of ice crystal growth. J. Atmos. Sci. 28, 728–736.

Jeffreys, H. (1916). Some problems of evaporation. Philos. Mag. 35, 270–280.

Jiang, H., and Feingold, G. (2006). Effect of aerosol on warm convective clouds: Aerosol-cloud-surface flux feedbacks in a new coupled large eddy model. J. Geophys. Res. 111, D01202. doi:10.1029/2005JD006138.

Johnson, D. B. (1976). Ultragiant urban aerosol particles. Science 194, 941–942.

Johnson, D. B. (1980). The influence of cloud-base temperature and pressure on droplet concentration. J. Atmos. Sci. 37, 2079–2085.

Johnson, D. B. (1982). The role ofgiant and ultragiant aerosol particles in warm rain initiation. J. Atmos. Sci. 39, 448–460.

Johnson, D. E., Wang, P. K., and Straka, J. M. (1993). Numerical simulations of the 2 August 1981 CCOPE Supercell storm with and without ice microphysics. J. Appl. Met. 32, 745–759.

Jonas, P., and Goldsmith, P. (1972). The collection efficiencies of small droplets falling through a sheared air flow. J. Fluid Mech. 52, 593–608.

Keller, V., and Hallett, J. (1982). Influence of air velocity on the habit of ice crystal growth from the vapor. J. Crystal Growth 60, 91–106.

Kessler III, E. (1969). On the distribution and continuity of water substance in atmospheric circulation. Meteorol. Monogr. 10.

Khairoutdinov, M., and Kogan, Y. (2000). A new cloud physics parameterization in a large-eddy simulation model of marine stratocumulus. Mon. Wea. Rev. 128, 229–243.

Khain, A. P., and Pinsky, M. B. (1997). Turbulence effects on the collision kernel. II: Increase of the swept volume of colliding drops. Q. J. R. Meteorol. Soc. 123, 1543–1560.

Khain, A. P., and Sednev, I. L. (1995). Simulations of hydrometeor size spectra evolution by water-water, ice-water and ice-ice interactions. Atmos. Res. 36, 107–138.

Khain, A. P., Ovtchinnikov, M., Pinsky, M., Pokrovsky, A., and Kugliak, H. (2000). Notes on the state-of-the-art numerical modeling of cloud microphysics. Atmos. Res. 55, 159–224.

Khain, A. P., Pokrovsky, A., and Sednev, I. (1999). Some effects of cloud-aerosol interaction on cloud microphysics structure and precipitation formation: Numerical experiments with a spectral microphysics cloud ensemble model. Atmos. Res. 52, 195–220.

Khain, A., Pokrovsky, A., Pinsky, M., Seifert, A., and Phillips, V. (2004). Simulation of effects of atmospheric aerosols on deep turbulent convective clouds using a spectral microphysics mixed-phase cumulus cloud model. Part I: Model description and possible applications. J. Atmos. Sci. 61, 2963–2982.

Klett, J. D., and Davis, M. H. (1973). Theoretical collision efficiencies of cloud droplets at small Reynolds number. J. Atmos. Sci. 30, 107–117.

Koenig, L. R., (1962). Ice in the summer atmosphere. Ph.D. Thesis, University of Chicago.

Koenig, L. R. (1963). The glaciating behavior of small cumulonimbus clouds. J. Atmos. Sci. 20, 29–47.

Koenig, L. R. (1971). Numerical modeling of ice deposition. J. Atmos. Sci. 28, 226–237.

Kovetz, A., and Olund, B. (1969). The effect of coalescence and condensation on rain formulation in a cloud of finite venical extent. J. Atmos. Sci. 26, 1060–1065.

Knupp, Kevin Robert, (1985). Precipitation convective downdraft structure: A synthesis of observations and modeling. Ph.D. dissertation, Colorado State University, Dept. of Atmospheric Science, Fort Collins, CO 80523, 196 pp.

Koziol, A. S., and Leighton, H. G. (1996). The effect of turbulence on the collision rates of small cloud drops. J. Atmos. Sci. 53 (13), 1910–1920.

Kulmala, M., Laaksonen, A., Korhonen, P., Vesala, T., Ahonen, T., and Barrett, J. C. (1993). The effect of atmospheric nitric acid vapor on cloud condensation nucleus activation. J. Geophys. Res. 98, 22949–22958.

Langmuir, I. (1948). The production of rain by a chain reaction in cumulus clouds at temperatures above freezing. J. Meteorol. 5, 175–192.

Latham, J., and Saunders, C. P. R. (1971). Experimental measurements of the collection efficiencies of ice crystals in electric fields. Q. J. R. Meteorol. Soc. 96, 257–265.

Levin Z., and Cotton W. R. (2007). Aerosol pollution impact on precipitation: A scientific review. In "Report from the WMO/IUGG International Aerosol Precipitation Science Assessment Group (IAPSAG), World Meteorological Organization, Geneva, Switzerland." 482 pp.

Levin, Z., Ganor, E., and Gladstein, V. (1996). The effects of desert particles coated with sulfate on rain formation in the eastern Mediterranean. J. Appl. Meteorol. 35, 1511–1523.

Levkov, L. (1971). Congelation de gouttes d'eau au contact de particules Cu. J. Rech. Atmos. 5, 133–136.

Lin, Y.-L., Farley, R. D., and Orville, H. D. (1983). Bulk parameterization of the snow field in a cloud model. J. Climate Appl. Meteorol. 22, 1065–1092.
List, R. (1959). Z. Angew. Math. Phys. 10, 143.
List, R. (1960). Z. Angew. Math. Phys. 11, 273.
List, (1965). "Proc. Cloud Phys. Conf., Tokyo, May 1965." p. 481. Meteorol. Soc. of Japan, Tokyo.
Liu, Y., and Daum, P. H. (2004). Parameterization of the autoconversion process. Part I: Analytical formulation of the Kessler-type parameterizations. J. Atmos. Sci. 61, 1539–1548.
Liu, J. Y., and Orville, H. D. (1969). Numerical modeling of precipitation and cloud shadow effects on mountain-induced cumuli. J. Atmos. Sci. 26, 1283–1298.
Locatelli, J. D., and Hobbs, P. (1974). Fall speeds and masses of solid precipitation particles. J. Geophys. Res. 79, 2185–2197.
Lohmann, U. (2004). Can anthropogenic aerosols decrease the snowfall rate?. J. Atmos. Sci. 61, 2457–2468.
Lord, S. J., Willoughby, H. E., and Piotrovicz, J. M. (1984). Role of a parameterized ice-phase microphysics in an axisymmetric nonhydrostatic tropical cyclone model. J. Atmos. Sci. 41, 2836–2848.
Ludlam, F. H. (1951). The production of showers by the coalescence of cloud droplets. Q. J. R. Meteorol. Soc 77, 402–417.
Ludlam, F. H. (1958). The hail problem. Nublia, 1, 12.
Lynn, B. H., Khain, A. P., Dudhia, J., Rosenfeld, D., Pokrovsky, A., and Seifert, A. (2005). Spectral (bin) microphysics coupled with a mesoscale model (MM5). Part I: Model description and first results. Mon. Wea. Rev. 133, 44–58.
Macklin, W. C. (1961). Accretion in mixed clouds. Q. J. R. Meteorol. Soc. 87, 413–424.
Magarvey, R. H., and Geldhart, J. W. (1962). Drop collisions under conditions of free fall. J. Atmos. Sci. 19, 107–113.
Magono, C. (1953). On the growth of snow flake and graupel. Sci. Rep. Yokohama Nat. University, Sect. 72, 321–335.
Magono, C, and Lee, C. W. (1973). The vertical structure of snow clouds as revealed by snow crystal sondes, Part II. J. Meteorol. Soc. Jpn. 51, 176–190.
Manton, M. J., and Cotton, W. R. (1977). Parameterization of the atmospheric surface layer. J. Atmos. Sci. 34, 331–334.
Marshall, J. S., and Palmer, W. M. (1948). The distribution of raindrops with size. J. Meteorol. 5, 165–166.
Marwitz, J. D. (1983). The kinematics of orographic airflow during Sierra storms. J. Atmos. Sci. 40, 1218–1227.
Mason, B. J. (1956). On the melting of hailstones. Q. J. R. Meteorol. Soc. 82, 209–216.
Mason, B. J. (1971). "The Physics of Clouds." Oxford University Press, Clarendon, London.

Mason, B. J., and Maybank, J. (1960). The fragmentation and electrification of freezing water drops. Q. J. R. Meteorol. Soc. 86, 176–186.

McCumber, M., Tao, W.-K., Simpson, J., Penc, R., and Soong, S.-T. (1991). Comparison of ice-phase microphysical parameterization schemes using numerical simulations of tropical convection. J. Appl. Met. 30, 985–1004.

McTaggert-Cowan, J. D., and List, R. (1975). Collision and breakup of water drops at terminal velocity. J. Atmos. Sci. 32, 1401–1411.

Mee, T. R., Takeuchi, D. M., (1968). Natural glaciation and particle size distribution in marine tropical cumuli. MR1 Final Rep. MR 168, FR-823 Contract No. E22-30-68(N), Exp. Meteorol. Branch, ESSA, University of Miami.

Meyers, M. P., DeMott, P. J., and Cotton, W. R. (1992). New primary ice-nucleation parameterizations in an explicit cloud model. J. Appl. Met. 31, 708–721.

Meyers, M. P., Walko, R. L., Harrington, J. Y., and Cotton, W. R. (1997). NewRAMS cloud microphysics prameterization. Part II: The two-momentscheme. Atmos. Res 45, 3–39.

Milbrandt, J. A., and Yau, M. K. (2005a). A multimoment bulk microphysics parameterization. Part I: Analysis of the role of the spectral shape parameter. J. Atmos. Sci. 62, 3051–3064.

Milbrandt, J. A., and Yau, M. K. (2005b). A multimoment bulk microphysics parameterization. Part II: A proposed three-moment closure and scheme description. J. Atmos. Sci. 62, 3065–3081.

Mitchell, D. L., Zhang, R., and Pitter, R. L. (1990). Mass-dimensional relationships for ice particles and the influence of riming on snowfall rates. J. Appl. Met. 29, 153–163.

Mordy, W. A. (1959). Computations of the growth by condensation of a population of cloud droplets. Tellus 11, 16–44.

Morrison, H., Curry, J. A., and Kovorostyanov, V. I. (2005a). A new double-moment microphysics parameterization for application in cloud and climate models. Part I: Description. J. Atmos. Sci. 62, 1665–1677.

Morrison, H., Curry, J. A., Shupe, M. D., and Zuidema, P. (2005b). A new double-moment microphysics parameterization for application in cloud and climate models. Part II: Single-column modeling of Arctic clouds. J. Atmos. Sci. 62, 1678–1693.

Mossop, S. C., and Hallett, J. (1974). Ice crystal concentration in cumulus clouds: Influence of the drop spectrum. Science 186, 632–633.

Mossop, S. C., and Ono, A. (1969). Measurements of ice crystal concentrations in clouds. J. Atmos. Sci. 26, 130–137.

Mossop, S. C., Ono, A., and Heffernan, K. J. (1967). Studies of ice crystals in natural clouds. J. Rech. Atmos. 3, 45–64.

Mossop, S. C., Ruskin, R. E., and Heffernan, K. J. (1968). Glaciation of a cumulus at approximately −4°C. J. Atmos. Sci. 25, 889–899.

Mossop, S. C., Ono, A., and Wishart, E. R. (1970). Ice particles in maritime clouds near Tasmania. Q. J. R. Meteorol. Soc. 96, 487–508.

Mossop, S. C., Cottis, R. E., and Bartlett, B. M. (1972). Ice crystal concentrations in cumulus and stratocumulus clouds. Q. J. R. Meteorol. Soc. 98, 105–123.

Murray, F. W., and Koenig, L. R. (1972). Numerical experiments on the relation between microphysics and dynamics in cumulus convection. Mon. Weather Rev. 100, 717–732.

Nakaya, U., and Terada Jr., T. (1935). Simultaneous observations of the mass falling velocity and form of individual snow crystals. J. Fac. Sci., Hokkaido Imp. University, Ser. 2 (4).

Neiburger, M., and Chien, C. W. (1960). Computations of the growth of cloud drops by condensation using an electronic digital computer. Geophys. Monogr, Am. Geophys. 5, 191–208.

Nelson, R. T., and Gokhale, N. R. (1968). Concentration of giant particles below cloud base. In "Nat. Conf. Weather Modif., 1st Albany, NY." pp. 89–98. Am. Meteorol. Soc, Boston, Massachusetts. Preprint.

Nickerson, E. C., Richard, E., Rosset, R., and Smith, D. R. (1986). The numerical simulation of clouds, rain, and airflow over the Vosges and Black Forest Mountains: A meso-$\beta$ model with parameterized microphysics. Mon. Weather Rev. 114, 398–414.

Oraltay, R. G., and Hallett, J. (1989). Evaporation and melting of ice crystals: A laboratory study. Atmos. Res. 24, 169–189.

Passarelli, R. E., and Srivastava, R. C. (1979). A new aspect of snowflake aggregation theory. J. Atmos. Sci. 36, 484–493.

Pawlowska, H., and Brenguier, J. L. (2000). Microphysical properties of stratocumulus clouds during ACE-2. Tellus. 52B, 867–886.

Pinsky, M. B., and Khain, A. P. (1997). Turbulence effects on droplet growth and size distributions in cloudsa review. J. Aerosol Sci. 28, 11271214.

Pinsky, M. B., and Khain, A. P. (2002). Effects of in-cloud nucleation and turbulence on droplet spectraum formation in cumulus clouds. Q. J. R. Meteorol. Soc. 128, 501–533.

Pitter, R. L., and Pruppacher, H. R. (1973). A wind tunnel investigation of freezing of small water drops falling at terminal velocity in air. Q. J. R. Meteorol. Soc. 99, 540–550.

Pruppacher, H. R., and Klett, J. D. (1978). "Microphysics of Clouds and Precipitation." Reidel, Boston, Massachusetts.

Rangno, A. L., and Hobbs, P. V. (1994). Ice particle concentrations and precipitation development in small continental cumuliform clouds. Q. J. R. Meteorol. Soc. 120, 573–601.

Rau, W. (1950). Uber die Wirkungsweise der Gefrierkeime im unterkuhlten Wasser. Z. Naturforsch., A 5A, 667–675.

Rauber, R.M., (1985). Physical structure of northern Colorado River basin cloud systems. Ph. D. Thesis, Atmos. Sci. Pap. No. 390, Dep. Atmos. Sci., Colorado State University.

Reinking, R. F. (1975). Formation of graupel. J. Appl. Meteorol. 14, 745–754.

Reisin, T., Levin, Z., and Tzivion, S. (1996a). Rain production in convective clouds as similated in an axisymmetric model with detailed microphysics. Part I: Description of the model. J. Atmos. Sci. 53, 497–519.

Reisin, T., Levin, Z., and Tzivion, S. (1996b). Rain production in convective clouds as similated in an axisymmetric model with detailed microphysics. Part II: Effects of varying drops and ice initiation. J. Atmos. Sci. 53, 1815–1837.

Reisner, J., Rasmussen, R. M., and Bruintjes, R. T. (1998). Explicit forecasting of supercooled liquid water in winter storms using the MM5 mesoscale model. Q. J. R. Met. Soc. 124, 1071–1107.

Roach, W. T. (1976). On the effect of radiative exchange on the growth by condensation of a cloud or fog droplet. Q. J. R. Meteorol. Soc. 102, 361–372.

Rogers, D. C., (1974). The aggregation of natural ice crystals. M.S. Thesis, Dep. Atmos. Resour., University of Wyoming.

Rogers, D. C. (1982). Field and laboratory studies of ice nucleation in winter orographic clouds. Ph.D. Dissertation, Univ. of Wyoming, Laramie, WY.

Rutledge, S. A., and Hobbs, P. V. (1984). The mesoscale and microscale structure and organization of clouds and precipitation in midlatitude cyclones. XII: A diagnostic modeling study of precipitation development in narrow cold-frontal rainbands. J. Atmos. Sci. 41, 2949–2972.

Saleeby, S. M., and Cotton, W. R. (2004). A large-droplet mode and prognostic number concentration of cloud droplets in the Colorado State University Regional Atmospheric Modeling System (RAMS). Part I: Module descriptions and Supercell test simulations. J. Appl. Met. 43, 182–195.

Saleeby, S. M., and Cotton, W. R. (2005). A large-droplet mode and prognostic number concentration of cloud droplets in the Colorado State University Regional Atmospheric Modeling System (RAMS). Part II: Sensitivity to a Colorado winter snowfall event. J. Appl. Meteorol. 44, 1912–1929.

Saleeby, S. M., Cheng, W. Y. Y., and Cotton, W. R. (2007). New developments in the regional atmospheric modeling system suitable for simulations of snowpack augmentation over complex terrain. J. Wea. Mod. 39, 37–49.

Sasyo, Y. (1971). Study of the formation of precipitation by the aggregation of snow particles and the accretion of cloud droplets on snowflakes. Pap. Meteorol. Geophys. 22, 69–142.

Schumann, (1938). Q. J. R. Meteorol. Soc. 63, 3.

Scott, B. D., and Hobbs, P. V. (1977). A theoretical study of the evolution of mixed-phase cumulus clouds. J. Atmos. Sci. 34, 812–826.

Seifert, A., Khain, A., Pokrovsky, A., and Beheng, K. D. (2006). A comparison of spectral bin and two-moment bulk mixed-phase cloud microphysics. Atmos. Res. 80, 46–66.

Shaw, R. A., Reade, W. C., Collins, L. R., and Verlinde, J. (1998). Preferential concentration of cloud droplets by turbulence: Effects on the early evolution of cumulus cloud droplet spectra. J. Atmos. Sci. 55, 1965–1976.

Spilhaus, A. F. (1948). Drop size intensity and radar echo in rain. J. Meleorol. 5, 161–164.

Squires, P. (1952). The growth of cloud drops by condensation. Aust. J. Sci. Res., Ser. A 5, 59–86.

Squires, P. (1956). The microstructure of cumuli in maritime and continental air. Tellus 8, 443–444.

Squires, P. (1958). The microstructure and colloidal stability ofwarm clouds. Tellus 10, 256–271.

Srivastava, R. C. (1978). Parameterization of raindrop size distributions. J. Atmos. Sci. 35, 108–117.

Stevens, B., Cotton, W. R., and Feingold, G. (1998). A critique of one- and two-dimensional models of boundary layer clouds with a binned representations of drop microphysics. Atmos. Res. 47-48, 529–553.

Stith, J. L., Burrows, D. A., and DeMott, P. J. (1994). Initiation of ice: Comparison of numerical model results with observations of ice development in a cumulu cloud. Atmos. Environ. 32, 13–30.

Straka, J. M., and Mansell, E. R. (2005). A bulk microphysics parameterization with multiple ice precipitation categories. J. Appl. Met. 44, 445–466.

Telford, J. (1955). A new aspect of coalescence theory. J. Meteorol. 12, 436–444.

Telford, J. W., and Chai, S. K. (1980). A new aspect of condensation theory. Pure Appl. Geophys. 118, 720–742.

Telford, J. W., Keck, T. S., and Chai, S. K. (1984). Entrainment at cloud tops and the droplet spectra. J. Atmos. Sci. 41, 3170–3179.

Tripoli, G. J., and Cotton, W. R. (1980). A numerical investigation of several factors contributing to the observed variable intensity of deep convection over South Florida. J. Appl. Meteorol. 19, 1037–1063.

Twomey, S. (1964). Statistical effect in the evolution of a distribution of cloud. J. Meteorol. 12, 436.

Twomey, S. (1966). Computations of rain formation by coalescence. J. Atmos. Sci. 23, 405–411.

Twomey, S. (1976). Computations of the absorption of solar radiation by clouds. J. Atmos. Sci. 33, 1087–1091.

Twomey, S., and Squires, P. (1959). The influence of cloud nucleus population on microstructure and stability of convective clouds. Tellus 11, 408–411.

Twomey, S., and Warner, J. (1967). Comparison of measurements of cloud droplets and cloud nuclei. J. Atmos. Sci. 24, 702–703.

Tzivion, S., Feingold, G., and Levin, Z. (1987). An efficient numerical solution to the stochastic collection equation. J. Atmos. Sci. 44, 3139–3149.

Vaillancourt, P. A., Yau, M. K., Bartello, P., and Grabowski, W. W. (2002). Microscopic approach to cloud droplet growth by condensation. Part II: Turbulence, clustering, and condensational growth. J. Atmos. Sci. 59, 3421–3435.

Vali, G. (1974). Contact ice nucleation by natural and artificial aerosols. In "Conf. on Cloud Physics." pp. 34–37. Amer. Meteorol. Soc, Tucson.

Vali, G., (1976). Contact-freezing nucleation measured by the DFC instrument. In "Third International Workshop on Ice Nucleus Measurements. Laramie, University of Wyoming." pp. 159–178.

Vali, G., and Stansbury, E. J. (1966). Time dependent characteristics of the heteorogeneous nucleation of ice. Can. J. Phys. 44, 477–502.

van den Heever, S., and Cotton, W. R. (2007). Urban aerosol impacts on downwind convective storms. J. Appl. Meteorol. 46, 828–850.

van den Heever, S. C., Carrio, G. G., Cotton, W. R., DeMott, P. J., and Prenni, A. J. (2006). Impacts of nucleating aerosol on Florida convection. Part I: Mesoscale Simulations. J. Atmos. Sci. 63, 1752–1775.

van den Heever, S. C., and Cotton, W. R. (2004). The impact of hail size on simulated supercell storms. J. Atmos. Sci. 61, 1596–1609.

Vardiman, L. (1978). The generation of secondary ice particles in cloud crystal-crystal collisions. J. Atmos. Sci. 35, 21682180.

Verlinde, J., Flatau, P. J., and Cotton, W. R. (1990). Analytical solutions to the collection growth equation: Comparison with approximate methods and application to cloud microphysics parameterization schemes. J. Atmos. Sci. 47, 2871–2880.

Warner, J. (1969). The micro-structure of cumulus clouds. I. General features of the droplet spectrum. J. Atmos. Sci. 26, 1049–1059.

Walko, R. L., Cotton, W. R., Meyers, M. P., and Harrington, J. Y. (1995). New RAMS cloud microphysics parameterization. Part I: The single-moment scheme. Atmos. Res. 38, 29–62.

Walko, R. L., Cotton, W. R., Feingold, G., and Stevens, B. (2000). Efficient computation of vaport and heat diffusion between hydrometeors in a numerical model. Atmos. Res. 53, 171–183.

Warshaw, M. (1967). Cloud-droplet coalescence: Statistical foundations and a one-dimensional sedimentation model. J. Atmos. Sci. 24, 278–286.

Wegener, A. (1911). Kerne derkristallbildung. In "Thermodynamik der Atmosphare, Barth, Leipzig." pp. 94–98.

Wisner, C., Orville, H. D., and Myers, C. (1972). A numerical model of a hail-bearing cloud. J. Atmos. Sci. 29, 1160–1181.

Woodcock, A. H. (1953). Salt nuclei in marine air as a function of altitude and wind force. J. Meteorol. 10, 362–371.

Xue, H., and Feingold, G. (2006). Large eddy simulations of trade-wind cumuli: Investigation of aerosol indirect effects. J. Atmos. Sci. 63, 1605–1622.

Yin, Y., Levin, Z., Reisin, T. G., and Tzivion, S. (2000). The effects of giant condensation nuclei on the development of precipitation in convective clouds a numerical study. Atmos. Res. 53, 91–116.

Young, K. C. (1974a). The role of contact nucleation in ice phase initiation in clouds. J. Atmos. Sci. 31, 768–776.

Young, K. C. (1974b). A numerical simulation of wintertime, orographic precipitation. I: Description of model microphysics and numerical techniques. J. Atmos. Sci. 31, 1735–1748.

Young, K. C. (1974c). A numerical simulation of wintertime, orographic precipitation. II: Comparison of natural and AgI-seeded conditions. J. Atmos. Sci. 31, 1749–1767.

Young, K. C. (1975). The evoiution of drop spectra due to condensation, coalescence and breakup. J. Atmos. Sci. 32, 965–973.

Young, K. C. (1978). On the Role of Mixing in Promoting Competition between Growing Hailstones. J. Atmos. Sci. 35, 2190–2193.

Ziegler, C. L. (1985). Retrieval of thermal and microphysical variables in observed convective storms. Part 1: Model development and preliminary testing. J. Atmos. Sci. 42, 1487–1509.

Chapter 5

# Radiative Transfer in a Cloudy Atmosphere and Its Parameterization

## 5.1. INTRODUCTION

In the early stages of cloud modeling, modelers ignored the effects of radiative transfer. This is largely because the emphasis was on the simulation of individual convective clouds. For convective time scales of the order of 30 minutes to 1 hour, radiative heating rates are of little importance. However, as cloud modeling moved to the simulation of stratocumulus clouds, fogs, middle and high clouds, and to cloud processes on the mesoscale, where time scales are on the order of a day, cloud modelers have begun to consider radiative transfer processes. Moreover, with recent emphasis on global climate change, interest in cloud radiation feedback increased dramatically because it is widely recognized that clouds influence global albedo and the longwave radiation budget, and contribute to global temperature changes. Even tradewind cumulus, which has a short cloud life, exhibits, on average, an important global albedo effect. Another impetus for cloud radiation studies is related to remote sensing in which instantaneous radiances are observed.

As with turbulent transport and cloud microphysical processes, cloud modelers must make a number of compromises in the design of a radiative transfer model (i.e. they must formulate the radiative transfer equations in a simplified parameterized form). Similar to atmospheric motions that span a broad range of eddy sizes and cloud hydrometeors that span a broad range of particle sizes, atmospheric radiation covers a broad spectrum of radiation frequencies, wavelengths, or wave numbers. The sun, for example, emits radiation approximately as a blackbody having temperatures between 6000 and 5700 K, which peaks in intensity at a wavelength of 0.470 μm but which spans the range from less than 0.2 μm to greater than 1.8 μm, but 50% of all solar energy is between 0.3 and 0.7 μ (see Fig. 5.1). In contrast, the radiation energy emitted by the earth corresponds approximately to blackbody radiation at a temperature of about 250 K. Thus, a combination of radiation emitted by the sun and the earth spans a range from less than 0.2 μm to greater than 50 μm. However, the spectrum of radiation emitted by the sun and by the earth exhibits

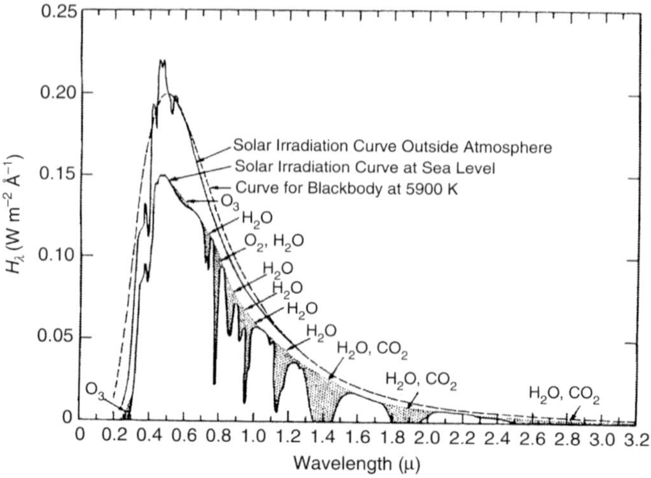


**FIGURE 5.1 Spectral distribution curves related to the sun; shaded areas indicate absorption, at sea level, due to the atmospheric constituents shown.** *(From Gast et al. (1965). Copyright © 1965 by McGraw-Hill. Reprinted by permission)*

very little overlap. For this reason we refer to radiation emitted by the sun as *shortwave radiation* and radiation emitted by the earth and its atmosphere as *longwave radiation*. The two regions are separated arbitrarily at around 4 μm which is sufficient when considering the dynamics of clouds where energy transfer is of primary importance. As we shall see later, this distinction allows some simplifications in the formulation of radiative transfer theories.

The most important consideration, as far as the dynamics of clouds is concerned, is that the net heating rate at various levels in a cloud is a consequence of the attenuation of atmospheric radiation. The net heating rate is a function of the net radiative flux divergence,

$$\partial\theta/\partial t = (1/\rho_0 c_p)(\partial F_N/\partial z), \tag{5.1}$$

where $F_N$ is the difference between downward and upward fluxes and has units of watts per square meter. In Eq. (5.1), for simplification, we consider only vertical flux divergences. In some cloud modeling problems, however, it may be desirable to consider horizontal radiative flux divergence as well (Marshak and Davis, 2005). An example is a valley fog located between two radiating valley sides, or radiative fluxes passing through a field of cumuli. Nonetheless, for most cloud modeling applications, consideration of vertical radiative flux divergences is sufficient. The net vertical flux $F_N$ is the difference between downward and upward fluxes, or

$$F_N = F\downarrow - F\uparrow. \tag{5.2}$$

The upward and downward fluxes, in turn, represent the fluxes integrated over all wavelengths and averaged over upward and downward looking hemispheres.

A divergence of radiative flux is caused by a combination of differential extinction of radiation and thermal emission. Extinction of radiation is a result of absorption and of scattering of radiant energy. The absorptance $A_\lambda$ represents the fraction of incoming radiation absorbed in a layer of the atmosphere. The reflectance $\mathrm{Re}_\lambda$ and transmittance $\mathrm{Tr}_\lambda$ simply represent the fractions of incoming radiation which are scattered out of the primary beam, and transmitted through a layer of the atmosphere, respectively. Note that all three processes vary with the wavelength of radiation. As a result of conservation of energy,

$$A_\lambda + \mathrm{Re}_\lambda + \mathrm{Tr}_\lambda = 1. \tag{5.3}$$

Light scattering without absorption is associated with redistribution of energy, a process which leads to a diffuse radiation field and is also called conservative scattering.

Absorption results in a change in the internal energy or temperature of the medium. In the atmosphere this change is usually a change in internal energy or temperature. If the incident energy as well as the scattered and transmitted energy remain constant for a time, the internal energy of the system will remain unchanged. A radiative equilibrium is established in which as much radiation is being emitted as is being absorbed. However, in conditions of local thermodynamic equilibrium, the emitted radiation is in thermal equilibrium with the source level. If the atmosphere were a pure blackbody (i.e. $A_\mathrm{A} = 1$ for all wavelengths), then the total emitted radiative flux would be

$$F = \sigma T^4, \tag{5.4}$$

where $\sigma$ is the Stefan-Boltzmann constant. According to Blevin and Brown (1971), $\sigma = (5.66961 \pm 0.0075) \times 10^{-8}$ W m$^{-2}$ K$^{-4}$. Equation (5.4) can be obtained by integrating the Planck radiation distribution function over all wavelengths (see any basic radiation physics text for a definition of the Planck function). The atmosphere does not, however, behave as a blackbody. Therefore, the amount of energy emitted by the absorbing atmosphere is given by

$$F = \varepsilon\sigma T^4, \tag{5.5}$$

where $\varepsilon$ represents the emittance of the atmosphere. The emittance is the ratio of the flux emitted by a body to the flux emitted by a blackbody at the same temperature. We shall see later that the effective emittance of a cloudy atmosphere varies with the liquid-water content and particle spectra in clouds.

## 5.2. ABSORPTANCE, REFLECTANCE, TRANSMITTANCE, AND EMITTANCE IN THE CLEAR ATMOSPHERE

As can be seen in Fig. 5.1, in a cloud-free and aerosol-free atmosphere, the primary absorbers of shortwave radiation are ozone and water vapor. Aerosols,

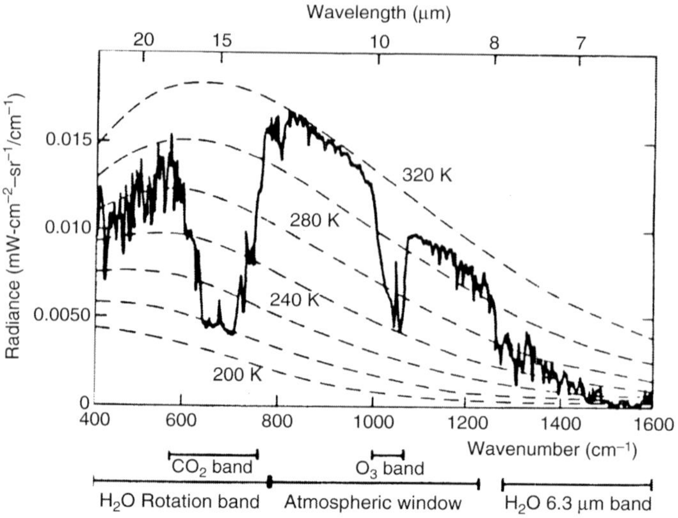


**FIGURE 5.2 Atmospheric spectrum obtained with a scanning interferometer on board the Nimbus 4 satellite.** The interferometer viewed the earth vertically as the satellite was passing over the North African desert. *(After Hanel et al. (1972), cited in Paltridge and Platt (1976))*

particularly soot or carbonaceous particles, also contribute to a lesser extent to absorption. At wavelengths shorter than 0.3 μm, oxygen and nitrogen absorb nearly all incoming solar radiation in the upper atmosphere. However, in the important region of visible radiation between 0.3 and 0.7 μm, little gaseous absorption occurs. Only weak absorption by ozone takes place in this spectral range. This is fortunate because these are the wavelengths in which the solar radiation peaks. At wavelengths of less than 0.7 μm, Rayleigh scattering of shortwave radiation back to space depletes the available flux. At longer wavelengths, absorption in various water vapor bands is quite pronounced. Some weak absorption by carbon dioxide and ozone also occurs at wavelengths greater than 0.7 μm. Absorption by carbon dioxide is a subject of considerable attention because of its importance for global warming. Cloud processes are important to climate change because positive or negative temperature changes may alter cloud cover and cloud microphysics, which can feedback on greenhouse gas warming.

Absorption of longwave radiation also occurs mainly in a series of bands. The principal absorbers of longwave or infrared (IR) radiation are water vapor, carbon dioxide, and ozone. Figure 5.2 illustrates the banded character of absorption in the IR region. Figure 5.2 represents the IR spectrum obtained by a scanning interferometer looking downward from a satellite over a desert region. Strong absorption by $CO_2$ at a wavelength band centered at 14.7 μm is shown by the emission of radiance at the temperature of 220 K. The stratosphere contributes mainly to the peak of this absorption band, with warmer tropospheric contributions occurring across the broader part of the absorption band. Water vapor absorption bands at 1.4, 1.9, 2.7, and 6.3 μm, and greater than 20 μm, cause emissions corresponding to mid-tropospheric temperatures. Little absorption is evident in the region called the "atmospheric window" between 8 and 14 μm. Here the radiance corresponds to the surface temperature of the desert, except for a slight depression in magnitude due to departures of the emittance of sand from unity. A distinct ozone absorption band is evident in the region of 9.6 μm in the middle of the window. Although not very evident in the figure, weak continuous absorption also occurs across the "window." The intensity of the continuum absorption depends on water vapor pressure. This continuous absorption is due to the presence of clusters or dimers $(H_2O)_2$ of water vapor molecules.

A detailed knowledge of the absorption spectra in the IR region is essential to predicting the rate of cooling due to longwave radiative transfer. In the following sections, we first examine the interactions among cloud particles and radiation. Then we review some of the techniques for calculating radiative transfer in a cloudy atmosphere.

## 5.3. SHORTWAVE RADIATIVE TRANSFER IN A CLOUDY ATMOSPHERE

The interaction of solar radiation incident upon a cloud is complicated by the fact that not only must we concern ourselves with the impact of a spectrum of radiative frequencies on cloud absorption, reflection, and transmission, but also with the consequences of a spectrum of droplets or ice crystals on radiative transfer. In the absence of emission, the azimuthally averaged, horizontally homogeneous plane parallel, time-independent radiative transfer equation appropriate to a cloudy medium is as follows:

$$\mu \frac{dI(\tau,\mu)}{d\tau} = -I(\tau,\mu) + \frac{\tilde{\omega}_0}{2}\int_{-1}^{+1} \overline{p}(\tau,\mu,\mu')I(\tau,\mu')d\mu' + \frac{S_0}{4\pi}\overline{p}(\tau,\mu,\mu_0)e^{-\tau/\mu_0}, \quad (5.6)$$

where $\tau$ is the optical depth, $\omega_0$ is the single-scattering albedo, $\overline{p}$ is the scattering phase function, and $S_0$ is the solar flux associated with a collimated beam incident on the cloud top. All parameters in Eq. (5.6) are functions of the frequency of radiation $\nu$ with the exception of $\mu$ and $\mu_0$, where $\mu$ is a function of the cosine of the zenith angle. The quantity $I(\tau,\mu)$ is the radiance along an angle given by $\mu$ through a cloudy layer defined by the optical depth $\tau$. The extinction optical thickness, $\tau$, is the sum of droplet scattering, droplet absorption, and gaseous absorption. Gaseous absorption is primarily due to water vapor in both a clear and a cloudy atmosphere. In fact, for some time it was thought that the major effect of clouds on shortwave radiation was to

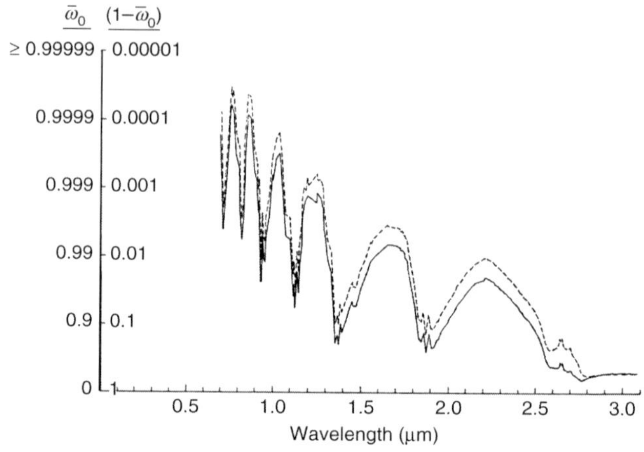


**FIGURE 5.3 Spectra of the single-scattering albedo $\omega_0$ as a function of wavelength.** *(From Twomey (1976))*

increase the optical path length of water vapor absorption as a result of multiple scattering by cloud droplets. We shall see that absorption by liquid droplets plays an important role in the overall absorption in clouds. Cloud droplet absorption is described in terms of $\tau_a$, the droplet absorption optical thickness. If we define $\tau_s$, the scattering optical thickness, and $\tau_g$, the gaseous absorption optical thickness, then the extinction optical thickness is given by

$$\tau = \tau_s + \tau_a + \tau_g. \tag{5.7}$$

For any frequency one can calculate the optical thickness $\tau(\nu)$ and thus obtain a *single-scattering albedo*

$$\tilde{\omega}_0 = \tau_s/\tau. \tag{5.8}$$

Thus, $\tilde{\omega}_0 = 1$ for a nonabsorbing cloud and $\tilde{\omega}_0 = 0$ when scattering is negligible. The optical thickness varies only slightly with $\nu$ and the average size of a droplet. The single-scattering albedo, on the other hand, varies strongly with both frequency (or wavelength) and droplet size. Figure 5.3 illustrates the variation in single-scattering albedo as a function of wavelength for a maritime cloud with a droplet concentration of 25 $cm^{-3}$ (solid curve) and a mildly continental cloud with a droplet concentration of 200 $cm^{-3}$ (dashed curve) for a given temperature and a liquid-water content of 0.33 $gm^{-3}$. In a continental cloud, such as over the High Plains of the United States, where droplet concentrations are approximately 700 $cm^{-3}$, or in polluted air masses, where droplet concentrations are approximately 2500 $cm^{-3}$, the single-scattering albedo would be much greater than in a clean maritime air mass. This is because for a given liquid-water content, the higher droplet concentrations result in much smaller droplet sizes.

The scattering phase function $\overline{p}(\tau, \mu, \mu')$ in Eq. (5.6) characterizes the angular distribution of the scattered radiation field. For spherical droplets, the function exhibits a strong peak in the forward direction and produces rainbow and glory effects in the backscattering directions; the sun has to be behind the observer to see a rainbow. As noted by Joseph et al. (1976), for some applications the phase function can be conveniently expressed as

$$\overline{p}(\tau, \mu, \mu') = (1 - g^2)/(1 + g^2 - 2g\mu\mu'), \tag{5.9}$$

which was formulated by Henyey and Greenstein (1941). The parameter $g$ in Eq. (5.9) is called the *asymmetry parameter*, which is the average value of the cosine of the scattering angle (weighted according to the probability of scattering in the various directions). For symmetric scattering, which puts equal amounts of energy in the forward and backward directions, $g$ is zero. This is the case for optically small droplets, ice crystals, aerosol particles, as well as air molecules; this regime is called Rayleigh scattering. As more energy is scattered in the forward hemisphere, $g$ increases toward unity. Conversely as more energy is scattered in backward, $g$ tends toward $-1$. For 2-stream models, $g$ describes the partitioning of energy between the upper and lower hemispheres.

Another parameter defining the radiative properties of clouds is the *optical thickness*. Excluding the effects of water vapor, the formal definition of the optical thickness is

$$\tau = \int_0^z \int_0^\infty f(r) Q_e(x, n_\lambda) \pi r^2 \mathrm{d}r \mathrm{d}z, \tag{5.10}$$

where $Q_e(x, n_\lambda)$ is the extinction efficiency, and $f(r)$ is the spectral density of droplets of radius $r$. The first integral is over the cloud depth $z$. The second integral is over the cloud droplet radius $r$. Evaluation of Eq. (5.10) requires a knowledge of both the cloud droplet distribution and the behavior of $Q_e$. The extinction efficiency is defined as the ratio of the extinction to the cross-sectional areas of cloud droplets. For spherical cloud droplets, $Q_e$ can be evaluated from the Mie solution to the light scattering problem; this is an exact solution to the Maxwell equations. It is a function of particle radius through the size parameter $x = 2\pi r/\lambda$ and the refractive index of the particle $n_\lambda$.

Figure 5.4 illustrates the variation of the scattering efficiency factor for nonabsorbing water drops ($n_i = 0$) as calculated by Hansen and Travis (1974) using Mie theory. The extinction efficiency is the sum of the scattering and absorption efficiencies. The variation in the absorption efficiency as a function of $x$ also resembles Fig. 5.4. The efficiency was calculated assuming droplets were distributed in a gamma-type size distribution function that is determined by two parameters: $a = r_e$, the effective radius defined later in Eq. (5.15), and a coefficient of dispersion ($b$) about the effective radius $a$. The coefficient of dispersion is a measure of the droplet spectral width, with $b = 0$ being a monodisperse distribution and larger values representing broader spectra. Typical observed droplet-size distributions in cumuli exhibit dispersion values

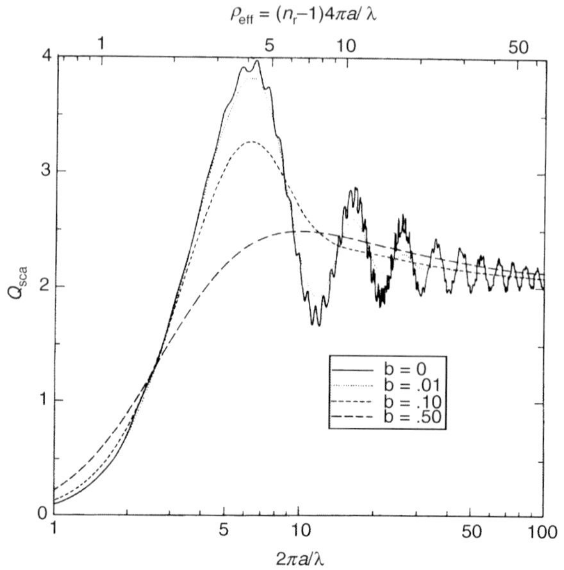


**FIGURE 5.4 Efficiency for scattering, $Q_{sca}$, as a function of the effective size parameter, $2\pi a/\lambda$.** A $\gamma$-type size distribution is used with effective radius $a = r_e$ and coefficient of dispersion $b$. For the case $b = 0$, $2\pi a/\lambda = 2\pi r/\lambda \equiv x$. The refractive index is $n_r = 1.33$, $n_i = 0$. *(From Hansen and Travis (1974). Copyright © 1974 by D. Reidel Publishing Company. Reprinted by permission)*

slightly less than 0.2. Remember that the parameter $x$ is a function of both the size of the droplets and the wavelength of electromagnetic radiation. Because the effective radius of most cloud droplet distributions is of the order of 10 μm, $x$ is typically greater than 10 for the visible spectrum. For large values of $x$, Fig. 5.4 illustrates (1) a series of regularly spaced broad maxima and minima called the interference structure, which oscillates about the approximate value of 2, and (2) irregular fine structure called the ripple structure. For broader droplet spectra, this fine structure is smeared out, but the feature of $Q_e$ approaching a limiting value of 2 for large $x$ remains.

Because for large values of $x$ (i.e. for shortwave radiation and typical cloud droplet-size distribution), $Q_e$ tends to be an almost constant value of 2, Eq. (5.10) can be written as

$$\tau = 2\pi \int_0^z \int_0^\infty f(r) r^2 \mathrm{d}r \mathrm{d}z. \tag{5.11}$$

Thus Eq. (5.11) is only a function of the drop-size distribution and the depth of the cloud. To a good approximation we can write

$$\tau \simeq 2\pi N_c \bar{r}^2 h, \tag{5.12}$$

where $N_c$ is the droplet concentration, $\bar{r}$ is the mean droplet radius, and $h$ is the geometric thickness of a cloud layer. Using the notation in Chapter 4, we can express Eq. (5.12) in terms of the cloud droplet mixing ratio $r_c$ as

$$\tau \simeq 2\pi(3\rho_0/\pi\rho_w)^{2/3} h N_c^{1/3} r_c^{2/3}, \tag{5.13}$$

where $\rho_0$ is the dry air density and $\rho_w$ is the density of water. The cloud droplet optical thickness is therefore a function of the cloud properties (liquid-water mixing ratio and cloud depth) as well as droplet concentration. The cloud droplet concentration, in turn, is largely determined by the aerosol content of the air mass or, in particular, the concentration of cloud condensation nuclei. For typical peak supersaturations in clouds, $N_c$ varies from as low as 10-50 $cm^{-3}$ for maritime clouds to 1000 $cm^{-3}$ for continental clouds. An important effect of changing the optical thickness of a cloud layer is that it changes the amount of reflected radiation and thereby alters the energy reaching the earth's surface and the atmosphere below a cloud layer. Twomey suggests that if the CCN concentration in the cleaner parts of the atmosphere, such as oceanic regions, were raised to continental atmospheric values, about 10% more energy would be reflected to space by relatively thin cloud layers. He also points out that an increase in cloud reflectivity by 10% is of more consequence than a similar increase in global cloudiness. This is because, while an increase in cloudiness reduces the incoming solar energy flux, it also reduces the outgoing infrared radiation flux. Thus both cooling and heating effects occur when global cloudiness is increased. In contrast, an increase in cloud reflectance due to enhanced CCN concentrations does not appreciably affect infrared radiation but does reflect more incoming solar radiation, which results in a net cooling effect.

Stephens (1978a) has shown that another approximation to Eq. (5.10) is

$$\tau \simeq \frac{3}{2} W/r_e, \tag{5.14}$$

where $W$ is the liquid-water path ($gm^{-2}$) and $r_e$ is an effective radius defined as

$$r_e = \int_0^\infty f(r) r^3 dr \triangle \int_0^\infty f(r) r^2 \, dr. \tag{5.15}$$

The liquid-water path represents the integrated liquid-water through a cloud of depth $h$,

$$W = \int_0^h \rho_0 r_c dz. \tag{5.16}$$

Note that the effective radius is the ratio of the third moment of the droplet size distribution to the second moment. In other words it is proportional to the water content of a cloud divided by the total surface area of droplets. If one imagines the same amount of water content of a cloud in a few big droplets, its effective radius will be larger than if the same water content were distributed on many small droplets. This is because, for the same water content, total surface area of many small droplets is greater than a few large droplets. Thus, effective radius conveniently describes two of the most important radiative cloud microphysical properties: volume and surface area.

It should be stressed that there is no instrument that directly measures effective radius. Nonetheless, effective radius has become an indispensable, if not abused, quantity in cloud remote sensing.

### 5.3.1. Absorption by Clouds

Absorption of solar radiation by clouds is generally quite small. Several investigators have estimated that precipitation-sized drops can appreciably increase cloud absorptance (Manton, 1980; Welch et al., 1980; Wiscombe et al., 1984). Because precipitation-sized drops form by collecting smaller cloud drops, they typically represent a miniscule percentage of the total drop concentration. Drizzle-sized drops have a concentration usually not exceeding 100 $m^{-3}$, compared to cloud droplet concentrations of the order of $10^8$ to $10^9$ $m^{-3}$. Thus precipitation-sized drops have little direct impact upon the reflectance of a cloud. On the other hand, because precipitation-sized drops can contribute significantly to the liquid-water content and, hence, to the liquid-water path, they can affect cloud absorption appreciably. Wiscombe et al. (1984) calculated that for a given liquid-water path, the absorptance is enhanced by 2-3% by the presence of large drops in a 4-km-deep cloud. Note, however, this is a small enhancement of a small amount of the total energy absorbed by clouds.

For a time it was thought that there was an appreciable difference, of as much as 40%, between the measured and the modeled cloud absorptance. This difference was called anomalous absorption (Cess et al., 1995; Ramanathan et al., 2003; Pilewskie and Valero, 1995). However, recent calculations using more sophisticated models and advanced measurement systems suggest that the difference between modeled and observed absorption is less than 10% which is within the range of instrument error (Ackerman et al., 2003; Sengupta and Ackerman, 2003).

### 5.3.2. Ice Clouds

The optical properties of ice clouds are complicated by the geometries of the ice particles, the uncertainties in ice crystal concentration, and their size spectra. We have seen in Chapter 4, that the habit of vapor-grown ice crystals varies with temperature and supersaturation with respect to ice. Moreover, as ice crystals grow by riming cloud droplets or aggregation, the geometrical and surface characteristics of the ice particles vary as do their optical properties. Important to any realistic assessment of the reflectance, transmittance, or absorptance of an ice or mixed-phase cloud system is the concentration of ice particles and their size spectra, this is especially true of thin cirrus clouds. Unfortunately, as we have pointed out in Chapter 4, we do not have a reliable way of diagnosing or predicting ice crystal concentrations. Equation (5.12) suggests that errors in ice crystal concentrations can lead to large errors in estimating the shortwave optical thickness ($\tau$) for an ice cloud.

One further complication of ice clouds is that ice crystals are non-spherical and they are generally not randomly oriented in space. Instead, large ice crystals fall preferentially with their major axis oriented horizontal (Ono, 1969; Jayaweera and Mason, 1965; Platt, 1978), while smaller crystals tend to fall randomly oriented, depending on their Reynolds number (Sassen, 1980). Some small ice crystals with diameters of about 30 μm also exhibit preferred orientations but with large tilt angles (Klett, 1995).

Evidence of the complexity of the shortwave radiative properties of ice clouds can be seen from the variety of optical phenomena that are frequently observed. Features such as halos, sundogs, and pillars result from the interaction of shortwave radiation with ice crystals having a particular crystal habit, spatial orientation, size spectra, and a growth history that often includes a relatively turbulent-free, slowly rising environment (Hallett, 1987; Tricker, 1970; Greenler, 1980).

## 5.4. LONGWAVE RADIATIVE TRANSFER IN A CLOUDY ATMOSPHERE

In this discussion, we refer to longwave radiation as radiation emitted by the earth's surface, or the atmosphere, having wavelengths greater than about 4 μm. The effect of clouds on longwave radiation is quite different from that of for shortwave radiation. In the case of shortwave radiation, we find that cloud droplets are strong scatterers of incident radiation. Absorption of solar radiation by cloud droplets and ice crystals is small. By contrast, longwave radiation is strongly absorbed by cloud droplets in optically thick clouds. As much as 90% of incident longwave radiation can be absorbed in less than 50 m pathlengths in a cloud with high liquid-water content. Scattering of longwave radiation in clouds is secondary to absorption. Thus optically thick clouds are often considered to be blackbodies with respect to longwave radiation. Yamamoto et al. (1970) suggested that cumulonimbus clouds could be considered blackbodies after a pathlength of only 12 m. By contrast, the blackbody depth of thin cirrus ice clouds may be greater than several kilometers (Stephens, 1983), which is greater than the depths of those clouds. Thus, cirrus clouds, thin stratus, and many fogs do not behave as blackbodies over the infrared range.

We noted previously that, in a cloud-free atmosphere, little gaseous absorption takes place between 8 and 14 μm, a band which is commonly referred to as the "atmospheric window." An exception is in the deep, maritime tropics where high values of low-level water vapor content can result in large amounts of gaseous absorption. In an optically-thick cloudy atmosphere, on the other hand, there are no spectral regions where absorption of longwave radiation is small. Clouds therefore have a major impact on the amount of longwave radiation emitted to space. Thus clouds are extremely important to the earths climate.

The behavior of the extinction efficiency $Q_e$ is quite different in the infrared region than it is over visible wavelengths for cloud particles. The extinction efficiency varies similarly to the behavior of the scattering efficiency shown in Fig. 5.4. Thus, for small values of $x$, $Q_e$ increases monotonically with $x$. Moreover, at wavelengths corresponding to the peak in the spectral density of terrestrial radiative flux ($10\ \mu m < \lambda < 20\ \mu m$), $Q_e$ varies almost linearly with droplet radius for droplets with radii less than 20 μm. Substitution of $Q_e = kr$ into Eq. (5.10) yields

$$\tau = \int_0^{\delta z} \int_0^{\infty} \pi k r^3 f(r)\, dr dz. \tag{5.17}$$

Because the liquid-water content is

$$\mathrm{LWC} = \int_0^{\infty} \frac{3\pi}{\rho_\ell} r^3 f(r)\, dr, \tag{5.18}$$

Eq. (5.18) shows that the optical thickness over the infrared range is principally a function of the liquid-water content of a cloud and is not strongly dependent upon the details of the cloud droplet spectrum. This greatly simplifies the parameterization of longwave radiative transfer through clouds.

The calculations by Wiscombe and Welch (1986), however, suggest that predictions of infrared cooling rates can be significantly affected by the presence of drizzle or raindrops near the tops of optically thick clouds. Estimated cooling rates for a cloud containing cloud droplets only, and for one containing precipitation, may differ by a factor of 4 in the topmost 50 m of a cloud. The differences between cooling rates for the two cloud types are reduced substantially at greater penetration distances in the cloud, although for an 8-km-deep cloud there is still nearly a factor of 2 difference in cooling rates between the two cloud types.

Paltridge and Platt (1976) noted that a commonly used approximation for estimating $Q_e$ for complex-shaped crystals is to use the equivalent sphere approximation, Eq. (5.17). Mie theory can then be used to calculate the variation of $Q_e$ as a function of $x$. The result is similar to Fig. 5.4, for water drops, with differences due to the variation of refractive indices between water and ice.

Although the longwave radiative reflectance is small, Stephens (1980) calculated that it can have a significant impact on the flux profiles in the cloud and thus on the cloud-heating profiles. This is true when the upward flux from the earth's surface is quite large. A longwave reflectance at the cloud base of only a few percent can thus significantly affect the upwelling fluxes at colder cloud temperatures. This can substantially alter the strength of flux divergence and, hence, the rate of radiational cooling near the cloud top. This effect is most pronounced in the tropics.

A commonly used concept in longwave radiation diagnostic studies as well as parameterizations is the effective emittance concept. Cox (1976) determined cloud emittance values from measurements of broadband radiative flux profiles through clouds. The emittance can thus be defined

$$\varepsilon(\uparrow) = \frac{F_B(\uparrow) - F_T(\uparrow)}{F_B(\uparrow) - \sigma T_T^4} \quad \text{for the upward irradiance,} \tag{5.19}$$

$$\varepsilon(\downarrow) = \frac{F_B(\downarrow) - F_T(\downarrow)}{\sigma T_B^4 - F_T(\downarrow)} \quad \text{for the downward irradiance.} \tag{5.20}$$

$F(\uparrow)$ and $F(\downarrow)$ refer to the upward and downward measured infrared irradiances, respectively. The subscripts T and B refer to the top and bottom of the cloud layer, respectively, and $\sigma$ is the Stefan-Boltzmann constant. The definition of effective emittance combines the effects of reflection, emission, and transmission by cloud droplets as well as gas molecules. The effective emittance is therefore not a scalar but a directionally dependent vector, since the emissivity is dependent upon the particular path the radiation takes through the atmosphere. As noted by Stephens (1980), when a cold cloud overlies a warm surface and reflects some longwave radiation, it can exhibit values of emittance considerably greater than unity.

## 5.5. RADIATIVE CHARACTERISTICS OF CLOUDS OF HORIZONTALLY FINITE EXTENT

Thus far we have considered only the radiative properties of horizontally infinite cloud layers. However, few cloud systems are horizontally homogeneous; a population of cumulus clouds is just one example. Even the tops of stratocumulus clouds are undulating, thus altering both the shortwave and longwave properties of those clouds.

Consideration of the finite geometries of clouds is quite complex. Looking out of the window at a few cumulus clouds, we observe the complicated shapes these clouds assume, sometimes identifying cloud shapes with animals or other familiar objects. Needless to say, describing such complicated shapes with mathematical functions or computer algorithms can be quite difficult. Often the cloud shapes are approximated as simple cubes (McKee and Cox, 1974, 1976; Davis et al., 1979a,b), while a few attempts have been made to simulate more complex shapes using superimposed sinusoidal functions (Takeuchi, 1986). But more recently the science of 3D cloud radiative transfer has made great advances as summarized in the book edited by Marshak and Davis (2005). Analytical modeling techniques such as the Fourier-Ricati approach of Gabriel et al. (1993) and the analytical-numerical methods of Evans (1998) have greatly advanced our understanding of 3D radiative transfer as well as many recent observational programs. Still at the time of this writing the state-of-the-art has not advanced to the point where 3D cloud resolving models interact explicitly with 3D radiative transfer. On the mesoscale and larger, a stochastic approach to parameterizing the 3D radiative properties of clouds may be needed as suggested by Gabriel et al. (1993).

In general, it appears that the finite geometrical properties of clouds are more important to the bulk radiative properties than are variations in the cloud microstructure.

## 5.6. RADIATIVE INFLUENCES ON CLOUD PARTICLE GROWTH

Traditionally, cloud physicists have ignored the effects of radiation (Mason, 1971; Byers, 1965). It is generally argued that because the temperature difference between cloud droplets or ice crystals and their immediate surroundings is so small, nearly as much radiation is emitted from the cloud particle as is absorbed. This view is probably valid for a cloud particle that resides in the middle of an optically thick cloud. However, a cloud particle that resides at the top of a cloud layer or in an optically thin cloud such as a cirrus cloud essentially "sees" outer space, especially in the 8- to 12-μm spectral window. As a consequence, the surface temperature of the cloud particle will be cooler. As a result, the droplet or ice crystal will experience a higher supersaturation, or, in a subsaturated environment, a lesser subsaturation. This can be more readily seen by considering the rate of mass change due to vapor deposition or evaporation of an ice crystal or cloud droplet of mass $M$,

$$\mathrm{d}M/\mathrm{d}t = 4\pi C D f_1 f_2[\rho_\mathrm{v}(T_\infty) - \rho_\mathrm{S}(T_\mathrm{s})], \tag{5.21}$$

where $C$ is the capacitance of an ice crystal as defined in Chapter 4. For a spherical droplet or ice crystal, $C = r$. The coefficient $D$ is the diffusivity of water vapor, $f_1$ is a factor that includes the accommodation coefficient for water molecules, $f_2$ is a ventilation function, $\rho_\mathrm{v}(T_\infty)$ is the vapor density some distance from the particle surface, and $\rho_\mathrm{S}(T_\mathrm{s})$ is the saturation vapor density with respect to ice or water at the surface temperature of the ice crystal. In order to estimate $T_\mathrm{s}$, we normally assume that a cloud droplet or ice crystal is in thermal equilibrium,

$$L(\mathrm{d}M/\mathrm{d}t) + Q_r = 4\pi C K f_1 f_2 \boldsymbol{\Delta} T, \tag{5.22}$$

where the first term on the left-hand side of Eq. (5.22) is the latent heat liberated in the growing cloud particle, $L$ is either the latent heat of sublimation or vaporization, and the second term on the left-hand side is the net radiative heating of the particle. The right-hand side represents the rate of heat diffusion away from the particle, where $K$ is the thermal diffusivity and $\boldsymbol{\Delta} T$ is the temperature difference between a cloud particle and its environment. Roach

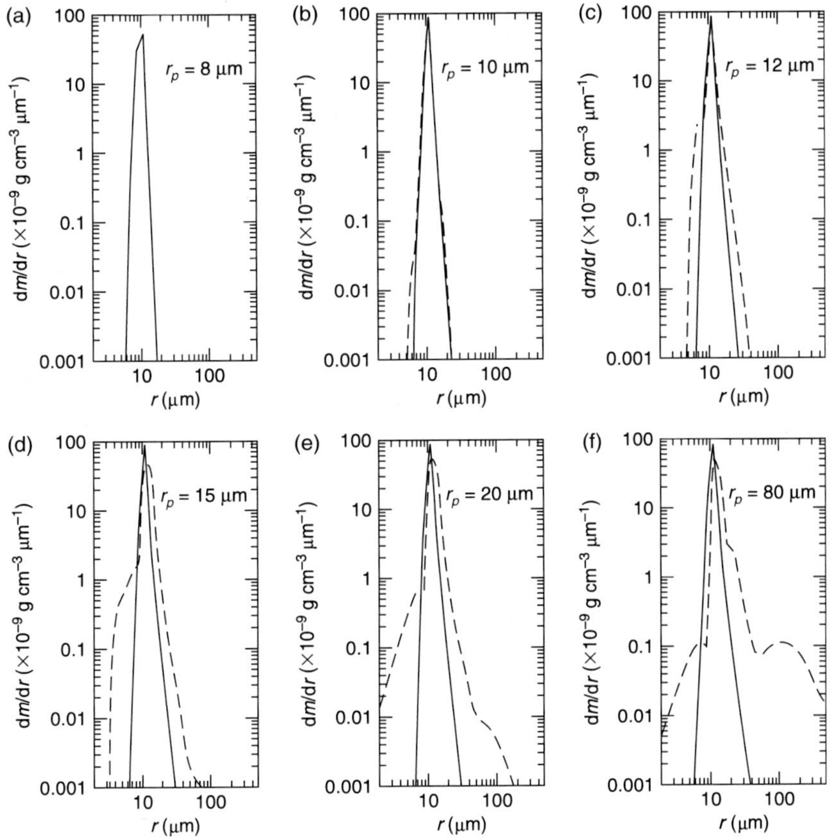


**FIGURE 5.5 Plots of d$m$/d$r$ for selected times during the control run (TAP). For (a) 4.1 h, (b) 4.2 h, (c) 4.4 h, (d) 4.6 h, (e) 4.8 h, and (f) 5.0 h.** The dashed line and solid line are for runs with (TAP) and with (TNR) radiation, respectively. Value or $r_p$ are given in the plot. *(From Harrington et al. (2000))*

(1976) and Barkstrom (1978) considered the radiative effects on cloud droplet growth while Stephens (1983), Hallett (1987), and Wu et al. (2000) considered the radiative influences on ice crystal growth. Barkstrom showed that for optically thick clouds radiation can be important to droplet condensation for those droplets residing within 20 m of cloud top. For optically thin clouds, such as fogs and thin stratus, cloud droplets throughout the cloud may be affected by radiation. Harrington et al. (2000) examined the radiative effect on warm season Arctic stratus by using a trajectory ensemble model (TEM) driven by the flow fields produced by a two-dimensional eddy resolving model. They found that the longwave radiative (LW) effect reduced the time-scale for the onset of drizzle formation for up to 30 minutes. As shown in Fig. 5.5 droplet radiative cooling caused a broadening of the droplet spectrum such that more larger droplets formed, and smaller droplets less than 10 μm evaporated as they lost out in the competition for water vapor. The effect is most important for parcels of air residing near cloud top for 12 minutes or more.

Hartman and Harrington (2005a,b) examined radiative influences on the initiation of drizzle drops further by considering the competitive influences of LW and shortwave (SW) radiation using a TEM driven by a large eddy simulation model (LES). In contrast to LW radiation cooling, which is confined to layers of the cloud near cloud top, SW heating is effective throughout the depth of shallow clouds and acts to suppress droplet growth. However, because SW heating stabilizes the cloud layer, cloud parcel lifetimes are increased, resulting in longer cloud residence times. But because SW heating occurs throughout the cloud layer, droplet growth is retarded compared to cases with only LW radiation cooling. Hartman and Harrington (2005a) also examined the influence of droplet concentration on the cloud response to LW radiation cooling. They found that for low droplet concentrations LW cooling had little influence on the initiation of collection as collection was very active without LW radiation cooling. However, for droplet concentrations greater about 200 $cm^{-3}$ LW cooling accelerated the onset of collection. Hartman and Harrington (2005b) also examined the influence of solar zenith angle $\theta_0$. They found that at small $\theta_0$ solar heating dominates over LW cooling causing a suppression of collection for smaller droplet concentrations. For larger droplet concentrations LW cooling dominates over SW heating even at small $\theta_0$. At large $\theta_0$, SW heating does not alter the initiation of collection, thus LW cooling enhances collection for all droplet concentrations.

For the case of thin ice crystal clouds such as some cirrus clouds, Stephens also concluded that (1) because radiation can enhance (suppress) particle growth (evaporation), radiative cooling at cloud top and warming at cloud base tend to broaden and narrow the spectrum, respectively, and (2) the influence of radiation on the survival distance of falling ice particles is most significant in air having a relative humidity greater than 70%. At lower relative humidities evaporation is so strong that survival distances are altered little by radiation.

Wu et al. (2000) further examined the radiative influences on cirrus ice crystal growth using a 2D cloud resolving model. Both SW and LW radiation were considered. They found that with radiative feedback the cloud was optically thinner, allowing SW radiation to penetrate deeper into the cloud layer. Smaller crystals were little affected by radiational heating while larger crystals, having a larger radiation cross sectional area, experienced substantially reduced vapor deposition growth. Thus the size of the ice crystals was limited by radiational heating. It is interesting that, because the crystals were limited in size, they did not precipitate out of the cloud so much, and contributed to a longer-lived cirrus cloud.

It is clear that radiative effects can be important to cloud particle growth and evaporation.

## 5.7. AEROSOL EFFECTS ON THE RADIATIVE PROPERTIES OF CLOUDS

We have already seen that the concentration and the size spectra of cloud droplets have an important influence upon the shortwave radiative properties of clouds. The cloud droplet concentration, in turn, is largely a function of the concentration of cloud condensation nuclei activated in typical cloud supersaturations. The width of the droplet size spectra is to some extent also a consequence of the aerosol spectrum, although, as noted in Chapter 4, the width of the droplet spectrum is influenced by other cloud macroscopic parameters. We have also seen that the concentration and size spectra of ice crystals have a strong impact upon the radiative properties of clouds. Furthermore, the concentrations of aerosols active as ice nuclei are believed to play an important role in determining the ice crystal concentration, particularly for cold clouds such as altostratus and cirrus. Also, in those clouds having weak vertical motions, the size spectrum of ice crystals is largely determined by the competition for vapor among the ice crystals nucleated on ice nuclei. Thus the size spectra and chemical composition of aerosols have important controlling influences on the concentration of cloud droplets and ice crystals as well as on their size spectra, which, in turn, have important impacts upon the radiative properties of clouds.

In addition to affecting cloud radiative properties indirectly by influencing the cloud microstructure, aerosols can directly affect the radiative properties of clear as well as cloudy air. Assessment of the radiative effects of aerosols requires an estimate of the single scattering properties and, just as with cloud particles, a knowledge of the aerosol optical thickness. The optical thickness is determined by Eq. (5.10), which, like cloud droplets, requires a knowledge of the size distribution of the aerosol particles. Estimates of the extinction efficiency $Q_e$ are complicated because aerosols are nonspherical, and, moreover, their variable chemical composition results in variability in their complex indices of refraction. Thus, the extinction efficiency varies with the source and life history of the aerosol. The life history is important because, as the aerosol ages, the particles coagulate with each other and form particles of mixed chemical composition. Furthermore, the extinction efficiency and the size spectra of the aerosol population change with relative humidity. At relative humidities greater than 70%, the hygroscopic aerosols take on water to become haze particles. As the relative humidity increases toward 100%, the aerosol particles swell in size and their complex indices of refraction change as the water-solution/particle mixture changes in relative amounts.

Computations of the radiative effects of natural dry aerosols suggest that polluted boundary layer air can result in shortwave radiative heating rates of the order of a few tenths of a degree to several degrees per hour (Braslau and Dave, 1975; Welch and Zdunkowski, 1976). Above the boundary layer, the aerosol shortwave radiative heating rates are much less. Even in the Saharan dust layer, heating rates are only of the order of 1-2 °C per day (Carlson and Benjamin,

1980; Ackerman and Cox, 1982). Nonetheless, such heating rates strengthen the overlying inversion, which further concentrates the pollutants in the lower troposphere. Absorption of solar radiation by aerosols can result in stabilization of a moist moderately stable layer and weaken convection and precipitation. This effect of aerosols has been termed the semi-direct effect (Grassl, 1975; Hansen et al., 1997). The reduction in cloud cover associated with this effect can alter the surface energy budget significantly. If the aerosol contains a large fraction of soot, such as the south Asian haze, then warming of the aerosol layer can desiccate stratocumulus cloud layers and alter the properties of the trade-wind cumulus layer (Ackerman et al., 2003). The influence of black carbon has a dominant effect on absorption of solar radiation within the atmosphere, which leads to lower surface temperatures (Ramanathan et al., 2001; Lohmann and Feichter, 2001), and reduces outgoing fluxes.

The impact of aerosols on infrared radiative transfer is usually less than it is for shortwave radiation. The effect of aerosols is greatest in the atmospheric window, where gaseous absorption of infrared radiation is least (Welch and Zdunkowski, 1976; Carlson and Benjamin, 1980; Ackerman et al., 1976). Before longwave radiative cooling effects can be detected, rather large concentrations of aerosols through a deep layer must be present. Ackerman and Cox (1982) could not detect any significant change in longwave fluxes due to dust over Saudi Arabia. Carlson and Benjamin (1980) calculated that the Saharan dust layer can affect longwave radiative fluxes if the dust layer is sufficiently deep. Several researchers have found that longwave radiative cooling can be appreciable in polluted boundary layer air (Welch and Zdunkowski, 1976; Saito, 1981). Andreyev and Ivlev (1980) investigated the radiative properties of various organic and inorganic natural aerosols. They found that organic aerosols are typically less than 0.5 μm in radius and affect infrared radiation little, but have a significant impact on shortwave radiative fluxes. Infrared radiative fluxes were mainly affected by the presence of large ($r > 0.15$ μm) mineral substances. Andreyev and Ivlev's evaluations did not include exposure of the aerosols to increasing relative humidity. Some of the small inorganic aerosols may be activated as haze particles at higher relative humidities. As a result, once the small aerosols have swollen in size, they may alter longwave radiative fluxes appreciably. Welch and Zdunkowski calculated that net longwave radiative fluxes changed by more than 25% from a dry polluted boundary layer compared to a moist polluted boundary layer.

The effect of increasing humidity on the radiative properties of aerosols is most pronounced in the shortwave spectrum. The swelling of aerosol particles or activation of haze particles has an appreciable impact on local visibility (Kasten, 1969; Takeda et al., 1986) and on the global albedo (Zdunkowski and Liou, 1976). Zdunkowski and Liou also calculate that the swelling of aerosol particles in humid atmospheres can alter the local albedo by as much as 5% relative to a dry atmosphere.

Aerosols can modify the radiative properties of clouds when they are mixed into a cloud system. The most hygroscopic of the aerosol particles participate in the nucleation of cloud droplets. Large and ultra-giant aerosols (greater than 1 μm in radius) rapidly become wetted regardless of their chemical composition, and become engulfed in cloud droplets or raindrops as a result of hydrodynamic capture. No longer functioning as aerosols, they still influence the radiative properties of the cloud system. Smaller, submicrometer-sized aerosols, however, can remain outside of cloud droplets (interstitial). Those that are non-hygroscopic may remain dry with little change in size. Hygroscopic aerosols, while not being activated as cloud droplets, will swell in size in the slightly sub-saturated or supersaturated environment. Thus, the interstitial aerosol population will resemble a cloud-free haze population, with the exception of the removal of the largest, most hygroscopic components by nucleation scavenging and hydrodynamic capture of the particles greater than 1 μm. After a time, the remainder of the submicrometer particles may also be scavenged by cloud droplets. The major scavenging processes for submicrometer aerosols are Brownian diffusion and thermophoretic/diffusiophoretic scavenging. As noted in Chapter 4, Brownian diffusion is quite slow, and because thermophoresis predominates over diffusiophoresis, for a submicrometer-sized aerosol, phoretic scavenging does not enhance scavenging of submicrometer aerosols in a supersaturated cloud. In subsaturated regions of a cloud, however, evaporating cloud droplets can be very effective at scavenging submicrometer aerosols by phoretic processes. The droplet spectrum has to be broad enough to allow the largest droplets to survive evaporation in local subsaturated regions, however. This means that in the least dilute regions of convective storm updrafts, the time scale is short owing to intense updraft speeds, and because little evaporation occurs in the weakly mixed regions, many submicrometer particles will survive the ascent into the upper troposphere. At least for some time, submicrometer aerosol particles may remain interstitial and affect the radiative properties of the stratiform region of those clouds.

The presence of the ice phase also contributes to scavenging of interstitial aerosols. Slowly settling, branched snowflakes can be very effective removers of any remaining particles greater than 1 μm. Likewise, as ice crystals sublimate in sub-ice-saturated environments, they can be effective scavengers of submicrometer aerosols by phoretic processes. Furthermore, as we noted in Chapter 4, when ice crystals grow by vapor deposition, they do so at the expense of cloud droplets, causing their evaporation. Again, if the droplet spectrum is broad enough, the larger droplets may survive evaporation and contribute to the scavenging of submicrometer aerosols. Still, these processes are slow enough that the anvils or stratiform regions of deep convective systems, especially in polluted environments, contain a substantial interstitial aerosol population for some time. A significant interstitial aerosol population also exists in the tops of boundary layer stratocumuli and relatively young middle-tropospheric stratus clouds.

Assuming that the interstitial aerosol size spectrum is unchanged by the presence of cloud, Newiger and Bahnke (1981) calculated that aerosols can enhance absorption in a 4-km-thick horizontally uniform cloud by factors of 1.3 to 2.4.

The next question concerns how the scavenged aerosol particles affect the radiative properties of clouds. If the particles are soluble, they will dissolve once they have become embedded in water droplets for a time. Only in very small droplets will the solution be concentrated enough to affect the refractive index of the droplet. If the droplet totally evaporates, however, they become aerosol particles again (perhaps altered in size and chemical composition) and affect radiative transfer again. If the particles, however, are insoluble, as are graphitic carbon or soot, they can remain embedded in droplets and alter the radiative properties of the droplets. If the insoluble particles are greater than 1 μm in radius, they could be scavenged by hydrodynamic capture. As noted previously, submicrometer-sized insoluble particles can be scavenged by Brownian or phoretic scavenging processes, or they can coagulate with hygroscopic particles and be removed by nucleation scavenging. Chylek et al. (1984) calculated the radiative properties of soot particles embedded in droplets. Assuming that the soot particles were randomly distributed throughout the droplets, they showed that the absorption efficiency at short wavelengths of graphite carbon particles embedded in droplets is more than twice the efficiency of the same particles freely suspended in air, as long as the volume fraction of the particles in water is small. Figure 5.6 illustrates the calculated change in cloud reflectance and absorptance as a function of the volume fraction of graphite carbon for optically thick clouds having three different droplet size spectra. The cloud absorptance is substantially enhanced while the reflectance is reduced for soot volume fractions greater than $10^{-5}$. Whether this amount of graphitic carbon occurs in cloud droplets in nature is unknown at this time. Certainly in the case of the hypothesized nuclear winter scenario (see, e.g. Pittock et al., 1986) one would expect to find volume fractions of graphitic carbon in excess of those shown in Fig. 5.6.

What would be the effects of graphitic carbon particles attached to ice crystals? Again, little is known about the range of volume fractions of carbon that become attached to ice crystals. One would expect that graphitic carbon particles attached in sufficiently high numbers to ice crystals would decrease the reflectance of ice clouds substantially. There is certainly evidence that soot embedded in surface snow significantly affects the albedo of the snow surface (Warren, 1982).

## 5.8. PARAMETERIZATION OF RADIATIVE TRANSFER IN CLOUDS

### 5.8.1. Introduction

We have seen in previous sections that clouds have an important impact upon radiative transfer. Because divergence of radiative fluxes contributes

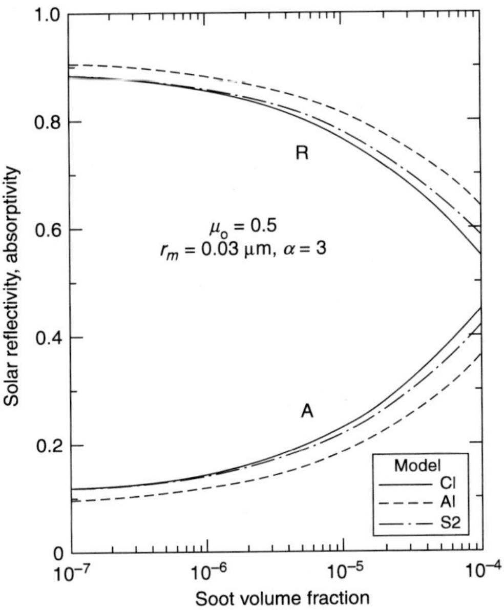


**FIGURE 5.6 Absorptance and reflectance for an optically thick cloud at $\lambda = 0.5\ \mu m$ as a function of soot volume fraction.** The cosine of the solar zenith angle taken to be 0.5. An amount of carbon between $5 \times 10^{-6}$ and $1 \times 10^{-5}$ by volume is required, depending on the cloud drop-size distribution, to obtain reflectance of 0.8 for thick clouds. *(From Chylek et al. (1984))*

to heating/cooling in a cloud system, radiative transfer processes can alter the thermodynamic stability of a cloud system and thereby contribute to the dynamics of the system. Because of the complexity of other cloud processes and their computational demands, cloud dynamicists and mesoscale modelers must seek simplifications in the formulation of radiative processes in models. Much of the research has been aimed at formulating radiative transfer parameterization schemes suitable for use in general circulation models. General circulation models typically do not have enough vertical resolution to make realistic estimates of the liquid-water path, let alone the other cloud properties such as hydrometeor type and spectra, and the finite dimensions of clouds.

## 5.8.2. Parameterization of Shortwave and Longwave Radiation in Clouds

One general approach to reducing the complexities associated with the solution to (5.6) is to introduce the so-called "two-stream approximation." In this approach the total radiation field is hemispherically-averaged, and represented by two streams: one in the upward direction (↑) and one in the downward direction (↓). The radiation intensity $I(\tau, \mu)$ is then integrated over the upward

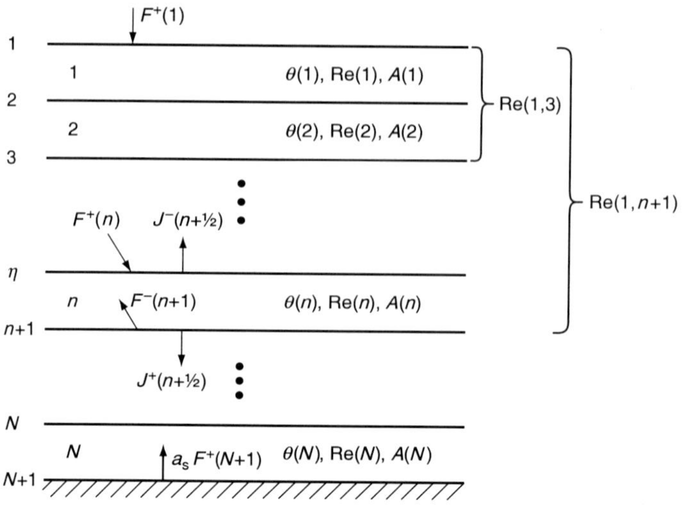


**FIGURE 5.7** **The shortwave radiative transfer model;** $F \downarrow$ and $F \uparrow$ denote the downward and upward flux. Re($n$) and $A(n)$ represent, respectively, the reflectance and absorptance at the $n$th layer. Re($1, n+1$) is the multiple reflectance from all layers above ($n+1$)th layer. *(Adapted by Chen and Cotton (1983); from Stephens and Webster (1979))*

and downward hemispheres to define the fluxes

$$F \uparrow\downarrow (\tau) = \int_0^1 \mu I(\tau, \pm\mu) \mathrm{d}\mu, \tag{5.23}$$

which, as Eqs (5.1) and (5.2) show, are the quantities needed to determine solar heating. Meador and Weaver (1980) showed that by assuming that $I$ is dependent on $\mu$, the hemispheric integral of Eq. (5.6) reduces to standard ordinary differential equations of the form

$$\frac{\mathrm{d}F \uparrow}{\mathrm{d}\tau} = \gamma_1 F \uparrow - \gamma_2 F \downarrow + (F_0/4)\tilde{\omega}_0 \gamma_3 \mathrm{e}^{-\tau/\mu_0}, \tag{5.24}$$

$$\frac{\mathrm{d}F \downarrow}{\mathrm{d}\tau} = \gamma_2 F \uparrow - \gamma_1 F \downarrow + (F_0/4)\tilde{\omega}_0 \gamma_4 \mathrm{e}^{-\tau/\mu_0}. \tag{5.25}$$

The $\gamma_i$ values are determined by the approximations used and are independent of $\tau$. Solutions to Eqs (5.25) and (5.26) can be obtained by standard techniques for specified boundary conditions. Meador and Weaver (1980) showed that several standard solution techniques, such as the Eddington approximation and the quadrature methods, could be transformed into the standard form of Eqs (5.25) and (5.26), with the appropriate specification of the $\gamma_i$ values. An illustration of a simple layered two-stream model is given in Fig. 5.7, where Re($n$) and $A(n)$ represent the reflectances and absorptances of each layer, respectively.

A few cloud resolving models (i.e. Fu et al., 1995) have used four-stream approximations as developed by Liou et al. (1988), Fu and Liou (1992). Although computationally more demanding than the two-stream approximation it offers more accuracy in radiative heating estimations particularly at higher latitudes where solar zenith angles are large.

Application of the two-stream or four-stream approximations to clouds requires the determination of the cloud properties $\tau, \tilde{\omega}$, and $g$ in terms of modeled or specified microphysical and macrophysical variables. For spherical droplets, these properties can be obtained using Mie theory (see van de Hulst, 1957). If, however, the particle shape is complex, such as is the case for ice crystals or aerosols, solutions are not easily obtainable using Mie theory. In this case approximate solutions are necessary. For very large particles ($x = 2\pi r/\lambda \gg 1$), it has become customary to employ geometrical optics, in which the paths of individual rays traveling through a droplet or ice particle are traced. Rays passing through a particle and those not interacting with the particle are not allowed to interact or interfere with each other. Figure 5.8 illustrates an application of geometrical optics to the rainbow problem.

A somewhat less restrictive technique for obtaining the extinction and absorption efficiency factors is the so-called anomalous diffraction theory (van de Hulst, 1957; Ackerman and Stephens, 1987; Mitchell, 2000; Mitchell et al., 2006). Like geometrical optics, it is also valid for particles much larger than the wavelength of radiation ($x \gg 1$), and for which the index of refraction is $n \sim 1$, these are often referred to as soft particles. The anomalous diffraction approximation is based on the premise that the extinction of light is primarily a result of the interference between the rays that pass through a particle and those rays that are not influenced by a particle (see Fig. 5.9).

### *5.8.2.1. Cloud Optical Depth*

Using Mie theory, or a suitable approximation to the interaction of radiation and droplets or ice particles, one can develop a parameterization of the cloud optical depth Eq. (5.10). Stephens (1978b) used eight different cloud droplet-size distributions to illustrate that the cloud optical depth, as calculated with Mie theory, is primarily a function of the cloud liquid-water path. He showed that $\tau$ could be approximated by Eq. (5.14). This is a useful approximation, because all one needs to do is calculate the liquid-water path, Eq. (5.16), and, given a climatologically derived effective radius, the optical thickness is easily determined.

A few cloud models using the two-stream approximation like Harrington (1997), Harrington et al. (1999) and Cotton et al. (2003) actually use the explicitly-resolved hydrometeor size distributions predicted with bin-microphysics models or diagnosed from two moment bulk schemes. This has the advantage that scattering and absorption by large raindrops and ice particles can be explicitly represented whereas use of the effective radius approximation masks the influence of these fewer but larger hydrometeors.

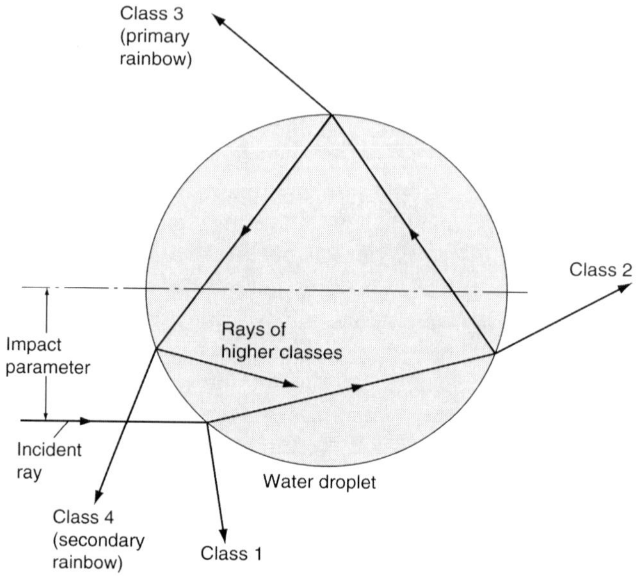


**FIGURE 5.8 Path of light through a droplet can be determined by applying the laws of geometrical optics.** Each time the beam strikes the surface, part of the light is reflected and part is refracted. Rays reflected directly from the surface are labeled rays of Class 1; those transmitted directly through the droplet are designated Class 2. The Class 3 rays emerge after one internal reflection; it is these that give rise to the primary rainbow. The secondary bow is made up of Class 4 rays, which have undergone two internal reflections. For rays of each class, only one factor determines the value of the scattering angle. That factor is the impact parameter: the displacement of the incident ray from an axis that passes through the center of the droplet. *(From Nussenzveig (1977). Copyright © 1977 by Scientific American, Inc. All rights reserved)*

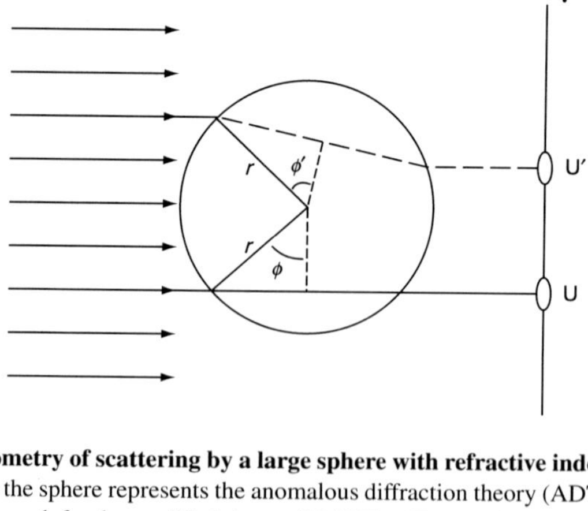


**FIGURE 5.9 Geometry of scattering by a large sphere with refractive index near 1.** The solid ray passing through the sphere represents the anomalous diffraction theory (ADT), while the broken ray describes the ray path for the modified theory (MADT). *(From Ackerman and Stephens (1987))*

#### 5.8.2.2. *Single-Scattering Albedo*

The single-scattering albedo also has been parameterized in terms of the effective radius. Based on geometrical optics, Liou (1980) developed the following parameterization of single-scattering albedo

$$\tilde{\omega}_0 = 1 - 1.7k' r_e, \tag{5.26}$$

where $k'$ is the complex part of the index of refraction ($n_\lambda = n_r - ik'$). Fouquart and Bonnel (1980) developed the expression

$$\tilde{\omega}_0 = 1 + \exp(-2k' r_e), \tag{5.27}$$

which is valid when liquid-water absorption is weak in the solar region. Using anomalous diffraction theory, van de Hulst (1957) derived a more general expression for $\tilde{\omega}_0$,

$$\tilde{\omega}_0 = 1 - \frac{1}{2}\left(\frac{4}{3}\rho \tan\Gamma - \rho^2 \tan^2\Gamma\right),$$

where $\rho = 2x(n_r - 1)$ and

$$\Gamma = \arctan(k'/n_r - 1). \tag{5.28}$$

Equation (5.28) is valid for small values of $4xk'$ and for $Q_e = 2$. Better estimates of $\tilde{\omega}_0$ can be obtained by using the tabulated values reported by Stephens et al. (1984). Stephens (1978b) showed that it is possible to remove the dependence of $\tilde{\omega}_0$ on $r_e$ by empirically tuning $\tilde{\omega}_0$ as well as the integrated phase function parameter using accurate numerical solutions. Fouquan and Bonnel also used more accurate calculations to derive the spectrally-averaged, single-scattering albedo

$$\tilde{\omega}_0 = 0.9989 - 0.0004 \exp(-0.15\tau). \tag{5.29}$$

Here again, the single scattering albedo can be calculated from explicitly represented hydrometeor size distributions rather than using effective radius approximations.

#### 5.8.2.3. *Implementation in Cloud Resolving Models and Larger-Scale Models*

In the past, computational efficiency has been gained in models by use of broad-band approximations wherein the radiances are averaged across rather large wavelength bands (i.e. Stephens (1978b, 1984b), Harrington (1997)). However, many operational and research groups have engineered there radiation codes to include detailed line, two-stream models by using neural network techniques (i.e. Key and Schweiger, 1998; Chevallier et al., 2000; Krasnopololsky et al., 2005) or look-up table methodologies (Pielke et al., 2006; Matsui et al., 2004; Leoncini and Pielke, 2007).

### *5.8.2.4. Radiative Transfer through a Multilayered Atmosphere*

Our discussion thus far has only dealt with a single homogeneous cloud layer with fixed values of $\tau$, $\tilde{\omega}_0$, and $g$. We now consider an atmosphere composed of a number of layers, each having different optical properties. The most commonly used approach is the so-called "adding" method. Grant and Hunt (1969a,b) considered an atmosphere composed of $n$ homogeneous layers, each with their respective reflective (Re), transmissive (Tr), and absorptive ($A$) properties. For such an atmosphere, the reflectance $\mathrm{Re}(1, n + 1)$ represents the combined multiple reflectance contributed by all layers above the $(n + 1)$th layer. It may be defined as

$$\mathrm{Re}(1, n+1) = \mathrm{Re}(n) + \frac{\mathrm{Tr}\downarrow(n)\mathrm{Tr}\uparrow(n)\mathrm{Re}(1, n)}{1 - \mathrm{Re}(1, n)\mathrm{Re}(n)}, \qquad (5.30)$$

where the reflection from a composite of all layers above the $(n + 1)$th layer is obtained by adding the reflectance from two layers whose reflectance is $\mathrm{Re}(n)$ and $\mathrm{Re}(1, n)$.

The flux transmitted through the upper layer is represented by $V\downarrow(n+1/2)$ which may be computed as

$$V\downarrow(n+1/2) = \frac{\mathrm{Tr}\downarrow(n)V\downarrow(n-1/2)}{1 - \mathrm{Re}(1, n)\mathrm{Re}(n)}. \qquad (5.31)$$

Similarly, the flux transmitted from the lower layer $V\uparrow(n + 1/2)$ is calculated as

$$V\uparrow(n+1/2) = \frac{\mathrm{Re}(n)V\downarrow(n-1/2)}{1 - \mathrm{Re}(1, n)\mathrm{Re}(n)}. \qquad (5.32)$$

The denominators in Eqs. (5.30)–(5.31) account for multiple reflections between layers. As noted by Stephens (1984a), this factor is especially large when the reflection between layers is large. This occurs when a dense cloud overlaps a bright surface such as another cloud or a snow-covered surface.

To close such a layered model, boundary conditions must be supplied. At the upper boundary, one can assume that the reflectance is zero and the downward-transmitted radiation corresponds to the flux coming into the atmosphere, or

$$\mathrm{Re}(1,\ 1) = 0,$$
$$V\downarrow(1/2) = F\downarrow(1).$$

If the model top were placed at the tropopause, then $F\downarrow(1)$ would correspond to the downward flux from stratospheric levels and above. At the lower boundary the upward flux is given by

$$F \uparrow (n+1) = a_s F \downarrow (n+1),$$

where $a_s$ is the albedo of the earth's surface. The value of $a_s$ varies depending on whether the surface is dry land, vegetated, snow covered, or a sea surface. It also varies depending on solar elevation. Typical values of $a_s$ for solar elevations of 45° are 7% for a water surface, 20-35% for dry grass lands, 30-40% for sand, and 80-85% for fresh snow.

This layered model illustrates the fact that the net radiative flux divergence is influenced by radiative transfer through all layers of the atmosphere above and below the level under consideration. To compute the transmittance and reflectance, we note that Eq. (5.3) can be integrated over all wavelengths shorter than 4 μm to give

$$\mathrm{Re} + \mathrm{Tr} = 1 - A. \qquad (5.33)$$

Thus, if a parameterization of absorptance and reflectance is available, one can compute the transmittance by using Eq. (5.33).

#### 5.8.2.5. *Partial Cloudiness*

We have noted that the radiative properties of a population of finite clouds can differ substantially from that of a horizontally homogeneous cloud system. In the case of a region covered by partial cloudiness, the normal procedure is to weight the reflectance, transmittance, and absorptance calculated for a cloudy atmosphere by the cloud fractional coverage, and the corresponding clear-air properties by the clear-air coverage. This ignores contributions from radiation emitted from the sides of clouds and the interaction of radiation among neighboring clouds. As noted above, the full 3D radiative interaction among clouds has not yet been parameterized for use in cloud resolving or larger-scale models. A statistical approach is probably needed.

### 5.8.3. Parameterization of Longwave Radiative Transfer in Clouds

We have seen in Section 5.4 that clouds are effective absorbers of longwave radiation. A cloud of only modest liquid-water content of 0.2 g kg$^{-1}$ may absorb up to 90% of the upwelling infrared radiant energy within a depth of only 50 m. By contrast, an equivalent penetration distance for shortwave radiation is at least 600 m for a cloud having the same liquid-water content. As a consequence, it is often assumed that optically thick clouds behave as blackbodies. Thus, all upwelling radiation is absorbed at the base of the cloud and is re-emitted in the upward and downward directions with a flux equal to $\sigma T_b^4$, where $T_b$ is the cloud-base temperature. At the cloud top, the upward flux emitted by an assumed blackbody cloud will simply be $\sigma T_T^4$, where $T_T$ is the temperature at the cloud top. Many clouds, such as cirrus, stratus, and stratocumulus, as well as fogs, are not optically thick and, therefore, the blackbody approximation is a poor one. One must then seek alternate parameterizations of the longwave properties of such clouds.

A common approach used in many models until recently was to use the mixed emittance concept introduced by Herman and Goody (1976). However, it has become standard practice in most models to apply the two-stream or four-stream approximation methodology to longwave radiation as well.

## 5.9. SUMMARY

In this chapter we have reviewed the interaction between the cloud microphysical and macrophysical structure and radiative transfer processes. We have also presented some of the concepts and approaches to parameterizing radiative transfer through clouds. In Part II of this book we will examine the effects of radiative processes on the dynamics and precipitation processes in several different cloud systems.

## REFERENCES

Ackerman, S. A., and Cox, S. K. (1982). The Saudi Arabian heat low: Aerosol distributions and thermodynamic structure. J. Geophys. Res. 87, 8991–9002.

Ackerman, S. A., and Stephens, G. L. (1987). The absorption of solar radiation by cloud droplets: An application of anomalous diffraction theory. J. Atmos. Sci. 44, 1574–1588.

Ackerman, T. P., Liou, K.-N., and Leovy, C. B. (1976). Infrared radiative transfer in polluted atmospheres. J. Appl. Meteorol. 15, 28–35.

Ackerman, T. P., Flynn, D. M., and Marchand, R. T. (2003). Quantifying the magnitude of anomalous solar absorption. J. Geophys. Res. 108, 4273. doi:10.1029/2002JD002674.

Andreyev, S. D., and Ivlev, L. S. (1980). Infrared radiation absorption by various atmospheric aerosol fractions. Izv. Acad. Sci. USSR Atmos. Oceanic Phys. (Engl. Transl.) 16, 663–669.

Barkstrom, B. R. (1978). Some effects of 8–12μm radiant energy transfer on the mass and heat budgets of cloud droplets. J. Atmos. Sci. 35, 665–673.

Blevin, W. R., and Brown, W. J. (1971). A precise measurement of the Stefan–Boltzmann constant. Metrologia 7, 15–29.

Braslau, N., and Dave, J. V. (1975). Atmospheric heating rates due to solar radiation for several aerosol-laden cloudy and cloud-free models. J. Appl. Meteorol. 14, 396–399.

Byers, H. R. (1965). "Elements of Cloud Physics." University of Chicago Press, Chicago, Illinois.

Carlson, T. N., and Benjamin, S. G. (1980). Radiative heating rates for Saharan dust. J. Atmos. Sci. 37, 193–213.

Cess, et al. (1995). Absorption of solar radiation by clouds: Observations versus models. Science 267, 496–499.

Chen, C., and Cotton, W. R. (1983). A one-dimensional simulation of the stratocumulus-capped mixed layer. Bound.-Layer Meteorol. 25, 289–321.

Chevallier, F., Morcrette, J.-J., Cheruy, F., and Scott, N. A. (2000). Use of a neural-network-based long-wave radiative-transfer scheme in the ECMWF atmospheric model. Q. J. R. Meteorol. Soc. 126, 761–776.

Chylek, P., Gupta, B. R. D., Knight, N. C., and Knight, C. A. (1984). Distribution of water in hailstones. J. Clim. Appl. Meteorol. 23, 1469–1472.

Cotton, W. R., Pielke Sr., R. A., Walko, R. L., Liston, G. E., Tremback, C. J., Jiang, H., McAnelly, R. L., Harrington, J. Y., Nicholls, M. E., Carri, G. G., and McFadden, J. P. (2003). RAMS 2001: Current status and future directions. Meteor. Atmos. Phys. 82, 5–29.

Cox, S. K. (1976). Observations of cloud infrared effective emissivity. J. Atmos. Sci. 33, 287–289.

Davis, J. M., Cox, S. K., and McKee, T. B. (1979a). Total shortwave radiative characteristics of absorbing finite clouds. J. Atmos. Sci. 36, 508–518.

Davis, J. M., Cox, S. K., and McKee, T. B. (1979b). Vertical and horizontal distribution of solar absorption in finite clouds. J. Atmos. Sci. 36, 1976–1984.

Evans, K. F. (1998). The spherical harmonics discrete ordinate method for three-dimensional atmospheric radiative transfer. J. Atmos. Sci. 55, 429–446.

Fouquart, Y., and Bonnel, B. (1980). Computations of solar heating of the earth's atmosphere: A new parameterization. Atmos. Phys. 53, 35–62.

Fu, Q., and Liou, K. N. (1992). On the correlated k-distribution method for radiative transfer in nonhomogeneous atmospheres. J. Atmos. Sci. 49, 2139–2156.

Fu, Q., Krueger, S. K., and Liou, K. N. (1995). Interactions of radiation and convection in simulated tropical cloud clusters. J. Atmos. Sci. 52, 1310–1328.

Gabriel, P. M., Tsay, S.-C., and Stephens, G. L. (1993). A Fourier-Riccati approach to radiative transfer. Part I: Foundations. J. Atmos. Sci. 50, 3125–3147.

Gast, P. R., Jursa, A. S., Castelli, J., Basu, S., and Aarons, J. (1965). Solar electromagnetic radiation. In "Handbook of Geophysics and Space Environments" (S. L. Valley, Ed.), pp. 16-1–16-38. McGraw-Hill, New York.

Grant, I. P., and Hunt, G. E. (1969a). Discrete space theory of radiative transfer. I. Fundamentals. Proc. R. Soc. London, Ser. A 313, 183–197.

Grant, I. P., and Hunt, G. E. (1969b). Discrete space theory of radiative transfer. II. Stability and non-negativity. Proc. R. Soc. London, Ser. A 313, 199–216.

Grassl, H. (1975). Albedo reduction and radiative heating of clouds by absorbing aerosol particles. Contrib. Atmos. Phys. 48, 199–210. Oxford.

Greenler, R. G. (1980). "Rainbows, Halos and Glories." Cambridge University Press, London.

Hallett, J. (1987). Faceted snow crystals. J. Opt. Soc. Am., A 4, 581–588.

Hanel, R. A., Conrath, B. J., Kunde, V. G., Prabhakara, C., Revah, I., Salomonson, V. V., and Wolford, G. (1972). The Nimbus 4 infrared spectroscopy experiment, 1. Calibrated thermal emission spectra. J. Geophys. Res. 11, 2629–2641.

Hansen, J. E., and Travis, L. D. (1974). Light scattering in planetary atmospheres. Space Sci. Rev. 16, 527–610.

Hansen, J., Sato, M., and Ruedy, R. (1997). Radiative forcing and climate response. J. Geophys. Res. 102, 6831–6864.

Harrington, J.Y. 1997. The effects of radiative and microphysical processes on simulated warm and transition season Arctic stratus. Ph.D. Dissertation, Colorado State University, 289 pp. [Available from Colorado State University, Dept. of Atmospheric, Fort Collins, CO 80523.].

Harrington, J. Y., Reisin, T., Cotton, W. R., and Kreidenweis, S. M. (1999). Cloud resolving simulations of Arctic stratus. Part II:Transition-season clouds. Atmos. Res. 55, 45–75.

Harrington, J. Y., Feingold, G., and Cotton, W. R. (2000). Radiative impacts on the growth of a population of drops within simulated summertime arctic stratus. J. Atmos. Sci. 57, 766–785.

Hartman, C. M., and Harrington, J. Y. (2005a). Radiative impacts on the growth o drops within simulated marine stratocumulus. Part I: Maximum solar heating. J. Atmos. Sci. 62, 2339–2351.

Hartman, C. M., and Harrington, J. Y. (2005b). Radiative impacts on the growth o drops within simulated marine stratocumulus. Part II: Solar zenith angle variations. J. Atmos. Sci. 62, 2323–2338.

Henyey, L. G., and Greenstein, J. L. (1941). Diffuse radiation in the galaxy. Astrophys. J 112, 445–463.

Herman, G. F., and Goody, R. (1976). Formation and persistence of summertime arctic stratus clouds. J. Atmos. Sci. 33, 1537–1553.

Jayaweera, D. O., and Mason, B. J. (1965). The behaviour of freely falling cylinders and cones in a viscous fluid. J. Fluid Mech 22, 709–720.

Joseph, J. H., Wiscombe, W. J., and Weinman, J. A. (1976). The delta-Eddington approximation for radiative transfer. J. Atmos. Sci 33, 2452–2459.

Kasten, F. (1969). Visibility forecast in the phase of pre-condensation. Tellus 5, 631–635.

Key, J. R., and Schweiger, A. J. (1998). Tools for atmospheric radiative transfer: Streamer and fluxnet. Computers & Geosci. 24, 443–451.

Klett, J. D. (1995). Orientation Model for Particles in Turbulence. J. Atmos. Sci. 52, 2276–2285.

Krasnopololsky, V. M., Fox-Rabinovitz, M. S., and Chalikov, D. V. (2005). New approach to calculation of atmospheric model physics: Accurate and fast neural network emulation of longwve radiation in a climate model. Mon. Wea. Rev. 133, 1370–1383.

Leoncini G., Pielke Sr. R. A. (2007). From model based parameterizations to Look Up Tables: An EOF approach. Wea. Forecasting (submitted for publication).

Liou, K.-N. (1980). "An Introduction to Atmospheric Radiation." Academic Press, New York.

Liou, K.-N., Fu, Q., and Ackerman, T. P. (1988). A simple formulation of the Delta-Four-stream approximations for radiative transfer parameterizaitons. J. Atmos. Sci. 45, 1940–1947.

Lohmann, U., and Feichter, J. (2001). Can the direct and semi-direct aerosol effect compete with the indirect effect on a global scale? Geophys. Res. Lett. 28, 159–161.

Manton, M. J. (1980). Computations of the effect of cloud properties on solar radiation. J. Rech. Atmos. 14, 1–16.

Marshak, A., and Davis, A. B. (2005). "3D Radiative Transfer in Cloudy Atmospheres." Springer-Verlag, Berlin, 686 pp.

Mason, B. J. (1971). "The Physics of Clouds." 2nd Ed., Oxford University Press, Clarendon, Oxford.

Matsui T., Leoncini G., Pielke Sr. R.A., Nair U.S., 2004. A new paradigm for parameterization in atmospheric models: Application to the new Fu-Liou radiation code. Atmospheric Science Paper No. 747, Colorado State University, Fort Collins, CO 80523, 32 pp.

McKee, T. B., and Cox, S. K. (1974). Scattering of visible radiation by finite clouds. J. Atmos. Sci. 31, 1885–1892.

McKee, T. B., and Cox, S. K. (1976). Simulated radiance patterns for finite cubic clouds. J. Atmos. Sci. 33, 2014–2020.

Meador, W. E., and Weaver, W. R. (1980). Two-stream approximations to radiative transfer in planetary atmospheres: A unified description of existing methods and a new improvement. J. Atmos. Sci. 37, 630–643.

Mitchell, D. L. (2000). Parameterization of the Mie extinction and absorption coefficients for water clouds. J. Atmos. Sci. 57, 1311–1326.

Mitchell, D. L., Baran, A. J., Arnott, W. P., and Schmitt, C. (2006). Testing and comparing the modified anomalous diffraction approximation. J. Atmos. Sci. 63, 2948–2962.

Newiger, M., and Bahnke, K. (1981). Influence of cloud composition and cloud geometry on the absorption of solar radiation. Contrib. Atmos. Phys. 54, 370–382.

Nussenzveig, H. M. (1977). The theory of the rainbow. Sci. Am. 236, 116–127.

Ono, A. (1969). The shape and riming properties of ice crystals in natural clouds. J. Atmos. Sci. 26, 138–147.

Paltridge, G. W., and Platt, C. M. R. (1976). "Radiative Processes in Meteorology and Climatology." In Developments in Atmospheric Science, vol. 5. Elsevier, New York.

Pielke Sr., R. A., Matsui, T., Leoncini, G., Nobis, T., Nair, U., Lu, E., Eastman, J., Kumar, S., Peters-Lidard, C., Tian, Y., and Walko, R. (2006). A new paradigm for parameterizations in numerical weather prediction and other atmospheric models. National Wea. Digest 30, 93–99.

Pilewskie, P., and Valero, F. P. J. (1995). Direct observations of excess solar absorption by clouds. Science 267, 1626–1629.

Pittock, A. B., Ackerman, T. P., Crutzen, P. J., MacCracken, M. C., Shapiro, C. S., and Turco, R. P. (1986). Environmental Consequences of Nuclear War, vol. 1.

Platt, C. M. R. (1978). Lidar backscatter from horizontal ice crystal plates. J. Appl. Meteorol. 17, 482–488.

Ramanathan, V., Crutzen, P. J., Lelieveld, J., Mitra, A. P., Althausen, D., Anderson, J., Andreae, M. O., Cantrell, W., Cass, G. R., Chung, C. E., Clarke, A. D., Coakley, J. A., Collins, W. D., Conant, W. C., Dulac, F., Heintzenberg, J., Heymsfield, A. J., Holben, B., Howell, S., Hudson, J., Jayaraman, A., Kiehl, J. T., Krishnamurti, T. N., Lubin, D., McFarquhar, G., Novakov, T., Ogren, J. A., Podgorny, I. A., Prather, K., Priestley, K., Prospero, J. M., Quinn, P. K., Rajeev, K., Rasch, P., Rupert, S., Sadourny, R., Satheesh, S. K., Shaw, G. E., Sheridan, P., and Valero, F. P. J. (2001). Indian Ocean Experiment: An integrated analysis of the climate forcing and effects of the great Indo-Asian haze. J. Geophys. Res. 106, 28371–28398.

Ramanathan, V., Subasilar, B., Zhang, G. J., Conant, W., Cess, R. D., Kiehl, J. T., Grassl, H., and Shi, L. (2003). Warm pool heat budget and shortwave cloud forcing: A missing physics?. Science 267, 499–503.

Roach, W. T. R. (1976). On the effect of radiative exchange on the growth by condensation of a cloud or fog droplet. Q. J. R. Meteorol. Soc. 102, 361.

Saito, T. (1981). The relationship between the increase rate of downward longwave radiation by atmospheric pollution and the visibility. J. Meteorol. Soc. 59, 254–261.

Sassen, K. (1980). Remote sensing of planar ice crystal fall attitudes. J. Meteorol. Soc. Japan 58, 422–429.

Sengupta, M., and Ackerman, T. P. (2003). Investigating anomalous absorption using surface measurements. J. Geophys. Res. 108, 4761. doi:10.1029/2003JD003411.

Stephens, G. L. (1978a). Radiation profiles in extended water clouds. I: Theory. J. Atmos. Sci. 35, 2111–2122.

Stephens, G. L. (1978b). Radiation profiles in extended water clouds. II: Parameterization schemes. J. Atmos. Sci. 35, 2123–2132.

Stephens, G. L. (1980). Radiative properties of cirrus clouds in the infrared region. J. Atmos. Sci. 37, 435–446.

Stephens, G. L. (1983). The influence of radiative transfer on the mass and heat budgets of ice crystals falling in the atmosphere. Meteorol. Monogr. 40, 1729–1739.

Stephens, G. L. (1984a). Scattering of plane waves by soft obstacles: Anomalous diffraction theory for circular cylinders. Appl. Opt. 23, 954–959.

Stephens, G. L. (1984b). The parameterization of radiation for numerical weather prediction and climate models. Mon. Weather Rev. 112, 826–862.

Stephens, G. L., and Webster, P. J. (1979). Sensitivity of radiative forcing to variable cloud and moisture. J. Atmos. Sci. 36, 1542–1556.

Stephens, G. L., Ackerman, S., and Smith, E. A. (1984). A shortwave parameterization revised to improve cloud absorption. Meteorol. Monogr. 41, 687–690.

Takeda, T., Pei-ming, W., and Okada, K. (1986). Dependence of light scattering coefficient of aerosols on relative humidity in the atmosphere of Nagoya. J. Meteorol. Soc. Japan 64, 957–966.

Takeuchi, Y. (1986). Effects of cloud shape on the light scattering. J. Meteorol. Soc. Jpn 64, 95–107.

Tricker, R. A. R. (1970). A note on the Lowitz and associated arcs. Weather 25, 503.

Twomey, S. (1976). Computations of the absorption of solar radiation by clouds. J. Atmos. Sci. 33, 1087–1091.

van de Hulst, H. C. (1957). "Light Scattering by Small Particles." Wiley, New York.

Warren, S. G. (1982). Optical properties of snow. Rev. Geophys. Space Phys. 20, 67–89.

Welch, R., and Zdunkowski, W. (1976). A radiation model ofthe polluted atmospheric boundary layer. J. Atmos. Sci. 33, 2170–2184.

Welch, R. M., Cox, S. K., and Davis, J. M. (1980). "Solar Radiation and Clouds", In Meteorol. Monograph, Vol. 17. 96 pp. Boston, Massachusetts.

Wiscombe, W. J., and Welch, R. (1986). Reply. J. Atmos. Sci. 43, 401–407.

Wiscombe, W. J., Welch, R. M., and Hall, W. D. (1984). The effects of very large drops on cloud absorption. Part I: Parcel models. J. Atmos. Sci. 41, 1336–1355.

Wu, T., Cotton, W. R., and Cheng, W. Y. Y. (2000). Radiative effects on the diffusional growth of ice particles in cirrus clouds. J. Atmos. Sci. 57, 2892–2904.

Yamamoto, G., Tanaka, M., and Asano, S. (1970). Radiative transfer in water clouds in the infrared region. J. Atmos. Sci. 27, 282–292.

Zdunkowski, W. G., and Liou, K.-N. (1976). Humidity effects on the radiative properties of a hazy atmosphere in the visible spectrum. Tellus 28, 31–36.

Part II

# The Dynamics of Clouds

Chapter 6

# Fogs and Stratocumulus Clouds

## 6.1. INTRODUCTION

We begin our discussion of the dynamics of clouds and cloud systems by discussing what may be the least dynamic of all cloud phenomena, fog. Cloud dynamicists often classify fog as a micrometeorological phenomenon because it forms next to the earth in the atmospheric boundary layer, a domain traditionally covered by micrometeorologists. Lecturers in micrometeorology, however, often consider fog to be in the discipline of cloud dynamics or, perhaps, mesoscale meteorology. Fog does, in fact, span all these disciplines (as do many cloud systems); it occurs in the atmospheric boundary layer; it is a well-defined cloud in most instances, and it exhibits horizontal and temporal variability on scales normally thought to be the domain of mesoscale meteorology. We discuss fog as a cloud system that may be considered part of the general class of boundary layer stratiform clouds.

## 6.2. TYPES OF FOG AND FORMATION MECHANISMS

Fog is normally categorized into four main types (see Willit, 1928; Byers, 1959; Jiusto, 1980):

A. Radiation fog
   (a) ground fog
   (b) high inversion fog
   (c) advection-radiation fog
   (d) upslope fog
   (e) mountain-valley fog
B. Frontal fog
   (a) prefrontal (warm front)
   (b) postfrontal (cold front)
   (c) frontal passage
C. Advection (mixing) fog
   (a) sea fog
   (b) tropical air fog
   (c) land and sea-breeze fog
   (d) steam fog ("Arctic sea smoke")

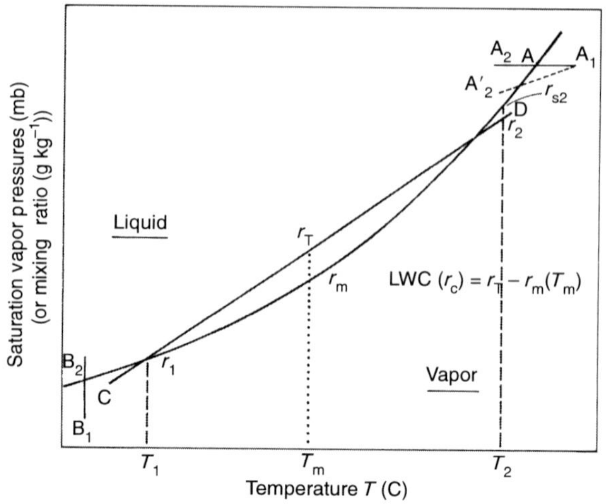


**FIGURE 6.1** **Phase diagram and fog formation processes.** *(Adapted from Jiusto (1980))*

D. Other
   (a) ice fog
   (b) snow fog

The physical mechanisms responsible for the formation of fog involve three primary processes: (1) cooling of air to its dewpoint, (2) addition of water vapor to the air, and (3) vertical mixing of moist air parcels having different temperatures.

The first mechanism generally explains radiation fogs, while the second causes frontal fogs and the third produces advection fog. A combination of all three mechanisms affects most fogs, though one mechanism may dominate.

To illustrate the processes further, consider the Clausius-Clapeyron diagram shown in Fig. 6.1. The formation of radiation fog may be illustrated by the line ($A_1$-$A_2$) where fog forms by the steady, isobaric cooling of air by contact with the cold ground and by radiative flux divergence in the moist or cloudy air until the air is cooled to saturation. According to Jiusto (1980), the line $A_1$-$A_2'$ actually better represents radiational fog formation than does $A_1$-$A_2$, because the water vapor content of the air is not fully conserved; some of the moisture is lost by dew deposition on the earth's surface. Thus, additional cooling of the air is needed for fog to form.

Frontal fogs often involve the addition of moisture by falling precipitation from relatively warm layers aloft into underlying cooler, subsaturated air. This is illustrated in Fig. 6.1 as line $B_1$-$B_2$, where the addition of water vapor causes the mixing ratio in the air to exceed the saturation mixing ratio at point $B_2$. Donaldson and Stewart (1993) pointed out that mixed phase precipitation from frontal clouds can produce fog-forming supersaturations by heat and vapor flows from particles that are warmer than the air, or by particles that are considerably colder than the air. Owing to the fact that they remain near 0 °C, freezing particles exhibit large temperature differences between the particles and the surrounding air near the ground, thus contributing to supersaturations with respect to water.

Advection or mixing fogs were first described by Taylor (1917), who observed them from a whaling ship off the Grand Banks of Newfoundland. Advection fogs form when near-saturated air parcels of different temperatures mix vertically. Consider, for example, the formation of a sea fog as warm, moist air flows over a cooler ocean surface. Suppose that the ocean surface has a temperature $T_1$, which is saturated at the mixing ratio $r_1$. Now a warm, moist air mass characterized by point D in Fig. 6.1, having a temperature $T_2$, flows over the cooler ocean surface. If the low-level air is well mixed in the vertical, the temperature will vary nearly linearly through the mixed layer. Likewise, the water vapor mixing ratio will vary linearly through the mixed layer along line $r_1 \rightarrow r_T \rightarrow r_2$ in Fig. 6.1. Because the saturation vapor pressure or the saturation mixing ratio varies nonlinearly along the line $r_1 \rightarrow r_m \rightarrow r_{s2}$, a region of the mixed layer becomes supersaturated with respect to water. A cloud thus forms which has a peak liquid-water content in the middle of the mixed layer equal to

$$r_c = r_T - r_m(T_m). \tag{6.1}$$

## 6.3. RADIATION FOG PHYSICS AND DYNAMICS

### 6.3.1. The Role of Radiative Cooling

Radiation fog commences with strong radiative cooling of the earth's surface. According to Taylor (1917), nights with clear skies, light winds, and high relative humidities favor radiation fog. Radiation cools the earth's surface, which then cools the air close to the earth's surface by conduction. In addition, radiative flux divergence in the moist atmosphere is also important (Zdunkowski and Nielsen, 1969; Zdunkowski and Barr, 1972; Brown and Roach, 1976). Brown and Roach (1976) concluded that gaseous radiative cooling is necessary to account for fog formation on the observed time scale of a few hours.

Once the fog forms aloft, radiative flux divergence at the fog top (Fleagle et al., 1952; Korb and Zdunkowski, 1970; Pilie et al., 1975) increases the stability at, and immediately above, the fog top and destabilizes the lapse rate within and below the fog. The resultant vertical mixing of the cold foggy air with clear, nearly saturated air below causes the fog to propagate downward. Radiative cooling at the fog top also increases the liquid-water content and decreases the visibility in the fog, often contributing to its upward propagation as well.

Radiation is also important on the scale of individual droplets in fogs. As noted in Chapter 5, Roach (1976) and Barkstrom (1978) include a radiative transfer term in the equations for heat and water mass budgets of a spherical droplet. When combined with the Clausius-Clapeyron equation, they produce an equation for droplet growth. The resulting expression for the saturation vapor pressure at the droplet surface, including the effects of radiative loss, indicates droplet growth can occur in a slightly subsaturated environment. This process allows for fog formation with consequent radiation-induced changes in the stability of the entire cloudy layer, even though the environment may never become supersaturated with respect to water on the average. Brown (1980) and Mason (1982) conclude that the importance of the radiation term in the droplet growth equation varies with the concentration of activated cloud condensation nuclei. With high CCN concentrations, the radiative term has little influence on either the mean droplet size or liquid-water content. They suggest this is due to the fact that the resultant more numerous droplets have small values of absorption efficiencies. They conclude that radiative exchange between droplets and their environment will be greatest in clean fogs and maritime layer clouds, and will be less in heavily polluted air.

### 6.3.2. The Role of Dew

A number of investigators have emphasized the role of dew deposition to the formation of fog (see Wells, 1838; Geiger, 1965; Pilie et al., 1975; Lala et al., 1975; Brown and Roach, 1976). The deposition of dew at the surface is responsible for the development of a downward transport of moisture and the formation of a nocturnal dew-point inversion. Dew-point inversions have been observed to extend to between 40 and 200 m above the surface.

Some of the moisture supplied to dew deposition may come from the underlying soil. Most, however, is extracted from the overlying air mass. Thus, Lala et al. (1975) and Brown and Roach (1976) view dew deposition as a "governor" on fog formation. For a given rate of radiative cooling, which drives the air toward saturation, if the dew deposition rate and accompanying downward transport of moisture is large, then fog formation may be inhibited. If the dew deposition at the surface is somewhat less, radiative cooling may be sufficient to initiate the formation of fog.

Pilie et al. (1975) also attribute the observed initial formation of fog aloft to the development of a dew-point inversion as a consequence of dew deposition. However, Jiusto (1980) has noted that many inland radiation fogs first develop at the surface and then build upward. The conditions under which fog formation occurs at the surface versus aloft are not well known.

From the time of fog formation and until sunrise, dew does not appear to serve any major function other than to maintain a saturated lower boundary (Pilie et al., 1975). After sunrise, however, the surface temperature begins to rise and evaporation of dew commences. As the fog layer warms, a supply of water vapor is needed in order to maintain saturation. Pilie et al. (1975) estimate that the dew evaporation rate is sufficient to allow persistence of a fog for several

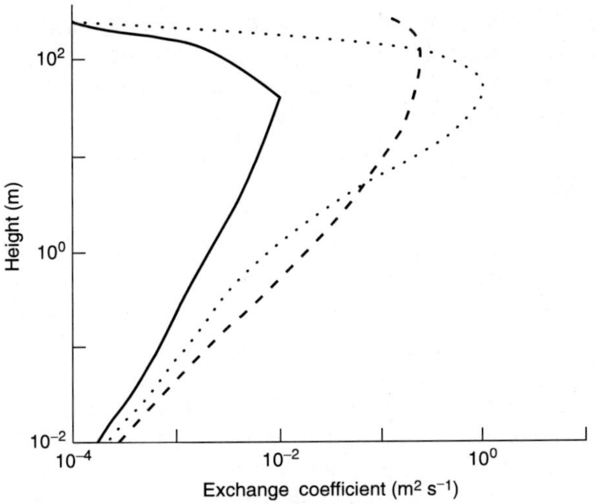


**FIGURE 6.2 Exchange coefficient regimes described in the text: dotted line, model I; dashed line, model II; solid line, model III.** *(From Brown and Roach (1976). Reproduced with the permission of the Controller of Her Britannic Majesty's Stationery Office)*

hours longer than over a dew-free surface. Eventually solar heating causes the saturation vapor pressure to increase above the actual vapor pressure, in spite of the evaporation of dew. This leads to fog dissipation, first at the surface and then propagating upward. Thus originates the term "fog lifting."

## 6.3.3. The Role of Turbulence

Turbulent transport of heat and moisture plays an important role in the evolution of a fog. However, there is not a consensus as to whether the role of turbulence is primarily constructive, contributing to the formation of fog, or destructive, contributing primarily to the dissipation of fog. Brown and Roach (1976) conclude that turbulence inhibits the formation of radiation fog. They base their conclusion partially on a fog model in which turbulence is modeled with an eddy viscosity closure. The magnitude of eddy viscosity was specified by various profiles shown in Fig. 6.2. They conclude that numerical experiments with model II provide the most realistic results when compared to observations reported by Roach (1976). Brown and Roach also experimented with a stability-dependent exchange coefficient formulation, which was derived using adiabatic similarity theory. As they point out, however, the existence of a constant-flux layer a few meters above the surface is questionable in such a stable environment as a radiation fog.

In a companion paper Roach (1976) infers from observations that turbulence hinders fog formation. They formed this conclusion because lulls in wind were accompanied by maximum cooling, and major lulls were accompanied by periods of significant fog development. Conversely, increases in wind (to $> 2\ m\ s^{-1}$) were associated with fog dispersal. They infer that as the wind speed decreases, turbulent transfer of moisture to the surface to form dew ceases. As a result, the moisture remains in the atmosphere, and as radiation cools the air, fog is formed. Alternately, at higher wind speeds, vertical mixing of drier air may inhibit fog formation.

Jiusto and Lala (1980) infer from observations that if radiative cooling and higher humidities extend to a greater depth, vertical turbulent mixing can contribute positively to the growth of the fog when it occurs in the presence of radiation cooling. Using various surface-layer turbulence exchange formulations, Welch et al. (1986) conclude that increased turbulence and reduced stability contribute to fog formation. They also suggest that fog intensification after sunrise is caused by increased turbulence generation and the resultant downward mixing of liquid in the upper part of the fog to the surface.

Lala et al. (1982) suggest that turbulence in the early evening may inhibit fog, whereas later in the evening turbulent mixing can intensify fog. Lala et al. (1975), Brown and Roach (1976), Zdunkowski and Barr (1972), and Welch et al. (1986) concur that the structure of fog and the occurrence and nonoccurrence of fog are strongly dependent upon the particular profile of eddy viscosity or turbulence models employed. Moreover, Brown and Roach (1976) conclude that a more realistic treatment of turbulence is needed, especially in the region beneath the fog top. They suggest that the top of deep fogs may behave somewhat like the ground, not only with respect to radiative processes, but also to turbulent transport. Comparison of more complicated turbulence models with observed fog structure shows that realistic simulations of fog life cycles can be obtained (Musson-Genon, 1987). This work further emphasizes the importance of turbulent transport to fog evolution.

The sensitivity of fog to turbulence, and specifically to wind speed, makes forecasting of fog quite challenging. Because radiation fog forms under conditions of weak large-scale pressure gradients, local winds are driven by very weak and small scale pressure gradient forcing by terrain and land-surface properties. Thus an increase in wind speed from say $1.5\ m\ s^{-1}$ to $2.5\ m\ s^{-1}$ can lead to the dissipation or no formation of fogs (Chibe, 2003).

### 6.3.4. The Role of Drop Settling

In Chapter 4 we discussed the partitioning of the droplet distribution in clouds into raindrops having significant terminal velocities, and into cloud droplets which are assumed to have negligible fall velocities. In those models, cloud droplets are as large as 40-50 μm in radius. In fogs, however, only a few droplets exceed 20 μm in radius. Nonetheless, Brown and Roach (1976) have concluded that settling of droplets can play an important role in the evolution of fog structure. This reflects the fact that vertical velocities in fogs are quite small, much smaller than in cumulus clouds for which the models in Chapter 4 were developed.

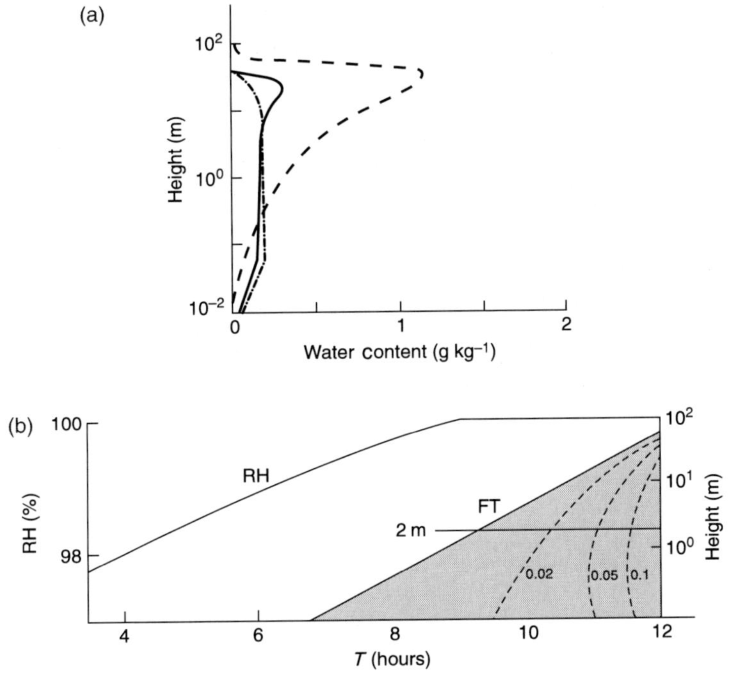


**FIGURE 6.3** **(a) Liquid-water content profiles after 5 h of integration with gravitational droplet settling excluded (dashed line) and included using $\overline{v} = 6.25w$ (solid line) or droplet settling included using $\overline{v} = 10w$ (dot-dash line). (b) Plot of relative humidity (RH) at 2 m, height of fog top (FT), and heights of liquid-water mixing ratio isopleths (dashed lines, in units of g kg$^{-1}$) against time ($T$) since beginning of integration of model II.** *(From Brown and Roach (1976). Reproduced with the permission of the Controller of Her Britannic Majesty's Stationery Office)*

Brown and Roach introduced drop settling into their model because, otherwise, unrealistically high liquid-water contents were predicted. Figure 6.3 illustrates the sensitivity of the model to several simple models of drop settling. Brown and Roach used the following feedback process to explain the sensitivity of the model to drop settling: the direct removal of liquid water by droplet settling causes a reduction of radiation cooling due to cloud droplets; lower radiation cooling, in turn, leads to a reduction in the rate of condensation and liquid-water contents. In a more detailed microphysical simulation of fog, Brown (1980) found that the simulated fog liquid-water content was sensitive to CCN concentrations. With lower concentrations of CCN, he found that the fog liquid-water content was about 20% less than with higher CCN concentrations; the difference was attributed to reduced drop settling in the high CCN case. Thus polluted air high in CCN concentrations can not only reduce visibility because for a given liquid-water content, a high concentration of small droplets scatter radiation more efficiently than fewer larger droplets, but also by reducing droplet sedimentation which contributes to higher liquid-water content, optically thicker fogs.

### 6.3.5. The Role of Vegetative Cover

Brown and Roach also investigated the role of surface vegetation on the rate of cooling at the surface. They argued that a grass surface will cool to a lower surface temperature than bare ground because of its small thermal capacity. Furthermore, grass will partially shield the soil from radiative loss. Thus, the air over a grass-covered surface will radiate to a colder surface than that of bare soil, and greater cooling of the air will occur. Using a single-column model Siebert et al. (1992) examined the influence of vegetation cover to the simulation of radiation fog. They found that for increasing vegetation coverage the fog height and fog water content increase appreciably.

### 6.3.6. Advection of Cloud Cover

The advection of a cloud layer over a fog can alter the net divergence of longwave radiation from the top of the fog and thereby lead to an alteration in the fog structure or perhaps even the destruction of the fog. Saunders (1957) noted that a large fraction of fog cases cleared following the arrival of a cloud deck. He also observed a rise in temperature near the ground. Subsequently, Saunders (1961) observed that the net outgoing radiation at fog top decreased substantially compared to a clear sky, depending on the height and temperature of the overlying cloud layer. The lower the height of the overlying cloud layer, the greater was the reduction in net outgoing longwave radiation. Saunders also concluded that the observed rise in temperature near the ground following the advection of a cloud deck over an overlying fog was a consequence of the reduction of outgoing radiation of the fog top. The temperature rise was also due in part to sustained upward heat flux from the ground. Because the upward heat flux from the ground is important in dissipating fog, another important controlling factor is the gradient in temperature between the soil and the air. If the soil-air temperature gradient is small, fog clearance is unlikely. If it is large, fog clearance is greatly enhanced when a cloud layer moves over the fog. Brown and Roach (1976) simulated the effects of advection of cloud cover over a fog by increasing the downward longwave radiation flux incident at the model top boundary from 30 to 95 W $m^{-2}$. This value, they argued, was appropriate to a stratocumulus of 600 (m) thickness and a base at 700 m in an atmosphere similar to one of the cases they had observed. Figure 6.4 illustrates the model-predicted changes in temperature and liquid-water content through the fog layer. Over a period of 2 hours the layer warmed and the liquid-water content decreased at all levels except near the top. They attributed the failure of the model to completely dissipate the fog, to weaknesses in their turbulence model. Nonetheless, their model shows quite clearly that the overlying cloud deck reduces the net radiative

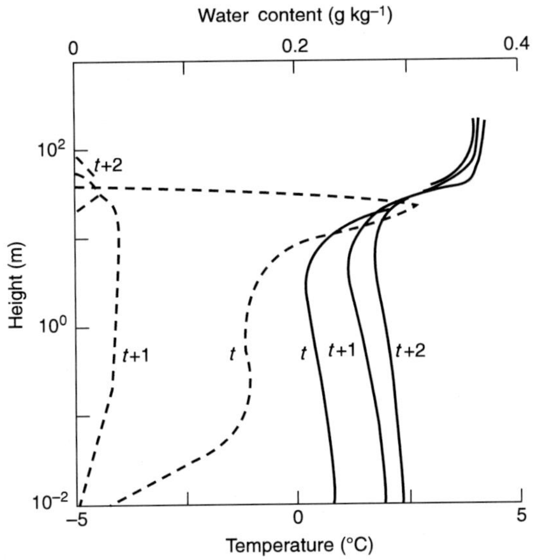


**FIGURE 6.4 Liquid-water content and temperature at times $t + x$ hours, where $t$ is the time of advection of cloud cover: dashed lines, liquid-water content; solid lines, temperature.** *(From Brown and Roach (1976). Reproduced with the permission of the Controller of Her Britannic Majesty's Stationery Office)*

cooling at the fog top to such an extent that the eddy transport of heat from the surface will drive the layer above water saturation.

## 6.3.7. Variability in Nocturnal Fogs

Once a nocturnal fog reaches maturity, the fog properties do not necessarily remain constant. Choularton et al. (1981) described periodic fluctuations in fog liquid-water content observed in Meppen, Germany. They noted that the liquid water fluctuated with periods of 51, 31, and 96 s, respectively, for 15- to 20-min intervals. They speculated that the periodicities were due to the passage of Bernard cell-type convection past the observation site.

Longer period oscillations in fog liquid-water content of the order of 45 min to 1 h were found by Lala et al. (1982) in observations of fog in Albany, New York. The source of the periodicity was not determined, however. In their simulations of the Albany fog cases, Welch et al. (1986) simulated a series of pronounced oscillations in the fog parameters, including liquid-water content. The period of the oscillation found in the model was on the order of 30-40 min, although the length of the period varied considerably with the particular formation of the turbulence closure model. The oscillation was associated with a surge in eddy mixing, which promotes entrainment of warm, dry air from the inversion top into the fog layer, and an upward mixing of moist air from near the surface. The result is a rapid decrease in liquid-water content in the lower 20 m.

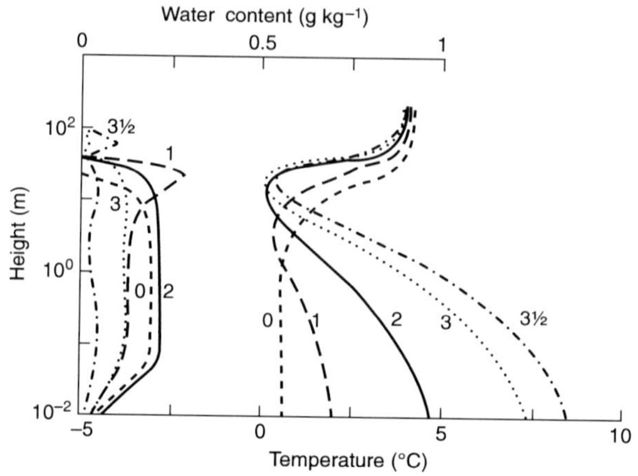


**FIGURE 6.5 Development of the model-predicted liquid-water and temperature profiles after sunrise: (- - -) sunrise; (– – –) 1 h; (——) 2 h; (. . .) 3 h; (– · –·) 4 h.** *(From Brown and Roach (1976). Reproduced with the permission of the Controller of Her Britannic Majesty's Stationery Office)*

Such longer period oscillations may be similar to those observed and modeled in stratocumulus clouds, as described later. Chen and Cotton (1987) attributed the oscillations in stratocumulus cloud simulations to interactions among radiation, turbulence, and droplet settling.

## 6.3.8. Dispersal of Fog by Solar Insolation

Brown and Roach (1976) investigated, with a numerical model, the dispersal of fog by solar heating. They introduced a simple shortwave radiative transfer model in which solar absorption in the fog was omitted. Fogs are so optically thin, they argued, that the solar heating rate is inconsequential relative to longwave radiative cooling and turbulent transport of heat. The primary feature modeled was the absorption of solar radiation at the earth's surface with about 30% of the incident solar beam reflected by the fog layer and 20% by the earth's surface. The predicted evolution of the liquid-water and temperature profiles after sunrise are shown in Fig. 6.5. A major difference between the solar radiation case and the case with overlying cloud cover is that longwave radiative cooling is not diminished within the fog as a consequence of the sunrise. Thus, for a while, longwave radiative cooling prevails over solar heating and the fog liquid-water content continues to rise after sunrise. After 3.5 h following sunrise, heat flux driven by solar heating of the surface prevails and the simulated fog is completely dissipated except for levels above 22 m. This deficiency of the model was again attributed to the inadequacy of the turbulence model. Brown and Roach also felt that their observations did not support the simulated mode of dissipation, namely from the surface upward. They argued that in this case the fog appeared to clear from above. Thus, their neglect of solar absorption in the fog layer may not have been justified. Welch et al. (1986) conclude that fog intensification after sunrise is caused by enhanced mixing of upper-level liquid to the surface as a result of increased turbulence due to surface heating. Evaporation of dew and soil moisture also supplies additional moisture. Eventual dissipation of the fog occurs when the effect of surface heat flux on the relative humidity exceeds the effect of the supply of moisture from the surface or mixed downward from the fog top. Using a two-dimensional model, Forkel et al. (1987) predicted that fog dissipation occurs more quickly in a polluted atmosphere due to larger solar heating rates within the fog caused by the larger concentrations of aerosol.

## 6.4. VALLEY FOG

Valley fogs are also radiation fogs and are therefore regulated by all the physical processes discussed in the preceding section. In addition, valley fogs are influenced by organized slope circulations that develop in response to radiative cooling and drainage of cooler air down the slopes of the valley, where the cool air is pooled into the valley basins. Pilie et al. (1975) and Pilie (1975) reported on an extensive study of the micrometeorological and microphysical characteristics of valley fogs near Elmira, New York. They interpreted their observations with respect to Defant's (1951) conceptual model of nocturnal mountain valley circulations shown in Fig. 6.6. According to their interpretation, the life cycle of a valley fog is as follows. Radiative cooling over the slope of the valley initiates a downslope wind and an upward return flow in the valley center region, as illustrated in Fig. 6.6a.

This process contributes to the formation of a deep nocturnal inversion. Several hours after sunset the mountain wind is established by drainage of cool air down the axis of the valley. The speed of the wind is a maximum at levels of 40-200 m, with speeds of 2-4 m $s^{-1}$. As the mountain wind matures, downslope winds prevail throughout the valley (see Fig. 6.6c) Finally, in the late night hours, the mountain wind occupies the entire valley and continues until after sunrise (see Fig. 6.6d).

Pilie et al. (1975) then explain the formation of deep valley fogs as follows:

1. Nocturnal radiation from the surface and subsequent turbulent heat transfer from air to ground, which produces an initial low-level temperature inversion, stimulates the downslope wind and upward return flow near the valley center. During the period, dew deposition at the cold surface creates the low-level dew-point inversion. The upward motion at the valley center carries the cool and somewhat dry air aloft to cause the inversion to deepen.
2. Approximately 3 h before fog formation, the mountain wind forms, providing continuity for the downslope wind while restricting the upward motion of air near the valley center. Cooling is therefore restricted to low and middle levels of the valley, i.e. those levels in which fog will eventually

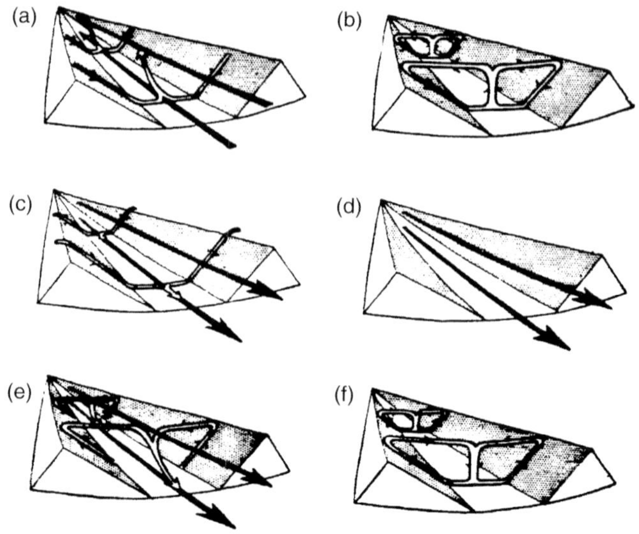


**FIGURE 6.6** **(a) Downslope wind begins shortly after sunset before up-valley mountain wind dies. (b) In late evening, up-valley wind dies and only downslope wind and return flow at center of valley exist. (c) Return flow at center of valley ceases and down-valley mountain wind becomes established. (d) Late at night the downslope wind ceases and the down-valley mountain wind persists. (e) Sunrise; onset of upslope winds (white arrows), continuation of mountain wind (black arrows). Valley cold, plains warm. (f) Forenoon (about 0900); strong slope winds, transition from mountain wind to valley wind. Valley temperature same as plains.** *(From Defant (1951) and Pilie et al. (1975))*

form. The continuing downslope wind mixes with warmer air at midlevels in the valley and causes the cooling rate to maximize in that region. Through this period, the dew-point inversion persists. Temperature and dew point, therefore, converge at midlevels, and a thin layer of fog forms aloft.

The subsequent downward propagation of the fog is then envisaged to occur in a similar manner to that of a "pure" radiation fog as discussed in Section 6.3.

Pilie et al. (1975) also noted that on several occasions fog formed following sunrise. On these occasions they suggested that surface warming and dew evaporation following sunrise cause vertical mixing of moist low-level air with cooler air in the valley center aloft. This suggests that the vertical mixing mechanisms discussed in Section 6.1 and the adiabatic cooling of rising air parcels contribute to the fog formation.

Slope/mountain or valley circulations that develop in response to solar heating of exposed valley walls or hills, such as illustrated in Fig. 6.6d, f, contribute to the eventual dissipation of the valley fog. As subsidence develops in the valley center in response to the slope circulation, the resultant adiabatic warming evaporates the fog. At the same time, slightly elevated hills in the valley center are the first regions where fog dissipates, because the optical depth of the fog is least over the tops of the hills. The solar insolation is able to penetrate the thinner fog layer, thus warming the higher ground. As the hills become exposed, slope circulations develop, with corresponding compensating sinking motions over the neighboring foggy lower terrain. Dissipation of the fog may then proceed quite rapidly. This is an area ripe for quantitative study that uses turbulence/radiation models coupled to mesoscale flow models.

## 6.5. MARINE FOG

Marine fog differs from the radiation fogs we have discussed in that radiation does not rapidly alter the surface of the ocean as it does the land. Surface fluxes of moisture and heat are important in marine fog formation, but their fluxes are not affected significantly (on the time scale of the diurnal cycle) by variations in solar and terrestrial radiation. Another factor of importance to marine fog is that the air mass, often of maritime origin, contains fewer active cloud condensation nuclei. Giant sea salt nuclei also may represent a substantial fraction of the activated aerosol in the low supersaturations encountered in fogs. Compared to the generally more "continental" radiation fogs, marine fogs are more prone to drizzle formation with the potential for radiation/drop settling influences on the fog structure, as discussed in Section 6.3.4.

### 6.5.1. Fog Formation in a Turbulence-Dominated Marine Boundary Layer

Because the surface of the ocean provides a source of moisture, fog may form over the ocean surface without significant radiative cooling. That is, the fog formation mechanisms are purely a function of the surface fluxes of heat, moisture, and momentum, and of the gradients of temperature, moisture, and wind in the overlying air mass. (Oliver et al., 1978; hereafter referred to as OLW) examined the existence criteria for a turbulence-dominated fog in the surface layer. The surface layer is a thin (~10 m) layer next to the ground in which turbulent fluxes of water, momentum, and heat are assumed to be constant. It is also a region in which the Monin-Obukhov similarity theory, or adiabatic similarity theory, is applied. That is, the fluxes and mean gradients within the surface layer can be related to distance away from the surface layer and to surface fluxes of heat, moisture, and momentum.

In their analysis, OLW employ the conservative thermodynamic variable $\theta_s$ which they define as

$$\theta_s = T - T_0 + \boldsymbol{\Gamma} z + (L_c/C_{pa})r_v, \tag{6.2}$$

where $\boldsymbol{\Gamma}$ represents the dry adiabatic lapse rate ($\boldsymbol{\Gamma} = g/C_{pa}$) and $T_0$ is the reference state temperature. The behavior of $\theta_s$ is similar to the behavior of $\theta_e$.

They also define a virtual potential temperature $\theta_v$ as

$$\theta_v = T - T_0 + \boldsymbol{\Gamma} z + (0.61r_v - r_c)T_0. \tag{6.3}$$

To denote surface properties, we use a double zero subscript, so surface temperature is denoted as $T_{00}$. The surface friction velocity and nondimensional surface heat and moisture fluxes are defined as

$$U_* = \overline{u''w''^{1/2}}|_{00}; \qquad \theta_{v}* = \overline{w''\theta_v''}|_{00}/U_*; \qquad r_* = \overline{w''r''}|_{00}/U_*. \quad (6.4)$$

In the surface layer the gradients of the mean variables can be expressed in terms of the diabatic profile functions: $\phi_u(\xi)$, $\phi_\theta(\xi)$, $\phi_r(\xi)$ as

$$\partial\overline{u}/\partial z = (u_*/kz)\phi_u(\xi), \quad (6.5)$$
$$\overline{\partial\theta_v}/\partial z = (\partial\theta_{v*}/kz)\phi_\theta(\xi), \quad (6.6)$$
$$\partial\overline{r}/\partial z = (r_*/kz)\phi_r(\xi), \quad (6.7)$$

where $k$ is the von Karman constant and $\xi = Z/L_m$. The parameter $L_{rn}$ is a generalized form of the Monin-Obukhov length defined as

$$L_m = -u_*^2 T_0/kg\theta_{v*}. \quad (6.8)$$

OLW generalized $L$ to include the virtual heat flux from the ocean surface. The original definition of $L$ by Obukhov (1946) included only the buoyancy effects of surface heat flux.

The behavior of the parameter $\xi = Z/L_m$ is analogous to the behavior of the Richardson number, which is the ratio of the rate of buoyant production of turbulence to shear production. Thus, if $\xi < 0$, the surface is characterized by a positive surface buoyancy flux, and buoyant production of turbulence dominates for large negative values of $\xi$. By contrast, a stably stratified surface layer will be characterized by $\xi > 0$. A large and positive value of $\xi$ implies that buoyancy is dominant over the mechanical production of turbulence. Because buoyancy accelerations damp the generation of turbulence in a stable atmosphere, large positive values of $\xi$ imply that turbulence will not be very active. However, small positive values of $\xi$ imply that mechanically generated turbulence will dominate the buoyant damping of turbulence.

In the absence of radiative effects in a surface-layer-bound fog, the turbulent correlations of $\theta_s$ and $r$ are identical. The distributions of $r - r_{00}$ as well as their turbulent correlations are similar to one another and can be represented as a function of $z/L_m$ and $z/z_{00}$, where $z_{00}$ is the surface friction height.

According to OLW, similarity theory leads to the following theorem as noted by Taylor (1917):

*"All fluid states in a surface layer in turbulent interaction must map onto a straight line in the r-θ plane which passes through the surface state ($r_{00}$, $\theta_{s00}$) and has a slope determined by the surface humidity to heat flux ratio $r^*/\theta_{s^*}$."*

Thus, OLW say that the existence criteria for a fog in the surface layer dominated by turbulence can be determined from Fig. 6.7, which is a

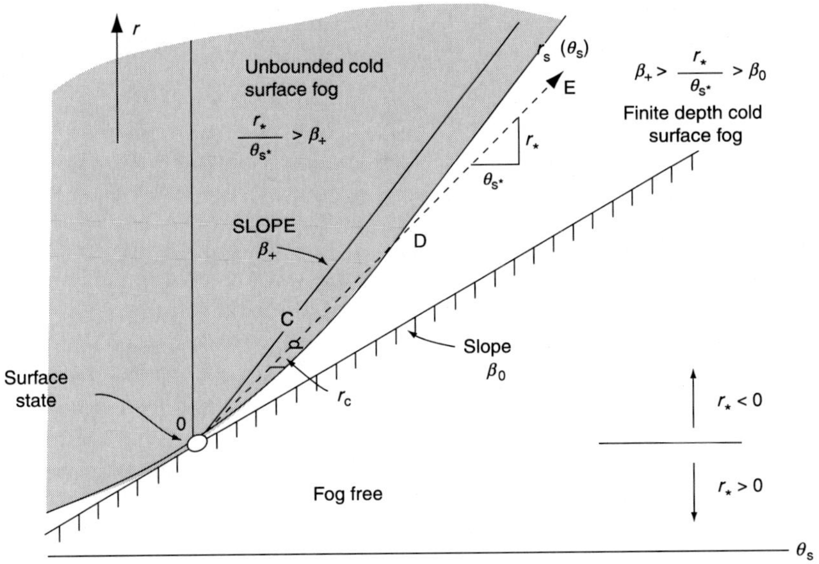


**FIGURE 6.7 Existence diagram for turbulence-dominated fog.** Slope $\rho_0$ of the saturation function $r_s$ in $\theta_s$ space determines the critical $r_*/\theta_{s*}$ for surface-layer distributions OCDE, which will be foggy or fog free. For a typical foggy layer, fog bank height is at D and maximum liquid-water content at C. *(From Oliver et al. (1978))*

generalization of Fig. 6.1. The surface state is denoted by the point "O" and the straight line OCDE represents the distribution of $r$ and $\theta_s$ throughout the surface layer. The variation of the saturation mixing ratio with $\theta_s$ derived from the Clausius-Clapyron equation is also shown in Fig. 6.7. (Note: OLW neglected variations in $r_s$ with pressure for shallow surface-layer applications.) The tangent to the saturation curve at the surface is denoted $\beta_{00}$ and defined as

$$\beta_{00} = (\partial r_s/\partial \theta_s)_p. \tag{6.9}$$

One can also relate $\beta_{00}$ to the logarithmic derivative of the saturation mixing ratio

$$\beta_T = (\partial \ln r_s/\partial \ln T)_p \tag{6.10}$$

as

$$\beta_{00} = (r_s/\tilde{\mu} T_0)\beta_T, \tag{6.11}$$

where

$$\tilde{\mu} = 1 + (R_v/C_{pa}) r_s \beta_T^2. \tag{6.12}$$

For saturated conditions at the surface, any parcel trajectory will be a straight line emanating from the surface point $r_{00}, \theta_{s00}$. Any parcel trajectory moving below the line $\beta_0$ will be a fog-free trajectory.

Thus, surface fog is *not possible* if

$$\theta_{s*} < 0, \qquad r_*/\theta_{s*} < \beta_0, \tag{6.13}$$

$$\theta_{s*} > 0, \qquad r_*/\theta_{s*} > \beta_0. \tag{6.14}$$

The saturation mixing ratio $r_s(\theta_s)$ becomes asymptotic to a straight line with slope $\beta_+ = r_{00}/\theta_{s00}$ for large $\theta_s$ (see upper part of diagram).

In the region between $\beta_+$ and $\beta_0$, surface fog may exist next to a cold surface and through the depth OCD on ray OCDE, with the top being at D. The maximum liquid-water content is at point C. Above the line $\beta_+$ cold surface fog exists, but it extends beyond the surface layer.

In the region $\theta_{s*} > 0$, warm surface fog exists. Such a fog may be bounded as OCDE or unbounded within the surface layer. OLW showed that surface fog existence criteria could be assessed in terms of the surface flux ratio $r_*/\theta_{s*}$, the saturation curve at the surface:

$$(r_*/\theta_{s*})_{\text{crit}} = \beta_0. \tag{6.15}$$

According to OLW the existence criteria may be summarized:

*"For a cold surface fog ($\theta_{s*} < 0$) a large humidity flux and small heat flux promote fog formation. Thus, dry nearly adiabatic air overrunning a cold water surface forms fog easily. On the other hand, for a warm surface fog ($\theta_{s*} > 0$), the flux ratio must be smaller in magnitude than the critical value. Hence, cold air with high relative humidity overrunning a warm surface will tend to fog."*

OLW also examined conditions when turbulence will dominate over radiation in fog formation. They concluded that radiation plays a significant role in most surface fogs of depths greater than a few meters. Nonetheless, these existence criteria can be useful for diagnosing the initiation of a radiation/turbulence fog.

## 6.5.2. The Advection-Radiative Fog

In this section we are concerned with a deeper type of fog in which turbulence and radiative processes play an important role. Pilie et al. (1979) hereafter referred to as P79 describe a case of marine fog formation off the western coast of California on 30 August 1972. The fog occurred as air blew over cold water to a patch of relatively warm water. Figure 6.8 illustrates the changes in the low-level temperature profiles in the clear region over cool water, at the edge of the fog, and in the fog layer.

As can be inferred from Fig. 6.8, upwind of the fog, heat was being transferred from the warmer air to the cooler ocean surface. Within the fog

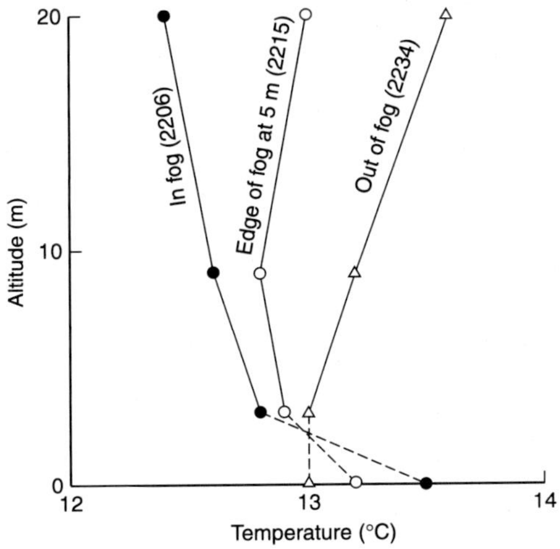


**FIGURE 6.8 Selected vertical profiles at indicated positions relative to fog edge, 30 August 1972.** *(From Pilie et al. (1979))*

layer, however, heat was being transferred from the warmer ocean surface to a "cooled" fog layer. P79 suggested that resultant mixing of the warm surface air over the warmer ocean and the near-saturated cool air produced the initial fog in accordance with the principles outlined in Sections 6.1 and 6.4 (i.e. Figs 6.1 and 6.7). Radiative cooling of the top of the fog layer enhances the low-level instability and further promotes the turbulent transport of heat and moisture. The cool fog represents a sink of moisture which enhances evaporation from the warm sea surface. This is an excellent example of the interaction between turbulence and radiative processes. The fog grows progressively deeper in the downwind direction as vertical eddy transport provides the moisture for further fog formation and radiation cooling lifts the inversion and encourages further eddy transport.

OLW developed a turbulence model of the growth of such an advective-radiative fog. Their model is a second-order transport turbulence model with radiative transfer processes included. The second-order transport model is closed with an eddy diffusion model on the third-order terms. The model also allows for cloud fractional coverage less than unity using a simple scheme. A simple shortwave and longwave radiative transfer model is employed which is based on the transmittance approach as described in Chapter 5. In their simulation of an advective-radiative fog with neglection of solar absorption, OLW simulated an air mass advecting over a warmer water surface by heating the surface by 0.5 °C while the air mass moves about 3 km. The predicted changes in liquid-water content and temperature are shown in Fig. 6.9. The fog first forms at the surface and builds upward. When the fog has deepened to a depth of 100 m, radiative cooling becomes significant enough to generate a secondary maximum of liquid-water content.

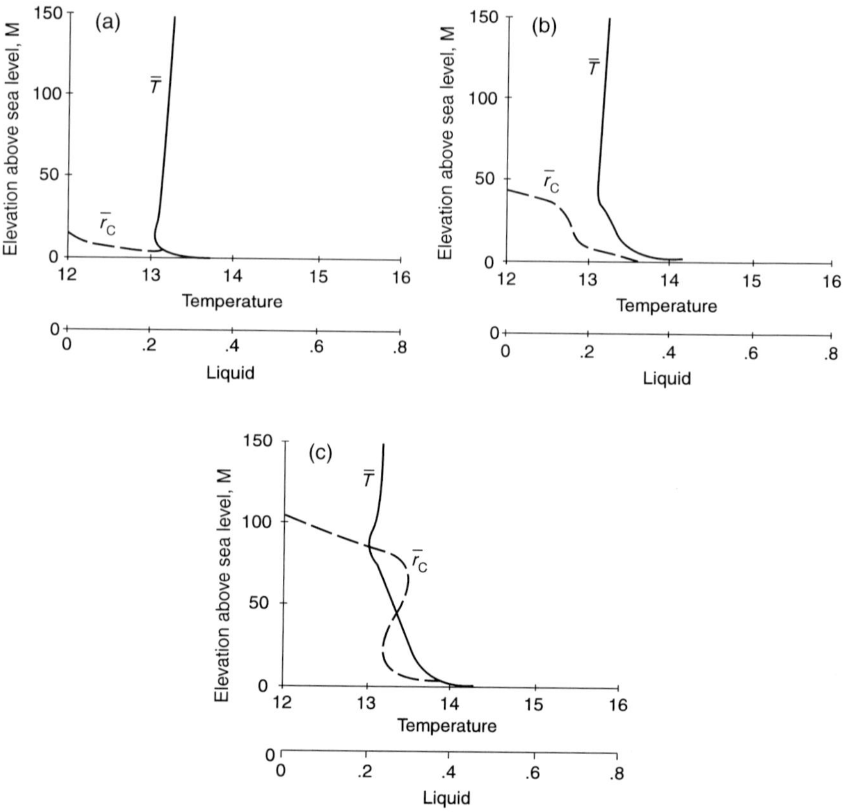


**FIGURE 6.9** **Temperature and liquid-water content profiles in an advective-radiative fog at $x = 100$ m (a), $x = 1000$ m (b), and $x = 3000$ m (c). $\overline{T}$, temperature (°C) ; $\overline{r}_c$, liquid water (g kg$^{-1}$).** *(From Oliver et al. (1978))*

## 6.5.3. Marine Fog Formation by a Stratus-Lowering Process

We shall reserve our discussion of the stratus-forming mechanisms until a later section. Before we leave the subject of fogs, however, we should note that, on occasion, fog forms as a result of the lowering of the base of stratus clouds. Anderson (1931) suggested that fog can form as a consequence of cloud-top radiation cooling, which causes instability in the cloud layer. This instability results in the downward development of the cloud base, in some cases to the surface. It has been noted by several authors that a condition necessary for stratus formation and lowering of the base requires that the inversion base must rise above the lifting condensation level.

P79 postulated the following conceptual model of the stratus-lowering process, shown schematically in Fig. 6.10. Net radiation from the stratus top causes rapid cooling that generates instability beneath the inversion. This causes

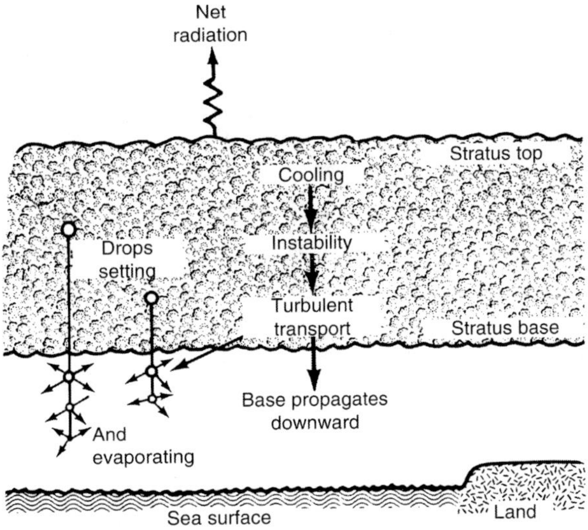


**FIGURE 6.10 Schematic representation of fog formation through the stratus-lowering process.** *(From Pilie et al. (1979))*

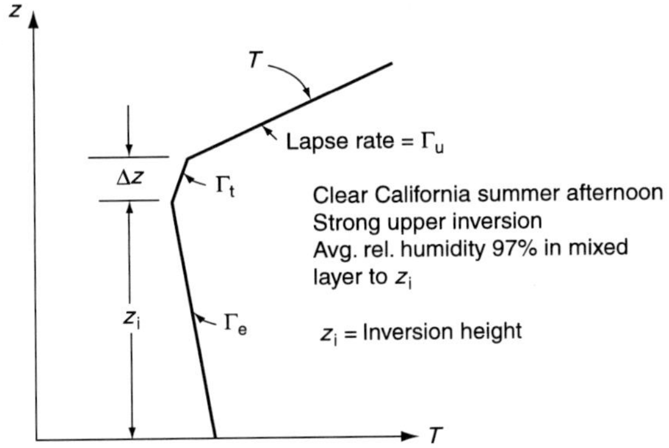


**FIGURE 6.11 Typical late afternoon temperature profile measured by Mack et al. (1974).** *(From Oliver et al. (1978))*

a turbulent transport downward of cool air and cloud droplets. Evaporation of droplets beneath cloud base causes cooling and an increase in humidity. This leads to a further lowering of cloud base.

Additional insight into the stratus-lowering process can be derived from the numerical experiments described by OLW. They commence a simulation with a sounding such as shown in Fig. 6.11 derived from data reported by Mack et al. (1974). The height of the mixed layer $z_i$ was varied from experiment to experiment. The experiments were all begun at 1900 local time. For $z_i$ equal to 300 m, the cloud base propagated downward until 0500, where it reached

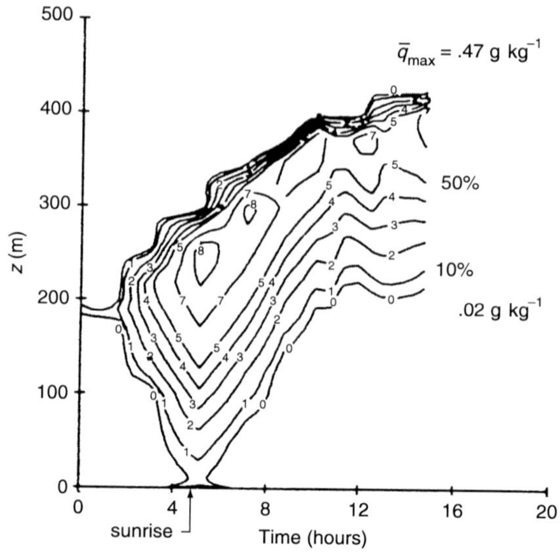


**FIGURE 6.12 Evolution of fog resulting from the lowering of stratus.** Fog shows some tendency to form at surface just before downward-propagating stratus reaches the surface. *(From Oliver et al. (1978))*

150 m above the surface. During the same period the top propagated upward to 450 m.

With $z_i$ initially at 200 m, the stratus base descended to the surface by 0500, and there it remained until 0700. Figure 6.12 illustrates the predicted time-height evolution of liquid-water content. The results are consistent with observations reported by Leipper (1948) and P79 which indicated that stratus lowering did not occur until the later afternoon or early evening inversion height was 400 m or lower.

A feature of the OLW simulation is that the turbulent moisture flux profiles are positive at all levels and at all times. Moreover, the model does not contain a formulation of precipitation processes. Thus, the results of the model suggested by P79 are not in accordance with OLW's model results. Perhaps the conceptual model of marine stratocumuli suggested by Schubert et al. (1979a, see Fig. 6.18) is applicable to the stratus-lowering process. We shall discuss this model in a later section.

The dissipation of surface fog following sunrise in OLW's model is a result of the radiative heating/cooling profiles predicted by the models shown in Figs 6.13 and 6.14. They note that direct solar heating is absorbed well into the interior of the fog, and that evaporation caused by solar heating occurs rather uniformly throughout the cloud because of the long absorption length for solar radiation (see Chapter 5). At the same time, infrared radiative cooling near cloud top causes instability and turbulence, which transports the shortwave warming throughout the cloud. As noted in Chapter 5, the magnitude of the solar

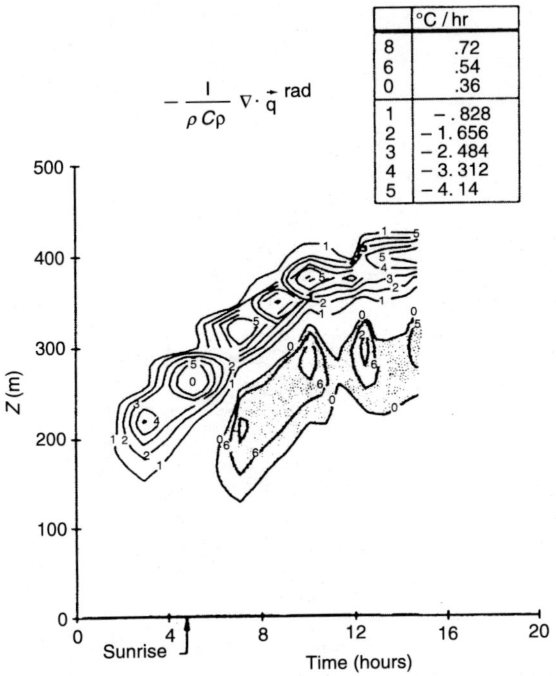


**FIGURE 6.13 Radiative cooling/heating distribution in a stratus-lowering fog.** Maximum cooling is concentrated at cloud top. After sunrise, net heating occurs, but deep within the cloud. *(From Oliver et al. (1978))*

absorption length is dependent upon the particular radiation parameterization used. It appears that OLW's parameterization exaggerates fog dissipation by solar heating. The drizzle process, however, may accelerate fog dissipation by removing the LWC near cloud top. This would allow shortwave radiation to penetrate more deeply into the interior of the fog where the radiation is absorbed more effectively due to the presence of large drops (see Chapter 5). This process should be studied quantitatively, however.

## 6.5.4. Fog Streets and Low-level Convergence

Marine fog does not always have a horizontally homogeneous structure. P79 refer to observations of what they call *fog streets*. That is, the fog is organized in alternating parallel lines of foggy and clear air. Figure 6.15 illustrates the observed changes in visibility as the ship *Acania* cruised crosswind through a region characterized by fog streets. They noted that the individual fog patches range in width from 0.5 to 2 km. Along the direction of the wind, the fog patches appeared with the upwind edges of the fog touching the surface. At distances of several hundred meters to several kilometers from the upwind edge, the fog base lifted from the surface and persisted as a stratus deck aloft. An example

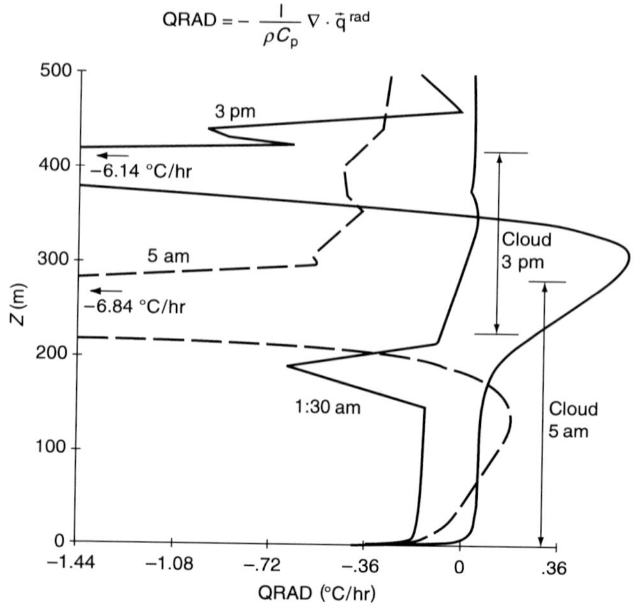


**FIGURE 6.14 Profiles of radiative cooling/heating at 0130 (stratus just forming), 0500 (stratus on the surface), and 1500 (stratus base lifted).** *(From Oliver et al. (1978))*

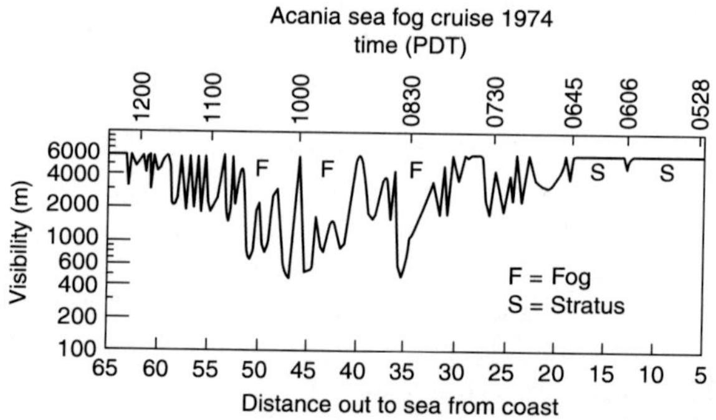


**FIGURE 6.15 Visibility data as a function of distance obtained while cruising crosswind through a series of fog patches, 23 August 1974.** *(From Pilie et al. (1979))*

of fog streets described by Walter and Overland (1984) is given in Chapter 7, Section 7.2.6. Mechanisms exhibiting a multiplicity of band scales responsible for the formation of such bands are also given in that section.

P79 also describe a case in which surface convergence appeared to be instrumental in fog formation. Upwind of the fog, the air was warmer than the sea, while within the fog it was cooler than the sea. Winds upwind of the fog were out of the northwest at 4-8 m $s^{-1}$, while within the fog the speeds were consistently lower and, in some locations, were reversed in direction.

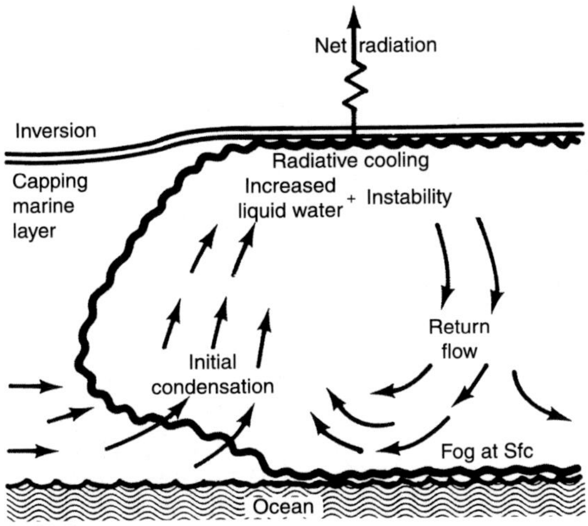


**FIGURE 6.16 Schematic representation of the vertical cross section of fog formed as a result of low-level convergence and radiative cooling.** *(From Pilie et al. (1979))*

Divergence values ranged from $-0.7 \times 10^{-4}$ to $-2.7 \times 10^{-4}$ $s^{-1}$ and persisted for at least 20 h over a region of approximately 2500 $km^2$. P79 estimated average vertical velocities over the foggy region to be 1-2 cm $s^{-1}$. However, they suggested that organized patterns of updrafts and downdrafts Fig. 6.16 existed within the area having significantly stronger magnitudes than the average. They postulated that adiabatic cooling within the local updrafts contributed to the fog liquid-water. Radiative cooling at fog top further contributed to fog liquid-water near the top, thus producing sufficient condensate to survive evaporative warming in descending drafts to the surface. Their conceptual model of such a convergence-driven, convective radiation fog is illustrated in Fig. 6.16.

This case is an excellent example of the interaction of mesoscale circulations, turbulence, and radiative processes in the initiation and maintenance of marine fogs.

### 6.5.5. Numerical Prediction of Fogs

Numerical weather prediction (NWP) guidance to fog forecasting has mainly been limited to the use of single-column models (Bergot and Guedalia, 1994; Bott and Trautman, 2002; Duynkerke, 1991; Teixera, 1999). This is due to the fact very fine vertical resolution is required for fog prediction, particularly radiation fog. Moreover, the physics required to simulate radiation fog, including detailed microphysics with radiation influences on droplet vapor deposition growth and the sedimentation of fog droplets, is not available in most cloud microphysics parameterizations in NWP models. The single-column models are either initialized with local soundings or from large-scale forecast fields such as done by Teixera (1999) using ECMWF

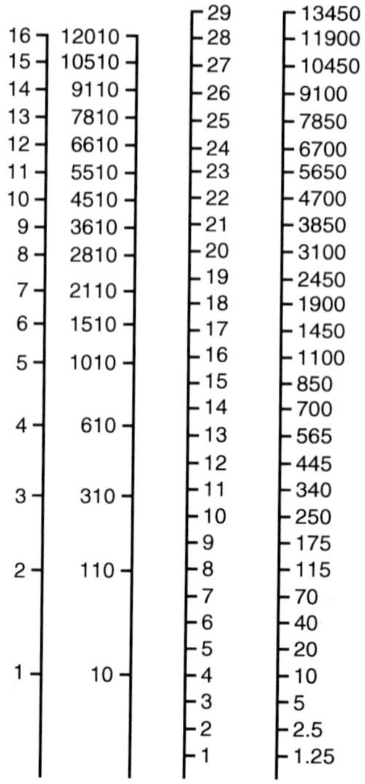


**FIGURE 6.17 Heights (m) of mesoscale model levels.** *(From Ballard et al. (1991))*

forecast fields. Single-column model forecasts are only useful in rather flat, homogeneous terrain. They are most useful when a forecaster experienced in local micrometeorological circulations interprets the single-column model predictions in relation to local phenomena.

There have been only a few prototype attempts to forecast fog with three dimensional NWP models. Ballard et al. (1991) describe the use of the United Kingdom Meteorological Office (UKMO) mesoscale model to fog forecasting. The model was setup with 15 km horizontal grid spacing and with either 16 or 29 vertical levels as shown in Fig. 6.17. The 29-level version has much finer vertical resolution than is typically used in most NWP models. Boundary conditions were obtained from the operational UKMO model with 75 km grid spacing. They found that the use of higher vertical resolution greatly improved the forecasts. Not surprisingly, the quality of the forecasts depended on the quality of the initial humidity and cloud water profiles, as well as the accuracy of the synoptic forecasts.

In another study Chibe (2003) applied the Regional Atmospheric Modeling System (RAMS; Cotton et al., 2003) to the simulation of several fog events. RAMS was set up using interactive nested grids with three grids, with a coarse grid having 50 km spacing in which the synoptic fields were initialized, a second grid with 10 km spacing, and a fog grid with 2 km spacing. The fog grid had

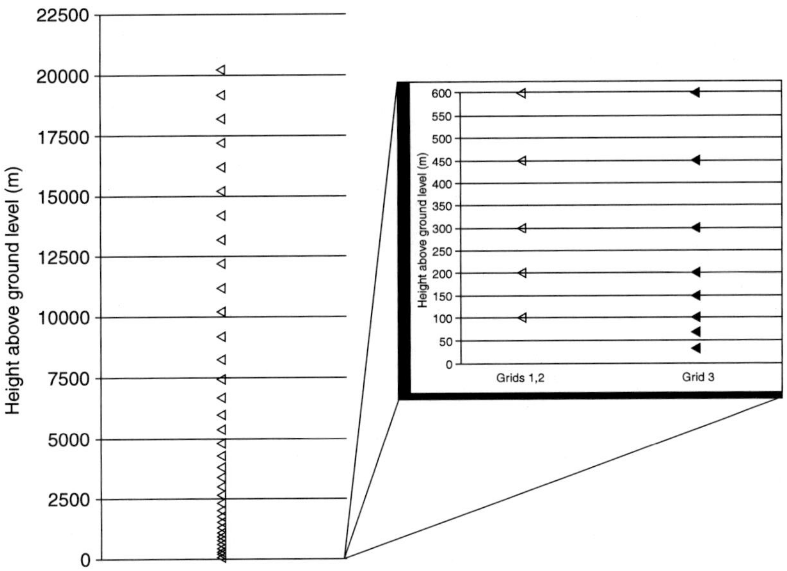


**FIGURE 6.18 The 34 vertical levels for grids one and two are depicted by the white arrows in the left-hand column. The insert show the lowest 600 m of the model, with the configuration for grids 1 and 2 illustrated by the white arrows on the left-hand side of the insert, and the insert, and the nested configuration for grid 3 illustrated by the black arrows on the right-hand side of the insert.** *(From Chibe (2003))*

refined vertical grid spacing as shown in Fig. 6.18, although not nearly as fine as used by Ballard et al. (1991). The cases simulated included a valley fog in central California and radiation fog in eastern Wisconsin. Errors in timing up of to an hour were found for the valley fog case. For the Wisconsin radiation fog, patches of fog were simulated while the observed fog was rather homogeneous in the region. The patchy nature of the forecast fog was attributed to errors in wind speed of about 0.5 m $s^{-1}$, where solid fog coverage was predicted if winds were below 2 m $s^{-1}$, and cleared regions where it exceeded it. This illustrates just how challenging fog forecasting is as very slight changes in fog speeds, and in some locations with local moisture sources, wind direction, can have a large influence on fog coverage and visibility.

## 6.6. STRATOCUMULUS CLOUDS

### 6.6.1. The Stratus-topped Boundary Layer

Low-level marine stratocumulus clouds occupy large portions of the eastern Pacific and eastern Atlantic oceans and small portions of the western Indian Ocean in regions of climatologically preferred positions of subtropical high pressure. Figure 6.19 illustrates a typical profile of temperature, moisture, and winds in an eastern North Pacific stratocumulus regime. The layer is

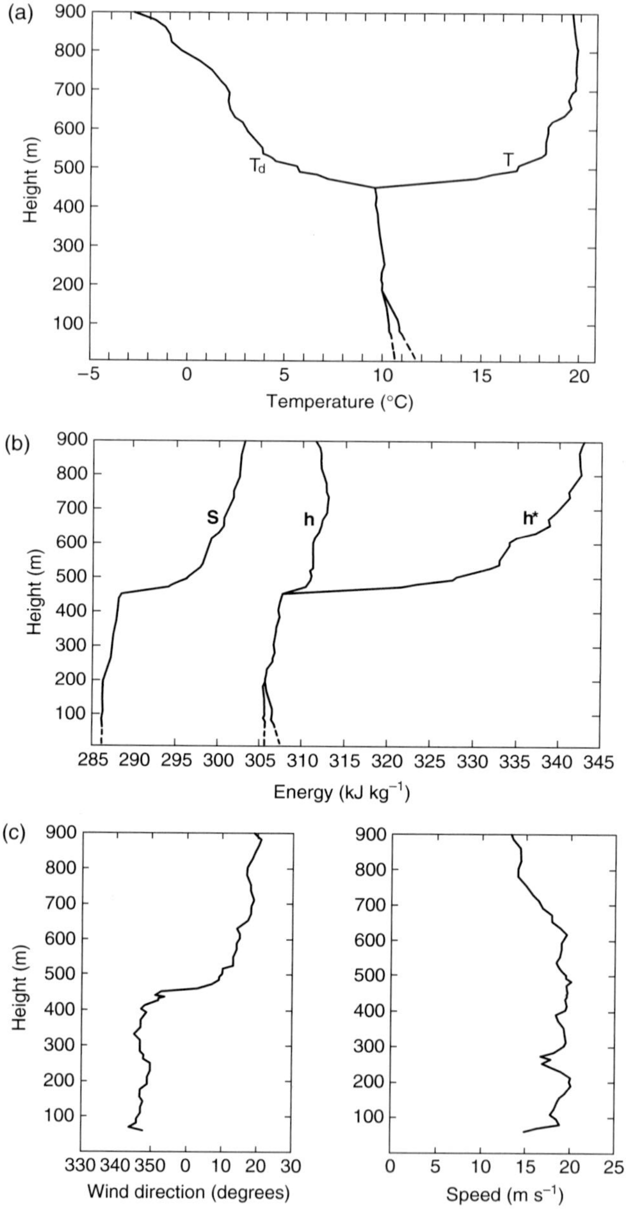


**FIGURE 6.19 Temperature, moisture, and wind data from an NCAR Electra sounding at 37.8°N, 125.0°W and between 1522 and 1526 GMT; (a) temperature and dew point; (b) dry static energy, moist static energy, and saturation static energy; (c) wind direction and speed. Dashed lines below 50 m are extrapolations.** *(From Schubert et al. (1979a))*

characterized by a nearly adiabatic, well-mixed subcloud layer (dry static energy, $s = c_pT + gz$, is constant, or $\theta$ is constant) and by a wet adiabatic cloud layer (moist static energy, $h = s + Lr_T$, or $\theta_e$, is constant), capped by a strong temperature inversion and drop in dew point. Winds through the entire mixed layer are nearly uniform. There are many questions that must be answered regarding the behavior and structure of stratocumulus clouds. Some of these are as follows:

1. How does a stratocumulus layer maintain a steady depth against the effects of subsidence which cause a cloud layer to become shallower?
2. What is the nature of entrainment into the top of a stratocumulus layer?
3. What are the relative roles of buoyancy and shear in generating turbulent kinetic energy in a cloud layer?
4. How important is cloud-top radiative cooling to the maintenance of a stratocumulus layer?
5. How important is drizzle to the dynamics and radiative properties of a stratocumulus layer?
6. What causes the breakup of a solid stratocumulus layer to form a field of cumuli?

In order to examine questions such as these as well as others, a hierarchy of models has been proposed to describe and predict the structure of marine stratocumulus clouds. These include one-dimensional layer-averaged or mixed-layer models, entity-type or plume models, higher ordered closure models, and large-eddy simulation models. Each of these modeling approaches offers certain advantages and disadvantages.

### 6.6.2. Layer-averaged Models

The primary advantage of layer-averaged models is their simplicity. They do not consume a great deal of computer time and, moreover, they yield a clear "signal" of the response of the simulated cloud layer to various physical processes. This is in contrast to higher ordered closure models or LES models, where the response of entrainment to changes in environmental parameters, for example, may involve a complex chain of events through triple correlation products or the ensemble-averaged effects of a large number of explicitly simulated convective cells. The primary disadvantage of the layer-averaged models is their lack of versatility. They are generally restricted to well-mixed layers where vertical shear of the horizontal wind is weak. Departures from a well-mixed state violate the fundamental premise of such models. The presence of drizzle, for example, was found by Nicholls (1984) to violate the fundamental mixed-layer hypothesis in some cases. Moreover, extension of layer-averaged models to include the effects of wind shear or the presence of a broken cloud field requires parameterizations or additional modeling assumptions.

The pioneering paper on this topic was published by Lilly (1968). He considered the effects of condensation and evaporation, large-scale vertical motion, and divergence of net radiation at the cloud top. The basic assumptions in his model are as follows:

(1) Below the base of the capping inversion, the boundary layer is well mixed, or uniform, in mean values of semiconservative properties such as total-moisture specific humidity ($q_W$) and wet-bulb potential temperature $\theta_W$ or equivalent-potential temperature $\theta_e$.
(2) The capping inversion is of negligible thickness.
(3) Turbulence in the mixed layer is generated entirely by buoyant production (the effects of wind shear are ignored).
(4) The upper cloudy portion of the mixed layer is entirely saturated, and the buoyant flux of entrainment occurs entirely in the saturated air.
(5) There is no precipitation or drizzle.
(6) The divergence of net radiation occurs entirely within the capping inversion and not at all within the upper mixed layer.
(7) The jump in $\theta_W$ or $\theta_e$ across the capping inversion ($\Delta\theta_W$ or $\Delta\theta_e$) must be positive for parcel stability and maintenance of the cloud layer.
(8) The entrainment rate or growth rate relative to any mean subsidence of the mixed layer is bounded on the upper side by that which can be deduced if there were no dissipation of kinetic energy (Ball, 1960; maximum entrainment assumption), and on the lower side by the value that can be deduced if buoyancy flux were just zero at some height within the mixed layer and positive at all other heights within the mixed layer (minimum entrainment rate assumption).

Entrainment here is viewed solely as cloud-top entrainment. Lilly viewed the *maximum entrainment* assumption as occurring when the dissipation and transport terms in the vertically integrated TKE equation are small compared to opposing positive and negative contributions due to buoyancy generation of TKE. The minimum entrainment is seen to occur when dissipation is so strong that a region of negative heat flux cannot be supported.

Schubert (1976) extended Lilly's model by retaining assumptions (1)-(6) but estimated the entrainment rate as the weighted average of the maximum and minimum entrainment assumptions.

The most controversial aspect of Lilly's model is the relationship between radiative and convective fluxes and entrainment at the top of the mixed layer. Lilly assumed that the radiative cooling was confined to the cloud-top jump region so that radiative cooling did not appear in the mixed-layer heat budget. Deardorff (1976; hereafter referred to as D76) allowed only a fraction of the cloud-top radiation divergence to occur within the capping inversion and the remaining fraction to occur within the uppermost mixed layer just below the capping inversion. By having radiative cooling extend over a finite depth, enhanced TKE and cooling occurred within the mixed layer.

Kahn and Businger (1979) took an even stronger stand and argued that essentially all the cloud-top divergence of net radiation should be placed within

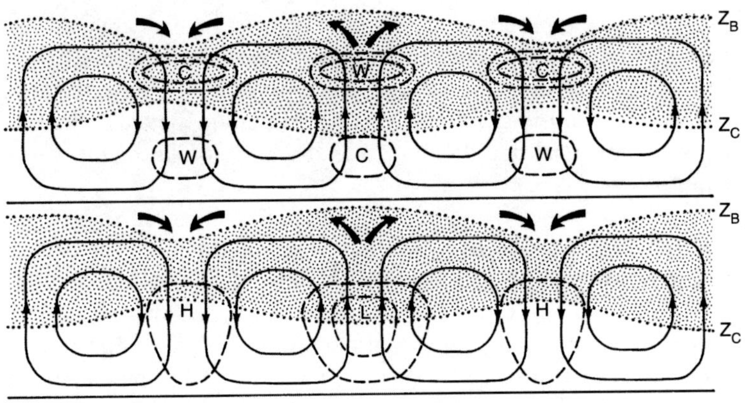


**FIGURE 6.20 Schematic depiction of the motion field of a convective element along with its associated cloud base, cloud top, temperature (top), and non-hydrostatic pressure (bottom) fields.** Note that the updraft has positive buoyancy in the cloud layer and negative buoyancy below. To accelerate surface air upward into the updraft requires the non-hydrostatic pressure shown. Since $\overline{w'p'} < 0$ at cloud base, the cloud layer does work on the subcloud layer. *(From Schubert et al. (1979a))*

the mixed layer and none within the capping inversion. They argued that all the cloudy air lies within the mixed layer and that the zone of longwave radiation cooling should extend below cloud top to the order of 100 m depth.

Schubert et al. (1979a) proposed a simple conceptual model of a stratocumulus cloud. In this model, shown in Fig. 6.20, air near the surface is accelerated toward the updraft and upward against negative buoyancy by lower pressure in the updraft near cloud base. If one computes the correlation of this pressure pattern with the convective-scale vertical motion field, one finds $(\overline{w''p''})_{z=z_c} < 0$. In other words, the work done on the subcloud layer by the cloud layer maintains the convection motions of the subcloud layer.

This conceptual model may also help explain the stratus-lowering process discussed in Section 6.5.3. That is, as radiative cooling destabilizes the stratus layer, convective fluxes can propagate downward by building a lower pressure below the regions of active updrafts. Such a lowering process would not require drizzle to moisten the air, because convection would transport water vapor from the ocean surface to the low-pressure region near cloud base.

This brief summary does not come close to reviewing all the various approaches to layer-averaged models. Our purpose is to give the reader a basic understanding of some of the major concepts involved in their formulation. For a comprehensive review on the subject the reader is referred to Stevens (2006).

### 6.6.3. Entity-type Models

Entity-type models are a convenient vehicle for expressing concepts about the structure of a stratocumulus field. For example, the stratocumulus field can be viewed as being composed of a field of thermal-like elements or vertically elongated plumes. Justification for the particular form of entity structure hypothesized comes from observations. Entity-type models have the advantage that they provide a convenient framework for examining both a variety of cloud microphysical processes and the internal consistency of entrainment processes or the effects of radiative cooling.

In an entity-type model, a stratocumulus field is viewed to be composed of a field of distinct convective elements which exhibit well-defined shapes, entrainment, or mixing laws, and so forth. The conceptual model illustrated in Fig. 6.20 is, in a sense, an entity-type model of a stratocumulus field in which the entities are regular thermal-like cells having comparable updraft and downdraft areas. Chai and Telford (1983) adopted a plume model to a stratocumulus field. They view a stratocumulus cloud as being composed of a population of plumelike updrafts and downdrafts. The ascending plumes originate at the top of a superadiabatic layer above the sea surface. The convective plume ascends, eventually becoming saturated, and gains buoyancy from latent heat release. Eventually the plume encounters a strong capping inversion which the rising plume cannot penetrate. The plume then turns over at the inversion base and descends toward the surface. Chai and Telford do not consider entrainment of air residing above the capping inversion into the descending plumes. They argue that entrainment effects happen rapidly and do not need to be modeled when seeking approximate time estimates. The descending plume is viewed to undergo horizontal motion at the bottom of its descent and re-enter an adjacent rising plume. As the air re-enters a plume, some surface air with surface properties will be mixed in. Thus, the plume becomes warmer and wetter. Because vapor is continuously added to the plumes near the sea surface, the cloud base descends and could eventually form sea fog. Only the erosion of the base of the capping inversion (a process called *encroachment*) by the rising plumes reduces the total moisture flux in the convective layer and inhibits the lowering of cloud base. Encroachment occurs when the whole layer is warmed up to the height of the inversion. When the air at greater heights is no longer buoyant relative to the air underneath, the air is incorporated into the boundary layer and the layer deepens. Mathematically the process is modeled by following the ascending and descending plumes which continually recycle air through the boundary layer. The plumes are assumed to remain well mixed internally and entrain air horizontally through the plume boundaries at rates proportional to the root-mean-square turbulent velocities. Chai and Telford also argue that radiative cooling is not necessary to model the time scales of overturning in the cloud layer ($\sim$15 min). While this may be valid from the perspective of the individual plumes, it is not necessarily valid with respect to time scales affecting the stability of the mixed layer as a whole (approximately several hours). Chai and Telford did not integrate through enough plume recycling to examine the importance of radiation to the longer term thermodynamic structure of the boundary layer.

### 6.6.4. Higher Order Closure Models

Higher order closure models, when applied as ensemble-averaged models, are more general than layer-averaged models in their formulation, and as such are more versatile in principle. That is, stable and unstable cloud-capped boundary layers as well as boundary layers with shear are all possible atmospheric states which a higher ordered closure model may be able to simulate. Moreover, they can be applied to mesoscale problems where strict horizontal homogeneity is not present. They also lend themselves to inclusion of complex physical models such as detailed bin models of cloud microphysics (Ackerman et al., 2004). Higher ordered closure models, however, are computationally demanding. Furthermore, important physical processes, such as entrainment, depend strongly on the details of the parameterization of poorly understood processes represented by terms such as triple-correlation terms and pressure-velocity correlations.

Higher order closure models of a stratocumulus layer involve the explicit prediction of turbulent kinetic energy, variances of the various quantities, and a number of covariances such as eddy fluxes. The general procedures involved in developing closure models are discussed in Chapter 3. Instead of specifying flux profiles through a mixed layer or entrainment rates at the top of a mixed layer, these quantities are explicitly predicted on a finite mesh. The validity of the predicted entrainment rates and the flux profiles depends on the vertical resolution of the model and on the closure assumptions. Examples of the application of closure models to the simulation of stratocumuli includes OLW's study of marine fog and the stratus-lowering process that we examined earlier. In addition, OLW examined the diurnal variation of a stratocumulus layer. Chen and Cotton (1983a) also examined the diurnal variation in a stratocumulus layer with a closure model; Bougeault (1985) studied the diurnal cycle of the marine stratocumulus layer with a higher order cloud model. We will discuss the results of more recent simulations of diurnal variations in a stratocumulus layer in another section. Chen and Cotton (1983b) and Moeng and Arakawa (1980) used higher order closure models to examine the onset of "entrainment instability" and the breakup of stratocumulus clouds.

Another approach to higher-order closure modeling is the so-called PDF approach. Joint PDFs of subgrid quantities, such as vertical velocity, temperature and moisture, are determined from prescribed basis functions in which various moments of the basis functions are predicted in the models (Pincus and Klein, 2000; Golaz et al., 2002a,b; Larson et al., 2005). Integration over the prescribed PDF basis functions permits closure of higher-order moments, buoyancy terms, diagnosis of cloud fraction and liquid-water contents. Using this approach Golaz et al. (2002b) showed that the scheme can predict a variety of boundary layer cloud regimes including solid stratus, trade-wind cumulus, and even cumulus-under-stratus. Larson et al. (2005) demonstrated that the PDF approach can be extended to precipitating cloud layers in which subgrid information is supplied to microphysics parameterizations using Latin-Hypercube sampling strategies.

### 6.6.5. Large-eddy Simulation Models

By far the most fundamental approach to modeling a cloud-capped boundary layer is the large-eddy simulation approach. The LES approach involves numerical integration of the equations of motion within a shallow boundary layer. LES offers the advantage of being able to simulate explicitly the detailed circulations and properties of the cloud layer. LES models are generally quite versatile, being able to simulate both a solid stratocumulus deck and a trade-wind cumulus field. The major deficiency of LES models is that they are computationally demanding. They also generate large volumes of data which require considerable analysis, also at large computational expense. Moreover, LES models can only simulate a limited range of scales of motion. Typically the largest scale resolved by a LES model is of the order of 5 km while the smallest is of the order of 50 m. This limitation becomes quite severe when the atmosphere is stably stratified and the dominant scales of motion may be of the order of 5 m. Observations also suggest that horizontal scales of the order of 5 m are important to cloud-top entrainment processes.

The pioneering work in the application of LES models to stratocumulus-topped mixed layers was performed by Deardorff (1980b). He used a primitive equation model valid for shallow convection. Chosen in the cloud simulations was $\mathbf{\Delta} y = \mathbf{\Delta} x = \mathbf{\Delta} z = 50$ m over a domain which was $2 \times 2 \times 2$ km. Explicit predictions were made for $\overline{u}$, $\overline{v}$, $\overline{w}$, $\overline{\theta}_1$, and $\overline{q}_{\mathrm{w}}$ where the total specific humidity is

$$\overline{q}_{\mathrm{w}} = q + q_1 \tag{6.16}$$

For eddies smaller than the grid scale, the model was closed with an eddy viscosity closure assumption in which the eddy viscosity was formulated to be a function of a predicted turbulent kinetic energy (TKE) [i.e. Eq. (3.31)]. The TKE was closed with a simple model of dissipation and with a down-gradient diffusion term on the triple-correlation terms and the pressure-velocity correlation terms. It should be noted that turbulence is relegated to scales less than 50. In 2007 the state-of-the-art is LES with horizontal grid spacing of 35 m and vertical spacing of 5 m (Stevens et al., 2005; Moeng, 2000) although some simulations have used as fine as 8 m in the horizontal and 4 m in the vertical (Stevens et al., 1999). The greatest challenge for LES models is to simulate cloud top entrainment properly. Stevens and Bretherton (1999) and Lewellen and Lewellen (1998) found that the simulated entrainment rates are much less sensitive to horizontal than to vertical resolution. Bretherton et al. (1999) concluded that accurate simulation of entrainment rates requires vertical resolution that is fine enough to resolve the horizontal variability of cloud top undulations. Some LES have been found to be very sensitive to

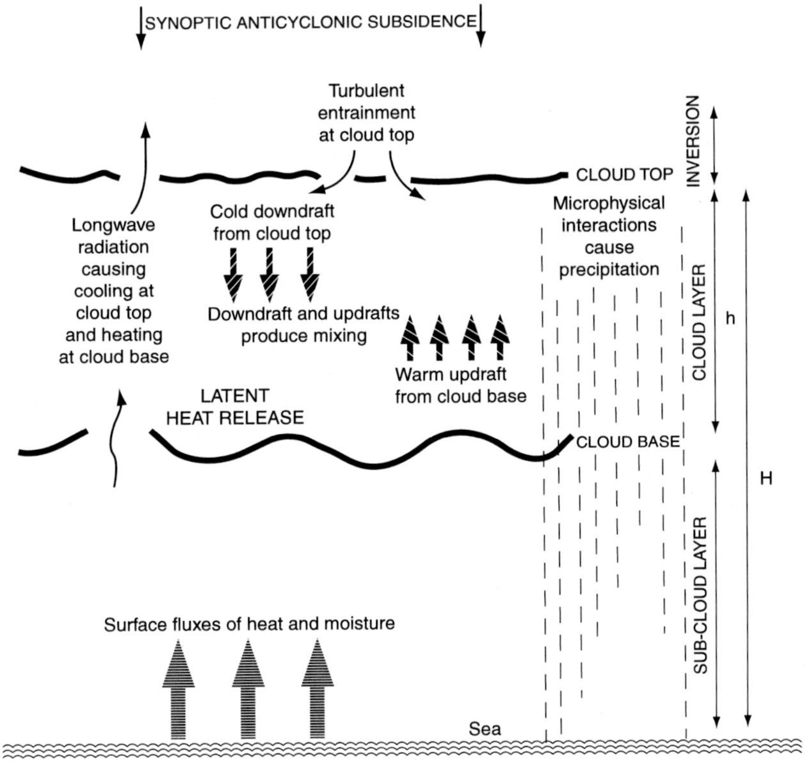


**FIGURE 6.21 Summary of physical process important for the development of nocturnal stratocumulus (after Bennetts et al. (1986)).** *(From Duynkerke et al. (1995))*

vertical grid spacing while others have not (Moeng, 1986). Stevens et al. (1999) conclude that this difference in sensitivity is due to differences in the sub-grid models used. Those that used the Smagorinsky sub-grid model were very sensitive to vertical grid spacing while those using the Deardorff turbulent kinetic energy (TKE) closure model were less sensitive to vertical grid spacing in their entrainment calculations. The Deardorff closure model allows compensating behavior between sub-grid-scale and resolved entrainment fluxes, yielding entrainment rates that are less sensitive to grid spacing and sub-grid-scale motions.

### 6.6.6. The Evolving Conceptual Model of the Stratus-topped Boundary Layer

Fig. 6.21 illustrates the physical processes important to the development of nocturnal stratus. At the surface there is a flux of heat and moisture into the subcloud layer. This leads to the formation of a cloud layer where latent heat of condensation contributes to buoyancy and enhancement of kinetic energy in the cloud layer. As the cloud layer thickens strong divergence of longwave radiation occurs at cloud top. The radiatively-cooled air descends, generating kinetic energy thereby contributing to the kinetic energy of the cloud layer. Near the base of the cloud convergence of longwave radiation produces a weak warming, also contributing to the kinetic energy of the cloud layer. Large scale subsidence, often in regions of subtropical high pressure, warms and dries the cloud top forming the strong inversion seen in Fig. 6.19. The thermal-like eddies in the cloudy layer penetrate into the inversion, engulfing warm, dry air and forcing it into the cloud layer. This process of engulfing mass into the cloud layer is called "entrainment." It should be distinguished from mixing, in which equal masses of cloudy air and above inversion become intermixed. The entrainment process, on the other hand, adds mass to the boundary layer which causes a deepening of the boundary layer working against sinking by large-scale subsidence. Drizzle or precipitation is also illustrated in the figure which we will discuss in a later section. Not shown in the figure is wind shear which can be another source of entrainment into the cloudy boundary layer. This will also be discussed in a later section.

The stratus-topped boundary layer can be distinguished from the dry boundary layer by the location of the main source of kinetic energy driving the layer. In the case of the dry boundary layer the main source of kinetic energy is the flux of heat and moisture from the surface. Figure 6.22 from Moeng's (1986) LES of an unstable cloud-capped boundary layer nicely illustrate the differences between dry and cloudy boundary layers. Contours of virtual dry static energy in the dry case illustrate that the plume-like thermals have their origin near the earth's surface. By contrast, contours of liquid-water static energy in the cloud-topped boundary layer illustrate that the boundary layer is dominated by descending plume-like convective elements. These descending elements obtain their negative buoyancy from radiative and evaporative cooling arising from entrainment of dry air at the capping inversion. Thus, the cloud-topped boundary layers are driven, not by heating from below, but by radiative and evaporative cooling near cloud top.

Because the energetics of the dry boundary layer are driven by heat fluxes from the surface, the buoyancy flux due to entrainment at the capping inversion ($\overline{w's'}|_e$) is approximated as a fraction of the surface buoyancy flux ($\overline{w's'}|_s$). Thus:

$$\overline{w's'}|_e = -A_e \overline{w's'}|_s$$

where $A_e$ is generally approximated as 0.2 (Stull, 1976).

The entrainment process is more complicated for a cloudy boundary layer, as the latent heat of condensation near cloud base and evaporation near cloud top, and radiative cooling near cloud top (plus a minor heating near cloud base) contribute to the kinetic energy of eddies impinging on the inversion layer and thereby being a source of entrainment. Moreover, as dry above-inversion-level

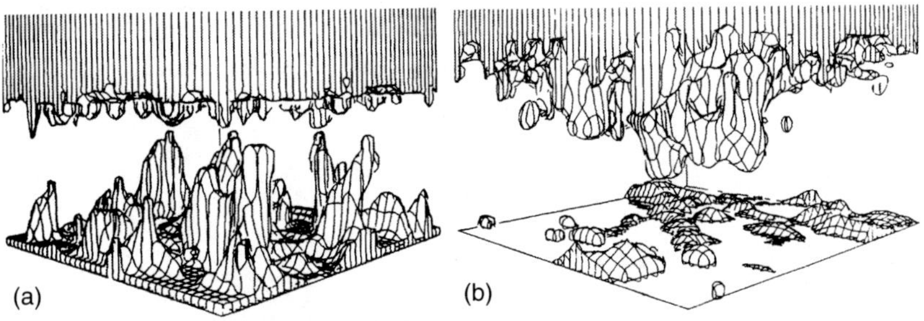


**FIGURE 6.22 Three-dimensional plots of a constant surface of (a) the virtual dry static energy from a clear convective PBL simulation and (b) the liquid-water static energy from a CTBL simulation.** *(From Moeng (1986))*

air is entrained into the cloudy layer liquid water is evaporated which reduces cloud top radiative cooling. Furthermore, if the evaporated cloud water chills the air sufficiently to make the entrained air dense enough it can penetratively descend into the cloud. The accelerating parcels of evaporatively-chilled air can lead to further entrainment of above-inversion-level air, sometimes leading to cloud top entrainment instability (CTEI) which we discuss more fully in the next section. When we consider the added complexity of entrainment for a cloudy boundary layer, it is not surprising that so many formulas describing cloud top entrainment have been proposed (Lilly, 1968, 2002a,b; Duynkerke, 1993; Lewellen and Lewellen, 1998; Deardorff, 1980a; Stage and Businger, 1981a,b; Lock, 1998), just to name a few.

An additional complication of a cloud-topped boundary layer over a dry convective boundary layer, is that the cloud layer can become decoupled from the surface forcing (Nicholls, 1984). Under this situation the buoyancy fluxes can reach a minimum near cloud base. Turbulent kinetic energy is supplied from the surface heat and moisture fluxes, while radiative cooling drives turbulent kinetic energy near cloud top. Nicholls (1984) noted that the factors contributing to decoupling or separation between the subcloud layer and the cloud layer are (1) a decrease in surface buoyancy fluxes, (2) a decrease in radiation flux divergence near cloud top, (3) an increase in evaporation in the subcloud layer by precipitation, (4) an increase in the entrainment of potentially warmer and drier air. He noted that decoupling can lead to the demise of the cloud layer because it will cut off the supply of moisture to the cloud layer. This will occur unless the cloud layer is supplied by moisture either by horizontal advection of moisture or by cumulus-under-stratus. Tjernström and Rogers (1996) also noted that during the daytime, solar heating can also contribute to decoupling when the surface buoyancy fluxes are not particularly strong. Using a mixed-layer model, Bretherton and Wyant (1997) developed decoupling criteria that is based on the flux ratio of boundary layer diabatic forcing to the surface latent heat flux. The threshold for decoupling is proportional to the fraction of the boundary layer filled by cloud and their model entrainment closure parameter.

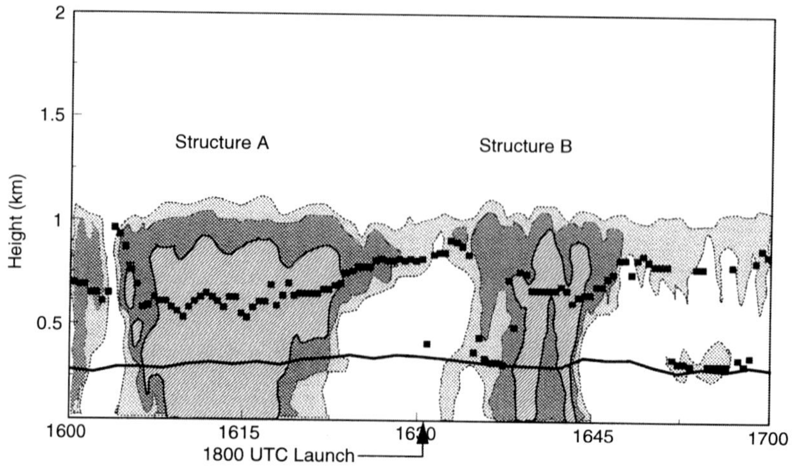


**FIGURE 6.23 Two cumulus-stratocumulus interaction events (marked A and B) shown by radar returns from an upward-pointing 94 GHz radar operating from the island of Santa Maria in the Azores.** The observations were made between 1600 and 1700 UTC on 15 June 1992. The return signal is indicated at three levels: the stippled light gray indicates low returns, the darker cross hatching indicates medium returns, and hatched areas outlined with a solid line represent saturation of the analog-digital converter. The filled squares represent cloud base heights from a laser ceilometer and the near-surface LCL is represented by the solid line. The clouds observed between the two events are decoupled from surface processes. *(From Miller and Albrecht (1995))*

Once a cumulus-under-stratus decoupled boundary layer forms the cloud layer looks quite different and behaves quite differently. Figure 6.23 (Miller and Albrecht, 1995) illustrates an upward pointing 94 GHz radar depiction of a cumulus-under-stratus regime, and Fig. 6.24 (Kropfli and Orr, 1993) is a schematic depiction of a radar-observed cloud during the Atlantic stratocumulus transition Experiment (ASTEX). They show that cumulus-under-stratus can resemble mini-cumulonimbus in structure including a convective core and an anvil. The anvil-like structure is feeding the stratiform layer with water substance. Tjernström and Rogers (1996) noted that the cumulus-under-stratus regime is characterized by more episodic moisture transport and entrainment than occurs in a fully coupled cloud top boundary layer. The cumulus depletes moisture from the layer near the surface and thereby impedes the cumulus activity, sometimes stopping altogether. At the top of the cloud layer the pulse of kinetic energy near the inversion locally enhances turbulent entrainment. After a time the moisture in the lower boundary layer will increase leading to another cumulus pulse (Rogers et al., 1995), and the process repeats itself.

## 6.6.7. Cloud Top Entrainment Instability

One of the most important questions that modelers of stratocumulus clouds must answer is, "at what point does a solid stratocumulus cloud layer break up into a

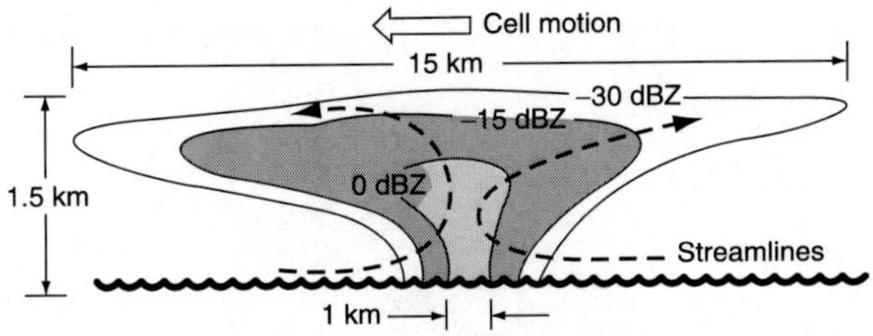


**FIGURE 6.24 Idealized representation of a microcell based on RHI and PPI scans through several microcells during ASTEX.** *(From Kropfli and Orr (1993))*

broken cumulus field?" Lilly (1968) suggested that whenever the jump $\boldsymbol{\Delta}\theta_e < 0$, where $\boldsymbol{\Delta}\theta_e$ is the above cloud value of $\theta_e$ minus the in-cloud value, the cloud top will become unstable to entrainment. Lilly argues that

*"If a parcel of the upper air is introduced into the cloud layer and mixed by turbulence, evaporation of cloud droplets into the dry parcel will reduce its temperature. If the mixed parcel reaches saturation at a colder temperature than that of the cloud top it will be negatively buoyant and can then penetrate freely into the cloud mass. In such a case the evaporation and penetration processes will occur spontaneously and increase unstably until the cloud is evaporated."*

Subsequently, Randall (1980) and Deardorff (1980a) pointed out that when water loading effects are considered, the criteria for cloud-top entrainment instability become $\boldsymbol{\Delta}\theta_e < -1$ to $-2$ K. This process is called cloud top entrainment instability (CTEI).

Kuo and Schubert (1988) summarized the thermodynamic criteria for entrainment instability. Consider that $X$ mass units of warm dry air just above the capping inversion have mixed with $1 - X$ mass units of cool moist air just below the inversion. If the subscripts a and b denote parcels originating above the inversion and below the inversion, respectively, then the equivalent potential temperature $\theta_e$ and total mixing ratio $r_T$ of the mixed parcel can be expressed as

$$\theta_e = \theta_{e_b} + X\boldsymbol{\Delta}\theta_e \tag{6.17}$$

and

$$r_T = r_b + X\boldsymbol{\Delta}r. \tag{6.18}$$

The resultant virtual potential temperature of the mixed parcel can be expressed as

$$\theta_v = \theta + \theta_0(0.608r_v - r_l), \tag{6.19}$$

where $r_v$ and $r_l$ are the vapor and liquid-water mixing ratios of the mixture, respectively, and $\theta_0$ is a reference potential temperature which he took to be 15 °C. If we denote $\tilde{X}$ as the mass units of dry air required to evaporate all the

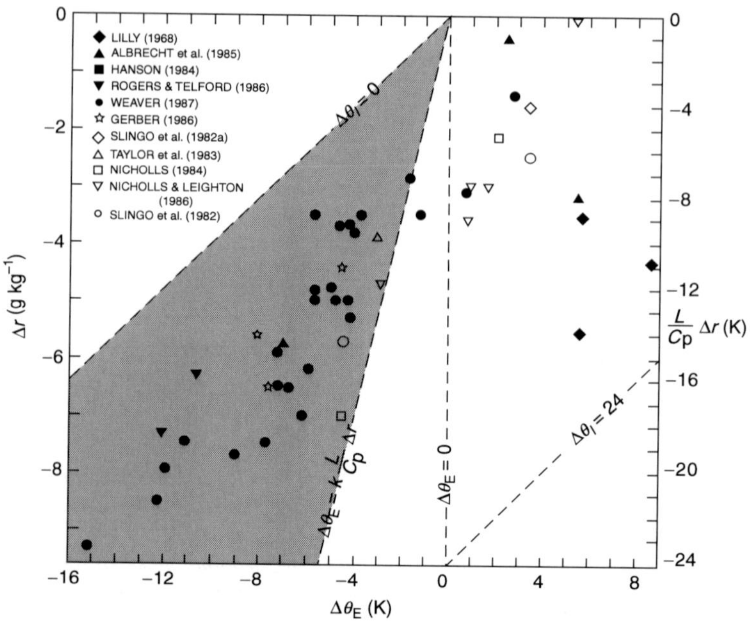


**FIGURE 6.25 The $\Delta\theta$, $\Delta r$ plane, with the Lilly critical curve ($\Delta\theta_e = 0$) and the Randall-Deardorff critical curve ($\Delta\theta_e = k(L/C_p)\Delta r$).** Observational data are indicated by the coded symbols, with solid symbols for subtropical cases and open symbols for mid-latitude cases. About two-thirds of the observed solid stratocumulus clouds occur in the shaded region, where the thermodynamic theory predicts stratocumulus breakup. *(Adapted from Kuo (1987))*

liquid water of the mixed parcel, the buoyancy of the mixed parcel when $X \leq \tilde{X}$ can be expressed as

$$\left(\frac{\theta_v - \theta_{vb}}{\theta_0}\right) g = X\left(\frac{C_p}{Lk}\boldsymbol{\Delta}\theta_e - \boldsymbol{\Delta}r\right) g, \tag{6.20}$$

where $k \simeq 0.23$. If the term within parentheses on the right-hand side of Eq. (6.20) is negative, all mixtures with $0 < X \leq \tilde{X}$ will be negatively buoyant. The thermodynamic criteria for the onset of entrainment instability are

$$\boldsymbol{\Delta}\theta_e < kL\boldsymbol{\Delta}r/C_p. \tag{6.21}$$

Figure 6.25 illustrates observations of solid-covered stratocumulus decks mapped relative to Lilly's ($\boldsymbol{\Delta}\theta_e = 0$) and the Randall-Deardoff [Eq. (6.22)] thermodynamic criteria for the onset of entrainment instability. About two-thirds of the observations occur to the left of the Randall-Deardorff critical line, suggesting that the thermodynamic criteria for stratocumulus breakup are not sufficient. A similar conclusion was reached by Nicholls and Turton (1986), Hanson (1984a,b), Rogers and Telford (1986), Weaver and Pearson

(1990), Albrecht (1991), Khalsa (1993), Wang and Albrecht (1994). Several alterative stability criteria have also been proposed (MacVean and Mason, 1990; Duynkerke, 1993).

It was suggested by Albrecht et al. (1985), Nicholls and Turton (1986), Rogers and Telford (1986), and Kuo and Schubert (1988), that the amount of liquid water available in the tops of many stratocumuli may be so small at times that the maximum negative virtual temperature difference that can be generated will not allow deep downward penetration of dry air into the cloud interior. This idea is supported by the LES simulations by Yamaguchi and Randall (2008) where it was shown that a cloud with larger liquid-water contents experienced more rapid entrainment and stronger turbulent kinetic energy. Thus the strength of CTEI is sensitive to the available liquid-water contents in the tops of clouds. Yamaguchi and Randall (2008) argue that CTEI can also be masked by strong cloud building processes such as strong cloud top radiative cooling, and surface evaporation.

Clearly the processes involved in the breakup of stratocumulus clouds are complicated. Not only is the thermodynamic state of the cloud-top layer important, but so also is the magnitude of moisture and heat fluxes from the underlying sea surface, cloud top radiative cooling, the turbulent kinetic energy in the bulk of the cloud layer, the liquid-water content, and, as suggested by Rogers and Telford (1986), perhaps even the drop-size spectrum. Moreover, Moeng and Arakawa (1980) and Kuo (1987) suggest that the magnitude of large-scale subsidence may also be important, with lesser subsidence favoring stratocumulus cloud breakup. We will examine this further in the next section when considering the transition from solid stratus to trade-wind cumulus.

### 6.6.8. Transition from Solid Stratus to Trade-wind Cumulus

It is well known that off the west coasts of the major continents a semi-permanent subtropical high pressure system exists. Within eastern parts of these high pressure systems air moves equatorward with sea surface temperature (SST) increasing. In response to the equatorward movement of the air over a distance of about 1000 km, there is a transition in cloudiness from solid stratus, to cumulus-under-stratus, to gradual dissipation of the stratocumulus layer, to a trade-wind cumulus regime (Klein et al., 1995). As the air column moves equatorward, it experiences changes in large-scale subsidence from a divergence of about $6 \times 10^{-6}\ s^{-1}$ to near zero (Neiburger et al., 1961; Klein and Hartman, 1993). As noted by Schubert et al. (1979b) this variation in subsidence increases the depth of the marine boundary layer. A schematic diagram illustrating the transition in cloud properties along the equatorward trajectory is shown in Fig. 6.26.

A number of models have been applied to the simulation of the evolution of the cloudy marine boundary layer along trajectories going from higher latitudes towards the tropics in which varying SSTs are prescribed. These models have

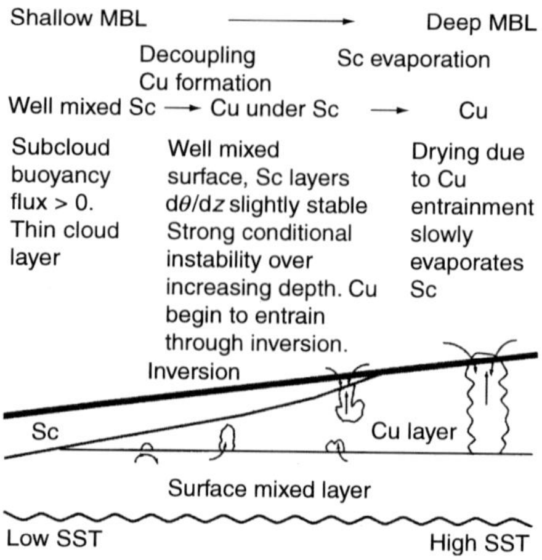


**FIGURE 6.26 A conceptual diagram of the STCT.** *(From Wyant et al. (1997))*

ranged from mixed-layer models (Wakefield and Schubert, 1981; Wyant and Bretherton, 1992) to higher-order closure models (Moeng and Arakawa, 1980; Bretherton, 1990, 1992) to two-dimensional (2-D) cloud-resolving models (CRMs; Krueger et al., 1995a,b; Wyant et al., 1997). The 2-D CRM simulations have been most instructive, as they show that as the air moves from cooler to warmer SSTs the boundary layer deepens to the point where the lower boundary layer becomes decoupled from the stratocumulus cloud layer above. At this point the marine boundary layer can be characterized as cumulus-under-stratus. This occurs as the increasingly vigorous convective eddies, in response to warming SSTs, increase entrainment at the capping inversion. Associated with the downward flux of entrained warm air, buoyancy fluxes below cloud base become increasingly negative. This disrupts the mixed layer and forms a stable layer below cloud base. As noted by Wyant et al. (1997) this stable layer acts as a valve which allows only the most powerful subcloud updrafts to ascend into the stratocumulus layer above. These more powerful updrafts begin to resemble cumuli. The detrained water substance from the cumulus eddies helps sustain the stratocumulus layer. But as the air column moves further equatorward where SSTs become warmer, the cumulus eddies increase in strength and thereby entrain larger amounts of dry above-inversion-level air. Wyant et al. (1997) argue that, as the strength of the cumulus eddies becomes greater, the ratio of penetratively entrained mass flux of dry air to upward cumulus mass flux of moist surface layer air increases, which leads to a drying of the cloud layer. This at first leads to mesoscale patches of stratocumulus free air and eventually to full transition to a trade-wind cumulus regime. Another interpretation is that as the cumulus-under-stratus eddies become more intense, the moist upward cumulus mass flux is compensated by increasingly strong subsidence which warms and dries the environment, leading to the demise of the stratocumulus layer.

Both Krueger et al. (1995b) and Wyant et al. (1997) concluded that CTEI could not explain the breakup of the stratocumulus layer. Krueger et al. (1995a,b) found that when their simulated stratocumulus layer was breaking up, negatively buoyant downdrafts were much less vigorous than updrafts, which is in contradiction to the predictions of CTEI. It is, however, consistent with the idea that weak widespread compensating subsidence could be the primary factor in dissipating the stratocumulus layer.

## 6.6.9. The Role of Vertical Shear of the Horizontal Wind

In our discussion of convectively unstable stratocumulus cloud layers, vertical shear of the horizontal wind is typically ignored as a contributing factor to the generation of kinetic energy in the cloud layer. This is particularly true of a number of the layer-averaged models such as Lilly (1968), Schubert (1976), and Deardorff (1976). The higher order closure models and 3-D large-eddy simulation models are inherently capable of including the effects of wind shear, though their application to the more stable cases with stronger wind shear has been quite limited.

Based on field observations, Mahrt and Paumier (1982) postulated a conceptual model of the entrainment process that differs little from the conceptual model of entrainment in isolated cumuli to be discussed in the following chapter. They observed that the coldest air near cloud top occurs on the downshear side of penetrative convective elements and on the upshear side of engulfed wisps or pockets of free-flow air. Mean vertical shear of the horizontal wind contributes to enhanced mixing in the downshear region of penetrative convective elements. Thus evaporative cooling contributes to the production of penetrative downdrafts on the downshear edge of the penetrating updrafts. If we also consider the observed and modeled flow about isolated penetrative towers described in Chapter 7, a schematic illustration of the interactions between penetrative updrafts, shear, and penetrative downdrafts is postulated as shown in Fig. 6.27. We will now examine the effects of vertical shear of the horizontal wind on the properties of stratocumulus clouds more fully.

Brost et al. (1982a,b) (hereafter referred to as BWL) presented the results of extensive analysis of several marine stratocumulus cases in which wind shear played an important role in the energetics of the cloud layer. Near the ocean surface, they characterized the cases as stable or slightly stable. In a horizontally homogeneous boundary layer, the time rate of change and advection of turbulent kinetic energy $\overline{e}$ are small; therefore Eq. (3.50) can be written as

$$0 = \underbrace{-\rho_0 \overline{u''w''}\frac{\partial \overline{u}}{\partial z}}_{(a)} \underbrace{- \rho_0 \overline{v''w''}\frac{\partial \overline{v}}{\partial z}}_{(b)} + \underbrace{\rho_0 \left( \frac{\overline{w''\alpha''_{\mathrm{m}}}}{\alpha_0} - \overline{w''r''_{\mathrm{w}}} \right) g}_{(c)}$$

$$\underbrace{- \frac{\partial}{\partial z}(\overline{w''e})}_{(d)} + \underbrace{\frac{\partial}{\partial z}(\overline{w''p''})}_{(e)} \underbrace{- \rho_0 \varepsilon}_{(f)} . \qquad (6.22)$$

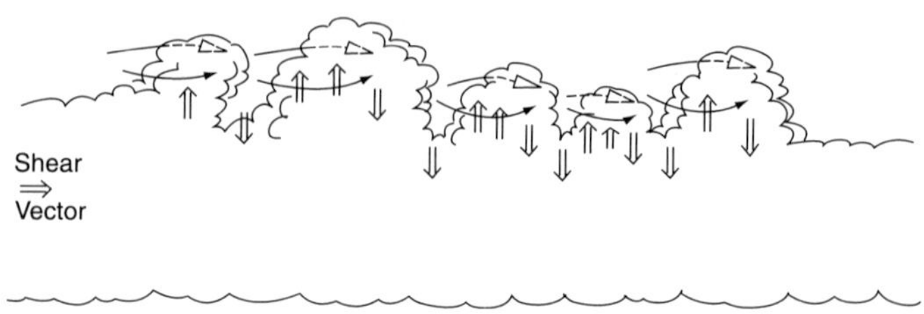


**FIGURE 6.27 Schematic model of a stratocumulus layer in sheared flow.** Double arrows represent updrafts and downdrafts. Higher speed flow above the capping inversion is shown diverging around the emerging towers and entraining downshear of the updrafts. Evaporation in the downshear regions is shown to be coupled with the formation of downdrafts.

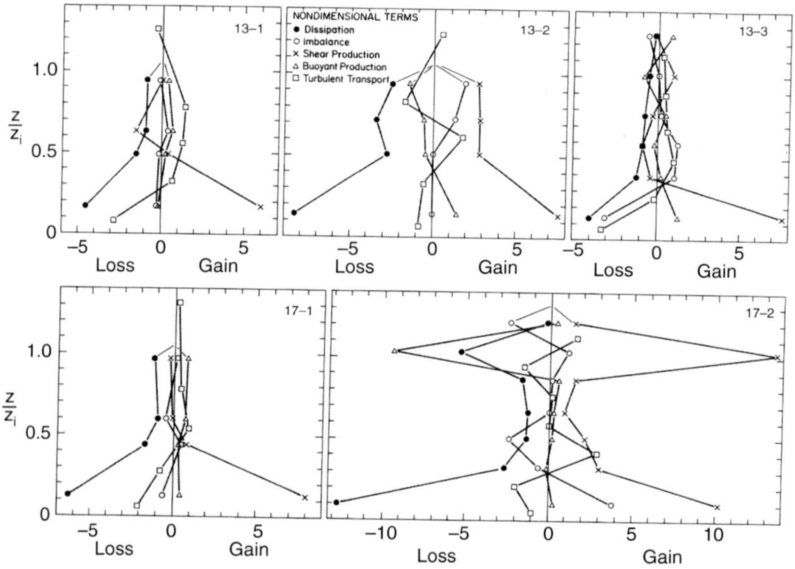


**FIGURE 6.28 Turbulent kinetic energy budgets. All terms are multiplied by $kz_i/u_*^3$.** The terms as identified in the figure are dissipation $\varepsilon$, shear production $-\overline{u''w''}(\partial\bar{u}/\partial z)-\overline{v''w''}(\partial\bar{v}/\partial z)$, buoyant production $(\overline{w''\alpha_m}/\alpha_0-\overline{w''r_w})g$, turbulent transport $-(1/\rho_0)(\partial/\partial z)(\overline{w''e})$, and pressure transport $-(\partial/\partial z)(\overline{w''p''}/\rho_0)$, which is calculated as a residual. *(From Brost et al. (1982b))*

Using a gust-probe-equipped aircraft, BWL measured the mechanical production terms (a and b), buoyancy production terms (c), and turbulent transport term (d). The dissipation rate $\varepsilon$ was estimated from the average slopes of the individual spectra for $u''$, $v''$, and $w''$. The pressure transport term (e) was evaluated as a residual from Eq. (6.22). Figure 6.28 illustrates the profiles of the various terms in Eq. (6.22) which have been made nondimensional by multiplying them by $kz_i/\rho_0 u_*^3$. They have been calculated for several cases analyzed by BWL.

The outstanding feature in Fig. 6.28 is that shear production dominates buoyancy production of turbulent kinetic energy in almost every case. It is especially pronounced in case 17-2. Shear production can be expected to be large near the surface, but in more convectively unstable cases, buoyancy production is usually thought to be dominant throughout the cloud layer (see Deardorff, 1980b). This is only evident in case 17-1. Both shear production and buoyancy production (suppression) are quite large in the capping inversion for case 17-2. Dissipation shows the expected behavior of decreasing away from the earth's surface, but occasionally shows a peak in the inversion, especially in case 17-2. Deardorff (1980b) also found a peak in $\varepsilon$ near the inversion in his simulations of convectively unstable stratocumuli. Deardorff found that pressure and turbulent transports act in concert near cloud base. They supply turbulent energy generated in the clear convective layer below. In the stable cases studied by BWL, pressure transport was quite small and not consistently related to turbulent transport.

One consequence of the greater role of shear over buoyant production of turbulence is that the nature of the entrainment processes appears to be altered. In their observational study of nocturnal stratocumuli over land, Caughey et al. (1982) noted that acoustic sounder records suggested the absence of thermal plumes that could penetrate into the inversion layer and mix dry air well downward into the cloud and subcloud layer. However, the region near cloud top was fully turbulent and exhibited vertical velocity fluctuations of the order of 1 m $s^{-1}$, typical of the convective boundary layer. The length scale of the entraining turbulent elements was of the order of 10 m, which is quite small. Caughey et al. (1982) and also Caughey and Kitchen (1984) suggested the entrainment mechanism occurred intermittently, perhaps triggered by the formation of Kelvin-Helmholtz instability waves which break down into turbulence. Circumstantial evidence supporting their hypothesis was provided by the observation of stronger wind shear in the inversion at the top of the cloud layer. They suggested that the stronger wind shear at the top of the cloud layer occasionally reduced the local gradient Richardson number below critical values (generally thought to be 0.25) such that turbulent breakdown could occur. On occasion they observed the turbulent layer near cloud top to be much thicker, suggesting that turbulent breakdown had occurred, which resulted in locally weaker shear and the return of Richardson numbers at the top of the cloud layer to supercritical values. It was hypothesized that the turbulence would then decay, causing the interface layer near cloud top to thin and return to near laminar flow, at which point the shear would intensify.

In their observation of a weaker, unstable, or stable cloud layer, Brost et al. (1982b) also found that the scale of the turbulent eddies contributing to cloud-top entrainment was only a few tens of meters.

These observations suggest a different link between cloud-top radiative cooling and entrainment in the stable or weakly convective stratocumulus cloud. It is generally believed that the local instability created by cloud-top radiative cooling is transmitted through the depth of the cloud layer and subcloud layer by strong vertical mixing caused by penetrative plumes. This is fundamental to the layer-averaged models such as Lilly's (1968), Schubert's (1976), and Stage and Businger's (1981a; 1981b). It is also a predicted response in the higher order closure models of OLW and Chen and Cotton (1983a,b). Thus, radiative cooling causes enhanced mixed-layer turbulence in the form of penetrative plumes, which, in turn, causes greater rates of cloud-top entrainment.

In the case of more stable stratocumulus, it appears that the local instability caused by cloud-top radiative cooling generates small-scale turbulence. The small-scale turbulence interacts with local wind shear at the top of the radiatively cooled layer. This causes sporadic turbulent breakdown or shear-driven entrainment. In some cases, there does not appear to be any direct communication between radiation cooling at cloud top and the energetics of the entire depth of the cloud layer. As noted by Nicholls and Leighton (1986), almost all the cloud-top radiative cooling is balanced locally by entrainment. As a result there is no net generation of positive buoyancy and associated convective transport. Such clouds may be more properly called stratus rather than stratocumulus and exhibit properties more similar to altostratus and cirrus than to boundary layer stratus.

### 6.6.10. The Role of Drizzle

In several of the cases observed by Brost et al. (1982b), drizzle drops were present. The vertical water flux by drizzle represented in some cases a significant fraction of the water flux in the cloud layer. Duynkerke et al. (1995) also diagnosed that drizzle was an important component of the water budget of stratocumulus clouds they observed.

Brost et al. (1982b) also suggested that drizzle can affect significantly the stability of a cloud layer by altering the vertical distribution of latent heating. That is, as drizzle settles from the top of the cloud layer, it removes water from that level which cannot be evaporated. As a consequence, drizzle contributes to a net latent heating in the upper part of the cloud layer. However, as the drizzle settles into the subcloud layer, it evaporates and causes evaporative cooling of the subcloud layer. In this process, the layer between the heating aloft and cooling below would be stabilized, whereas the shallow layers above the heating zones or below the evaporatively cooled zone would be destabilized. They hypothesized that this would tend to form two shallow unstable layers decoupled by an intermediate stable layer. Paluch and Lenschow (1991) concluded from their observational study that when drizzle largely evaporates mid-way through the subcloud layer it has a destabilizing effect. However, when drizzle settles through the entire subcloud layer, it tends to stabilize the subcloud layer. Modeling studies by Feingold et al. (1996) also show such a response. As a result of the stabilizing influence of drizzle, moisture tends to build up near the surface until sufficient CAPE is built up. At that point cumulus-under-stratus can form (Nicholls, 1984). This has been shown in LES modeling studies by Stevens et al. (1998). It is interesting to note that in an LES modeling study of the effects of enhanced CCN on stratocumulus optical properties Jiang et al. (2002) found a similar response. For the clean case where small drizzle drops evaporated in the mid-subcloud layer thus destabilizing the layer, they found that increasing CCN concentrations suppressed drizzle which led to a more stable boundary layer. As a consequence, the cloud liquid-water contents were less and the cloud albedo was either reduced or remained the same. This response is opposite to that found by Albrecht (1989) using a simple layer-averaged model. They found that increased CCN concentrations would reduce drizzle formation which would lead to clouds with higher liquid-water contents, greater cloud coverage, and therefore greater cloud albedo.

It was noted earlier that under conditions supporting a cumulus-under-stratus regime, the clouds can resemble mini-cumulonimbus clouds. Jensen et al. (2000) go further and note that in a drizzling boundary layer, the evaporating drizzle can form cold-pools which produce propagating cells that resemble miniature squall lines.

Regarding the nonlinear response of stratocumulus clouds to varying CCN concentrations, Ackerman et al. (2004) found, using a one-dimensional turbulent closure model, that increasing CCN in some situations also resulted in lowered cloud albedo in contrast to the Albrecht (1989) hypothesis. Only when the layer above the boundary layer was moist did they find that increasing CCN resulting in greater liquid water paths and higher cloud albedo. When the layer was dry, however, higher CCN concentrations led to enhanced entrainment of dry air into the boundary layer and thus the liquid-water paths and cloud albedo were reduced. To understand the relationship between CCN concentration and entrainment, we have to realize that for the same liquid-water content increasing CCN results in higher concentrations of smaller cloud droplets. Thus for a given liquid-water content, higher concentration smaller droplets expose a larger surface area to the subsaturated air and thereby evaporate more quickly than a few larger droplets. This is because the surface to volume ratio is larger for small spherical droplets than for larger droplets. Now we speculate if the boundary layer meets the conditions supporting CTEI then the higher concentration CCN clouds will produce more vigorous evaporatively-chilled downdrafts following entrainment of dry air, which will contribute to greater depletion of cloud liquid-water in the cloud and reduced cloud albedo. One can imagine that if the liquid-water content in the cloud is large enough, higher CCN clouds could be transformed into broken cumulus clouds following entrainment of dry air.

We have noted above that drizzling regions of stratocumuli often appear like mini-cumulonimbi and even squall lines. Other observations have associated drizzle to the formation of anomalous regions of cloud organization that have been called pockets of open cells (POCs; Stevens et al., 2005; van Zanten and Stevens, 2005) or rifts (Sharon et al., 2006). The POCs are regions surrounded by solid stratocumulus, that have cloud patterning similar to open cellular

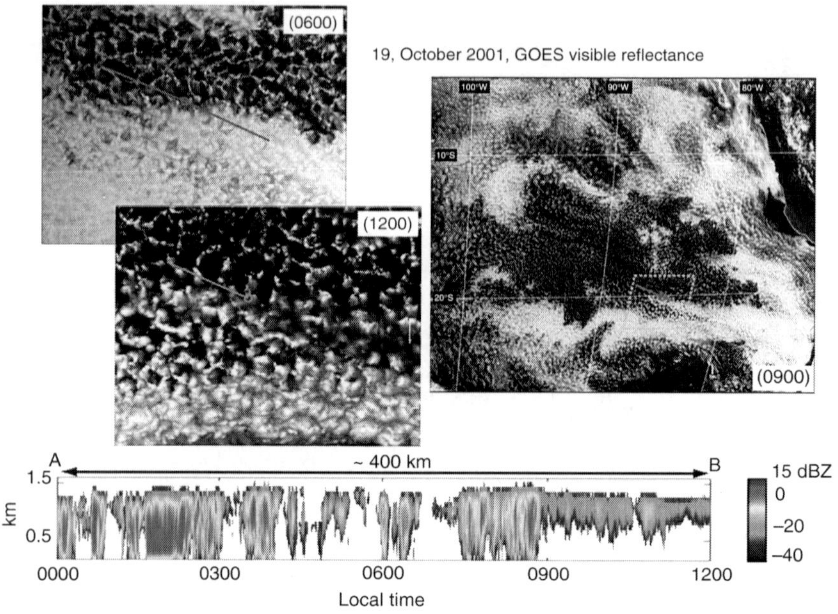


**FIGURE 6.29** **(top) Channel I (0.6 μm) reflectance from the southeast Pacific from GOES-8 on 19 Oct 2001. On the right is a large areal view at 0900 LT, on the left, zoomed images of cloud features in the region directly over the ship are shown for 0600 and 1200 LT. (bottom) Radar reflectivity data is taken from an upward-pointing cloud radar operated from the NOAA RV *Ron H. Brown*, which was on station at 20°S, 85°W as part of the EPIC experiment.** In each satellite image the position of the it Ron H. Brown is indicated by the orange open-circle marker, and the approximate trajectory of the cloud field, as estimated from surface wind measurements, is indicated by the orange line. (For interpretation of the references to color in this figure legend, the reader is referred to the web version of this book.) *(From Stevens et al. (2005))*

convection (Agee, 1984) as described in the next chapter. Figure 6.29 shows a satellite depiction of a region characterized by POCs. Also shown in the figure are cloud radar reflectivities which show the cellular higher reflectivity or higher precipitation amounts in the POC region. Figure 6.30 is a conceptual model showing the differences between the POC region which is characterized by heavier drizzle amounts and higher values of $\theta_e$, and the solid stratus region where little precipitation is observed and $\theta_e$ values are lower and more uniform. Likewise the rift regions are characterized by greater amounts of precipitation and very low CCN concentrations. In fact, Fig. 6.31 shows these are regions where ship tracks are most evident. It is commonly observed that the ship tracks are regions of enhanced albedo where ships are producing enhanced CCN concentrations (Durkee et al., 2000a,b). Because the rifts are regions where CCN concentrations are low and drizzle is heavy, they are also regions where cloud albedo is low. Thus they provide a sharp contrast to the ship-track regions.

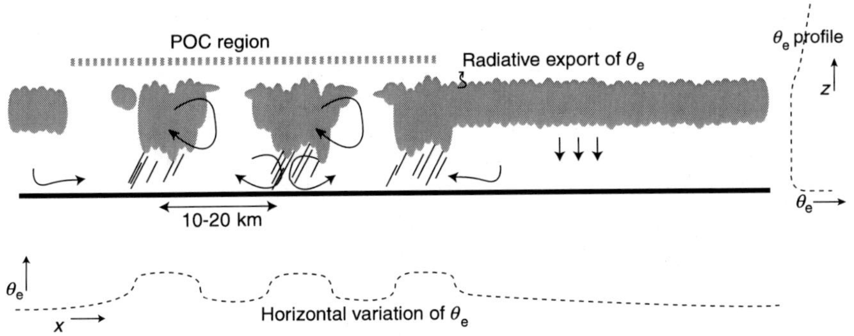


**FIGURE 6.30 Conceptual rendering of POC and neighboring non-POC or stratiform region.** Also shown is a schematic of the horizontal and vertical variations in $\theta_e$ and inferred mesoscale circulations. Note that the net upward motion in the POC is left out. *(From van Zanten and Stevens (2005))*

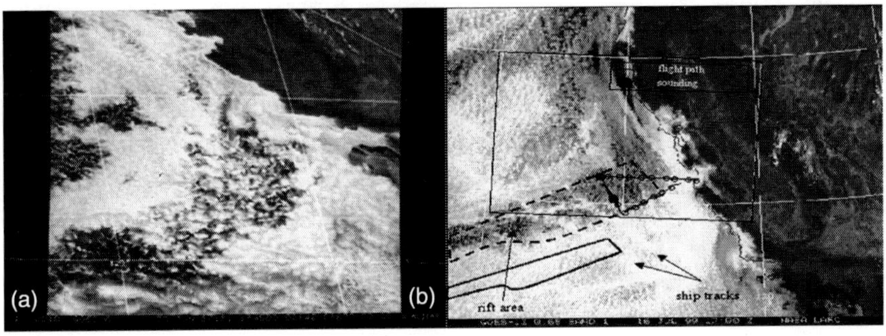


**FIGURE 6.31 (a) GOES-10 satellite image of stratocumulus clouds and embedded rift observed off the coast of California, 1845 UTC 30 Jun 1987. (b) GOES-10 visible satellite image of stratocumulus clouds and embedded rift observed off the coast of California, 1700 UTC 16 Jul 1999 with progression of the rift shown by the dashed line (corresponds to 1700 UTC) and the solid thick line (corresponds to 2300 UTC).** The flight patch is also overlaid to identify location of soundings and path of aircraft. The circles indicate the location of the soundings taken during the flight (the open circles were not analyzed in detail in this study, the solid circles indicate the three soundings analyzed in detail; the two circles in the rift area almost overlap completely, appearing as only one circle). *(From Sharon et al. (2006))*

The basic causes of POCs and rifts are still being debated but it seems apparent that the drizzle process is a key factor.

In summary, settling of drizzle and evaporation of drops can be important to the water budget and energetics of stratocumulus clouds. This is especially true of stratocumulus clouds occurring in the presence of weak surface fluxes. Its role in the more strongly forced cases has not been so clearly identified. It is also important to recognize that drizzle may be more important to the dynamics of nocturnal stratocumuli. This is because it is frequently observed (see Kraus, 1963, for example) that there is a well-defined nighttime maximum in maritime precipitation. Kraus found the largest diurnal amplitude to occur in mid-latitude regions where the rain is mostly nonconvective. He speculated that absorption of solar radiation during the daytime resulted in less liquid-water production than at night. Thus, drizzle effects could be more important in the somewhat wetter and deeper, nocturnal stratocumulus clouds. Finally because drizzle alters the thermodynamics of the cloud and subcloud layer substantially, it also alters the dynamics of stratocumulus clouds in a very nonlinear way. This nonlinear response to precipitation is not unlike that which we will see is characteristic of deep convective cloud systems.

### 6.6.11. Role of Large-scale Subsidence

It is generally recognized that large-scale subsidence plays an important role in establishing the environmental conditions favorable for formation of marine stratocumulus. That is, large-scale subsidence establishes the pronounced capping inversion, which serves as an upper lid to the atmospheric boundary layer and confines the moisture and heat fluxes from the ocean surface in a shallow layer. The overlying air mass is also dried out by the sinking motion. While subsidence may establish environmental conditions favorable for maintaining a solid stratus deck, too much subsidence may be responsible for the breakup of stratocumulus clouds. Roach et al. (1982) examined whether subsidence could account for the observed dispersal of a 300-m-thick cloud in 2-4 h at night. From the analysis of the vertical thermodynamic structure upwind of the cloud, they determined that a downward displacement of the air mass of 220 m in 2-4 h would result in the evaporation of the entire cloud. This corresponds to a subsidence rate of 2-4 cm $s^{-1}$. Subsidence rates of this magnitude cannot be directly measured but can be diagnosed by integrating the divergence of the horizontal wind field. They estimated that such a magnitude of subsidence was plausible, but the wind observations were too sparse to obtain a definitive mesoscale analysis of the local subsidence field. Thus, future observational studies of stratocumulus clouds should include the measurement of winds with sufficient accuracy and spatial resolution to calculate subsidence rates.

Chen and Cotton (1987) performed some sensitivity experiments with their higher order closure model to determine if large-scale subsidence played a more active role in the determination of cloud structure. In their first experiment they imposed a divergence of $5.0 \times 10^{-6}$ $s^{-1}$, which produces a 0.25-cm $s^{-1}$ subsidence rate at a height of 550 m. This corresponds to estimates for typical large-scale subsidence off the coast of California in the United States. If the intensity of the capping inversion is 10 K over a depth of 25 m, this subsidence rate will warm the capping inversion by 190 K per day. Thus a significant fraction of the longwave radiative cooling at cloud top is balanced by subsidence warming. As a consequence, the upward heat flux in the cloud layer and buoyancy production of turbulence are reduced. Relative to an environment with

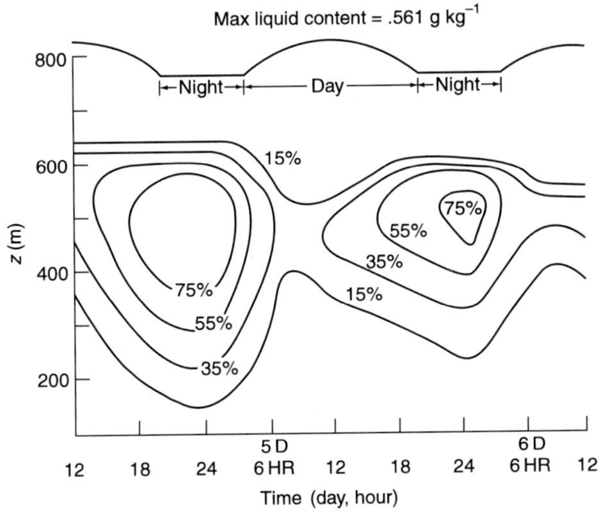


**FIGURE 6.32 Diurnal variation of liquid water in stratus cloud formed in a subsidence-capped boundary layer.** *(From Oliver et al. (1978))*

no vertical motion, this imposed sinking reduces the liquid-water content by 11.4%. If the subsidence is further increased by a factor of 2 (i.e. to 0.5 cm $s^{-1}$ at 550 m), the liquid-water content is lowered to 45% of the value obtained in a zero-subsidence calculation. We shall see below that large-scale subsidence has an important influence on the response of the cloud layer to solar heating.

## 6.6.12. Diurnal Variations in Marine Stratocumuli

Using a second-order turbulent transport model, OLW simulated a diurnally varying stratocumulus cloud structure. As shown in Fig. 6.32, the stratus top rises and the cloud base lowers during the night and thickens until sunrise, when the stratus begins to dissipate due to solar insolation. As noted previously, however, the location of maximum solar heating well into the interior of the cloud layer shown in Fig. 6.33 may only be possible in clouds that are optically thin (i.e. contain little liquid-water content). The corresponding distribution of turbulent kinetic energy predicted by OLW is shown in Fig. 6.34. This shows a maximum in TKE in the interior of the cloud during the nighttime with a more vertically uniform profile of TKE from the surface to near cloud top during the daytime.

Other higher order closure models (Bougeault, 1985; Chen and Cotton, 1987) simulate a diurnal variation in cloud base height but not cloud top. Only when large-scale subsidence was imposed did Chen and Cotton find both cloud top descent and cloud base ascent after sunrise. Smith and Kao (1996) also found that large-scale subsidence was needed to simulate the observed diurnal variation in cloud top height.

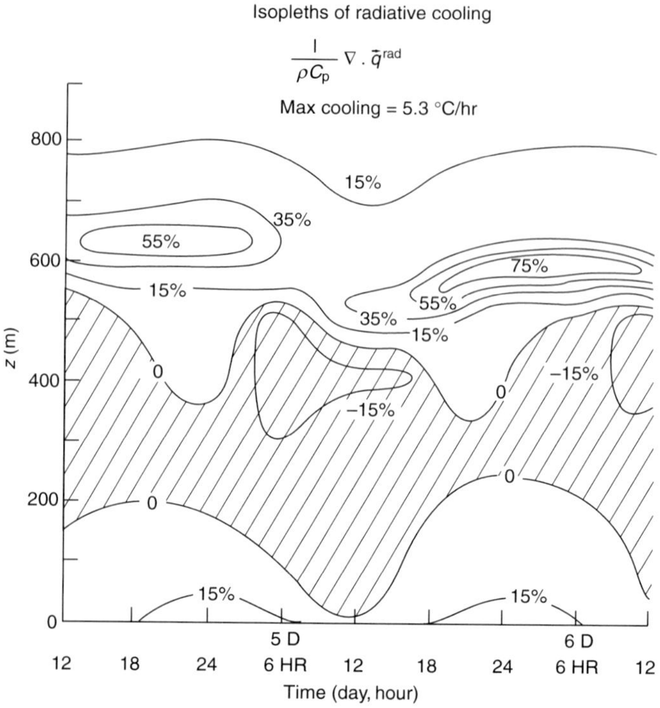


**FIGURE 6.33 Distribution of radiative cooling/heating in stratus cloud. Note that net heating occurs deep within the cloud interior.** *(From Oliver et al. (1978))*

Nicholls (1984) inferred from aircraft observations around the United Kingdom that absorption of solar radiation tends to establish a more stable stratification within the boundary layer which can lead to decoupling between the near surface mixed layer and the cloudy layer above.

Some observations suggest that if the clouds are thin enough solar heating may contribute to a reduction in cloud fraction (Rozendall et al., 1995). Pincus and Baker (1997) analyzed Lagrangian-type measurements taken during the Atlantic Stratocumulus Transition Experiment (Albrecht et al., 1995). They found that clouds that were optically thick in the morning hours were more likely to exhibit large diurnal variations in optical depth but remain unbroken. On the other hand, clouds that were optically thin in the morning were likely to break up without showing large variations in optical thickness. They concluded that a threshold optical depth of 10 demarked clouds that were likely to breakup versus those that remained solid stratus, with clouds having an optical depth greater than 10 unlikely to exhibit a cloud fraction of less than 80% during the day.

Both observations (Twomey, 1983) and modeling studies (Chen and Cotton, 1987) suggest that the humidity above the capping inversion top has a significant

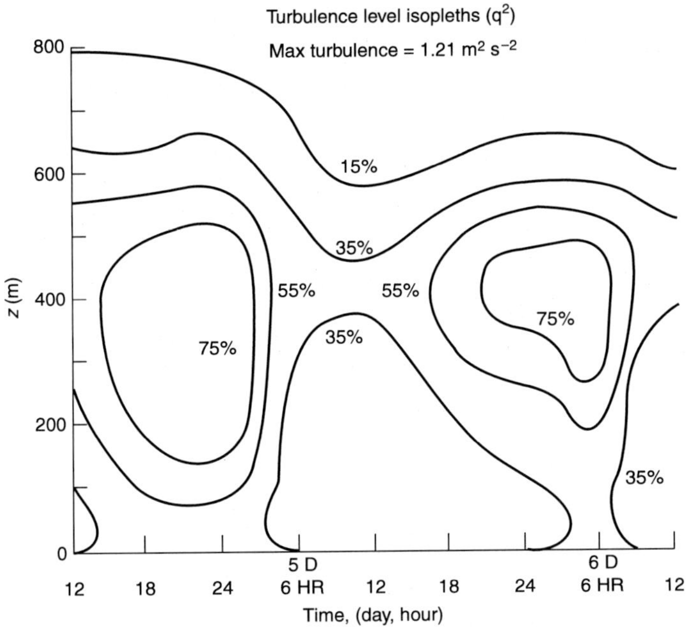


**FIGURE 6.34** **Distribution of turbulence in diurnally varying stratus cloud.** *(From Oliver et al. (1978))*

control on the response of a cloud layer to solar heating. The presence of a moist layer weakens longwave radiative cooling at cloud top. Instead of the maximum radiation cooling being localized near cloud top, it becomes distributed between cloud top and the top of the overlying moist layer. Because the moist layer is transparent to solar radiation, solar heating is unaffected. As a result, a cloud layer becomes far more responsive to solar heating, with cloud top descending and cloud base rising rapidly following sunrise.

## 6.6.13. Influence of Mid- and High-level Clouds

In our discussion of radiation fogs we noted that the advection of mid-and high-level clouds over a ground-based fog layer caused the dissipation of the fog by the reduction of longwave radiation cooling at fog top. One might expect a similar phenomenon to occur with marine stratocumulus clouds. Chen and Cotton (1987) simulated the response of a marine stratocumulus layer to upper level cloud layers with a second-order closure model. They imposed a cloud layer at a height of 5 km with a thickness of 50 m and a liquid-water content of 0.2 g kg$^{-1}$ over the marine stratocumulus layer previously described. The upper-level cloud layer reduced the longwave radiative cooling rate by 170° day$^{-1}$ and, as a consequence, lowered the maximum liquid-water content by 0.1 g kg$^{-1}$. In another experiment, Chen and Cotton introduced a cloud at a height of 3 km which was nearly 10 °C warmer than the first overlying cloud layer. As a result of the warmer radiating temperature of the overlying cloud, longwave radiative cooling was reduced by about 232 °C day$^{-1}$. The weaker longwave radiative cooling reduced the buoyancy production of turbulence and further reduced the cloud liquid-water content. While the response of a marine stratocumulus cloud to overlying cloud layers is more subtle than in a ground-based, radiative fog, the consequences are nonetheless quite important to the overall internal structure and dynamics of the stratocumulus layer.

## 6.7. ARCTIC STRATUS CLOUDS

Low-level stratiform clouds are a common feature in the central Arctic region in the summertime. Monthly average cloud cover amounts are nearly 70% for the months of May through September (Tsay and Jayaweera, 1984; Curry et al., 1996). They form in the boundary layer and in the free atmosphere, typically at heights below 2000 m. They are normally rather tenuous clouds with thicknesses of a few hundred meters and are frequently found to be composed of two or more well-defined layers (Jayaweera and Ohtake, 1973; Herman, 1977). The presence or absence of clouds, and their optical thickness has a large impact on the climate system by modulating the surface energy budget (e.g. Royer et al., 1990; Curry and Ebert, 1990; Ebert and Curry, 1993; Curry et al., 1993, 1995), and the surface heat and momentum fluxes (Maykut and Untersteiner, 1971). In particular, low-level cloudiness has a strong influence on the surface energy budget over the Arctic basin (Curry et al. 2000; Perovich et al., 1999; Carrió et al., 2005a,b).

According to Herman and Goody (1976) and Tsay and Jayaweera (1984), Arctic stratus clouds can form either as a convectively unstable cloud layer when cold polar air flows over a warmer sea surface, or as a stable cloud layer when warm, moist air flows over a cold sea surface. Tsay and Jayaweera note that clouds that form in a stable air mass are relatively thin and have bases low enough to frequently reach the sea surface. The clouds forming over the cold ice surface are for the most part what we have called decoupled clouds. The moisture supply to these low level clouds either comes from open leads (Burk et al., 1997) or horizontal advection of moist air from open ocean regions. Upper level stratiform cloud layers probably result from horizontal advection of moist air undergoing weak ascent.

The most unique aspect of summertime Arctic stratus clouds is that they occur with a large solar zenith angle (~74°), and at 80°N the sun remains above the horizon for 24 h day$^{-1}$ from May through August. Diurnal variations in cloud properties can therefore be expected to be slight. Herman and Goody (1976) concluded from their model calculations that the layered structure of Arctic stratus clouds was due to the persistent solar heating in the cloud interior.

Curry (1986) calculated shortwave and longwave radiative heating rates based on cloud liquid-water contents and droplet-size distributions observed in Arctic stratus. As seen in Fig. 6.35, owing to the large solar zenith angle,

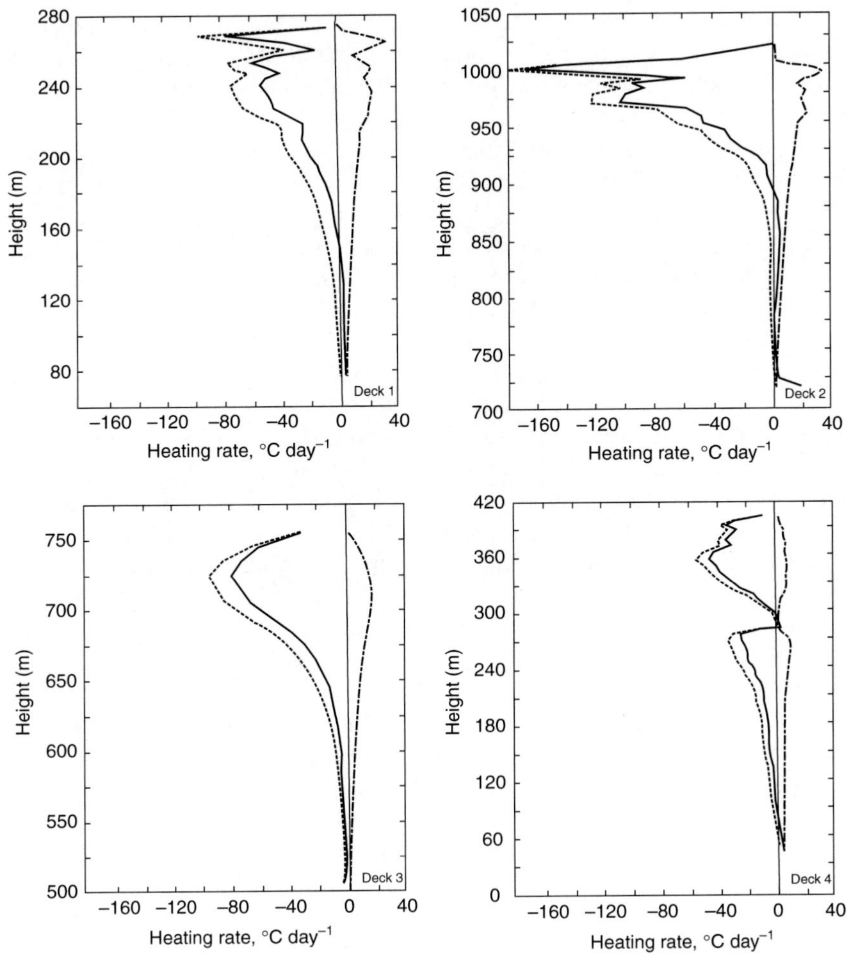


**FIGURE 6.35 Vertical profiles of heating rates: thermal radiation (dotted lines), solar radiation (dash-dot lines), and net radiation (solid lines).** *(From Curry (1986))*

longwave radiative cooling predominates over shortwave radiative heating, and the cloud layers exhibit a net radiative cooling. It is therefore common to find that the warmest days in the Arctic are cloudy days even during the summer months. This is particularly evident in the layered cloud (Deck 4) shown in Fig. 6.35. Of course, it is not essential that solar heating results in a net radiative warming for it to contribute to the dissipation of a layer. If radiative cooling maintains a cloud against the warming and drying effects of subsidence, then a reduced rate of net radiative cooling could lead to dissipation of a cloud layer. Curry et al. (1988) pointed out that the air above cloud top is often quite moist, thus the drying effects of subsidence are reduced.

Tsay and Jayaweera's (1984) analysis of the large-scale processes responsible for the formation of Arctic stratus clouds suggests that the layered structure may be a result of several simultaneous large-scale processes contributing to cloud formation.

In many respects the dynamics of layered Arctic stratus clouds appears to resemble the dynamics of weakly unstable stratocumuli observed over the United Kingdom (e.g. Nicholls, 1984; Nicholls and Leighton, 1986). In particular, an analysis of eddy fluxes reveals that many of the cloud layers are not coupled to surface fluxes. The dynamics of Arctic stratus, however, resemble more the dynamics of middle and high clouds discussed in Chapter 10 than ordinary stratocumulus.

Another distinquishing feature of Arctic stratus clouds is that except for mid-summer they are dominated by the ice-phase. Furthermore because they are often quite tenuous, relatively modest variations in the concentrations of CCN and IN can have large consequences. It has long been known that there are major intrusions of polluted air into the Arctic basin (Shaw, 1983, 1986, 1995) and that they often contain high concentrations of CCN (Borys and Rahn, 1981; Rahn, 1981; Patterson et al., 1982). More recently, during the recent SHEBA-FIRE spring field campaign, the airmass was found to be moderately polluted in terms of both cloud condensation nuclei (CCN) and ice nuclei (IN) concentrations immediately above the boundary layer on several occasions (Curry et al. 2000; Rogers et al., 2001; Carrió et al., 2005a,b).

Several modeling studies suggested that variations in CCN and IN concentrations associated with pollution intrusions into the pristine Arctic environment can affect the microphysical, dynamical and radiative properties of Arctic boundary layer clouds in several ways. Moreover, an observational study (Lubin and Vogelmann, 2006) concluded that for mostly liquid Arctic clouds enhanced CCN concentrations can increase downward longwave radiation to the sea ice that is climatologically significant ($\sim$3.4 W $m^{-2}$) and comparable to a warming effect from established greenhouse gases. Girard and Blanchet (2001a,b) investigated the effect of pollution-derived sulphuric acid aerosols for wintertime cloud systems. They hypothesized that IN concentrations are significantly reduced when the sulphuric acid aerosol concentration is high, as the sulphate coating deactivates the IN (Borys, 1989). Consequently, ice crystal number concentrations decrease and their size increases, increasing the sedimentation rates and the dehydration rate of the lower troposphere. As a result, the infrared radiation flux reaching the surface and the greenhouse effect are decreased (Girard and Blanchet, 2001b). For a largely liquid stratus deck, Harrington et al. (1999) and Harrington and Olsson (2001) used cloud-resolving model (CRM) simulations to show that a largely liquid stratus deck can be transformed into a broken optically-thin ice cloud by modest increases of IN concentrations.

CRM and LES studies of cloud-aerosol interactions for Arctic boundary layer clouds (Jiang et al., 2000, 2001; Carrió et al., 2005a,b) indicate an important response to varying amounts of CCN and IN. The latter two studies used data obtained during the Surface Heat Budget of the Arctic Ocean (SHEBA) field campaign. They found that enhanced IN concentrations increased simulated infrared emissivity due to higher ice crystal number concentrations and ice water contents due to lower sedimentation rates and higher ice particle residence times (ie. the clouds were over-seeded). In addition, Carrió et al. (2005b) extended the study to the entire melting season with the CRM coupled to the Los Alamos National Laboratory sea ice model (CICE). That study showed that cloud properties are modified to such an extent that the underlying sea-ice is affected.

In summary we have much to learn about Arctic stratus clouds and the extent to which they are modified by aerosol pollution intrusions in the Arctic basin. Because clouds in this region are so important to both regional and global climate, furthering that understanding is paramount to understanding our climate system.

## REFERENCES

Ackerman, A. S., Kirkpatrick, M. P., Stevens, D. E., and Toon, O. B. (2004). The impact of humidity above stratiform clouds on indirect aerosol climate forcing. Nature 432, 1014–1017.

Agee, E. M. (1984). Observations from space and thermal convection: A historial perspective. Bull. Amer. Meteorol. Soc. 65, 938–949.

Albrecht, B. A. (1989). Aerosols, cloud microphysics, and fractional cloudiness. Science 245, 1227–1230.

Albrecht, B. A. (1991). Fractional cloudiness and cloud-top entrainment instability. J. Atmos. Sci. 48, 1519–1525.

Albrecht, B. A., Penc, R. S., and Schubert, W. H. (1985). An observational study of cloud-topped mixed layers. J. Atmos. Sci. 42, 800–822.

Albrecht, Bruce A., Bretherton, Christopher S., Johnson, Doug, Scubert, Wayne H., and Shelby Frisch, A. (1995). The Atlantic Stratocumulus Transition Experiment—ASTEX. Bull. Amer. Meteor. Soc. 76, 889–904.

Anderson, J. B. (1931). Observations from air planes of cloud and fog conditions along the southern California Coast. Mon. Weather Rev. 59, 264–270.

Ball, F. K. (1960). Control of inversion height by surface heating. Q. J. R. Meteorol. Soc. 86, 483–494.

Ballard, S. P., Golding, B. W., and Smith, R. N. B. (1991). Mesoscale model experimental forecasts of the Haar of the Northeast Scotland. Mon. Weather Rev. 119, 2107–2123.

Barkstrom, B. R. (1978). Some effects of 8–12 μm radiant energy transfer on the mass and heat budgets of cloud droplets. J. Atmos. Sci. 35, 665–673.

Bennetts, D. A., McCallum, E., and Nicholls, S. (1986). Stratocumulus: an introductory account. Meteorol. Mag. 115, 65–76.

Bergot, T., and Guedalia, D. (1994a). Numerical forecasting of radiation fog. Part I: Numerical model and sensivity tests. Mon. Weather Rev. 142, 1218–1230.

Borys, R. D. (1989). Studies of ice nucleation by Arctic aerosol on AGASP-II. J. Atmos. Chem. 9, 169–185.

Borys, R. D., and Rahn, K. A. (1981). Long-range atmospheric transport of cloud-active aerosol to Iceland. Atmos. Env. 15 (8), 1491–1501.

Bott, A., and Trautman, T. (2002). FAFOG new efficient forecast model of radiation fog and low-level striform clouds. Atmos. Res. 64, 191–203.

Bougeault, P. (1985). The diurnal cycle of the marine stratocumulus layer: A higher-order model study. J. Atmos. Sci. 42, 2826–2843.

Bretherton, C. S., and Wyant, M. C. (1997). Moisture transport, lower-tropospheric stability, and decoupling of cloud-topped boundary layers. J. Atmos. Sci. 54, 148–167.

Bretherton, C. S., and coauthors, (1999). An intercomparison of radiatively driven entrainment and turbulence in a smoke cloud, as simulated by different numerical models. Q. J. R. Meterol. Soc. 125, 391–423.

Bretherton, C. S. (1990). Lagrangian development of a cloud-topped boundary layer in a turbulence closure model. In "Conf. on Cloud Physics, San Francisco, CA." pp. 48–55, Amer. Meteorol. Soc. Preprints.

Bretherton, C. S. (1992). A conceptual model of the stratocumulus-trade-cumulus transition in the subtropical oceans. In "Proc., 11th Int. Conf. on Clouds and Precipitation, Vol. 1, Montreal, Quebec, Canada." pp. 374–377, International Commission of Clouds and Precipitation and International Association of Meteorology and Atmospheric Physics.

Brost, R. A., Lenschow, D. H., and Wyngaard, J. C. (1982a). Marine stratocumulus layers. Part I: Mean conditions. J. Atmos. Sci. 39, 800–817.

Brost, R. A., Wyngaard, J. C., and Lenschow, D. H. (1982b). Marine stratocumulus layers. Part II: Turbulence budgets. J. Atmos. Sci. 39, 818–836.

Brown, R. (1980). A numerical study of radiation fog with an explicit formulation of the microphysics. Q. J. R. Meteorol. Soc. 106, 781–802.

Brown, R., and Roach, W. T. (1976). The physics of radiation fog: II—A numerical study. Q. J. R. Meteorol. Soc. 102, 335–354.

Burk, S. D., Fett, R. W., and Englebretson, R. E. (1997). Numerical simulation of cloud plumes emanating from Arctic leads. J. Geophys. Res. 102, 16,529–16,544.

Byers, H. R. (1959). "General Meteorology." McGraw-Hill, New York.

Carrió, G. G., Jiang, H., and Cotton, W. R. (2005a). Impact of aerosol intrusions on Arctic boundary layer clouds. Part I: 4 May 1998 case. J. Atmos. Sci. 62, 3082–3093.

Carrió, G. G., Jiang, H., and Cotton, W. R. (2005b). Impact of aerosol intrusions on Arctic boundary layer clouds. Part II: Multi-month simulations. J. Atmos. Sci. 62, 3094–3105.

Caughey, S. J., and Kitchen, M. (1984). Simultaneous measurements of the turbulent and microphysical structure of nocturnal stratocumulus cloud. Q. J. R. Meteorol. Soc. 110, 13–34.
Caughey, S. J., Crease, B. A., and Roach, W. T. (1982). A field study of nocturnal stratocumulus. II: Turbulence structure and entrainment. Q. J. R. Meteorol. Soc. 108, 125–144.
Chai, S. K., and Telford, J. W. (1983). Convection model for stratus cloud over a warm water surface. Boundary-Layer Meteorol. 26, 25–49.
Chen, C., and Cotton, W. R. (1983a). A one-dimensional simulation of the stratocumulus-capped mixed layer. Boundary-Layer Meteorol. 25, 289–321.
Chen, C, and Cotton, W. R. (1983b). Numerical experiments with a one-dimensional higher order turbulence model: Simulation of the Wangara day 33 case. Boundary-Layer Meteorol. 25, 375–404.
Chen, C., and Cotton, W. R. (1987). The physics of the marine stratocumulus-capped mixed layer. J. Atmos. Sci. 44, 2951–2977.
Chibe, R. J. (2003). The numerical simulation of fog with the RAMS@CSU cloud-resolving mesoscale forecast model. M.S. Thesis, Colorado State University, Dept. of Atmospheric Science, Fort Collins, CO 100 pp. [Available as Atmospheric Science Paper No. 741.].
Choularton, T. W., Fullarton, G., Latham, J., Mill, C. S., Smith, M. H., and Stromberg, I. M. (1981). A field study of radiation fog in Meppen, West Germany. Q. J. R. Meteorol. Soc. 107, 381–394.
Cotton, W. R., Pielke Sr., R. A., Walko, R. L., Liston, G. E., Tremback, C. J., Jiang, H., McAnelly, R. L., Harrington, J. Y., Nicholls, M. E., Carrió, G. G., and McFadden, J. P. (2003). RAMS 2001: Current status and future directions. Meteorol. Atmos. Phys. 82, 5–29.
Curry, J. A. (1986). Interactions among turbulence, radiation and microphysics in Arctic stratus clouds. J. Atmos. Sci. 43, 90–106.
Curry, J. A., and Ebert, E. E. (1990). Sensitivity of the thickness of Arctic sea ice to the optical properties of clouds. Ann. Glaciol. 14, 43–46.
Curry, J. A., Schramm, J. L., and Ebert, E. E. (1995). Sea ice-albedo climate feedback mechanism. J. Climate 8, 240–247.
Curry, J. A., Rossow, W. B., Randall, D., and Schramm, J. L. (1996). Overview of Arctic cloud and radiation characteristics. J. Climate 9, 1731–1764.
Curry, J. A., Ebert, E. E., and Herman, G. F. (1988). Mean and turbulence structure of the summertime Arctic cloudy boundary layer. Q. J. R. Meteorol. Soc. 114, 715–746.
Curry, J. A., Schramm, J. L., and Ebert, E. E. (1993). Impact of clouds on the surface radiation balance of the Arctic Ocean. Meteorol. Atmos. Phys. 51, 197–217.
Deardorff, J. W. (1976). On the entrainment rate of a stratocumulus-topped mixed layer. Q. J. R. Meteorol. Soc. 102, 563–582.
Deardorff, J. W. (1980a). Cloud top entrainment instability. J. Atmos. Sci. 37, 131–147.

Deardorff, J. W. (1980b). Stratocumulus-capped mixed layers derived from a three-dimensional model. Bound Layer Meteorol. 18, 495–527.

Defant, F. (1951). Local winds. In "Compendium of Meteorology" (T. F. Malone, Ed.), pp. 655–672. Am. Meteorol. Soc, Boston, Massachusetts.

Donaldson, N. R., and Stewart, R. E. (1993). Fog induced by mixed-phase precipitation. Atmos. Res. 29, 9–25.

Durkee, P. A., and coauthors, (2000a). Composite ship track characteristics. J. Atmos. Sci. 57, 2542–2553.

Durkee, P. A., coauthors, K. J., Noone, R. T., and Bluth, (2000b). The Monterey Area Ship Track experiment. J. Atmos. Sci. 57, 2523–2728.

Duynkerke, P. G. (1991). Radiation fog: A comparison of model simulation with detailed observations. Mon. Weather Rev. 119, 324–341.

Duynkerke, P. G. (1993). The stability of cloud top with regard to entrainment: Amendment of the theory of cloud-top entrainment instability. J. Atmos. Sci. 50, 495–502.

Duynkerke, P. G., Zhang, H., and Jonker, P. J. (1995). Microphysical and turbulent structure of nocturnal stratocumulus as observed during ASTEX. J. Atmos. Sci. 52, 2763–2777.

Ebert, E. E., and Curry, J. A. (1993). An intermediate one-dimensional thermodynamic sea ice model for investigating ice-atmosphere interactions. J. Geophys. Res. 98, 10,085–10,109.

Feingold, G., Stevens, B., Cotton, W. R., and Frisch, A. S. (1996). The relationship between drop in-cloud residence time and drizzle production in numerically simulated stratocumulus clouds. J. Atmos. Sci. 53, 1108–1122.

Fleagle, R. G., Parrott, W. H., and Barad, M. L. (1952). Theory and effects ofvertical temperature distribution in turbid air. J. Meteorol. 9, 53–60.

Forkel, R., Sievers, U., and Zdunkowski, W. (1987). Fog modelling with a new treatment of the chemical equilibrium condition. Contrib. Atmos. Phys. 60, 340–360.

Geiger, R. (1965). "The Climate Near the Ground." Harvard University Press, Cambridge, Massachusetts.

Girard, E., and Blanchet, J.-P. (2001a). Microphysical parameterization of Arctic diamond dust, ice forg, and thin stratus for climate models. J. Atmos. Sci. 58, 1181–1198.

Girard, E., and Blanchet, J.-P. (2001b). Simulation of Arctic diamond dust, ice fog, and thin stratus using an explicit aerosol-cloud-radiation fog. J. Atmos. Sci. 58, 1199–1221.

Golaz, J.-C., Larson, V. E., and Cotton, W. R. (2002a). A PDF-based model for boundary layer clouds. Part I: Method and model description. J. Atmos. Sci. 59, 3540–3551.

Golaz, J.-C., Larson, V. E., and Cotton, W. R. (2002b). A PDF-based model for boundary layer clouds. Part II: Model results. J. Atmos. Sci. 59, 3552–3571.

Hanson, H. P. (1984a). Stratocumulus instability reconsidered: A search for physical mechanisms. Tellus 36A, 355–368.

Hanson, H. P. (1984b). On mixed-layer modeling of the stratocumulus-topped marine boundary layer. J. Atmos. Sci. 41, 1226–1234.

Harrington, J. Y., and Olsson, P. Q. (2001). On the potential influence of ice nuclei on surface-forced marine stratocumulus clouds dynamics. J. Geophys. Res. 106 (D21), 27–473, 27–484.

Harrington, J. Y., Reisin, T., Cotton, W. R., and Kreidenweis, S. M. (1999). Cloud resolving simulations of Arctic stratus. Part II: Transition-season clouds. Atmos. Res. 55, 45–75.

Herman, G. F. (1977). Solar radiation in summertime Arctic stratus clouds. Q. J. R. Meteorol. Soc. 34, 1423–1432.

Herman, G. F., and Goody, R. (1976). Formation and persistence of summertime Arctic stratus clouds. J. Atmos. Sci. 33, 1537–1553.

Jayaweera, K. O., and Ohtake, T. (1973). Concentrations of ice crystals in Arctic stratus clouds. J. Rech. Atmos 7, 199–207.

Jensen, J. B., Lee, S., Krummel, P. B., Katzfey, J., and Gogoasa, D. (2000). Precipitation in marine cumulus and stratocumulus. Part I: Thermodynamic and dynamic observations of closed cell circulations and cumulus bands. Atmos. Res. 54, 117–155.

Jiang, H., Cotton, W. R., Pinto, J. O., Curry, J. A., and Weissbluth, M. J. (2000). Cloud resolving simulations of mixed-phase Arctic stratus observed during BASE: Sensitivity to concentration of ice crystals and large-scale heat and moisture advection. J. Atmos. Sci. 57, 2105–2117.

Jiang, H., Feingold, G., Cotton, W. R., and Duynkerke, P. G. (2001). Large-eddy simulations of entrainment of cloud condensation nuclei into the Arctic boundary layer: 18 May 1998 FIRE/SHEBA case study. J. Geophys. Res. 106 (D14), 15,113–15,122.

Jiang, H., Feingold, G., and Cotton, W. R. (2002). Simulations of aerosol-cloud-dynamical feedbacks resulting from entrainment of aerosol into the marine boundary layer during the Atlantic Stratocumulus Transition Experiment. J. Geophys. Res. 107 (D24), 4813. doi:10.1029/2001JD001502.

Jiusto, J. E. (1980). Fog structure. lnvited Rev. Pap., Symp. Workshop Clouds. Their Form., Opt. Prop. Eff., IFA ORS, Williamsburg, Va.

Jiusto, J. E., and Lala, G. G. (1980). Thermodynamics of radiation fog formation and dissipation—A case study. In "Int. Cloud Phys. Conf., 8th, Clermont-Ferrand, Fr." pp. 333–335. Preprints.

Kahn, P. H., and Businger, J. A. (1979). The effect of radiative flux divergence on entrainment of a saturated convective boundary layer. Q. J. R. Meteorol. Soc. 105, 303–304.

Khalsa, S. J. S. (1993). Direct sampling of entrainment events in a marine stratocumulus layer. J. Atmos. Sci. 50, 1734–1750.

Klein, S. A, and Hartman, D. L. (1993). The seasonal cycle of low stratiform clouds. J. Climate 6, 1587–1606.

Klein, S. A, Hartman, D. L., and Norris, J. R. (1995). On the relationships among low cloud structure, sea surface temperature, and atmospheric circulation in the summertime northeast Pacific. J. Climate 8, 1140–1155.

Korb, G., and Zdunkowski, W. (1970). Distribution of radiative energy in ground fog. Tellus 22, 298–320.

Kraus, E. B. (1963). The diurnal precipitation change over the sea. J. Atmos. Sci. 20, 551–556.

Kropfli, R. A., and Orr, B. W. (1993). Observations of microcells in the marine boundary layer with 8-mm wavelength Doppler radar. In "Proceedings, 26th International Conference on Radar Meteorology, 24-28 May 1993." pp. 492–494. Amer. Met. Soc., Norman, OK.

Krueger, S. K., McLean, G. T., and Fu, Q. (1995a). Numerical simulation of the stratus-to-cumulus transition in the subtropical marine boundary layer. Part I: Boundary-layer circulation. J. Atmos. Sci. 52, 2839–2850.

Krueger, S. K., McLean, G. T., and Fu, Q. (1995b). Numerical simulation of the stratus-to-cumulus transition in the subtropical marine boundary layer. Part II: Boundary-layer structure. J. Atmos. Sci. 52, 2851–2868.

Kuo, H.-C. (1987). Dynamical modeling of marine boundary layer convection. Thesis, Dep. Atmos. Sci., Colorado State University.

Kuo, H.-C., and Schubert, W. H. (1988). Stability of cloud-topped boundary layers. Q. J. R. Meteorol. Soc. 114, 887–916.

Lala, G. G., Mandel, E., and Jiusto, J. E. (1975). A numerical evaluation of radiation fog variables. J. Atmos. Sci. 32, 720–728.

Lala, G. G., Jiusto, J. E., Meyer, M. B., and Kornfein, M. (1982). Mechanisms of radiation fog formation on four consecutive nights. In "Conf. Cloud Phys., Chicago, Ill".

Larson, V. E., Golaz, J.-C., Jiang, H., and Cotton, W. R. (2005). Supplying local microphysics parameterizations with information about subgrid variability: Latin hypercube sampling. J. Atmos. Sci. 62, 4010–4026.

Leipper, D. F. (1948). Fog development at San Diego, California. J. Mar. Res. 7, 337–346.

Lewellen, D. C., and Lewellen, W. S. (1998). Large-eddy boundary layer entrainment. J. Atmos. Sci. 55, 2645–2665.

Lilly, D. K. (1968). Models of cloud-topped mixed layers under a strong inversion. Q. J. R. Meteorol. Soc. 94, 292–309.

Lilly, D. K. (2002a). Entrainment into mixed layers. Part I: Sharp-edged and smoothed tops. J. Atmos. Sci. 59, 3340–3352.

Lilly, D. K. (2002b). Entrainment into mixed layers. Part II: A new closure. J. Atmos. Sci. 59, 3353–3361.

Lock, A. P. (1998). The parameterization of entrainment in cloudy boundary layers. Q. J. R. Meteorol. Soc. 124, 2729–2753.

Lubin, D., and Vogelmann, A. (2006). A climatologically significant aerosol longwave indrect effect in the Arctic. Nature 439, 453–456.

Mack, E. J., Katz, U., Rogers, C., and Pilie, R. (1974). The microstructure of California coastal stratus and fog at sea. Rep. CJ-5405-M-1, Calspan Corp., Buffalo, New York.

MacVean, M. K., and Mason, P. J. (1990). Cloud top entrainment instability through small-scale mixing and its parameterization in numerical models. J. Atmos. Sci. 47, 1012–1030.

Mahrt, L., and Paumier, J. (1982). Cloud-top entrainment instability observed in AMTEX. J. Atmos. Sci. 39, 622–634.

Mason, J. (1982). The physics of radiation fog. J. Meteorol. Soc. Jpn 60, 486–498.

Maykut, G. A., and Untersteiner, N. (1971). Some results from a time dependent thermodynamic model of sea ice. J. Geophys. Res. 76, 1550–1575.

Miller, M. A., and Albrecht, B. A. (1995). Surface-based observations of mesoscale cumulus-stratocumulus interaction during ASTEX. J. Atmos. Sci. 52, 2809–2826.

Moeng, C.-H. (1986). Large-eddy simulation of a stratus-topped boundary layer. Part I: Structure and budgets. J. Atmos. Sci. 43, 2886–2900.

Moeng, C.-H., and Arakawa, A. (1980). A numerical study of a marine subtropical stratus cloud layer and its stability. J. Atmos. Sci. 37, 2661–2676.

Moeng, C.-H. (2000). Entrainment rate, cloud fraction, and liquid water path of PBL stratocumulus clouds. J. Atmos. Sci. 57, 3627–3643.

Musson-Genon, L. (1987). Numerical simulation of a fog event with a one-dimensional boundary layer model. Mon. Weather Rev. 115, 592–607.

Neiburger, M., Johnson, D. S., and Chen, C.-W. (1961). "The Inversion over the Eastern North Pacific Ocean." In Studies of the Structure of the Amosphere over the Eastern Pacific Ocean in Summer, vol. 1. University of Chicago Press, 94 pp.

Nicholls, S. (1984). The dynamics of stratocumulus: Aircraft observations and comparisons with a mixed-layer model. Q. J. R. Meteorol. Soc. 110, 783–820.

Nicholls, S., and Leighton, J. (1986). An observational study of the structure of stratiform cloud sheets: Part 1. Structure. Q. J. R. Meteorol. Soc. 112, 431–460.

Nicholls, S., and Turton, J. D. (1986). An observational study of the structure of stratiform cloud sheets. Part 11. Entrainment. Q. J. R. Meteorol. Soc. 112, 461–480.

Obukhov, A. M. (1946). Turbulence in an atmosphere with non-uniform temperature. Tr. Inst. Teor. Geofiz., Akad. Nauk SSSR 1, 95–115. Engl. transl., Boundary-Layer Meteorol. 2, 7–29 (1971).

Oliver, D. A., Lewellen, W. S., and Williamson, G. G. (1978). The interaction between turbulent and radiative transport in the development of fog and low-level straius. J. Atmos. Sci. 35, 301–316.

Paluch, I. R., and Lenschow, D. H. (1991). Stratiform cloud formation in the marine boundary layer. J. Atmos. Sci. 48, 2141–2158.

Patterson, E. M., Marshall, B. T., and Rahn, K. A. (1982). Radiative properties of the arctic aerosol. Atmos. Environ. 16, 2967–2677.

Perovich, D. K., Grenfell, T. C., Light, B., Richter-Menge, J. A., Sturm, M., Tucker III, W. B., Eicken, H., Maykut, G. A., and Elder, B. (1999). SHEBA: Snow and ice studies, CD-ROM, October 1999.

Pilie, R. J. (1975). The life cycle of valley fog. Part 11: Fog microphysics. J. Appl. Meteorol. 14, 364–374.

Pilie, R. J., Mack, E. J., Kocmond, W. C., Eadie, W. J., and Rogers, C. W. (1975). The life cycle of valley fog, I, Micrometeorological characteristics. J. Appl. Meteorol. 14, 357–363.

Pilie, R. J., Mack, E. J., Rogers, C. W., Katz, U., and Kocmond, W. C. (1979). The formation of marine fog and the development of fog-stratus systems along the California coast. J. Appl. Meteorol. 18, 1275–1286.

Pincus, R., and Klein, S. A. (2000). Unresolved spatial varability and microphysical process rates in large-scale models. J. Geophys. Res. 105 (D22), 27,059–27,065.

Pincus, R., and Baker, M. B. (1997). What controls stratocumulus radiative properties? Lagrangian observations of cloud evolution. J. Atmos. Sci. 54, 2215–2236.

Rahn, K. A. (1981). Relative importances of North America and Eurasia as sources of arctic aerosol. Atmos. Environ. 15, 1447–1455.

Randall, D. A. (1980). Entrainment into a stratocumulus layer with distributed radiative cooling. J. Atmos. Sci. 37, 148–159.

Roach, W. T. (1976). On the effect of radiative exchange on the growth by condensation of a cloud or fog droplet. Q. J. R. Meteorol. Soc. 102, 361–372.

Roach, W. T., Brown, R., Caughey, S. J., Crease, B. A., and Slingo, A. (1982). A field study of nocturnal stratocumulus: I. Mean structure and budgets. Q. J. R. Meteorol. Soc. 108, 103–123.

Rogers, D. P., and Telford, J. W. (1986). Metastable stratus tops. Q. J. R. Meteorol. Soc. 112, 481–500.

Rogers, D. C., DeMott, P. J., and Kreidenweis, S. M. (2001). Airborne measurements of tropospheric ice-nucleating aerosol particles in the Arctic spring. J. Geophys. Res. 106, 15,053–15,063.

Rogers, D. P., Yang, X., Norris, P. M., Johnson, D. W., and Martin, G. M. (1995). Diurnal evolution of the cloud-topped marine boundary layer. Part I: Nocturnal stratocumulus development. J. Atmos. Sci. 52, 2953–2966.

Royer, J. R., Planton, S., and Deque, M. (1990). A sensitivity experiment for the removal of Arctic sea ice with the French specitral general circulation model. Clim. Dyn. 5, 1–17.

Rozendall, M. A., Leovy, C. B., and Klein, S. A. (1995). An observational study of diurnal variations of marine stratiform cloud. J. Climate 8, 1795–1809.

Saunders, P. M. (1957). The thermodynamics of saturated air: A contribution to the classical theory. Q. J. R. Meteorol. Soc. 83, 342–350.

Saunders, P. M. (1961). An observational study of cumulus. J. Atmos. Sci. 18, 451–467.

Schubert, W. H. (1976). Experiments with Lilly's cloud-topped model. J. Atmos. Sci. 33, 436–446.

Schubert, W. H., Wakefield, J. S., Steiner, E. J., and Cox, S. K. (1979a). Marine stratocumulus convection. Part 1: Governing equations and horizontally homogeneous solutions. J. Atmos. Sci. 33, 1286–1307.

Schubert, W. H., Wakefield, J. S., Steiner, E. J., and Cox, S. K. (1979b). Marine stratocumulus convection. Part II: Horizontally homogeneous solutions. J. Atmos. Sci. 36, 1308–1324.

Sharon, T. M., Albrecht, B. A., Jonsson, H. H., Minnis, P., Khaiyer, M. M., van Reken, T. M., Seinfeld, J., and Flagan, R. (2006). Aerosol and cloud microphysical chacteristics of rifts and gradients in maritime stratocumulus clouds. J. Atmos. Sci. 63, 983–997.

Shaw, G. E. (1983). On the aerosol particle size distribution spectrum in Alaskan air mass systems: Arctic haze and non-haze episodes. J. Atmos. Sci. 40, 1313–1320.

Shaw, G. E. (1986). Cloud condensation nuclei associated with Arctic haze. Atmos. Environ. 20 (7), 1453–1456.

Shaw, G. E. (1995). The Arctic haze phenomenon. Bull. Amer. Meteorol. Soc. 76, 2403–2413.

Siebert, J., Bott, A., and Zdunkowski, W. (1992). Influence of a vegetation-soil model on the simulation of radiation fog. Beitr. Phys. Atmos. 65, 93–106.

Smith, W. S., and Kao, C.-Y. J. (1996). Numerical simulations of the marine stratocumulus-capped boundary layer and its diurnal variation. Mon. Weather Rev. 124, 1803–1816.

Stage, S. A., and Businger, J. A. (1981a). A model for entrainment into a cloud-topped marine boundary layer. Part I: Model description and application to a cold-air outbreak episode. J. Atmos. Sci. 38, 2213–2229.

Stage, S. A., and Businger, J. A. (1981b). A model for entrainment into a cloud-topped marine boundary layer. Part II: Discussion of model behavior and comparison with other models. J. Atmos. Sci. 38, 2230–2242.

Stevens, B. (2006). Bulk boundary-layer concepts for simplified models of tropical dynamics. Theor. Comput. Fluid Dyn. doi:10.1007/s00162-006-0032-z.

Stevens, B., Cotton, W. R., Feingold, G., and Moeng, C.-H. (1998). Large-eddy simulations of strongly precipitating, shallow, stratocumulus-topped boundary layers. J. Atmos. Sci. 55, 3616–3638.

Stevens, B., Moeng, C.-H., and Sullivan, P. P. (1999). Large-eddy simulations of radiatively driven convection: Sensitivities to the representation of small scales. J. Atmos. Sci. 56, 3963–3984.

Stevens, B., Vali, G., Comstock, K., Wood, R., van Zanten, M. C., Austin, P. H., Bretherton, C. S., and Lenschow, D. H. (2005). Pockets of open cells and drizzle in marine stratocumulus. Bull. Amer. Meteorol. Soc. 86, 51–57.

Stevens, D., and Bretherton, C. S. (1999). Effects of resolution on the simulation of stratocumulus entrainment. Q. J. R. Meteorol. Soc. 125, 425–439.

Stull, R. B. (1976). The Energetics of Entrainment Across a Density Interface. J. Atmos. Sci. 33, 1260–1267.

Taylor, G. I. (1917). The formation of fog and mist. Q. J. R. Meteorol. Soc. 43, 241–268.

Teixera, J. (1999). Simulation of fog with the ECMWF prognostic cloud scheme. Q. J. R. Meteorol. Soc. 125, 529–552.

Tjernström, M., and Rogers, D. P. (1996). Turbulence structure in decoupled marine stratocumulus: A case study from the ASTEX field experiment. J. Atmos. Sci. 53, 598–619.

Tsay, S.-C., and Jayaweera, K. (1984). Physical characteristics of arctic stratus clouds. J. Climate Appl. Meteorol. 23, 584–596.

Twomey, S. (1983). Radiative effects in California stratus. Contrib. Atmos. Phys. 56, 429–439.

van Zanten, M. C., and Stevens, B. (2005). Observations of the structure of heavily precipitating marine stratocumulus. J. Atmos. Sci. 62, 4327–4342.

Wakefield, J. S., and Schubert, W. H. (1981). Mixed-layer model simulation of eastern Northe Pacific stratocumulus. Mon. Weather Rev. 109, 1952–1968.

Walter Jr., B. A., and Overland, J. E. (1984). Observations of longitudinal rolls in a near neutral atmosphere. Mon. Weather Rev. 112, 200–208.

Wang, Q., and Albrecht, B. A. (1994). Observations of cloud-top entrainment in marine stratocumulus clouds. J. Atmos. Sci. 51, 1530–1547.

Weaver, C. J., and Pearson, R. (1990). Entrainment instability and vertical motion as causes of stratocumulus breakup. Q. J. R. Meteorol. Soc. 116, 1359–1388.

Welch, R. M., Ravichandran, M. G., and Cox, S. K. (1986). Prediction of quasi-periodic oscillations in radiation fogs. Part I: Comparison of simple similarity approaches. J. Atmos. Sci. 43, 633–651.

Wells, W. C. (1838). An essay on dew and several appearances connected with it. In "On the Influence of Physical Agents on Life" (W. F. Edwards., Ed.), Haswell, Barrington & Haswell, Philadelphia, Pennsylvania.

Willit, H. C. (1928). Fog and haze. Mon. Weather Rev. 56, 435.

Wyant, M. C., and Bretherton, C. S. (1992). The dynamics of decoupling in a cloud-topped boundary layer. In "Proc., 11th Conf. on Clouds and Precipitation, Vol. 1, Montreal, Quebec, Canada." pp. 417–420. International Comission on Clouds and Precipitation and International Association of Meteorology and Atmospheric Physics.

Wyant, M. C., Bretherton, C. S., Rand, H. A., and Stevens, D. E. (1997). Numerical simulation and a conceptual model of the stratocumulus to trade cumulus transition. J. Atmos. Sci. 54, 168–192.

Yamaguchi, Takanobu, and Randall, David A. (2008). Large-eddy simulation of evaporatively driven entrainment in cloud-topped mixed layers. J. Atmos. Sci. 65, 1481–1504.

Zdunkowski, W. G., and Barr, A. E. (1972). A radiative conductive model for the prediction of radiative fog. Boundary-Layer Meteorol. 2, 152–177.

Zdunkowski, W. B., and Nielsen, B. C. (1969). A preliminary prediction analysis of radiation fog. Pure Appl. Geophys. 75, 278–299.

Chapter 7

# Cumulus Clouds

## 7.1. INTRODUCTION

Cumulus clouds take on a variety of forms and sizes ranging from non-precipitating fair-weather cumuli to heavily precipitating thunderstorms. In this chapter we shall discuss the dynamic characteristics of cumuli, ranging from boundary layer cumuli to towering cumuli or cumulus congestus. Stull (1985) proposed a classification of fair-weather cumulus clouds according to their interaction with the atmospheric boundary layer. He considered three categories: forced, active, and passive clouds. Figure 7.1 illustrates the differences among the three cloud categories. Forced cumulus clouds form at the tops of boundary layer thermals that overshoot into the stable layer that caps the ABL. The thermals rise above the lifting condensation level but because they are unable to rise above the level of free convection (LFC), they remain negatively buoyant during overshoot. Active fair-weather cumulus clouds ascend above the LFC and therefore become positively buoyant. As a consequence of gaining positive buoyancy, they ascend to greater heights than forced cumuli, and develop circulations that depart from those characteristic of dry ABL thermals. Passive clouds are the decaying remnants of formerly active clouds. They can be readily identified by the absence of a flat base. Their importance is mainly due to the fact that decaying clouds may account for a significant fraction of the total cloud coverage (Albrecht, 1981). This is especially true on those days in which the free atmosphere is humid and cloud evaporation is slow. Thus, passive clouds can significantly affect the amount of radiative heating/cooling in the ABL. Over land, passive clouds mainly affect the ABL by shading the ground and thereby reducing surface heating. We begin this chapter by examining boundary layer or fair-weather cumuli as an ensemble of cloud elements that can be thought of as an extension of the cloud-free boundary layer. We then examine the organization of cumulus clouds. Finally, we focus on the properties of individual cumuli by examining the processes of entrainment in cumuli and the initiation and maintenance of convective-scale updrafts and downdrafts, and the interaction of clouds.

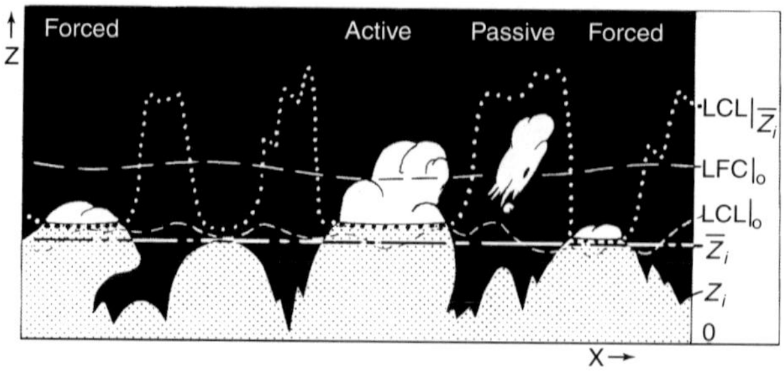


**FIGURE 7.1 Schematic of the relationship of cumulus clouds to various fair-weather mixed-layer (ML) characteristics.** The lightly shaded region denotes ML air, the black region denotes free-atmosphere air, and the white regions denote clouds. The horizontal average ML depth is indicated by $\bar{Z}_i$, while $Z_i$ is used for the local ML top. The dotted line shows the local lifting condensation level (LCL) for air measured at $\bar{Z}_i$; the short-dash line shows the local LCL for air measured in the surface layer; the long-dash line shows the local level of free convection (LFC) for air measured in the surface layer. Cloud classes are labeled above their respective sketches. *(From Stull (1985))*

## 7.2. BOUNDARY LAYER CUMULI—AN ENSEMBLE VIEW

Boundary layer cumuli are cumulus clouds whose vertical extent is limited by a pronounced capping inversion, often viewed as the top of the atmospheric boundary layer. As such, they do not markedly differ from stratocumulus clouds, except that the coverage of cumulus clouds remains considerably less than 100%. In fact, the breakup of a stratocumulus layer often results in the formation of a layer of boundary-layer-confined cumuli. Such clouds are often non-precipitating, and are usually referred to as fair-weather cumulus, cumulus humilus, or trade-wind cumulus. However it is not uncommon that warm-based, maritime boundary layer cumuli may precipitate, especially during high wind situations. Boundary layer cumuli are often composed of an ensemble of forced, active, and passive clouds. The active cells, however, do not ascend more than a few kilometers above the capping inversion.

There is some evidence that boundary-layer-confined cumuli differ significantly from more active, inversion-penetrating cumulus clouds such as towering cumuli or cumulonimbi. In a diagnostic study of tropical cumuli, Esbensen (1978) concluded that the time scale for boundary layer cumuli to reach a state of quasi-equilibrium with the thermodynamic structure of their environment is significantly longer than the adjustment time for deeper clouds. This has led to the idea that it may be necessary to parameterize separately the effects of deep and shallow cumuli for many applications. The observational study of the distribution of cloud cover by Plank (1969) also suggests that boundary layer cumuli and deep cumuli differ markedly. Plank noted that the morning populations of cumuli over Florida were essentially unimodal, whereas the distribution in the afternoon was bimodal. The morning cumuli were relatively small, so they were probably boundary-layer-confined cumuli. In the afternoon, the cloud populations could be considered to be a mix of

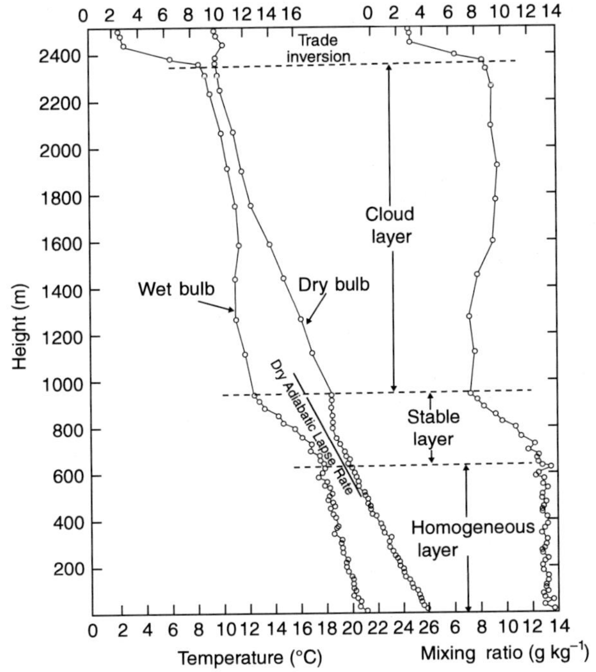


**FIGURE 7.2 Structure of the undisturbed tropical atmosphere based upon aircraft soundings.** *(After Malkus (1956); from Garstang and Betts (1974))*

boundary layer cumuli, towering cumuli, and cumulonimbi. We shall show later that the dominant mechanisms of entrainment in boundary layer cumuli and deep cumuli may also be quite different.

Since boundary layer cumuli may be an intimate component affecting the overall structure of the ABL, let us examine the characteristic thermodynamic structure of a cumulus-capped ABL. An example is the layered structure of the undisturbed trade-wind atmosphere. Similar structures are observed in middle latitudes in postfrontal air and in the early morning hours over land before the onset of deep convective disturbances. As summarized by Garstang and Betts (1974), the undisturbed trade-wind ABL can be characterized by five distinct layers (see Fig. 7.2):

(1) The surface layer, which may be adiabatic over water or super-adiabatic over land and which may exhibit a decrease in specific humidity or mixing ratio. The surface layer is generally less than a few hundred meters in depth.
(2) A mixed layer, which is nearly adiabatic throughout and is nearly constant in specific humidity. Its depth may extend to 900 mbar in the undisturbed trades and much higher over continental regions.
(3) A transition layer, which is nearly isothermal but has sharply decreased moisture and a thickness less than 100 m. This layer separates the cloud

layer above from the dry mixed layer below.

(4) A cloud layer, which is conditionally unstable and extends from the transition layer to the base of the capping inversion and has a temperature gradient somewhat greater than wet adiabatic.
(5) The trade-wind inversion (or, in general, the capping inversion), which caps the ABL and shows a strong increase in temperature and decrease in specific humidity with height.

The depths and structure of these layers are controlled by the fluxes of heat, moisture, and momentum at the surface, the strength of subsidence in the lower atmosphere, and the intensity of cumulus convection. In an analysis of data obtained from multiple-level aircraft flights during GATE, Nicholls and LeMone (1980) noted that a significant change in the distribution of heating and moistening throughout the depth of the mixed layer or subcloud layer was associated with the onset of cumulus convection. The virtual heat fluxes or buoyancy fluxes in the subcloud layer did not vary with the presence or absence of cumuli. Sommeria (1976) obtained a similar result in his 3-D large-eddy simulation of the trade-wind boundary layer.

There have been few attempts to obtain direct measurements of fluxes and energy budgets through the cloud layer because of two major problems. First, the cloud layer is too high for tethered balloon-borne sensors. Furthermore, most researchers agree that it is difficult to obtain reliable, high-frequency sampling of temperature and total water (water vapor plus condensed water) from high-speed airplanes. Direct (in situ) probes of air temperature either become wet or respond too slowly for accurate calculations of fluxes using eddy-correlation analyses. Likewise, total water content samplers either have too slow a response or are costly and unreliable.

The second problem arises from the marked difference in the character of eddies in a cumulus layer as opposed to those in the subcloud layer. In the cloud layer, the eddies composing the cumuli are clustered in widely spaced localized regions and separated by cloud-free, relatively quiescent regions. In order to obtain an adequate sampling of an ensemble of cloud and cloud-free eddies in the cloud layer, an aircraft must traverse a large volume of air. Unfortunately, cumuli are rarely horizontally homogeneous over long distances, so one can seldom obtain meaningful flux measurements by simply flying an airplane over long distances. Furthermore, because clusters of cumuli are rarely steady for periods of several hours, one cannot always obtain a meaningful ensemble average by flying an aircraft back and forth through a particular cluster. The expense of using multiple aircraft at different levels, and the present lack of appropriate cloud and temperature sensors, usually force us to resort to cloud models or diagnostic studies in which cloud-layer fluxes are inferred as a residual from total moisture and heat budget Analyses (i.e. Nitta and Esbensen, 1974; Holland and Rasmusson, 1973; Augstein et al., 1974).

Pennell and LeMone (1974) obtained some direct estimates of momentum fluxes and the kinetic energy budget of a trade-wind cloud layer under relatively

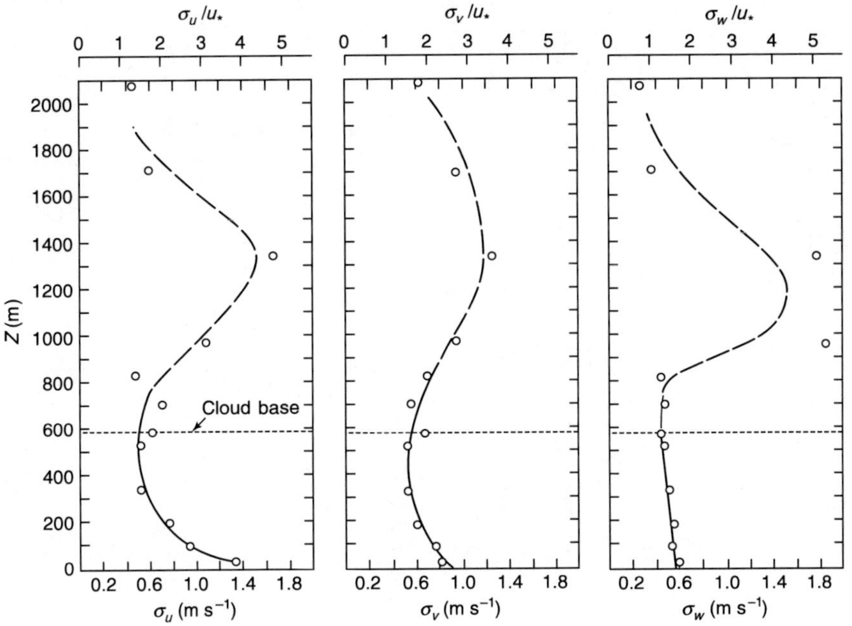


**FIGURE 7.3 Standard deviations of $u'$, $v'$, and $w'$ as a function of height for the cloudy region ($u_* = 46$ cm s$^{-1}$).** The sampling time is probably not sufficient to establish the profiles above cloud base. *(From Pennell and LeMone (1974))*

high wind conditions. Using an aircraft equipped with a gust probe and an Inertial Navigation System, they obtained accurate measurements of the three wind components. Their measurements were, however, subject to the sampling problem noted above. Figure 7.3 illustrates the measured profiles of the standard deviations of the horizontal winds ($\sigma_u$ and $\sigma_v$) and vertical velocity ($\sigma_w$) obtained through the depth of the boundary layer with winds as large as 16 m s$^{-1}$ during a period of enhanced cloud activity. Below cloud base, the profiles of $\sigma_u$, $\sigma_v$, and $\sigma_w$ are reasonably consistent with observations of the ABL under cloud-free conditions (Lenschow, 1970; Kaimal et al., 1976). In particular, $\sigma_u$ and $\sigma_v\sigma_u$ and $\sigma_v$ exhibit a sharp decrease from a maximum value near the surface. The standard deviation of vertical velocity, however, exhibits a slight linear decay with height, whereas some other observations and laboratory models suggest (Willis and Deardorff, 1974; Nicholls and LeMone, 1980) that $\sigma_w$ should exhibit a maximum in the subcloud layer at about one-half the mixed layer depth. Above cloud base, where the release of latent heat is expected to be important, there is the suggestion that $\sigma_u$, $\sigma_v$, and $\sigma_w$ reach a maximum. Likewise, the eddy flux of horizontal momentum ($-u'w'$), shown in Fig. 7.4, exhibits a maximum in the cloud layer. Clearly, the vertical transport of momentum by clouds is important.

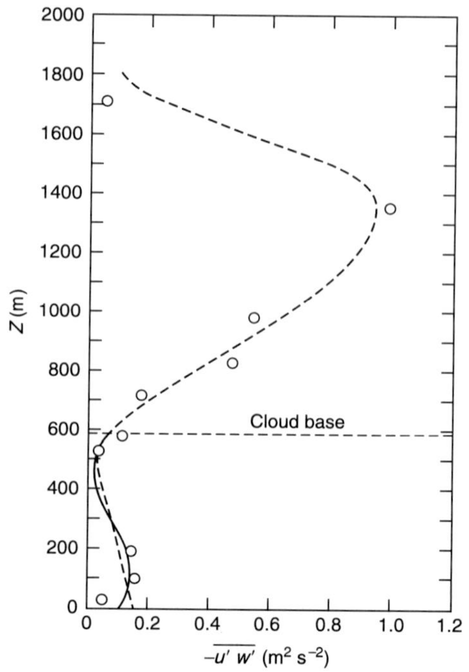


**FIGURE 7.4 Momentum flux as a function of height for the cloudy region.** *(From Pennell and LeMone (1974))*

A three-dimensional large-eddy simulation (LES) and a 2-D cloud-ensemble model (CEM), have been applied to the trade-wind boundary layer (Sommeria, 1976; Sommeria and LeMone, 1978; Nicholls et al., 1982; Cuijpers and Duynkerke, 1993; Krueger and Bergeron, 1994; Abel and Shipway, 2007). Only the latter considered the influence of precipitation on boundary layer structure (we shall examine this in a later section). The Sommeria (1976) and Sommeria and LeMone (1978) model is an extension of Deardorff's (1972) boundary layer model to a non-precipitating cumulus layer. Using a grid resolution of 50 m, the model covered a horizontal domain of 2 × 2 km. The model data were compared with the observations during suppressed convective activity reported by Pennell and LeMone (1974). Figures 7.5 and 7.6 illustrate the simulated and observed profiles of the variances of the wind components $u$ and $w$. The observed profiles are shown for raw variances and variances calculated after the wind field is filtered to remove wavelengths greater than 2.2 km. The model variances are shown for the case of weak cloud activity, which corresponds to the observed case, relatively strong cloud activity, and for the subgrid-scale contribution to the variances. The agreement between simulated and observed horizontal wind variances is quite good when compared to the filtered wind data. However, the unfiltered $\sigma_u^2$ are considerably greater than the observed, suggesting that contribution to the wind fluctuations on scales greater than

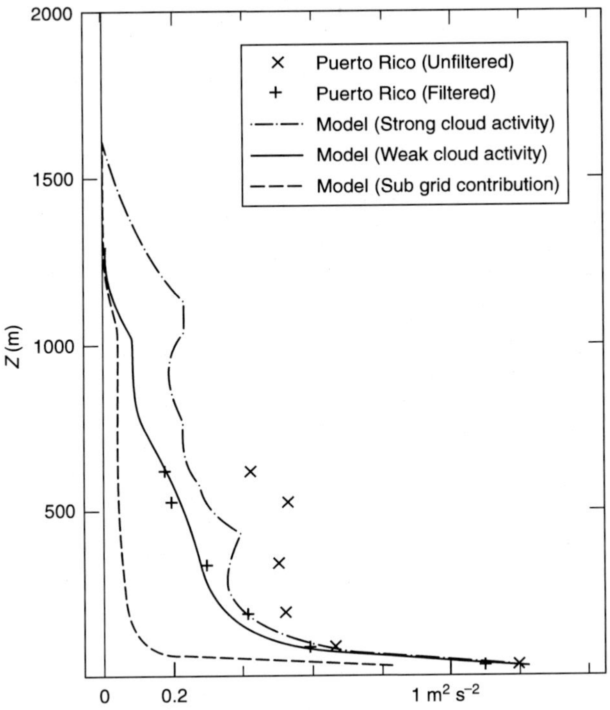


**FIGURE 7.5 Comparison of the variances of the *u* component of the velocity in the model and in the Puerto Rico experiment.** *(From Sommeria and LeMone (1978))*

2 km is quite significant. Both the filtered and unfiltered observed $\sigma_w^2$ data are consistent with the simulated data. This suggests that eddy scales larger than 2 km are not very important for generating vertical velocity fluctuations. In the case of the simulated enhanced cloud activity, $\sigma_w^2$ exhibits a relative maximum in the cloud layer. This is consistent with the observed profile of $\sigma_w^2$ during enhanced cloud activity shown in Fig. 7.3, except that the observed values are quite a bit larger.

As illustrated in Fig. 7.7, the model underestimates the variance in specific humidity. The authors suggested that this may be due to the neglect of anomalies in sea surface temperature which would generate local anomalies in surface evaporation rates. However, the errors are greatest relative to the unfiltered data; this suggests that scales larger than the domain of the model are contributing substantially to moisture fluctuations and perhaps to low-level moisture convergence.

The vertical transport of horizontal momentum, simulated by the model shown in Fig. 7.8, agrees with observations in the subcloud layer. In the cloud layer, the observed fluxes are positive, especially under enhanced cloud activity

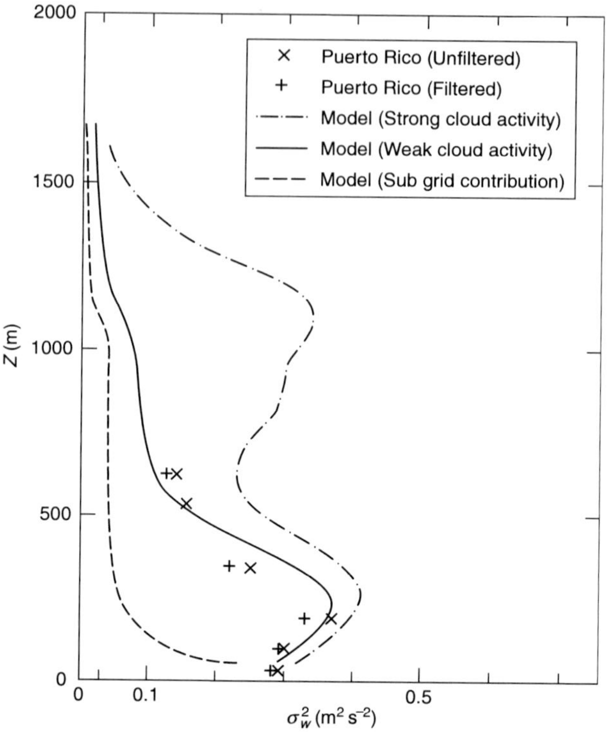


**FIGURE 7.6 Comparison of the variances of the vertical component of the velocity in the model and Puerto Rico experiment.** *(From Sommeria and LeMone (1978))*

(see Fig. 7.4), whereas the simulated fluxes are negative. The differences, however, may be due to the somewhat different wind profiles used in the model.

Figures 7.9 and 7.10 illustrate the observed and simulated moisture flux and the flux of virtual potential temperature. The observed and simulated fluxes of these quantities generally agree with observations in the subcloud layer. As convective activity increases, the moisture flux also increases substantially in the subcloud layer, whereas the buoyancy flux remains unchanged. In the cloud layer, however, both moisture and buoyancy fluxes are substantially modulated by the vigor of convective activity. The various terms contributing to turbulent kinetic energy [see Eq. (3.31)] emulated by the model, are illustrated in Fig. 7.11, for the case of strong cloud activity. The terms in the kinetic energy equation calculated from the somewhat more suppressed observed case are also shown in Fig. 7.11. As far as the kinetic energy budget is concerned, Sommeria and LeMone (1978) suggested that both the subcloud and the cloud layer could be divided into two layers. In the lower mixed layer up to approximately 200 m, shear production (and, secondarily, buoyancy) generates kinetic energy which is locally removed by viscous dissipation and vertical turbulent transport.

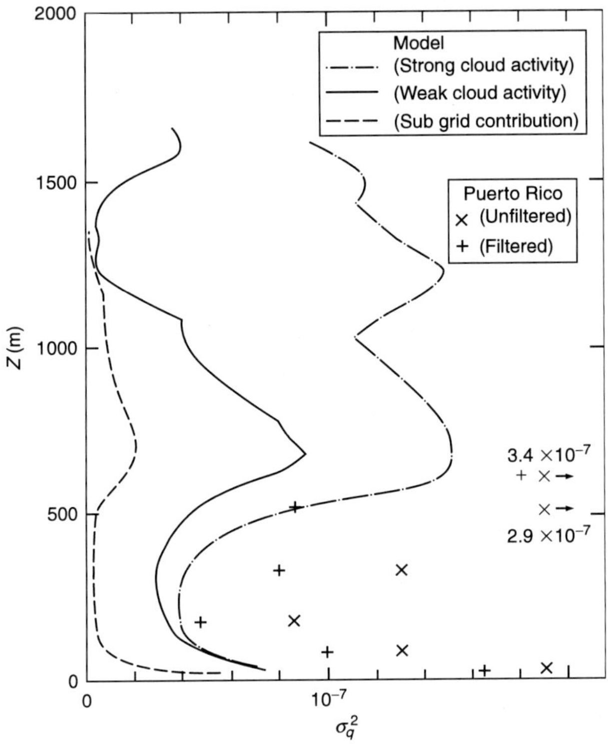


**FIGURE 7.7 Comparison of the variances of the specific humidity in the model and in the Puerto Rico experiment.** Units are mass of water vapor per mass of air squared. *(From Sommeria and LeMone (1978))*

In the upper mixed layer (up to about 700 m), turbulent transport of kinetic energy (and to a lesser extent, shear and buoyancy production) is balanced by dissipation. In the lower part of the cloud layer (approximately 700 to 1050 m), kinetic energy is produced by the buoyancy generated by condensation, or, secondarily, by shear and pressure transport. Local molecular dissipation and vertical turbulent transport share equally in balancing the production terms. Kinetic energy production in the upper part of the cloud layer is dominated by turbulent transport from lower levels along with smaller contributions from shear and buoyancy. Along with molecular viscosity, pressure transport plays a significant part in the dissipation of kinetic energy.

Nicholls et al. (1982) applied Sommeria's model to a low-wind-speed case observed in the trade-wind region during GATE. Comparison with observations revealed that the model does quite well in predicting the observed heights of cloud base and cloud top, cloud cover, vertical fluxes of heat, moisture, and momentum, and vertical velocity variance. Again, the model underestimated the horizontal wind variances and the variances of moisture and temperature.

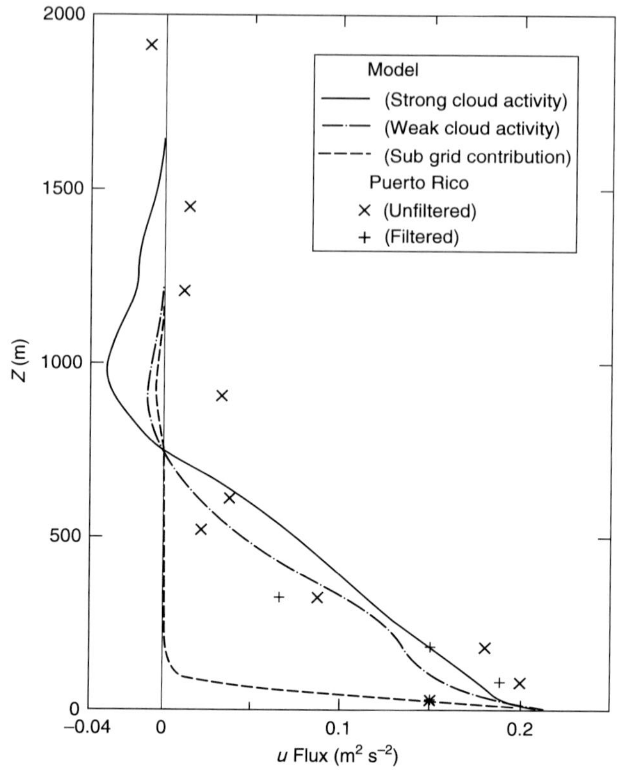


**FIGURE 7.8 Comparison of the flux of the *u* component of velocity in the model and in the Puerto Rico experiment.** *(From Sommeria and LeMone (1978))*

The reason for the underestimation of moisture variance can be seen in Fig. 7.12, which illustrates a composite spectrum of the variations in specific humidity $q$, vertical velocity $w$, and vertical moisture flux. The interesting feature of the moisture spectrum is that it exhibits a peak at a wavelength of about 10 km. This implies that the largest contribution to the total moisture variance comes from mesoscale eddies having wavelengths of the order of 10 km. Nicholls and LeMone (1980) found that the horizontal wind variances also exhibited spectral peaks in wavelengths of the order of 10 km. However, as shown in b and c in Fig. 7.12, the composite vertical velocity spectrum and the co-spectrum between $w$ and $q$ (which represents vertical moisture flux) exhibit peaks at wavelengths of about 700 m. The vertical eddy fluxes of $u$, $v$, and $T$ also exhibit peaks at approximately 700 m. The results of this study and the earlier work of Sommeria and LeMone (1978) suggest that vertical velocity variance and vertical eddy fluxes scale with either (1) the boundary layer depth or (2) the depth of the subcloud layer. Total wind variances and total variances of such scalar properties as specific humidity (and perhaps temperature) scale with mesoscale eddies,

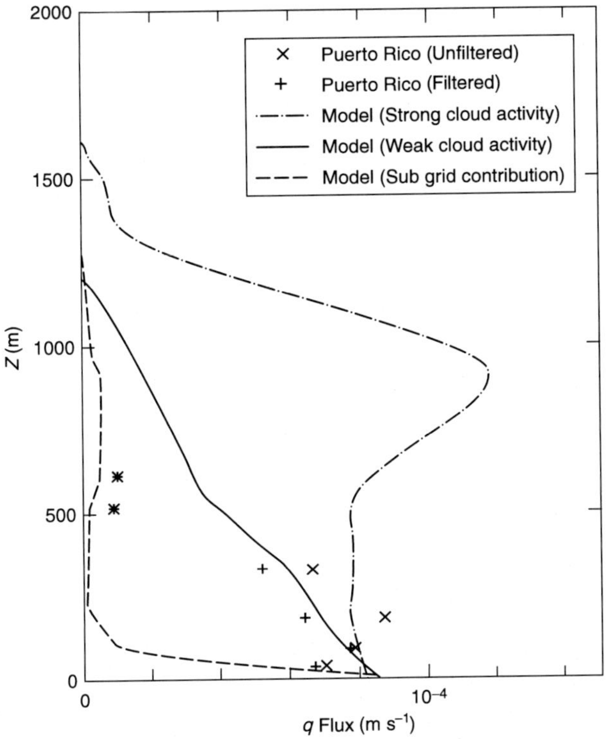


**FIGURE 7.9 Comparison of the flux of specific humidity in the model and in the Puerto Rico experiment.** *(From Sommeria and LeMone (1978))*

whose wavelength contributing to the maximum variance of a quantity may vary depending on the meteorological situation. Thus, a large-eddy simulation model of limited horizontal extent such as Sommeria's appears to be quite useful for simulating eddies contributing to vertical fluxes.

If the horizontal wind fluctuations or variances in scalar properties are important to a particular problem, then the inability of a model to simulate mesoscale contributions to the total variances can lead to serious errors. One example is the prediction of cloud cover. Nicholls et al. (1982) noted that the model underpredicted cloud cover; where the observed coverage was between 5 and 10%, the modeled coverage was only 4 to 5%. We shall examine later the important role of precipitation in controlling mesoscale organization of clouds and cloud cover. The earlier work of Sommeria (1976) and Sommeria and LeMone (1978) also indicated that the vertical fluxes of moisture from the sea surface were insufficient to maintain relatively steady moist convection. It is likely that this deficiency was caused not by an inadequate parameterization of surface fluxes but by a lack of resolution of mesoscale eddies contributing to low-level moisture convergence.

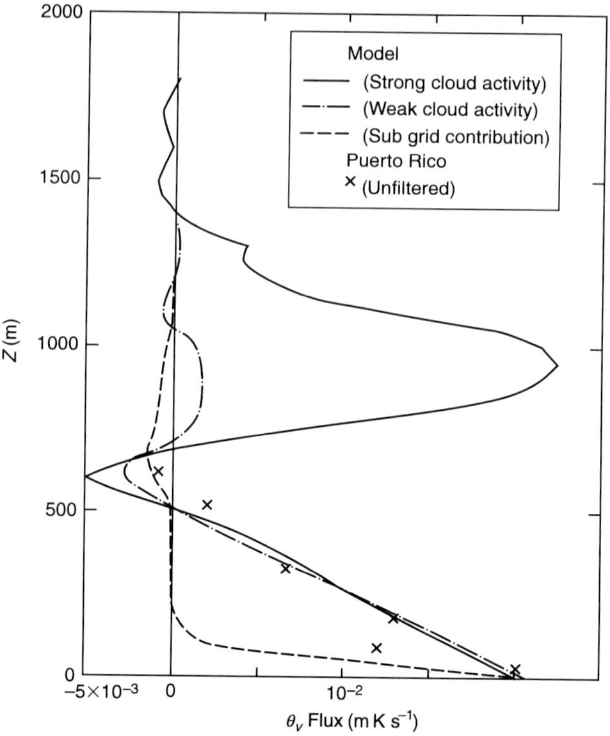


**FIGURE 7.10 Comparison of the flux of virtual potential temperature in the model and in the Puerto Rico experiment.** *(From Sommeria and LeMone (1978))*

An example of advances in LES as well as some of their deficiencies can be found in the model intercomparison study of Stevens et al. (2001). The largest differences amongst the models was in the simulation of cloud amount. The models exhibited large variability of cloud amount, even those with very fine resolution. The models were very sensitive to the details of the numerical algorithms. Moreover, part of the sensitivity was attributed to a physically realistic feedback through cloud top radiative cooling, higher cloud fractions yield larger amounts of cloud top radiative cooling, which then feeds back into the cloud layer circulations.

Despite these deficiencies, LES models have been very useful in providing new insights into the behavior of boundary layer clouds. The Stevens et al. (2001) simulations revealed a convective circulation that is distinctly organized into two main layers; a cumulus layer of about 700 m depth that is deeply rooted not only in the subcloud layer but in the surface layer as well. Several sublayers were also identified including a detrainment layer around 1400 m in which stratiform clouds develop, a highly variable transition layer, and a surface layer. The simulations indicate that the transition layer is best characterized

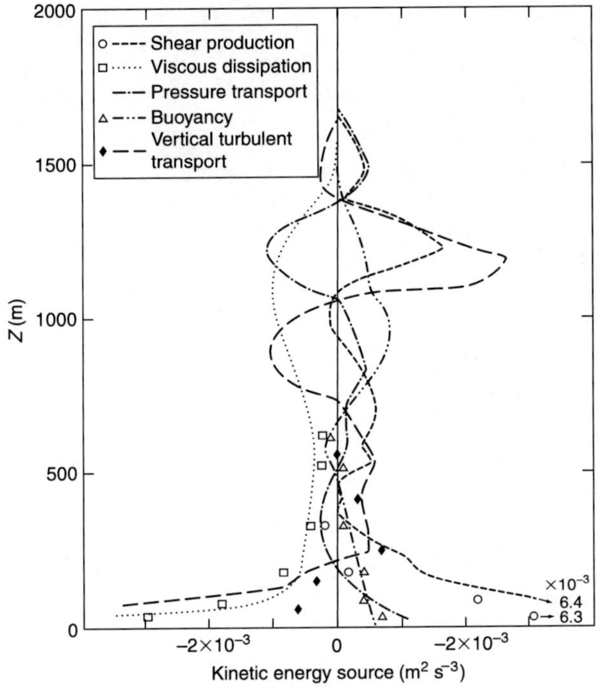


**FIGURE 7.11 Terms of the kinetic energy budget in the Puerto Rico experiment and the model.** The model results are for the time of strong cloud activity, from model times 2.2 to 2.48 h. The signs denote observational data; the lines denote model results. *(From Sommeria and LeMone (1978))*

by moisture variances and gradients. Furthermore the transition layer is by no means a stable layer, but it undergoes marked variations in height of over 400 m. Finally, the LES models illustrate how the circulations in the cloud layer and subcloud layer differ. While the subcloud layer is characterized by a relatively homogeneous ensemble of numerous weak updrafts and downdrafts, the cloud layer is dominated by intermittent structures (i.e. cells). These entities exhibit large vertical velocities, have a narrow region of compensating downward motion in a sheath surrounding the clouds, and are surrounded by a large region of weak compensating subsidence. Figure 7.13 is a cartoon that nicely illustrates these features. LES models can guide us in formulating and testing simpler models of boundary layer clouds. Bougeault (1981a,b) and Golaz et al. (2002a,b) used LES models to develop and test one-dimensional higher order closure models of the trade-wind cumulus layer. Likewise, Siebesma and Cuijpers (1995), De Roode and Bretherton (2003), and Stevens et al. (2001) use LES to evaluate and test parameters and assumptions in mass-flux models of the cumulus topped boundary layer.

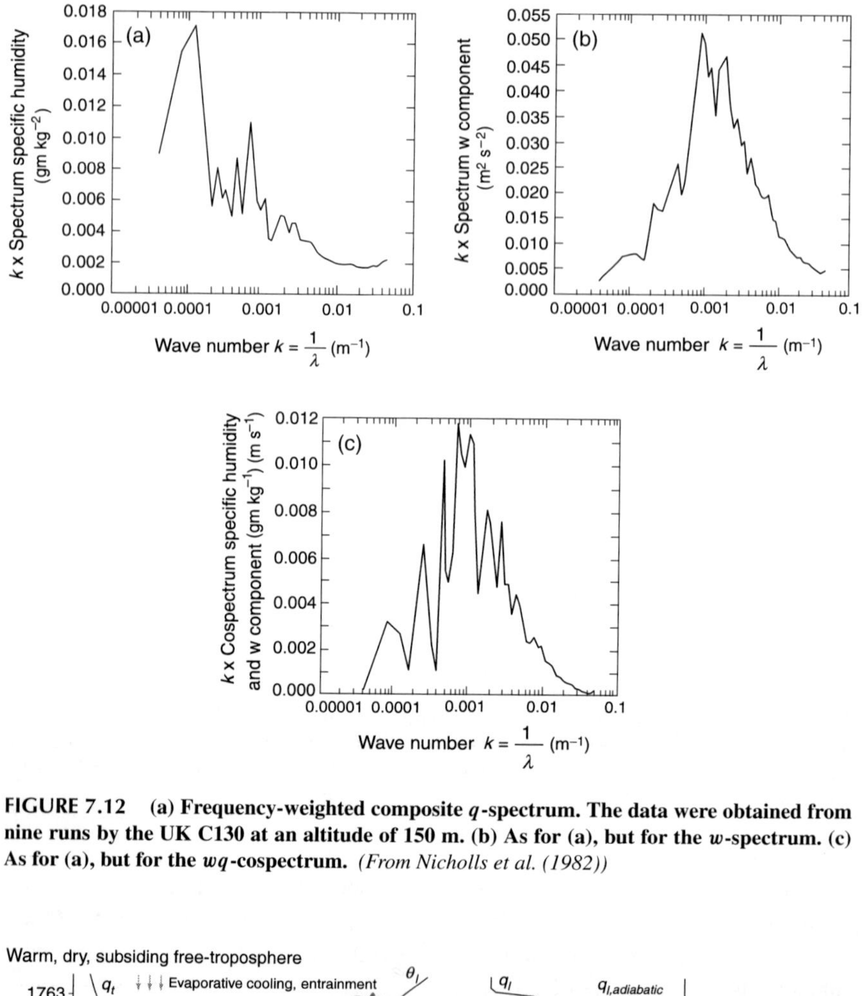


**FIGURE 7.12** **(a) Frequency-weighted composite $q$-spectrum. The data were obtained from nine runs by the UK C130 at an altitude of 150 m. (b) As for (a), but for the $w$-spectrum. (c) As for (a), but for the $wq$-cospectrum.** *(From Nicholls et al. (1982))*

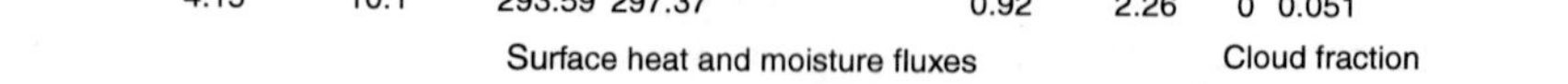


**FIGURE 7.13** **Cartoon of trade-wind boundary layer from large-eddy simulation.** Heights of cloud base, level of maximum $\theta_l$ gradient (inversion height), and maximum cloud penetration depth are indicated, as are subcloud layer and inversion-level values of thermodynamic quantities. Cloud water contents are averaged over cloudy points only, with adiabatic liquid water contents indicated by the dash-dot line. The far right panel shows cloud fraction, which maximizes near cloud base at just over 5%. *(From Stevens (2005))*

The layer-averaged models of the trade-wind boundary layer using the mass-flux approach, such as Betts' (1975), Nitta's (1975), Esbensen's (1976), and Albrecht et al.'s (1979), which have been successful in simulating the bulk properties of the cloud layer, consider explicitly lateral entrainment in cumuli. Beniston and Sommeria (1981), in fact, tested Betts' model against the data simulated by Sommeria's 3-D model, and found that the thermodynamic fluxes calculated by Betts' scheme were consistent with the 3-D model data.

Betts formulated his model in terms of the thermodynamic quantities' moist and dry static energies. The dry static energy $s$ is defined as

$$s = c_{\mathrm{p}}T + gz, \tag{7.1}$$

and moist static energy $h$ is defined as

$$h = c_{\mathrm{p}}T + gz + Lq. \tag{7.2}$$

In a water-saturated environment they may be defined as

$$S_{\mathrm{L}} = c_{\mathrm{p}}T + gz - Lq_{\mathrm{L}} \tag{7.3}$$

and

$$H_{\mathrm{s}} = c_{\mathrm{p}}T + qz + Lq_{\mathrm{S}}. \tag{7.4}$$

The behavior of the quantities s and h may be considered to be analogous to the behavior of $\theta$ and $\theta_{\mathrm{e}}$, respectively.

The fundamental premise in Betts' model is that the eddy fluxes of heat and moisture can be represented as the product of a single convective mass flux ($\omega^*$) and a perturbation quantity derived from a single entraining cloud parcel (subscript c) rising through a known mean environment. Thus,

$$F_{\mathrm{X}} = -\overline{\omega' x'} \equiv \omega^*(X_{\mathrm{c}} - X_{\mathrm{e}}), \tag{7.5}$$

where $X$ represents $h, q$, or $S_{\mathrm{L}}$. The convective mass flux represents the cloud area-averaged flux due to clouds having active updrafts with velocity $w_c$ and having a fractional coverage $\sigma$. Thus,

$$\omega^* = \sigma\rho w_{\mathrm{c}}. \tag{7.6}$$

We write the budget equations for a saturated cloud layer in pressure coordinates,

$$Q_1 = \frac{\partial\bar{s}}{\partial t} + \mathrm{V}\cdot\nabla\bar{s} + \overline{\omega}\frac{\partial\bar{s}}{\partial p} = Q_{\mathrm{R}} - \frac{\partial}{\partial p}(\overline{\omega' S_{\mathrm{L}}'}), \tag{7.7}$$

$$Q_2 = L\left(\frac{\partial\overline{q_{\mathrm{v}}}}{\partial l} + \mathrm{V}\cdot\nabla\overline{q_{\mathrm{v}}} + \overline{\omega}\frac{\partial\overline{q_{\mathrm{v}}}}{\partial p}\right) = -L\frac{\partial}{\partial p}(\overline{\omega' q_{\mathrm{v}}'}), \tag{7.8}$$

$$Q_1 - Q_2 = \frac{\partial \bar{h}}{\partial t} + \mathrm{V} \cdot \nabla \bar{h} + \bar{\omega} \frac{\partial \bar{h}}{\partial p} = Q_\mathrm{R} - \frac{\partial}{\partial p} (\overline{\omega' h'}), \tag{7.9}$$

where $Q_R$ is the radiative heating and $Q_1$ and $Q_2$ are the apparent heat source and apparent water vapor sink, respectively. Substitution of Eq. (7.5) into Eq. (7.8), for example, gives

$$\begin{aligned} Q_1 - Q_2 - Q_\mathrm{R} &= \frac{\partial}{\partial p} \left[ w^* (h_\mathrm{sc} - \overline{h_e}) \right] \\ &= -\omega^* \frac{\partial h_\mathrm{e}}{\partial p} + \omega^* \frac{\partial h_\mathrm{sc}}{\partial p} + (h_\mathrm{sc} - h_{\bar{\mathrm{e}}}) \frac{\partial \omega^*}{\partial p}. \end{aligned} \tag{7.10}$$

Betts defined the lateral entrainment rate for an ascending cloud parcel as $\lambda (> 0)$, where

$$\partial h_\mathrm{sc} / \partial p = \lambda (h_\mathrm{sc} - \bar{h}). \tag{7.11}$$

Substitution of Eq. (7.10) into Eq. (7.9) yields the budget equation

$$Q_1 - Q_2 - Q_\mathrm{R} = -\omega^* \frac{\partial h_\mathrm{e}}{\partial p} + \omega^* (h_\mathrm{sc} - h_\mathrm{e}) \left( \lambda + \frac{1}{\omega^*} \frac{\partial \omega^*}{\partial p} \right). \tag{7.12}$$

The term $\lambda + (1/\omega^*)(\partial \omega^* / \partial p)$ represents two sources, namely, the entrainment of dry environmental air into clouds at the rate $\lambda$, and the net detrainment of cloud mass having properties differing from the environment. As noted by Betts, entrainment affects the properties of a cloud, while detrainment affects the environment on the time scale of the response of the environment. As a consequence, when Betts (1975) applied his model to the diagnostic interpretation of trade-wind cumulus budgets, he found that the diagnosed cloud transports were relatively insensitive to variations in entrainment rates. Even complete neglect of entrainment did not affect the results substantially. By contrast, the diagnosed detrainment rates were much larger and nearly independent of $\lambda$. In another diagnostic study, Esbensen (1978) concluded that detrainment dominates the mass budgets of shallow cumulus clouds. These diagnostic studies resolve the apparent paradox of how 1-D higher order transport models of the cumulus-capped boundary layer (i.e. Bougeault, 1981b; Golaz et al., 2002a,b), which only calculate the net flux divergences (or detrainment rates) that affect the cloud environment and have no explicit representation of entrainment, can represent the cumulus layer well. Because detrainment predominates over entrainment in determining the budgets of a shallow cumulus layer, a lack of consideration of entrainment in such higher order models does not appear to be a serious defect.

A consistent result from LES models, higher order transport models, and diagnostic models, is that boundary layer cumuli moisten the cloud layer. That is, the detrainment of water from clouds dominates the drying effect of cumulus-induced subsidence. Furthermore, the lower part of the cloud layer is warmed and the upper part cooled. The evaporative cooling at the top of the cloud layer and the radiative cooling maintain the trade inversion against the warming produced by large-scale subsidence. The moistened, destabilized cloud layer, however, favors the development of towering cumuli that can penetrate well above the capping inversion.

### 7.2.1. Towering Cumuli in the Tradewind Environment

The classic structure of the trade-wind regime shown in Fig. 7.2 is based on soundings for the undisturbed trade-wind environment. However, one does not have to reside in the trade-wind environment long to realize that this simple structure does not prevail under more disturbed conditions, such as when easterly waves move through the area. Moreover, as one moves towards the equator the presence of deep convective clouds becomes more prevalent (Simpson, 1992). There the emphasis shifts to cumulonimbi that extend through the depth of the troposphere and sometimes penetrate into the lower stratosphere. Although early observational studies in the tropics indicated that, along with cumulonimbi and boundary layer cumuli, cumulus congestus are quite prevalent (Malkus, 1962), cumulus congestus have often been overlooked. Johnson et al. (1999) argue that cumulus congestus, along with boundary layer cumuli and cumulonimbi, are important to the tropospheric moisture distribution, heating, divergence, cloud detrainment, and fractional cloudiness in the tropics. They note that cumulus congestus with tops below 10 km contribute 28% of the convective rainfall observed over the tropical western Pacific warm pool during TOGA-COARE (Tropical Ocean-Global Atmospheric Coupled Ocean-Atmospheric Response Experiment; (Webster and Lukas, 1992). If one excludes the heavily raining MCS category, the fraction increases to over 50% of the convective rainfall in that region. Figure 7.14 is a cartoon that illustrates the prevalence of cumulus congestus in the deep tropics.

Johnson et al. (1999) note that the peaks in the distribution of boundary cumuli and cumulonimbi in the tropics correspond to prominent stable layers, like the trade-wind inversion (~2 km) and the tropical tropopause (~15-16 km). Likewise, the peak in the distribution of cumulus congestus corresponds to a middle-troposphere stable layer near the (0°C) level (~5 km). This stable layer, although relatively weak, can inhibit the further growth of cumulus congestus and lead to the prevalence of detrainment just above it or at 7.5 km (400 mb). Not only is the further growth of cumulus congestus inhibited, but cumulonimbi are often observed to detrain cloud substance above the (0°C) level creating what is often called *shelf clouds*. Those congestus that do penetrate mid-troposheric stable layers often glaciate and thereby experience a boost in buoyancy from the additional latent heats of freezing and sublimation, and may penetrate to the tropopause (Johnson et al., 1999; Zipser, 2003). Figure 7.15 illustrates some soundings in which inversions corresponding to the melting level can be

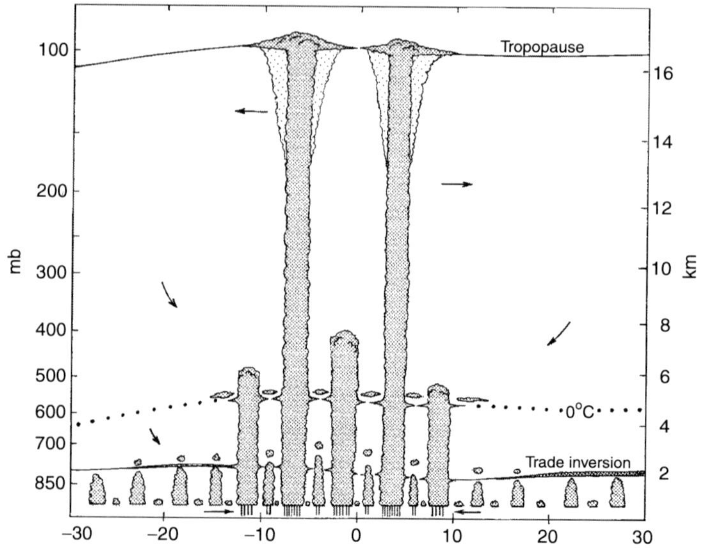


**FIGURE 7.14 Conceptual model of tropical cumulus cloud distributions from 308N to 308S based on IOP-mean radar data and thermal stratification.** Three main cloud types are indicated: shallow cumulus, cumulus congestus, and cumulonimbus. Within the shallow cumulus classification, there are two subdivisions: forced and active cumuli. Three stable layers are indicated: the trade inversion, the 08C layer, and the tropopause. Shelf clouds and cloud debris near the trade and 08C stable layers represent detrainment there. Cirrus anvils occur near the tropopause. Considerable overshooting of the trade and 08C stable layers occurs in the equatorial trough zone. Arrows indicate meridional circulation. Although double ITCZ is indicated, representing IOP-mean, this structure is transient over the warm pool and a single ITCZ often exists. *(From Johnson et al. (1999))*

discerned. The causes of these stable layers have not been fully identified, but Johnson et al. (1999) suggest that melting by snow and ice may be a causal factor. They note that these inversions are strongest in the vicinity of MSCs and cumulonimbus activity, and thus are remnants of melting in pre-existing convection. Another possibility is that these stable layers may be the result of gravity waves or "buoyancy bores" emanating from deep convection (Mapes, 1993; Mapes and Houze, 1995). These waves leave in their wake a region of enhanced stability in the middle troposphere over a several hundred kilometer region. These waves could be focused near the melting level by the stable layers produced by melting precipitation near deep convective systems, and thus communicate the stabilizing effects of melting over a much larger area.

## 7.2.2. Role of Radiation

The importance of shortwave and longwave radiative transfer to the ensemble structure of a boundary layer cumulus layer has been investigated using 3-D,

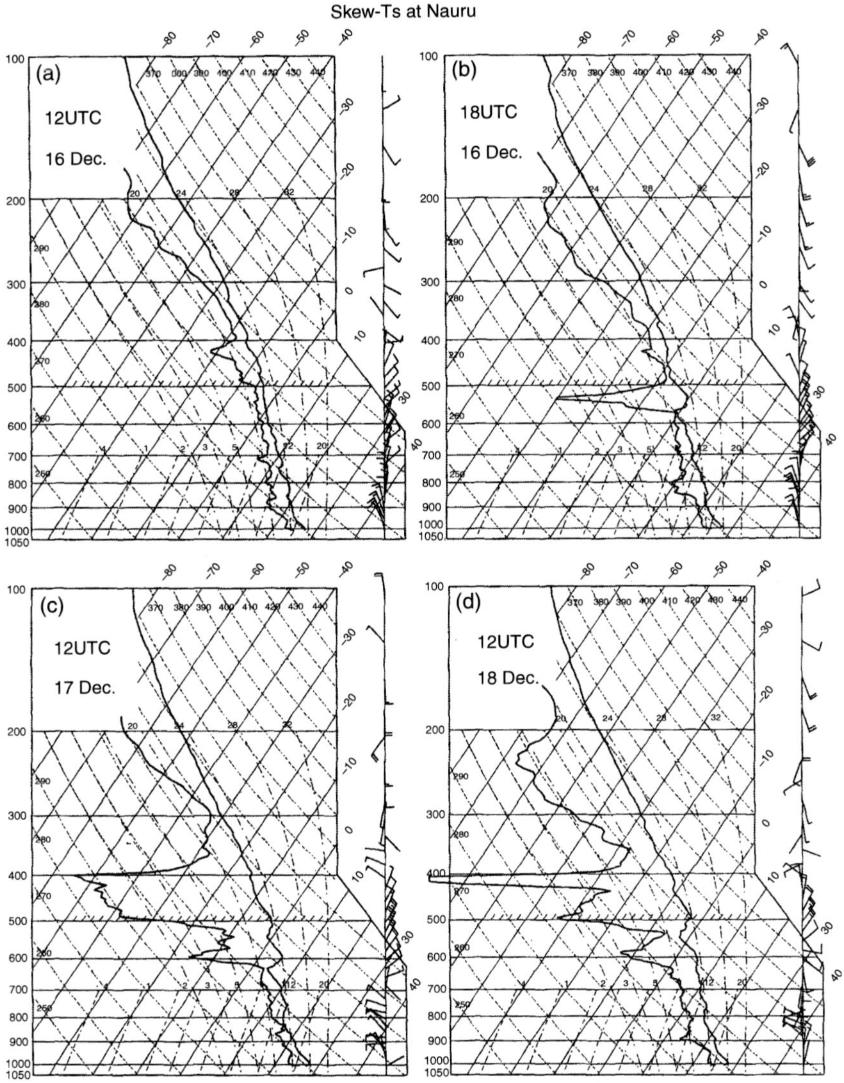


**FIGURE 7.15 Skew *T* diagrams for Nauru at (a) 1200 UTC 16 December, (b) 1800 UTC 16 December, (c) 1200 UTC 17 December, and (d) 1200 UTC 18 December 1992, depicting a sequence of soundings from a 2-day period that was characterized by a sudden onset of dry, stable layers near the 0°C level with associated acceleration of the northeasterly flow, suggesting dry intrusions from the subtropics.** *(From Johnson et al. (1996))*

1-D, observations and diagnostic models. These studies primarily focus on moist tropical or subtropical cumulus layers over the ocean.

Like stratocumulus clouds, we expect longwave radiative cooling to be primarily at cloud top and shortwave heating to be distributed deeper into the cloud layer, the difference, however, is that owing to the small cloud fractions the impact of cloud radiative heating on boundary layer stability is much less than for a solid stratus deck. As noted by Stevens et al. (2001) the response to radiation can be much greater if a more solid cloud deck forms at the level where strong detrainment of moisture substance occurs. Using a diagnostic mass flux model Albrecht et al. (1979) found that clouds modified the vertical distribution of longwave radiational cooling, but did not change substantially the net cooling through the depth of the boundary layer. Under cloudy conditions, the clear-sky radiative cooling rate was confined to a thin layer at the top of the cloud layer. Likewise, they found that shortwave solar heating rates were confined to a thin layer at cloud top or the capping inversion. Calculations with their model further demonstrated that longwave radiative cooling and evaporation of cloud water at the inversion, balances warming due to subsidence.

In a subsequent study, Brill and Albrecht (1982) modified the assumed vertical distribution of solar heating in the cloudy boundary layer. This change was motivated by studies by Welch et al. (1976), Grassl (1977), Tanaka et al. (1977), and Davis et al. (1979) which suggested that significant solar heating occurs below cloud top, particularly when the zenith angle is small, or at larger zenith angles when the cloud fraction is small. Brill and Albrecht (1982) used the results of Davis et al. (1979) to parameterize the distribution of heating into a subcloud heating rate $Q_M$, a lower half of the cloud-layer heating rate $Q_{CB}$, and a top-half heating rate $Q_{CT}$. In the top half of the cloud layer, a variable fraction of the heating rate was placed in the inversion. Placing all the top-half heating in the inversion resulted in a much greater diurnal variation of the inversion than placing it below. However, when all the top-half solar heating is distributed below the top of the inversion, the cloud layer is stabilized by warming the top half of the cloud layer more than the bottom half.

There is evidence that boundary layer cumulus can play a role in observed variations of sea-surface temperature. It is observed that strong diurnal variations in sea surface temperature occur when strong large-scale subsidence is present where the boundary layer is either cloud-free or composed of small cloud coverage boundary layer cumulus clouds (Sui et al., 1997; Rickenbach and Rutledge, 1997). This diurnal signal disappears when larger cloud coverage, deep convection is prevalent.

One can expect very important interactions between radiative transfer processes and the cloud layer over land, where surface heating/cooling can vary substantially over the diurnal cycle. This is illustrated by the LES experiments performed by Golaz et al. (2001) in which the impact of varying amounts of initial soil moisture to the development of a cumulus-topped boundary layer were examined. They found that drier soil moisture conditions led to clouds with higher cloud bases, but little change in cloud top. With larger amounts of soil moisture, CAPE was larger, which lead to more active cumulus clouds, but the total cloud fraction was little changed, staying about 12% during midday.

The liquid water path increased with greater amounts of soil moisture, owing to the greater cloud depths associated with the lowering of the cloud base. It was found that the cloud fraction was not correlated with relative humidity.

### 7.2.3. Cloud Coverage

As noted above, the amount of cloudiness has a significant impact upon the distribution of radiative heating in the boundary layer. In addition, the amount of cloud cover controls heating at both the ocean and land surfaces, and effects the net latent heating in the boundary layer and hence the resultant thermodynamic and dynamic impacts of the clouds on the environment.

The prediction or diagnosis of fractional cloud cover is important to a number of meteorological problems. For example, the problem of predicting conditions favorable for severe aircraft icing in supercooled boundary layer clouds depends upon both the liquid-water content and the fractional area of cloud coverage.

Unfortunately, our ability to predict cloud cover using the physical parameterization and resolution of current forecast models and observations is severely limited. LES models seem to adequately predict cloud cover induced by boundary layer eddies having a horizontal scale on the order of the boundary layer depth. However, we have noted previously that these models might underestimate total cloud cover. We indicated that this may be because the eddies that contribute the most to moisture variance have scales of 10 km or more; larger than the LES model domain. These mesoscale eddies contribute to pockets of open cells (POCs) and rifts, as discussed in preceding sections. It is our concern that this same problem may apply equally to higher order closure boundary layer models, wherein cloud coverage is diagnosed as a function of model-simulated variances and covariances of thermodynamic variables Bougeault's (1981a), or derived from the diagnosed PDFs of thermodynamic variables (Golaz et al., 2002a,b). Likewise, models based on mass-flux approaches in which cloud cover is either predicted (Tiedke, 1993) or diagnosed (Albrecht, 1981; Lappen and Randall, 2001a,b,c) only consider cloud cover contributions by eddies on the cumulus scale and not by mesoscale eddies.

As noted in Chapter 6, drizzle and precipitation processes can have an important impact on the dynamics of boundary layer clouds and thereby impact cloud cover in a very significant way. The response, however, is very nonlinear, such that modest increases in CCN suppress drizzle formation and lead to larger cloud fractions (Petters et al., 2006) in accordance with Albrecht (1989) hypothesis. On the other hand, LES modeling studies by Xue et al. (2008) show that the increase in cloud fraction with CCN occurs only for modest increases in CCN concentrations. When CCN concentrations exceed 100/cc, further increases in CCN lead to a reduction in cloud fraction. Clearly we are only beginning to understand the complex relationships between drizzle processes and cloud fraction. We shall see in the next section that this conclusion applies to cloud organization as well.

### 7.2.4. Organization of Cumuli

It is generally agreed that cumuli originate in boundary layer eddies. The association between subcloud eddies and cumulus clouds is most obvious over land, where thermals associated with mountains (Orville, 1965) or hot spots over flat terrain (Woodward, 1959) have been found to be the sources of cumuli. In the case of the marine boundary layer, however, the association between subcloud eddies and cumuli is much less distinct. In fact, early observations by Bunker et al. (1949) suggested that the origin of non-precipitating trade cumuli was not in the subcloud layer. More recent observations using modern airborne instrumentation reported by LeMone and Pennell (1976), as well as 3-D modeling studies (Sommeria, 1976), indicate that cumuli have distinct (though not very intense) roots in the marine subcloud layer. It is not surprising, then, that the organization of cumuli often reflects the organization of eddies in the subcloud layer. In fact, boundary layer cumuli are frequently used as a means of detecting subcloud eddies (Kuettner, 1959; Plank, 1966; Walter, 1980).

Interest in the study of factors affecting cloud organization has been stimulated by the availability of satellites, which provide a perspective of cloud systems that has previously been unavailable. Furthermore, high-altitude aircraft have become available to the meteorological community. Early studies using these observational platforms suggested that there often exists an analogous behavior between the organization of atmospheric cloud systems and convective circulations observed in shallow laboratory fluid experiments. This observation inspired more laboratory studies of convection (e.g. Krishnamurti, 1968a,b, 1975b; Faller, 1965) as well as a proliferation of theoretical analyses of laboratory analogs to atmospheric flows (Kuettner, 1971; Krishnamurti, 1975a; Agee and Chen, 1973; Asai, 1970, 1972; Sun, 1978; Shirer, 1980). The latter studies can be traced to the classical convection theory developed by Rayleigh (1916).

Three different forms of convection that commonly appear in laboratory experiments as well as in the atmosphere have been identified. First, there is a form of cellular convection that occurs in laboratory fluid experiments that are initially at rest. One form of convection exhibits ascending motion at the center of a cell and descending motion at the edge of the cell and is commonly referred to as *closed cellular convection*. Figure 7.16 provides an example of closed cellular convection in the upper right panel. *Open cellular convection*, in contrast, refers to cells that have downward motion at their centers and thin regions of ascent at their boundaries. Figure 7.17 and the lower right panel of Fig. 7.16 provide examples of open cellular convection in the atmosphere.

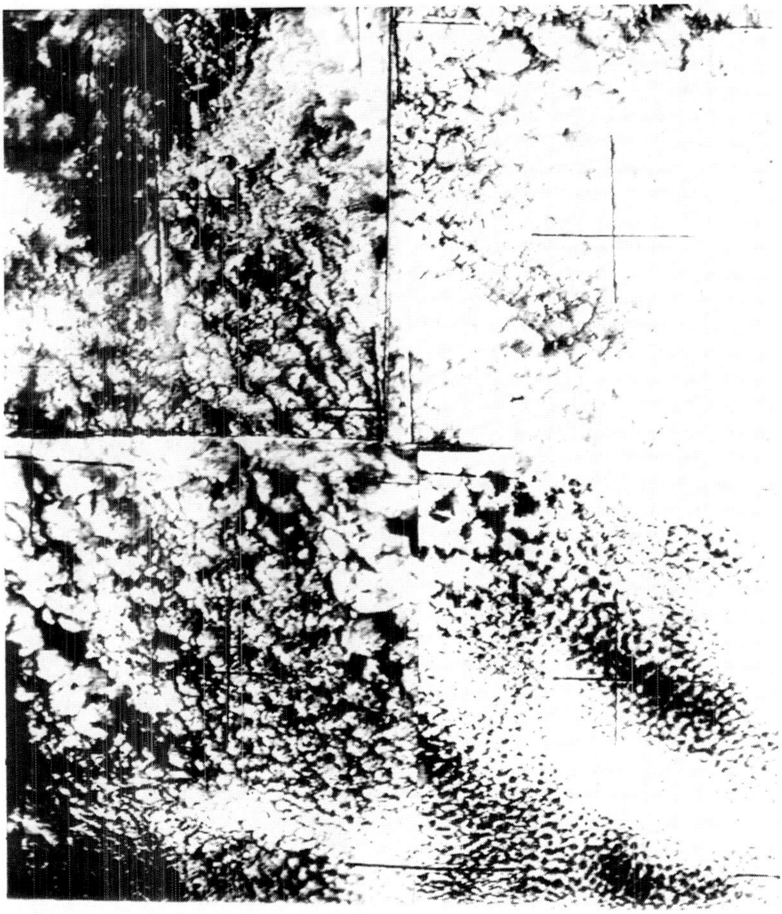

**FIGURE 7.16 Photograph of closed mesoscale cellular convection taken by COSMOS-144 at 0800 GMT, 20 April 1967.** (Provided through the courtesy of Dr. V. A. Bugaev, Director, Hydrometeorological Research Centre of the USSR.) *(From Agee and Chen (1973))*

The third form of convection is represented by cloud bands that are typically oriented parallel to the wind shear vector in the boundary layer or the average boundary layer wind. Wind shear is most important to cloud organization but the average boundary layer wind is usually in the same direction. Such clouds, often called *cloud streets*, are illustrated in Fig. 7.18. Because cloud streets are aligned parallel to the boundary layer wind shear, they are often called *longitudinal roll clouds*. Clouds are also observed to align in parallel bands that are perpendicular to the wind shear vector. Such clouds bands are called *transverse cloud bands*. Malkus and Riehl (1964) presented an example of a case in which both transverse and longitudinal cloud bands coexisted in the

FIGURE 7.17 **Hexagonal "open" cells north of Cuba, photographed from Gemini 5 at 18.27 GMT, 23 August 1965.** Convective enhancement at the vertices of the hexagons can also be noted. *(From Agee and Asai (1982))*

same environment. Figure 7.19 illustrates this case: to the left of a tropical wave trough only longitudinal roll clouds are present, whereas to the right of the trough, both longitudinal rolls aligned parallel to the low-level wind and transverse cloud bands aligned perpendicular to the low-level wind are evident.

In the first edition of this book we reviewed the linear theoretical models that simulate laboratory convection experiments. However, because more recent research is revealing that precipitation contributes significantly to the organization of cellular convection in the atmosphere, we will not include that overview in this edition. Furthermore, while there exists a qualitative similarity between laboratory convection experiments and atmospheric open and closed cellular convection, some substantial differences also exist. If we define $\lambda$ as the diameter of a convective cell and $h$ the depth of the convective layer, typical aspect ratios ($\lambda/h$) in laboratory experiments range from 2:1 to 4:1. In the atmosphere, however, the convective cells are extremely flat with aspect ratios of the order of 30:1. Fiedler (1985) noted that while individual cumulus cells display aspect ratios of 3:1, the atmospheric open and closed cellular convection is composed of many cumulus-scale elements and thus is distinct from both the cumulus-scale and laboratory convection experiments.

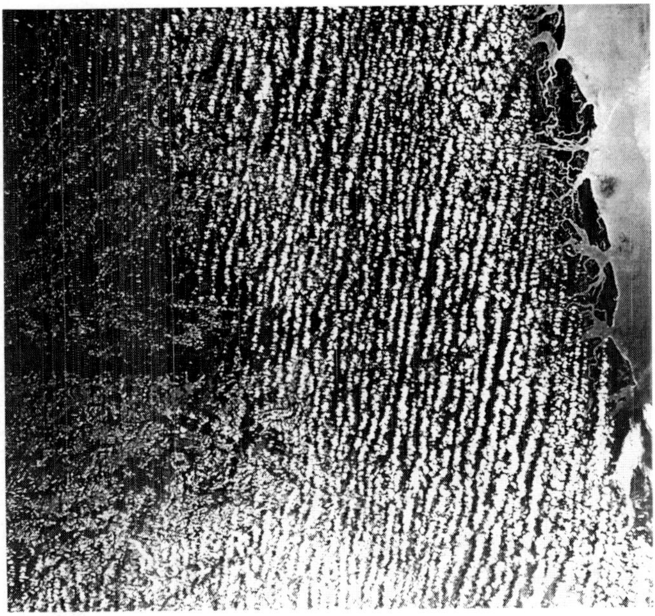

**FIGURE 7.18 Cloud streets over Georgia developing near the coastline in a southerly flow on 4 April 1968, as seen from Apollo 6.** Maximum length of bands: over 100 km; spacing 2 to 2.5 km. (Photo from NASA.) *(From Kuettner (1971))*

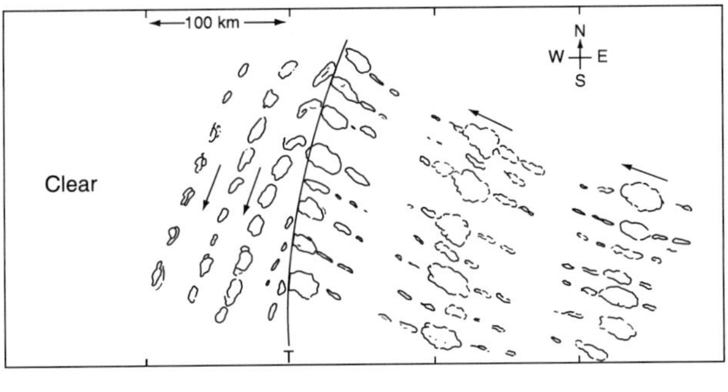


**FIGURE 7.19 Sketch of cloud observations near wave trough.** *(From Malkus and Riehl (1964))*

## 7.2.5. Cellular Convection

As mentioned in preceding sections and Chapter 6, it is now emerging that the formation of anomalous regions of cloud organization called pockets of open cells (POCs; Stevens et al., 2005; van Zanten et al., 2005) are associated with drizzle formation. We suspect that closed cellular convection is also associated with precipitation, or perhaps more vigorous convection without precipitation, and that no current theory can explain the different regimes of cellular convection in the atmosphere. POCs are observed to occur in regions where CCN concentrations are quite low (Petters et al., 2006). Petters et al. (2006) speculate that cloud processing of aerosol via nucleation scavenging, coalescence, and precipitation, perturb the abundance of CCN in the atmosphere and thereby play a role in cloud organization and cloud fractional coverage. The LES of Xue et al. (2008) suggests that "precipitation promotes the development of open cells through a precipitation-dynamical feedback. In the middle of a precipitating cell, air motions are downward and hence divergent at the surface. At the edges of these cells air motions are upward and convergent, leading to the formation of new convective clouds, some of which evolve into new zones of precipitation. These new precipitation events, in turn, generate divergence zones. Divergent regions are shown to be cooler and drier than the convergent zones near the surface." These emerging concepts clearly deviate from those derived from laboratory convection experiments and linear models.

### 7.2.6. Roll Convection

As mentioned above, it is commonly observed that boundary layer cumuli organize into longitudinal rolls or cloud streets. Clouds so organized are commonly observed under relatively high wind conditions (Kuettner, 1959, 1971). Using free convection scaling arguments, Deardorff (1972, 1976) concluded that roll-type convection will predominate whenever $-Z_i/L \leq 4.5$ to 25, where $L$ is the Monin-Obukhov length defined as $L = U^3T/kg\overline{w''T''}$ and $Z_i$ is the boundary layer depth. For large values of $-Z_i/L$, more or less random convection can be expected. LeMone (1973) observed rolls over the range $3 \leq -Z_i/L \leq 10$. Grossman (1982) determined from BOMEX data that for small values of $-Z_i/L$ roll convection dominates, while for larger $-Z_i/L$ values free convection (or random cells) dominates. Table 7.1 shows the different regimes determined by Grossman (1982). Christian and Wakimoto (1989) found from radar studies over the Colorado High Plains that roll structures occurred for very large values of $Z_i/L$ where random free cells should dominate. They speculated that their result could be due to the low observed wind speeds and that the roll structure was enhanced by other factors such as terrain. Thus a precise value of $-Z_i/L$ which separates the two regimes is still debatable.

Kuettner (1959, 1971) also points out that cloud streets are preferred in an environment exhibiting a relatively unidirectional wind profile that is strongly curved, such as that illustrated in Fig. 7.20. Kuettner also concludes that, typically, the cloud layer is convectively unstable. Other features of cloud streets: the lengths are 20 to 500 km, spacing between bands is 2 to 8 km, the depth of the boundary layer is 0.8 to 2 km, the width-to-height ratio varies from 2 to 4, and the vertical shear of the horizontal wind is $10^{-7}$ to $10^{-6}$. LeMone (1973) observed width-to-height ratios ranging from 2.2 to 6.5. The orientation

**TABLE 7.1** Categories of dominant eddy structure.

| Category | $Z_I/L$ parameter | Remarks |
|---|---|---|
| 1 | $-Z_I/L < 5.0$ | Only roll vortex motion. |
| 2 | $-Z_I/L < 7.3$ | Rolls coexist with convective cells and are necessary for their maintenance; rolls dominate. |
| 3 | $7.3 < -Z_I/L < 21.4$ | Rolls coexist with random cells but are not necessary for their maintenance; random cells dominate. |
| 4 | $21.4 < -Z_I/L < 30^4$ | Not only random cells; shear important to cell structure and morphology. |
| 5 | $-Z_I/L > 30$ | Random cells only; shear unimportant to cell structure and morphology. |

**Source**: From Christian and Wakimoto (1989)

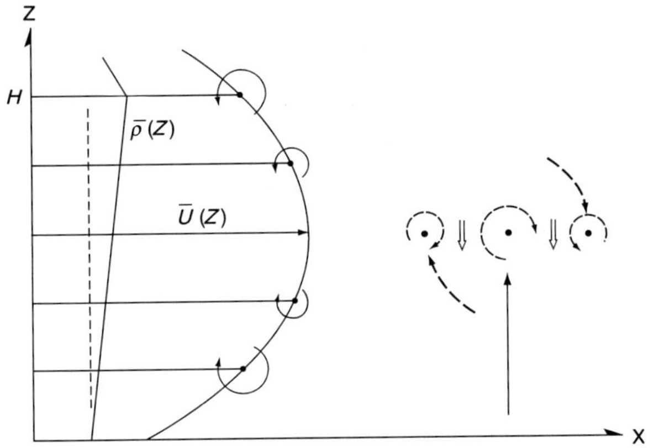


**FIGURE 7.20 Schematic presentation of basic flow and density profile (left), relative vorticity of vertically displaced convective element, and resulting restoring force (right).** *(From Kuettner (1971))*

of rolls is roughly parallel to the winds at the top of the convective layer. Kelley (1984) observed roll orientations from 2° to 15° to the right of the geostrophic wind at the top of the convective layer. Walter and Overland (1984), however, found fog streets to be oriented 16°-18° to the left of the geostrophic wind in a near-neutral boundary layer. LeMone (1973) obtained a similar result.

Cloud streets, roll vortices, or longitudinal roll clouds have been the subject of intensive theoretical and laboratory investigations. Laboratory experiments and theoretical studies by Faller (1965) and Lilly (1966) have predicted the formation of roll vortices in a neutrally stratified, viscous boundary layer having a wind profile that varies both in direction and in speed following an Ekman profile (Blackadar, 1962). The mechanism of instability leading to roll vortex formation that they identified is generally referred to as parallel instability. It is dependent upon shear along the roll axis, the earth's rotation (or Coriolis effect), and viscosity, but is independent of thermal instability.

A related mechanism for roll formation is commonly called inflection point instability. This mechanism of instability occurs in inviscid models in which the wind profile exhibits an inflection point or shear in the cross-roll direction (Brown, 1961; Barcilon, 1965; Faller, 1965). Lilly (1966) argued that in an unstable boundary layer, the inflection point instability mechanism would be suppressed, whereas the parallel instability mechanism would continue to operate. LeMone (1973) concluded that her observations of roll structure were consistent with the predictions of inflection point instability. She noted, however, that the neglect of buoyancy contributions may have contributed to an underestimate of the cross-roll component of the wind variance in those models.

Several investigators have used linear theory to examine roll vortex formation in a thermally unstable boundary with wind shear but no change in wind direction (Kuo, 1963; Kuettner, 1971; Asai, 1970, 1972). These studies show that the most unstable mode is the longitudinal mode in which the bands are oriented parallel to the wind shear and are stationary relative to the mean flow. However, in an unstable boundary layer in which both the speed and direction of the wind vary with height, Asai and Nakasuji (1973) found no preferred orientation of the rolls. Sun (1978) further extended the linear models to include latent heat effects. Following the wave-CISK hypothesis (Lindzen, 1974), latent heat was released in vertical columns in proportion to low-level moisture convergence, and evaporative cooling occurred when the low-level moisture was divergent. Sun found that the orientation and phase speed of propagation varied depending on whether or not buoyancy generated by latent heat release is dominant. If buoyancy dominated, the cloud bands were parallel to the wind shear and remained stationary relative to the mean wind. If the buoyancy force was weak and conversion of kinetic energy from the mean flow dominated, then the most unstable mode was a transverse mode (perpendicular to wind shear), and the cloud bands propagated relative to the mean wind.

Shirer (1980, 1982, 1986) also included latent heat effects in a nonlinear spectral model of cloud bands. In his model, latent heat is released in regions of upward motion above cloud base while air descends dry adiabatically. Because of the model's general formulation, it can be used to investigate all the previously described instability mechanisms believed to be important to cloud street formation. Like Sun (1978), Shirer (1980) found that both longitudinal and transverse roll clouds could coexist in a given environment. However, the particular character of the wind profile was the major determining influence. If the wind direction was invariant with height, longitudinal rolls were the predominant mode. However, if the basic wind changed direction as well as speed, some wind profiles resulted in longitudinal rolls, some in transverse rolls, and others led to two or three coexisting bands of differing alignments.

The linear models mentioned above elucidated three instability mechanisms leading to cloud street formation: inflection point instability, parallel instability, and thermal instability. Because the first two occur in a neutral atmosphere with shear, they can be referred to as dynamic instability mechanisms. Using a 3-D truncated spectral model, Shirer (1982, 1986) concluded that the thermal and parallel instability mechanisms are linked, such that only one convective mode develops when both instability mechanisms are operating. This result is consistent with Lilly's earlier conclusions. As shown by Kelley (1984) and Shirer and Brummer (1986), Shirer's model yields good predictions of observed roll cloud orientation angles and wavelengths.

The emphasis in most of the theoretical studies as well as observational investigations is on roll structures having width-to-height ratios in the range 2 to 4. Occasionally, width-to-height ratios as large as 20 to 30 are reported (Ogura, 1985). In one case, Walter and Overland (1984) observed a multiplicity of band scales coexisting over a region. Band scales of 1.3-1.7, 5-6, 12-15, and 25-30 km were superimposed over a broad-scale fog bank. They suggested that the bands spaced at 12-15 km could be traced upstream to individual mountain peaks, which may have played a role in their genesis. They noted that upstream topographic influences on cloud street spacing has also been reported by Higuchi (1963), Asai (1966), and Tsuchiya and Fujita (1967) over the Sea of Japan. Walter and Overland also suggested that Fiedler's mesoscale entrainment instability mechanism, or a resonant subharmonic to the basic boundary layer instability, may also be causing the larger scale bands. It is also possible that the larger scale bands may be a result of symmetric instability (Emanuel, 1979). If we define the vertical component of absolute vorticity as $\eta$, and the gradient Richardson number $R_i$ as $R_i = N^2/(\partial U/\partial Z)^2$, where $N$ is the Brunt-Väisälä frequency, then the criterion for symmetric instability is $R_i < f/\eta$. Thus, symmetric instability will occur if the vertical shear is large or the absolute vorticity small. The resultant rolls are aligned parallel to the vertical shear vector.

It should be remembered that in all these laboratory and analytic studies of factors affecting cloud organization, idealized boundary layer models are assumed. In many of the analytic theories, a constant eddy viscosity model is specified that is analogous to molecular viscosity in the laboratory experiments. This assumption affects the quantitative results of the models. Sun (1978), for example, assumed that the constant vertical eddy diffusivities for heat and momentum differed from the horizontal viscosities;this assumption improved his predictions of the spacing of the cloud bands. We emphasize that we must be cautious in our interpretations of the results of these analytic and laboratory studies when applying them quantitatively to the atmosphere.

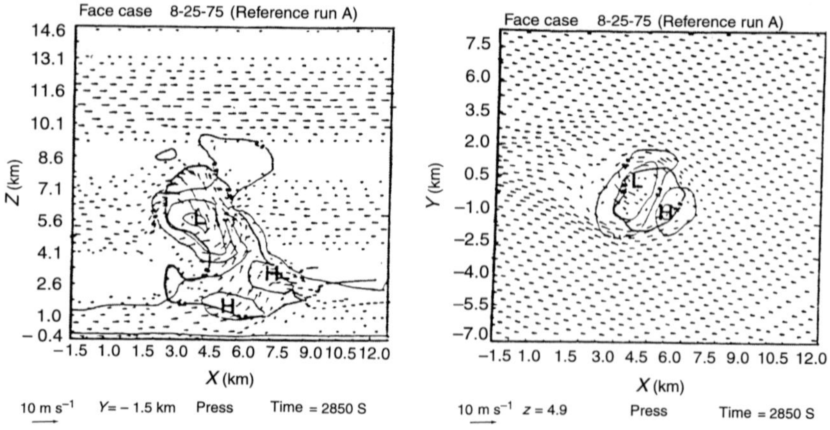


**FIGURE 7.21** **Pressure analysis for natural cloud A.** Contours are every 0.01 kPa, minima and maxima are denoted by L and H, respectively. *(From Cotton et al. (1981))*

## 7.2.7. The Role of Gravity Waves in Cloud Organization

In the studies of cloud organization examined thus far, the emphasis has been on instabilities in the boundary layer caused by shear and thermodynamic stratification. Clark et al. (1986) and Kuettner et al. (1987) have argued that gravity waves in the stably stratified atmosphere above the atmospheric boundary layer play a role in cloud organization and spacing. They suggest that the gravity waves are excited by what they call "thermal forcing," in which boundary layer thermals deform the capping inversion or interface region, causing ripples which produce vertically propagating gravity waves. They also suggest that boundary layer eddies and cumulus clouds can act in an analogous way to obstacles in the flow in the presence of mean environmental shear. As we shall see in subsequent sections, as cumulus clouds penetrate into a sheared environment, pressure perturbations develop about the rising cloud towers with positive pressure anomalies on the upshear flanks of the updraft and at cloud top, and negative pressure perturbations on the downshear flanks of the updraft (see Fig. 7.21). The positive pressure perturbations tend to divert flow in the stably stratified environment over and around the updrafts of the convective towers. Thus the diversion of flow about the actively rising cumulus towers in a sheared environment qualitatively resembles flow about a solid obstacle. The resultant flow perturbations can excite gravity waves in the stably stratified environment, which can propagate vertically and horizontally. That boundary layer eddies or cumulus clouds can excite gravity waves is not too surprising. As Clark et al. and Kuettner et al. pointed out, sailplane pilots have known for some time that so-called thermal waves can frequently be found above the tops of cumulus clouds. What is surprising, however, is that Clark et al. showed, with a two-dimensional numerical prediction model, that the gravity waves excited by clouds and boundary layer thermals can feed back on the boundary layer eddies, causing a change in the spacing of cloud lines. They speculate that such tropospheric/boundary layer interactions can result in a weak resonant response that may at times involve the entire depth of the troposphere. This work introduces an entirely new perspective to the problem of cloud organization. It suggests that changes in vertical wind shear and thermodynamic stratification in the overlying stably stratified troposphere may be important to the organization of boundary-layer-confined cumulus clouds. Such changes in environmental shear and stratification will alter the vertical propagation of gravity waves, causing reflection and refraction of gravity wave energy into the boundary layer. Petters et al. (2006) hinted that gravity waves may be participating in the formation of POCs.

The numerical experiments reported by Clark et al. (1986) and Clark and Hauf (1986) suggest that not only is the interaction between cumulus clouds and the gravity wave field important to the spacing of cumulus clouds, but it is also important to the commonly observed phenomenon that new cloud turret growth occurs preferentially on the upshear flank of a cloud system (e.g. (Malkus, 1952)). Clark and Hauf showed that in a sheared environment, as the cumulus clouds penetrate the stably stratified environment, gravity waves are excited and exhibit downward motion on the downshear side of the cloud tower and upward motion both on the upshear flank of the cloud and over the top of the turret. This they attribute to the faster propagation of gravity waves upstream than the thermals, which are emitted from the boundary layer.

A clear implication of the Clark et al. and Clark and Hauf studies is that the dynamics of cumulus clouds cannot be modeled properly by confining the cloud model to simulating motions only within the visible portion of the cloud. A cumulus cloud is clearly the product of interactions between the gravity wave field and convective thermals, which generally have their origins in the atmospheric boundary layer. We shall see in subsequent chapters that the interaction between gravity waves and convection is also important to the propagation and behavior of individual cumulonimbus clouds and mesoscale convective systems.

### 7.2.8. The Observed Structure of Individual Cumuli

In this section we focus on a few observations of the liquid-water content and velocity structure of non-precipitating or lightly precipitating cumuli to illustrate the complexity of structure of these seemingly simple clouds. In spite of being taken over 50 years ago, observations of cumuli, obtained by the Australian group (Warner, 1955, 1970, 1978; Squires, 1958a,b) using an instrumented aircraft capable of simultaneously measuring cloud vertical velocities and cloud micro-and macrostructure, remain as some of the best sets of data. Advances in electronic instrumentation have served to provide great details of droplet size-spectra, and even inhomogeneities in droplets on scales of a few meters, but the Australian observations still stand the test of

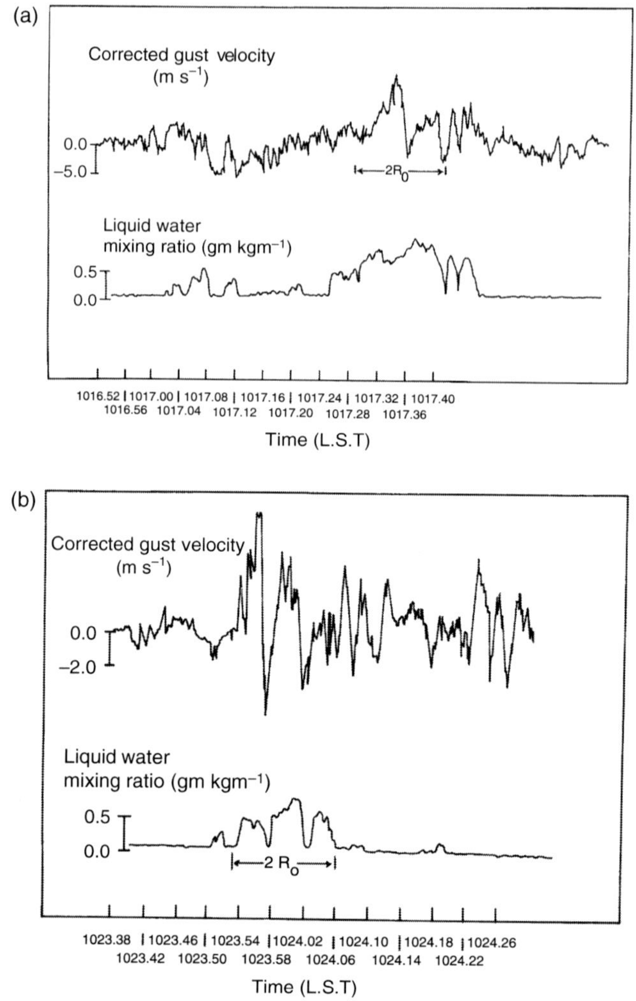


**FIGURE 7.22 Observed vertical velocity and cloud liquid-water content at 2.77 km (a), 2.34 km (b), 1.9 km (c), and 1.5 km (d).** All heights mean sea level (MSL). *(From Cotton (1975a))*

time. Figure 7.22 illustrates a set of observations obtained by the cloud physics group of CSIRO under the direction of J. Warner near Bundaberg, Queensland, Australia. The observations were performed in an aircraft descending from 2.77 km MSL to 1.5 km MSL. Most noteworthy in these figures is the extreme variability of vertical velocity as a cloud is transected. In some cases (see Fig. 7.22b), the vertical velocity varies from updrafts about 6 m $s^{-1}$ to downdrafts of more than 4 m $s^{-1}$ in less than 50 m across the cloud. Upon

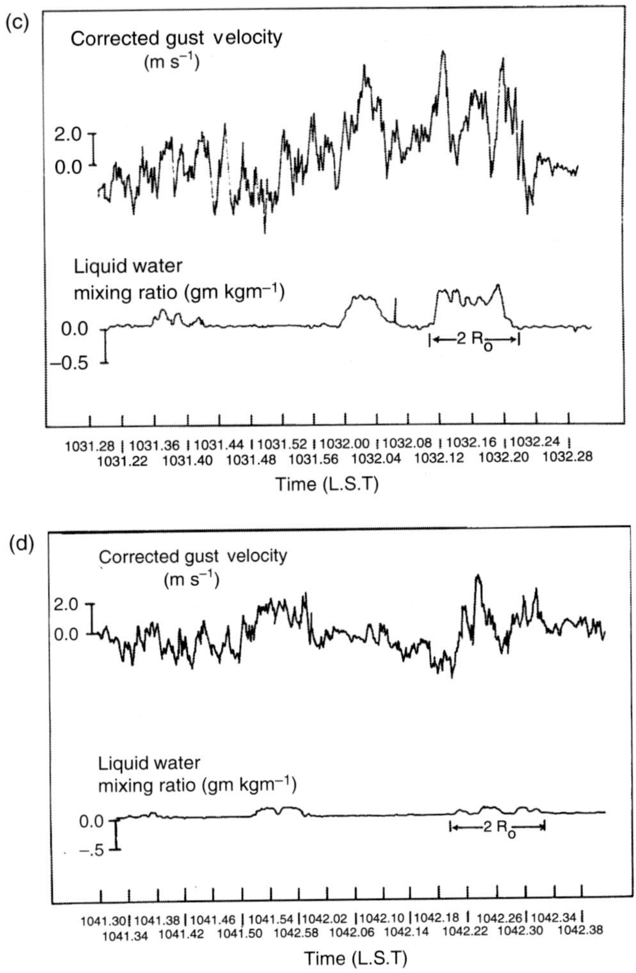


**FIGURE 7.22** *(continued)*

entering the cloud, the LWC jumps from zero to substantial values. Through most of the cloud, the LWC fluctuates slightly in response to vertical velocity fluctuations while maintaining high values, on the average. Only in regions of vigorous downdrafts does the LWC plunge to near-zero values in the cloud interior.

Warner (1955) examined the vertical variation in LWC from cloud base to near cloud top. Figure 7.23 illustrates that the peak LWC monotonically rises from cloud base to within 300 m of cloud top, where it rapidly falls to zero. If parcels of air entered the base of a cloud and then rose through the depth of the

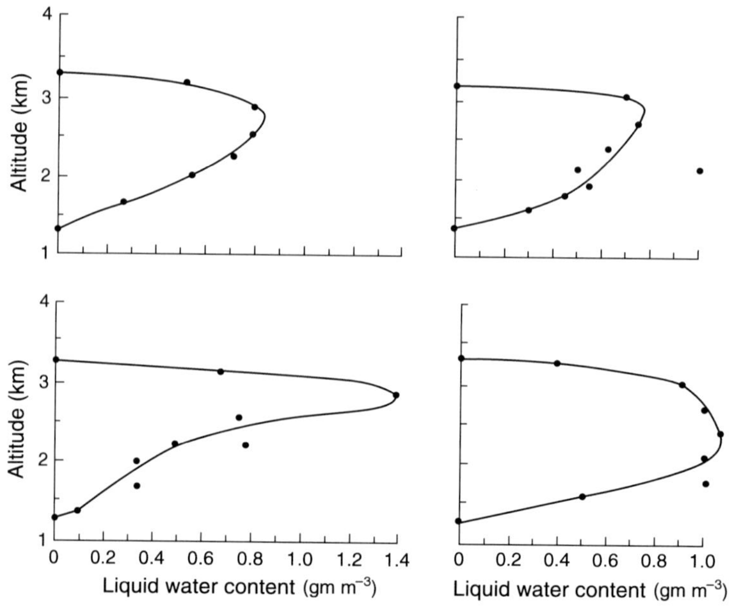


**FIGURE 7.23** **Variation of peak water content with height.** *(From Warner (1955))*

cloud without mixing with their environment, then the LWC would be given by

$$r_{1a} = r_T - r_s(T, P), \tag{7.13}$$

where $r_T$ represents the total mixing ratio of the air that would be conserved in the absence of mixing, and $r_s(T, P)$ represents the saturation mixing ratio that varies with pressure $P$ and temperature $T$. For an unmixed parcel of air, the temperature $T$ would be given by the rate of wet adiabatic cooling as the parcel ascends from cloud base. We define the quantity $r_{1a}$ as the adiabatic liquid-water mixing ratio. The ratio $(r_1/r_{1a})$ represents the departure of the LWC from adiabatic values due to the effects of mixing. Only if the cloud base temperature varies significantly from the estimated value, or if precipitation allows water drops to settle to lower levels in a cloud, will $r_1/r_{1a}$ exceed unity. Figure 7.24 illustrates observed profiles of $r_1/r_{1a}$ computed by Warner (1955), Squires (1958a), Ackerman (1959), and Skatskii (1965). Warner (1955) computed $r_{1a}$ as the average of the peaks in LWC shown in Fig. 7.24, whereas Squires (1958a) computed the average of a set of randomly distributed sampling points. Thus, Warner's (1955) averaging procedure results in slightly higher average values of LWC. Why Skatskii's average values of $r_1/r_{1a}$ are higher than Warner's is not known. The consistent feature of these observations is that the liquid-water content in cumuli departs substantially from adiabatic values within 500 m

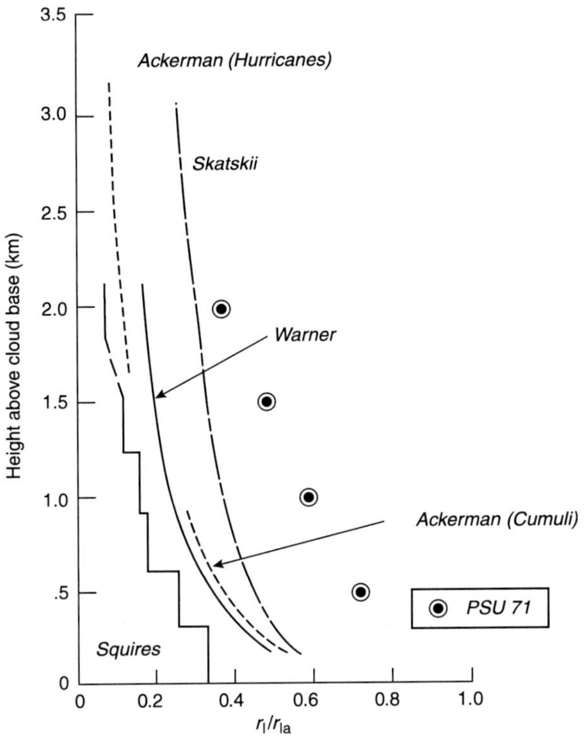


**FIGURE 7.24 The ratio of the mean liquid-water content at a given height above cloud base to the adiabatic value.** The circular symbol represents calculated values of $r_l/r_{la}$ using a lateral entrainment model. *(From Cotton (1975a))*

of cloud base. Furthermore, the average value of $r_l/r_{la}$ rapidly approaches asymptotic values of the order 0.2. Squires (1958a) noted that the adiabatic LWC was attained in only 3% of his samples. The observations reported by Squires and Warner, however, were randomly oriented with respect to the wind shear vector. As we shall see, the asymmetry in cloud structure caused by wind shear can influence the likelihood of sampling near-adiabatic values of LWC. Raga et al. (1990) computed profiles of $r_l/r_{la}$ from observations in warm trade-wind cumuli over Hawaii and found the same shape as shown in Fig. 7.24 but the shifted to higher values implying the Hawaiian clouds were less diluted.

As Fig. 7.22 shows, the most noteworthy feature of the structure of the vertical cross sections is its extreme variability. Warner (1970, 1977) examined the velocity structure of cumuli to determine the updraft features that vary most consistently with height or with the lifetime of the cloud. First of all, Warner (1970) found that the vertical velocity averaged over the width of the cloud showed no consistent variations with height. The root-mean-square vertical velocity $\omega$ ($\omega = \overline{(w''^2)}^{1/2}$) averaged over the width of the cloud, however,

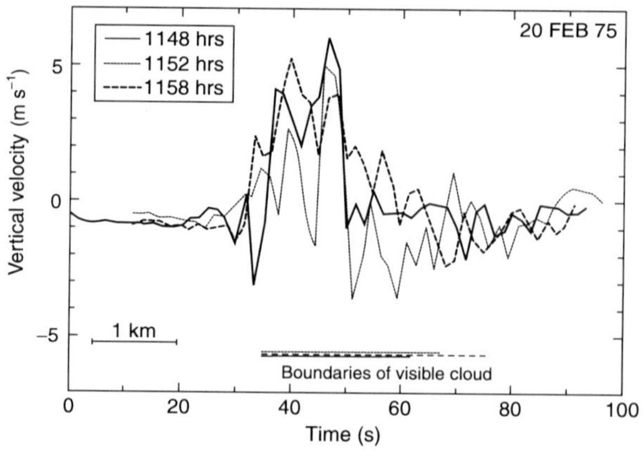


**FIGURE 7.25 Filtered velocity data from successive penetrations of a cumulus cloud showing the persistence of updraft structure near the upshear cloud boundary (left-hand side of diagram).** *(From Warner (1977))*

exhibited a consistent, monotonic increase with height. He found that the root-mean-square velocity could be described by the regression equation

$$\omega = a_1 + b_1 z, \tag{7.14}$$

where $z$ represents the height above cloud base in kilometers. The coefficient $a_1 = 1.26$ for observations over Cofl's Harbour, Australia, and $a_1 = 1.13$ for observations over Bundaberg, Australia. Likewise $b_1$ varied from 0.76 over Coff's Harbour to 0.69 over Bundaberg.

The peak updrafts and downdrafts also showed consistent variations with height, both increasing with height above cloud base. The maximum updraft Warner encountered was 12.7 m s$^{-1}$ at 2 km above the base of a 1.7-km-deep cloud. Warner (1970) noted that neither peak, average, nor root-mean-square velocity varied significantly with cloud width.

Examining the time variation of these properties, Warner (1977) found that the root-mean-square vertical velocity at a given level varied little throughout the lifetime of a cloud even for sampling periods as long as 20 minutes. Warner (1977) did find that a coherent, persistent updraft structure often occupies a small fraction of the cross-sectional area of the visible cloud. Figure 7.25 shows that a persistent updraft exists on the upshear side of a cloud for successive penetrations. Also, while fluctuations in vertical velocity on the upshear side of the cloud remain relatively small, they are more pronounced on the downshear side and extend a considerable distance from the cloud.

The appearance of an asymmetric structure of cumuli growing in an environment with wind shear was first noted by Malkus (1949), and was elaborated upon more fully by Heymsfield et al. (1978). Heymsfield et al. used an instrumented sailplane that allowed the simultaneous observation of cloud

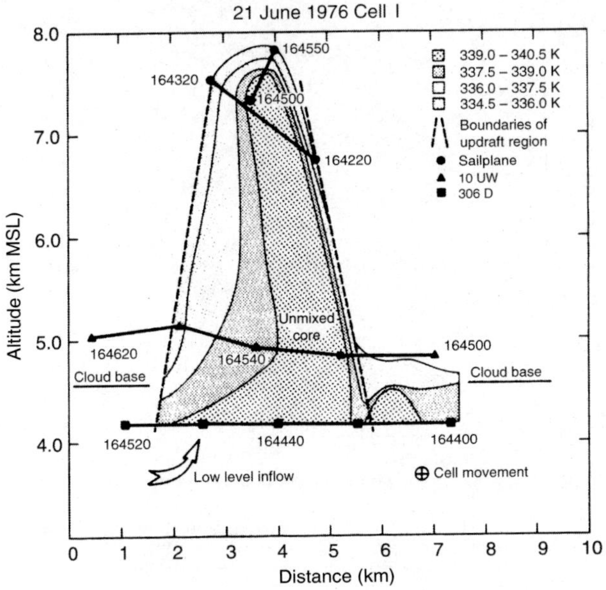


**FIGURE 7.26 Vertical section of $\theta_e$ data based on measurements from three aircraft plotted nearly perpendicular to the cell motion.** Low-level inflow region is positioned on left side of cell. Times of aircraft penetrations are indicated. *(From Heymsfield et al. (1978))*

microstructure, temperature, and rise rate of a cloud (while spiraling upward in cloud updrafts). They used $\theta_e$ and $r_l/r_{la}$ as two indicators of the degree of mixing in the cloud updrafts. Figure 7.26 illustrates an inferred vertical cross section of $\theta_e$ in a towering cumuli observed over northeastern Colorado. A core of constant $\theta_e$ can be seen nearly throughout the vertical extent of the cloud on its upshear flank. Over a much greater area of the cloud than the "unmixed" core, values of $\theta_e$ are smaller than those found at cloud base. Figure 7.27 illustrates a horizontal cross section of $\theta_e$ and $r_l/r_{la}$ observed by the sailplane. A consistent pattern of an unmixed core on the upshear side of the cloud and vigorous mixing downshear can be seen in this figure. In another study, Ramond (1978) inferred, from aircraft observations, the pressure field about growing cumuli. He diagnosed regions of positive pressure anomalies upwind of actively growing cumuli. He suggested that the positive pressure anomaly forms a protective zone against entrainment, which provides a mechanism for forming upshear undiluted cores.

Raga et al. (1990) also found regions of Hawaii cumuli to be composed of essentially undiluted cloud base air. They found these regions to be generally located on the upshear side of the cloud or near the center of a particular cross section. The scales of these features ranged from 100 m to 400 m and these undiluted regions often comprised a significant fraction of the cloud at mid-levels.

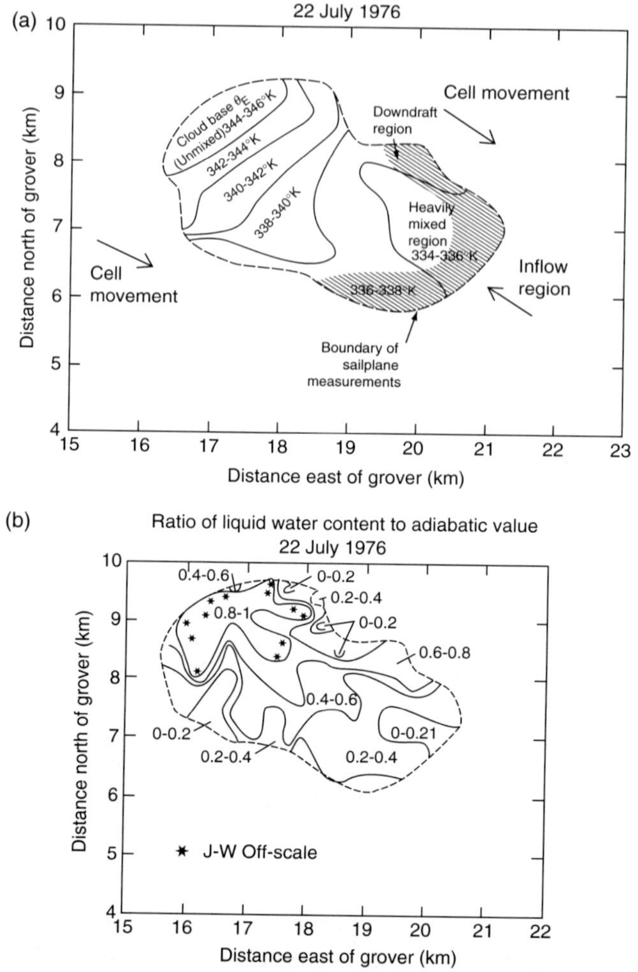


**FIGURE 7.27 Horizontal projection of data obtained from the sailplane on 22 July 1976 over an altitude range of 3.1 km. Data have been plotted relative to the cell motion. (a) $\tau_e$ data; (b) LWC/LWC$_A$ data.** *(From Heymsfield et al. (1978))*

Using a tethered balloon system with multilevel turbulence probes, Kitchen and Caughey (1981) inferred a P-shaped circulation in shallow cumuli. Figures 7.28–7.30 illustrate the inferred circulation field for three different cumuli. The feature of an updraft overturning at cloud top in the direction of the mean wind shear forming a reversed letter P pattern is consistent with the circulations inferred by Heymsfield et al. (1978) for larger towering cumuli. A feature of the circulation depicted in Figs 7.28–7.30 that was not inferred by Heymsfield et al. is the strong downdraft at the upwind boundary of the cloud. This, they suggested, is part of the return flow of the main updraft.

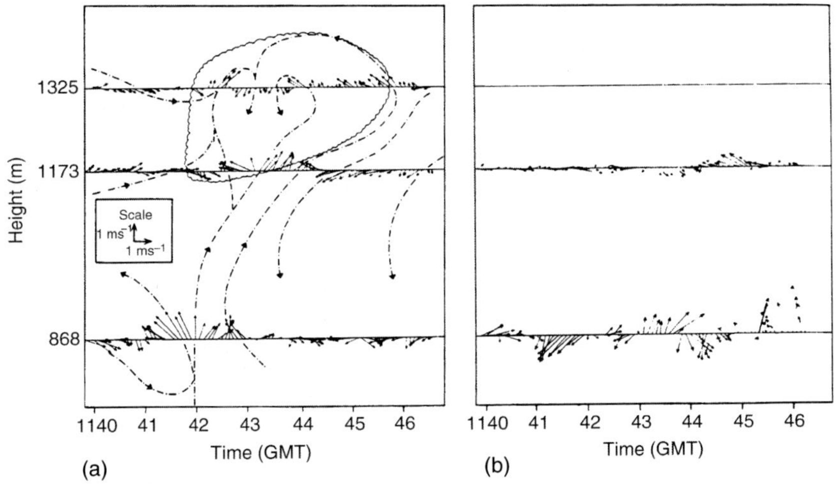


**FIGURE 7.28 Gust vectors (5s) constructed from the $u$, $v$, and $w$ wind components (see text) for cloud 1 in the vertical (a) and horizontal (b) planes.** A suggested flow pattern (dot-dash lines with directional arrows) has an inclined updraft overturning at cloud top in the direction of the mean wind shear; this circulation assumes the shape of a reversed letter P. Schematic cloud boundaries (wavy lines) and possible horizontal rotations (large curved arrows) are marked. *(From Kitchen and Caughey (1981))*

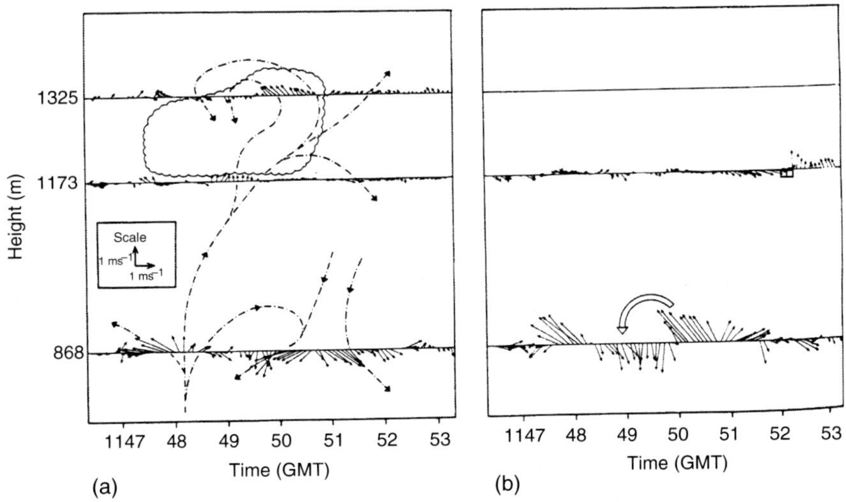


**FIGURE 7.29 As in Fig. 7.28, but for cloud 2, a weak dissolving cloud.** *(From Kitchen and Caughey (1981))*

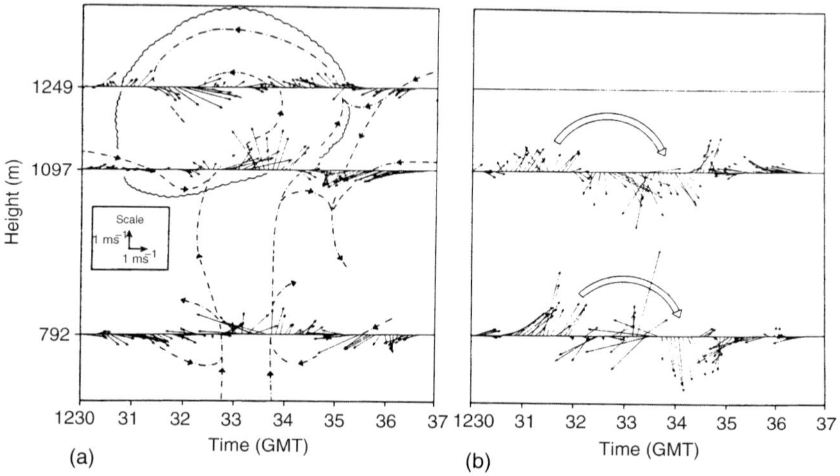


**FIGURE 7.30 As in Fig. 7.28, but for cloud 6, a larger cumulus cloud in the mature stage of development.** *(FromKitchen and Caughey (1981))*

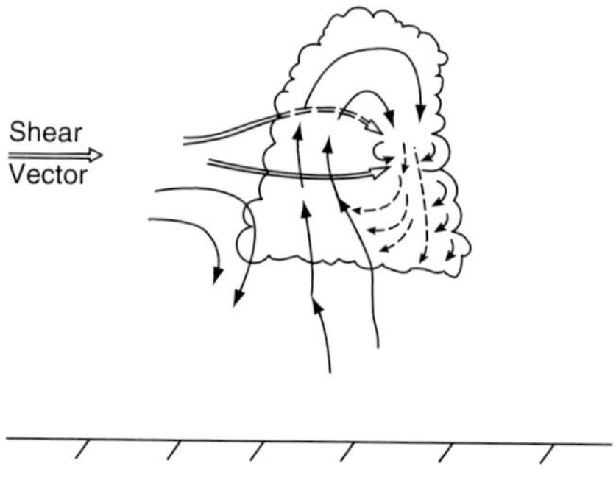


**FIGURE 7.31 Illustration of circulations in shallow cumuli.** Dashed lines depict parcel motions following mixing in the downshear flank of the cloud.

A conceptual model of the circulation in cumuli is illustrated in Fig. 7.31. Air rises through the cloud in an updraft that overturns downshear near cloud top, and thus forms a downdraft. Also illustrated are downdrafts along the upshear flank of the cloud. Superimposed on the P-shaped circulation is a horizontal flow that diverges about the main updraft and converges in the descending branch of the P-shaped circulation. The notion that downdrafts tend to form downshear of a cumuli is also consistent with the observationally-based inferences reported by Raymond et al. (1991), Telford and Wagner (1980) and Rogers et al. (1985).

Keeping these observed structures in mind, we now examine the theories for entrainment, detrainment, and downdraft initiation in cumuli.

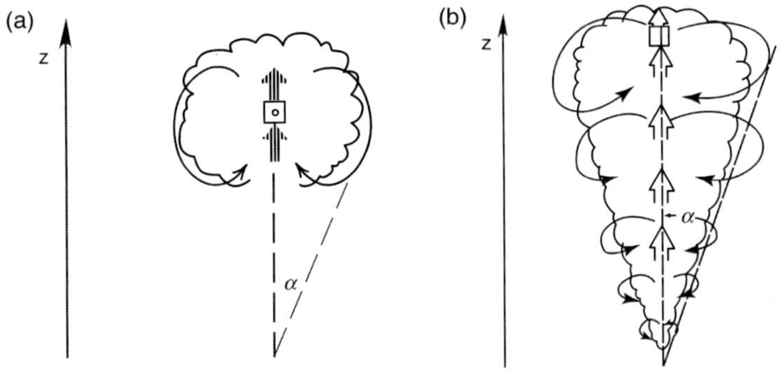


**FIGURE 7.32** **(a) Schematic view of the "bubble" or "thermal" model of lateral entrainment in cumuli. (b) Schematic view of the "steady-state jet" model of lateral entrainment in cumuli.**

## 7.3. THEORIES OF ENTRAINMENT, DETRAINMENT, AND DOWNDRAFT INITIATION IN CUMULI

As seen in the preceding section, there is considerable evidence that vigorous entrainment of dry environmental air takes place in cumuli leading to average LWCs well below adiabatic values. Entrainment also leads to a vertical velocity structure noted for its variability rather than to a distinct organized structure on the scale of the cloud. Early concepts of entrainment of dry environmental air into clouds viewed entrainment as occurring primarily through the cloud sides. That is, clouds were thought to be composed of a principal buoyant updraft having a "jetlike" or "bubblelike" structure. Entrainment was thus seen to be generated by the shear between the principle updraft and a stagnant environment (Ludlam and Scorer, 1953; Malkus and Scorer, 1955; Woodward, 1959; Levine, 1959; Malkus, 1960; Schmidt, 1949; Stommel, 1947; Squires and Turner, 1962). The laterally entrained air was then thought to mix homogeneously across the width of the updraft. As a result, the thermodynamic properties of updraft air at any level in the cloud should be a mixture of the properties of air entering cloud base and air entrained into the updraft at all levels below that level. Also, one would expect that a distinct gradient in cloud properties should exist between the central interior of the cloud and near the cloud edge. Thus, vertical velocity should be greatest near the middle of the cloud, and the LWC should be highest in the middle. The early concepts of entrainment were derived from laboratory experiments in which a denser fluid, like saline water was placed into a tank and then the expansion of the fluid was evaluated photogrammetrically. Figure 7.32a and b illustrate schematic models of the "bubble" and "jet" concepts of lateral entrainment. A corollary of the lateral entrainment theories is that the fractional rate of entrainment of environmental air into the updraft should vary inversely with the cloud radius (Malkus, 1960). Malkus (1960) formulated the rate of entrainment ($\mu_C$) for the bubble model as

$$\mu_C = (1/M_c)(dM_c/dz) = b/R, \tag{7.15}$$

where $M_c$ is the cloud mass, $b$ is a dimensionless coefficient, and $R$ is the radius of a cloud. Based on laboratory experiments (Turner, 1962, 1963), the rate of entrainment for a thermal is $b = 3\alpha = 0.6$, where $\alpha$ is the half-broadening angle of the rising bubble illustrated in Fig. 7.32a. In the case of a steady jet, Squires and Turner (1962) formulated the lateral entrainment rate ($\mu$) as

$$\mu = (1/F_m)(dF_m/dz) = b/R, \tag{7.16}$$

where $F_m$ is the vertical mass flux, $b = 2\alpha = 0.2$, and again $\alpha$ is the half-broadening angle shown in Fig. 7.32b.

Aside from the differences in entrainment coefficients, these two models share much in common. Both models contain the same form of vertical rise rate equation (Morton et al., 1956; Malkus and Williams, 1963), both view the entrainment process in cumuli to be primarily lateral, and both sets of equations can be integrated in a Lagrangian marching-type of solution technique. That is, one can vertically integrate the equations governing the rise rate and buoyancy by following the center of mass of a bubble or a characteristic parcel moving through a steady-state jet-like cloud (see a and b in Fig. 7.32). The lateral entrainment concepts have been generalized to models of a sequence of parcels released from the ground or the atmosphere (Danielsen et al., 1972; Lopez, 1973), and to one-dimensional, time-dependent numerical models where the governing equations are cast in Eulerian form (Weinstein, 1970; Wisner et al., 1972; Asai and Kasahara, 1967).

Perhaps the first challenge to the lateral entrainment concept came from Warner (1955), who observed that the ratio $r_l/r_{la}$ at a given height above cloud base varied little with the horizontal dimension of a cloud. As illustrated in Fig. 7.22, Warner also noted that the LWC varies sharply across the cloud boundary, with the edges of the cloud being nearly as wet as the center. Subsequently, Squires (1958a) hypothesized that the observed structure of cumuli can be better explained by the entrainment of dry air into the tops of cumuli. The mechanism as envisaged is similar to the theory for entrainment instability (CTEI) in stratocumulus clouds described in Chapter 6. That is, tongues or plumes of dry environmental air engulfed into the tops of cumuli will cause the evaporation of neighboring cloudy air, thereby chilling the air and resulting in penetrating downdrafts. The downdrafts, bringing air having above-cloud-top-level environmental properties, will penetrate well into the interior of the cloud. A property of this theory is that the vigor of entrainment will increase in proportion to the liquid water produced by the cloud and by the dryness of the cloud environment.

A further deficiency of the lateral entrainment theory was pointed out by Warner (1970) and verified by Cotton (1975b). Warner showed that if

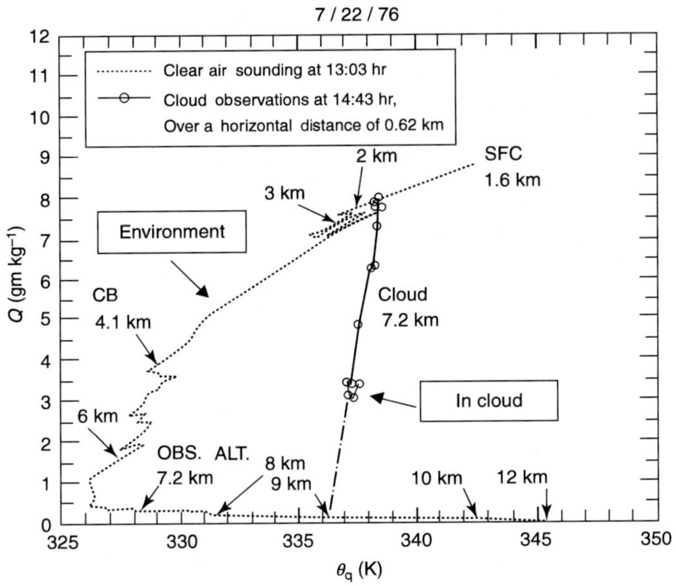


**FIGURE 7.33** **Data collected on 22 July 1976 at 7.2 km (−15 °C); Air with the observed properties could have been formed by mixing air from the surface levels with air from 9 km.** *(From Paluch (1979))*

the lateral entrainment rate was adjusted to enable the steady-state model to predict the observed cloud-top height, the LWC predicted by the model (or the profile of $r_l/r_{la}$) exceeded observed average peak values at all levels in the cloud. Fig. 7.24 illustrates this result; the figure shows Cotton's (1975a) calculation of $r_l/r_{la}$ exceeding observed values at all heights. In addition to the predicted magnitudes being too large, the predicted profile slope does not have the marked drop-off within the first few hundred meters above cloud base. Cotton (1975a) also showed that a time-dependent, one-dimensional model in which entrainment occurs principally laterally shares the same deficiency with predicted profiles of $r_c/r_a$, exceeding observed values.

Thermodynamic analysis of aircraft observations by Paluch (1979) further supports a CTEI concept of entrainment. She used the total mixing ratio $r_T$ and wet equivalent potential temperature $\theta_q$ as tracers of cloud motions. The wet equivalent potential temperature is similar to $\theta_e$, except it is conservative for reversible adiabatic ascent and descent rather than for pseudo-adiabatic motions. The selected parameters have the property of mixing in a nearly linear way, which simplifies interpretation of the data. Paluch's analysis was applied to data collected by an instrumented sailplane ascending in developing cumuli in northeast Colorado. Figure 7.33 illustrates a linear mixing line. She concluded that air at a given observation level in an updraft could not have originated as a mixture of environmental air at that level or below and air originating near the

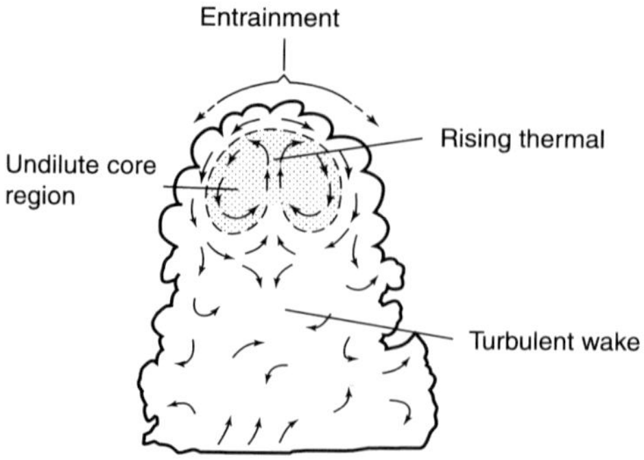


**FIGURE 7.34 Schematic model of a cumulus cloud showing a shedding thermal that has ascended from cloud base.** Continuous entrainment into the surface of the thermal erodes the core, and the remaining undiluted core region continues its ascent, leaving a turbulent wake of mixed air behind it. *(From Blyth et al. (1988))*

earth's surface. Instead, most of the air entrained in updrafts originated several kilometers above the observation level near cloud-top height. Paluch's results have been supported by similar thermodynamic analyses reported by Boatman and Auer (1983), Jensen et al. (1985) and Blyth and Latham (1985), all of which showed that mixing occurs predominantly between two levels, with one level being near cloud top. This is curious indeed, as how would a cloud know of the existence of its eventual cloud top? Gardiner and Rogers (1987) argued that the clouds selected for analysis in those studies all were of limited vertical extent, and were capped by very stable, dry air which the clouds could not penetrate. Thus, the cloud-top remained at a given level for an extended period of the cloud lifetime. This enhances the likelihood that entrainment will take place from a single level. Using his saturation point analysis technique, Betts (1982) also concluded that mixing between cloud base air and air near the maximum cloud top could only explain Warner's liquid water content measurements. Blyth et al. (1988), on the other hand, interpreted observations from 80 summertime cumulus clouds using Paluch's analysis, as indicating that the source of the entrained air was near the aircraft penetration level or slightly above. Thus environmental air enters a cloud at all levels. They interpreted their results as indicating that cumulus clouds were fundamentally like shedding thermals. A schematic model of this concept can be seen in Fig. 7.34. It shows that entrainment occurs near a rising cloud top and mixed volumes of air descend along the advancing edge of the thermal core into a trailing wake where they mix with cloudy and dry air in that wake region.

This conceptual model may be considered a limiting case for an environment without shear. In the case of sheared flow, Betts (1982) saturation point analysis and Gardiner and Rogers (1987) interpretation of Paluch's analysis technique suggests that the conceptual model of entrainment in cumuli in sheared flow, as proposed by Heymsfield et al. (1978) and modeled numerically by Cotton and Tripoli (1979) and Tripoli and Cotton (1980) and illustrated in Fig. 7.31, is appropriate. Gardiner and Rogers (1987) concluded that the updrafts located on the upshear side of the cloud constantly mix with the environment as the air parcels rise. Air residing in the downshear portion of the cloud experiences considerable cloud-top mixing, which favors the formation of negatively buoyant downdrafts. The rather coarse resolution modeling studies reported by Tripoli and Cotton (1980), and observational studies of towering cumuli flanking cumulonimbi (Knupp and Cotton, 1982a,b), suggest that the entrainment rate is modulated by (1) the difference between the updraft horizontal momentum and environmental horizontal momentum, (2) the updraft momentum flux and its vertical divergence, and (3) the saturation point or $\theta_{ES}$ of the environment. That work suggests that entrainment is controlled by both cloud-scale dynamics and the thermodynamics of evaporation processes.

The rather coarse resolution model sensitivity experiments reported by Tripoli and Cotton (1980) suggested that the vertical momentum flux carried by the updraft and its vertical divergence have a significant regulating influence on the intensity of large-eddy entrainment. If, for a given environmental stability in the cloud layer, the vertical momentum flux entering cloud base is relatively strong (weak), then the vertical divergence of the updraft momentum flux will be weak (strong) or the rate of large-eddy entrainment will be weak (strong). Thus, small-scale mesoscale convergence fields,local boundary features such as small hills or surface "hot spots" can have an important modulating influence upon the intensity of large-eddy entrainment in towering cumuli. Likewise, wherever there is strong divergence of updraft momentum flux, or wherever parcel buoyancy decreases with height, *detrainment* should be expected (Bretherton and Smolarkiewicz, 1989).

Finally, the drier the cloud environment (as evidenced by the minimum value of $\theta_{ES}$, the greater will be the tendency for air entrained on the downshear side of the updraft to generate penetrative downdrafts. The modeling studies of Cotton et al. (1981) suggest that the penetrative downdrafts intensify the downshear relative pressure low, which further enhances the intensity of large-eddy entrainment. This positive-feedback loop between penetrative downdrafts and entrainment is analogous to "pulling the plug in a bathtub."

In the above discussion we referred to the entrainment process by eddies on the scale of a cloud tower radius as *large-eddy entrainment.* Such eddies are readily observed by multiple-Doppler radar (e.g. Knupp and Cotton (1982a,b)) and are modeled by numerical prediction models. Entrainment by large eddies should be distinguished from entrainment by small eddies or turbulent eddies having horizontal scales of a few tens of meters.

As noted previously, Kitchen and Caughey (1981) observed the characteristics of eddies in boundary layer cumuli using multilevel sensors on a tethered balloon system. Using power spectrum analysis, they found that the kinetic energy of the cloud resided on two distinct scales, one being on the scale of the major updrafts and downdrafts (i.e. 0.5 km) and the second on a scale less than or equal to 10 m. They noted that the data suggested a shift in both peaks to shorter scales as the top is approached. They inferred that this was due to the influence of the capping inversion. Thus, one would anticipate that a small-eddy, cloud-top entrainment process would be more prevalent near an inversion in boundary layer cumuli.

There seem to be two schools of entrainment. One school can be thought of as the dynamic school of entrainment. In this school entrainment is driven primarily by the bulk dynamics of the cloud and dynamical instabilities that develop at shear interfaces on the bulk cloud circulations. This concept is represented in the two and three dimensional idealized simulations by Klaassen and Clark (1985), Clark et al. (1988), Grabowski and Clark (1991, 1993a,b), Grabowski (1995). In this school of thought Kelvin-Helmhotz type instabilities generated by local and thermal-scale shears at the cloud-clear air interface, and Rayleigh-Taylor type of instabilities in which small scale vortical circulations are generated by baroclinic torques generated by vertical and horizontal density gradients, are the principle drivers of entrainment. Grabowski (1995) for example argues that entrainment is *caused* by these interfacial instabilities while buoyancy reversal is an *effect* of entrainment. He argues that CTEI is not the primary driver of entrainment.

The other school argues that entrainment is induced by buoyancy differences associated with phase changes as cloudy air mixes with dry environmental air, or is based on the CTEI concept. Origins of this school of thought can be found in the papers by Squires (1958a,b) and Telford (1975) in which entrainment was described as being due to penetrative downdrafts originating near the tops of growing cumuli. This concept has lead to the formulation of what are called buoyancy-sorting models (Raymond, 1979; Raymond and Blyth, 1986; Emanuel, 1991) in which parcels of air gain negative buoyancy and aggregate at their level of neutral buoyancy.

Carpenter et al. (1998a,b,c) describe simulations of New Mexico cumuli using 50 m grid spacing. The clouds were initiated with an applied surface heat source as an idealization of thermal mountain slope flow convection. No environmental winds or wind shear was simulated. Relatively undiluted boundary layer air could be found in rising thermals at all levels of the simulated clouds. They interpreted their results in terms of the shedding thermal model illustrated in Fig. 7.34. Strong narrow penetrative downdrafts associated with the collapse of overshooting turrents were responsible for much of the mixing as well as detrainment of mass and moisture in narrow layers. Figure 7.35 illustrates a conceptual model of a collapsing turrent, followed by detrainment. Some aspects of these model results support both schools of thought in that the shedding thermal model can be thought of as a *bulk* idealization of the initiation of entrainment by dynamical instabilities along the boundaries of the descending branch of the toroidal circulation. We suspect that Grabowski and Clark would argue that at 50 m grid spacing the actual dynamics of small scale eddies were not explicitly represented but instead small-scale eddy mixing, modeled by an

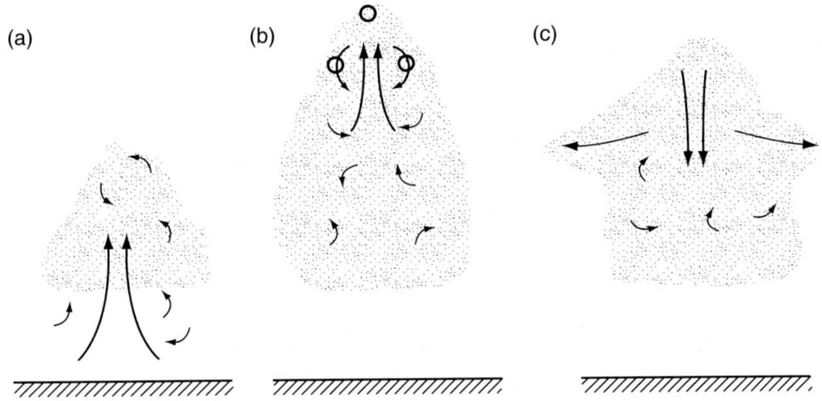


**FIGURE 7.35 Schematic diagram of the life cycle of a modeled cloud turret.** (a) Convection begins when a boundary layer updraft becomes organized and penetrates a weak inversion near cloud base. (b) A toroidal circulation develops as the turret ascends and accelerates. Environmental air entrained in this manner is deposited in a trailing wake. Strong downdrafts associated with the turrets collapse originate either near the top of the turret or in the descending branch of the toroidal circulation (open circles). (c) Downdrafts develop in response to overshooting and are further fueled by evaporative cooling. They descend to their level of neutral buoyancy, which is typically located at mid levels within the cloud. Parcels within strong downdrafts are detrained away from the cloud in a narrow layer. *(From Carpenter et al. (1998c))*

eddy viscosity parameterization, suppressed the production of vorticity along the cloud boundaries which are the major contributors to entrainment.

Recent LES modeling results of simulations of trade-wind cumuli by Jiang and Feingold (2006), Xue and Feingold (2006), and (Xue et al., 2008) suggest that variations in CCN concentration effect entrainment. That is, owing to the large net surface area (for a given LWC) of cloud droplets when CCN concentrations are high, compared to when CCN concentrations are low, faster evaporation rates are simulated in entrained eddies when CCN concentrations are high. As a consequence penetrative downdrafts are enhanced and greater amounts of entrainment are simulated. This leads to clouds with lower cloud tops and lower areal coverage. This work supports the idea that entrainment is controlled in part by the CTEI concept. That is not to say that entrainment is not initiated by dynamic instabilities associated with shears along the cloud boundaries but that, at the very least, evaporation of drops mixing with the entrained air amplifies the entrainment process. Likewise, the importance of vertical shear of environmental winds as a mechanism of enhancing dynamic forcing of entrainment, relative to a no-sheared environment, versus CTEI type forcing of entrainment, has yet to be quantified. This is important to understanding of the potential impacts of aerosols on entrainment and cloud characteristics in a climate sense. As environmental wind shear is increased, will dynamically-driven entrainment be enhanced to such an extent that aerosol effects on entrainment will be masked or relegated to secondary importance?

Likewise we need to quantify the relative roles of large eddies versus small eddies to the entrainment process. The evidence is compelling that small eddies, less than 10 m scales, initiate the entrainment process, but is it necessary to explicitly resolve eddies on such small scales in order to represent the bulk thermodynamic and dynamical properties of cumuli properly?

## 7.4. THE ROLE OF PRECIPITATION

In Chapter 4 we noted that the presence of condensed water affects the vertical accelerations in a cloud by generating a downward-directed drag force equivalent to the weight of the suspended water. Thus, one immediate consequence of a precipitation process is that it unloads an updraft from the weight of the condensed water. Several modeling studies (e.g. Simpson et al., 1965; Simpson and Wiggert, 1969; Weinstein and Davis, 1968) have shown that, as a result of precipitation unloading, a cumulus tower can penetrate to greater heights than its non-precipitating counterpart. Additional modeling studies (Das, 1964; Takeda, 1965, 1966a,b; Srivastava, 1967) demonstrated that, in a nonshearing environment, the water which is removed from the upper parts of the updraft accumulates at lower levels, where it can eventually lead to the decay of the updraft.

In the case of a convective tower growing in a sheared environment, however, the precipitation may not settle into the updraft at lower levels. Instead, precipitation falls downshear of the updraft. Precipitation particles, therefore, typically settle in the region that, we have seen, is already preferred for downdraft formation as a consequence of the interaction of cloud vertical motions with environmental shear. Precipitation can thus enhance the negative buoyancy in that region. Moreover, as cloud water is depleted by evaporation in downdrafts, the settling of precipitation in downdrafts provides an additional supply of water to be evaporated in those regions. Thus, we see that there exists an intimate coupling between entrainment and precipitation processes in the initiation and maintenance of downdrafts in towering cumuli.

A number of modeling studies (Liu and Orville, 1969; Murray and Koenig, 1972; Yau, 1980) have shown that the thermodynamic consequences of the precipitation process are far more important than the water loading alterations. This is particularly true of the latter stages of the cloud lifetime, in which the evaporation of rainwater below the cloud is important in stimulating and intensifying downdrafts. As the downdraft air approaches the earth's surface, it spreads laterally. The evaporatively-chilled, dense downdraft air can undercut surrounding moist, potentially warm (high $\theta_e$) low-level air, thereby lifting it to the lifting condensation level (or saturation point). In some cases, the lifting by the downdraft outflow may be great enough to lift the air to the level of free convection (i.e. the LFC, the height at which an air parcel becomes buoyant), thus initiating new convective towers. We shall see, however, that a distinguishing feature between a towering cumulus and a cumulonimbus cloud is that the downdraft outflow from a towering cumulus is frequently not vigorous enough to lift the low-level air to the level of free convection (LFC). Moreover, the profile of the horizontal wind or wind shear profile has a profound influence upon whether the precipitating downdraft from a towering cumulus can constructively or destructively influence the subsequent propagation of the cloud (i.e. create new convective towers).

Consider, for example, the three-dimensional simulations of towering cumuli over Florida reported by Levy and Cotton (1984). Low-level flow in this case was westerly while mid-tropospheric flow was easterly. As a consequence, Levy found that the precipitation falling out of the cumulus tower fell into the relative inflow flank of the cloud updraft. As a result, the low-level updraft was quenched by the diverging rain-chilled air. This resulted in the dissipation of the parent cumulus tower. New cell growth occurred on the flanks of the diverging evaporatively-cooled air, but these cells were relatively weak in intensity and shallow in depth. By contrast, when the low-level flow was altered from westerly to easterly, the evaporating rain did not cut off the low-level updraft inflow air; this caused a longer-lived cloud circulation and a 32% increase in rainfall. Therefore, depending upon the environmental wind field, evaporation of precipitation can constructively or destructively contribute to the propagation and longevity of a towering cumulus cloud. As we shall see later, however, the volume of rainfall available for evaporation in the subcloud layer is also important to the subsequent propagation of convective towers. If the volume of rainfall from an individual cumulus tower or from several neighboring cumulus towers is large enough, the diverging rain-chilled air may initiate new cloud growth of sufficient depth to contribute further to the propagation of the cloud system, even in the case of westerly low-level flow illustrated above. Likewise, the size spectrum of raindrops is importance to the amount of evaporation in the subcloud layer and the intensity of low-level cold pools. If, for a given rainwater content, many small raindrops are present, as opposed to a few large drops, subcloud evaporative cooling will be greater. This is a consequence of the greater surface to volume ratio of small drops versus large drops.

### 7.4.1. The Role of the Ice Phase

Since the early cloud-seeding experiments reported by Kraus and Squires (1947), wherein it was hypothesized that the observed "explosive growth" of cumuli was the result of artificially induced glaciation of the cloud, there has been intensive interest and research in the relationships between evolution of the ice phase and the dynamics of cumuli. As we have seen in previous sections, mixing or entrainment processes in cumuli act as a brake against cumulus growth (Malkus and Simpson, 1964). There often exists a delicate balance between the production of cloud buoyancy forces and their destruction by entrainment. Simple inspection of a thermodynamic diagram (e.g. skew T log P, or tephigram) reveals that the latent heat liberated during the growth of cloud droplets is quite large in the lower troposphere. However, in the middle to upper troposphere, the moist adiabat becomes more and more parallel to the dry adiabat; this illustrates the fact that cloud buoyancy production by condensation growth of cloud droplets becomes less important. By contrast, at these same levels in a typical tropical and mid-latitude environment, the formation of the ice phase becomes most active. Thus, the latent heat liberated during the freezing of supercooled droplets and the vapor deposition growth of ice particles can augment the diminishing cloud buoyancy production during condensation growth of cloud droplets. In some cases, the additional buoyancy may be sufficient to allow a cloud to penetrate weak stable layers or to survive the entrainment of relatively dry environmental air. The results of calculations of cloud growth with simple one-dimensional cloud models (Simpson et al., 1965; Weinstein and Davis, 1968; Simpson and Wiggert, 1969; Cotton, 1972) suggests that the additional buoyancy liberated by the growth of the ice phase can, in some cases, promote the vertical growth of convective towers several kilometers above their non-glaciated counterparts.

Zipser (2003) elaborated on these concepts more fully with specific focus on tropical convective clouds. He noted that there is virtually no evidence of unmixed updrafts in tropical oceanic cumuli. Instead he argues that their convective cores are highly diluted by entrainment in the lower troposphere. He argues that vigorous oceanic cumulonimbus are able to reach the tropical tropopause as a result of the combined effects of the freezing of condensate, and the shifting from the condensation to sublimation latent heating rates. He illustrated this with Figs 7.36 and 7.37. First he argued that the pseudo-adiabatic curve and associated CAPE is not representative of tropical oceanic convection as much of the low-level buoyancy in those clouds is depleted by entrainment. In fact he argues that CAPE is overrated as a predictor of the behavior of convection as it does not take into account entrainment, water loading, the vertical distribution of buoyancy or ice latent heat effects. Zipser made a rough estimate of the effects of depletion of buoyancy by entrainment and water loading, by reducing buoyancy by 1/3 in Figs 7.36 and 7.37. By doing so, clouds do not reach the tropical tropopause yet cloud updraft velocities are more in accord with observations. He then assumed that all condensate will freeze between (−4 °C) and (−15 °C) and that ice adiabatic ascent will occur above the (−15 °C) level. This is in accordance with observations in tropical cumulus (Black and Hallett, 1986; Willis and Hallett, 1991). This illustrates that tropical convection can then reach the tropical tropopause owing to the additional latent heating from the ice phase.

The formation of frozen precipitation can also affect cloud circulations by the redistributions of total water or water loading caused by the differing terminal velocities and resultant settling of the various frozen species through updrafts and downdrafts. In some cold cloud-base continental cumuli, precipitation may only be initiated by ice-phase precipitation processes. Furthermore, the formation of rapidly falling hailstones can unload a vigorous

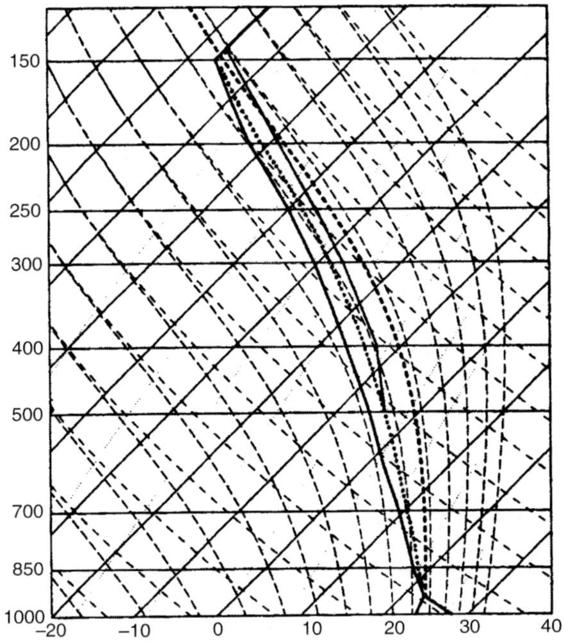


**FIGURE 7.36 Sounding adapted from Riehl and Malkus (1958) representing the properties of the environment in the equatorial trough zone (solid) and hypothetical pseudo-adiabatic ascent from cloud base (dashed).** The dotted curve represents the actual temperature of the ascent at one-third pseudo-adiabatic buoyancy. The thin solid curve shows the temperature increase from that of the dotted curve from freezing one-third the adiabatic water load between 500 and 400 hPa, and subsequent ascent along the ice adiabat. *(From Zipser (2003))*

updraft that is otherwise burdened by the weight of large quantities of slowly settling supercooled raindrops.

The buoyancy released during the formation of the ice phase is greatly dependent upon the prior rate of formation of liquid precipitation. For example, using a one-dimensional cloud model, Cotton (1972) found that the suppression of the warm-rain process could, in some circumstances, favor explosive cloud development. That is, suppose there exists a cumulus tower that is sufficiently vigorous (i.e. thermally buoyant) to overcome the negative buoyancy due to water loading. If the tower then penetrates into supercooled levels, the amount of condensed liquid water available for freezing is much greater than if the cloud had formed precipitation. It can easily be shown that the buoyancy gained by the freezing of condensed water exceeds the negative buoyancy contribution of the weight of the condensed water. Thus, if freezing commences before entrainment processes erode the cloud buoyancy, the additional condensed liquid water (relative to a raining cumuli) can contribute to explosive growth of the cloud tower.

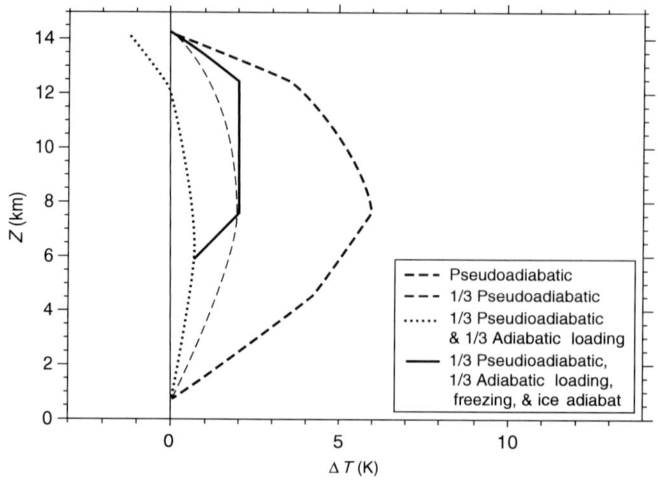


**FIGURE 7.37 Effective buoyancy of updraft from Fig. 7.36 according to various assumptions: standard pseudo-adiabatic ascent (heavy dashes), one-third pseudo-adiabatic buoyancy (light dashes), one-third pseudo-adiabatic buoyancy but adding drag from one-third adiabatic water loading (dots), and the effect on the dotted curve of freezing one-third adiabatic water load and subsequently following the ice adiabat (solid).** *(From Zipser (2003))*

However, the speed of glaciation of a cloud is also highly dependent upon the prior history of warm-rain processes. Several modeling studies (Cotton, 1972; Koenig and Murray, 1976; Scott and Hobbs, 1977) have shown that the coexistence of large, supercooled raindrops and small ice crystals nucleated by deposition, condensation-freezing, or Brownian contact nucleation, favors the rapid conversion of a cloud from the liquid phase to the ice phase. This is because, in the absence of supercooled raindrops, small ice crystals first grow by vapor deposition until they become large enough to commence riming or accreting small cloud droplets. The riming process then proceeds relatively slowly until they have grown to millimeter-sized graupel particles. Thereafter, the conversion of the cloud to the ice phase can proceed relatively quickly. However, if supercooled raindrops are present, the slow-growth period can be circumvented. The large raindrops then quickly collide with small ice crystals; they immediately freeze to become frozen raindrops. The frozen raindrops can rapidly collect small supercooled cloud droplets, which further enhance the rate of conversion of a cloud to the ice phase. Secondary ice-crystal production by the rime-splinter mechanism (Hallett and Mossop, 1974; Mossop and Hallett, 1974) further accelerates the glaciation rate of the cloud. Several modeling studies (Chisnell and Latham, 1976a,b; Koenig, 1977; Lamb et al., 1981) have shown that the presence of supercooled raindrops accelerates the cloud into a mature riming stage wherein large quantities of secondary ice crystals can be produced in the temperature range −3 to −8 °C. The small secondary ice crystals collide with any remaining supercooled raindrops, causing them to freeze and further accelerate the glaciation process.

However, as noted by Keller and Sax (1981), in broad, sustained rapid-updraft regions, even when the criteria for rime-splinter secondary production are met, the secondary crystals and graupel will be swept upward and removed from the generation zone. Until the updraft weakens and graupel particles settle back into the generation zone, the positive-feedback aspect of the multiplication mechanism is broken. Therefore, the opportunities are greatest for rapid and complete glaciation of a single steady updraft if the updraft velocity is relatively weak. In contrast, Keller and Sax (1981) observed high concentrations of ice particles in the active updraft portion of a pulsating convective tower. They postulated that the graupel particles swept aloft in the first bubble of a pulsating convective tower settled downward into the secondary ice-particle production zone (−3 to −8 °C), wherein they became incorporated into a new convective bubble and contributed to a prolific production of secondary ice crystals by the rime-splinter mechanism. This demonstrates that there exists a very intimate, nonlinear coupling between buoyancy production by glaciation of a cloud and the evolution of the microstructure of the cloud, and the evolving cloud motion field.

This is illustrated quite dramatically in modeling studies of the effects of pollution aerosols on cloud dynamics and precipitation processes. Seifert and Beheng (2006), Khain et al. (2004), Zhang et al. (2005), van den Heever et al. (2006), and van den Heever and Cotton (2007) all found that enhanced pollution-induced higher CCN concentrations produced smaller cloud droplets, which reduced the production of drizzle drops or supercooled raindrops. If the convection was strong enough (i.e. CAPE was large enough) this resulted in more supercooled liquid water being thrust into higher levels of the cloud. The associated latent heat release results in more vigorous convection when these droplets freeze. In contrast, in a clean cloud where CCN concentrations are low, vigorous warm-cloud precipitation processes deplete the amount of liquid water lofted into sub-freezing temperatures and less latent heat is released when clouds glaciate, resulting in less vigorous convection. In some cases this leads to the formation of more vigorous squall line convective systems in polluted environments but not in clean air (Khain et al., 2004; Zhang et al., 2005). In other cases, while convection is initially intensified in a polluted environment the amount of precipitation on the ground is reduced in a polluted environment (van den Heever et al., 2006; van den Heever and Cotton, 2007).

In our discussion of the role of the ice phase in cumuli, we have not yet considered the influence of melting. Actually, there have been very few quantitative investigations of the role of melting in ordinary cumulus clouds. We will examine the influence of melting in larger-scale cloud systems such as cumulonimbi and mesoscale convective systems more fully in subsequent chapters.

## 7.5. CLOUD MERGER AND LARGER SCALE CONVERGENCE

A fundamental characteristic of the upscale growth of convective cloud systems is that individual convective elements merge together to form larger convective clouds. The merging of convective clouds has been identified observationally, both visually and by radar (Byers and Braham, 1949), and from the analysis of aircraft-measured updraft structures (Malkus, 1954). As pointed out by Westcott (1984) there are many definitions of cloud merger. The most commonly used definition is that obtained from the analysis of radar data. Wiggert and Ostlund (1975), for example, defined a merger as the combination of two previously separate radar echoes at the 1 mm $hr^{-1}$ isopleth of rainfall rate. A first-order merger was the result of the joining of two or more previously independent single echoes, and a second order merger was the result of the merger of two or more first- or second-order echoes.

The importance of the merger of convective clouds to convective precipitation was emphasized by Simpson et al. (1971), who noted that the merger of two moderately sized cumulonimbi produced a 10- to 20-fold increase of rainfall. In a subsequent study, Simpson et al. (1980) found that merged convective systems are responsible for 86% of the rainfall over an area under surveillance in south Florida, even though only 10% of the cells were merged. Simpson et al. (1993) reported on a similar study over northern Australia for so-called Hectors. They found that during monsoon break periods, 70% of the rainfall comes from second-order mergers, which represent less than 5% of the total number of convective systems. The more numerous single cells contribute less than 10% to the observed precipitation.

Using inferences drawn from radar observations of convective clouds, Simpson et al. (1980) postulated that the approach or collision of downdraft-induced gust fronts from adjacent cumulus clouds is the primary mechanism of shower merger. Figures 7.38 and 7.39 illustrate a conceptual model of the merger process in a light wind and weak shear case (Fig. 7.38) and in a moderate shear case (Fig. 7.39). Illustrated is a precursor "bridge" between cloud towers which, they noted, virtually always precedes radar echo merger. As the downdraft outflows approach and collide, new towers surge upward from the bridge filling the gap. In a sheared environment, such as often occurs over Florida in the summer, east winds decrease in intensity with height and return to stronger northerlies just below the tropopause. In such an environment a young cloud (illustrated on the right side of Fig. 7.39) contains mostly updrafts and moves toward the west (to the left) faster than the winds in the cloud layer. The gust front spreads out nearly symmetrically from the base of the young cloud. In an older cloud (illustrated on the left side of Fig. 7.39) downdrafts are predominant. Since the downdrafts transport slower moving easterlies downward, the relative motion of the older cloud is slower than that of the young cloud, and of the wind in the subcloud layer. As a result, the downdraft spreads out mostly on the downshear (right side), where it collides with the ambient low-level easterly flow, setting off new towers. The new

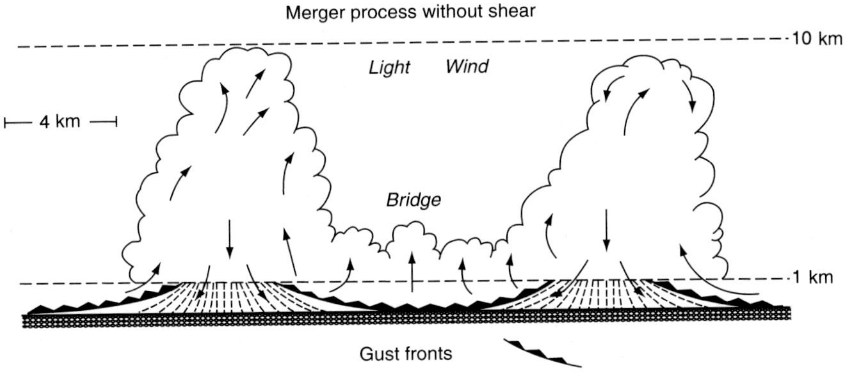


**FIGURE 7.38 Schematic illustration relating downdraft interaction to bridging and merger in case of light wind and weak shear.** *(From Simpson et al. (1980))*

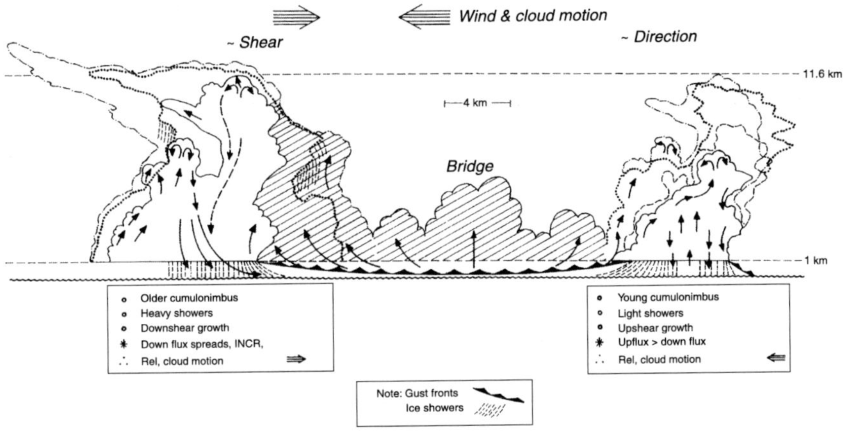


**FIGURE 7.39 Schematic illustration relating downdraft interaction to bridging and merger in case of moderate shear opposite to wind direction through most of the vertical extent of cloud layer.** Younger cumulonimbus on right has predominant up motions and moves faster than the wind. Older cumulonimbus on left has predominant down motions and moves more slowly than wind, so clouds move and propagate toward each other. Interaction of downdrafts enhances bridge development. *(From Simpson et al. (1980))*

towers may then serve as the bridge for new cloud growth. Therefore, they hypothesize that in a sheared environment, relative motion and propagation (by the downdraft and outflows) play a major role in the merger process.

Several different concepts of the merger process have been proposed. One view is that the merger process is a stochastic process in which larger aggregates of clouds arise spontaneously from the random clumping of smaller cloud elements. Ludlam and Scorer (1953), Lopez (1978), and Randall and Huffman

(1980) represent several advocates of such an approach. The idea is that cumulus clouds leave behind, following their decay, a local environment which is moister and more unstable (warm near cloud base and cool near cloud top). Thus, they provide an environment which is more favorable for further cloud growth. Given an initially random distribution of cloud elements, the locations where the greatest concentration of clouds initially occurred (by chance) would be able to support more vigorous, larger clouds. Thus, larger clumps of cumuli would be expected to arise in the more favored environment. Once a large clump of cumuli has formed, by virtue of its larger size and longer lifetime, the favored cumuli would be expected to aggregate randomly with smaller clumps. This would cause a still larger cloud system and subsequently a more favored environment for more cloud growth. This concept essentially views the merger process as a static process and ignores the dynamic properties of cumuli (i.e. their motion fields and pressure fields).

The role of cloud-induced pressure anomalies and circulations in the merger process has been emphasized by several investigators. In the numerical investigation of buoyant thermals, Wilkins et al. (1976) found that the overlapping buoyancy force fields between two neighboring thermals favors their coalescence. However, they also found that the velocity fields between the two rising thermals tend to interfere with each other and to suppress the circulation and velocity of rise. This causes a mutual repulsion between thermals. Orville et al. (1980) performed two-dimensional simulations of the interactions among precipitating cumuli. They varied both the spacing and the time of development of neighboring cumuli. They obtained results similar to Wilkins et al. (1976), in that two identical clouds initiated at the same time and height, generated circulation fields and pressure fields between them which inhibited cloud merger. When two clouds were started at different times, or if one cloud was substantially stronger than the other and their spacing was less than about 7 km, cloud merger resulted. In contrast to Orville et al. (1980), Turpeinen (1982) found in three-dimensional numerical simulations of cloud interactions that differences in the timing and intensity of two neighboring cloud impulses did not promote cloud merger. Instead, they found that the maximum center-to-center separation between cloud impulses could not be greater than 3.6 km for a cloud merger to occur. In agreement with Orville et al. (1980), Turpeinen found that the pressure perturbation field had the dominant influence on merger.

Further insight into the processes involved in merger can be gained by looking at the results of three-dimensional numerical simulations of cloud merger described by Turpeinen (1982). He performed two simulations of cloud merger, one with no ambient wind and the second with the wind field observed on Day 261 of GATE. Two cumulus clouds were initiated with impulses of radii 3.6 km and with a center-line separation distance of 3.2 km. As noted previously, this was the minimum separation distance for which merger was simulated. A warm-rain process was simulated in both cases. Figure 7.40

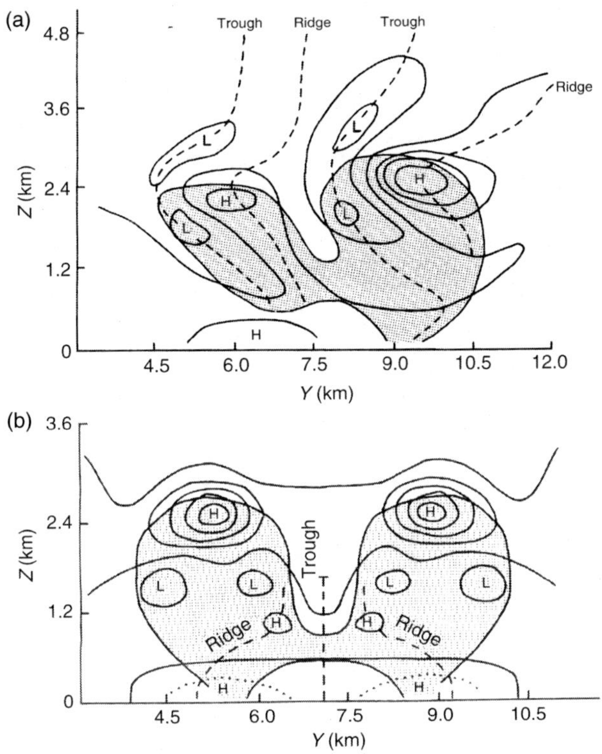


**FIGURE 7.40 Vertical section of nondimensional perturbation pressure (solid lines, in intervals of 1.5 x $10^{-5}$) and cloud water content (shaded areas, $r_c \geq 0.01$ g kg$^{-1}$) with locations of troughs and ridges (dashed lines) for the run with shear (top) and no-shear (bottom).** The dotted line indicates a perturbation pressure of $5 \times 10^{-6}$. In the shear run the downshear cell is the left and the upshear cell is to the right. *(Adapted from Turpeinen (1982))*

(bottom) illustrates a perturbation pressure field. A schematic illustration of the simulated perturbation pressure field for the no shear (bottom) and shear (top) cases is shown in Fig. 7.40. For the no-shear case, a symmetric pressure field is simulated with relative highs near cloud top and near the surface; the latter high is caused by the evaporatively-chilled downdraft outflow. At midlevels, relative low-pressure cells were simulated in the cloudy air. However, between the clouds a relative low-pressure trough is found which is caused by subsidence warming in the clear air, as in Orville et al. (1980). As in Simpson et al.'s (1980) conceptual model (Fig. 7.38), a cloudy bridge can be seen as a precursor to the simulated merger event. High pressure occurs in the cloudy bridge region. This Turpeinen explains in terms of the vertical gradient of buoyancy with condensation warming at the base of the cloudy bridge and evaporative cooling at its top. The combined effects of the cloudy bridge and the clear-air subsidence establish a vertical pressure gradient which favors accelerated cloud growth. Turpeinen admitted that the cloudy bridge may have been a result of his overlapping initial impulses rather than the forced uplifting caused by the outflows from two neighboring clouds, as Simpson et al. (1980) inferred.

The process of environmental wind shear complicates the picture by displacing the central trough upshear so that it resides well within the upshear cell (see Fig. 7.40, top panel). There also exists a surface pressure high below the downshear cell only, since the upshear cell is not precipitating at the time of merging. The resulting pressure field favors the intensification and extended lifetime of the upshear cell. Because the merger in this case takes place within the upshear cell, this differs from Simpson et al. (1980), who postulated that even in the presence of shear, merger should take place in the cloud bridge between the merging cells. Turpeinen suggested that the cases analyzed by Simpson et al. (1980) differed from his simulated cases in that mesoscale convergence could have played a role in the observed merger process.

Idealized 3-D simulations by Kogan and Shapiro (1996) suggests that low-level merger was a consequence of relative advection between isolated cells. That is, merger occurred as a result of a neighboring cell being caught in the radial inflow of its neighbor and advected inward. Likewise, Bennets et al. (1982) attributed merger in their 3-D simulations to the mutual attraction between neighboring cells rather than the collision between gust fronts formed by downdrafts.

Cloud ensemble simulations provide another perspective of merger. In the 2-D and 3-D cloud ensemble simulations reported by Tao and Simpson (1984, 1989) a random ensemble of clouds are initiated by introducing warm bubbles in the boundary layer and then, after many simulated hours, merging events are identified in the simulations. These merging events can be both first-order and second-order. In Tao and Simpson's (1984) 2-D experiments over 200 groups of cloud systems were simulated under different regimes of stratification and cloud forcing. In the 3-D simulations ten merged precipitating systems were identified. Their results support the concept that a shallow cloud bridge forms in response to colliding cold outflows. The 2-D simulations revealed that convective activity and the amount of precipitation increased with the strength of large-scale forcing or large-scale convergence. Likewise, they found a greater occurrence of cloud merger with increased intensity of large-scale lifting. The most favorable environmental conditions for cloud merger were (1) more unstable thermodynamic stratification and (2) stronger large-scale lifting. In the 3-D simulations they found that merger occurred most frequently when cells were roughly aligned parallel to the wind shear vector.

Orville et al. (1980) also suggested that mesoscale convergence in the lower levels would increase the number of merger events because the effect of convergence is to move cells closer together. Simpson et al. (1980) provided circumstantial evidence that mesoscale convergence aids the merger process. They showed that many of the observed radar echo merger events coincided with the zones of convergence predicted by the three-dimensional sea-breeze model developed by Pielke (1974). This model did not contain any parameterization of cloud-scale processes. Thus, any degree of association between the predictions by the model and the observed merger events could be attributed to the sea-breeze-generated convergence zones.

There are several reasons why one would expect areas of mesoscale convergence to be favored sites for cloud merger. As noted by Simpson et al. (1980), the horizontal influx of warm moist air provides a source for sustained buoyant ascent and the release of the slice method constraint (Bjerknes, 1938; Cressman, 1946), so that a larger fraction of the area can be filled by buoyant updrafts. This would also result in the more vigorous clouds being located closer together than in a divergent air mass.

Using a two-dimensional cloud model, Chen and Orville (1980) found that randomly initiated thermal eddies tended to merge together more frequently when a mesoscale convergence field was present. Also, the resultant clouds were broader and deeper in the presence of convergence than they were with no convergence or with divergence in the subcloud layer.

Tripoli and Cotton (1980) found that a three-dimensional cloud model predicted larger, more vigorous convective storms and greater precipitation intensities when low-level convergence was present. Also, Ulanski and Garstang (1978a,b) presented observational evidence that surface convergence patterns nearly always precede the development of radar echoes for periods as long as 90 minutes over southern Florida. They suggested that "the most crucial factor in determining the total amount of rainfall produced by a given storm is the size of the area of surface convergence." Because larger precipitation rates result in greater subcloud evaporation rates, one would expect that the vigor of downdraft outflows from neighboring cumuli would be greater in an environment with low-level convergence, which favors the downdraft-related mechanisms of merger postulated by Simpson et al. (1980).

In this chapter we have discussed the dynamics of cumuli, ranging in scale from shallow cumuli to towering cumuli to precipitating cumuli. Of these clouds, the shallow cumuli are the least dependent upon mesoscale moisture convergence to sustain the cloud population. Such clouds can be sustained by boundary layer fluxes of heat and moisture. However, larger cumuli which transport large amounts of moisture, often through the depth of the troposphere, become dependent upon a supply of moisture and moist static energy which is greater than can be supplied by vertical eddy transport in the boundary layer. As we shall see in the next chapter, some cumulonimbi in a strongly sheared environment may be able to obtain sufficient moisture and moist static energy to sustain them by virtue of the convergence fields which they induce through downdraft outflows and cloud-induced vertical pressure gradients.

We have also seen in this chapter that boundary layer thermals and shallow cumulus clouds can excite gravity waves in the overlying, stably stratified environment, which, in turn, affect subsequent cloud organization and new cell growth. The merger of neighboring cumulus clouds and up-scale growth of cloud systems may also be partly a result of the interactions of gravity waves excited by neighboring convective clouds. We shall see in subsequent chapters that the interaction between cumulonimbus clouds and gravity waves may be responsible for propagation of thunderstorms and the further up-scale growth of cloud systems into mesoscale convective systems.

## REFERENCES

Abel, S. J., and Shipway, B. J. (2007). A comparison of cloud-resolving model simulations of trade wind cumulus with aircraft observations taken during RICO. Q. J. R. Meteorol. Soc. 133, 781–794.

Ackerman, B. (1959). The variability of the water contents of tropical cumuli. J. Meteorol. 16, 191–198.

Agee, E. M. and Asai, T., (Eds.) (1982). Cloud dynamics: An introduction to shallow convective systems. In "Proc. Symp. Gen. Assem. IAMAP, 3rd, Hamburg, West Germany" Reidel, Boston, Massachusetts, 423 pp.

Agee, E. M., and Chen, T. S. (1973). A model for investigating eddy viscosity effects on mesoscale cellular convection. J. Atmos. Sci. 30, 180–189.

Albrecht, B. A. (1981). Parameterization of trade-cumulus cloud amounts. J. Atmos. Sci. 38, 97–105.

Albrecht, B. A. (1989). Aerosols, cloud microphysics, and fractional cloudiness. Science 245, 1227–1230.

Albrecht, B. A., Betts, A. K., Schubert, W. H., and Cox, S. K. (1979). A model of the thermodynamic structure of the trade-wind boundary layer, Part I. Theoretical formulation and sensitivity experiments. J. Atmos. Sci. 38, 73–89.

Asai, T. (1966). Cloud bands over the Japan Sea off the Hokuriku district during a cold air outburst. Pap. Meteorol. Geophys 16, 179–194.

Asai, T. (1970). Stability of a plane parallel flow with variable vertical shear and unstable stratification. J. Meteorol. Soc. Jpn 48, 129–139.

Asai, T. (1972). Thermal instability of a shear flow turning the direction with height. J. Meteorol. Soc. Jpn 50, 525–532.

Asai, T., and Kasahara, A. (1967). A theoretical study of the compensating downward motions associated with cumulus clouds. J. Atmos. Sci. 24, 487–497.

Asai, T., and Nakasuji, I. (1973). On the stability of Ekman boundary layer flow with thermally unstable stratification. J. Meteorol. Soc. Jpn 51, 29–42.

Augstein, A., Schmidt, H., and Ostapoff, F. (1974). The vertical structure of the atmospheric planetary boundary layer in undisturbed trade winds over the Atlantic ocean. Boundary-Layer Meteorol. 6, 129–150.

Barcilon, V. (1965). Stability of non-divergent Ekman layers. Tellus 17, 53–68.

Beniston, M. G., and Sommeria, G. (1981). Use of a detailed planetary boundary layer model for parameterization purposes. J. Atmos. Sci. 38, 780–797.

Bennets, D. A., Bader, M. J., and Marles, R. H. (1982). Convective cloud merging and its effect on rainfall. Nature 300, 42–45.

Betts, A. K. (1975). Parametric interpretation of trade-wind cumulus budget studies. J. Atmos. Sci. 32, 1934.

Betts, A. K. (1982). Cloud thermodynamic models in saturation point coordinates. J. Atmos. Sci. 39, 2182–2191.

Bjerknes, J. (1938). Saturated ascent of air through dry-adiabatically descending environment. Q. J. R. Meteorol. Soc. 64, 325–330.

Black, R. A., and Hallett, J. (1986). Observations of the distribution of ice in hurricanes. J. Atmos. Sci. 43, 802–822.

Blackadar, A. K. (1962). The vertical distribution of wind and turbulence in a neutral atmosphere. J. Geophys. Res 67, 3095–3102.

Blyth, A. M., and Latham, J. (1985). An airbome study of vertical structure and microphysical variability within a small cumulus. Q. J. R. Meteorol. Soc. 111, 773–792.

Blyth, A. M., Cooper, W. A., and Jensen, J. B. (1988). A study of the source of entrained air in Montana cumuli. J. Atmos. Sci. 45, 3944–3964.

Boatman, J. F., and Auer Jr., A. H. (1983). The role of cloud top entrainment in cumulus clouds. J. Atmos. Sci. 40, 1517–1534.

Bougeault, P. (1981a). Modeling the trade-wind cumulus boundary layer. Part I: Testing the ensemble cloud relations against numerical data. J. Atmos. Sci. 38, 2414–2428.

Bougeault, P. (1981b). Modeling the trade-wind cumulus boundary layer. Part II: A high-order one-dimensional model. J. Atmos. Sci. 38, 2429–2439.

Bretherton, C. S., and Smolarkiewicz, P. K. (1989). Gravity waves, compensating subsidence and detrainment around cumulus clouds. J. Atmos. Sci. 46, 740–759.

Brill, K., and Albrecht, B. (1982). Diurnal variation of the trade-wind boundary layer. Mon. Weather Rev. 110, 601–613.

Brown, W. B. (1961). A stability criterion for three-dimensional laminar boundary layers. In "Boundary Layer and Flow Control, Vol. 2." pp. 913–923. Pergamon, New York.

Bunker, A. F., Haurwitz, B., Malkus, J. S., and Stommel, H. (1949). Vertical distribution of temperature and humidity over the Caribbean sea. Pap. Phys. Oceanogr. Meteorol. 11.

Byers, H. R., and Braham, R. R. (1949). "The Thunderstorm." U.S. Weather Bur., Washington, DC.

Carpenter, R. L, Droegemeier, K. K., and Blyth, A. M. (1998a). Entrainment and detraiment in numerically simulated cumulus congestus clouds. Part I: General results. J. Atmos. Sci. 55, 3417–3432.

Carpenter, R. L, Droegemeier, K. K., and Blyth, A. M. (1998b). Entrainment and detraiment in numerically simulated cumulus congestus clouds. Part II: Cloud budgets. J. Atmos. Sci. 55, 3433–3439.

Carpenter, R. L, Droegemeier, K. K., and Blyth, A. M. (1998c). Entrainment and detraiment in numerically simulated cumulus congestus clouds. Part III: Parcel analysis. J. Atmos. Sci. 55, 3440–3455.

Chen, C. H., and Orville, H. D. (1980). Effects of mesoscale on cloud convection. J. Appl. Meteorol. 19, 256–274.

Chisnell, R. F., and Latham, J. (1976a). Ice multiplication in cumulus clouds. Q. J. R. Meteorol. Soc. 102, 133–156.

Chisnell, R. F., and Latham, J. (1976b). Comments on the paper by Mason, Production of ice crystals by riming in slightly supercooled cumulus. Q. J. R. Meteorol. Soc. 102, 713–715.

Christian, T. W., and Wakimoto, R. M. (1989). The relationship between radar reflectivities and clouds associated with horizontal roll convection on 8 August 1982. Mon. Weather Rev. 117, 1530–1544.

Clark, T. L., and Hauf, T. (1986). Upshear cumulus development: A result of boundary layer/free atmosphere interactions. In "Conf. Radar. MeteoroL Conf. Cloud Phys., 23rd, Snowmass, Colo." pp. J18–J21. Am. Meteorol. Soc., Boston, Massachusetts. Preprints.

Clark, T. L., Hauf, T., and Kuettner, J. (1986). Convectively forced internal gravity waves: Results from two-dimensional numerical experiments. Q. J. R. Meteorol. Soc. 112, 899–925.

Clark, T. L., Smolarkiewicz, P. K., and Hall, W. D. (1988). Three-dimensional cumulus entrainment studies. In "Tenth International Cloud Physics Conference, Bad Homburg (FRG)." pp. 88–90. Preprints.

Cotton, W. R. (1972). Numerical simulation of precipitation development in supercooled cumuli. Mon. Weather Rev. 100, 757–763.

Cotton, W. R. (1975a). On parameterization of turbulent transport in cumulus clouds. J. Atmos. Sci. 32, 548–564.

Cotton, W. R. (1975b). Theoretical cumulus dynamics. Rev. Geophys. Space Phys. 13, 419–448.

Cotton, W. R., and Tripoli, G. J. (1979). Reply. J. Atmos. Sci. 36, 1610–1611.

Cotton, W. R., Nehrkorn, T., and Hindman, E. E. (1981). The dynamic response of Florida cumulus to seeding. Final Rep. to NOAA, Environ. Res. Lab., Weather Modif. Program Off., Grant No. 04-78-B01-29.

Cressman, G. P. (1946). The influence of the field of horizontal divergence on convective cloudiness. J. Meteorol. 3, 85–88.

Cuijpers, J. W. M., and Duynkerke, P. G. (1993). Large eddy simulation of trade wind cumulus clouds. J. Atmos. Sci. 50, 3894–3908.

Danielsen, E. F., Bleck, R., and Morris, D. A. (1972). Hail growth by stochastic collection in a cumulus model. J. Atmos. Sci. 29, 135–155.

Das, P. (1964). Role of condensed water in the life cycle of a convective cloud. J. Atmos. Sci. 21, 404–418.

Davis, J. M., Cox, S. K., and McKee, T. B. (1979). Vertical and horizontal distributions of solar absorption in finite clouds. J. Atmos. Sci. 36, 1976–1984.

Deardorff, J. W. (1972). Numerical investigation of neutral and unstable planetary boundary layers. J. Atmos. Sci. 29, 91–115.

Deardorff, J. W. (1976). On the entrainment rate of a stratocumulus topped mixed layer. Q. J. R. Meteorol. Soc. 102, 563–583.

De Roode, S. R., and Bretherton, C. S. (2003). Mass-flux budget of shallow cumulus convection. J. Atmos. Sci. 60, 137–151.

Emanuel, K. A. (1979). Inertial instability and mesoscale convective systems. Part I: Linear theory of inertial instability in rotating viscous fluids. J. Atmos. Sci. 36, 2425–2449.

Emanuel, K. A. (1991). A scheme for representing cumulus convection in large-scale models. J. Atmos. Sci. 48, 2313–2335.

Esbensen, S. (1976). Thermodynamic effects of clouds in the trade wind planetary boundary layer. Ph.D. Thesis, University of California at Los Angeles.

Esbensen, S. (1978). Bulk thermodynamic effects and properties of small tropical cumuli. J. Atmos. Sci. 35, 826–387.

Faller, A. J. (1965). Large eddies in the atmospheric boundary layer and their possible role in the formation of cloud rows. J. Atmos. Sci. 22, 176–184.

Fiedler, B. H. (1985). Mesoscale cellular convection: Is it convection?. Tellus 37A, 163–175.

Gardiner, B. A., and Rogers, D. P. (1987). On mixing processes in continental cumulus clouds. J. Atmos. Sci. 44, 250–259.

Garstang, M., and Betts, A. K. (1974). A review of the tropical boundary layer and cumulus convection: Structure, parameterization and modeling. Bull. Am. Meteorol. Soc. 55, 1195–1205.

Golaz, J.-C., Jiang, H., and Cotton, W. R. (2001). A large-eddy simulation study of cumulus clouds over land and sensitivity to soil moisture. Atmos. Res. 59–60, 373–392.

Golaz, J.-C., Larson, V. E., and Cotton, W. R. (2002a). A PDF-based model for boundary layer clouds. Part I: Method and model description. J. Atmos. Sci. 59, 3540–3551.

Golaz, J.-C., Larson, V. E., and Cotton, W. R. (2002b). A PDF-based model for boundary layer clouds. Part II: Model results. J. Atmos. Sci. 59, 3552–3571.

Grabowski, W. W. (1995). Entrainment and mixing in buoyancy-reversing convection with application to cloud-top entrainment instability. Q. J. R. Meteorol. Soc. 121, 231–253.

Grabowski, W. W., and Clark, T. L. (1991). Cloud-environment interface instability: Rising thermal calculations in two spatial dimensions. J. Atmos. Sci. 48, 527–546.

Grabowski, W. W., and Clark, T. L. (1993a). Cloud-environment interface instability. Part II: Extension to three spatial dimensions. J. Atmos. Sci. 50, 555–573.

Grabowski, W. W., and Clark, T. L. (1993b). Cloud-environment interface instability. Part III: Direct influence on environmental shear. J. Atmos. Sci. 50, 3821–3828.

Grassl, H. (1977). Radiative effects of absorbing aerosol pa]ticles inside clouds. In "Radiation in the Atmosphere" (H. J. Bolle, Ed.), pp. N180–N182. Science Press, Princeton, New Jersey.

Grossman, R. L. (1982). An analysis of vertical velocity spectra obtained in the BOMEX fair-weather, trade-wind boundary layer. Bound. Layer Meteorol. 23, 323–357.

Hallett, J., and Mossop, S. C. (1974). Production of secondary ice particles during the riming process. Nature (London) 249, 26–28.

Heymsfield, A. J., Johnson, D. N., and Dye, J. E. (1978). Observations of moist adiabatic ascent in northeast Colorado cumulus congestus clouds. J. Atmos. Sci. 35, 1689–1703.

Higuchi, K. (1963). The band structure of snowfalls. J. Meteorol. Soc. Jpn 41, 53–70.

Holland, J. Z., and Rasmusson, E. M. (1973). Measurements of the atmospheric mass, energy, and momentum budgets over a 500-kilometer square of tropical ocean. Mon. Weather Rev. 101, 44–57.

Jensen, J. B., Austin, P. H., Baker, M. B., and Blyth, A. M. (1985). Turbulent mixing, spectral evolution and dynamics in a warm cumulus cloud. J. Atmos. Sci. 42, 173–192.

Jiang, H., and Feingold, G. (2006). Effect of aerosol on warm convective clouds: Aerosol-cloud-surface flux feedbacks in a new coupled large eddy model. J. Geophys. Res. 111, D01202. doi:10.1029/2005JD006138.

Johnson, R. H., Rickenbach, T. M., Rutledge, S. A., Ciesielski, P. E., and Schubert, W. H. (1999). Trimodal characteristics of tropical convection. J. Climate 12, 2397–2418.

Johnson, R. H., Ciesielski, P. E., and Hart, K. A. (1996). Tropical inversions near the 0 °C level. J. Atmos. Sci. 53, 1838–1855.

Kaimal, J. C., Wyngaard, J. C., Haugen, D. A., Cote, O. R., Izumi, Y., Caughey, S. J., and Readings, C. J. (1976). Turbulence structure in the convective boundary layer. J. Atmos. Sci. 33, 2152–2169.

Keller, V. W., and Sax, R. I. (1981). Microphysical development of a pulsating cumulus tower: A case study. Q. J. R. Meteorol. Soc. 107, 679–697.

Kelley, R. D. (1984). Horizontal roll and boundary-layer interrelationships observed over Lake Michigan. J. Atmos. Sci. 41, 1816–1826.

Khain, A., Pokrovsky, A., Pinsky, M., Seifert, A., and Phillips, V. (2004). Simulation of effects of atmospheric aerosols on deep turbulent convective clouds using a spectral microphysics mixed-phase cumulus cloud model. Part I: Model description and possible applications. J. Atmos. Sci. 61, 2963–2982.

Kitchen, M., and Caughey, S. J. (1981). Tethered-balloon observations of the structure of small cumulus clouds. Q. J. R. Meteorol. Soc. 107, 853–874.

Klaassen, G. P., and Clark, T. L. (1985). Dynamics of the cloud-environment interface and entrainment in small cumuli: Two-dimensional simulations in the absence of ambient shear. J. Atmos. Sci. 42, 2621–2642.

Knupp, K. R., and Cotton, W. R. (1982a). An intense, quasi-study thunderstorm over mountainous terrain. Part II: Doppler radar observations of the storm morphological structure. J. Atmos. Sci. 39, 343–358.

Knupp, K. R., and Cotton, W. R. (1982b). An intense, quasi-steady thunderstorm over mountainous terrain. Part III: Doppler radar observations of the turbulent structure. J. Atmos. Sci. 39, 359–368.

Koenig, L. R. (1977). The rime-splintering hypothesis of cumulus glaciation examined using a field-of-flow cloud model. Q. J. R. Meteorol. Soc. 103, 585–606.

Koenig, L. R., and Murray, F. W. (1976). Ice-bearing cumulus cloud evolution: Numerical simulations and general comparison against observations. J. Appl. Meteorol. 15, 747–762.

Kogan, Y. L., and Shapiro, A. (1996). The simulation of a convective cloud in a 3D model with explicit microphysics. Part II: Dynamical and microphysical aspects of cloud merger. J. Atmos. Sci. 53, 2525–2545.

Kraus, E. B., and Squires, P. (1947). Experiments on the stimulation of clouds to produce rain. Nature (London) 159, 489.

Krishnamurti, R. (1968a). Finite amplitude convection with changing mean temperature. Part 1. Theory. J. Fluid Mech. 33, 445–455.

Krishnamurti, R. (1968b). Finite amplitude convection with changing mean temperature. Part 2. An experimental test of the theory. J. Fluid Mech. 33, 457–463.

Krishnamurti, R. (1975a). On cellular cloud patterns. Part 1: Mathematical model. J. Atmos. Sci. 32, 1353–1363.

Krishnamurti, R. (1975b). On cellular cloud patterns. Part 2: Laboratory model. J. Atmos. Sci. 32, 1364–1372.

Krueger, S. K., and Bergeron, A. (1994). Modeling the trade cumulus boundary layer. Atmos. Res. 33, 169–192.

Kuettner, J. P. (1959). The band structure of the atmosphere. Tellus 11, 267–294.

Kuettner, J. P. (1971). Cloud bands in the earth's atmosphere: Observations and theory. Tellus 23, 404–425.

Kuettner, J. P., Hildebrand, P. A., and Clark, T. L. (1987). Convection waves: Observations of gravity wave systems over convectively active boundary layers. Q. J. R. Meteorol. Soc. 113, 445–467.

Kuo, H. L. (1963). Perturbations of plane couette flow in stratified fluid and origin of cloud streets. Phy. Fluids 6, 195–211.

Lamb, D., Hallet, J., and Sax, R. I. (1981). Mechanistic limitations to the release of latent heat during the natural and artificial glaciation of deep convective clouds. Q. J. R. Meteorol. Soc. 107, 935–954.

Lappen, C.-L, and Randall, D. A. (2001a). Towards a unified parameterization of the boundary layer and moist convection. Part I. A new type of mass-flux model. J. Atmos. Sci. 58, 2021–2036.

Lappen, C.-L., and Randall, D. A. (2001b). Towards a unified parameterization of the boundary layer and moist convection. Part II. Lateral mass exchanges and sub-plume-scale fluxes. J. Atmos. Sci. 58, 2037–2051.

Lappen, C.-L., and Randall, D. A. (2001c). Towards a unified parameterization of the boundary layer and moist convection. Part III. Simulations of multiple convective regimes. J. Atmos. Sci. 58, 2052–2072.

LeMone, M. A. (1973). The structure and dynamics of horizontal roll vortices in the planetary boundary layer. J. Atmos. Sci. 30, 1077–1091.

LeMone, M. A., and Pennell, W. T. (1976). The relationship of trade-wind cumulus distributions to subcloud layer fluxes and structure. Mon. Weather Rev. 104, 524–539.

Lenschow, D. H. (1970). Airplane measurements of planetary boundary structure. J. Appl. Meteorol. 9, 874–884.

Levine, J. (1959). Spherical vortex theory of bubble-like motion in cumulus clouds. J. Meteorol. 16, 653–662.

Levy, G., and Cotton, W. R. (1984). A numerical investigation of mechanisms linking glaciation of the ice-phase to the boundary layer. J. Clim. Appl. Meteorol. 23, 1505–1519.

Lilly, D. K. (1966). On the stability of Ekman boundary flow. J. Atmos. Sci. 23, 481–494.

Lindzen, R. S. (1974). Wave-CISK in the tropics. J. Atmos. Sci. 31, 156–179.

Liu, J. Y., and Orville, H. D. (1969). Numerical modeling of precipitation and cloud shadow effects on mountain-induced cumuli. J. Atmos. Sci. 26, 1283–1298.

Lopez, R. E. (1973). A parametric model of cumulus convection. J. Atmos. Sci. 30, 1354–1373.

Lopez, R. E. (1978). Internal structure and development processes of c-scale aggregates of cumulus clouds. Mon. Weather Rev. 108, 1488–1494.

Ludlam, F. H., and Scorer, R. S. (1953). Convection in the atmosphere. Q. J. R. Meteorol. Soc. 79, 94–103.

Malkus, J. S. (1949). Effects of wind shear on some aspects of convection. Trans. Am. Geophys. Union 30 (1).

Malkus, J. S. (1952). Recent advances in the study of convective clouds and their interaction with the environment. Tellus 2, 71–87.

Malkus, J. S. (1954). Some results of a trade cumulus cloud investigation. J. Meteorol. 11, 220–237.

Malkus, J. S. (1956). On the maintenance of the trade winds. Tellus 8, 335–350.

Malkus, J. S. (1960). Penetrative convection and an application to hurricane cumulonimbus towers. In "Cumulus Dynamics" (C. F. Anderson, Ed.), pp. 65–84. Pergamon, Oxford.

Malkus, J. S. (1962). Large-scale interactions. In "The Sea: Ideas and Observations on Progress in the Study of the Seas." pp. 88–294. John Wiley and Sons.

Malkus, J. S., and Riehl, H. (1964). "Cloud Structure and Distributions Over the Tropical Pacific Ocean." University of California Press, Berkeley.

Malkus, J. S., and Scorer, R. S. (1955). The erosion of cumulus towers. J. Meteorol. 12, 43–57.

Malkus, J. S., and Simpson, R. H. (1964). Modification experiments on tropical cumulus clouds. Science 145, 541–548.

Malkus, J. S., and Williams, R. T. (1963). On the interaction between severe storms and large cumulus clouds. Meteorol. Monogr. 5, 59–64.

Mapes, B. E. (1993). Gregarious tropical convection. J. Atmos. Sci. 50, 2026–2037.

Mapes, B. E., and Houze Jr., R. A. (1995). Diabatic divergence profiles in western Pacific mesoscale convective systems. J. Atmos. Sci. 52, 1807–1828.

Morton, B. R., Sir G. Taylor, F. R. S., and Turner, J. S. (1956). Turbulent gravitational convection from maintained and instantaneous sources. Proc. R. Soc. London, Ser. A 235, 1–23.

Mossop, S. C., and Hallett, J. (1974). Ice crystal concentration in cumulus clouds: Influence of the drop spectrum. Science 186, 632–634.

Murray, F. W., and Koenig, L. R. (1972). Numerical experiments on the relation between microphysics and dynamics in cumulus convection. Mon. Weather Rev. 100, 717–732.

Nicholls, S., and LeMone, M. A. (1980). The fair weather boundary layer in GATE: The relationship of subcloud fluxes and structure to the distribution and enhancement of cumulus clouds. J. Atmos. Sci. 37, 2051–2067.

Nicholls, S., LeMone, M. A., and Sommeria, G. (1982). The simulation of a fair weather marine boundary layer in GATE using a three-dimensional model. Q. J. R. Meteorol. Soc. 108, 167–190.

Nitta, T. (1975). Observational determination of cloud mass flux distributions. J. Atmos. Sci. 32, 73–91.

Nitta, T., and Esbensen, S. (1974). Heat and moisture budgets using BOMEX data. Mon. Weather Rev. 102, 17–28.

Ogura, Y. (1985). Modeling studies of convection. Adv. Geophys. 288, 387–421.

Orville, H. D. (1965). A numerical study of the initiation of cumulus clouds over mountainous terrain. J. Atmos. Sci. 22, 684–699.

Orville, Harold D., Kuo, Y.-H., Farley, R. D., and Hwang, C. S. (1980). Numerical simulation of cloud interactions. J. Rech. Atmos. 14, 499–516.

Paluch, I. R. (1979). The entrainment mechanism in Colorado cumuli. J. Atmos. Sci. 36, 2467–2478.

Pennell, W. T., and LeMone, M. A. (1974). An experimental study of turbulence structure in the fair-weather trade wind boundary layer. J. Atmos. Sci. 31, 1308–1323.

Petters, M. D., Snider, J. R., Stevens, B., Vali, G., Faloona, I., and Russell, L. M. (2006). Accumulation mode aerosol, pockets of open cells, and particle nucleation in the remote subtropical Pacific marine boundary layer. J. Geophys. Res. 111, D02206. doi:10.1029/2004JD005694.

Pielke, R. A. (1974). A three-dimensional numerical model of the sea breezes over South Florida. Mon. Weather Rev. 102, 115–134.

Plank, V. G. (1969). The size distribution of cumulus clouds in representative Florida populations. J. Appl. Meteorol. 8, 46–67.

Plank, V. G. (1966). Wind conditions in situations of pattern form and nonpattem form cumulus convection. Tellus 18, 1–12.

Raga, G. B., Jensen, J. B., and Baker, M. B. (1990). Characteristics of cumulus band clouds off the coast of Hawaii. J. Atmos. Sci. 47, 338–355.

Ramond, D. (1978). Pressure perturbations in deep convection. J. Atmos. Sci. 35, 1704–1711.

Randall, D. A., and Huffman, G. J. (1980). A stochastic model of cumulus clumping. J. Atmos. Sci. 37, 2068–2078.

Rayleigh, Lord O. M. (1916). On convection currents in a horizontal layer of fluid when the higher temperature is on the underside. Philos. Mag. 32, 529–546.

Raymond, D. J. (1979). A two-scale model of moist, non-precipitating convection. J. Atmos. Sci. 36, 816–831.

Raymond, D. J., and Blyth, A. M. (1986). A stochastic mixing model for nonprecipitating cumulus clouds. J. Atmos. Sci. 43, 2708–2718.

Raymond, D. J., Solomon, R., and Blyth, A. M. (1991). Mass fluxes in New Mexico mountain thunderstorms from radar and aircraft measurements. Q. J. R. Meteorol. Soc. 117, 587–621.

Rickenbach, T. M., and Rutledge, S. A. (1997). "The Diurnal Variation of Rainfall Over the Western Pacific Warm Pool: Dependence on Convective Organization. Proceedings, 22nd Conference on Hurricanes and Tropical Meteorology, Ft. Collins, CO, May 1997." Amer. Met. Soc.

Riehl, H., and Malkus, J. S. (1958). On the heat balance in the equatorial trough zone. Geophysica 6, 503–538.

Rogers, D. P., Telford, J. W., and Chai, S. K. (1985). Entrainment and the temporal development of the microphysics of convective clouds. J. Atmos. Sci. 42, 1846–1858.

Schmidt, F. H. (1949). Some speculation on the resistance to motion of cumuliform clouds. Meded. Ned. Meteorol. Inst. (b) Deel 1, Mr. 8.

Scott, B. C., and Hobbs, P. V. (1977). A theoretical study of the evolution of mixed-phase cumulus clouds. J. Atmos. Sci. 34, 812–826.

Seifert, A., and Beheng, K. D. (2006). A two-moment cloud microphysics parameterization for mixed-phase clouds. Part II: Maritime vs. continental deep convective storms. Meteorol. Atmos. Phys. 92, 67–82.

Shirer, H. N. (1980). Bifurcation and stability in a model of moist convection in a shearing environment. J. Atmos. Sci. 37, 1586–1602.

Shirer, H. N. (1982). Toward a unified theory of atmospheric convective instability. In "Cloud Dynamics" (E. M. Agee and T. Asai, Eds.), pp. 163–177. Reidel, Dordrecht, Netherlands.

Shirer, H. N. (1986). On cloud street development in three dimensions. Parallel and Rayleigh instabilities. Contrib. Atmos. Phys. 59, 126–149.
Shirer, H. N., and Brummer, B. (1986). Cloud streets during Kon Tur: A comparison of parallel/thermal instability modes with observations. Contrib. Atmos. Phys. 59, 150–161.
Siebesma, A. P., and Cuijpers, J. W. M. (1995). Evaluation of parametric assumptions for shallow cumulus convection. J. Atmos. Sci. 52, 650–666.
Simpson, J., and Wiggert, V. (1969). Models of precipitating cumulus tower. Mon. Weather Rev. 97, 471–489.
Simpson, J., Woodley, W. L., Miller, A. H., and Cotton, G. F. (1971). Precipitation results of two randomized pyrotechnic cumulus seeding experiments. J. Appl. Meteorol. 10, 526–544.
Simpson, J., Westcott, N. E., Clerman, R. J., and Pielke, R. A. (1980). On cumulus mergers. Arch. Meteorol. Geophys. Bioklimatol., Ser. A 29, 1–40.
Simpson, J., Keenan, Th. D., Ferrier, B., Simpson, R. H., and Holland, G. J. (1993). Cumulus mergers in the maritime continent region. Meteorol. Atmos. Phys. 51, 73–99.
Simpson, J., Simpson, R. H., Andrews, D. A., and Eaton, M. A. (1965). Experimental cumulus dynamics. Rev. Geophys. 3, 387–431.
Simpson, J. (1992). Global circulation and tropical cloud activity. In "The global role of tropical rainfall" (J. S. Theon et al., Eds.), pp. 77–92. AA Deepak Publishing.
Skatskii, V. I. (1965). Some results from experimental study of the liquid water content in cumulus clouds. Izv. Acad. Sci. USSR, Atmos. Oceanic Phys. (Engl. Transl.) 1, 479–487.
Sommeria, G. (1976). Three-dimensional simulation of turbulent processes in an undisturbed trade wind boundary layer. J. Atmos. Sci. 33, 216–241.
Sommeria, G., and LeMone, M. A. (1978). Direct testing of a three-dimensional model of the planetary boundary layer against experimental data. J. Atmos. Sci. 35, 25–39.
Squires, P. (1958a). Penetrative downdraughts in cumuli. Tellus 10, 381–389.
Squires, P. (1958b). The microstructure and colloidal stability of warm clouds. Tellus 10, 256–271.
Squires, P., and Turner, J. S. (1962). An entraining jet model for cumulonimbus updraughts. Tellus 14, 422–434.
Srivastava, R. C. (1967). A study of the effect of precipitation on cumulus dynamics. J. Atmos. Sci. 24, 36–45.
Stevens, B. (2005). Atmospheric moist convection. Ann. Rev. Earth Planet. Sci. 33, 605–643.
Stevens, B., Ackerman, A. A., Albrecht, B. A., Brown, A. R., and Chlond, A. (2001). Simulations of trade-wind cumuli under a strong inversion. J. Atmos. Sci. 58, 1870–1891.
Stevens, B., Vali, G., Comstock, K., Wood, R., and van Zanten, M. C. (2005). Pockets of open cells and drizzle in marine stratocumulus. Bull. Amer. Meteorol. Soc. 86, 51–57.

Stommel, H. (1947). Entrainment of air into a cumulus cloud. Part I. J. Appl Meteorol. 4, 91–94.

Stull, R. B. (1985). A fair-weather cumulus cloud classification scheme for mixed-layer studies. J. Clim. Appl. Meteorol. 24, 49–56.

Sui, C.-H., Lau, K.-M., Takayabu, Y. N., and Short, D. A. (1997). Diurnal variations in tropical oceanic cumulus convection during TOGA Coare. J. Atmos. Sci. 54, 639–655.

Sun, W. Y. (1978). Stability analysis of deep cloud streets. J. Atmos. Sci. 35, 466–483.

Takeda, T. (1965). The downdraft in convective shower-cloud under the vertical wind shear and its significance for the maintenance of convective system. J. Meteorol. Soc. Jpn 43, 302–309.

Takeda, T. (1966a). Effect of the prevailing wind with vertical shear on the convective cloud accompanied with heavy rainfall. J. Meteorol. Soc. Jpn 44, 129–143.

Takeda, T. (1966b). The downdraft in the convective cloud and raindrops: A numerical computation. J. Meteorol. Soc. Jpn 44, 1–11.

Tanaka, M., Asano, S., and Yamamoto, G. (1977). Transfer of solar radiation through water clouds. In "Radiation in the Atmosphere" (H.-J. Bolle, Ed.), pp. 177–179. Science Press, Princeton, New Jersey.

Tao, W.-K., and Simpson, J. (1984). Cloud interactions and merging: Numerical simulations. J. Atmos. Sci. 41, 2901–2917.

Tao, W. K., and Simpson, J. (1989). A further study of cumulus interactions and mergers: Three-dimensional simulation with trajectory analyses. J. Atmos. Sci. 46, 2974–3004.

Telford, J. W. (1975). Turbulence, entrainment and mixing in cloud dynamics. Pure Appl. Geophys. 113, 1067–1084.

Telford, J. W., and Wagner, P. B. (1980). The dynamical and liquid water structure of the small cumulus as determined from its environment. Pure Appl. Geophys. 118, 935–952.

Tiedke, M. (1993). Representation of clouds in large-scale models. Mon. Weather Rev. 121, 3040–3061.

Tripoli, G. J., and Cotton, W. R. (1980). A numerical investigation of several factors contributing to the observed variable intensity of deep convection over South Florida. J. Appl. Meteorol. 19, 1037–1063.

Tsuchiya, K., and Fujita, T. (1967). A satellite meteorological study of evaporation and cloud formation over the western Pacific under the influence of the winter monsoon. J. Meteorol. Soc. Jpn 45, 232–250.

Turner, J. S. (1962). "Starting plumes" in neutral surroundings. J. Fluid Mech. 13, 356–368.

Turner, J. S. (1963). Model experiments relating to thermals with increasing buoyancy. Q. J. R. Meteorol. Soc. 89, 62–74.

Turpeinen, O. (1982). Cloud interactions and merging on Day 261 of GATE. Mon. Weather Rev. 110, 1238–1254.

Ulanski, S., and Garstang, M. (1978a). The role of surface divergence and vorticity in the lifecycle of convective rainfall, Part I: Observation and analysis. J. Atmos. Sci. 35, 1047–1062.
Ulanski, S., and Garstang, M. (1978b). The role of surface divergence and vorticity in the life cycle of convective rainfall. Part II: Descriptive model. J. Atmos. Sci. 35, 1063–1069.
van den Heever, S., and Cotton, W. R. (2007). Urban aerosol impacts on downwind convective storms. J. Appl. Meteorol. 46, 828–850.
van den Heever, S. C., Carrio, G. G., Cotton, W. R., DeMott, P. J., and Prenni, A. J. (2006). Impacts of nucleating aerosol on Florida convection. Part I: Mesoscale Simulations. J. Atmos. Sci. 63, 1752–1775.
van Zanten, M. C., Stevens, B., Vali, G., and Lenschow, D. H. (2005). Observations of drizzle in nocturnal marine stratocumulus. J. Atmos. Sci. 62, 88–106. 2005.
Walter, B. A. (1980). Wintertime observations of roll clouds over the Bering Sea. Mon. Weather Rev. 108, 2024–2031.
Walter Jr., B. A., and Overland, J. E. (1984). Observations of longitudinal rolls in a near a neutral atmosphere. Mon. Weather Rev. 112, 200–208.
Warner, J. (1955). The water content of cumuliform cloud. Tellus 7, 449–457.
Warner, J. (1970). The microstructure of cumulus cloud. Part III. The nature of the updraft. J. Atmos. Sci. 27, 682–688.
Warner, J. (1977). Time variation of updraft and water content in small cumulus clouds. J. Atmos. Sci. 34, 1306–1312.
Warner, J. (1978). Physical aspects of the design of PEP. Rep. No.9, pp. 49–64. Precipitation Enhancement Proj., Geneva Weather Modif. Programme, World Meteorol. Organ.
Webster, P. J., and Lukas, R. (1992). TOGA COARE: The coupled ocean-atmosphere response experiment. Bull. Amer. Meteor. Soc. 73, 1377–1416.
Weinstein, A. T. (1970). A numerical model of cumulus dynamics and microphysics. J. Atmos. Sci. 27, 246–255.
Weinstein, A. T., and Davis, L. G. (1968). A parameterized numerical model of cumulus convection. Rep. II, GA-777, **43**, Natl. Sci. Found., Washington, DC.
Welch, R., Geleyn, J. F., Korb, G., and Zdunkowski, W. (1976). Radiative transfer of solar radiation in model clouds. Contrib. Atmos. Phys. 49, 128–146.
Westcott, N. (1984). A historical perspective on cloud mergers. Bull. Amer. Meterol. Soc. 65, 219–226.
Wiggert, V., and Ostlund, S. (1975). Computerized rain assessment and tracking of South Florida radar echoes. Bull. Amer. Meteor. Soc. 56, 17–26.
Wilkins, E. M., Sasaki, Y. K., Gerber, G. E., and Chaplin Jr., W. H. (1976). Numerical simulation of the lateral interactions between buoyant clouds. J. Atmos. Sci. 33, 1321–1329.

Willis, G. E., and Deardorff, J. W. (1974). A laboratory model of the unstable planetary boundary layer. J. Atmos. Sci. 31, 1297–1307.

Willis, P. T., and Hallett, J. (1991). Microphysical measurements from an aircraft ascending with a growing isolated maritime cumulus tower. J. Atmos. Sci. 48, 283–300.

Wisner, C., Orville, H. D., and Myers, C. (1972). A numerical model of a hail-bearing cloud. J. Atmos. Sci. 29, 1160–1181.

Woodward, E. B. (1959). The motion in and around isolated thermals. Q. J. R. Meteorol. Soc. 85, 144–151.

Xue, H., and Feingold, G. (2006). Large-eddy simulations of trade wind cumului: Investigation of aerosol indirect effects. J. Atmos. Sci. 63, 1605–1622.

Xue, H., Feingold, G., and Stevens, B. (2008). Aerosol effects on clouds, precipitation, and the organization of shallow cumulus convection. J. Atmos. Sci. 65, 392–406.

Yau, M. K. (1980). A two-cylinder model of cumulus cells and its application in computing cumulus transports. J. Atmos. Sci. 37, 488–494.

Zhang, J., Lohmann, U., and Stier, P. (2005). A microphysical parameterization for convective clouds in the ECHAM5 Climate Model: 1. Single column model results evaluated at the Oklahoma RM site. J. Geophys. Res. 110, D15S07. doi:10.1029/2004JD005128.

Zipser, E. J. (2003). Some views on "Hot Towers" after 50 years of tropical field programs and two years of TRMM data. In "Cloud Systems, Hurricanes, and the Tropical Rainfall Measuring Mission (TRMM)" (W.-K. Tao and R. Adler, Eds.), American Met. Soc., 29, No. 51, pp. 49-58.

Chapter 8

# Cumulonimbus Clouds and Severe Convective Storms

## 8.1. INTRODUCTION

The cumulonimbus cloud, or thunderstorm, is a convective cloud or cloud system that produces rainfall and lightning. It often produces large hail, severe wind gusts, tornadoes, and heavy rainfall. Many regions of the earth depend almost totally upon cumulonimbus clouds for rainfall. Cumulonimbus clouds also play an important role in the global energetics and the general circulation of the atmosphere by efficiently transporting moisture and sensible and latent heat into the upper portions of the troposphere and lower stratosphere. They also affect the radiative budgets of the troposphere. Moreover, cumulonimbus clouds influence tropospheric air quality and the chemistry of precipitation.

We begin this chapter by reviewing descriptive models of thunderstorms, and then we attempt to identify several storm types. This will provide the reader with a perspective on the variable nature of thunderstorms and an introduction to the terminology that is typically used in discussing thunderstorms.

## 8.2. DESCRIPTIVE STORM MODELS AND STORM TYPES

As defined by Byers and Braham (1949) and Browning (1977), the fundamental building block of a cumulonimbus cloud is the "cell." Normally identified by radar as a relatively intense volume of precipitation or a local, relative maximum in reflectivity, the cell can also be described as a region of relatively strong updrafts having spatial and temporal coherency. These updrafts give rise to local regions of intense precipitation, which may not be exactly co-located with the updrafts.

Byers and Braham (1949) identified three stages in the evolution of an ordinary cumulonimbus cloud: the *cumulus stage*, the *mature stage*, and the *dissipating stage*. During the *cumulus stage*, cloud towers, mainly with updrafts, characterize the system. The characteristics of the cumulus stage parallel those described in Chapter 7, except that the horizontal scale of the updrafts is generally larger than that of typical cumuli. As illustrated in Fig. 8.1a, the cumulus stage is characterized by one or more towering cumulus clouds that

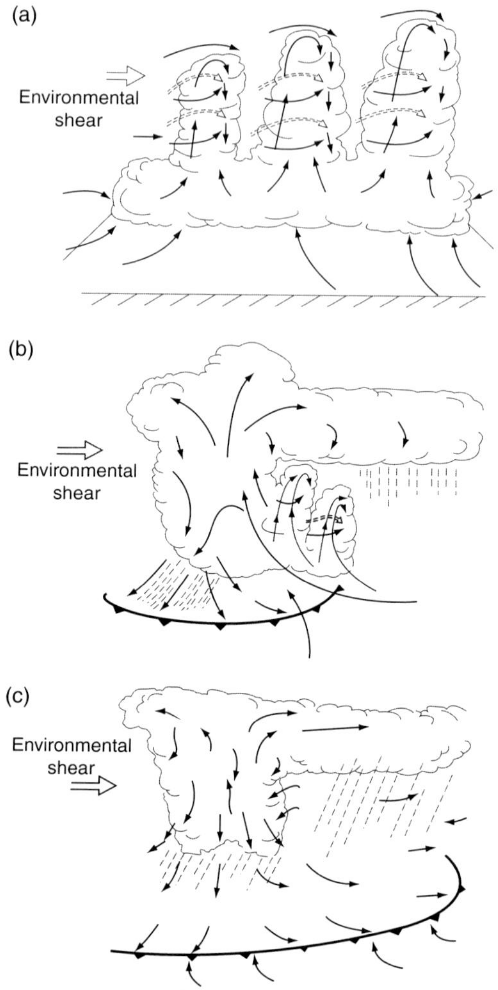


**FIGURE 8.1 Schematic model of the life cycle of an ordinary thunderstorm.** (a) The cumulus stage is characterized by one or more towers fed by low-level convergence of moist air. Air motions are primarily upward, with some lateral and cloud-top entrainment depicted. (b) The mature stage is characterized by both updrafts and downdrafts and rainfall. Evaporative cooling at low levels forms a cold-pool and a gust front that advances, lifting warm, moist, unstable air. An anvil at upper levels begins to form. (c) The dissipating stage is characterized by downdrafts and diminishing convective rainfall. Stratiform rainfall from the anvil cloud is also common. The gust front advances ahead of the storm, preventing air from being lifted at the gust front into the convective storm.

are fed by moisture convergence in the boundary layer. While updrafts prevail during this stage, penetrative downdrafts near cloud top and on the downshear flank of the cumuli can occur. Also during this stage, precipitation may form in the upper portion of the cumuli, but significant rainfall in the subcloud layer is unlikely.

The merger of the cumulus elements into a larger-scale convective system characterizes the transition to the *mature stage*. As noted in Chapter 7, the merger process is frequently associated with the collision of downdraft-induced gust fronts from adjacent cumulus clouds. Thus, the onset of precipitation into the subcloud layer is also characteristic of the transition from the cumulus to the mature phase. In Fig. 8.1b, both updrafts and downdrafts characterize the mature phase. Updrafts may extend through the depth of the troposphere. Divergence of the updrafts just below the tropopause results in the formation of the anvil cloud, and a cloud dome is often present. A cloud dome signifies updraft air overshooting into the stable stratosphere. Near the ground, the diverging downdraft air, chilled by the evaporation and melting of precipitation, spreads out to form a gust front. This front forces warm, moist air ahead of it, thus feeding the updrafts of new cumuli. Heavy localized rain showers also characterize this stage.

Downdrafts characterize the lower portions during the *dissipating stage* of a cumulonimbus. However, local pockets of convective updrafts can remain, as shown in Fig. 8.1c, especially in the upper half of the cloud. Entrainment through the sides of the cloud and turbulence also occur. The cloud dome, often visible during the mature phase, is absent. Near the ground, the diverging, evaporatively-chilled air feeds the gust front, and the front advances far away from the cloud; thus air lifted by the gust front can no longer feed the storm updrafts. Light but steady, stratiform precipitation prevails during the dissipating stages.

Byers and Braham (1949) attributed the demise of the ordinary thunderstorm to the advancing gust front moving so far away from the parent storm that inflow into the storm could no longer feed the updrafts. Another view of the demise of the storm can be seen from Bretherton and Smolarkiewicz (1989) hypothesis that gravity waves excited by the storm act to normalize buoyancy gradients or normalize mass fluxes in the storm. During the growth stage the convective cells exhibit large buoyancies. As a result gravity waves propagate from the storm and normalize the buoyancy gradients by inducing subsidence which enhances the buoyancy of the environment. As a result the buoyancy of the storm updrafts are reduced and the updrafts in the storm weaken. In response to the weakened updrafts precipitation settles faster toward the ground. In response to the displacement of water mass lower in the cloud, water loading, melting of hail and graupel, and enhanced entrainment into the cloud induces downdrafts lower in the storm. Entrainment would be enhanced by mass flux divergence or a reversal in buoyancy gradients associated with water loading and melting. Once the precipitation-loaded downdrafts are exposed to subcloud air, evaporation continues to accelerate the downdrafts until the surface is reached, whereupon gust fronts generate new lifting to fuel the storm. With time, gravity-wave induced subsidence, which normalizes buoyancy gradients between the storm and its environment, weakens the storm buoyancy to the point that updrafts cease and the storm decays.

In this concept vertical shear of the horizontal wind contributes to storm intensity and lifetimes in two ways. First, we have seen in Chapter 7 cumuli growing in shear, preferentially develop entrainment and downdrafts downshear. Furthermore, the preferred location for precipitation to settle is in the downshear flank of the cloud. Thus the precipitation-loaded downdrafts are less likely to interfere with the updrafts forming preferentially on the upshear flank (owing to Doppler-shifted gravity wave propagation from the storm, as noted in Chapter 7). Thus vertical shear of the horizontal winds favors a longer-lived ordinary thunderstorm. Second, from a gravity wave perspective, as gravity waves work to normalize buoyancy gradients between storm updrafts and their environment in a sheared environment, they do not do so symmetrically about the storm. Instead, waves propagating downshear are Doppler-shifted such that they propagate rapidly away from the storm downshear while remaining close to the storm upshear. This means that the near environment of the storm in the downshear flank experiences less subsidence warming than the upshear flank and favors the propagation of the storm in the downshear direction. It is likely that this favors greater longevity of the storm in a sheared environment than in a weakly sheared environment.

There have been numerous attempts to classify thunderstorms into various storm types (Browning, 1977; Chisholm, 1973; Marwitz, 1972a,b,c; Weisman and Klemp, 1984; Foote, 1984). Browning (1977) used the label "ordinary" to refer to a thunderstorm that undergoes the three stages in evolution in a period of 45-60 min and in which the mature stage lasts for only 15-30 min.

Byers and Braham (1949), Browning (1977), Marwitz (1972b), and Chisholm and Renick (1972), among others, have identified another form of Thunderstorm called the *multicell storm*, which, as its name implies, is typically composed of two to four cells. At any time, some cells may be in the cumulus stage, others in the mature stage, and others in the dissipating stage of the life cycle of a cumulonimbus cell. Figure 8.2 shows schematic horizontal and vertical radar depictions of a multicell storm existing at various times in the evolution of the storm system. New cumulus towers typically form on the right flank of the storm complex (i.e. cells at time 0 in Fig. 8.2). Some researchers have referred to the flanking cumulus towers as "feeder" clouds (Dennis et al., 1970) because the new cells move into the storm complex and merge with the parent cell. Others, such as Browning (1977), refer to the flanking line of cumuli as "daughter" cells, wherein the new cells do not merge with the parent circulation but grow rapidly to become the new storm center. The older parent cell then begins to decay. According to Browning, new cells typically form at intervals of 5-10 min and exhibit characteristic lifetimes of 20-30 min. Generally, multicell thunderstorms are somewhat less intense than supercell storms. As seen in Fig. 8.2, they frequently exhibit radar features such as hook echoes and weak-echo regions (WERs), although generally the WER is not fully bounded in the form of an echo-free vault. Occasionally we observe short-lived bounded weak echo regions (BWERs) or echo-free vaults in multicelled storms.

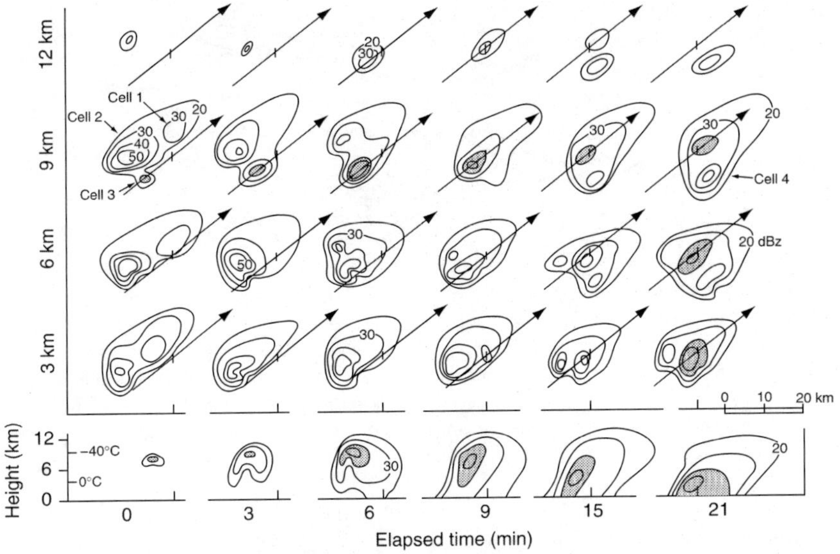


**FIGURE 8.2 Schematic horizontal and vertical radar sections for an ordinary multicell storm at various stages during its evolution showing reflectivity contours at 10-dBZ intervals.** Horizontal sections are illustrated for four altitudes (3, 6, 9, and 12 km AGL) at six different times. The arrow superimposed on each section depicts the direction of cell motion and is also a geographical reference line for the vertical sections at the bottom of the figure. Cell 3 is shaded to emphasize the history of an individual cell. *(From Chisholm and Renick (1972))*

Browning (1977) distinguished between a typical multicell storm and a supercell by the visual appearance of daughter cells (Fig. 8.3). However, some scientists argue that a supercell storm is nothing more than a multicell storm in which the daughter cells are embedded within the forward over-hanging anvil cloud and precipitation. In some cases multicell storms evolve into supercell storms (e.g. Vasiloff et al., 1986; Knupp and Cotton, 1982a,b).

Large size and intensity distinguish supercell storms, where the updraft and downdraft circulations coexist in a nearly steady-state form for periods of 30 minutes or longer. The model of a supercell thunderstorm has undergone a series of refinements and changes over the years (Browning and Ludlam, 1960, 1962; Browning and Donaldson, 1963; Browning, 1965, 1977; Marwitz, 1972a,b,c; Chisholm, 1973; Browning and Foote, 1976; Lemon and Doswell, 1979; Rotunno and Klemp, 1985). Researchers often refer to supercell storms as severe right (SR) storms because the major low-level inflow is on the right flank of the storm relative to its direction of motion. Therefore the SR storm propagates to the right of the mean tropospheric winds.

Figure 8.3 shows a schematic model of a supercell storm moving toward the east. The model illustrates a broad, intense updraft entering the southeast flank of the storm; the updraft rises vertically and then curves anticyclonically in the anvil outflow region. Figure 8.3 also illustrates a mid-level downdraft that

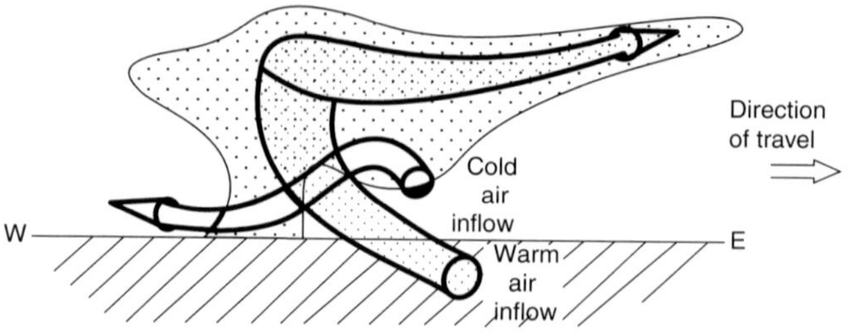


**FIGURE 8.3 Model showing the airflow within a three-dimensional severe right storm traveling to the right of the tropospheric winds.** The extent of precipitation is lightly stippled and the updraft and downdraft circulations are shown more heavily stippled. Air is shown entering and leaving the updraft with a component into the plane of the diagram. However, the principal difference of this organization is that cold-air inflow, entering from outside the plane of the vertical section, produces a downdraft ahead of the updraft rather than behind it. *(From Browning (1968))*

originates on the forward flank of the storm, curves to the north, and exits on the rear flank of the storm. Some supercell models (e.g. Lemon and Doswell, 1979) exhibit downdrafts originating in both the forward and rear flanks of the storm. We shall discuss the origin of such downdrafts in subsequent sections.

Rotating thunderstorms (or *mesocyclones*), typically associated with tornado-producing storms, and radar-echo characteristics also help identify quasi-steady supercell storms. Figure 8.4 illustrates plan views (a) and a vertical cross section (b) through a supercell storm. Particularly noteworthy is the "hook echo" that wraps around a so-called "vault" (Browning and Ludlam, 1960, 1962) or "bounded weak-echo region" (BWER) (Chisholm and Renick, 1972). Browning hypothesized that the vault, or BWER, is caused by air rising so rapidly in strong updrafts that insufficient time is available for the formation of radar-detectable precipitation elements. Weisman and Klemp (1984) have found in their numerical experiments that the BWER resides on the gradient of strong updrafts. They suggested that both the rotational character of the updraft and the updraft strength are important in producing the features of the BWER. Figure 8.4b shows that an intense precipitation region on the storm's rear flank, and an overhanging precipitating region on the storm's forward flank, bound the echo-free vault.

There is evidence of a continuum of storm types, ranging from the lesser organized multicell storms, to organized multicell storms, to the steady supercell storms (e.g. Foote and Wade, 1982). Vasiloff et al. (1986) suggest that the distance between successive updraft cells $L$ relative to updraft diameter $D$, can be used to identify storm type. When $L > D$, storms appear as multicell storms, and when $L \ll D$ they resemble supercell storms. When $L/D < 1$, they suggest that weak evolving multicell storms should prevail, in which individual updraft perturbations associated with cells are embedded within a larger scale region of background updraft.

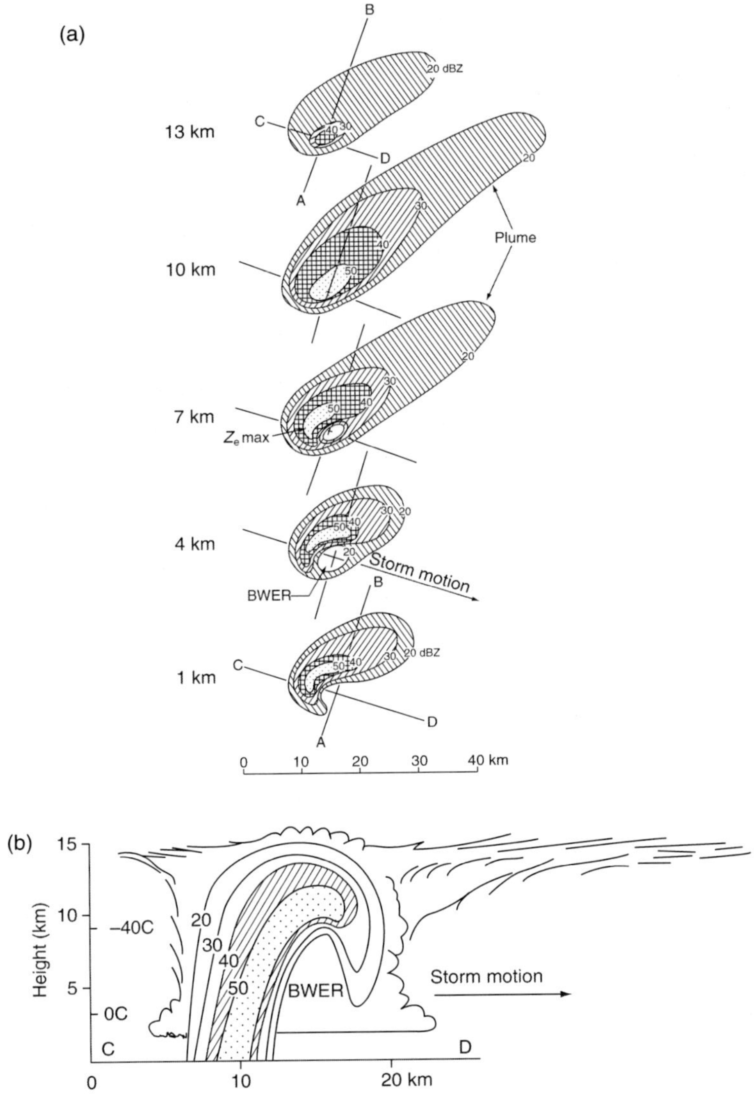


**FIGURE 8.4 (a) Schematic horizontal sections showing the radar structure of a unicellular supercell storm at altitudes of 1, 4, 7, 10, and 13 km AGL.** Reflectivity contours are labeled in dBZ. Note the indentation on the right front quadrant of the storm at 1 km which appears as a weak echo vault (or BWER, bounded weak echo region, as it is labeled here) at 4 and 7 km. On the left rear side of the vault is a reflectivity maximum extending from the top of the vault to the ground. **(b) Schematic vertical section through a unicellular supercell storm in the plane of storm motion [along CD in (a)].** Note the reflectivity maximum, referred to elsewhere as the hail cascade, which is situated on the (left) rear flank of the vault (or BWER, as it is labeled here). The overhanging region of echo bounding the other side of the vault is referred to as the embryo curtain, where it is shown to be due to millimeter-sized particles, some of which are recycled across the main updraft to grow into large hailstones. *(From Chisholm and Renick (1972))*

Moreover it has become common to sub-classify supercell thunderstorms into classic supercells, high-precipitation (HP) supercells, and low-precipitation (LP) supercells (Moller and Doswell, 1988; Moller et al., 1994; Doswell et al., 1990; Doswell and Burgess, 1993). The classic supercell is as described above. In addition the classic supercell typically has these features:

- Frequently develop well away from competing storms.
- Radar signature frequently shows hook echos.
- Large outbreaks of tornadoes are often associated with these storms.
- Moderate precipitation rates.
- Often produce large hail.

LP supercells (Burgess and Davies-Jones, 1979; Bluestein and Parks, 1983; Bluestein and Woodall, 1990) form little if any precipitation and, as a result, are difficult to detect with conventional radars. Other general characteristics are:

- Usually form along the surface dryline in the western plains.
- Produce little rain but often large hail.
- Exhibit visual evidence of rotation.
- Only produce occasional tornadoes.
- Tend to be smaller than classic supercells.
- No evidence of strong downdraft to the surface.
- Almost always isolated cells.
- Found to rotate both cyclonically and anticyclonically with cyclonic rotation most prevalent.

A conceptual model of an LP storm is shown in Fig. 8.5.

HP supercells occur most frequently in the eastern US and western High Plains (Doswell and Burgess, 1993). Their general characteristics (Doswell, 1985; Nelson, 1987; Moller and Doswell, 1988; Moller et al., 1990, 1994; Doswell et al., 1990; Doswell and Burgess, 1993) are as follows:

- Extensive precipitation along the right rear flank of the storm including torrential rain and hail.
- Mesocyclone embedded in extensive precipitation.
- Not clearly isolated from surrounding convection but are distinct in character.
- Often associated with widespread damaging hail or wind in form of long wide swaths; Derecho events (see Chapter 9) may have embedded HP cells.
- Usually larger than classic supercells.
- Updrafts take on arc shape as updrafts form on the southern end of a gust front.
- Tornadoes occur with the mesocyclone or along the leading edge of the gust front.
- Sometimes exhibit multicell characteristics including those such as several high reflectivity cores, multiple mesocyclones and multiple BWERs.

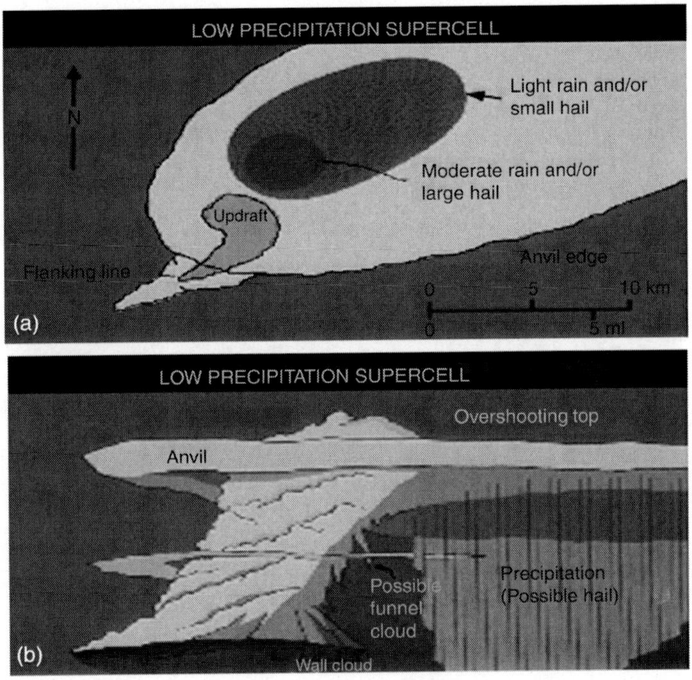


**FIGURE 8.5 A conceptual model of a low-precipitation supercell. (a) The low-level precipitation structure and cloud features looking down from above. (b) Visual structures from an observer's point of view to the east of the storm.** *(From Doswell and Burgess (1993))*

A conceptual model of an HP storm is shown in Fig. 8.6.

Weisman and Klemp (1984) propose a more dynamic classification of supercell versus ordinary multicell storms. They differentiate between supercell and multicell storms by noting such features as dynamically-induced low-level pressure minima, vertical pressure gradients that enhance updrafts, degree of correlation between updraft and vertical vorticity, and propagation characteristics of the storm. They argue that strong shear over the lowest 6 km, or a proper range of bulk Richardson number

$$R_i\,(\text{bulk}) = \frac{\text{CAPE}}{0.5 \times (\overline{u^2} + \overline{v^2})}, \tag{8.1}$$

where CAPE equals

$$\text{CAPE} = g \int_{\text{LFC}}^{\text{EL}} \frac{\theta' - \theta_0}{\theta_0} \tag{8.2}$$

are important.

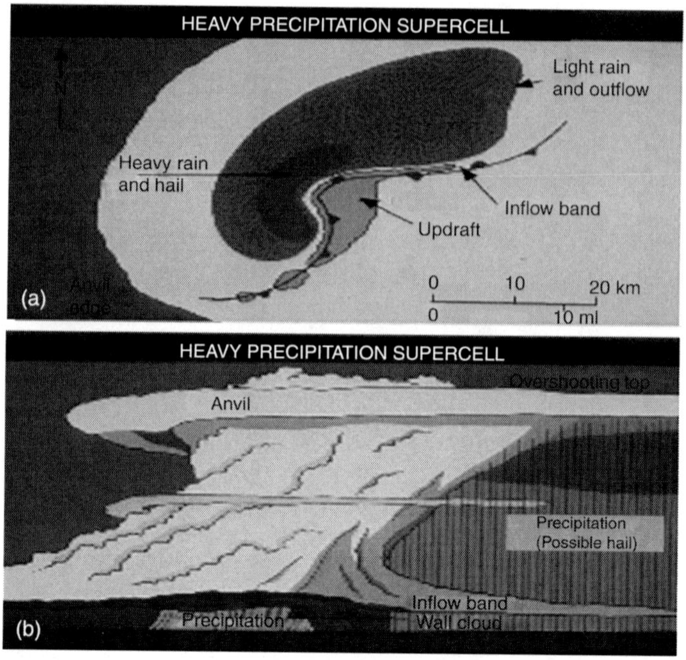


**FIGURE 8.6 A conceptual model of a high-precipitation supercell. (a) The low-level precipitation structure and cloud features looking down from above. (b) Visual structures from an observer's point of view to the east of the storm.** *(From Doswell and Burgess (1993))*

They found that when $R_i$ (bulk) lies between 15 and 45, supercell storms are favored, while multicell storms are favored for values greater than 45.

Occasionally thunderstorms organize into clusters of cumulonimbi called mesoscale convective systems (MCSs) that have maximum dimensions of 100 km or more (see Chapter 9). Often MCSs form major lines of thunderstorms called *squall lines*, in which the cells align in a direction perpendicular to the direction of movement of the storm system (Newton, 1963). Typically the thunderstorm building blocks of the squall-line system go through a multicellular life cycle with new cells forming on the southern flank of the squall-line system. One or several supercell storms may comprise the squall-line building blocks. More circular mesoscale convective systems, as viewed from satellites, are referred to as mesoscale convective complexes (Maddox, 1980) in mid-latitudes, or as cloud clusters in the tropics. Occasionally MCCs have been observed to contain simultaneously, smaller squall lines and multicellular thunderstorms (Wetzel et al., 1983).

## 8.3. UPDRAFTS AND TURBULENCE IN CUMULONIMBI

One of the fundamental properties of a cumulonimbus cloud is the intense, deep updraft associated with the cloud system. Many factors govern the intensity of updrafts in convective clouds. This can readily be seen by applying Eq. (3.14) to the updraft velocity $\overline{w}$ averaged across its width. Neglecting Coriolis effects, Eq. (3.14) becomes

$$\frac{d\overline{w}}{dt} = \underbrace{-\frac{1}{\rho_0}\frac{\partial p'}{\partial z}}_{(1)} + \underbrace{\left(\frac{\theta'_v}{\theta_0} - \frac{c_v}{c_p}\frac{p'}{\rho_0} - r'_w\right)g}_{(2)} \underbrace{-\frac{1}{\rho_0}\frac{\partial}{\partial x_j}(\rho_0\overline{w''u''_j})}_{(3)} + \underbrace{\text{viscous terms}}_{(4)}. \tag{8.3}$$

Thus we see that the mean updraft speed is controlled by (1) local vertical pressure gradients; (2) buoyancy due to virtual temperature anomalies, pressure anomalies, and the drag or loading due to the presence of liquid or frozen water; (3) turbulent Reynolds stresses; and (4) viscous diffusion and dissipation. For high-Reynolds-number flows characteristic of cumulonimbi, viscosity has a negligible influence on mean updraft speeds. The three remaining terms, however, all exert an important influence on the magnitude and scale of updrafts in cumulonimbi, although their relative importance varies with location and the lifetime of the cloud system.

Turbulence also distinguishes cumulonimbus clouds. One can apply Eq. (3.30) to form a turbulent kinetic energy equation averaged across the widths of updrafts (or downdrafts) of cumulonimbi (Eq. (8.4)),

$$\frac{d\overline{e}}{dt} = \underbrace{-\rho_0\overline{u''_i u''_j}\frac{\partial\overline{u}_i}{\partial x_j}}_{(a)} + \underbrace{\left(\frac{\overline{u''_i\theta''_v}}{\theta_0} - \frac{c_v}{c_p}\frac{\overline{u''_i p''}}{p_0} - \overline{u''_i r''_w}\right)g\delta_{i3}}_{(b)} \underbrace{-\frac{\partial}{\partial x_j}(\overline{eu''_j})}_{(c)} \underbrace{-\frac{\partial}{\partial x_j}(\overline{u''_j p''})}_{(d)} \underbrace{-\rho_0\varepsilon}_{(e)}, \tag{8.4}$$

where (a) represents the mechanical or shear production of turbulent kinetic energy, (b) represents buoyant production of TKE, (c) represents the transport of kinetic energy by turbulence, (d) represents the diffusion of TKE by pressure-velocity correlations, and (e) is the rate of molecular dissipation of TKE.

We next examine observational and modeling studies that shed light upon the characteristics of updrafts and turbulence in cumulonimbi.

## 8.4. UPDRAFT MAGNITUDES AND PROFILES

As can be seen from Eq. (8.3), a significant force in determining the magnitude of updrafts in cumulonimbi is the average buoyancy across the updraft width. The buoyancy, in turn, is regulated by the stability of the environment and by the amount of turbulent mixing between the updraft and the dry environment. As a rough estimate of the effects of environmental stability on updraft intensities,

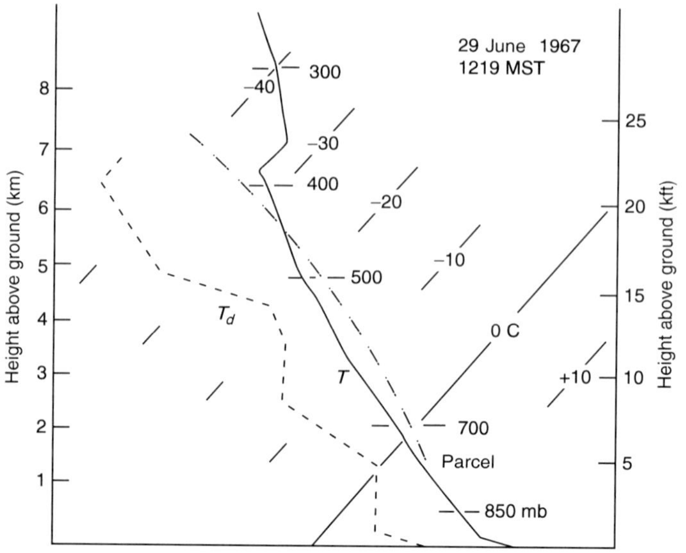


**FIGURE 8.7 Radiosonde sounding for 1219 MST 29 June 1967.** The dot-dashed line indicates a moist adiabatic parcel trajectory using representative cloud-based conditions. *(From Chisholm (1973))*

it is common practice to assume that a parcel of air lifted from the surface will rise dry adiabatically to the lifting condensation level and thereafter rise wet adiabatically until the air parcel attains substantial negative buoyancy. Ignoring all mixing processes and pressure-gradient influences, we can then use Eq. (8.3) to estimate the vertical profile of updraft velocity. Thus, for a hailstorm environment, such as that shown in Fig. 8.7, which exhibits a maximum buoyancy of 2.5-3 °C at 3.5 km above ground level (AGL), Chisholm (1973) calculated the corresponding vertical velocity and liquid-water content profiles shown in Fig. 8.8. The maximum vertical velocity is 25 m $s^{-1}$ near 6 km AGL.

Considerably more intense updraft velocities can be expected in supercell thunderstorms. Miller et al. (1988) diagnosed updraft magnitudes in excess of 40 m $s^{-1}$ in a supercell storm observed near Miles City, Montana (Fig. 8.9). At the other extreme, in a disturbed tropical environment, such as a hurricane, the latent heat released by large numbers of cumulonimbi drives the environmental sounding to nearly wet adiabatic, resulting in maximum parcel buoyancies of the order of 1 °C or less. As a result, typical updraft velocities are less than 6 m $s^{-1}$. Figure 8.10 shows this in the updraft data compiled by Zipser and LeMone (1980) and Jorgensen (1984). By contrast, observations made during the thunderstorm project of ordinary thunderstorms over Ohio and Florida in the United States, exhibit typical updraft magnitudes that exceed 10 m $s^{-1}$.

Severe mid-latitude cumulonimbi over land exhibit the highest updraft speeds; these exceed 40 m $s^{-1}$ (see Fig. 8.9). Also, the width of updrafts in severe continental updrafts is much larger. Lucas et al. (1994) note that

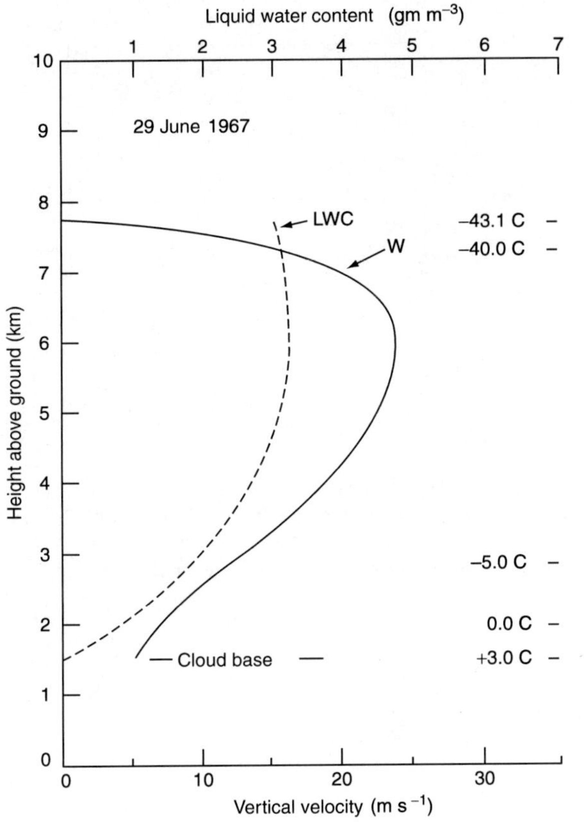


**FIGURE 8.8 Vertical velocity (W) and water content (LWC) profiles for 29 June 1967.** Note the maximum W value near 25 m $s^{-1}$ and the storm top at 7.8 km AGL. *(From Chisholm (1973))*

convective cores in oceanic cumulonimbi below 6 km are one half to one-third of the strength of corresponding continental storms. They note that this cannot be explained by differences in CAPE. They conclude that higher water loading and greater entrainment rates (owing to smaller width updrafts) reduce the buoyancy of oceanic updrafts relative to their continental counterparts. Note that most of the observations in Fig. 8.10 and those reported by Lucas et al. (1994) are below the freezing level owing to the height limitations of most direct cloud-probing aircraft.

Above the freezing level, updraft speeds may greatly exceed values attained in lower levels. This is because the largest virtual temperature anomalies usually occur in severe thunderstorm environments above the freezing level. Moreover, the additional latent heat released by the freezing of supercooled water drops and by the vapor deposition growth of ice crystals, further contributes to cloud buoyancy. Owing to the small differences between saturation mixing ratios and environmental mixing ratios at colder temperatures, entrainment has a less inhibiting influence on updraft intensities than it does in the lower troposphere.

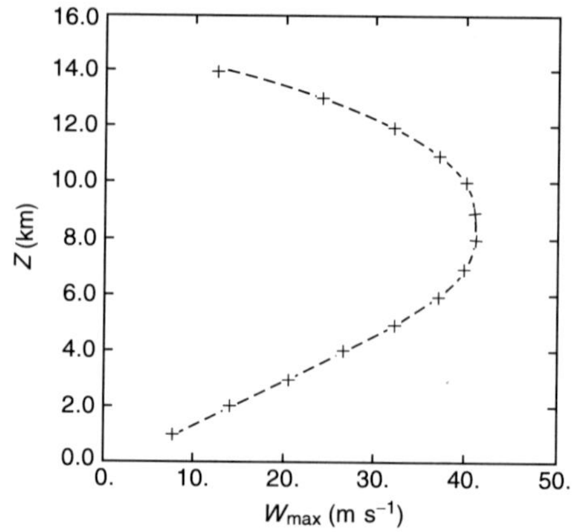


**FIGURE 8.9 Vertical profile of maximum updraft speed versus height obtained from multiple-Doppler radar from a supercell storm observed during the 1981 CCOPE near Miles City, Montana.** The maximum value of updraft speed through the entire depth was 40.88 m $s^{-1}$. *(Provided by L. Jay Miller for the storm described in Miller et al. (1988))*

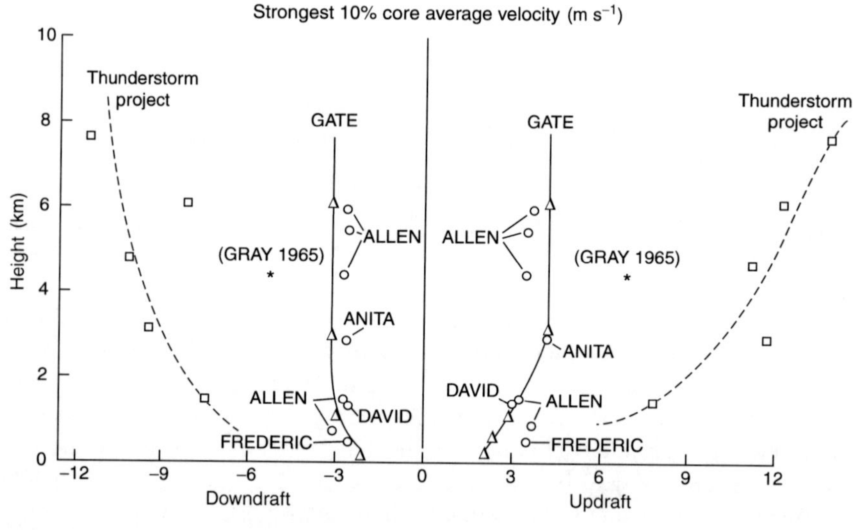


**FIGURE 8.10 Strongest 10% level average vertical velocity in updraft and downdraft cores as a function of height.** GATE values are denoted by triangles, Thunderstorm Project values by squares, and hurricane values by circles and storm name. Observations made by Gray (1965) are also indicated. *(From Jorgensen (1984))*

This was clearly illustrated in early one-dimensional modeling studies (Simpson et al., 1965; Simpson and Wiggert, 1969; Cotton, 1972). Figure 8.11 illustrates how different rates of activation of the ice phase affect updraft speeds aloft. Clearly, the ice phase can accelerate updrafts greatly in the

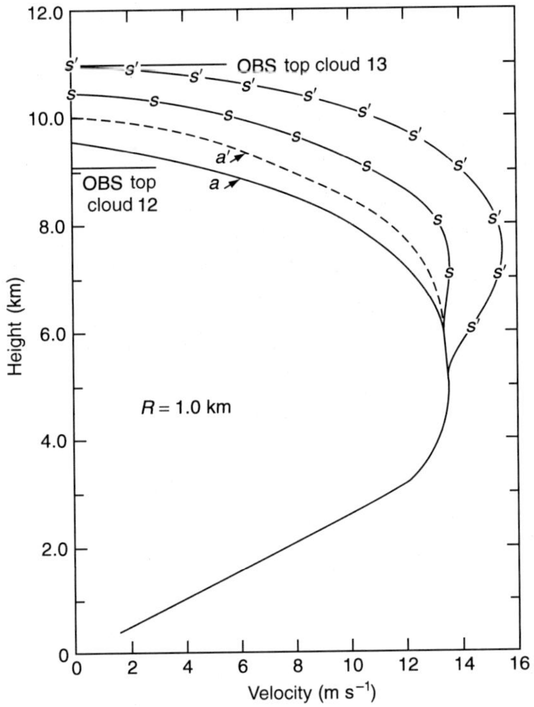


**FIGURE 8.11 Predicted cloud vertical velocity as a function of height for 27 May 1968, EML case study.** *(From Cotton (1972))*

upper troposphere. Vertically-pointing Doppler radar-estimated updraft speeds reported by Battan (1975) and shown in Fig. 8.12 also show updraft speeds reaching maximum values in the upper troposphere. Battan inferred updraft speeds greater than 24 m $s^{-1}$ at the 10.5-km level in ordinary cumulonimbi. Zipser (2003) referred to several observations in oceanic cumulonimbi using airborne Doppler radars which measured updrafts of over 20 m $s^{-1}$ in the upper troposphere. Zipser also mentioned several encounters with updrafts exceeding 20 m $s^{-1}$ in observations of oceanic convective storms. The author (Cotton) also remembers encountering hard bubble-like cores emanating out of soft low-level convection in the NASA 990 during GATE. Estimated updraft strengths were well over 20 m $s^{-1}$ at ~30,000 feet MSL.

Updrafts in cumulonimbi are not regulated solely by buoyancy [i.e. term (2) in Eq. (8.3)]. Analyses of updrafts below cloud base and in weak-echo regions of severe cumulonimbi (Auer and Marwitz, 1968; Marwitz, 1973; Grandia and Marwitz, 1975; Ellrod and Marwitz, 1976) have revealed that updrafts below cloud base are frequently negatively buoyant. Researchers inferred that the vertical pressure-gradient force [term (1) in Eq. (8.3)] accelerates air near the surface upward to cloud base. Three-dimensional thunderstorm models and

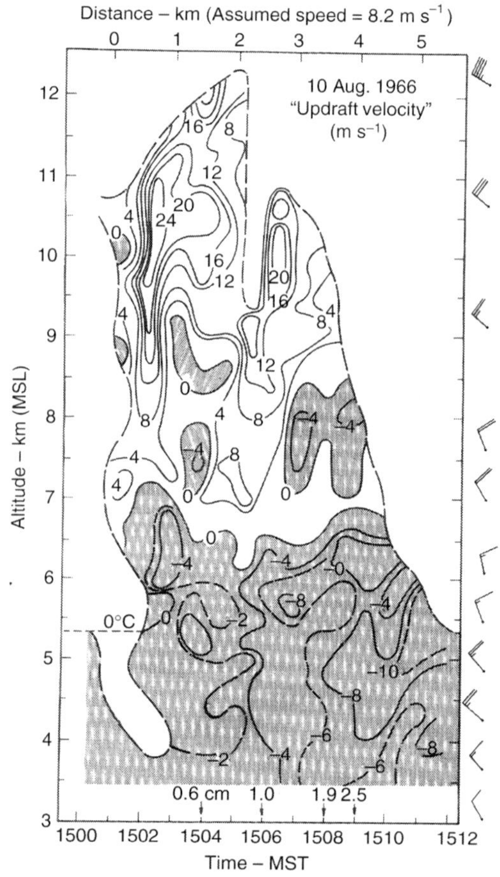


**FIGURE 8.12 An estimated updraft velocity calculated as $W_R = \overline{V} + 3.8Z^{0.072}$.** *(From Battan (1975))*

linear theory support this inference. Weisman and Klemp (1984) concluded from numerical experiments that, with a clockwise-curved wind hodograph, the updraft on the right flank of the storm system is locally induced by the lowering of pressure caused by the dynamic interaction of the storm with environmental wind shear. This dynamic forcing is often sufficiently strong to lift some of the negatively buoyant low-level air into the updraft circulation (Schlesinger, 1975, 1978, 1980; Rotunno and Klemp, 1982, 1985). Schlesinger (1978) also noted that a significant pressure high is simulated near the tops of growing cumulonimbi. We will examine the processes responsible for lowering of pressures in a cloud more fully in subsequent sections.

Marwitz (1973) and Sulakvelidze et al. (1967) also found that the height of the maximum updrafts were often quite low in the WER, i.e. 10-25 m s$^{-1}$ at some 1-4 km above cloud base. Presumably water loading and entrainment of low-valued $\theta_e$ air could have weakened updraft strengths at levels above the updraft maximum in the WER. However, because researchers could not reliably estimate $w$ by tracking chaff or balloons in supercooled clouds, these observations do not preclude the possibility that a primary updraft maximum existed at heights above the freezing level.

## 8.5. TURBULENCE

The observations reported by Marwitz (1973), Grandia and Marwitz (1975), and Ellrod and Marwitz (1976) indicate that the updraft air entering the base of cumulonimbi is smooth and relatively free of turbulence and remains so through a significant depth of the WER. This is consistent with the observation that the updrafts are negatively buoyant and accelerated by the vertical pressure-gradient force; negative buoyancy suppresses the production of turbulence.

The turbulent structure of cumulonimbi has been observed by aircraft (Steiner and Rhyne, 1962; Marwitz, 1973; Grandia and Marwitz, 1975; Ellrod and Marwitz, 1976) and Doppler radar (Frisch and Clifford, 1974; Frisch and Strauch, 1976; Battan, 1975, 1980; Donaldson and Wexler, 1969; Battan and Theiss, 1973). Turbulence levels can be estimated with a Doppler radar using two methods. The first method involves estimates of the energy dissipation rate [Eq. (8.4), term (e)] from the variance of the Doppler spectra obtained over a given radar pulse volume. The estimate of the energy dissipation rate assumes that the turbulence is homogeneous and isotropic over the radar pulse volume. This estimate of turbulence can be analogous to a subgrid-scale estimate of turbulence in a large-eddy simulation model. The radar pulse volume is a conical section having typical dimensions of 150 m in pulse length and 350 m in transverse beam width. The second method of estimating turbulence levels by Doppler radar is from the variability in mean radial Doppler velocities $V_R$ from radar range gate to range gate ($\Delta V_r/\Delta R$) or by directly calculating spectral energy using point estimates of $V_r$. This estimate is analogous to the explicitly predicted turbulence at grid points in a LES model.

Knupp and Cotton (1982b) synthesized both estimates of turbulence variations in a cumulonimbus cloud with multiple-Doppler estimated mean flow fields and radar reflectivities. The storm system they analyzed was a quasi-steady storm that moved to the left of the mean cloud-layer environmental winds. Figure 8.13 illustrates a conceptual model of the storm updraft and downdraft circulations as inferred from the Doppler analyses. Major features are a primary updraft located in the northwest (downshear) storm quadrant with peak speeds of approximately 25 m $s^{-1}$ and a secondary weaker updraft (10-20 m $s^{-1}$) located in the southern (upshear) quadrant. The most significant downdrafts were situated in the southeast-southwest storm quadrants, where relative inflow of low-valued $\theta_e$ air produced evaporation of cloud and precipitation particles. Figure 8.14b shows that a pronounced minimum in $\theta_e$ between 3 and 4 km AGL characterized the environment. Furthermore, the

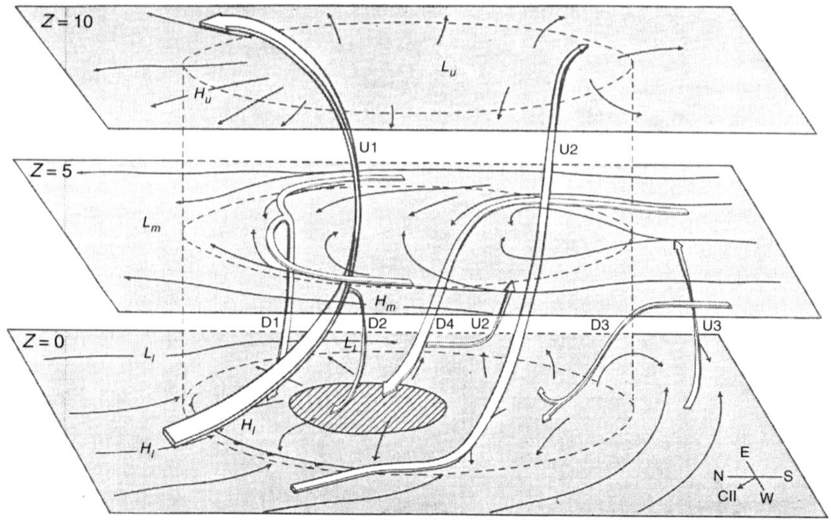


**FIGURE 8.13 Conceptual model of the flow patterns within C11 during its intense quasi-steady stage.** Streamlines depict airflow (storm relative) in the given horizontal planes. The arrowed ribbons represent updraft and downdraft circulations. Each H and L denotes regions of strong and weak flow, respectively, at lower (1), middle (m), and upper (u) levels. The hatched region denotes heavy rain. *(From Knupp and Cotton (1982a))*

wind hodograph shown in Fig. 8.15 exhibits strong shear at low levels near 4 km MSL. Figure 8.16 shows that the largest dissipation ($\varepsilon$) magnitudes at low levels (1 km AGL) were in the eastern storm quadrant, where the air mass was confluent and a downdraft was present. Within the primary updraft inflow sector (northwest quadrant), turbulence levels were quite low. Figures 8.17 and 8.18 illustrate vertical cross sections of $\varepsilon$, $\mathbf{\Delta} V_{\mathrm{R}}/\mathbf{\Delta} R$, mean wind vectors, and radar reflectivity, as well as horizontal sections at 4.0 and 7.0 km AGL. At 4.0 km, turbulence was most substantial along the western and southwestern regions of the storm. Turbulence in these regions appears to be generated by the inflow of dry environmental air and wind shear in the vicinity of the updraft. Both buoyant and shear production of turbulence appear to be important. Strong shear production is evident in the region (labeled SH in Fig. 8.17d) in which the horizontal flow carried by the updraft encounters the mid-level southerly flow diverging around the primary updraft. While the most intense levels of turbulence exist between 4 and 6 km AGL, the largest areal coverage of turbulence is in the upper portion of the storm in the form of small turbulence eddies, because large values of $\varepsilon$ prevail. In the 4- to 6-km layer, $\mathbf{\Delta} V_{\mathrm{r}}/\mathbf{\Delta} R$ has the largest areal coverage, which suggests that entrainment first occur in large eddies that then cascade to smaller scales as they are carried aloft in the updrafts. Moreover, the turbulence is initiated at levels just below the 4-km level, where the minimum values in environmental $\theta_{\mathrm{e}}$ reside (see Fig. 8.14b). At the same level, the greatest difference exists between the horizontal momentum

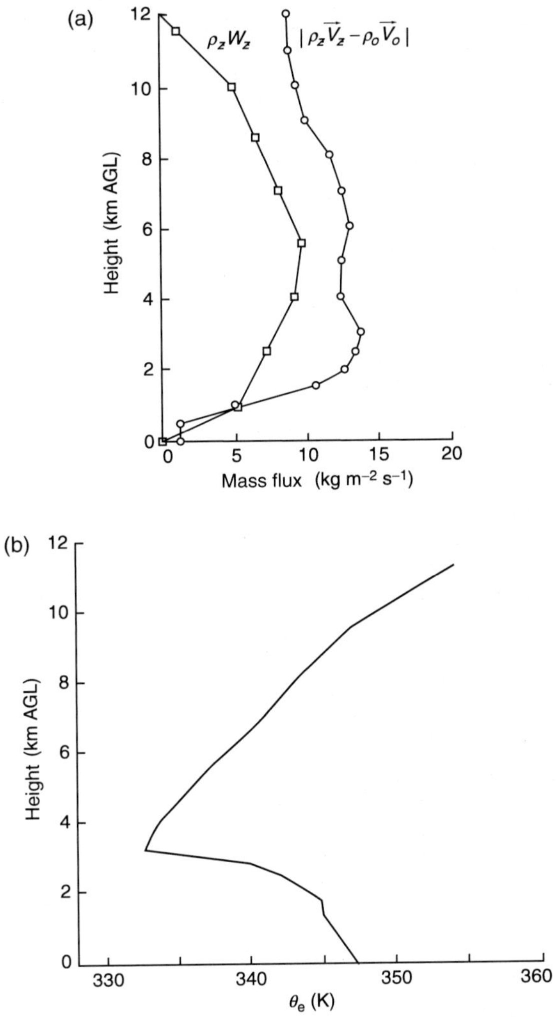


**FIGURE 8.14 Vertical profiles of (a) the magnitude of the difference between low-level momentum ($\rho_0 V_0$) and environmental momentum at cloud levels $\rho_z V_z$ and $|\rho_0 V_0 - \rho_z V_z|$, and the maximum updraft mass flux ($\rho_z w_z$) at 1927 MDT; (b) environmental equivalent potential temperature ($\theta_e$).** Add 3 km for height above MSL. *(From Knupp and Cotton (1982b))*

carried by the updraft ($\rho_z V_z$) and the environmental momentum ($\rho_0 V_0$) (see Fig. 8.14a). The differences in horizontal momentum create horizontal shears and also support shear and buoyant production of turbulence as well as the generation of entrainment in the downshear flank of the cloud, as discussed in Chapter 7. Also shown in Fig. 8.14a is the vertical profile of updraft mass flux ($\rho_z w_z$). Large values of ($\rho_z w_z$) imply the existence of substantial lateral shear between the updraft and a relatively quiescent environment. Moreover,

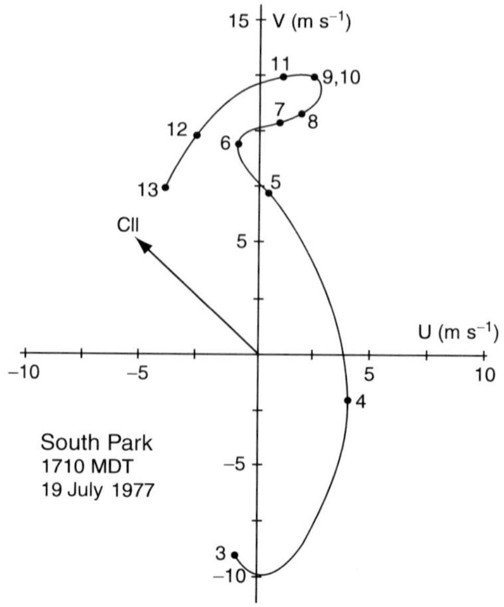


**FIGURE 8.15 Environmental wind hodograph derived from the 1710 MDT South Park sounding.** Winds above 11 km MSL are probably influenced by anvil outflow from storms to the east. Numbers adjacent to hodograph curve denote height (kilometers MSL). The motion of storm C11 is depicted by the arrow. *(From Knupp and Cotton (1982a))*

considering mass continuity, strong vertical divergence of $(\rho_z w_z)$ implies that a strong horizontal convergence (or dynamic entrainment) in the updraft must be present. Thus, above 4 km AGL, buoyancy and shear production of turbulence can be induced by a variety of cloud-environment interactions. In general, the intensity and spatial distribution of turbulence will vary, depending on many of the characteristics of the storm environment we have described above. As we shall see, the same environmental characteristics that enhance entrainment into the storm updrafts and turbulence generation also affect the genesis and intensity of storm downdrafts.

## 8.6. DOWNDRAFTS: ORIGIN AND INTENSITY

The same forces that affect updrafts initiate, maintain, or dissipate downdrafts in cumulonimbi. Thus Eq. (8.3) applies equally well to estimating downdraft speeds. Vertical pressure-gradient force, buoyancy, and turbulent Reynolds stresses initiate, maintain, or dissipate downdrafts. Similarly, Eq. (8.4) can describe the rate of change of turbulent kinetic energy in downdrafts. As in updrafts, both buoyant and shear production of turbulence are important. The relative importance of the various terms in Eqs (8.3) and (8.4), however, may differ in downdrafts as compared to updrafts. Water loading, for example, may play a critical role in the initiation and maintenance of downdrafts because

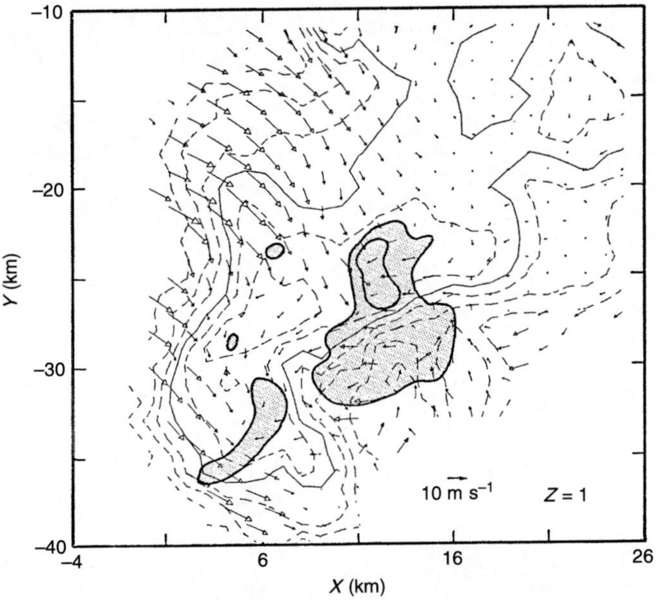


**FIGURE 8.16 Patterns of reflectivity factor, mean wind vectors (storm relative), and dissipation rate estimates at 1 km AGL at 1907 MDT.** Reflectivity contours (thin solid and dashed lines) are drawn every 5 dBZ, with 30 dBZ denoted by thick solid lines. Dissipation rate contours are drawn at 0.02 and 0.04 $m^2\ s^{-1}$. The NOAA-D radar is located at the coordinate origin. *(From Knupp and Cotton (1982b))*

precipitation is more likely to settle in downdraft regions. Likewise melting can play an important role in downdraft intensity.

Downdrafts within cumulonimbi exhibit a wide spectrum of magnitudes and sizes. Vertical velocity data tabulated from the Thunderstorm Project flights (Byers and Braham, 1949) reveal median downdraft speeds and widths of 5-6 m $s^{-1}$ and 1.2 km, respectively. Data acquired from intense northeastern Colorado cumulonimbi, as summarized by Musil et al. (1977), show a respective mean maximum downdraft speed and width of 8 m $s^{-1}$ and 2.5 km. Maximum measured downdraft gusts and widths have exceeded 20 m $s^{-1}$ and 7 km in several cases. While these peak values are typically measured at and above mid-levels, downdrafts of similar size and magnitude may also exist at low levels, because in situ measurements within and near low-level precipitation cores have been avoided. Sinclair (1973, 1979) has reported frequent occurrences of downdrafts at middle to upper cloud levels in both clear and cloudy portions of intense cumulonimbi. Other indirect evidence supporting the presence of clear-air downdrafts or larger regions of weaker subsidence adjacent to precipitating convection is summarized in Fritsch (1975) and in Hoxit et al. (1976).

Relative to mid-latitude continental cumulonimbi, tropical maritime cumulonimbi contain much weaker updrafts and downdrafts. LeMone and Zipser (1980) summarized vertical motions measured within GATE convective

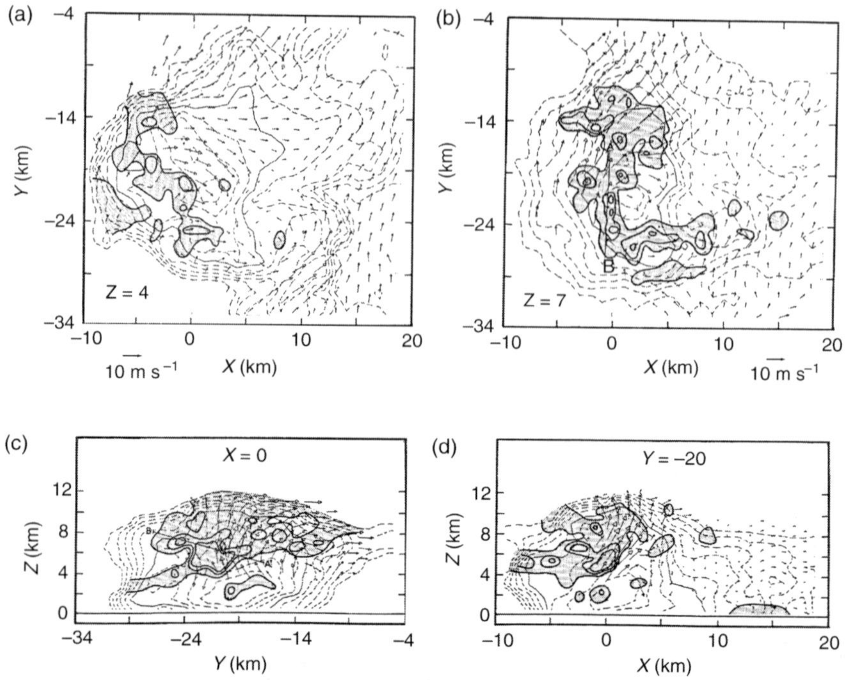


**FIGURE 8.17 Patterns of reflectivity factor, mean wind vectors (storm relative), and dissipation rate estimates at 1927 MDT for selected horizontal and vertical planes.** Contours of $\varepsilon$ are drawn for $\varepsilon = 0.02, 0.04$, and $0.07\ m^2\ s^{-3}$. The NOAA-D radar is located at the coordinate origin. *(From Knupp and Cotton (1982b))*

clouds, and they found a median value of 1.8 m $s^{-1}$ with maximum downdraft speeds of 10 m $s^{-1}$ in rare instances. A summary of these measurements and a comparison to Thunderstorm Project data are shown in Fig. 8.10. This shows that GATE draft magnitudes are one-third to one-half the Thunderstorm Project draft magnitudes. Figure 8.10 also shows the composite draft profiles Jorgensen (1984) obtained through hurricane convective bands and inner cores, that are similar to GATE profiles.

Also shown in Fig. 8.10 is an increase with height of both updraft and downdraft magnitudes for GATE, Thunderstorm Project, and hurricane drafts, a behavior similar to that measured in nonprecipitating cumulus congestus. Such a pattern may be biased due to the previously mentioned lack of penetrations through low-level precipitation cores, particularly over continents. Some measurements within mid-latitude precipitating convection indicate that low-level downdrafts associated with precipitation may attain intense magnitudes. For example, Rodi et al. (1983) measured 15 m $s^{-1}$ peak downdrafts within light precipitation beneath the bases of cumulus congestus clouds forming above deep, dry mixed layers in Colorado.

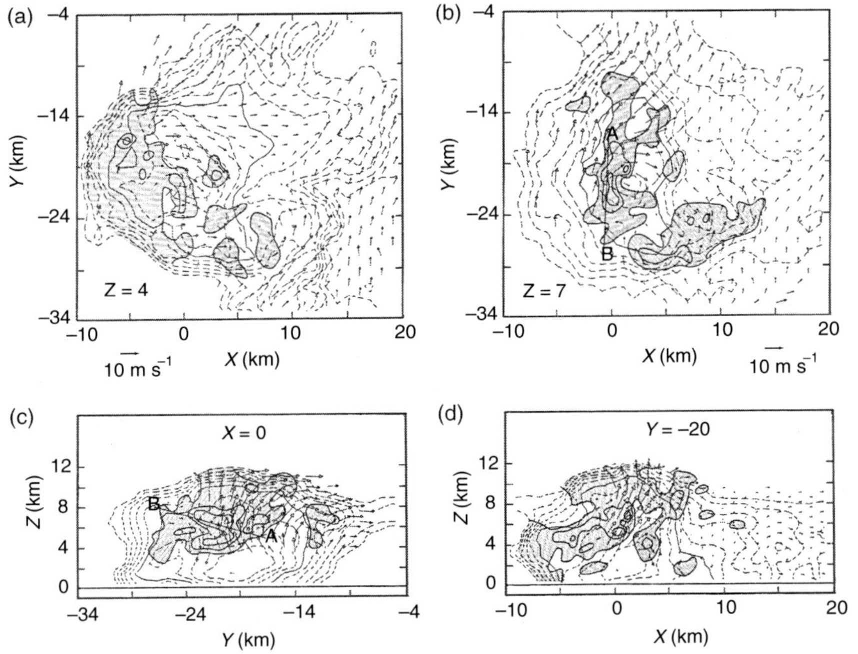


**FIGURE 8.18 Patterns of reflectivity factor, mean wind vectors (storm relative), and radial velocity differences ($\Delta V_r/\Delta R$) at 1927 MDT.** $\Delta V_r/\Delta R$ contours are drawn at $5 \times 10^{-3}$, $10 \times 10^{-3}$, and $15 \times 10^{-3}$ $s^{-1}$. The NOAA-D radar is located at the coordinate origin. *(From Knupp and Cotton (1982b))*

Vertically pointing Doppler (VPD) radar observations provide further evidence that strong downdrafts of continental cumulonimbi exist at low levels. Battan (1975, 1980) has presented the most comprehensive set (four cases) of VPD observations. These are supplemented by additional VPD data contained in Battan and Theiss (1970), Strauch and Merrem (1976), Wilson and Fujita (1979), and Mueller and Hildebrand (1983). Figure 8.12 typifies one of those presented in Battan's (1975; 1980) examples. The VPD observations generally reveal vertically continuous, large-scale downdrafts (6-12 m $s^{-1}$ maximum) in the lowest 3-4 km. Pockets of downdraft that typify the middle to upper levels represent small-scale drafts commonly measured by aircraft (e.g., Fig. 8.18). The observations of Wilson and Fujita (1979) show considerable variability in small-scale intense updrafts and downdrafts near the radar-echo top. The magnitude of such near-cloud-top (overshooting) downdrafts may approach 40 m $s^{-1}$, as Fujita (1974) determined from airborne photogrammetric analyses of intense cumulonimbi. Downdrafts exceeding 10 m $s^{-1}$ were even found in radar observations of Florida thunderstorms at 10 km altitude (Yuder and Houze, 1995). Cloud-top-overshooting downdrafts differ from those driven by evaporative cooling of cloud and precipitation at lower to middle levels.

Newton (1966) calculated downdraft strengths by integrating the vertical equation of motion, including buoyancy forces, modified by entrainment of momentum. He found upper-level updrafts ascend to heights exceeding (overshooting) the parcel equilibrium level, beyond which, rising air quickly becomes negatively buoyant. Very strong downward forces from negative buoyancy then lead to downdrafts that are negatively buoyant only in the upper levels. These initial overshooting downdrafts similarly overshoot their level of neutral buoyancy, leading to subsequent decaying buoyancy oscillations. This can be viewed as an extreme example of buoyancy sorting as discussed in Chapter 7.

In an analysis of tropical convective systems, Zipser (1969, 1977) distinguished 1-5 m $s^{-1}$ downdrafts of scale ~1 km associated with active convective cloud cores from weaker, 0.1-0.5 m $s^{-1}$ mesoscale downdrafts of scale 10-100 km associated with an extensive anvil cloud trailing the active convection. Other studies on both squall lines (Houze, 1977; Ogura and Liou, 1980) and cumulonimbus cloud clusters (Leary and Houze, 1979a) show similar scale separation between convective-scale and mesoscale downdrafts. Mesoscale downdrafts will be discussed more fully in Chapter 9.

Similar spatial-scale variations, greater than one order of magnitude, appear in convective downdrafts. As discussed previously, direct aircraft observations revealed nonprecipitating convective cloud downdrafts no greater than ~500 m, in contrast to ~10-km-wide downdrafts occasionally measured within precipitating cumulonimbi. Indirect observations suggest a similar range of scales in intense downdrafts (downbursts).

From their inspection of surface damage patterns that exhibited scales from a few hundred meters to greater than 10 km, Fujita and Wakimoto (1981) and Forbes and Wakimoto (1983) inferred a wide spectrum of downburst sizes. Fujita (1981) also found short time scales (~5 min) of low-level outflow wind associated with small downbursts.

A number of investigators have inferred downdraft source levels by analyzing thermodynamic tracers such as equivalent potential temperature ($\theta_e$), wet-bulb potential temperature ($\theta_w$), or moist static energy ($h = c_pT + Lq + gz$), all of which are conserved approximately for dry and moist adiabatic processes, assuming no mixing or ice-phase change. Vertical profiles of $\theta_e$ in the environment of cumulonimbi typically show a minimum near 500-600 mbar. Mal and Desai (1938), Normand (1946), and Newton (1950) were among the first to apply this principle in inferring that cold downdraft air measured at the surface originated several kilometers above.

Other investigators have subsequently indicated that low-valued mid-level $\theta_e$ air often reaches the surface within downdrafts. Using an analysis of $\theta_e$, Zipser (1969) inferred that mid-level air near the level of minimum $\theta_e$ descended approximately 500 mbar to the surface behind a tropical squall line. Similar inferences concerning the origin of downdraft air are made using thermodynamic analysis in many other cases, for example, in mid-latitude convective storms (Newton and Newton, 1959; Foote and Fankhauser, 1973;

Fankhauser, 1976; Lemon, 1976; Barnes, 1978a,b; Ogura and Liou, 1980). In other less intense cases, downdrafts apparently originate just above cloud base, significantly below the level of minimum $\theta_e$. Betts (1976) estimated that downdraft air descended about 100 mbar from just above the cloud base of Venezuelan storms. Barnes and Garstang (1982) and Johnson and Nicholls (1983) inferred downdraft source levels near 700-750 mbar for precipitating tropical convection of moderate intensity.

Many analyses indicate that low-level downdrafts are closely associated with precipitation falling beneath cloud base from convective clouds of weak to severe intensity. Byers and Braham (1949) demonstrated a close association between downdrafts and surface rainfall. They inferred that downdrafts were initiated by precipitation loading and were maintained by evaporation of cloud and precipitation. Other striking examples showing this relationship can be seen in the surface mesonet analyses of Foote and Fankhauser (1973), Fankhauser (1976, 1982), Holle and Maier (1980), and Wade and Foote (1982). Finally, Barnes and Garstang (1982) established a positive correlation between areal precipitation rate and downdraft transport of mass and low static energy ($h$) into the boundary layer. They determined that precipitation rates needed to exceed a threshold of $\sim$2 mm h$^{-1}$ (averaged over an area of $\sim$16 km$^2$) before air with low values of $h$ was transported into the subcloud layer.

In other cases, the lowest valued $\theta_e$ air is located just upshear of the downdraft and precipitation core (e.g. Barnes, 1978a,b; Nelson, 1977; Lemon, 1976. The multiple-Doppler radar presentations in Kropfli and Miller (1976), Ray et al. (1981), Foote and Frank (1983), and Wilson et al. (1984), among others, also illustrate that low-level downdrafts are either located within or along the upshear edge of heaviest low-level precipitation.

The relationship between subcloud precipitation and downdrafts appears to be especially strong in cases where precipitating cumulus congestus and cumulonimbi form above deep, dry boundary layers in the western United States. These are the very conditions that support the formation of dry downbursts or dry microbursts (Krumm, 1954; MacDonald, 1976; Brown et al., 1982; Hjelmfelt, 1987; Proctor, 1988, 1989). These intense downdrafts and outflows that generate hazardous wind shears for aviation arise from rather innocuous looking shallow cumuli producing virgae. In the presence of deep dry adiabatic layers even light amounts of precipitation can produce strong low-level downdrafts. Often sublimation and melting of ice crystals and evaporation of small drops (melted snow flakes) cools the air appreciably, giving rise to strong negative buoyancy unimpeded by dry adiabatic descent, owing to the dry adiabatic stability of the subcloud layer (Rodi et al., 1983; Proctor, 1989; Wakimoto et al., 1994). Such high-based cumuli forming over a dry adiabatic subcloud layer are quite common along the eastern slopes of the Rocky Mountains in Colorado and over the High Plains to the east. Wet downbursts on the other hand, often form from more intense cumulonimbi where greater amounts of precipitation produce strong water loading and melting of hail that

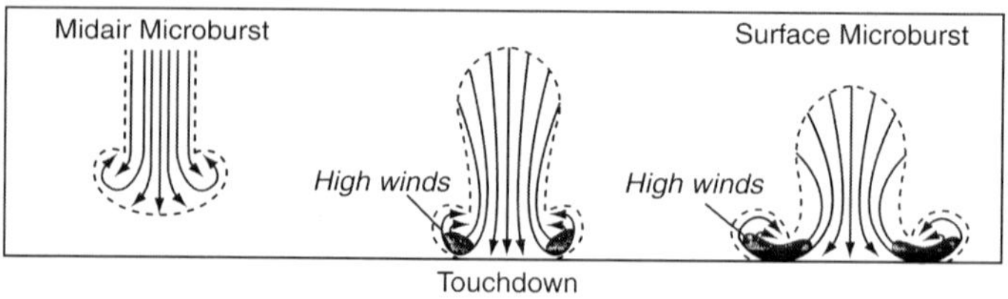


**FIGURE 8.19 Conceptual model of a microburst hypothesized to explain ground-damage patterns.** Three stages of development are shown. A midair microburst may or may not descend to the surface. If it doesn't, the outburst winds develop immediately after its touchdown. Based on a figure from Fujita (1985). *(From Wakimoto (2001))*

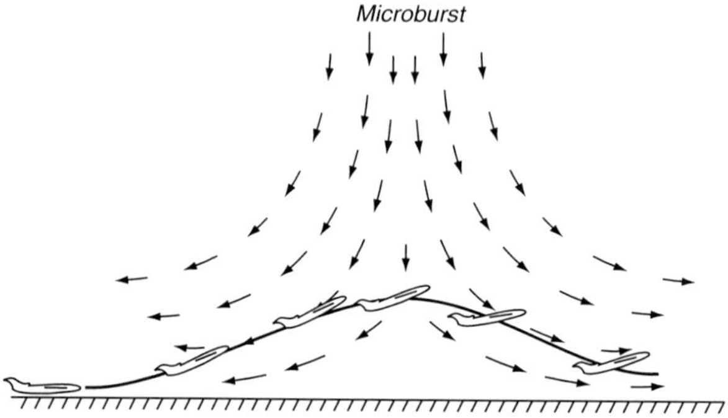


**FIGURE 8.20 Schematic diagram illustrating the impact of a microburst on aircraft performance during takeoff.** The airplane first encounters a headwind and first experiences added lift. This is followed in short succession by a decreasing headwind component, a downdraft, and finally a strong tailwind, which may lead to an impact with the ground. Composite drawing based on numerous studies of aircraft accidents by Fujita and Caracena (1977), Fujita and Byers (1977), and Fujita (1978, 1985, 1986). *(From Wakimoto (2001))*

contribute to strong downdrafts and microbursts (Wakimoto and Bringi, 1988). Figure 8.19 is a schematic diagram of a downburst descending to the surface and generating strong surface winds. Figure 8.20 illustrates how such a downburst can effect flight operations. It illustrates an aircraft taking off and finding itself in a strong downdraft with strong headwinds. Later the aircraft continues to be in a strong downdraft but the wind shifts from a strong headwind to a tailwind. As a result aircraft flight speeds can abruptly fall below stall speed and the aircraft descends without full flight control. Imagine further, a light aircraft like a glider making an approach to land, turning from base leg to final and encountering a downburst, and abruptly encountering not only a strong downdraft but a strong tail wind. The aircraft at this point would be flying slowly, perhaps 10 to 15 kts above stall speed. Upon encountering a downburst with say a 50 kt tailwind component, airspeed can drop from say 65 kts to more like 15 kts. If the sailplane is making a turn from baseleg to final, the pilot finds himself (herself) in one of the deadliest situations a pilot can encounter, a "stall-spin" situation with no chance to recover since the aircraft is close to the ground on final approach.

Numerous modeling studies have demonstrated the importance of the drop-size distribution on downdraft speeds. Hookings (1965) calculated that, with other factors remaining unchanged, more vigorous downdrafts were produced for (1) smaller droplet sizes, (2) greater liquid-water content, and (3) lower initial humidity at downdraft origin. Kamburova and Ludlam (1966) and Das and Subba Rao (1972) showed that for specified downdraft speeds, the rates of evaporation and associated cooling strongly depend on precipitation size and intensity. Knupp (1985) demonstrated in a two-dimensional cloud simulation that quartering the characteristic precipitation size for both rain and graupel particles increased maximum downdraft magnitudes over 2 m $s^{-1}$, enhanced low-level cooling by 3-5 °C, and increased downdraft depth by 0.8 km. The dependence of downdraft intensity on precipitation size is relatively easy to understand, because evaporation occurs at the surface of raindrops. Because the surface-to-volume ratio is greatest for smaller drops, a downdraft containing a given liquid-water content will expose a greater surface area to evaporation if the drops are small than if they are large. The same arguments hold for hail as melting of hail is controlled by heat exchanges at the hail surface. Thus van den Heever and Cotton (2004) showed that, other things being the same, downdrafts were stronger when small hail occurred as opposed to fewer large hailstones.

Occasionally, low-level downdrafts in severe storms appear to assume a two-celled pattern: one associated with heavy precipitation as described above, and another located on the upshear flank within lighter precipitation. The schematic in Fig. 8.21 illustrates the relative locations of what Lemon and Doswell (1979) term a forward-flank downdraft (FFD) located within the precipitation core downshear, and a colder rear-flank downdraft (RFD) within lighter precipitation on the upshear storm flank. Lemon and Doswell speculate that the RFD is initially dynamically forced by perturbation pressure gradients on the upshear flank at high levels (7-10 km), and is then maintained by loading and evaporation of anvil precipitation at middle to lower levels. The pressure gradients were assumed to be generated by high pressure, typically present within the upshear flank of updrafts in which perturbation pressure increases with height up to mid-levels. Other possible mechanisms will be discussed in subsequent sections.

Some investigations indicate that minimum-valued $\theta_e$ air lies within downdraft cores, while in others the $\theta_e$ minimum lies on the upshear edge of the downdraft core.

In a comprehensive analysis of low-level downdrafts in high-plains convective storms over the United States, Knupp (1985) included the synthesis of a number of cases observed by multiple-Doppler radar, two- and three-dimensional numerical simulations of the observed convective storms, model sensitivity studies, and the use of simple diagnostic models. Figure 8.22a illustrates a conceptual model of several predominant downdraft flow branches

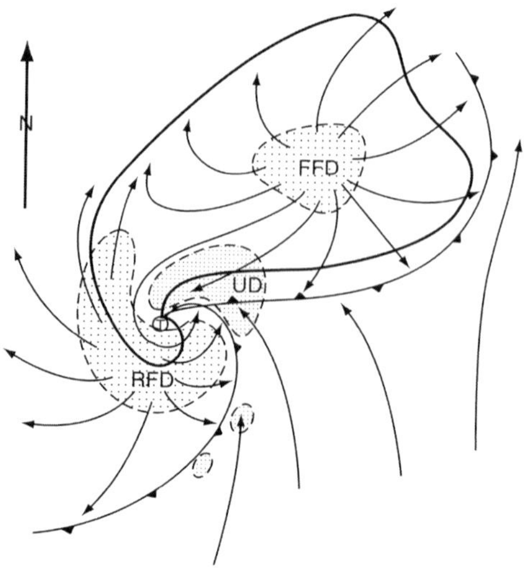


**FIGURE 8.21 Schematic plan view of surface features associated with a tornadic thunderstorm.** Gust fronts are depicted by barbed frontal symbols. Low level positions of draft features are denoted by stippled areas, where UD is updraft, RFD is rear-flanking downdraft, and FFD is forward-flanking downdraft. Streamlines denote storm-relative flow. *(From Lemon and Doswell (1979))*

that would occur in an environment having moderate shear, such as is illustrated in Fig. 8.22c. Figure 8.20c illustrates a sounding that has a planetary boundary layer depth of approximately 2 km, a conditionally unstable layer between 2 and 4 km, and an absolutely stable structure above 4 km. A projection of the downdraft branches onto an east-west plane is illustrated in Fig. 8.22b. Also shown in Fig. 8.22b are processes affecting downdraft intensities and the relationship between the sounding and those branches.

Knupp referred to those branches originating above the PBL as "mid-level" and those originating within the PBL he called "up-down." Originating within the updraft inflow sector, the "up-down" branch may rise up to 4 km before descending within the precipitation-laden primary downdraft region. Initially an upward-directed pressure gradient assists the lifting of the parcel until latent heating produces positive buoyancy along the upper portion of its path. Cooling produced by melting of precipitation along with "loading" by precipitation may provide sufficient negative buoyancy to initiate its transition to the down segment of this branch. Thus the "up-down" branch typically occurs during the mature storm phases. Both Doppler radar analyses and cloud-model results indicate that mixing occurs near the summit between moist air of this branch and drier, lower valued $\theta_e$ air flowing along the mid-level branch (Fig. 8.22a). Such mixing produces subsaturated, intermediate-valued $\theta_e$ air, which promotes increased evaporation rates along the descending portion of the "up-down" branch. The sudden decrease in buoyancy associated with the

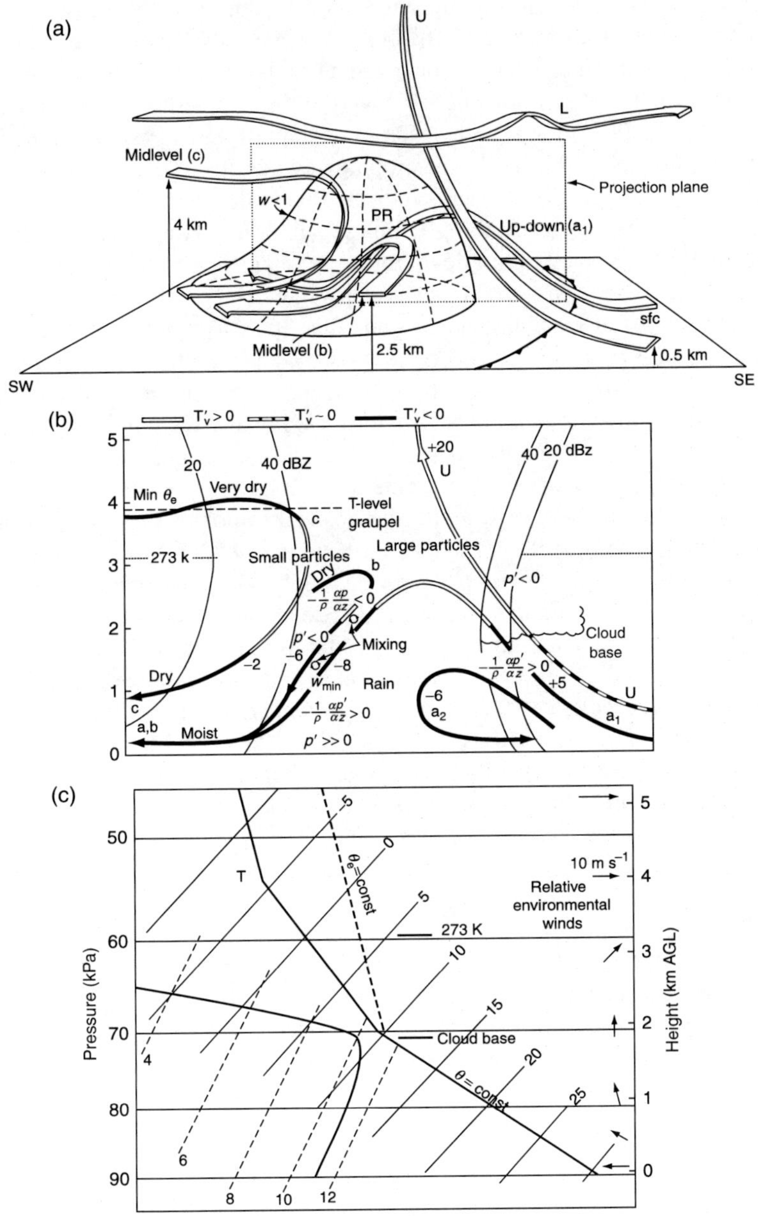


FIGURE 8.22 **(a) Schematic illustrating primary relative flow branches comprising the low-level precipitation-associated downdraft located along the upshear flank with respect to the updraft. (b) Projection of primary downdraft flow branches onto a vertical east-west plane. Physical processes along each branch are portrayed. (c) Sounding illustrating the relationship of downdraft properties to environmental structure.** *(From Knupp (1985))*

enhanced evaporation rates, along with water loading and, perhaps, a relative movement of this trajectory away from the region of favorable upward-directed pressure gradient forces, all contribute to the fast descent rates along this branch. The mid-level downdraft branches (b) and (c) originate within a 2-to 4-km AGL layer above the PBL in the southwestern quadrant of the storm. These branches are most pronounced during the developing downdraft stages.

Observations and model results indicate that both branches exhibit a period of weak ascent before descending within precipitation. Air with low-valued $\theta_e$ is transported to near the "up-down" branch where mixing occurs (trajectory b) and along trajectory c to the surface. In agreement with earlier diagnostic modeling studies of Kamburova and Ludlam (1966), Knupp found that descent rates along mid-level branches depend on the static stability of environmental air at the level of the origin of the branch, with greater stability favoring slower descent rates. Downdraft branches such as c typically descend at rates of 1-3 m $s^{-1}$ outside heaviest precipitation and reach the surface some distance behind the convective precipitation cores. This accounts for the lowest valued $\theta_e$ air at the surface being observed along the upshear sector of the storm (Lemon, 1976; Barnes, 1978a,b; Ray et al., 1981; Knupp, 1985). Knupp noted that the thermodynamic and kinematic properties along branch c resemble those found within mesoscale downdrafts (Zipser, 1977; Leary and Houze, 1979b).

Figure 8.22b indicates that as the precipitation-laden air along downdraft branches (b) and the "up-down" branch enters the subcloud layer, evaporation of precipitation accelerates the air downward and creates a downward-directed pressure-gradient force. This pressure-gradient force further enhances the downward acceleration of air and encourages the mixing between it and branch b containing lower valued $\theta_e$ air. This example demonstrates that the vertical pressure-gradient force contributes significantly to downdraft intensities and to the volume of downdraft air displaced into the subcloud layer.

An interesting finding by Knupp was the importance of melting to total cooling along downdraft trajectories. He found that melting can account for 10%-60% of the total cooling along certain downdraft trajectories. The largest contribution by melting occurred along the "up-down" trajectories in which parcels often pass through the melting zone twice or reside within the melting zone for considerable time at the apex of the "up-down" path. Precipitation evaporation accounts for the remainder of the total cooling and is typically greatest within drier air along mid-level trajectories. The contribution of melting to total downdraft cooling becomes greater in clouds having relatively moist subcloud layers and low cloud-base heights, such as exist in the maritime tropics.

The importance of the "up-down" downdraft branch is generally under appreciated. This is partly because it is not often observed by multiple Doppler radars as very short base lines are required. Moreover modeling studies may not recognize it as instantaneous snapshots of streamlines merge "up-down" downdrafts with mid-level downdrafts. Trajectories are required to

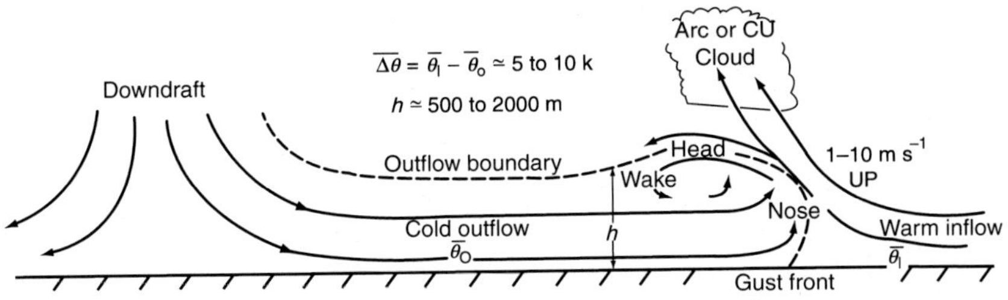


**FIGURE 8.23 Schematic structure of a gust front.** *(Adapted from Goff (1976), Fankhauser (1982), and Wakimoto (1982))*

readily identify them. Furthermore, models with ice phase microphysics are more likely to simulate them owing to the importance of melting. In spite of the tendency to ignore "up-down" downdrafts several modeling studies (Schmidt, 1991; Bernardet and Cotton, 1998) have implicated them in the formation of extreme surface wind events called derechos (see Chapter 9).

We can see from preceding sections that the intensity of updrafts and downdrafts in cumulonimbi is quite variable both in space and in time. Furthermore, the organization of updrafts and downdrafts in cumulonimbi is fundamental in determining the particular character of the storm system. We will now examine the downdraft outflows near the surface.

## 8.7. LOW-LEVEL OUTFLOWS AND GUST FRONTS

As the downdraft air approaches the surface, it diverges and forms a gust front that commonly produces significant convergence along its leading edge. Gust front properties and associated flows have been studied from analyses of observations (Charba, 1974; Goff, 1976; Wakimoto, 1982; Fankhauser, 1982; Sinclair and Purdom, 1983), laboratory experiments (Simpson, 1969; Simpson and Britter, 1980), and numerical modeling studies (Mitchell and Hovermale, 1977; Thorpe et al., 1980; Droegemeier and Wilhelmson, 1985a,b). Figure 8.23 illustrates some gust-front features as composited from observational and numerical studies. Major features include the nose, or elevated leading edge, the head, which marks the greatest height of the advancing system, and the turbulent wake. The gust-front system is usually 0.5-2 km deep, depending on strength and distance from the downdraft source. Bryan (2005) describe cold-pools associated with MCSs greater than 4 km in depth. The greatest vertical motions range from several to $\sim$10 m s$^{-1}$ near the upper portion of the front edge (e.g. Matthews, 1981; Wakimoto, 1982; Fankhauser, 1982), above which arc clouds, or deep convection may form. Downward motion characterizes the turbulent wake zone behind the head.

Wakimoto (1982) described four stages in the evolution of a gust front. During Stage I evaporatively-cooled downdraft air begins to diverge near the surface. During Stage II an advancing gust front forms; it exhibits a roll structure at its leading edge. The mature stage is similar to that illustrated in Fig. 8.23. During the dissipating stage (Stage IV), the gust front is no longer fed by evaporatively-chilled downdraft air, and the advancing gust front shrinks in vertical extent.

Movement speeds of the advancing front relative to ambient flow are often found to be close to the speed of density currents, where, if we approximate the density difference across the gust front as $\boldsymbol{\Delta}\rho/\rho_0 \simeq \overline{\boldsymbol{\Delta}\theta}/\theta_0$, we find

$$c = k[gh(\overline{\boldsymbol{\Delta}\theta}/\theta_0)]^{1/2}, \tag{8.5}$$

where $\overline{\boldsymbol{\Delta}\theta}$ is the average potential temperature deficit over the depth $h$ of the outflow, $\theta_0$ is the mean potential temperature, and $k$ is the Froude number, which is a constant that is determined empirically or theoretically. Typical values are about 0.8, ranging from 0.7 to 1.1 (Wakimoto, 1982). The typical speed of the gust front is 10 m s$^{-1}$, although speeds in excess of 20 m s$^{-1}$ have been observed (Wakimoto, 1982; Goff, 1975).

By introducing the equation of state, Seitter (1987) showed that the density difference across the gust front can be estimated from the surface hydrostatic pressure difference across the gust front, $\boldsymbol{\Delta}p = gh\boldsymbol{\Delta}\rho$, thus

$$c = k(\boldsymbol{\Delta}p/\rho_0)^{1/2}, \tag{8.6}$$

where $k$ is the Froude number and $\rho_0$ is the density of air. An advantage of Eq. (8.6) is that only surface pressure need be measured to estimate the speed of propagation of a gust front. Moreover, Nicholls (1987) found in two-dimensional numerical simulations of tropical squall lines that, whenever clouds occurred above the gust front, latent heat release and associated warming as well as water loading at levels above the depth of the outflow altered the surface pressure difference. This, in turn, altered the propagation speed of the gust front. These effects are not depicted in Eq. (8.5). Thus, both Seitter (1987) and Nicholls (1987) found that Eq. (8.6) yields a more consistent prediction of observed and modeled gust-front speeds than does Eq. (8.5), in which only temperature perturbations in the cold-pool affect gust-front movement.

Equation (8.6) is valid only when there is no ambient wind ahead of the gust front. Seitter noted that simply adding the wind component $\overline{U}$ parallel to the gust front motion to Eq. (8.6) does not yield satisfactory agreement with observations. Based on laboratory simulations of density currents reported by Simpson and Britter (1980), Seitter suggested that $\overline{U}$ must be multiplied by the factor 0.62 before being added to Eq. (8.6). Note that $\overline{U}$ is the wind component parallel to the gust-front motion that is averaged over the depth of the head of the gust front and is positive in the direction of the gust-front motion.

As noted by Bryan and Rotunno (2008), Eqs (8.5) and (8.6) are strictly valid only for shallow density currents, such as those studied in laboratory flows. In the case of deeper atmospheric cold-pools, where density varies appreciably with height, gust front propagation speeds are slower. Bryan and Rotunno (2008) derived more complicated equations for gust front propagation speeds for deep atmospheres. They showed that for atmospheric flows the estimated propagation speeds are as much as 25% slower than estimates based on incompressible equations.

While outflow $\theta$ profiles may assume constant mixed-layer values around the active turbulent portions, the coldest air may become stratified within the lower 500 m of less vigorous outflow air (e.g. Betts, 1984). The vertical motion over gust fronts is one of the primary mechanisms for triggering new clouds (e.g. (Purdom, 1982). Gust-front interactions with other meso-scale boundaries, such as cold fronts, dry lines, sea-breeze fronts, or other gust fronts, are important in the formation of convective clouds (e.g. Purdom, 1976, 1979; Sinclair and Purdom, 1983).

Purdom (1976) and Weaver and Nelson (1982) observed that new cloud growth is particularly explosive within regions where two outflow boundaries collide. The relative importance of such boundaries has been established by Purdom and Marcus (1982), who found that outflow interactions were present in 73% of all cumulonimbi initiated over the southeast United States during the 1979 summer months. Surface analyses presented in Cooper et al. (1982) also suggest that, in several cases examined, convergence produced by outflow from precipitating convection intensified and expanded cumulonimbi activity over southern Florida.

Droegemeier and Wilhelmson (1985a,b), simulated cloud development along intersecting thunderstorm outflow boundaries using the Klemp and Wilhelmson (1978a) three-dimensional cloud model. In the no-wind simulations (or those with weak wind shear), as the outflows from two convective storms collide, two symmetrical vertical velocity maxima are generated on the ends of the intersecting outflow boundaries. The structure of the convergence field driving the vertical velocity is shown schematically in Fig. 8.24. As the two outflows approach within several kilometers of each other, a zone of lifting is created between them. As the gust fronts come closer, air in the region labeled A is squeezed out laterally and vertically. The largest accelerations are horizontal because vertical accelerations require work against gravity. As a result, two regions of maximum convergence form on either side of the intersection point, as Fig. 8.24b shows (the center of $A$). If the low-level air is moist and the level of free convection is low enough, two clouds form over the two zones of maximum convergence.

Droegemeier and Wilhelmson (1985b) also simulated the effects of a unidirectional wind shear of varying depths and intensities on outflow interactions. The shear vector was aligned perpendicular to the center line of two initial outflow-producing clouds (i.e. oriented vertically in Fig. 8.24). The model results showed that in strong shear, the upshear member of the pair of

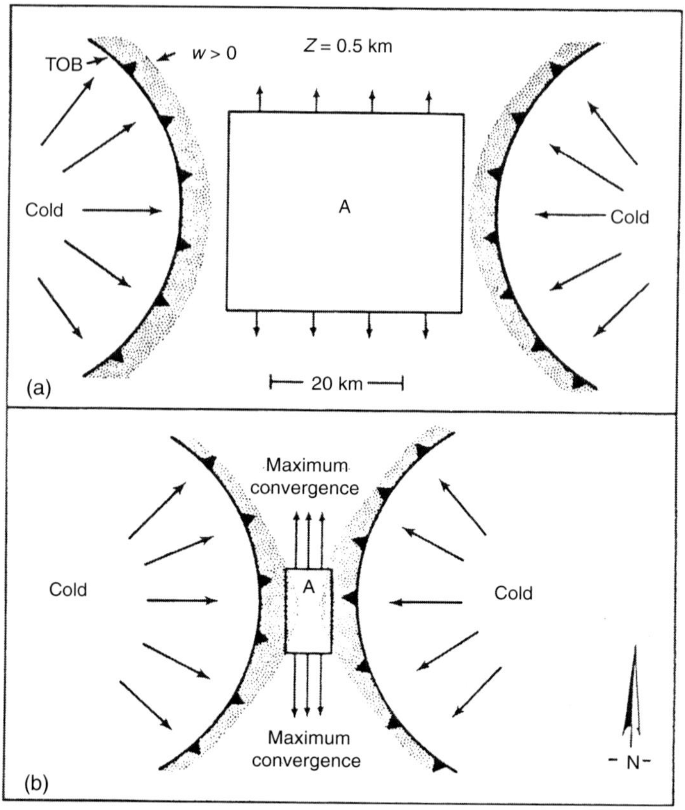


**FIGURE 8.24 Schematic diagram for the case of no wind shear in the model environment showing a plan view of (a) two outflows approximately 40 km apart and moving toward each other.** The gust front is indicated by the bold solid lines with barbs, and the regions of upward motion along the gust fronts are stippled. An arbitrary area A is shown by the box, and the small arrows indicate horizontal flow out of A as the outflows approach each other. **(b) The outflows are now approximately 10 km apart.** The regions of maximum horizontal convergence due to the rapid flow out of A from mass continuity are also indicated. *(From Droegemeier and Wilhelmson (1985a))*

clouds that formed in the maximum convergence zones became the stronger cell. This was because the upshear member had a head start on development and because the downdraft from the upshear member tilted downshear, thus suppressing the other cloud.

Downdraft outflows appear to exert primary influences on cumulonimbus maintenance and motion. Results of several three-dimensional cloud modeling studies indicate that convergence and associated uplift along downdraft-driven gust fronts may produce quasi-stationary storms (Miller, 1978), upshear propagation of tropical squall lines (Moncrieff and Miller, 1976), and storm regeneration (Thorpe and Miller, 1978; Klemp et al., 1981). Moncrieff and

Miller and Thorpe and Miller inferred that steady storm behavior requires zero relative motion between a storm and its gust front. This condition is satisfied when the low-level updraft inflow equals the outflow velocity of air behind the gust front. The importance of gust-front convergence in maintaining cumulonimbi and controlling storm motion has also been emphasized in the observational studies of Weaver and Nelson (1982), Knupp and Cotton (1982a) and Fankhauser (1982), and in the cloud model investigations of Klemp et al. (1981) andWilhelmson and Chen (1982).

As we have seen, downdraft outflows can also produce severe weather such as damaging winds and extreme wind shears. Wind shear associated with low-level downdraft divergence patterns is sometimes sufficiently extreme to cause aircraft accidents (e.g. Fujita and Caracena, 1977; National Research Council, 1983). Damaging surface winds are also directly responsible for deaths and numerous injuries. In some cases such damaging wind systems can be widespread and prolonged (Fujita, 1978; Fujita and Wakimoto, 1981; Johns and Hirt, 1983) contributing to derechos as discussed in Chapter 9.

Downdraft outflows and gust fronts can sometimes transform into bores and solitary waves (Knupp, 2006). Figure 8.25 illustrates the transition from a gust front, to a bore, to a solitary wave. Like gust fronts bores generate wind shifts in the direction of propagation, a pressure rise and strong updrafts. On the other hand, a solitary wave exhibits minor changes in surface winds in the form of a convergence/divergence couplet, and surface pressure oscillations associated with ascent and descent of stable air. As noted by Knupp (2006), the primary difference between a *solitary wave* and a *bore* is that upward displaced parcels are long lasting in the wave of a bore, whereas net parcel displacements are negligible for a solitary wave. Thus after bore passage, surface pressure remains high (Koch and Clark, 1999), but during a solitary wave passage, surface pressure oscillates to a maximum and then approaches preexisting pressure. Using hydraulic theory bore propagation speeds can be estimated from Eq. (8.7):

$$C_{\text{bore}} = c_w \left[ \frac{1}{2} \frac{h_1}{h_0} \left( 1 + \frac{h_a}{h_0} \right) \right]^{1/2}, \tag{8.7}$$

where $h_1$ is the depth of the bore, $h_0$ is the height of the surface-based inversion. The term $c_{gw}$ is the speed of propagation of a gravity wave given by Eq. (8.8),

$$C_{gw} = \left[ g \left( \frac{\Delta\theta_v}{\theta_v} \right) h_0 \right]^{1/2}, \tag{8.8}$$

where $\delta\theta_v$ is the difference in average $\theta_v$ across the inversion. Knupp (2006) observed the wave to propagate at 18 m s$^{-1}$ whereas Eq. (8.8) predicted 12 m s$^{-1}$. He noted that nonlinear effects could be important in his observed

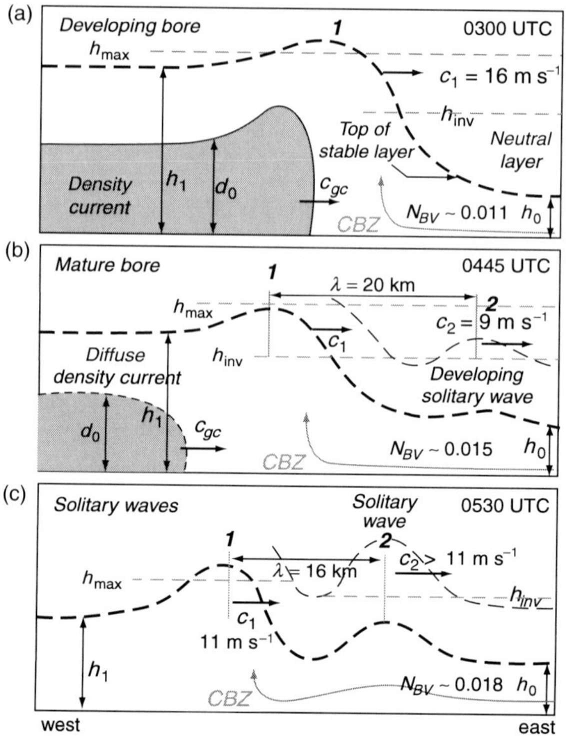


**FIGURE 8.25 Schematic vertical cross section showing the transition from a gust front/bore to an eventual solitary wave pair.** The three stages include the (a) developing bore, (b) mature bore, and (c) solitary waves 1 and 2. The NBL depth ($h_0$) and stability [expressed in terms of Brunt Väisälä frequency ($N_{BV}$, units of s$^{-1}$)] both increased during this sequence. The heavy dashed line represents the top of the stable layer, and h1 represents the mean bore depth. The thin dashed line ($\theta$ = const) is drawn to delineate the leading solitary wave, labeled 2. The shaded region in (a) and (b) denotes cool, dry air associated with the gravity current of depth $d_0$. The solid gray streamline represents low-level relative flow that defines the CBZ associated with the bore. The initial bore formed in advance of the gust front by 0300 UTC. A second solitary wave formed within the elevated inversion layer ($h_{inv}$) in advance of the bore about 30 min before the time in (b). By 0500 UTC, the gravity current disappeared. Propagation speeds of the gravity current and two buoyancy waves are represented by $C_{gc}$, $C_1$, and $C_2$. *(From Knupp (2006))*

large amplitude wave. But, Knupp (2006) found that Eq. (8.7) predicted the bore propagation speed of 16.3 to 16.9 m s$^{-1}$ which was close to the observed 15.9 m s$^{-1}$ speed. Knupp (2006) noted that the bore displaced air vertically about 2 km which should be sufficient to initiate convection.

Clearly, downdraft outflows, gust fronts, bores, and solitary waves play an important role in the behavior of convective storms. We will now examine theoretical models of storm propagation which include these as well as other mechanisms of storm propagation and movement.

## 8.8. THEORIES OF STORM MOVEMENT AND PROPAGATION

Storm movement and propagation can be classified into three different mechanisms: (1) translation or advection, (2) forced propagation, and (3) autopropagation. Translation or advection is the process whereby a storm is blown along by the mean wind as it evolves through its lifetime. Forced propagation refers to the sustained regeneration of a convective storm by some external forcing mechanism, usually larger in scale than the convective storm. Examples of external forcing mechanisms are fronts and rainband convergence associated with mid-latitude cyclones, sea-breeze fronts, convergence associated with mountains, convergence associated with variations in land-surface properties, like mesoscale patches of moist or dry soils either produced naturally through rainfall patterns or through irrigation, rainband convergence in tropical cyclones, frontal boundaries produced by the outflows from decayed convective storms, and convergence associated with gravity waves excited by external forcing mechanisms. Often the systems providing the forced propagation have lifetimes considerably longer than individual thunderstorms and are only mildly modulated by the presence of the storms.

Autopropagation refers to the process in which a thunderstorm can regenerate itself or cause the generation of similar storm elements (cells) within the same general system. Examples of autopropagation mechanisms are downdraft forcing and gust fronts, updraft forcing via warming aloft (causing enhanced inflow due to vertical pressure gradients), development of vertical pressure gradients due to storm rotation, and the triggering of gravity waves by a thunderstorm which generates areas of enhanced low-level convergence.

Many convective storm systems are affected by all three mechanisms of movement and propagation for at least some part of their lifetime. Some storm systems are primarily affected by autopropagation mechanisms for a major portion of their lifetime. An example of the latter is the Supercell thunderstorm. Squall-line thunderstorms can form by both forced propagation and autopropagation. The severe, prefrontal squall-line thunderstorm system (Newton, 1963), once initiated, is a classic example of an autopropagating system. Other squall-line thunderstorm systems are coupled with fronts throughout their lifetime. Such frontal squall lines are influenced primarily by forced propagation.

In this section we focus our attention on the theories of autopropagation of thunderstorms. We break these down into two classes: (1) propagation by gust fronts and cold-pools and (2) propagation by gravity waves. In Cotton and Anthes (1989) we reviewed two analytical theories of storm propagation in some detail. They were (1) the convective overturning models of Moncrieff and Green (1972; hereafter referred to as MG), Moncrieff and Miller (1976; hereafter referred to as MM), Moncrieff (1981; hereafter referred to as M), and Thorpe et al. (1982; hereafter referred to as TMM), and (2) the wave-CISK models of Lindzen (1974), Stevens and Lindzen (1978), Raymond (1975; 1976; 1983; hereafter referred to as R75, R76, R83, respectively), Davies (1979), and

Silva-Dias et al. 1984; hereafter referred to as SBS). These models represent two fundamentally different concepts of autopropagation of cumulonimbi, although they also exhibit some common features. We will not review these theories in detail here as both theories have been found to have major deficiencies when compared to observations. The Moncrieff-Green-Miller-Thorpe (MGMT) models are comprehensive theories of steady, convective, overturning storms. A common assumption used in the MGMT models is that the flow is steady in a reference frame that moves at the translation speed of the thunderstorm system. Each of the MGMT models requires the explicit prescription of "characteristic" updraft/downdraft streamline patterns. Miller and Moncrieff (1983) summarized the formulation of the MGMT models.

The other class of analytical model is called wave-CISK (conditional instability of the second kind). In this approach a convective system is viewed as a symbiotic interplay among the convective elements and a wave disturbance in the stably stratified, cloud-free environment. The convective elements provide the heating and, perhaps, momentum transports to initiate and drive the wave disturbance, while the wave provides the large-scale, moisture supply to the clouds. As seen in Chapter 7, even shallow cumulus clouds excite gravity waves in the stably stratified troposphere. The gravity waves, in turn, appear to exert an important influence on the resultant organization of clouds and propagation of clouds. In the wave-CISK concept simple linear models are used to describe the interaction between deep convection and gravity waves.

The wave-CISK concept is a derivative of the original CISK concept pioneered by Charney and Eliassen (1964) and Ooyama (1964) in its application to hurricanes. In those studies, cumulus convection was perceived to be driven by boundary layer frictional convergence associated with a tropical depression. Latent heat release and warming by convection then led to a deepening of the cyclone with a corresponding increase in boundary layer convergence.

Wave-CISK differs from CISK in that the forcing mechanism for cumulus convection is the convergence field associated with inviscid gravity-wave motions rather than boundary layer convergence. As in CISK, cumulus convection is introduced as a heat source with a specified distribution in the vertical. Figure 8.26 illustrates a schematic view of Raymond's (1975) wave-CISK representation of a convective system. There is no explicit scale separation specified in Raymond's wave-CISK models. Instead, a separation occurs between moist and dry processes. Moist processes are not a part of the solutions of the model; rather, they are specified in some parameterized form. The parameterized moist processes then serve as forcing functions to the set of linearized equations of motion. Predicted growth rates and phase speeds of the normal mode solutions to dry, inviscid wave motion are obtained as an eigenvalue problem. Omitted in the wave-CISK scheme is the possibility of moist, mesoscale ascent and descent, which are known to occur in tropical cyclones and other mesoscale convective systems such as squall lines and mesoscale convective complexes (see Chapter 9).

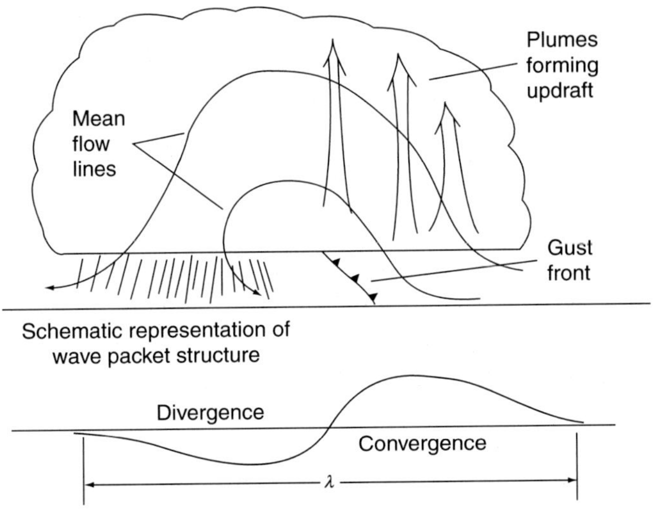


**FIGURE 8.26 Schematic diagram of the hypothesized relationship between a packet of gravity waves and the associated convective storm, subject to the limitations of a two-dimensional drawing.** The wave packet consists of one convergent and one divergent region with a dominant wavelength λ comparable to the diameter of the storm. Convective plumes develop in the convergent region, passing into the divergent region as they decay and produce rain. *(From Raymond (1975))*

As noted by Raymond (1983), it is the parameterization of these terms that is the Achilles heel of wave-CISK models. That is, the model predictions generally depend critically upon the details of the parameterization embedded in those terms. Particularly important is the parameterization of cumulus convective heating and fluxes.

Both the MGMT and wave-CISK models are extreme simplifications or idealizations of the behavior of wet convective systems. The formulations of both models have aspects in common with theoretical models of stably stratified flow over mountain barriers (see Chapter 11 for similarities between mountain wave theory and these models). Therefore, wave solutions are inherent in both approaches, as is the role of horizontal pressure gradients on storm propagation. The assumption of steady state is a severe limitation of the MGMT models, as is the assumption of linearity in the wave-CISK models. Comparison of model predictions with observations has generally been limited to a few case studies. Unfortunately, because the MGMT models require the updraft/downdraft profiles to be specified and the wave-CISK models are sensitive to the details of the convective parameterizations, each new development in model formulation requires a completely new evaluation of the model.

Rather than get bogged down in the details of these highly-simplified theories, let us overview the concepts of autopropagation of convective storms by downdrafts and gust fronts and by gravity waves.

### 8.8.1. Autopropagation by Gust Fronts and Cold-Pools

The conceptual model of ordinary thunderstorms proposed by Byers and Braham (1949) relies heavily on storm propagation by diverging, evaporatively-chilled air feeding the gust front near the ground, which then lifts moist conditionally unstable air along the gust front to feed the storm updrafts. We have seen in Section 8.7 that the speed of propagation of the gust front is proportional to the depth of the cold-pool and the density difference (or temperature contrast) across the gust front. If the depth of the cold-pool and the vertical ascent along the gust front is great enough to lift air to its level of free convection, then the advancing gust front is a major mechanism for autopropagation of thunderstorms.

Further clarification of the role of the cold-pool in storm propagation has been obtained in the two-dimensional numerical experiments by Thorpe et al. (1982) and the two-and three-dimensional numerical simulations by Rotunno et al. (1988; hereafter referred to as RKW). As we shall see, most of the research on autopropagation of convective storms has been directed towards understanding squall line propagation. Nonetheless, these concepts are also relevant to individual cumulonimbi. The upshear tilt of a convective system signals the increasing dominance of the cold-pool and its interaction with lower tropospheric shear. Thorpe et al. (1982) argued that a cold-pool acts as an obstacle to the flow. At the interface between the outflow from the cold-pool and the storm-relative inflow, steady convergence is established which prolongs the cell's life. If the storm-relative inflow is weak, $u_0$ reverses to less than or equal to 15 m s$^{-1}$ (e.g. the low-level shear is weak), the gust front is not stationary, and localized boundary layer convergence is not maintained and the convection dies. These results are consistent with earlier two-dimensional simulations reported by Takeda (1971), who concluded that a jet at low levels, particularly at 2.5 km AGL, is most favorable for producing a "long-lasting" cloud system. RKW interpreted the cold-pool and relative-inflow interaction in terms of the opposing shears present on each side of the cold-pool and inflow interface. Equation (8.9) shows that the time rate of change of vorticity is proportional to the horizontal buoyancy gradient:

$$\rho_o \frac{\mathrm{d}}{\mathrm{d}t}(\eta/\rho_o) = -\frac{\partial B}{\partial x}, \tag{8.9}$$

where $\eta$ represents the component of vorticity normal to a vertical plane, $\rho_o$ is the average density in the layer of the cold-pool, and $B$ is the buoyancy.

They noted that the circulation associated with a cold-pool is characterized by a negative horizontal component of vorticity at the cold-pool and inflow interface. In contrast a strong relative inflow, particularly one with a low-level jet, is characterized by a positive vorticity of the low-level shear. They argued that the optimum situation for maintaining steady, upshear-tilted convection is one in which the negative vorticity associated with the cold-pool and the positive

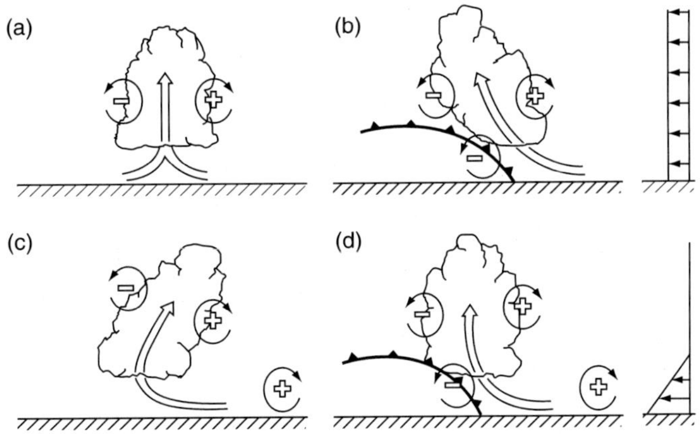


**FIGURE 8.27 Schematic diagram showing how a buoyant updraft may be influenced by wind shear and/or a cold-pool.** (a) With no shear and no cold-pool, the axis of the updraft produced by the thermally created, symmetric vorticity distribution is vertical. (b) With a cold-pool, the distribution is biased by the negative vorticity of the underlying cold-pool and causes the updraft to lean upshear. (c) With shear, the distribution is biased toward positive vorticity and this causes the updraft to lean back over the cold-pool. (d) With both a cold-pool and shear, the two effects may negate each other, and allow an erect updraft. *(From Rotunno et al. (1988))*

vorticity associated with the inflow are approximately balanced. Figure 8.27 illustrates this concept. Because cold-pools are generally shallow, shear can promote convection only when restricted to low levels. This conclusion is consistent with Bluestein and Jain (1985) composite study, which revealed that squall lines, in general, are oriented along the shear vector in the lowest level.

Owing to the finite extent of their cold-pools and curvature of the gust fronts, this theory is only relevant to small portions of individual cumulonimbi where the vorticity gradients are maximized. Nonetheless it highlights the importance of cold-pool interactions with environmental shear as being important to autopropagation of cumulonimbi.

Cold-pool propagation of storms may be considered a near-field propagation mechanism which is generally limited to distances of the order of 25-50 km from a storm. In contrast, gravity wave propagation can be considered a "far-field" propagation mechanism that can extend 50 to 100's of kilometers away from a given storm, although we shall see there are exceptions to that view when the boundary layer is stably stratified.

## 8.8.2. Autopropagation by Gravity Waves

We have seen that as cumulus clouds develop buoyancy (both positive and negative), gravity waves are excited, which tend to neutralize the buoyancy gradients (Bretherton and Smolarkiewicz, 1989). To a large extent the depth of the layer in which the wave is traveling determines the phase speed of the

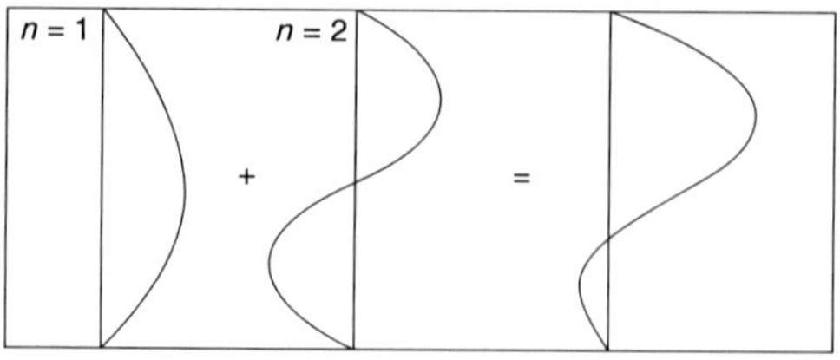


**FIGURE 8.28 Vertical distribution of the thermal forcing for $n = 1$ and $n = 2$ and their sum.** $Q_0$ is the same magnitude for each mode, positive for $n = 1$, negative for $n = 2$. *(From Nicholls et al. (1991))*

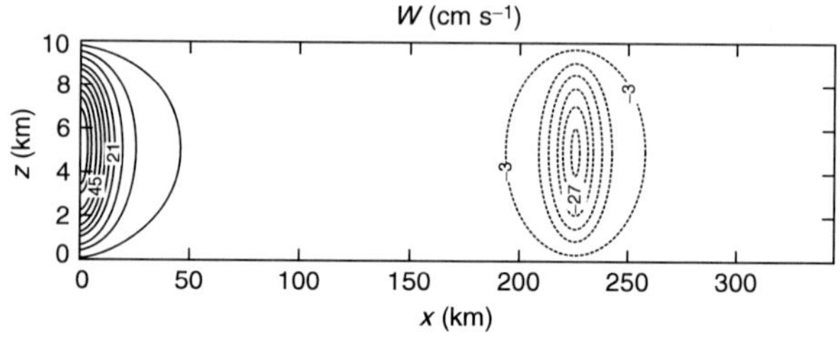


**FIGURE 8.29 Rigid-lid solution for the $n = 1$ mode. $Q_{mp} = 2.0$ J kg$^{-1}$ s$^{-1}$, $a = 10$ km, $N = 0.01$ s$^{-1}$ and $t = 2$ h.** Shows vertical velocity. The contour interval is 6 cm s$^{-1}$. *(From Nicholls et al. (1991))*

gravity wave. Thus Eq. (8.10) gives us:

$$C = \frac{NH}{\pi(1/2 + n)}, \quad n = 0, 1, 2, \tag{8.10}$$

where $N$ is the Brunt Väisälä, $H$ is the depth of the layer, and $n$ is the mode or vertical wavenumber.

The Brunt Väisälä frequency is given by:

$$N^2 = \frac{-g}{\rho_o}\frac{\partial \rho_o}{\partial z}. \tag{8.11}$$

Likewise the amplitude of the wave is proportional to the heating rate and inversely proportional to $N^2$. Now suppose that an idealized heating profile for a growing cumulonimbi resembles the profile shown in Fig. 8.28, labeled $n = 1$ for a deep troposphere heating mode. This mode produces fast-moving subsidence warming at large distances from the storm as illustrated in Fig. 8.29. Mapes (1993) noted this wave is not an ordinary wave, in the sense it does not have periodic structure in space and time. Instead he argues it is analogous to a *tidal bore* in water which propagates at the speed of an internal wave while

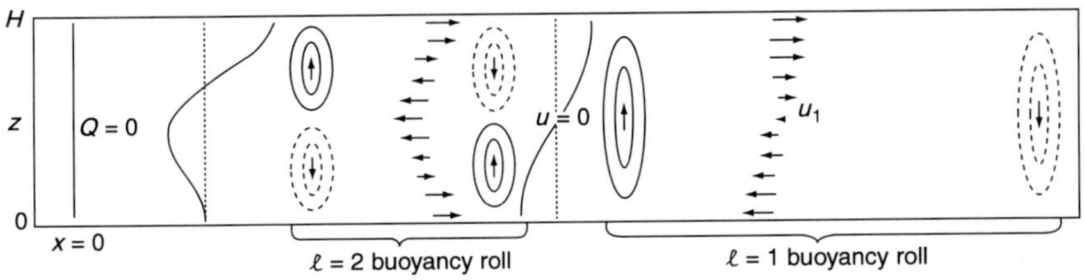


**FIGURE 8.30** **Two "buoyancy rolls."** Schematic of the buoyancy bores, horizontal winds, and horizontal displacements of material lines at a time $r$, after the two-mode heat source (3) acted near $x = 0$ from time 0 until time $r_2 < r/2$. The $\ell = 1$ "buoyancy rolls" (bounded on each end by $\ell = 1$ buoyancy bores) has completely outrun the $\ell = 2$ buoyancy rolls. *(From Mapes (1993))*

irrevocably changing the depth of the water as it passes. He thus refers to these waves as buoyancy bores.

Now suppose that the cumulonimbus has evolved into its mature phase, or an MCS has developed where low-level rain evaporation and melting of hail produces low level cooling. Likewise an anvil forms which in the case of an MCS can transform into a deep stratiform-anvil cloud with latent heating due to freezing and ice-deposition through a deep layer in the upper troposphere. The resultant idealized heating profile for this system can be represented by the $n = 2$ mode and the combined effects $n^1 + n^2$ are illustrated in the right-hand panel of Fig. 8.28. As shown in Fig. 8.30, initially this heating profile results in sinking motion at low levels and ascent aloft, but it reverts to a propagating wave with subsidence aloft and ascent at low levels. This ascent at low-levels with corresponding low-level convergence represents what Mapes (1993) calls *buoyancy bores* that favor the clustering of convection. He hypothesizes that these buoyancy bores are the mechanism for what he calls "*gregarious convection*" leading to the formation of "*superclusters*."

The wave-CISK models mentioned earlier represent a class of linear wave models in which convection is phase-locked to the convergence induced by a gravity wave. In the regions of convergence, a heating profile similar to that shown in Fig. 8.28 is imposed. The imposed heating then reinforces the gravity wave modes which then continue to propagate as a convectively-reinforced gravity wave. The problem is that numerical modeling studies of convective systems have shown that convection can excite many different scales and amplitudes of gravity waves, few of which will necessarily phase-lock with convection (Tripoli and Cotton, 1989a,b; Schmidt and Cotton, 1990). Thus the coupling of gravity waves and convection can only occur under very specific conditions.

The problem is that without phase-locking, gravity waves propagate energy vertically, and without a mechanism to reinforce the wave or reflect the wave, a wave will rapidly lose amplitude before traveling very far (Lindzen and Tung, 1976). It can be shown that reflection of gravity waves or wave trapping can occur when $l^2$ is less than $k^2$, where $l^2$ is the Scorer parameter given by

$$l^2 = \frac{N^2}{(U_0 - c)^2} - \frac{\partial^2 U_0/\partial z^2}{(U_0 - c)}, \tag{8.12}$$

$k^2$ is the horizontal wave number, $U_0$ is the ambient wind speed, and $c$ is the gravity wave speed. Wave trapping is likely to occur if $N^2$ is small or negative (low stability), $(U_0 - c)$ is large (strong shear) or $\partial^2 U_0/\partial z^2$ is large (a sharp jet).

There are several examples of where wave ducting can reinforce convective systems. Tripoli and Cotton (1989a,b) showed in their numerical simulations that radiative cooling at the top of a stratiform anvil layer of an MCS can lead to a layer of low stability which can serve to trap wave energy. This trapping layer can thus enhance low-level convergence and provide a mechanism for propagation of an MCS. Schmidt and Cotton (1990) simulated a squall line that was able to propagate over a stable layer such as occurs at night time over the plains of the US. The presence of such a nocturnal stable layer is not favorable for thunderstorm and squall line propagation by cold-pools or the RKW theory. Nonetheless, such nocturnal thunderstorm complexes are quite common over the High Plains of the US. Schmidt and Cotton (1990) showed that the presence of a low-level stable layer, with an overlaying layer of weak stability (sufficient to trap wave energy and support deep convection) and an upper tropospheric stable layer with both low-level shear and deep tropospheric shear, provided an ideal environment for sustaining a squall line by gravity wave propagation.

The debate about whether thunderstorms and MCSs (squall lines in particular) propagate primarily by gust front/cold-pool dynamics or gravity waves began over 60 years ago (Hamilton and Archbold, 1945; Tepper, 1950; Newton, 1950) and is sure to continue for some time to come. We will return to this discussion more when we focus on MCSs and Squall lines in Chapter 9.

### 8.8.3. Updraft Splitting and Storm Propagation

The process of storm splitting was first recognized from the behavior of radar reflectivity fields (Fujita and Grandoso, 1968; Achtemeier, 1975). More recently it has been identified by multiple-Doppler radar as a process of a splitting of the updraft in convective storms (Bluestein and Sohl, 1979; Knupp and Cotton, 1982a). Three-dimensional numerical models have been used extensively to analyze the splitting process (Wilhelmson and Klemp, 1978, 1981; Klemp and Wilhelmson, 1978a,b; Thorpe and Miller, 1978; Schlesinger, 1980; Clark, 1979; Rotunno and Klemp, 1982, 1985; Weisman and Klemp, 1984; Tripoli and Cotton, 1986).

Klemp and Wilhelmson (1978a) argued that downward drag due to strong rainwater loading in the cloud interior contributes to updraft splitting. Schlesinger (1980) concluded that the dynamic pressure perturbations associated with the entrainment of potentially cool air outside the storm also contributed strongly to storm splitting. Clark (1979) also concluded that the entrainment process contributed to storm splitting. He argued that the entrainment process and, hence, storm splitting were very sensitive to the details of the often unrealistic model initial conditions.

Vertical shear of the horizontal wind is clearly important in the splitting process. Thorpe and Miller (1978) concluded that strong shear contributes to a large downshear slope of the updraft core with rain falling into the inflow region, resulting in a split updraft cell. Klemp and Wilhelmson's (1978a) calculations indicate that strong low-level vertical shear or the presence of a low-level jet favors the updraft splitting process.

The directional shear of the horizontal wind has been found to be important in determining which of the two split-cell members predominates. If the wind shear is unidirectional (i.e. shows no change in direction with height), such as illustrated in Fig. 8.31, Wilhelmson and Klemp (1981) found that a vortex pair develops in a horizontal plane in which positive vertical vorticity occurs on the right-hand side of the updraft relative to the shear vector. On the left-hand side of the updraft, negative vorticity occurs. This vortex pair is essentially responsible for the production of entrainment on the down-shear flank of towering cumulus clouds as discussed in Chapter 7. Rotunno (1981) showed that the vortex pair is a direct consequence of the tilting of vortex tubes by rising convective updrafts. The vortex tubes, shown in Fig. 8.31a, are produced by the environmental shear. The downshear downdraft induced by entrainment and rainwater loading subsequently splits the original cell into two cells, one dominated by cyclonic vorticity at low levels propagating to the right of the shear vector; a second cell propagating to the left of the shear vector is dominated by anticyclonic vorticity. These two storms, illustrated in Fig. 8.31b, possess mirror-image symmetry. Furthermore, these storms possess properties similar to Browning's (1964) and Browning and Wexler's (1968) conceptual model of severe right (SR)- and severe left (SL)-moving storms.

Klemp and Wilhelmson (1978b) found that the relative strengths of the SR- and SL-moving storms varied with directional changes of the shear vector. They found that if the wind shear vector veers with height (i.e. turns clockwise in the northern hemisphere), the development of a cyclonic, right-moving storm is favored. By contrast, if the wind shear vector backs with height (i.e. turns counterclockwise in the northern hemisphere), then an anticyclonic, left-moving storm prevails. Rotunno and Klemp (1982) developed a simple linear model to explain these results. Figure 8.32 is a conceptual model illustrating the variations in pressure gradient with changes in the direction of the shear vector. Figure 8.32a illustrates the case of unidirectional shear flow where symmetric vortex pairs develop in the updraft with relative high-pressure upshear of the updraft and relative low-pressure downshear. With no change in the shear vector with height, the relative highs and lows are vertically aligned. Thus, the storm circulations develop symmetrically about the shear vector. When the shear vector changes direction with height, such symmetry is lost, however.

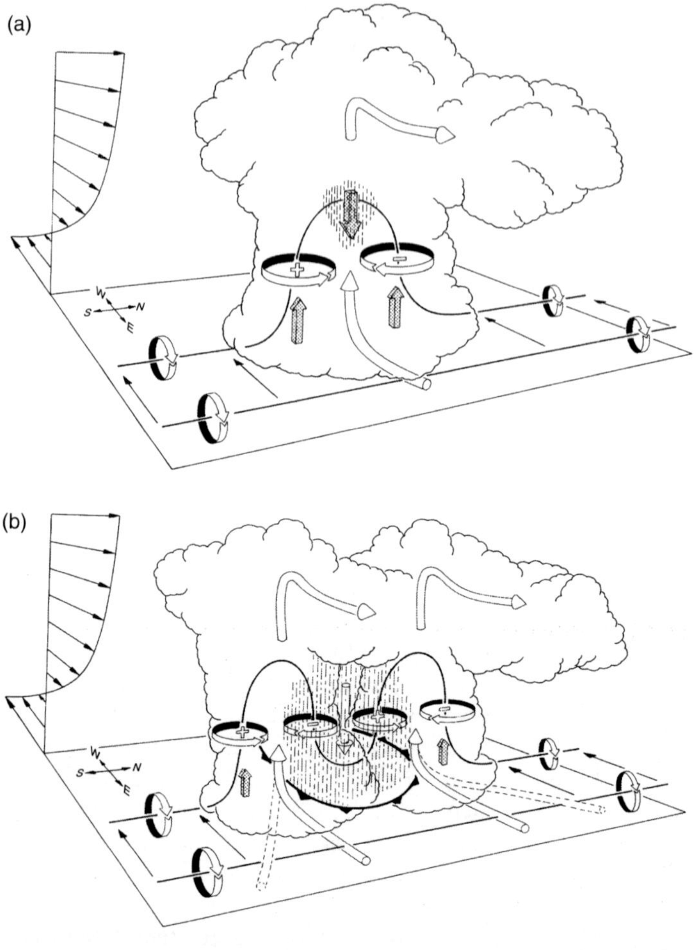


**FIGURE 8.31 Schematic depicting how a typical vortex tube contained within (westerly) environmental shear is deformed as it interacts with a convective cell (viewed from the southeast).** Cylindrical arrows show the direction of cloud-relative airflow, and heavy solid lines represent vortex lines with the sense of rotation indicated by circular arrows. Shaded arrows represent the forcing influences that promote new updraft and downdraft growth. Vertical dashed lines denote regions of precipitation. (a) Initial stage: vortex tube loops into the vertical as it is swept into the updraft. (b) Splitting stage: downdraft forming between the splitting updraft cells tilts vortex tubes downward, producing two vortex pairs. The barbed line at the surface marks the boundary of the cold air spreading out beneath the storm. *(Adapted by Klemp (1987), from Rotunno (1981). Copyright © 1987 by Annual Reviews, Inc. Reproduced with permission.)*

Figure 8.32b illustrates the case when a shear vector turns clockwise with height. This illustrates that a relative high forms to the south and a low to the north at low levels, and at a higher level a high is situated to the north and a low to the south. This creates a favorable vertical pressure gradient for updraft

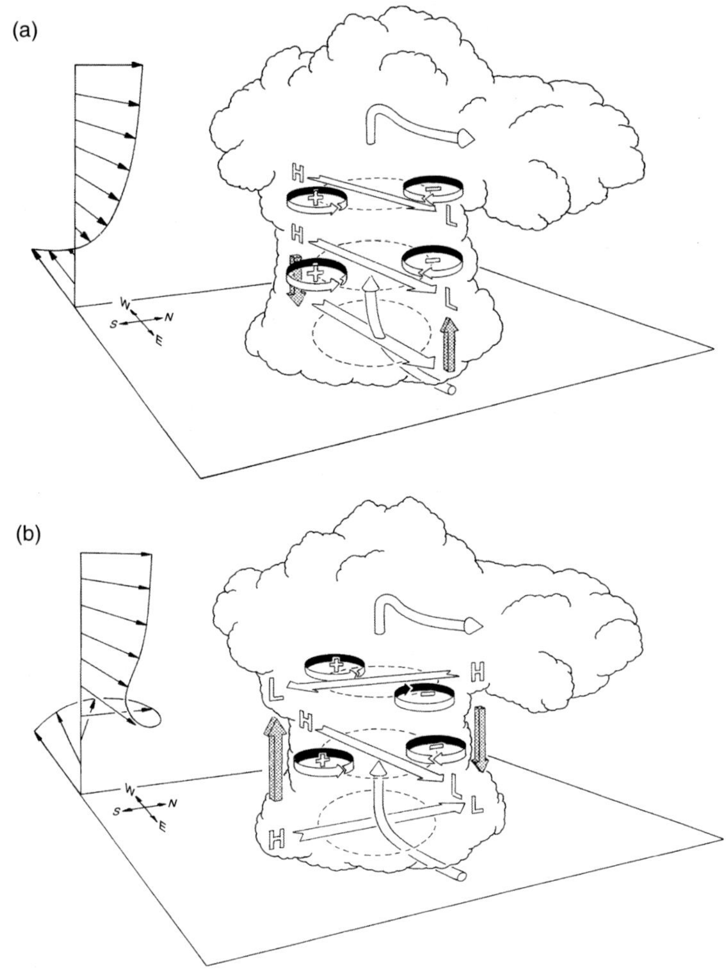


FIGURE 8.32 **Schematic illustrating the pressure and vertical vorticity perturbations arising as an updraft interacts with an environmental wind shear that (a) does not change direction with height and (b) turns clockwise with height.** The high (H) to low (L) horizontal pressure gradients parallel to the shear vectors (flat arrows) are labeled along with the preferred location of cyclonic (+) and anticyclonic (−) vorticity. The shaded arrows depict the orientation of the resulting vertical pressure gradients. *(Adapted by Klemp (1987), from Rotunno and Klemp (1982). Copyright © 1987 by Annual Reviews, Inc. Reproduced with permission.)*

intensification on the southern side of the updraft and an unfavorable one on the northern side. Therefore, such a wind hodograph favors the SR-moving storm.

Weisman and Klemp (1984) performed a detailed analysis of the pressure fields and convergence patterns simulated by a three-dimensional cloud model for wind profiles, in which the shear vector turns clockwise through 180° over the lowest 5 km. As noted previously, such profiles favor the formation of the SR storm system. As in earlier simulations, splitting of the updraft of the initial cell occurred which resulted in right-flank and left-flank updraft cells. Two distinct mechanisms forced the low-level convergence fields supporting the updrafts. In the case of the left-flank updraft, the strong convergence was driven by a cold-pool spreading against an incoming flow of potentially unstable environmental air. The convergence is thus forced by high pressure behind the gust front. This effect is strongest where the storm relative inflow and downdraft outflow most directly oppose each other. The simulated left-flank cells were multicellular, shorter lived, and less intense than the right-flank cells. However, the left-flank cells produced the most rainfall.

In the case of the right-flank cell, the low-level convergence field was locally induced by the lowering of pressure caused by the dynamic interaction of the storm with environmental wind shear, similar to that illustrated in Fig. 8.32b. The resultant mesolow produced low-level convergence of moist air into the updraft in regions where convergence caused by the spreading cold outflow is relatively weak. The strength of such a dynamically-induced mesolow increases with higher values of environmental wind shear. Weisman and Klemp found that a bulk Richardson number,

$$R_{\mathrm{i}} = B/\frac{1}{2}(\boldsymbol{\Delta}\overline{u})^2, \tag{8.13}$$

where $B$ represents the convective available potential energy in the storm's environment and $\boldsymbol{\Delta}\overline{u}$ represents a difference between environmental wind speeds at low levels and mid-levels, was a useful parameter for determining when such a mesolow would prevail. As in the MG definition, the Richardson number is a measure of the ratio of available potential to available kinetic energy. Low values of $R_{\mathrm{i}}$ correspond to an environment where the available kinetic energy is large. They found that the right-flank cells increase in strength with decreasing values of $R_{\mathrm{i}}$ until the shear becomes so strong that the initial convection was suppressed ($R_{\mathrm{i}} \sim 15$). The right-flank cells predominate over the left-flank cells (in terms of updraft intensities) for $R_{\mathrm{i}}$ less than about 6.0.

As noted by Rotunno and Klemp (1982), the majority of severe tornadic storms over the United States are SR-moving storms. This is consistent with Maddox's (1976) climatology of tornadic storms which reveals that they occur in a wind field in which the wind shear vector between the surface and 700 mbar turns clockwise. However, occasionally counterclockwise wind shear vectors are observed along with intense SL-moving storms. One such storm, described by Knupp and Cotton (1982a,b), was an intense, quasi-steady left-moving storm system which formed in an environment exhibiting a counterclockwise wind shear vector. This particular local environmental wind field was established by the interaction of the large-scale flow field with local topographically-induced circulations (Cotton et al., 1982). At first it was thought that the left-moving character of the storm system and its primary updraft/downdraft structure were a consequence of the complex mesoscale environment in which the storm formed

(Cotton et al., 1982). However, subsequent three-dimensional numerical experiments reported by Tripoli and Cotton (1986) revealed that a left-moving storm system, having the observed updraft/downdraft structure and intensity, was simulated with a horizontally homogeneous environment in which the low-level shear vector turned counterclockwise. The simulated steady left-moving storm formed as a result of splitting of the updraft of the parent cell. A less intense, multicellular right-moving cell was simulated which also exhibited characteristics similar to an observed right-moving cell. Consistent with Weisman and Kemp's modeling results, the simulated left-moving cell was supported by a dynamically-induced mesolow. It appears that the complicated early history of convection was only important in that it established the local environmental shear that was favorable for the dominance of the left-moving cell.

In summary, several mechanisms are involved in the propagation of convective storms. Which mechanism prevails depends strongly upon the environmental winds and stability. Updraft rotation also affects storm propagation. We now examine how updraft rotation affects storm dynamics as well as the formation of tornadoes.

## 8.9. MESOCYCLONES AND TORNADOS

In this section we concentrate on the processes involved in the generation of thunderstorm rotation and tornado-scale vorticities. The emphasis, however, is on the generation of rotation on the thunderstorm scale, with only a brief discussion of structure and vorticity in tornadoes. We begin by defining the various parameters and equations that are used to describe and measure rotation in thunderstorms and tornadoes. We then discuss the larger scale conditions most favorable for generating tornado-producing storms then discuss the processes involved in developing rotating thunderstorms (mesocyclones). We conclude with a brief summary of the characteristics of tornadoes and models of tornadoes.

### 8.9.1. Vorticity and Circulation Equations

One measure of rotation frequently used in meteorology is the vorticity which represents the local and instantaneous rate of rotation of the system. In the case of solid rotation, vorticity is twice the angular velocity. To obtain the vorticity of a system, one takes the curl or vector cross product of the equations of motion. As shown by Dutton (1976), this results in the general vorticity equation of the form:

$$\begin{aligned} \mathrm{d}\omega/\mathrm{d}t &= (\omega \cdot \nabla)V - \omega\nabla \cdot \mathbf{V} - \nabla\alpha \times \nabla p \\ &\quad + \gamma\nabla^2\boldsymbol{\xi} + \nabla\gamma \times \left[\nabla^2\mathbf{V} + \nabla(\nabla \cdot \mathbf{V})\right], \end{aligned} \tag{8.14}$$

where the absolute vorticity $\omega = \xi + 2\mathrm{r}$, $\xi$ is the relative vorticity, and $\mathbf{r}$ is the earth's rotation rate. Ignoring the effects of viscosity, Eq. (8.14) can be decomposed into its scalar components as follows:

$$\frac{\mathrm{d}}{\mathrm{d}t}\begin{bmatrix}\xi\\ \eta\\ \zeta\end{bmatrix} = \overset{\text{(a)}}{(\boldsymbol{\omega}\cdot\nabla)\begin{bmatrix}u\\ v\\ w\end{bmatrix}} - \overset{\text{(b)}}{\begin{bmatrix}\xi\\ \eta\\ \zeta\end{bmatrix}\nabla\cdot\mathbf{V}} + \overset{\text{(c)}}{\begin{bmatrix}g(\partial/\partial y)(\alpha'/\alpha_0)\\ -g(\partial/\partial x)(\alpha'/\alpha_0)\\ 0\end{bmatrix}}, \quad (8.15)$$

Terms (a) and (b) of Eq. (8.15) represent the so-called tilting and convergence terms, respectively. These terms act together to concentrate and transfer vorticity from one plane to another. An updraft, for example, can tilt a horizontally-oriented vortex tube into a vertically-oriented one. Term (b), the convergence term, represents the fluid analog to the angular acceleration of a solid rotating system due to change in the radius of rotation. Term (c) is called the solenoidal term and it represents the production of vorticity by baroclinicity. It acts predominantly in the horizontal, where large-scale baroclinic fronts or convective-scale gust fronts can generate vorticity about a horizontal axis. Figure 8.33 illustrates schematically the contributions by (a) convergence, (b) by tilting, and (c) by baroclinicity to vertical vorticity production. One must be cautious about focusing on one component of the vorticity equation, such as the vertical, since the vortex tube strength can increase due to the other two horizontal components. As pointed out by Grasso (1992) one might conclude, erroneously, that say, a thunderstorm is not rotating faster simply because the vertical component of vorticity is not changing. The thunderstorm may be rotating faster simply because it is leaning away from the vertical as most thunderstorms do. It is therefore important to consider the full three-dimensional vorticity equation.

As discussed by Davies-Jones (1984) tilting of horizontally-oriented vorticity can occur either when the environmental winds are perpendicular to the flow (called crosswise) or when parallel to the flow (called streamwise). Figure 8.34a (or top) shows the case when the environmental flow is perpendicular to the vortex lines induced by vertical shear of the horizontal wind. In this case tilting produces a cyclonic/anticyclonic pair in which the cyclonic (anticyclonic) vortex is on the right (left) resulting in no net updraft rotation. Now consider Figure 8.34b (or bottom) in which the environmental flow is parallel to the vortex lines induced by vertical shear of the horizontal winds. In this case the cyclonic vortex is on the rising side of the peak which produces a positive correlation between vertical velocity and vertical vorticity. Thus Davies-Jones (1984) identifies tilting of streamwise vorticity to be the origin of updraft rotation in supercell storms.

For a Boussinesq system (see Chapter 2), Eq. (8.15) reduces to

$$\frac{\mathrm{d}}{\mathrm{d}t}\begin{bmatrix}\xi\\ \eta\\ \zeta\end{bmatrix} = (\boldsymbol{\omega}\cdot\nabla)\begin{bmatrix}u\\ v\\ w\end{bmatrix} - \begin{bmatrix}\xi\\ \eta\\ \zeta\end{bmatrix}\nabla\cdot\mathbf{V} + \begin{bmatrix}g(\partial/\partial y)(\alpha'/\alpha_0)\\ -g(\partial/\partial x)(\alpha'/\alpha_0)\\ 0\end{bmatrix}, \quad (8.16)$$

where the solenoidal term affects only the horizontal components of vorticity.

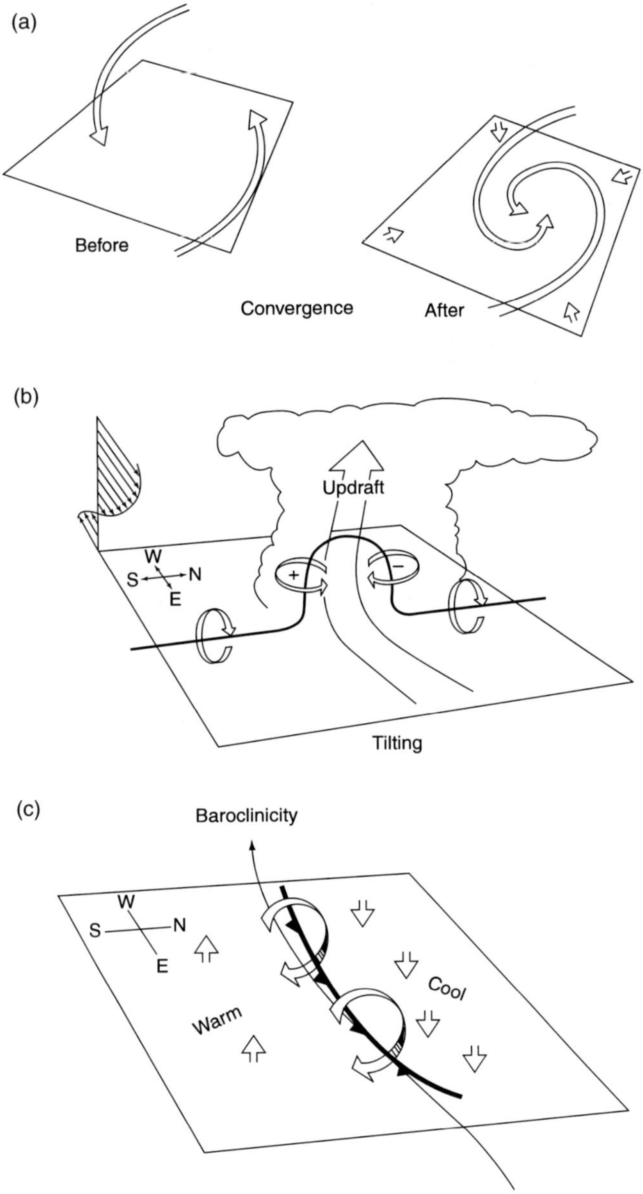


**FIGURE 8.33 Processes responsible for generating rotation or vorticity. (a) Convergence of air can strengthen vorticity. (b) Tilting of vorticity in the horizontal plane by updrafts can produce vorticity in the vertical plane. (c) Strong horizontal temperature gradient or baroclinity can be tilted into the vertical.**

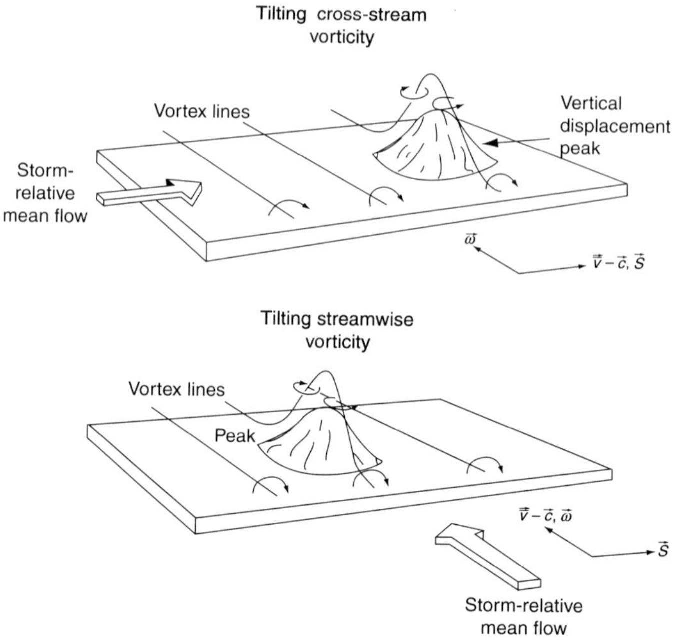


**FIGURE 8.34 (top) Case when the environmental flow is perpendicular to the vortex lines induced by vertical shear of the horizontal wind.** In this case tilting produces a cyclonic/anticyclonic pair in which the cyclonic (anticyclonic) vortex is on the right (left) resulting in no net updraft rotation. **(bottom) Case where the environmental flow is parallel to the vortex lines induced by vertical shear of the horizontal winds.** In this case the cyclonic vortex is on the rising side of the peak which produces a positive correlation between vertical velocity and vertical vorticity. *(Adapted from Davies-Jones (1984))*

Another useful quantity for investigating the rotational properties of thunderstorms is *equivalent potential vorticity*, defined as

$$\omega_{\theta e} = \alpha_0 \nabla\theta_e \cdot \boldsymbol{\omega}. \tag{8.17}$$

Rotunno and Klemp (1985) extended the concept of potential vorticity (see Dutton, 1976, p. 382) to a cloud system. Introducing the anelastic continuity equation (see Chapter 2) in Eq. (8.15) and defining $B = \alpha' g/\alpha_0$, we have

$$\frac{1}{\alpha_0}\frac{d(\alpha_0\boldsymbol{\omega})}{dt} = (\omega\cdot\nabla)\mathbf{V} + \nabla\times(B\mathbf{k}). \tag{8.18}$$

Making use of the identity

$$\boldsymbol{\omega}\cdot\frac{d}{dt}\nabla\theta_e = \omega\cdot\nabla\frac{d\theta_e}{dt} - \nabla\theta_e\cdot[(\omega\cdot\nabla)\mathbf{V}] \tag{8.19}$$

in the left-hand side of Eq. (8.18) yields

$$\alpha\nabla\theta_e\left[\frac{1}{\alpha_0}\frac{d(\alpha_0\omega)}{dt} - \omega\cdot\nabla\mathbf{V}\right] = \frac{d}{dt}(\alpha_0\nabla\theta_e - \omega) - \alpha\omega\cdot\nabla\frac{d\theta_e}{dt}, \quad (8.20)$$

or

$$\frac{d}{dt}(\alpha_0\nabla\theta_e\cdot\omega) - \alpha_0\omega\cdot\nabla\frac{d\theta_e}{dt} = -\alpha_0\nabla\theta_e\cdot\nabla\times(B\mathbf{k}). \quad (8.21)$$

Assuming that cloud updrafts and downdrafts are wet adiabatic, $d\theta_e/dt = 0$, then

$$\frac{d}{dt}(\omega_{\theta_e}) = \frac{d}{dt}(\alpha_0\nabla\theta_e\cdot\omega) = -\alpha_0\nabla\theta_e\cdot\nabla\times(B\mathbf{k}). \quad (8.22)$$

Ignoring the effects of water loading, in saturated air $B = B(\theta_e)$. Therefore, the right-hand side is zero and $\omega_{\theta_e}$ is conserved. In unsaturated downdrafts, Rotunno and Klemp argue by scale analysis that the right-hand side is small so that $\omega_{\theta_e}$ is nearly conserved. Thus, $\omega_{\theta_e}$ has the property of being nearly conserved along updraft and downdraft trajectories in convective storms. This useful property can help us understand the intimate relationship between the vorticity and thermodynamics of a convective storm.

Another parameter useful for studying storm rotation is the circulation $C(t)$ defined as

$$C(t) = \oint \mathbf{V}\cdot d\mathbf{l}, \quad (8.23)$$

where the integration is performed around a closed material surface. The change in circulation around the material surface for an inviscid, Boussinesq system is given by

$$dC/dt = \oint B\mathbf{k}\cdot d\mathbf{l}. \quad (8.24)$$

Lilly (1986a,b) has advocated the use of the "helicity" concept in studying thunderstorm rotation. The helicity $H$ is defined as

$$H = \mathbf{V}\cdot\boldsymbol{\omega}, \quad (8.25)$$

which is essentially a covariance of the velocity and vorticity vectors. Lilly also defined relative helicity $r_H$ as

$$r_H = H/V\omega, \quad (8.26)$$

where $r_H$ ranges in magnitude between $+1$ and $-1$.

It is also useful to consider the pressure diagnostic equation in terms of vorticity. In Chapter 2 we derived a diagnostic equation for pressure for an anelastic system in terms of buoyancy gradients and the momentum fields. However, one can also form a diagnostic pressure equation for an anelastic system in terms of relative vorticity. Grasso (1992) derived the full diagnostic equation but then showed that, to a good approximation, the diagnostic pressure equation given in Eq. (8.27) is useful near and within a supercell thunderstorm (see Brandes, 1984).

$$\frac{\partial}{\partial x_i}\frac{\partial p^*}{\partial x_i} \simeq -\rho_o(z)d_{ij}d_{ij} + \frac{\rho_o(z)}{2}\omega_j\omega_j. \tag{8.27}$$

The first term on the right-hand side represents the rate of strain. Both the convergence and divergence fields are included in this term and are correlated with a perturbation high pressure. The second term is the square of the magnitude of total relative vorticity. This term is always positive so that either cyclonic or anticyclonic vorticity contributes to a lowering of pressure.

We will subsequently use these principles in our examination of factors contributing to thunderstorm rotation and tornado formation.

### 8.9.2. Large-Scale Conditions Associated with Tornadoes

In this section we examine the large-scale conditions associated with the formation of tornadoes. As we shall see, forecasting tornado-producing storms involves forecasting intense storms that are also capable of producing hail, flash floods, and strong, nonrotational surface winds. Few severe storms produce only one form of severe event. However, we can identify conditions which favor the formation of the most intense tornadic outbreaks. These conditions usually occur in the springtime associated with vigorous baroclinically-driven mid-latitude cyclonic storms.

Figure 8.35 illustrates the synoptic-scale conditions often associated with major tornado outbreaks. Pertinent features are the wave cyclone with its attendant warm front (WF) and cold front (CF), along with the low-level jet (LJ) and upper-level jet (UJ) streams. Beebe and Bates (1955) noted that the region labeled A where the two jet streams cross is the most favored region for the formation of tornado-producing storms. This region is often characterized by the following conditions:

1. Strong low-level advection of heat and moist air, providing an adequate supply of moist static energy.
2. A conditionally unstable atmosphere through a deep layer (high CAPE).
3. A strong capping inversion, which inhibits the widespread outbreak of thunderstorms (high convective inhibition or CIN).
4. Moderate to strong vertical shear of the horizontal winds.

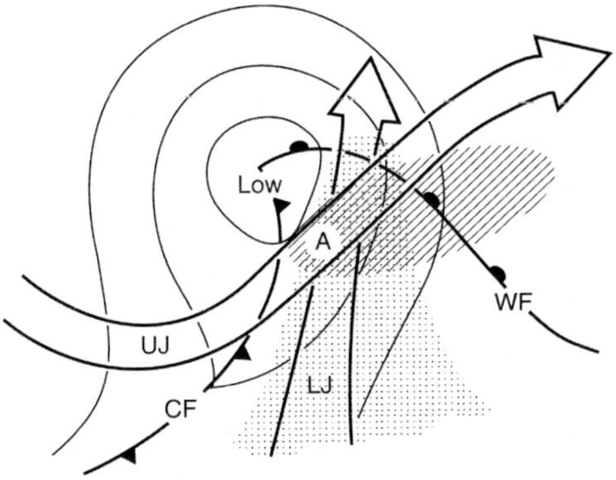


**FIGURE 8.35 Idealized sketch of a situation especially favorable for development of severe thunderstorms.** Thin lines denote sea-level isobars around a low-pressure center, CF and WF indicating cold and warm fronts at the ground. Broad arrows represent jet streams in low (LJ) and upper (UJ) levels of the troposphere. Region rich in water vapor at low levels is stippled. Severe storms are most likely to originate near A and, during the ensuing 6-12 h as the cyclone moves eastward and the severe storms move generally eastward at a faster speed, affect the hatched region (outside which thunderstorms may also occur that will probably be less intense). *(From Newton et al. (1978))*

Also necessary for severe storm occurrence is some mechanism for breaking the capping inversion. Some of the major mechanisms are frontal lifting, surface heating, drylines, gravity waves, and lifting due to terrain effects.

It has also been shown that major tornadic outbreaks are associated with jet streaks (speed maxima) in the upper-level jet stream. Riehl et al. (1952) concluded that jet streaks were associated with divergence fields which produce upward motion in the left front of the jet exit zone and in the right rear of the entrance zone of the jet streak. Subsequently, Uccellini and Johnson (1979) showed that the UJ and LJ jets are often coupled by mass adjustments associated with the propagation of geostrophically unbalanced flows associated with the jet streaks. Thus Kloth and Davies-Jones (1980) found that the left-exit and right-entrance zones are favored zones for upward motion and tornado occurrences.

The warm sector of the wave cyclone is often a region of strong positive vorticity advection (PVA), which, is associated with upward motion and is therefore a very important parameter in severe-storm forecasting (Koscielski, 1965; Miller, 1967). However, Maddox and Doswell (1982) showed several examples of tornadic outbreaks that were not associated with intense cyclonic storms and their attendant $PVA$ and UJ/LJ relationships. These were significant events;they had as many as 15 tornado reports and considerable loss of life and property damage. They noted that the most consistent large-scale feature associated with such storms was a pronounced low-level thermal advection field. It should be remembered that the omega equation (see Chapter 9) shows that for quasi-geostrophic motion, thermal advection contributes directly to large-scale vertical motion. Thus, Maddox and Doswell argue that in the absence of significant PVA and jet-streak dynamics, low-level warm advection in regions of strong conditional instability signals the presence of large-scale upward motion that can trigger significant severe weather occurrences.

Tornadoes are associated with land-falling tropical cyclones and can account for a significant fraction of the damage and loss of life in such storms. Novlan and Gray (1974) found that 25% of the hurricanes in the United States spawned tornadoes, with 10 as the average number of tornadoes per storm. This number excludes hurricane Beulah in 1967, which produced 141 tornadoes. Fujita et al. (1972) found the average number of tornadoes associated with typhoons in Japan to be somewhat less, that is, 2.3.

Both observational and modeling studies have documented the environmental and storm-scale characteristics associated with tornado formation in tropical cyclones (TCs) (Novlan and Gray, 1974; Gentry, 1983; McCaul, 1991; McCaul and Weisman, 1996; Spratt et al., 2000; McCaul et al., 2004; Curtis, 2004; Schneider and Sharp, 2007; Eastin et al., 2007). These studies indicate that tornadoes in TCs preferentially form in outer convective rainbands located in the on-shore flow of the right-front quadrant, usually within 100 to 400 km of the storm center. Favorable environmental characteristics include strong low-level wind shear (greater than 20 m $s^{-1}$ over the lowest kilometer), moderate CAPE (greater than 500 J $kg^{-1}$), strong low-level storm-relative helicity (greater than 100 $m^2$ $s^{-2}$), dry air at mid-levels adjacent to the rainband, a low level thermal boundary or convergence axis, a very moist low level air or a low LCL.

As early as Fujita et al. (1972), tornados have been associated with rotating thunderstorms embedded in the tropical cyclones. These are now generally called *mini-supercells* (Suzuki et al., 2000) owing to the fact they tend to be shallower (echo tops 8-12 km), smaller in diameter (less than 10 km), shorter lived (often less than 20 min) than Great Plains supercells. Figure 8.36 illustrates a major outbreak of tornadoes associated with landfall of hurricane Ivan.

Tornadoes are also associated with mesoscale convective complexes (see Chapter 9), a storm system that often develops in weakly baroclinic atmospheres.

In summary, we find that the major tornadic outbreaks generally occur with vigorous mid-latitude and land-falling TCs. However, any time the environment is able to support an intense thunderstorm system, tornadoes are possible, along with other forms of severe weather such as flash floods, hail, and severe winds, so forecasting tornadoes remains a major challenge to meteorologists. For a more extensive discussion about forecasting tornadoes and other severe storms the reader is referred to Moller (2001).

### 8.9.3. Rotating Thunderstorms or Mesocyclones

In this section we will examine the characteristics and dynamics of rotating thunderstorms, also known as mesocyclones. If a rotating thunderstorm produces a tornado, it is also referred to as a tornado-cyclone. Most severe tornadoes are associated with rotating thunderstorms.

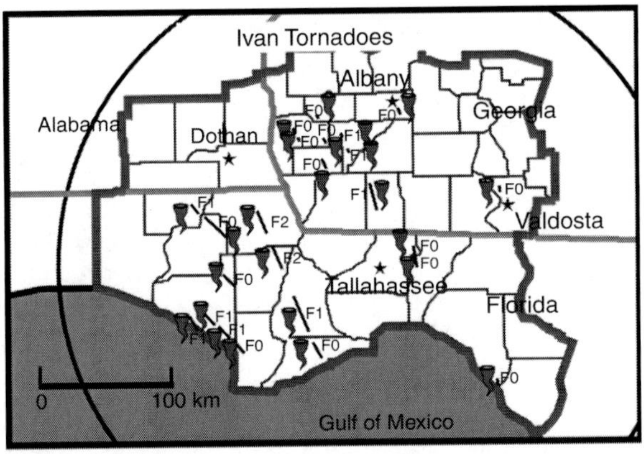


**FIGURE 8.36 Tornado tracks of F-scale of verified tornado touchdowns during Hurricane Ivan on 15-16 September 2004.** *(From Watson et al. (2005))*

The basic conceptual model of a rotating thunderstorm builds on the supercell storm model as discussed in Section 8.2. A characteristic of supercell storms is that they are rotating thunderstorms. However, one must not be misled into thinking that only supercell storms are rotating storms. There is considerable evidence that the mature cell in multicellular storms often exhibits the rotational characteristics of the supercell storm. Even cumulus congestus are often observed to be rotating, particularly when water spouts form.

A comprehensive conceptual model of a rotating thunderstorm has been derived by Lemon and Doswell (1979) from the synthesis of data obtained for aircraft, Doppler radar, and surface mesoscale networks, as well as from visual observations. Figure 8.21 illustrates the surface features of the mesocyclone at the time of tornado formation. Pertinent features are the locations of the storm updraft (UD), and the forward-flanking downdraft (FFD) and rearward flanking downdraft (RFD). The storm system resembles an occluding mid-latitude cyclonic storm. The most intense tornadoes (T) are often located near the tip of the "occlusion" close to the updraft/downdraft interface. The authors emphasize the importance of the rear-flanking downdraft in establishing the occlusion and the descent of the mesocyclone from mid-levels. Often associated with the descending mesocylone is a local region of intense horizontal shear, which can be detected by Doppler radar. This local horizontal shear region is referred to as the tornado vortex signature (TVS), and is often found to precede tornado touchdown at the surface by tens of minutes.

A three-dimensional depiction of the temporal evolution of Lemon and Doswell's model storm is shown in Fig. 8.37. The evolution of a tornado-producing storm is depicted in four separate stages.

***Stage 1***: At this stage the storm has a strong rotating updraft with a forward-flanking downdraft in the precipitation region. A gust front at the surface

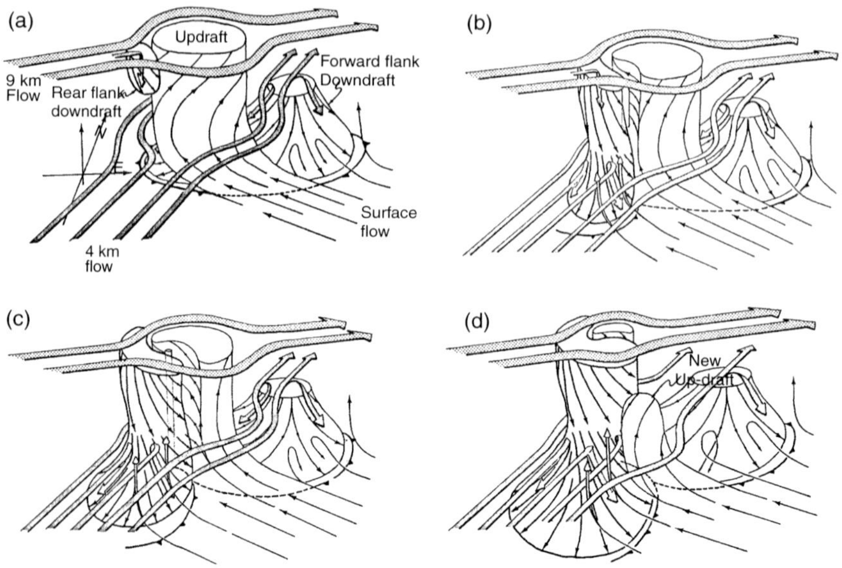


**FIGURE 8.37 Schematic three-dimensional depiction of evolution of the drafts, tornado, and mesocyclone in an evolving supercell storm.** The stippled flow line suggesting descent of air from the 9-km stagnation point has been omitted from (c) and (d), for simplicity. Fine stippling denotes the TVS. Flow lines throughout the figure are storm relative and conceptual only, and are not intended to represent flux, streamlines, or trajectories. Conventional frontal symbols are used to denote outflow boundaries at the surface, as in Fig. 8.21. Salient features are labeled on the figure. *(From Lemon and Doswell (1979))*

forms at the forward and right flanks of the radar echoes. Only an incipient rear-flanking downdraft is evident at mid-levels.

***Stage 2*:** As seen in Fig. 8.37b, the rear-flanking downdraft has now descended to the surface, initiating the first stages of the wave-cyclone-like features there. At the interface between the updraft and the RFD, a mid-level vortex or TVS forms at the rear flank of the storm. At this stage, a bounded weak-echo region or radar hook echo characteristically forms.

***Stage 3*:** This stage is identified by the formation of the low-level occlusion (see Fig. 8.37c), which is a result of the spreading gust front from the RFD catching up with the outflow from the FFD. The tornado reaches the surface near the tip of the occlusion. The bounded weak-echo region or radar hook echo region begins to fill or wrap up. Precipitation becomes widespread throughout the storm and the major tornado dissipates.

***Stage 4*:** The major updraft continues to weaken and downdrafts spread throughout the rotating storm. For the storm to persist, a new updraft must form on the right-hand side of the spreading surface outflow as shown in Fig. 8.37d. The sequence is then repeated, leading to the periodic occurrence of tornadoes.

Lemon and Doswell also performed an analysis of the vorticity production in such a rotating storm [see Eq. (8.16)] and concluded that the tilting term [term (a)] prevails. They also noted that the solenoidal term may be significant at the interface between the cold, dense, rear-flanking downdraft and the updraft.

A more extensive analysis of vorticity production in rotating thunderstorms has been done by Rotunno and Klemp (1985). They applied a three-dimensional cloud model to a case study of a tornado-producing storm described by Klemp et al. (1981). They used the actual thermodynamic sounding, but modified the wind profile to have the same magnitude of shear without directional change with height. The model responded to the initial thermal perturbation in a manner similar to the storm-splitting studies described previously. As described by Wilhelmson and Klemp (1981), a downdraft forms downshear which splits the original updraft into two cells, one dominated by cyclonic vorticity at mid-levels, which propagates to the right of the shear vector, and a second propagating to the left dominated by anticyclonic vorticity.

Focusing on the SR member of the split pair, Rotunno and Klemp integrated the model for 150 min. As shown in Fig. 8.38, the model simulated a surface occlusion similar to the conceptual model developed by Lemon and Doswell (1979). Note the well-defined rotation at both upper and lower levels. At the surface, the maximum vorticity is located at the updraft center at the position marked by a v at 90 min. By 150 min the maximum vorticity has shifted to the sharp gradient between updraft and downdraft, but now it is on the updraft side. At this time (150 min) the updraft strength has weakened sharply. The reduction in updraft strength occurs when the low-level rotation exceeds the mid-level rotation. Associated with the rotating updraft is a cyclostrophic reduction in pressure (see Eq. (8.27)) at the updraft center. When the low-level portion of the updraft rotates faster than the mid-level portion, an adverse vertical pressure gradient is established that retards the updraft strength. This process, called the "vortex valve" effect by Lemon et al. (1975), explained the decline in updraft strength as low-level rotation increased in the Union City, Oklahoma, tornado. Figure 8.39a illustrates schematically the downward-directed pressure gradient force.

### 8.9.4. The Genesis of Low-level Rotation and Tornadoes

In this section we discuss the theories for genesis of tornadoes in supercell and non-supercell storms. We begin by discussing non-supercell tornado (NST) formation. There have been several observational studies of tornadoes forming from cumulonimbi that do not exhibit the characteristics of rotation found in supercell storms (Wakimoto and Wilson, 1989; Brady and Szoke, 1989; Wilczak et al., 1992). A conceptual model that Wakimoto and Wilson (1989) derived

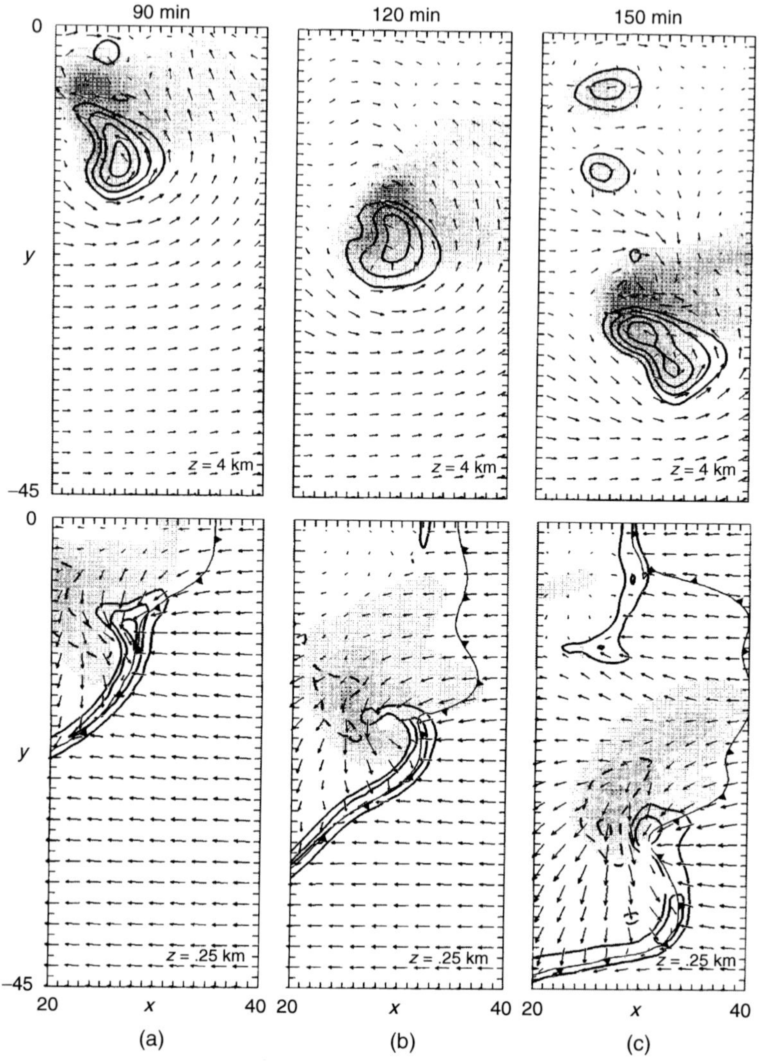


**FIGURE 8.38 Time sequence of the flow at $z = 4$ and 0.25 km over a 20× 45-km portion of the computational domain at (a) 90 min, (b) 120 min, and (c) 150 min.** The contour interval for $w$ is 5 m s$^{-1}$ at $z = 4$ km and is 1 m s$^{-1}$ at 0.25 km with the zero lines omitted. The rainwater field is shaded in 2 g kg$^{-1}$ increments beginning with 1 g kg$^{-1}$; horizontal wind vectors are represented as one grid length = 10 m s$^{-1}$ plotted at every other grid point. The locations of the low-level maxima of vertical vorticity are indicated by the $v$. *(From Rotunno and Klemp (1985))*

from observations is shown in Fig. 8.40. In their model misocyclones or small vortices originate at low levels by shearing instability along a convergence boundary such as a cold outflow from early storms. Relatively small cumuli form above the convergence boundary. They hypothesized that by chance, one

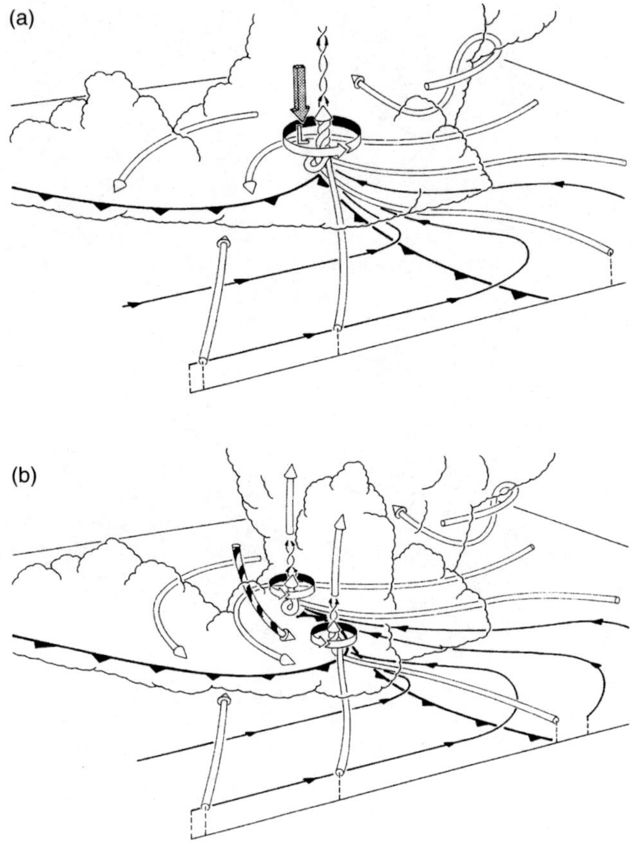


**FIGURE 8.39 Expanded three-dimensional perspective, viewed from the southeast, of the low-level flow (a) and (b) about 10 min later after the rear-flanking downdraft has intensified.** The cylindrical arrows depict the flow in and around the storm. The vector direction of vortex lines are indicated by arrows along the lines. The sense of rotation is indicated by the circular ribbon arrows. The heavy barbed line marks the boundary of the cold air beneath the storm. The shaded arrow in (a) represents the rotationally induced vertical pressure gradient, and the striped arrow in (b) denotes the rear-flanking downdraft. *(From Klemp (1987). Copyright © 1987 by Annual Reviews, Inc. Reproduced with permission.)*

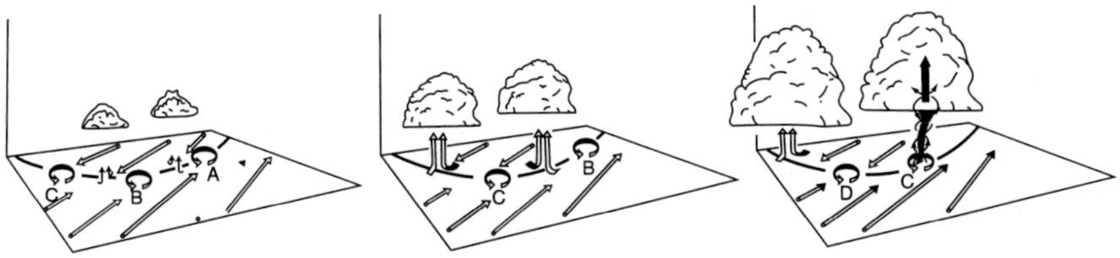


**FIGURE 8.40 Schematic model of the lifecycle of the non-supercell tornado.** The black line is the radar detectable convergence boundary. Low-level vortices are labeled with letters. *(From Wakimoto and Wilson (1989))*

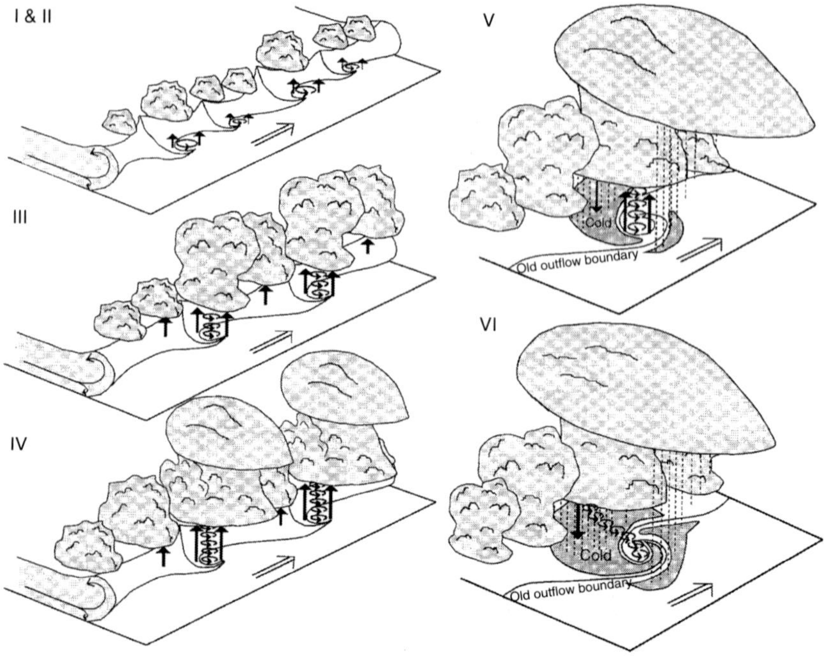


**FIGURE 8.41 Schematic presentation of the life cycle stages of NST evolution.** Stage I, vortex sheet development; stage II, vortex sheet rollup; stage III, misocyclone interaction and merger; stage IV, early mature NST; stage V, late morning NSF; and stage VI, dissipation. The diagrams in stages V and VI focus on just one member of the NSF family. The viewing perspective is from an elevated position looking northwest. *(From Lee and Wilhelmson (1997))*

of the low-level vortices along the boundary co-locates with a cumulus above. Then convergence and stretching by the towering cumulus generates a tornado of modest intensity.

Lee and Wilhelmson (1997) performed idealized three dimensional simulations of a NST with 60 m grid spacing. Their simulations produced a quite realistic looking line of vortices some of which evolved into tornadic vortices. Figure 8.41 portrays a conceptual model of NST evolution they derived from their simulations. During Stage I an outflow boundary encounters an air mass with a low-level wind field relatively parallel to the leading edge of the boundary. At the interface between these two air masses strong horizontal shear (i.e. a vertical vortex sheet) is found along the boundaries. Horizontal shearing instabilities develop along the boundary leading to the the formation of misoscale eddies (Stage II). During stage III the misoscale eddies merge and coalesce to form strong vortices which feed on the vorticity of neighboring eddies. These larger-scale vortices produce an asymmetric pattern of deep convection along the boundary which converges and stretches the vorticity in the low-level misocyclones. They conclude that the co-location of the deep convective cells and the larger surface misocyclones is not by chance, but that the misocyclones induce low-level convergence which favors the formation of the deep convective cells. Once the deep convective cells become coupled to the low-level vorticities (Stage IV) low-level stretching associated with friction-induced low-level inflow intensifies the misocyclone to tornadic intensity. The most intense phase (Stage V) occurs when precipitation-induced cold outflows increase low-level convergence and stretching of the vortex. The final dissipation stage (Stage VI) begins when the area beneath the cumulus is filled with negatively buoyant downdraft air near the surface. At this stage the tornadic vortices begins to tilt away from their associated convective towers as they are advected along with the advancing cold-pool. We call this type of tornado formation *bottom-up* tornado genesis.

In the case of tornado genesis in supercell storms, early views were based on a *top-down* concept in which the mesocyclone builds downward by a pressure-deficit tube (Smith and Leslie, 1978, 1979). The concept can be understood by examining Eq. (8.27). As vorticity in the mesocyclone intensifies, pressure will fall in the center of the vortex. Beneath the region of lowered pressure, a strong upward-directed pressure gradient will form which will accelerate the updrafts, causing enhanced convergence and stretching, leading to enhancement of vorticity at lower levels, further lowering of pressure and downward propagation of the vortex and associated pressure deficit tube. While this may occur in idealized axisymmetric models, it is not very likely that vorticity in the mesocyclone will become sufficiently localized and intense that a pressure-deficit tube can form. Grasso and Cotton (1995) simulated the formation of a pressure deficit tube in an idealized three dimensional supercell simulation. However, the simulated vortex did not build down from the center of the mesocyclone. Instead, the vortex was initiated at the periphery of the updraft in the region of strong horizontal gradient of updraft velocity. Furthermore, the pressure-deficit tube vortex did not directly build downward to the surface to form a surface-based tornado vortex. Instead the descending vortex coalesced with a pre-existing surface-based vortex, possibly associated with the downdraft. The resultant enriched vorticity allowed the pressure deficit tube to descend to the surface forming a tornado.

Several observations have shown that tornadoes form in the zone of the strong horizontal gradient of vertical velocity that resides between the main updraft of the storm and the rear-flanking downdraft. The existence of the rear-flanking downdraft has been attributed to evaporative cooling (Barnes, 1978a) or to dynamical interaction between the environment and the storm (Lemon and Doswell, 1979). Klemp and Rotunno (1983) argue that the rear-flanking downdraft is induced by intense rotation near the ground. The downward or adverse pressure gradient noted previously with the low-level rotating updraft is also thought to drive the rear-flanking downdraft. This is illustrated in Fig. 8.39b.

Rotunno and Klemp showed that the development of the mid-level rotation was fundamentally different than that at low levels in the simulated storm. At mid-levels the primary source of rotation is the vertical shear of the horizontal wind, which is tilted [term (a) of Eq. (8.15)] into the vertical. They used the concept of equivalent vorticity [see Eq. (8.22)] to illustrate this process. It should be noted that if equivalent vorticity is conserved, a vortex line must lie on a surface of constant $\theta_e$. Analysis of the various terms contributing to vorticity production in a Boussinesq system, Eq. (8.16), Rotunno and Klemp found that a major contribution to the production of horizontal vorticity at low levels was the baroclinicity along the gust front [term (c) in Eq. (8.16)] (see also the cartoon in Fig. 8.33). As a vortex line travels along the gust front, it mixes with low-valued $\theta_e$ downdraft air, where baroclinic-production of vorticity is also occurring. Rotunno and Klemp applied the circulation equation, Eq. (8.24), to the gust-front region to investigate further the process involved in the generation of vorticity or circulation in that region. That analysis showed the importance of the cold air in generating the low-level circulation. It also showed that the circulation produced along the gust front is a consequence of the vorticity production in all the air parcels that move along the front. This includes the mixing of air parcels possessing ambient vorticity and air parcels in which vorticity is produced baroclinically.

An obvious question concerns the relationship between the mid-level rotation and the low-level rotation of the storm. Rotunno and Klemp argue that the primary importance of the mid-level rotation is that it affects the transport of potentially cold air to the forward and left flanks of the storm. Upon being evaporatively-chilled by rain, it descends and forms a cold-pool on the left and forward flank. This location is the right place for baroclinically-produced horizontal vorticity of the proper sign to become tilted into positive vertical vorticity upon encountering updrafts.

A weakness in Klemp and Rotunno's argument is that little tilting of the baroclinically- produced horizontal vorticity can occur near the ground where vertical motion is nearly zero. If a tornado derives its vorticity from near-surface tilting of baroclinically-produced vorticity, how then can a tornado's vortex lines remain essentially vertical down close to the ground, turning horizontal in the friction layer? A number of researchers have shown that a downdraft is needed to produce strong rotation near the ground (Davies-Jones, 1982) (Davies-Jones and Brooks, 1993; Walko, 1993; Wicker and Wilhelmson, 1995; Grasso, 1996; Finley, 1997; Finley et al., 2001; Gaudet and Cotton, 2006). Thus in order for Klemp and Rotunno's concept to work, the baroclinically-produced low-level vortex must build downward by a pressure-deficit tube. Wicker and Wilhelmson's (1995) 3-D simulation of the genesis of two tornadoes in a supercell storm was interpreted as a low-level mesocyclone that built downwards to the surface by what we interpret as a pressure-deficit tube as they did not actually call the mechanism that. They concluded that the mesocyclone formed as in the Klemp-Rotunno conceptual model by tilting of

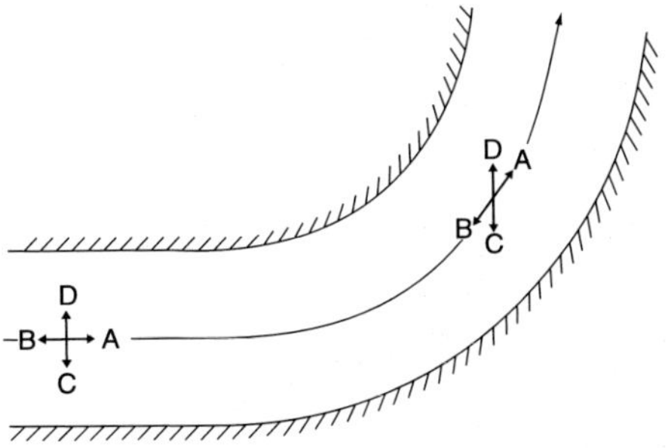


**FIGURE 8.42 Diagram of flow around a river bend, demonstrating the development of streamwise vorticity.** Upstream of the bend the flow is parallel with speed shear (crosswise vorticity) owing to friction at the river bottom. Consider the fluid cross ABCD with arm AB along a streamline and CD along a vortex line. Since flow around the bend generates no vertical vorticity to a first approximation, the arms of the cross must rotate in opposite directions. Thus the vortex line CD turns toward the streamwise direction AB. *(From Davies-Jones et al. (2001))*

baroclinically-produced vorticity along the flank of the FFD gust front. Even so, Wicker and Wilhelson did note that their simulations formed a variety of surface-based vortices which formed weaker tornadoes that they referred to being akin to 'gustnadoes'. It is not clear from their analysis if surface-based vortices preceded the surface penetration of the mesocyclone at which point tornado genesis occurred. In our opinion, it is likely that, in general, the low-level mesocyclone will have to couple with a pre-existing surface-based vortex to reach tornado intensity. Thus we suspect in general, as the mid-level mesocyclone and the low-level mesocyclone are generated by separate mechanisms, so also is the formation of the surface-based tornadic circulation different from the low-level mesocyclone. The key to forming intense tornadic vortices is the coupling between incipient surface-based tornado vortices, low-level mesocyclones, and mid-level cyclones. Once that coupling is established, an intense surface-based tornado is spawned.

A more likely scenario of tornado genesis, is that the surface-based tornadic intensity vortex is either formed by descending air in the RFD or that it forms by some mechanism of shearing instability along the gust front of the RFD similar perhaps to NSTs. Davies-Jones et al. (2001) provide a conceptual model of how near-surface vorticity develops in association with a downdraft. They argue that the mesocyclone aloft draws a thin curtain of rain around the rear side of the mesocyclone where it falls into dry air that is overtaking the storm and is diverted around the updraft. As the dry air enters and flows through the precipitation region, it is cooled and moistened by evaporating rain. The cooled, moistened air becomes negatively-buoyant and descends to near the surface. It is then argued that if the flow is cyclonically-curved rather than being straight, cross-wise vorticity will be converted into streamwise-vorticity by the river bend effect as shown in Fig. 8.42. They note that horizontal vorticity is generated baroclinically along the sides of the downdraft, where the vortex

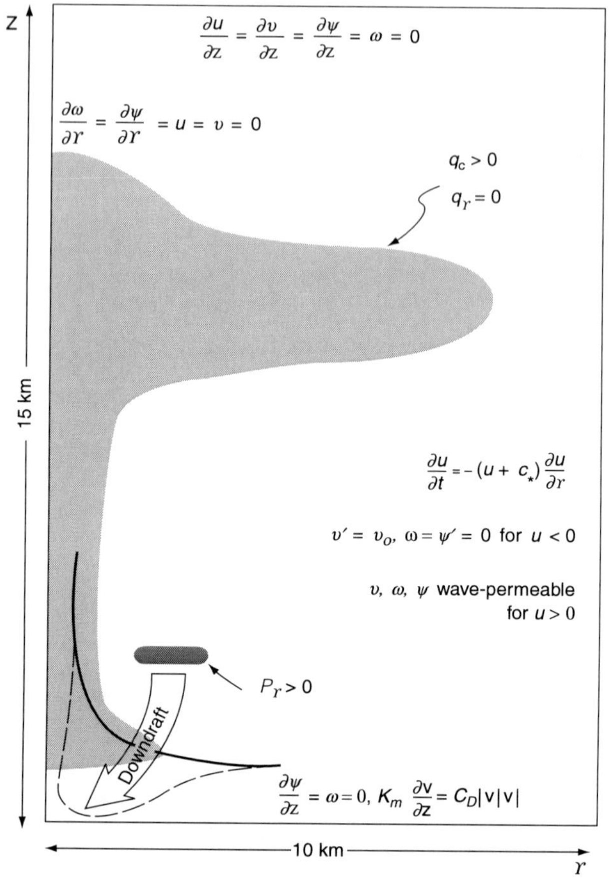


**FIGURE 8.43 Idealized axisymmetric simulations design. A moist, rotating updraft (rotation is imposed) free of rainwater is generated along the axis (light gray shading denotes the cloud water field).** Once an approximately steady state is achieved, rainwater is introduced aloft on the periphery of the updraft within the dark gray region (rainwater is imposed by way of a production term $P$). The resulting precipitation-driven downdraft transports angular momentum toward the ground. A schematic vortex line is shown before (black solid line) and after (black dashed line) for the formation of the downdraft. The boundary conditions are indicated at their respective locations, with $\psi$ indicating a scalar. All other variables are defined as in the appendix. As defined by Klemp and Wilhelmson (1978), $c_*$ is a assumed intrinsic phase speed (40 m $s^{-1}$ assumed) of the dominant gravity wave modes moving out through the lateral boundary. *(From Markowski et al. (2003))*

lines are tilted upward, producing a cyclonic/anticyclonic vortex pair on the sides of the downdraft. The main mesocyclone updraft aloft draws upward the almost saturated, rain-chilled cyclonic vortex into the storm, stretching it vertically and thereby intensifying the surface-based vortex into a mesocyclone. An axisymmetric representation of the final stages of tornado initiation by this concept can be seen in Fig. 8.43. This illustrates an initial horizontally-oriented vortex line that becomes tilted into the vertical by a downdraft.

Observational studies have shown that it is difficult to distinguish between low-level mesocyclones that produce strong tornadoes and those that do not (Blanchard and Straka, 1998; Trapp, 1999; Wakimoto and Cai, 2000). Some of the factors that favor tornado genesis that have been proposed: (1) the intersection of a mesoscale boundary with a supercell circulation (Grasso, 1996; Finley, 1997; Markowski et al., 1998; Wakimoto and Liu, 1998; Atkins et al., 1999; Rasmussen et al., 2000; Glimore and Wicker, 2002); (2) the presence of streamwise vorticity near the surface (Markowski et al., 2000), large values of shear/buoyancy parameters (Rasmussen and Blanchard, 1998); (3) the presence of low LFCs (Rasmussen and Blanchard, 1998) and/or relatively warm and unstable RFDs (Markowski et al., 2002) and (4) the breakdown of the low-level mesocyclone into secondary vortices (Wakimoto and Liu, 1998). Because tornadoes have the size of the order of 100 m and vorticity of 1 $s^{-1}$, while low level mesocyclones have the size of the order of 10 km and vorticity of 0.1 $s^{-1}$, some mechanism(s) of vorticity-enhancement by convergence is necessary for tornado genesis.

Gaudet and Cotton (2006) illustrate by three dimensional numerical simulation the amplification of vertical vorticity by collision of outflow boundaries. In their simulation, low-level vorticity first becomes maximized in a line along the RFD gust front, near the intersection of the warm inflow sector, the FFD gust front, and the RFD gust front, similar to the locations of low level mesocyclones in the literature (see Fig. 8.44). The vorticity line then begins to bow (see Fig. 8.45) and finally becomes concentrated in discrete centers that are connected by spiraling filaments (see Fig. 8.46). At the time that vorticity concentrated into a compact center, a sudden decrease in low-level pressure occurred and the vortex deepened and intensified to tornadic intensity. Eventually the vortex migrated away from the gust front and over the cold-pool. Gaudet et al. (2006) subsequently describe the mechanisms of transformation of the vortex line into a locally concentrated vortex using simple two-dimensional analytical models such as a Burgers vortex layer (Burgers, 1948).

The tornado simulated by Gaudet and Cotton (2006) was relatively weak with maximum surface winds of 50 m $s^{-1}$. They attributed this to the relatively coarse grid spacing (100 m) and the fact that mesoscale variability such as found by Markowski et al. (1998) and Rasmussen et al. (2000) was absent in the simulations. Grasso (1996), for example, simulated the formation of two different tornadic storms that formed along a dry-line. Using six levels of interactive nested grids with the finest grid having 111m, Grasso simulated the formation of a tornado using 26 April 1991 input data. The simulated tornado reached a maximum horizontal tangential wind speed of 102 m $s^{-1}$ and a pressure drop of 95 mb within the core. The intensification of the surface-based vortex followed a sequence similar to that simulated by Gaudet and Cotton (2006) but the tornado built out of a mesoscale field of vorticity associated with shear along a dry-line boundary. The ingestion of this pre-existing mesoscale vorticity was a contributing factor to such an intense vortex.

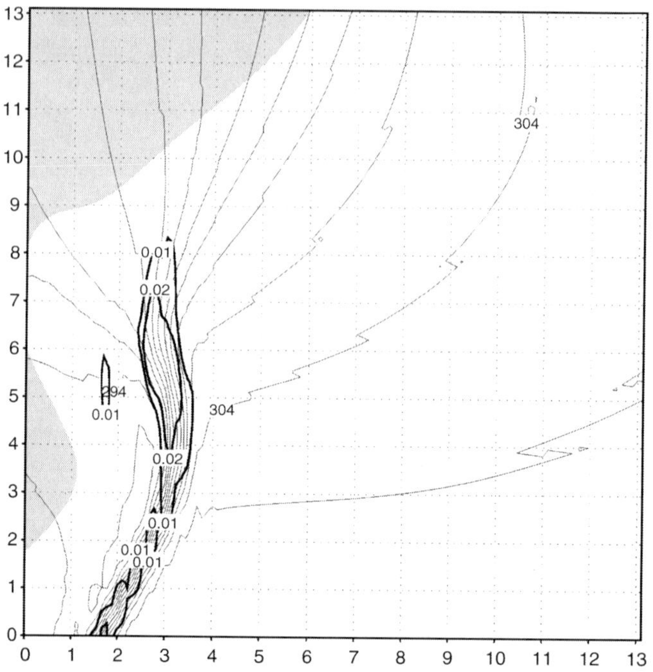


**FIGURE 8.44 Potential temperature and vertical velocity for grid 2 at 3300 s and 19 m above the surface.** Potential temperature is represented by thin contours in 1-K increments; vertical vorticity is represented by thick contours at 0.01, 0.02, and every 0.04 $s^{-1}$ thereafter. Axes labels represent distance in km. Shading represents the rain mixing ratio > 2 g $kg^{-1}$. The maximum vertical vorticity is 0.046 $s^{-1}$. *(From Gaudet and Cotton (2006))*

There is observational evidence that the air within RFDs of tornadic supercell storms is more buoyant and potentially buoyant (higher CAPE) than in non-tornadic supercell storms (Fujita et al., 1977; Brown and Knupp, 1980; Rasmussen and Straka, 1996; Markowski et al., 2002; Grzych et al., 2007). Markowski et al. (2002) also found that there was a high correlation between the coldness of the downdraft and the ambient (inflow) relative humidity. More buoyant low-level downdrafts were found in moist level environments than in dry. This is also consistent with low LCLs in TC tornado environments as noted earlier. Markowski et al. (2003) performed simulations examining the thermodynamic characteristics of downdrafts that were favorable for tornado genesis in an axisymmetric model. They found that for imposed downdrafts with low relative humidities and rain curtains having relatively low rainfall rates, the resultant downdrafts were warmer than when the relative humidity was small and the precipitation rates in the rain curtain large. When the downdrafts were relatively warm and moist, they are better able to recycle downdraft air into vigorous updrafts which can converge the low-level vorticity into tornado level intensity. As previously shown in Fig. 8.43, a downdraft that is warmer is more

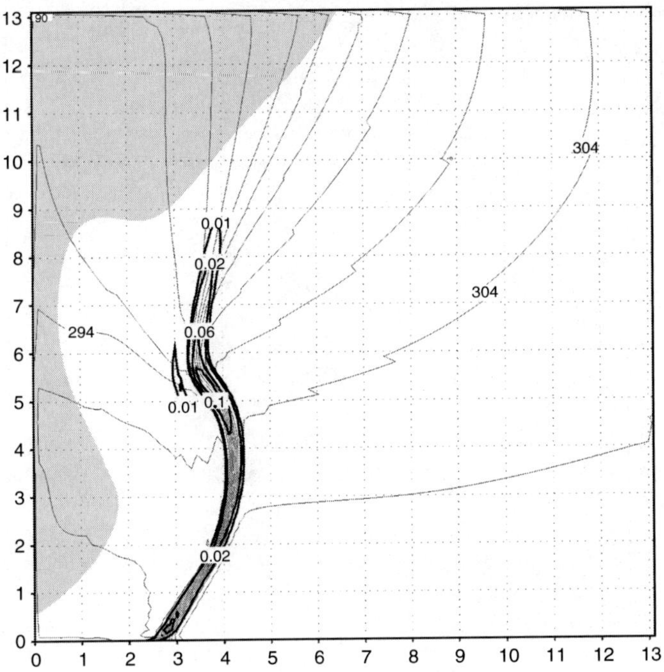


**FIGURE 8.45** **Same as Fig. 8.44, but at 3360 s. The maximum vertical vorticity is 0.12 $s^{-1}$.** *(From Gaudet and Cotton (2006))*

likely to tilt an initially horizontally-oriented vortex line into the vertical by a downdraft and converge the vorticity in the ascending former downdraft into tornado levels of intensity.

As well as the relative humidity at downdraft source levels and the magnitude of precipitation rates, van den Heever and Cotton (2004) and Gilmore et al. (2004) have shown that when hailstones and raindrops are relatively large, the evaporation rates and melting rates in downdrafts are less. This is a result of the fact that for a given hydrometeor water content, owing to the larger surface to volume ratio for small hailstones/raindrops relative to larger hydrometeors, and that both melting and evaporation occur primarily on the surface of a raindrop (hailstone), greater cooling will occur in downdrafts composed of small raindrops and hailstones than larger particles. Thus Snook and Xue (2008) found in their three dimensional simulations of tornado genesis, that RFD cold-pools were warmer and exhibited higher CAPE values when the simulated storm rainshafts were composed of larger-sized hailstones and raindrops. When the simulated cold-pools were warmer, the gust fronts moved more slowly which permitted vortices formed along the gust front to remain aligned with the mesocyclone aloft. On the other hand, colder cold-pools produced gust fronts that raced ahead of the mesocyclone aloft resulting in gentler, slanted

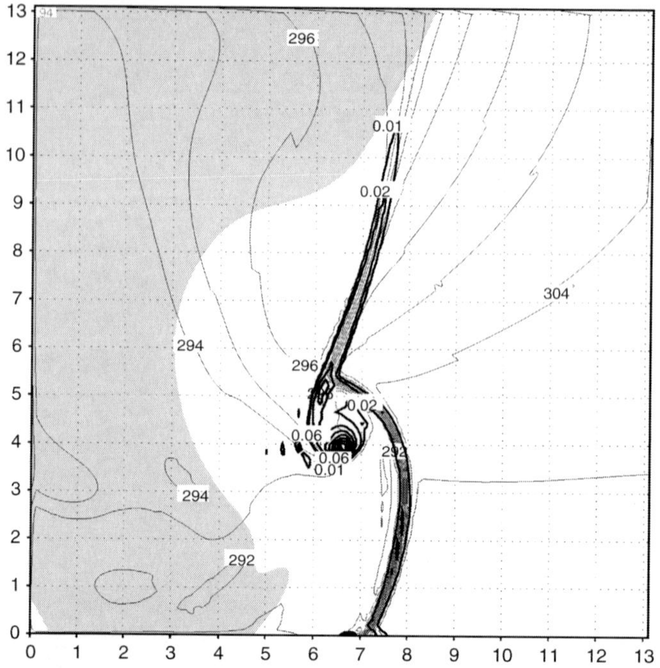


**FIGURE 8.46** **Same as Fig. 8.44, but at 3600 s. The maximum vertical vorticity is 0.32 $s^{-1}$.** *(From Gaudet and Cotton (2006))*

updrafts which did not favor a coupling between vortices formed along the gust front and a higher-level mesocyclone. The analysis of three dimensional simulations of supercell storms with varying CCN concentrations by Lerach et al. (2008) suggests that higher concentrations of CCN retard the warm-cloud collision-coalescence process which results in greater amounts of supercooled water being transported aloft. This results in reduced rainfall at lower levels and, as a consequence, reduced evaporation rates, which favors warmer cold-pools. We also expect that the greater amounts of supercooled water aloft would favor the growth of larger hailstones. The melting and evaporation of these larger hailstones favor the formation of warmer cold-pools than if more numerous smaller hailstones were present. However, this response is secondary to the change in low-level rainfall rates in response to CCN concentration variations for the one case simulated by Lerach et al. (2008). This results in thermodynamic properties of the cold-pool being more favorable for tornado genesis in polluted air, everything else being the same. Lerach et al's results suggest that the CCN ingested into a storm might be a contributing factor in determining if a supercell storm generates tornadoes or not.

Finally, the interpretation of observations and modeling studies has focused on the properties of the RFD as being controlled by air originating at mid-levels and descending while precipitation is evaporating and melting. As noted by Markowski et al. (2002) the thermodynamic properties of near-surface air having an origin at mid-levels is controlled in part by the amount of entrainment of dry (low-$\theta_e$) air that occurs aloft. Trajectory analysis of the source of surface air in the RFD region near the gust-front in three-dimensional simulations of tornadic supercells by Finley (1997) suggests that a source of the downdraft air was the under appreciated "up-down" downdraft component. Air in this downdraft component would be expected to be potentially warmer than mid-level downdraft source air as it starts in ascending warm, high CAPE air and remains in the interior of the storm where it is less likely to be exposed to entrained dry air. Thus this downdraft component, which is driven by water loading, melting of hail, and downward-directed pressure gradients, is more likely to contribute to cold-pool air that is more buoyant and potentially buoyant. More research is needed to determine if this downdraft component is a major contributor to the characteristics of RFD cold-pool characteristics in tornadic supercell storms.

In summary, it seems like every observation and modeling study of supercell tornado genesis comes up with a new or slightly different perspective. Perhaps it is naíve to expect that any one mechanism can prevail in tornado genesis in supercells. In a sense it is like trying to describe or predict a single boundary layer eddy in a convective boundary layer. We speculate that a strong supercell storm is like a giant vacuum cleaner of surface vortices. These surface-based-vortices can form by a variety of mechanisms discussed above. Like NSTs, these surface-based vortices can merge and coalesce to form more intense, deeper surface-based vortices. Wicker and Wilhelmson's (1995) simulation of the genesis of supercell tornadoes exhibited the formation of surface-based vortices which merged or coalesced into intense surface-based vortices. It was not clear from their analysis, however, if the merged surface-based vortices were important to the genesis of the two simulated intense tornadoes.

If RFD gust fronts are relatively warm and buoyant, the gust fronts move more slowly in a storm-relative sense, increasing the likelihood that an intensified and deepened surface-based vortex ahead of the gust front will become coupled to a low-level mesocyclone and the parent mesocyclone aloft much like NSTs. Once a surface-based vortex becomes coupled to the mesocyclone aloft, stretching and convergence of the surface-based vortex by the strong mesocyclone updrafts favors the intensification of the vortex into a tornado. Perhaps, as suggested by Greg Tripoli (personal communication), the main role of the low-level mesocyclone is not convergence and stretching of a surface-based vortex, but convergence of surface-based vortices which favors their merging and coalescence to form an intense tornadic vortex. From this perspective the tornado genesis process is quite stochastic with many processes contributing to the formation of surface-based vortices. The strong dynamics of the supercell including its downdrafts and gust fronts, develops strong low-level shears which favors generation of surface vortices by shearing instabilities and vortex breakdown. But, it is the overall intensity of the supercell mesocyclone and its associated updrafts, and the strength of the underlying cold-pool that determines the probability that surface-based vortices (initiated by a variety of mechanisms) will be intensified to tornadic strength.

### 8.9.5. Tornado Features

As can be seen from the preceding section, there is considerable experimental as well as numerical modeling evidence that tornadoes form in a highly baroclinic environment of mesocyclones. The extent to which this baroclinicity is present in or affects the tornado circulation itself remains under debate. There is some evidence from the analysis of debris and cloud-tag trajectories (Golden and Purcell, 1978) that the circulation in tornadoes is strongly asymmetric in both the rotational and vertical wind components. Most theoretical, numerical, and laboratory models of tornadoes do not include the effects of this baroclinicity nor asymmetric characteristics of the tornado circulation. It remains to be seen how well the axisymmetric concepts of tornado rotation relate to the asymmetric, baroclinic concepts of the parent rotating thunderstorm. Clearly the biggest gap in our understanding of tornado formation occurs on scales between the low-level mesocyclone and the tornado vortex. The scale of a tornado is normally only a few hundred meters and rarely more than 1 km. The important scales of motion within the tornado itself are of the order of tens of meters. Thus, to model or observe a tornado, scales of the order of tens of meters must be resolved.

We will briefly review some of the theories and concepts of the dynamics of tornado vorticities in this section. For more complete reviews the reader is referred to Morton (1966), Davies-Jones and Kessler (1974), Davies-Jones (1982), Lewellen (1976), Snow (1982), Lewellen (1993), and Davies-Jones et al. (2001).

Current concepts governing the dynamics of tornadoes have been derived from a few photogrammetric studies of tornado motions (Fujita, 1960; Hoecker, 1960a,b; Golden and Purcell, 1978), some axisymmetric, numerical model simulations and analytic models, axisymmetric, barotropic laboratory vortex experiments and LES models of tornadoes (Lewellen and Lewellen, 2007a,b; Lewellen et al., 2000). To illustrate the characteristics of the flow field in steady, fully developed tornadoes, we shall partition the flow into five regions, following Morton (1970), Lewellen (1976), and Snow (1982) as shown in Fig. 8.47. Some of the characteristics of these regions described by Snow (1982) are given below.

#### 8.9.5.1. *Region I: The Outer Flow*

The flow in region I responds to the concentrated vorticity in the core (region II) and to the positive buoyancy and vertical pressure gradients associated with the cloud aloft (region V). The flow in this region is expected to approximately conserve its angular momentum. The flow therefore spins faster as it approaches the central core.

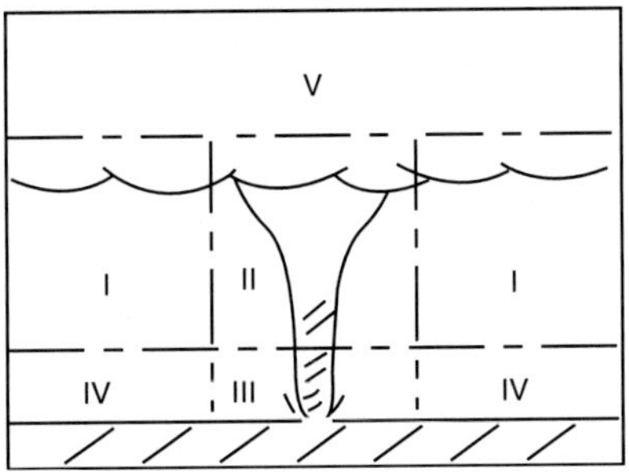


**FIGURE 8.47 Sketch of an idealized tornado vortex showing the five regions of the flow discussed in the text: region I, outer flow; region II, core; region III, corner; region IV, inflow layer; and region *V*, convective plume.** *(From Snow (1982). Copyright by the American Geophysical Union)*

### *8.9.5.2. Region II: The Core*

The core region surrounds the central axis and extends outward to the radius of maximum tangential winds. It often contains a visible condensation funnel which extends from cloud base. Dust and debris raised from the surface may further outline the region. Owing to the high rotation speeds in this region, the flow is in approximate cyclostrophic balance. Cyclostrophic balance and a radial increase of angular momentum suppress radial motions or entrainment. Centrifugal effects also contribute to a lowering of the central pressure below that occurring at the same level in the outer-flow region. These conditions have led to the application of a variety of simplified flow models in this region (Lewellen, 1976).

### *8.9.5.3. Region III: The Corner*

This region represents that part of the boundary layer where the flow changes from primarily horizontal flow (region IV) to upward motion into the core. Wilson and Rotunno (1982) suggest that the outer boundary of this region should be identified with the radius of maximum tangential winds. Vertical pressure-gradient forces are responsible for deflecting the flow from the horizontal into the vertical. This is the region where the maximum velocities occur near the top of the ground boundary layer ($\sim$100 m) and can be significantly higher than predicted by cyclostrophic balance (Lewellen, 1976, 1993). Turbulence in this region is critical in determining both the maximum velocity and the detailed structure of the tornado.

### *8.9.5.4. Region IV: Boundary Layer Flow*

As the low-level inflow air interacts with the earth's surface, a turbulent boundary layer is created of the order of tens of meters in depth, and occasionally, in the largest tornadoes, a few hundred meters. The dynamics of

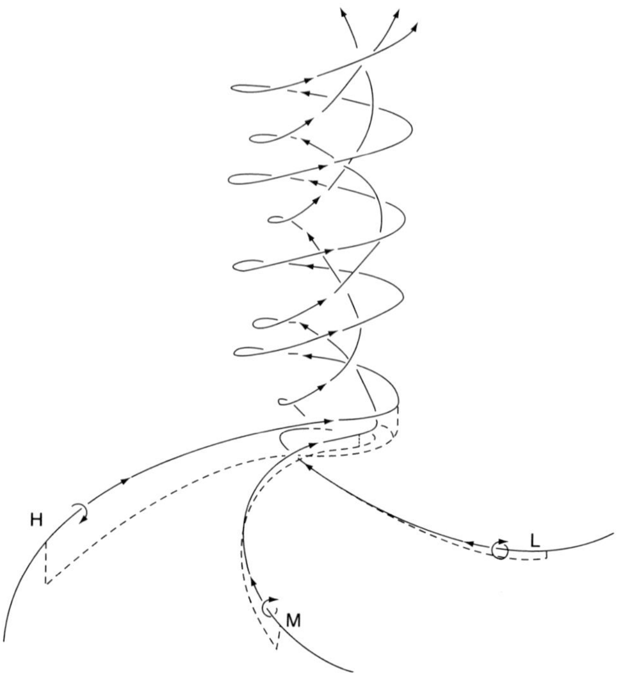


**FIGURE 8.48 Schematic interpreting the near-surface vorticity distribution.** Initially, all vortex lines spiral in the same manner. As the radial flow accelerates, the vortex lines at the top of the boundary layer (H) are turned to spiral opposite to those at lower levels. As the radius decreases, the inflection point (change in direction of spiraling) moves inward and downward, affecting next those at mid-levels (M). Lines at the very bottom (L) are not modified. The figure is based on work by Rotunno (1980). *(From Snow (1982). Copyright by the American Geophysical Union)*

this layer has been discussed by Rotunno (1980), Baker (1981), and Wilson (1981). These studies suggest that the cyclostrophic balance that is characteristic of region II is disrupted by frictional effects. As a result the flow is retarded and a net inward-directed force is present. This results in an inward acceleration of the flow toward the core, with the rate of inflow being limited by the eddy stresses near the surface and inertial forces. Figure 8.48 is a schematic illustration of several vortex lines at different levels in the boundary layer synthesized by Snow (1982). The schematic also includes an illustration of the corner region.

### *8.9.5.5. Region V: The Rotating Updraft*

This region is the storm-scale rotating updraft region discussed in Section 8.9.3. It is also the region where vertically-oriented vortex lines in the tornado spread laterally outward, eventually becoming horizontal. As mentioned earlier, many unanswered questions remain with regard to the interactions between this region and the tornado-scale regions above.

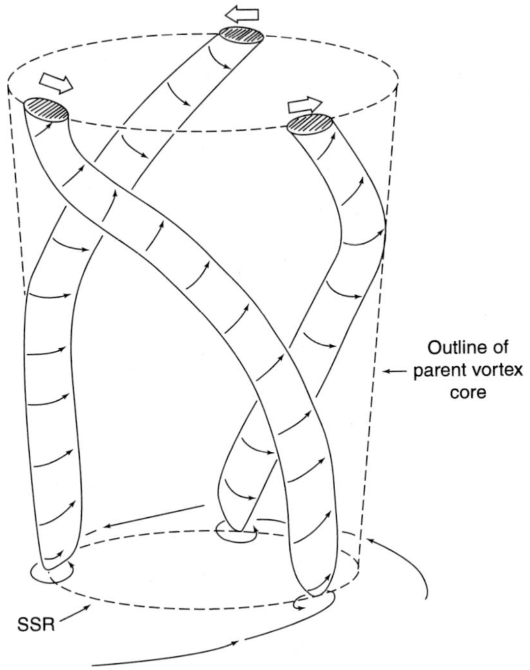


**FIGURE 8.49** **Sketch of a vortex containing three subsidiary vortices.** These are in rapid rotation about their spiral axes, while moving about the central core. *(From Snow (1982). Copyright by the American Geophysical Union)*

## 8.9.6. Secondary Vortices

Tornadoes do not necessarily consist of a single vortex. Fujita (1970) and Agee et al. (1977) describe secondary vortices or suction spots existing within the main tornado circulation. Fujita (1970) has noted that much of the damage associated with tornadoes occurs within cycloidal streaks within the main track. Photographs of tornadoes reported by Agee et al. (1977) show multiple vortices that apparently rotate around the tornado axis and have diameters considerably less than the parent tornado. Investigations into the structure and causes of multiple vortices have involved laboratory experiments (Ward, 1972; Church et al., 1979; Church and Snow, 1993), analytic theory and numerical simulations (Snow, 1978; Gall and Staley, 1981; Gall, 1983; Rotunno, 1977, 1981, 1984; Rotunno and Lilly, 1981; Walko and Gall, 1984; Finley, 1997). Figure 8.49 is a schematic drawing of three secondary vortices by Snow (1982).

The single most important parameter describing the potential for a tornadic or any intense vortex to break down into a secondary vortex is the swirl ratio

$$S = \frac{v_0}{\bar{w}} \tag{8.28}$$

where $v_0$ represents the vertically-averaged tangential velocity and $\bar{w}$ is the

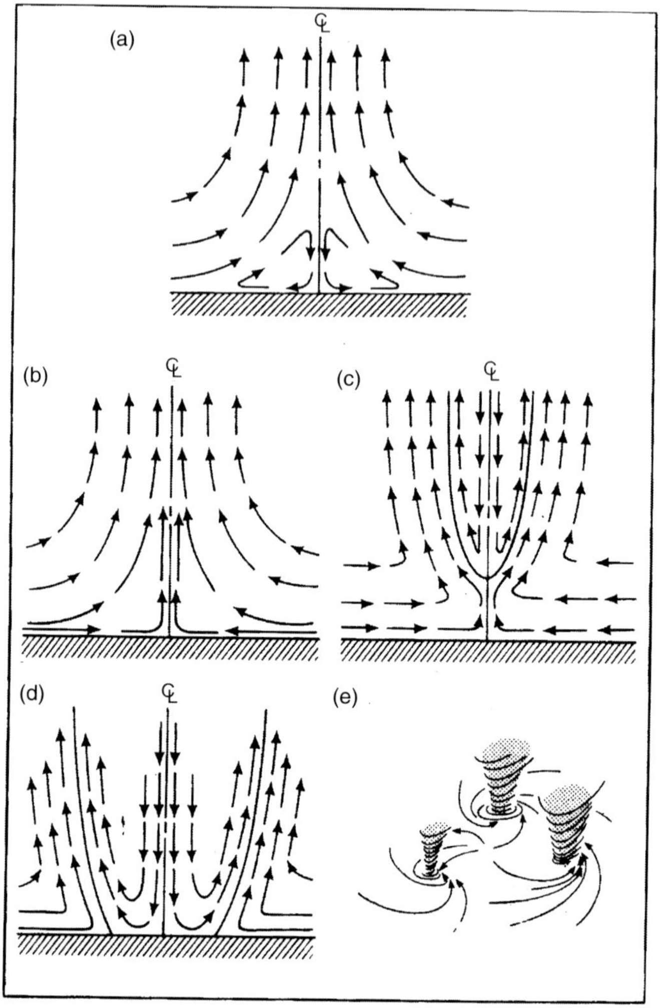


**FIGURE 8.50** **Effect of increasing swirl ratio on vortex flow. (a) Weak swirl-flow in the boundary layer separates and passes around the corner region; (b) one-cell vortex; (c) vortex breakdown; (d) two-cell vortex with downdraft penetrating to the surface; (e) multiple vortices.** *(From Davies-Jones (1983))*

average vertical velocity inside the updraft radius $r_0$. Figure 8.50, illustrates the breakdown of a single primary tornado vortex into secondary vortices. For small values of $S$ ($S$ less than or equal to 0.1) a ring-like zone of separated flow develops along the lower boundary in the inflow region. This prevents low-level angular momentum from reaching the center of the vortex and no concentrated vortex forms at the surface (Fig. 8.50a). As S increases slowly the vortex develops aloft and builds downward. When $S = 0.1$, as shown in

Fig. 8.50b, the vortex is a one-celled vortex with ascending motion everywhere within the core radius and maximum upward velocity at its center. As $S$ is increased further the core structure undergoes a dramatic change called vortex breakdown. The core radius abruptly increases as upward air diverges and flows around the breakdown bubble. As $S$ increases further the vortex breakdown penetrates to the surface, followed by radial expansion of the core, and penetration of the central downdraft to the surface. At this point the vortex has a two-celled structure in which the central downdraft is surrounded by an annular updraft region through the depth of the core (Fig. 8.50d). For large values of $S$ ($S = 2 - 2.5$), the core expands until it fills the updraft hole. At this stage the updraft is confined to a narrow annulus, and the downdraft becomes stronger and larger until it occupies much of the convective region. After breakdown occurs secondary vortices form which circle the center of the parent vortex. These secondary vortices are observed to form in the laboratory around $S = 0.5$.

To her surprise Finley (1997) found that in a rather crude simulation (111 m grid spacing) of a long-lasting tornado (over 50 minutes), six secondary vortices formed during a ten minute period toward the end of the simulation. The following is a summary from her results:

- The onset of secondary vortex formation occurred shortly after the parent vortex developed a two-celled structure throughout the boundary layer.
- The secondary vortices developed where the tangential wind, vertical vorticity, vertical velocity were generally the largest in the parent vortex.
- The secondary vortices wrapped around the parent vortex cyclonically with height.
- The calculated swirl ratio was 1.1-2.3 which is in the range for secondary vortices observed in laboratory and field studies.
- The region in the parent vortex where the secondary vortices developed coincided with the position of downdrafts that develop on the outside of the parent vortex and wrap around the southern and eastern side of the vortex with time.
- It was proposed that the downdrafts enhance vertical and tangential velocities within the vortex thereby destabilizing the flow in those regions.

It appears that the downdrafts served as a trigger for the onset of secondary vortex formation. Much higher resolution simulations having grid spacing of a few meters are needed to further explore the factors leading to secondary vortex formation in tornadoes.

## 8.10. HAILSTORMS

The formation of large hail is a result of a broad range of interacting scales of motion and physical processes. However, the size of the hailstones is also strongly dependent on the strength of updrafts and the microstructure of the storm system. This includes the source, location, and size of the ice particles (hail embryos) suitable for initiation of the rapid stages of hail particle growth; the time needed to grow large hailstones relative to the time constraints provided by the storm system; the nature of hail particle growth in the optimum temperature, liquid-water content, and updraft velocity regions; and the rapidity of hailstone melting processes. Danielsen (1977) also argued that the concentrations of cloud condensation nuclei and ice nuclei and, perhaps, the aerosol size distribution, are important to the production of hail embryos. In this section we will not concentrate on the detailed physics of hailstone growth processes. For reviews of the properties and growth of hailstones, the reader is referred to Mason (1971), Pruppacher and Klett (1980), and List (1982).

### 8.10.1. Synoptic and Mesoscale Conditions Suitable for the Formation of Hailstones

Forecasting hailstorms is similar to the forecasting of the occurrence of severe convective storms. Beyond that, it is difficult to distinguish between a hail-producing storm and a storm that produces severe winds and tornadoes. This is not too surprising, because often a severe convective storm that produces large hail also produces tornadoes and strong winds. In general, flash-flood storms prevail in conditions of low wind and low wind shear, whereas most hailstorms prevail in stronger wind shear environments. The Cheyenne, Wyoming, 1985 flash flood and hailstorm, however, represents an exception.

One of the most important factors affecting the formation of hailstorms is the thermodynamic instability of the atmosphere. The more unstable the atmosphere, the more likely that thunderstorms will form with updraft strengths capable of supporting large hailstones. Most of the hail forecast techniques and indices use this concept as the basis of the forecast scheme (Humphreys, 1928; Showalter, 1953; Fawbush and Miller, 1953; Foster and Bates, 1956; Galway, 1956; Boyden, 1963; Miller, 1972; Zverev, 1972; Haagenson and Danielsen, 1972; Maxwell, 1974). The most sophisticated of the thermodynamic stability forecast schemes use simple one-dimensional cloud models to predict maximum updraft velocity and cloud temperature (Fawbush and Miller, 1953; Miller, 1972; Maxwell, 1974; Zverev, 1972; Foster and Bates, 1956; Haagenson and Danielsen, 1972). Danielsen (1977) used such a model to predict hail growth. Renick and Maxwell (1977) summarized the hail forecast scheme using the nomogram applied to Alberta, Canada, thunderstorms shown in Fig. 8.51. Due to deficiencies in the one-dimensional model, such as the neglect of vertical pressure-gradient forces, the peak updraft velocity predicted by the model was adjusted to a lower value. Using observations as a basis, English (1973) adjusted the height of maximum updraft velocity to cloud-base height plus 0.61 times cloud-top height. With these adjustments, the nomogram is then used to predict the likely size of hailstones that will form. For example, if the model-predicted maximum $W$ is between 35 and 44 m s$^{-1}$ and the temperature (at the adjusted

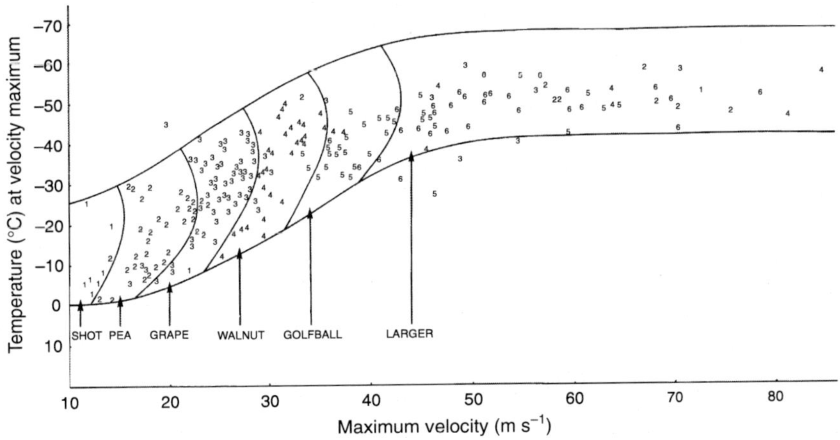


**FIGURE 8.51 Results and proposed nomogram for forecasting maximum hail size from model diagnostic reruns of 1969-1973 hail days.** Hail size categories are (1) shot, 1-3 mm; (2) pea, 4-12 mm; (3) grape, 13-20 mm; (4) walnut, 21-30 mm; (5) golfball, 33-52 mm; (6) larger than golfball. *(From Renick and Maxwell (1977))*

height) is between about −32 and −38 °C, then hail between 33 and 52 mm is forecast.

Hailstorms are frequently associated with an upper-level jet stream and strong wind shear (Ludlam, 1963; Schleusener, 1962; Modahl, 1969; Das, 1962; Frisby, 1962; Fawbush and Miller, 1953). Often the wind veers with height from a low-level southeasterly jet to a mid-level to upper-level westerly jet in the northern hemisphere. Wind shear aids the formation of a sustained updraft/downdraft couplet and a long-lived self-propagating storm system. However, damaging hail can also form in shorter lived, intense single cells that develop in relatively weakly sheared environments (Battan, 1964; Battan and Theiss, 1973; Renick, 1971). Thus, strong wind shear is not a necessary condition for the formation of damaging hail.

The height of the melting level is also important in determining the amount, if any, of hail that will reach the surface. Foote (1984) illustrated the importance of melting by modifying the temperature and humidity profiles for a Colorado hailstorm to representative profiles of Alberta, Canada, and southern Arizona. For Colorado conditions, he calculated that 74% of the hail falling through the 0 °C level melts before reaching the surface. For Alberta the amount was 42%, while it was 90% for southern Arizona. This result is consistent with the observation that hailstorms are more frequent at higher latitudes. In fact, G. B. Foote (personal communication) has noted that any convective storm with a radar-echo top over 8 km above ground level in Alberta is likely to produce significant hailfall. This probably explains the usefulness of the height at which the wet-bulb temperature is 0 °C (often referred to as the wet-bulb zero height, or WBZ) as a predictor for hail formation (Miller, 1972). Large hail is generally associated with a WBZ between 2.1 and 2.7 km, while only small hailstones are typically found when the WBZ is below 1.5 km or above 3.4 km. Miller and McGinley (1977) argue that the WBZ represents the minimum depth that a hailstone will fall without melting appreciably, after allowing for evaporative cooling of the stone. One would, therefore, expect that there should not be a lower limit on the WBZ height. However, Morgan (1970) found the WBZ is correlated with environmental low-level moisture in Italy's Po Valley. Thus, if the WBZ is too close to the ground, environmental moisture is not great enough to support intense convection.

As will be discussed below, Danielsen (1977) argued that the distributions of cloud condensation nuclei and ice nuclei of an air mass are also important to the formation of hail, and that variations in these quantities contribute to the difficulty in forecasting hail. As noted in Chapter 4, cloud-base temperature also influences the microphysical evolution of the storm system. Thus, if the cloud-base temperature is relatively warm, warm-cloud collision and coalescence processes are more likely to prevail. A vigorous warm-cloud precipitation process below the freezing level is likely to deprive the hail generation levels of supercooled liquid-water as well as to introduce a large number of hail embryos that can compete for the available water at freezing temperatures. Thus, both the greater prominence of warm-rain processes in warm-based cumulonimbi as well as the greater depths available for melting of hailstones may contribute to infrequent observations of hailfalls in the tropics.

The climatology of the frequency of hailstorms suggests that thermally-driven mesoscale circulations, such as mountain ridge/valley circulations and sea-breeze and land-breeze circulations, aid in the formation of hail-stones. This is dramatically illustrated in the July climatology of hailfalls shown in Fig. 8.52. To the east of the central Rocky Mountains, three predominant regions of frequent hailfall can be seen over southern Wyoming, central Colorado, and northern New Mexico. Each of these hail/frequency maxima correspond to elevated ridges that extend into the High Plains. Also, the maxima near Rapid City, South Dakota, correspond to the location of the Black Hills, and the maxima near Cody, Wyoming, correspond to the Big Horn Mountains. Presumably ridge/valley circulations and flow about local hills establish low-level convergence zones that enhance the formation of the most intense, hail-producing storms. Likewise, the maxima along the shores of the Great Lakes may be associated with sea-breeze and land-breeze circulations in those regions.

It is clear from the foregoing discussion that a large range of scales of motion and a number of physical processes contribute to the formation of significant hailfalls. Using the 30 July 1979 hailstorm over Fort Collins, Colorado, as an example, Fritsch and Rodgers (1981) argued that successful prediction of hailstorms requires the quantitative assessment of a multitude of smaller scale or mesoscale meteorological processes. They argue that quantification of these processes will require advances in (1) observational capability and (2) increased computer power.

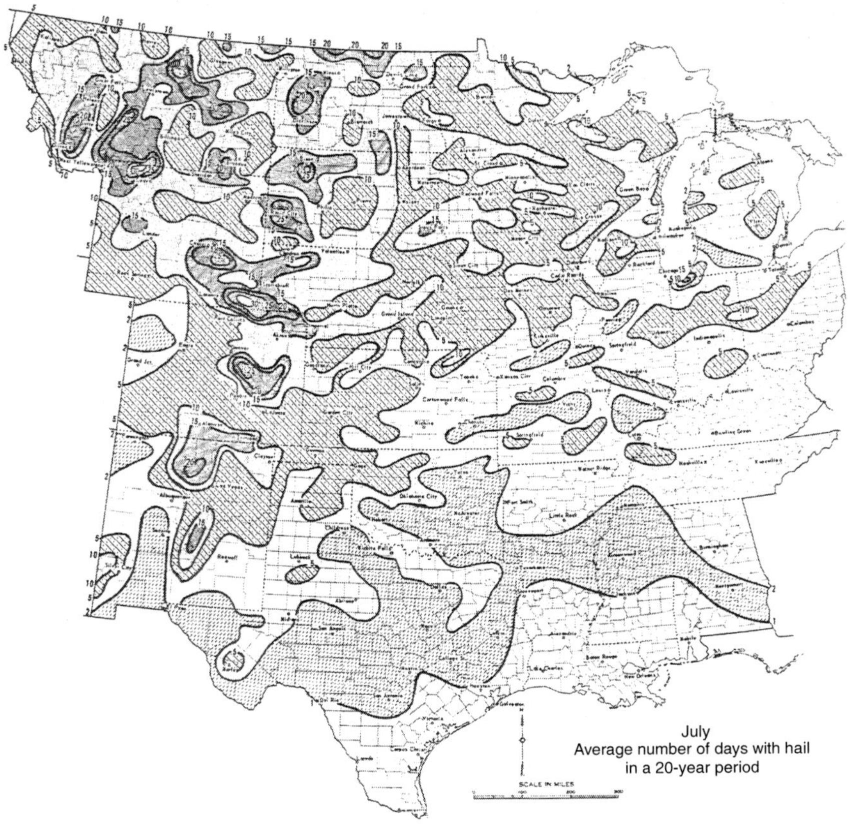


**FIGURE 8.52** **Average number of days with hail in July in a 20-yr period.** *(From Stout and Changnon (1968))*

## 8.11. MODELS OF HAILSTORMS AND HAIL FORMATION PROCESSES

In this section we will not attempt to review the extensive literature on conceptual models and concepts of hail growth. We will focus instead on a few key papers illustrating the differing concepts of hailstorm structure and hailstone growth. Nor will we concentrate on the detailed particle-growth physics such as the mathematical models of hailstone growth by collection or accretion, wet and dry growth, and melting processes. The reader is instead referred to the summary of microphysical processes in Chapter 4 as well as to the articles mentioned in the introduction to this section.

There is a great diversity of views, concepts, and models of the formation and growth of hail. Some of the diversity is due to the evolution in our ability to observe quantitatively convective storms as well as to model them numerically. Early concepts and models were essentially one-dimensional. As we obtained the ability to observe convective storms with multiple-Doppler radar and aircraft, and as our numerical models evolved from one dimension to two dimensions (see review in Orville, 1978) and three dimensions (Xu, 1983), the models of hail growth have evolved into complex interactions between particle-growth physics and a time-evolving three-dimensional flow structure of the storm system. Some of the diversity in concepts may result from geographical variations in the structure of hailstorm motion fields, water contents, and microstructure. Because only a few hailstorms have been simulated numerically in three dimensions, and because combined multiple-Doppler radar and aircraft observations of hailstorms have been limited to a few experimental cases and geographical areas, it is possible that other storm types and hail growth models may apply to storms that have not yet been observed with modern instrumentation or modeled on computers.

### 8.11.1. The Soviet Hail Model

The Soviet hail model is described by Sulakvelidze et al. (1967). It formed the basis of the Soviet hail-suppression techniques (Sulakvelidze et al., 1967) as well as being the model tested in the United States in the National Hail Research Project (Foote and Knight, 1979). Because this model was developed prior to modern multiple-Doppler radar and multidimensional numerical models, it is vague in its depiction of the flow structure of a hailstorm. In this model, the updraft of the storm must be evolving slowly in time and nearly erect in its vertical extent. Moreover, the updraft velocity should increase with height and exhibit a maximum ($W_{\mathrm{m}}$) somewhere in the warmer portion ($-5$ to $-15$ °C) of the supercooled liquid-water zone. Another aspect of the Soviet model is that a warm-rain, collision, and coalescence process takes place at levels below the level of maximum velocity. This allows the formation of large, supercooled raindrops whose terminal velocities are comparable to $W_{\mathrm{m}}$. Thus, the growing raindrops will ascend to near $W_{\mathrm{m}}$ where they will be trapped in what the Soviets call an "accumulation zone." In this region the liquid-water content of the raindrops will increase to greater than 10-15 g m$^{-3}$. For this large water content to remain suspended, the temperature excess ($\Delta T$) in the updraft must exceed 3-5 °C.

If some of the large supercooled drops reach a temperature range of $-15$ to $-22$ °C and freeze by immersion freezing, or if smaller ice particles form by some type of primary and secondary nucleation and are rapidly swept up by the supercooled raindrops, millimeter-sized hail embryos form in a water-rich environment. The Soviets calculate that hail will grow by accretion of cloud droplets from 0.1 to 1-3 cm in just 2-4 min as the hailstone settles from the accumulation zone to cloud base. This model is in accord with radar observations of a rapidly developing reflectivity maximum aloft, which then descends rapidly to the surface. A crucial aspect of the Soviet model is that the warm-rain process be present to feed the accumulation zone. If the warm-rain process takes place too quickly, so much water will be depleted in the warmer portions of the cloud that an insufficient amount will be available to accumulate in the supercooled zones, thus limiting hail growth. We suggest that this is one factor contributing to the low frequency of hail occurrence in the tropics. However, if a warm-rain process is not present, supercooled water cannot accumulate in a relatively narrow zone and the rapid generation of hail will not occur. If the Soviet model is valid, it should apply in clouds whose bases are moderately warm (10-15 °C), where the air mass exhibits moderate CCN concentrations (e.g. 300-600 $cm^{-3}$), and in multicellular storms whose peak updrafts are relatively low in height and low in magnitude ($W_m \sim$ 12-20 m $s^{-1}$).

In the Soviet hail model, supercooled raindrops serve as millimeter-sized embryos for the growth of hailstones. Many hailstorms exhibit steady updrafts of the magnitude of 25-50 m $s^{-1}$. With such strong updrafts, the time available for broadening the droplet spectrum to precipitation-size raindrops is extremely short. Danielsen et al. (1972) calculated that the initial droplet spectrum had to be quite broad before hail could form and grow entirely within such strong updrafts. Presumably, this motivated Danielsen's (1977) hypothesis that details of the air mass aerosol population are important to hail growth. Low concentrations of CCN or large concentrations of ultragiant aerosol particles can result in rapid initial broadening of the cloud droplet spectrum and enhance the formation of supercooled raindrops as hailstone embryos. In the case of hailstorms over the High Plains of the United States and Canada, there is considerable evidence that only 20% of the hailstone embryos are frozen raindrops (e.g. Knight et al., 1974). Many of those may actually be ice particles that have melted and have been swept up by the vigorous updrafts. Thus, for cold cloud-base, continental hailstorms, the millimeter-sized embryos for hailstone growth generally form by primary and/or secondary nucleation of ice crystals, followed by slow growth by vapor deposition, riming, and aggregation. The time required for the growth of millimeter-sized particles in 35-40 m $s^{-1}$ steady updrafts is longer than the time necessary to eject such particles in the anvil region of the cloud system. The hail growth process in such storms is thus viewed as a multiple-staged process in which hail embryos first form in relatively weak updrafts and then are transported into the more vigorous, water-rich updrafts. Such a multistaged process depends on the detailed temporal and spatial characteristics of the storm circulations. We will thus examine the hail growth processes in several storm types, namely, multi-celled storms, supercell storms, and a more hybrid storm system that Foote and Wade (1982) have labeled "organized" multicelled storms. We shall see that there exists a continuum of storm types, ranging from lesser organized multicell storms to the classical steady, dominant updraft/downdraft model of a supercell storm.

At one extreme of the continuum resides the classic multicell storm illustrated in Fig. 8.2. Individual cells in the storm system evolve through the ordinary storm life cycle of growth stage and mature stage, followed by decay stage. Using the "daughter" cell concept proposed by Browning (1977), millimeter-sized hail embryos first form during the cumulus stage in flanking cumulus towers. As the towers grow into the mature stage, these daughter cells become the "parent" cell of the storm complex. The hail embryos grow rapidly in the water-rich environment of the upper portions of the evolving cell and then descend through the storm system as the cell evolves into its dissipating stage.

### 8.11.1.1. *Ordinary Multicell Storms*

Battan's (1975) single-Doppler radar observations of a multicellular hail storm reveal a structure quite different from that of the Soviet model. Instead of a quasi-steady updraft and an accumulation zone, his observations reveal that the cloud system is composed of a series of turbulent thermals that are some 1-2 km in diameter. Figure 8.12 illustrates that some of the thermals exhibit updraft speeds greater than 24 m $s^{-1}$. Battan hypothesized that hail embryos form in the pulsating updrafts and are carried aloft while growing into hailstones. The larger hailstones may fall out of the thermal and settle through weak updrafts or downdrafts before reaching the ground. In some cases, the descending hailstones may encounter several vigorous updraft cells ascending and descending accordingly as they eventually find their way to the ground. Battan's concept of the hail growth process is certainly more complex than the Soviet model and, as we shall see, the models for hail growth in more organized multicellular storms and supercell storms.

### 8.11.1.2. *Organized Multicellular Storms*

To illustrate the concept of hail growth in an organized multicellular storm, we will use the 22 July 1976 case study observed during the National Hail Research Experiment (NHRE). This storm has been described extensively (Foote and Wade, 1982; Heymsfield and Musil, 1982; Foote and Frank, 1983; Jameson and Heymsfield, 1980; Heymsfield et al., 1980; Foote, 1984). According to Foote and Wade (1982) the early history of the storm system was characterized as an organized multicellular storm system in which new cells periodically formed on the right flank (relative to the storm's direction of motion) of the older parent cells. Following the merger of several weaker cells, an intense cell formed which exhibited a persistent intense updraft. During the latter portion of the storm's lifetime, this steady flow persisted, while a series of reflectivity and updraft perturbations could be seen super-imposed on the steady circulation. Figure 8.53 illustrates the persistent updraft/downdraft structure derived from multiple-Doppler radar data. Major updraft components are labeled A, B, F, and G. Updraft A is the dominant updraft feature; it originates at low levels in the south-southeast, rises abruptly as it is lifted over the gust front, then passes through cloud base, ascends through cloud levels, and turns to form the anvil streaming to the northeast. On the flanks of A the updraft streamlines labeled B tilt more to the back of the storm and enter the lower part of the anvil. Updraft streamlines F and G can be identified with the "up-down", downdraft circulation identified by Knupp (1985).

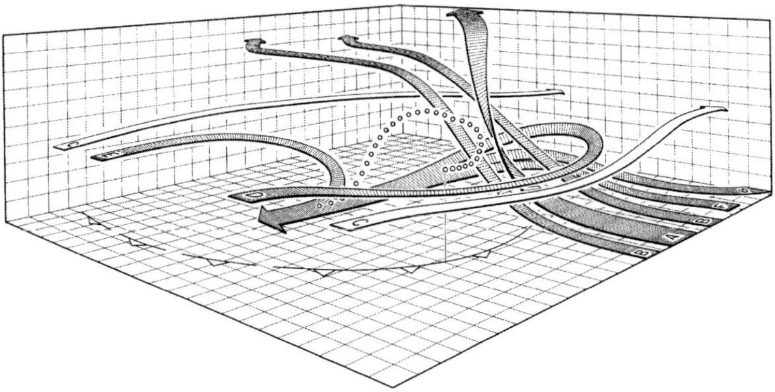

**FIGURE 8.53** **Major components of the airflow in the Westplains storm.** The strong updraft is depicted by the ribbon labeled A, which starts in the low levels to the south-southeast of the storm, rises sharply in the storm interior, and leaves the storm toward the northeast to form the anvil outflow. On the flanks of the strong updraft the air rises more slowly and penetrates farther to the rear of the storm before also turning to the northeast. In the middle levels there is a tendency for the westerly environmental flow to be diverted around the sides of the storm (streamlines labeled C), but some air also enters the storm (streamlines D and E) and contributes to the downdraft. A contribution to the downdraft flux is also made by air originally in the low levels to the southeast and east of the storm (streamlines F and G), which then rises several kilometers before turning downward in the vicinity of the echo core. The various streamlines are depicted relative to the storm, which is moving toward the south-southeast as shown, rather than relative to the ground. The small circles indicate the possible trajectory of a hailstone. *(From Foote and Frank (1983))*

At mid-levels, the flow is from the west. Streamlines labeled C encounter positive pressure anomalies on the upshear side of the updraft and are diverted around the storm. Streamline D is also diverted around the storm, but along the east side of the storm it encounters lower pressure, causing it to divert close to the updraft, where precipitation loading, evaporation, and melting cause it to descend as a major component of the storm downdraft and surface outflow. The flow along streamline E is substantially weaker than along D and enters the rear of the storm to become a rear-flanking downdraft. Extension of Knupp's (1985) analysis of thunderstorm downdrafts to this case suggests that air along this streamline is forced to descend by a downward pressure-gradient force as well as by negative buoyancy created by melting and mixing of low-valued $\theta_e$ air with cloudy air.

Particularly important to the formation of hail is the fact that the updraft streamlines are horizontally convergent. Frank and Foote (1982) found that the convergence in the updrafts extends from the surface to the 6- to 7-km levels. Foote (1984) simulated hail growth in a model of the Westplains, Colorado storm by releasing hail embryos of 0.1, 0.2, 0.4, 0.6, and 0.8 cm in diameter in the observed flow fields. The particles were released at altitudes of 5, 6, 7, and 8 km MSL in the shaded region shown in Fig. 8.54. He then computed the

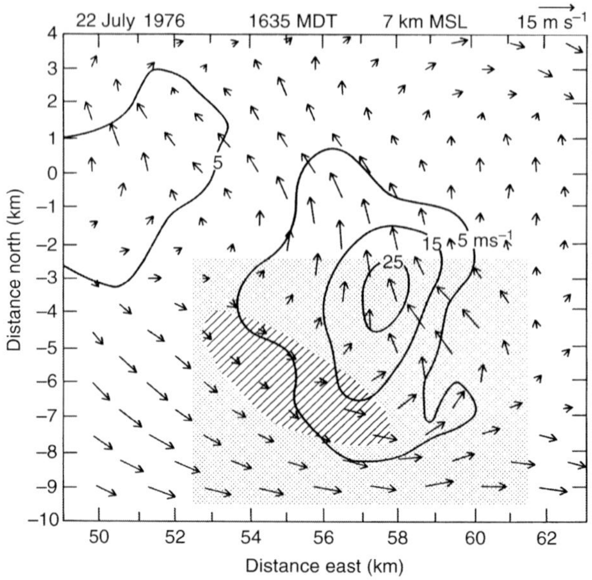


**FIGURE 8.54 Embryo starting positions relative to the airflow at 7 km MSL.** Embryos of various sizes are inserted into the flow over a 1-km grid defined within the shaded region. Horizontal wind vectors are shown (scale in upper right), and updraft contours are indicated. The largest hail originated within the hatched region, where embryos were in the best position to be transported into the updraft core. *(From Foote (1984))*

growth of the hail embryos into hailstones by continuous accretion using the growth equations developed by Paluch (1978). Foote found that with the right combination of initial particle size, injection altitude, and particle drag law, hail could originate at any position in the shaded region except near the southern corners. Figures 8.55 and 8.56 illustrate a plan view and vertical cross section of three selected hailstone trajectories that produced the largest simulated hailstones. Note that the embryos enter the strong updraft region and then turn northward and move along the long dimension of the updraft. It is important to hail growth in the early period that the particle fall speed and updraft speed are nearly balanced so that the hailstone moves less than 2.0 km vertically. During this period the hailstone accretes cloud liquid-water. Subsequently, it passes to the north of the updraft core where it falls to the ground. Foote noted that the hailstones falling into strong downdrafts are more likely to fall to the ground as large hail, since they spend less time in warm air. He calculated that a 1.0 cm hailstone particle above the melting level can lose 73% of its mass due to melting.

Overall, Foote concluded that (1) hail is grown during a single pass through the sloping updraft, passing generally from south to north; (2) embryos of a single size can produce a wide range of hail sizes, and the range of sizes is

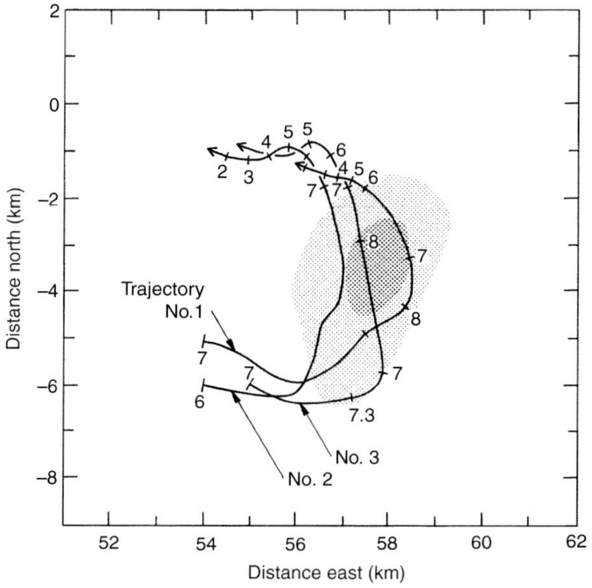


**FIGURE 8.55 Plan view of trajectories 1, 2, and 3 relative to the 15- and 25 m s$^{-1}$ updraft contours as in Fig. 8.54.** These are examples of trajectories that produced the largest hail. Heights are indicated alongside the tracks in units of kilometers above mean sea level. *(From Foote (1984))*

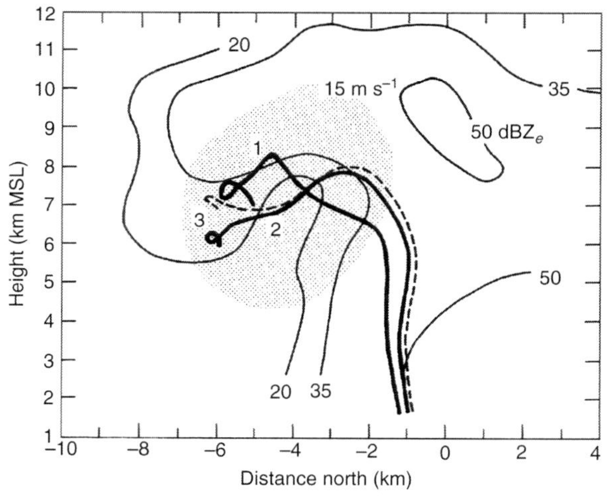


**FIGURE 8.56 Trajectories 1, 2, and 3 projected onto a north-south vertical section.** Reflectivities are shown at three levels of intensity for a north-south plane passing through the updraft maximum. The region of updraft exceeding 15 m s$^{-1}$ in the same plane is also indicated. After entering the updraft from the west, the hailstones grow while making a simple traverse to the north, falling out in the vicinity of the radar echo core. *(From Foote (1984))*

insensitive to initial embryos size; (3) most of the hailstone growth occurs in the temperature range −10 to −25 °C; and (4) larger hail tends to originate from embryos that find themselves in a region on the southwest side of the large updraft.

Not fully resolved is the question of the origin of the embryos of the hailstones. Heymsfield et al. (1980) argued that hail embryos originate in feeder cells which were either small-scale, embedded updrafts and flanking cumulus towers, or upwind mature cells. They concluded that aggregates of dendrites and smaller single dendrites were the predominant type of particle found within the feeder cells, and these were the embryos for most graupel particles and hailstones. They also suggested that embryo production would be enhanced if new turrets injected large aggregates which settle in the precipitation debris region associated with the forward overhang of the mature cell. Once the embryos are formed, they are carried into the mature cell by the storm circulation described by Foote and Frank (1983) and Foote (1984). Heymsfield et al., also concluded that the feeder cells have to be relatively close to the parent updraft to be effective suppliers of hail embryos. The actual effective distance separating the feeder from the parent updraft depends on the strength of the relative winds and the height at which the embryos are detrained from the feeder cells. As we shall see, the relationship of feeder cells to hail growth is an important difference between hail growth in multicellular storms and in supercell storms.

#### *8.11.1.3. Supercell Storms*

Earlier in this chapter we identified the supercell thunderstorm as a quasi-steady storm system exhibiting a persistent, primary updraft/downdraft circulation that generally travels to the right of the mean tropospheric winds in the northern hemisphere. Such a storm system is characterized by a persistent, bounded weak echo region or echo-free vault. Supercell hail-storms have been described by Browning and Ludlam (1962), Browning and Donaldson (1963), Browning (1962, 1965), Marwitz (1972a), Chisholm (1973), and Chisholm and Renick (1972). While supercell storms are rather infrequent in terms of the total number of thunderstorms or even hail-producing storms, they generally produce the largest hailstones and hail-swaths which are quite long and wide. Therefore, they are a major contributor to hail damage (Summers, 1972).

In the context of hail growth, the best example of a supercell storm is the conceptual model developed by Browning and Foote (1976). Their model was derived from calibrated radar and multiple aircraft observations of the Fleming, Colorado, hailstorm, which occurred on 21 June 1972 over northeast Colorado. Figure 8.57 illustrates the evolution of the radar reflectivity field and hailswath for the storm system. Note that the hailswath was approximately 300 km long and 15-20 km wide, with baseball-sized hail falling near the town of Fleming. Figures 8.58 and 8.59 illustrate horizontal maps of the radar reflectivity fields and a vertical cross section along a northwest to southeast line through the

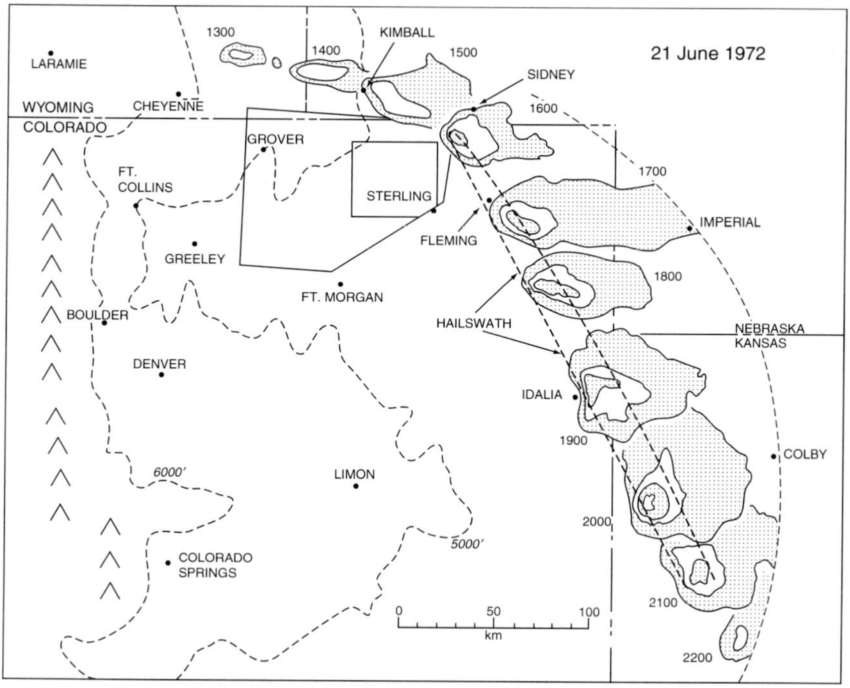


**FIGURE 8.57 Hourly positions of the Fleming hailstorm as determined by the NWS Limon radar (CHILL radar data used, 1300-1500 MDT).** The approximate limits of the hailswath are indicated by the bold dashed line. Continuity of the swath is not well established, but the total extent is. Special rawinsonde sites were located near the towns of Grover, Ft. Morgan, Sterling, and Kimball. Contour intervals are roughly 12 dB above 20 dBZ. *(From Browning and Foote (1976))*

bounded weak echo region during the time when large hail began reaching the ground. Figure 8.58a shows the weak echo region which is bounded by the large forward overhang or embryo curtain on its southeast quadrant and by the main hail precipitation region on its northern quadrant. The wind field at mid-levels observed by aircraft (see Fig. 8.60) suggests a flow field similar to that obtained by Doppler radar in the organized multicellular storm discussed previously. The flow diverges about the main updraft region and converges downshear of it. Figure 8.58 shows that superimposed on the quasi-steady storm structure were a number of local reflectivity maxima, which Browning and Foote labeled "hot spots." They did not regard the hot spots as being particularly important in themselves, but only used them as tracers of air motions. The hot spots were observed to form along the western flank of the storm and more cyclonically around the main embryo curtain.

Browning and Foote visualized hail growth in such a storm system as a three-stage process, as shown schematically on Fig. 8.61. During stage 1, hail embryos form in a relatively narrow region on the edge of the main updraft,

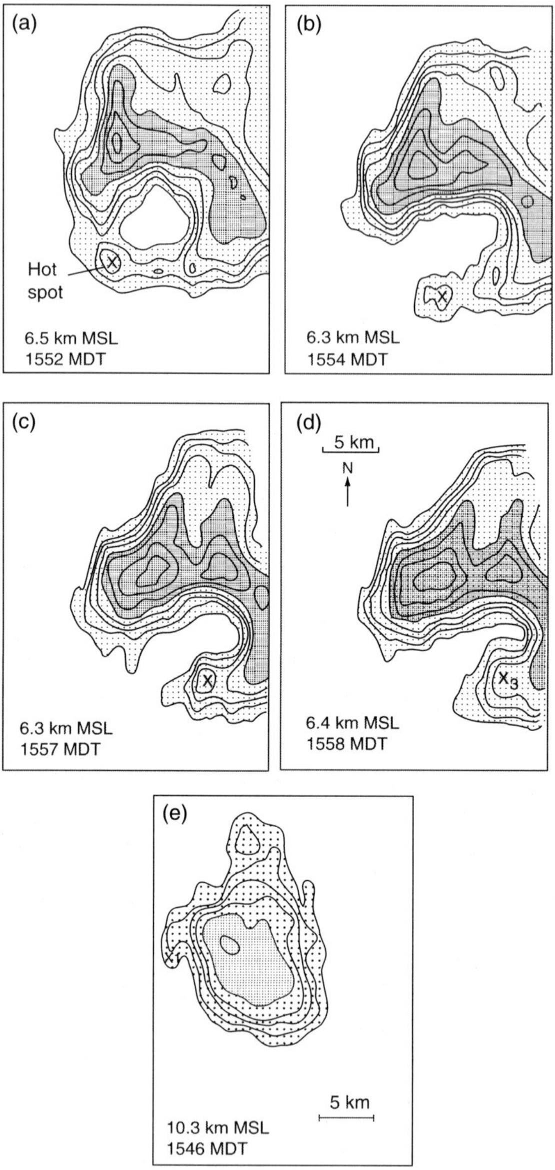


**FIGURE 8.58 Quasi-horizontal sections at four altitudes showing three-dimensional pattern of radar reflectivity for the Fleming storm at 1552-1553 MDT.** Reflectivity contours are at 5-dBZ intervals. Areas in excess of 30 and 50 dBZ, respectively, are stippled thinly and thickly; X is a fiducial mark. *(From Browning and Foote (1976))*

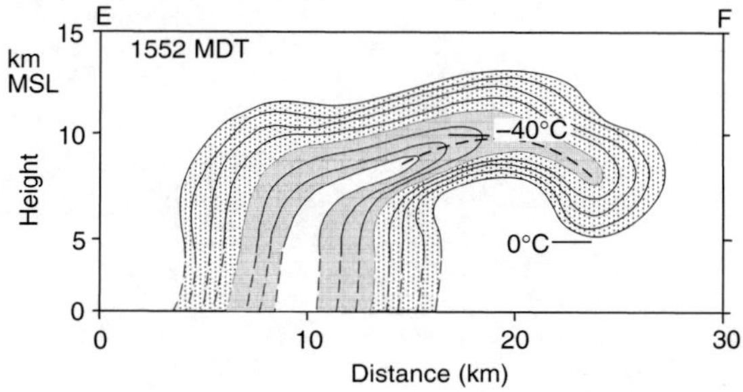


**FIGURE 8.59 Pattern of radar reflectivity for the Fleming storm in a vertical section along a NW to SE line through the bounded weak echo region in Fig. 8.58.** Contours and shading are as in Fig. 8.58. The resolution of this figure, as in Fig. 8.58, is limited by the 1° beamwidth and by the 1-s time integration while scanning in azimuth at 1° $s^{-1}$. The Grover radar was located about 95 km west of the storm. *(From Browning and Foote (1976))*

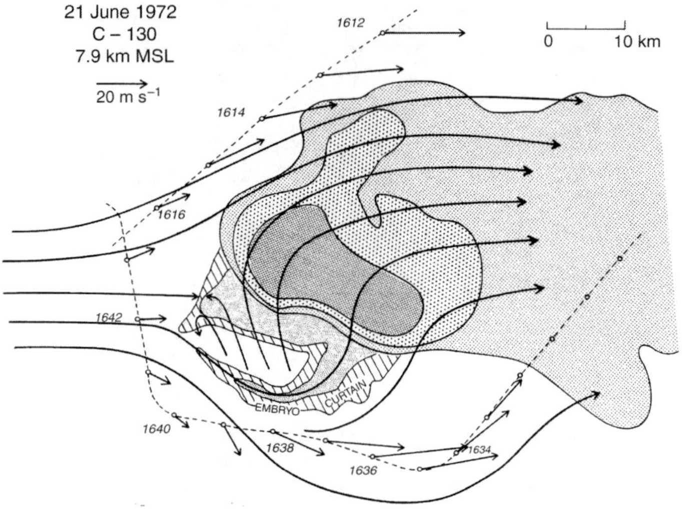


**FIGURE 8.60 Winds measured by the C-130 aircraft as it flew around the Fleming storm at 7.9 km MSL (track and winds shown relative to the storm).** The radar data consist of a low-level PPI from the CHILL system, and a horizontal composite from the DC-6 radar showing the overhanging echo at an altitude of 7.5 km. The streamline analysis emphasizes the blocking flow with a forward stagnation point. North is toward the top of the figure. *(From Browning and Foote (1976))*

where speeds are typically 10 m $s^{-1}$, allowing time for growth to millimeter size. Those particles forming on the western edge of the updraft have a good chance of entering the embryo curtain region, and they follow the trajectory labeled 1 in Fig. 8.61b. The particles labeled trajectory 0 ascend through the core of the updraft and do not have sufficient time to grow to large size. They

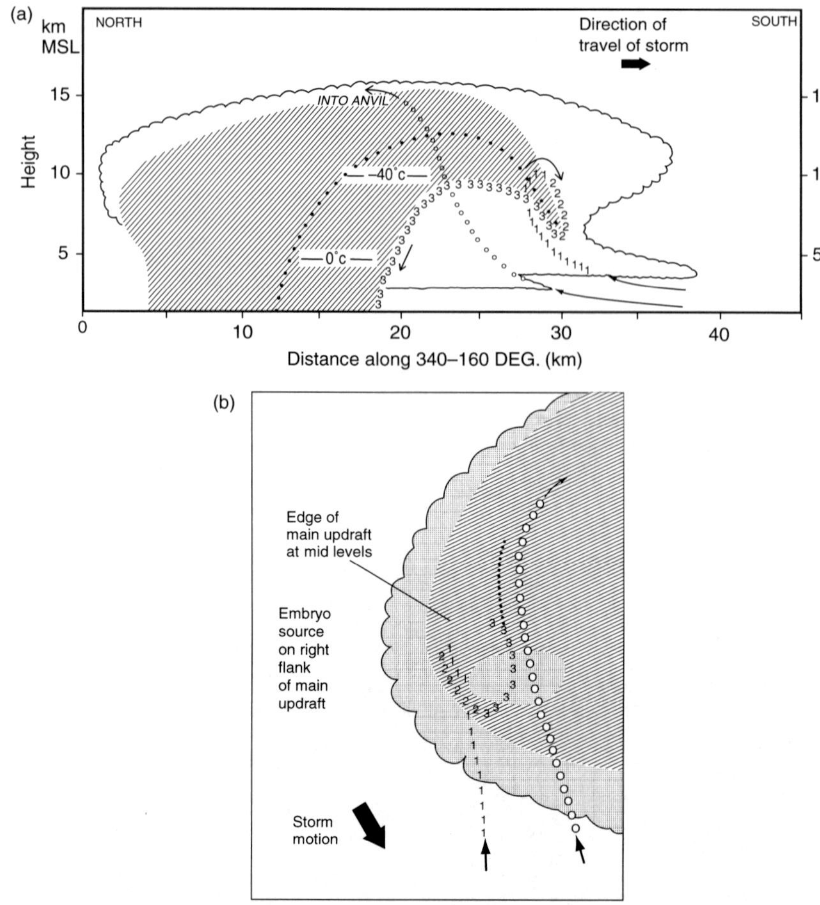


**FIGURE 8.61 (a) and (b) Schematic model of hailstone trajectories within a supercell storm based upon the airflow model inferred by Browning and Foote (1976). (a) Hail trajectories in a vertical section along the direction of travel in the storm; (b) these same trajectories in plan view.** Trajectories 1, 2, and 3 represent the three stages in the growth of large hailstones discussed in the text. The transition from stage 2 to stage 3 corresponds to the re-entry of a hailstone embryo into the main updraft, prior to a final up-and-down trajectory during which the hailstone may grow large, especially if it grows close to the boundary of the vault. Other, slightly less favored, hailstones will grow a little farther away from the edge of the vault and will follow trajectories resembling the dotted trajectory. Cloud particles growing "from scratch" within the updraft core are carried rapidly up and out into the anvil along trajectory 0 before they can attain precipitation size. *(From Browning and Foote (1976))*

form the weak echo region and are exhausted into the anvil outflow. During stage 2, the embryos formed on the western edge of the main updraft are carried along the southern flank of the storm by the diverging flow field. Some of the large hail embryos settle into the region of weak updrafts that characterizes the embryo curtain. The trajectory labeled 2 also illustrates particles that, as

Browning and Foote hypothesize, find their way into the embryo curtain by erosion of updraft air and particles, as the environmental flow meets the updraft air on its western flank. Some further growth of the embryos is likely as they descend in the embryo curtain. Some of the particles settle out of the lower tip of the curtain and re-enter the foot of the main updraft, commencing stage 3.

Stage 3 represents the mature and final stage of hail growth in which the hailstones experience nearly adiabatic liquid-water contents during their ascent in the main updraft. Similar to Browning (1963) and English (1973), Browning and Foote visualize the growth of hail from embryos as a single up-and-down cycle. Those embryos re-entering the main updraft at the lowest levels, where the updraft is weakest, are likely to have their fall speed nearly balanced by the updraft speed. As a result of their slow rise rate, they will have plenty of time to accrete the abundant liquid water. Browning and Foote (1976) and Browning et al. (1963) emphasize that the requirement that the particle fall speed nearly balances the updraft speed accounts for the infrequent occurrence of large hail. The most favored particles remain near the lower boundary of the embryo curtain as they slowly ascend and rapidly accrete cloud droplets. Eventually, they reach the apex of their ascent where, according to Atlas (1966), they spend a large amount of time accreting cloud droplets in a region called the "balance level." Their fall velocity will eventually become large enough to overcome the intense updraft speeds and/or they will settle into the downdraft region as they travel to the northern flank of the storm in a tilted updraft.

Nelson (1983) computed hail growth in flow fields derived by multiple-Doppler radar in a supercell storm. Like Browning and Foote, he calculated that major hail growth occurred in a single up-and-down trajectory. His calculations suggested that a factor more important for hail growth than maximum updraft speed is the presence of a broad region of moderate updraft (20-40 m $s^{-1}$), because hail cannot grow in the strongest part of the updraft. He concludes that the flow field of the storm, not the lack of embryos, is a major modulating factor in hail growth.

A major point of contention with the Browning and Foote model is the assumption that embryos are formed at the edges of the main updraft or are eroded from the main updraft by the interaction between the environmental flow and the updraft. A competing hypothesis is that supercell storms contain flanking, towering cumulus elements or feeder clouds that are embedded within the overall supercell precipitation field. Krauss and Marwitz (1984) provide evidence for such embedded feeder clouds in a supercell storm. They hypothesize that the feeder cells serve as a source region for hailstone embryoes, as they do in the conceptual models of multicell storms. The supercell storm observed by Nelson (1983) also contained local updraft regions that could be interpreted as embedded feeder cells. One could argue that the supercell storms observed by Krauss and Marwitz (1984) and Nelson (1983) represent different positions in the continuum of storm types residing between the classic supercell storm of Browning and Foote and the organized multicell storm described previously. Alternatively, one could interpret the hot spots described by Browning and Foote as actual embedded feeder cells that represent the primary sources for embryos to the main updraft.

### 8.11.2. The role of aerosols in hail formation

There has been surprisingly little research on the impacts of natural aerosol variability on hailstone sizes and storm damage. Danielsen (1977) argued that the concentrations of cloud condensation nuclei and ice nuclei and, perhaps, the aerosol size distribution are important to the production of hail embryos. However, most of the studies of the influence of aerosols on hail formation have been done in the context of cloud seeding for hail suppression. Furthermore, most of the cloud seeding concepts focus on seeding with IN. An overview of hail suppression concepts can be found in Cotton and Pielke (2007). Briefly, the concepts include the embryo competition concept first introduced by Iribarne and de Pena (1962). It involves the introduction of modest concentrations of hailstone embryos (of the order of 10 per cubic meter) in the regions of major hailstone growth. This is done by seeding clouds with IN such as AgI in flanking cumulus congestus which serve as a source region for hailstone embryos. If, indeed, this hypothesis is viable, then one would expect that natural variations in IN could also enhance or reduce the likelihood of beneficial competition, and thereby have a modest regulating influence on hail production.

Another idea for hail suppression is called early rainout. The idea behind early rainout is to initiate ice-phase precipitation lower into the feeder or daughter cell clouds where temperatures are in the range −5 to −15 °C. If the prematurely initiated precipitation settles below an otherwise rain-free base, it could fall into the inflow of the storm and impede the flow of moisture into the storm, which, in turn, would reduce supercooled liquid-water contents deeper in the storm. In addition, initiation of ice lower in the smaller turrets has the potential of reducing supercooled liquid-water available for hail growth in the larger turrets, where updrafts are stronger and more conducive to the growth of larger hailstones. Again, if this hypothesis for hail suppression is viable, one would expect that natural variations in IN concentrations will either enhance or suppress hail formation.

Another strategy for hail suppression is called trajectory lowering. The basis of this concept is to seed the base of clouds with salt particles, or some other hygroscopic material, and thereby initiate a warm rain process in the lower levels of the cloud. According to this concept, precipitation settling out of the lower part of the cloud will deplete the liquid-water in the cloud and therefore limit hailstone growth. It is based on hygroscopic seeding strategies in which salt particles or some other hygroscopic material is introduced into the base of flanking clouds, thereby initiating a more vigorous warm rain process in the lower levels of the cloud. Some cloud modeling studies (Young, 1977) suggest that this may be a feasible approach in regions such as the High Plains of the

United States or Canada where cloud base temperatures are cold and cloud droplet concentrations are large. Applying this concept to naturally varying CCN concentrations, we would expect that ingestion of low CCN concentrations or giant CCN into the base of hailstorms would deplete the liquid-water in the storms thus reducing the potential for the formation of large hailstones. On the other hand, if the storm ingests polluted air with high CCN concentrations, greater liquid-water contents would be expected in the hail forming regions thus increasing the chance for large hailstone formation.

We have seen that hailstorms are a result of a complex series of scale interactions, ranging from the synoptic scale down to the scale of hailstones and their embryos. We now examine the processes involved in the formation of rainfall in cumulonimbus clouds.

## 8.12. RAINFALL FROM CUMULONIMBUS CLOUDS

In many parts of the world, rainfall from cumulonimbus clouds is the dominant contributor to crop-growing season rainfall and to the supply of water for people and livestock. Simultaneously, cumulonimbi can produce heavy rainfall and flash floods, which can kill people and livestock and cause millions of dollars worth of property damage and losses.

The amount of rainfall produced by cumulonimbi depends on the organization and structure of weather systems over a broad range of scales, ranging from the synoptic scales of motion, through the mesoscale, and down to the scale of individual cumulonimbi. This environmental organization can result in rainfall from cumulonimbi ranging from less than 1 cm for ordinary storms (Byers and Braham, 1949) to 38 cm for the more severe flash-flood-producing storms (Maddox et al., 1978).

It is not surprising that the rainfall from cumulonimbus clouds depends strongly on the moisture content of the air mass near the surface as well as the moisture content of the air through the depth of the troposphere. The flash-flood events over the United States analyzed by Maddox et al. (1979) exhibited typical surface mixing ratios of 10 g $kg^{-1}$ to greater than 14 g $kg^{-1}$ and precipitable water contents ranging from 2.58 to 4.16 cm.

Often, heavy rainfall occurs during periods when the large-scale flow pattern appears innocuous, typically exhibiting a large-scale ridge pattern (Maddox et al., 1979). They noted that in many flash floods a shortwave trough at the 500-mbar level was evident to the west or northwest (upstream) of the heaviest rain areas. The heavy rain events over the western United States, however, were often more directly associated with a shortwave trough at the 500-mbar level. This results from the more complicated interactions between large-scale weather disturbances and the complex topography of that region.

A common characteristic of the flow patterns at the 850-mbar level associated with heavy convective rainfall over the United States was the presence of a pronounced low-level jet. The LLJ is effective in fueling convective storms with low-level moisture.

Maddox et al. (1979) noted that there exist several characteristic surface weather patterns associated with flash floods over the United States. Some events are associated with active cyclonic storms which exhibit slowly moving north-and south-oriented fronts. Others are associated with a nearly stationary surface front oriented in an east-west direction. Still other flash-flood events were not associated with a synoptic-scale surface front but were linked to a nearly stationary thunderstorm outflow boundary. Presumably such thunderstorm outflow boundaries could be associated with large complexes of thunderstorms or mesoscale convective systems described by Maddox (1980), Bosart and Sanders (1981), and Wetzel et al. (1982). We will describe these systems more fully in Chapter 9.

The mesoscale processes associated with heavy thunderstorm rainfall events are variable and complex. Often the heavy rainfall is associated with the co-location of thermally-driven mesoscale circulations such as sea-breeze fronts (Cotton et al., 1974), mountain ridge/valley circulations (Maddox et al., 1978) upper-level large-scale troughs, or surface fronts. In some cases slope circulations associated with small hills or urban heat-island circulations are contributing factors to flash-flood events (Miller, 1978). Some are associated with MCCs that survive for several days and locally interact with complex terrain to yield heavy rainfall (Bosart and Sanders, 1981).

The presence of weak winds aloft and weak-to-moderate vertical shear of the horizontal wind is often a characteristic of heavy convective rainfall events (Maddox et al., 1979). Weak winds aloft result in reduced storm motion and more localized heavy rainfall, compared to strong winds aloft in which rain is spread over a large area as the storm moves with stronger winds. The role of weak-to-moderate wind shear in producing heavy convective rainfall is more complicated. Figure 8.62 illustrates the variation of precipitation efficiency with vertical shear of the horizontal winds for High Plains thunderstorms over North America as compiled by Marwitz (1972d). Precipitation efficiency (PE) is defined as the ratio of the measured precipitation rate at the surface to the water vapor flux though the base of a cloud system. The figure suggests that cumulonimbi residing in high wind shears have low precipitation efficiencies, whereas clouds existing in an environment of low wind shear exhibit high precipitation efficiencies.

There are many definitions of PE in the literature but it generally refers to the fraction of either the water vapor input at cloud base or condensed water in cloud that falls out as precipitation. Instantaneous values of PE vary from near zero early in the lifetime of a cloud before precipitation has commenced to values exceeding 100% during the dissipation stages of a storm, when cloud-base moisture fluxes are near zero (Market et al., 2003). Doswell et al. (1996) suggest that PE is most meaningful when averaged over the storm lifetime. Market et al. (2003) propose that it is best to define a volume around a moving system and employ storm-relative winds in evaluating PE. In this way, storm-averaged PE can be obtained for a moving system. One must be cautious in

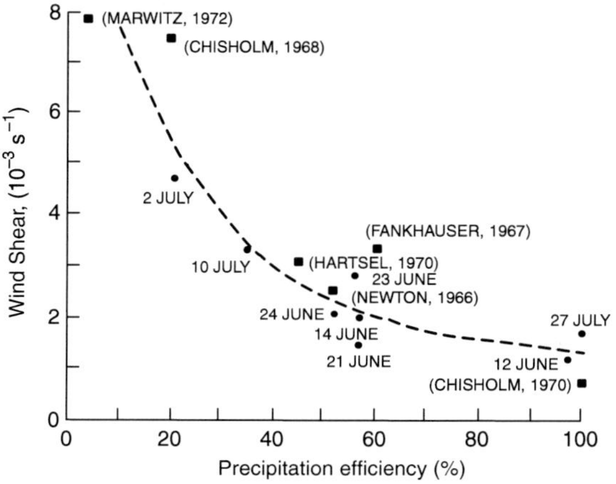


**FIGURE 8.62 Scatter diagram of wind shear versus precipitation efficiency for 14 thunderstorms which occurred on the High Plains of North America.** *(From Marwitz (1972d))*

comparing PE estimates from different studies because of differing definitions of PE, different data sources such as aircraft in situ measurements versus radar measurements, and different time and spatial averaging (Hobbs et al., 1980).

Fankhauser (1982) presented one of the most detailed studies of thunderstorm precipitation efficiency. Data were taken from seven storms during the Cooperative Convective Precipitation Experiment (CCOPE) using aircraft, rawinsondes, a surface mesonetwork and radar histories, and calculated values ranging from 19% to 47%. Various environmental quantities, such as kinetic energy, cloud base area and height, and cloud base mixing ratio were found to be factors that had a strong positive correlation to PE levels. He found that variables such as the bulk Richardson number, the ratio of buoyant energy to the amount of wind sheer, and convective available potential energy (CAPE), a measure of potential updraft strength, did not correlate well with PE.

Another strategy for estimating precipitation efficiencies is to use drying ratio (DR) as proposed by Smith et al. (2003). This method is described in Section 11.9.1.

Let us consider the water budget of a typical cumulonimbus cloud. The primary source of water for a cumulonimbus cloud is the flow of water vapor into the base of the cloud. As the air ascends and cools, the vapor is converted into liquid cloud droplets and some is converted into liquid or frozen precipitation. A portion of the water rapidly falls out as surface rainfall, while some of the water is injected into the anvil portion of the cloud, where it eventually evaporates or slowly settles out as steady precipitation. Some of the water is evaporated from the sides of the cloud due to entrainment processes. As the cloud decays, some of the cloud water and smaller precipitation elements also evaporate. Another portion evaporates in the dry subcloud layer in low-level down-drafts. As an indication of the relative contribution of the various water sinks for a severe squall-line thunderstorm, Newton (1966) estimated that between 45% and 53% of the water vapor entering the updraft reaches the ground as precipitation, while about 40% evaporates in downdrafts and about 10% is injected into the anvil portion of the cloud. As noted by Fujita (1959), the loss of water by subcloud evaporation is appreciable and increases with the height of cloud base. In one example, he showed that the amount of rain that evaporated as it descended from a cloud-base height of approximately 3.0 km equaled the amount of rain reaching the ground.

As we have seen, vertical shear of the horizontal wind increases the rates of entrainment into the cloud system and aids the organization of the storm into a vigorous updraft/downdraft couplet. With larger values of wind shear, one would expect greater water losses in downdrafts and greater rates of transport of water into the upper troposphere, thus lowering the precipitation efficiency of storms.

In contrast, wind shear may increase the storm-relative inflow of warm moist air into the storm system, and sustain the cloud lifetime such that, even with reduced precipitation efficiencies, greater amounts of precipitation are produced than in low-shear environments. Wind shear, however, is clearly not the only important factor influencing the precipitation efficiency of a cumulonimbus. For thunderstorms observed over Florida and Ohio during the Thunderstorm Project, Braham (1952) estimated a precipitation efficiency of only 10%. He attributed the low precipitation efficiency of these storms to the loss of water by entrainment, which is greater in smaller, ordinary cumulonimbi.

Not only is the magnitude of wind shear important to the efficiency of precipitation production, but its directional variation is also critical to rainfall production. Miller (1978) simulated a localized heavy rainfall event over London, called the Hampstead storm. The model for this study was a three-dimensional, nonhydrostatic numerical model described by Miller and Pearce (1974). An important feature of the environment of the storm was that the wind veered with height. Veering of environmental winds is also a characteristic of the flash-flood events analyzed by Maddox et al. (1979). Miller demonstrated that the veering wind profile was important in producing a localized heavy rainfall event.

The storm simulated by Miller was a multicellular storm system in which new cells repeatedly formed on the southeast flank of the spreading low-level outflow. It was on this flank that the spreading outflow most directly opposed the low-level winds. It is commonly observed that flash-flood-producing storms are multicellular (e.g. Caracena et al., 1979; Dennis et al., 1973; Hoxit et al., 1978). This is consistent with Weisman and Klemp's (1982) model results which suggested that multicellular storms prevail in relatively low-sheared environments in which the bulk Richardson number exceeds 50. Miller noted that part of the mature cell and the decaying older cell merged to form an elongated raining anvil. The elongated rainfall pattern created a similarly elongated mesohigh that, in turn, assisted in the persistence and regeneration of the storm system as a whole. The mesoscale circulation of the storm, as

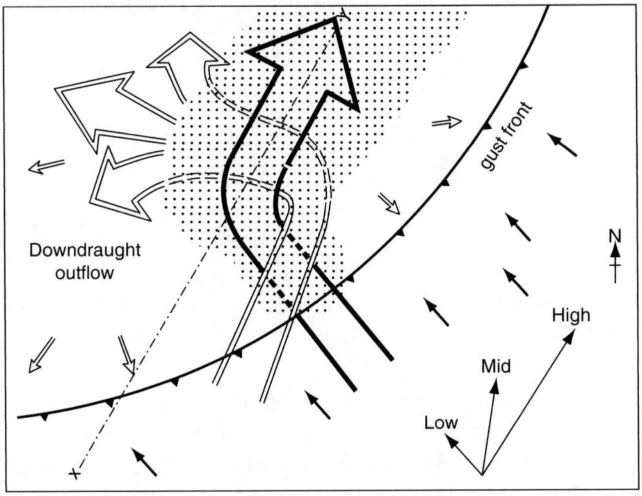


**FIGURE 8.63 A schematic of the primary features of the storm model deduced from the simulation.** *(From Miller (1978))*

characterized by the low-level inflow, anvil outflow, and downdraft outflow, was quite persistent. Figure 8.63 illustrates a conceptual model of the interlocking nature of the persistent updraft/downdraft circulation. The advance of the gust front toward the southeast was nearly matched by the prevailing southeasterly flow, so the gust front remained nearly stationary relative to the ground. Consequently, new cell development occurred repeatedly at the point of maximum convergence between the outflow and inflow at a nearly constant geographical position.

Miller demonstrated further that the veering of the winds through 90° between the surface and 400 mbar was instrumental in maximizing the convergence of air approaching the storm. He did so by repeating the simulation with middle-level and upper-level winds rotated to a southeasterly direction but with no speed change. The storm system elongated toward the northwest and hence in the direction of the low-level relative flow. As a result, the basic stationary character of the storm circulation was lost. Figure 8.64 is a schematic diagram of the results of the two numerical experiments. Miller's results are consistent with the rule of thumb that the precipitation from a storm system increases with the horizontal area of the storm. The elongated precipitation/convergence pattern would be expected to enhance the total volume of rainfall from the storm system. Miller's simulation is an ideal illustration of the importance of the speed and direction of environmental winds on the rainfall production of a storm system.

Not only are synoptic-scale, mesoscale, and storm-scale features important to rainfall production, but the microphysical characteristics of the storm are also important. Generally, heavy rain-producing storms are warm-based storms in an

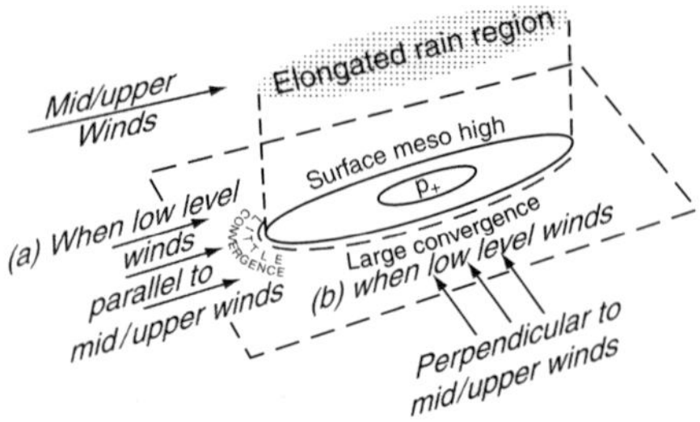


**FIGURE 8.64 A schematic of the proposed mechanism by which vector shear assists the regeneration or maintenance of a storm system.** *(From Miller (1978))*

environment with a high mixing ratio, which afford the opportunity for warm-rain processes to be very active. If the air mass is maritime (low in CCN) or is cleansed by earlier precipitation scavenging of aerosol, then the opportunity for dominance of the warm-rain process results in the precipitation process being concentrated at low levels. This was apparently true for the Big Thompson Storm (Caracena et al., 1979). Modeling studies indicate that a cloud system in which the warm-rain process is predominant is a more efficient rain-producing cloud than is a cloud system dominated by ice processes (Levy and Cotton, 1985; Tripoli and Cotton, 1982). As the ice phase becomes more predominant in a large convective storm, more total water is thrust upward into the anvil region.

This brings us to the question of the relationship between the production of hail versus rainfall in cumulonimbi. As we have seen, the storms producing the largest hailstones occur in strongly sheared environments. Thus, in general, we should not expect that the storm systems producing the largest hailstones are also heavy rain-producing storms. However, there are many storm systems that produce large quantities of smaller yet damaging hailstones. Such storms occur in less severely sheared environments and can thus be prolific rain producers as well. For example, calculations by Crow et al. (1976) suggest that for northeastern Colorado hailstorms, the hail contribution to the total precipitation mass is typically less than 4%. Rainfall is the dominant form of precipitation even in some of the most severe hailstorms. Dennis (1980) has noted that cloud seeders are concerned about whether seeding clouds to reduce hail increases, decreases, or has no effect upon rainfall production by those storm systems. When one considers the small fraction of the water budget that hail represents, it is not surprising that this problem has not been resolved at this time.

## 8.13. AEROSOL IMPACTS ON CONVECTIVE PRECIPITATION

Aerosols can influence convective precipitation by direct radiative heating affects, semi-direct affects, and indirect affects. Absorption of solar radiation by absorbing aerosols such as black carbon can heat the air layer which can stabilize an air mass making it less likely to produce convective precipitation. This process is called the direct heating affect of aerosols (Grassl, 1975; Hansen et al., 1997; Koren et al., 2005). In some instances the absorption of solar radiation can lead to the complete desiccation of clouds by warming the air mass and lowering the relative humidity of the cloud layer. This process is called the semi-direct effect (Ackerman et al., 2000). The complete desiccation of a stratocumulus or stratus layer over land can increase surface heating and enhance the opportunity for deep convective precipitation. However, in this section we concentrate on the indirect affects of aerosol serving as CCN. As we have seen earlier, enhanced CCN concentrations suppress the rate of formation of precipitation sized drops by collision and coalescence. As we shall see, this does not always mean that precipitation will be decreased.

In fact, Seifert and Beheng (2006) showed that the effect of changes in CCN on mixed phase convective clouds is quite dependent on cloud type. They found that for small convective storms, an increase in CCN decreases precipitation and the maximum updraft velocities. For multicellular storms, the increase in CCN has the opposite effect namely, promoting secondary convection, and increasing maximum updrafts and total precipitation. Supercell storms were the least sensitive to CCN. Their study also showed that the most important pathway for feedbacks from microphysics to dynamics is via the release of latent heat of freezing. That is, higher CCN concentrations produced higher amounts of supercooled liquid-water and thus greater amounts of latent heat of freezing. The added freezing led to explosive growth of the simulated clouds and enhanced precipitation.

However, other modeling studies by Khain et al. (2005) show complex dynamical responses to aerosols, sometimes leading to greater precipitation amounts and other times less. In one simulation ordinary thunderstorms were transformed into a long-lived squall line system following the explosive growth of convective cells. Zhang et al. (2005) came to similar conclusions in their model simulations for different three-week periods over the ARM site in Oklahoma. Similarly, mesoscale simulations of deep convection over Florida by Lynn et al. (2005) showed that higher CCN concentrations delayed the onset of precipitation but led to more intense convective storms with higher peak precipitation rates. However, the accumulated precipitation was largest for the cleaner atmosphere.

In mesoscale simulations of entrainment of Saharan dust into Florida thunderstorms with bin-emulating bulk microphysics, van den Heever et al. (2006) found that dust not only impacts cloud microphysical processes but also the dynamical characteristics of convective storms. Dust may serve as CCN, GCCN, and IN. The effect of dust on cloud microstructure and storm dynamics in turn alters the accumulated surface precipitation and the radiative properties of anvils. These results suggest that the dynamic structure of the storms is influenced by varying dust concentrations. In particular, the updrafts were consistently stronger and more numerous when Saharan dust was present compared with a clean air mass. Like Seifert and Beheng (2006), they found that dust results in enhanced glaciation of convective clouds, which then leads to dynamical invigoration of the clouds, larger amounts of processed water, and thereby enhanced rainfall at the ground. However, Van den Heever et al's. simulations suggested that rainfall is enhanced by dust ingestion only during the first two hours of the formation of deep convective cells, and it is reduced on the ground later in the day. Thus the clean aerosol simulations produced the largest surface rain volume at the end of the day. This is a result of complex dynamical responses of clouds to aerosol changes associated with subcloud evaporation of rain, in which low-level cold-pools influence storm propagation and to scavenging of dust, so that few GCCN and IN remained late in the day.

Another study that further illustrates the complexity of aerosol interactions with convective storms is van den Heever and Cotton's (2007) examination of the impacts of urban-enhanced aerosol concentrations on convective storm development and precipitation over and downwind of St. Louis, MO. In the van den Heever and Cotton (2007) study RAMS was set up as a cloud-resolving, mesoscale model with both sophisticated land-use processes and aerosol microphysics using a bin-model emulation approach. The results indicate that urban land-use forced convergence downwind of the city, rather than the presence of greater aerosol concentrations, is the dominant control on the locations and amounts of precipitation in the vicinity of an urban complex. Once convection is initiated, urban-enhanced aerosols can exert a significant effect on the dynamics, microphysics and precipitation produced by these storms. The model results indicate, however, that the response to urban-enhanced aerosol depends on the background concentrations of aerosols; a weaker response occurs with increasing background aerosol concentrations. It was found that when aerosol concentrations were enhanced, cloud water was enhanced aloft, rather than in the clean control simulation in which only observed rural aerosol concentrations were utilized. The updrafts were also stronger initially, and the downdrafts developed more quickly. The larger amounts of supercooled liquid-water available, together with the stronger updrafts, led to the generation of greater ice mixing ratios earlier in the storm development. Greater amounts of surface precipitation were also produced in this case during the first hour and a quarter to hour and a half of convective storm formation. However, the greater and more rapid production of surface precipitation generates stronger downdrafts and more intense cold-pools earlier in the storm life cycle than in the clean control simulation. This is detrimental to the updraft development and strength, the evidence of which is the earlier demise of the storm closest to the urban region following storm splitting. In the clean control simulation, the updrafts develop later in association with the delayed hydrometeor development, but they are eventually stronger than those in the simulation in which aerosol concentrations are enhanced. The storms last longer following storm splitting, and new storm development occurs downwind of the city later on in the simulation. This results in increased amounts of accumulated surface precipitation during this time. The variations in storm dynamics in response to variations in aerosol concentrations, result in the greatest accumulated surface precipitation when aerosols are increased early in the afternoon. However as the simulation progresses, this trend reverses, and later in the afternoon, the largest accumulated precipitation occurs in the clean control case.

In conclusion, in contrast to the assertions of Rosenfeld et al. (2008), the indirect affects of aerosol on deep convective clouds does not necessarily lead to an increase in precipitation on the ground in a climatological sense. Instead, the extremely complex, non-linear relationships between the microphysics and dynamics of storms, particularly cold-pool interactions, make it difficult to make absolute statements regarding the impacts of aerosol pollution on precipitation.

## 8.14. THUNDERSTORM ELECTRIFICATION AND STORM DYNAMICS

In this section we review thunderstorm electrification processes with a focus on the possible impact of storm electrical processes on storm dynamics and the impact of storm dynamics on charge separation processes. We begin by discussing the possible role of storm electrification processes on the storm thermodynamics and dynamics.

### 8.14.1. Influence of Storm Electrification on Cloud Dynamics

Vonnegut (1960) hypothesized that lightning discharges in the core of the tornado vortex would generate sufficient heating to become a significant energy source in driving the tornado. In order for lightning to be a significant energy source, the frequency of lightning flashes in the vortex must be very high, of the order of 1000 $km^{-2}$ $min^{-1}$. While there are occasional reports of electrical glows in and near tornadoes (Vaughan and Vonnegut, 1976), there is little confirmation evidence of vigorous lightning activity such as strong radio sferics (Davies-Jones and Golden, 1975a), or other evidence of electromagnetic disturbances near tornadoes (Zrnic, 1976). As noted by Davies-Jones (1982), the importance of electrical heating in tornadoes is by no means fully resolved; the pros and cons of the theory are still being debated (Davies-Jones and Golden, 1975a,b,c; Vonnegut, 1975; Colgate, 1975; Watkins et al., 1978).

Electrical heating, however, is not the only way in which cloud electrification processes can affect storm dynamics. It is possible that the electrical fields and high space charge densities in thunderstorms could accelerate air parcels directly. However, numerical calculations (Chiu, 1978) and simple order-of-magnitude estimates (Vonnegut, 1963) suggest that such electrical forces are small in comparison to buoyancy accelerations. Cloud electrification may also affect the dynamics of clouds by altering the terminal velocities of precipitation elements. Numerical calculations by Levin and Ziv (1974) and Chiu (1978) suggest that when the electrical field strengths approach breakdown potential, the terminal velocities of precipitation particles can be appreciably modified. The precipitation particles are then levitated by the electric fields. The alteration of fall velocities of precipitation particles then influences the distribution of water substances; the distribution, in turn, alters the buoyancy of the cloud by changing its water loading. The numerical experiments by Chiu (1978) suggest that these effects can significantly alter the subsequent dynamics of a cloud. Rawlins (1982) found in a three-dimensional cloud model that levitation of hail had a negligible influence on the early development of electrification but had a modest influence at large field strengths, suggesting that levitation altered the water distribution of the cloud.

Schonland (1950) and Levin and Ziv (1974) suggested that the cessation of levitation of precipitation elements immediately following a lightning discharge results in the occurrence of the so-called rain gush. Moore et al. (1964) observed as much as a 10-fold increase in precipitation content of a storm following a lightning discharge. If levitation of precipitation particles is important, then one should observe substantial changes in particle motions by vertically pointing Doppler radar immediately following lightning discharges. Williams and Lhermitte (1983) attempted to examine such a response in Florida thunderstorms. They observed infrequent Doppler velocity changes at high levels in the cloud where the radar reflectivities and precipitation particle sizes were small. In general, however, they found little correspondence between velocity changes of the precipitation and lightning discharges in regions of high reflectivity. This suggests that levitation does occur, at least on smaller particles, but there is little evidence that it is strong enough to substantially alter the motions of larger precipitation particles.

Cloud electrification processes can also alter the dynamics of clouds in a more subtle way. There is considerable evidence suggesting that the presence of strong electric fields and charged drops can enhance the collection efficiency among cloud and precipitation elements (Sartor and Miller, 1965; Davis, 1961, 1964a,b; Lindblad and Semonin, 1963; Semonin and Plumlee, 1966; Schlamp et al., 1976, 1979; Latham, 1969; Saunders and Wahab, 1975). As a result of enhanced collection processes, larger, faster falling precipitation elements will more readily form, resulting in a redistribution in condensed water with its subsequent impact upon cloud buoyancy and dynamics. Indeed, Moore et al. (1964) suggested that the rain gush is a result of increased coalescence of precipitation elements in the strong electric fields prior to the lightning stroke. The Doppler radar observations reported by Williams and Lhermitte (1983) are consistent with Moore's hypothesis. They noted that gradual changes in downward particle velocity were well correlated with electric field changes. One would expect a more gradual response in particle motions by enhanced collection than by levitation effects.

In conclusion, there is considerable evidence that cloud electrification processes can influence the dynamics of clouds. It appears, however, that these effects are localized, and they occur at electric field strengths approaching breakdown potential.

## 8.14.2. Influence of Cloud Dynamics on Cloud-Charging Processes

To begin our discussion of the influence of cloud dynamics on cloud-charging processes, we shall follow the lead of Mason (1971) by first listing the characteristics of a satisfactory charge separation theory. In so doing we will update Mason's characteristics with the results of more recent observations. The updated characteristics are as follows:

1. The average duration of precipitation and electrical activity from a single thunderstorm cell is about 30 min.
2. The electric field strength destroyed in a lightning flash is about 3-4 kV $cm^{-1}$; the breakdown field in clear air is much higher (30 kV $cm^{-1}$).
3. In a large, extensive cumulonimbus cloud, this charge is generated and separated in a volume bounded by the $-5$ and $-40$ °C levels and having a radius of approximately 2 km.
4. Negative charges are usually centered between the $-10$ and $-20$ °C levels with the positive charge several kilometers above, and a secondary pocket of positive charge is occasionally found near cloud base in precipitation; the center of negative space charge may be somewhat lower in mesoscale systems, closer to the freezing level.
5. The charge generation and separation processes are closely associated with the development of precipitation, although the space charge center appears to be displaced both vertically and horizontally from the main precipitation core.
6. Sufficient charge must be generated and separated to supply the first lightning flash within about 20 min of the appearance of precipitation particles of radar-detectable size.

The charge generation theories consistent with most of these characteristics can be classified as being either a precipitation-related theory or a convection theory. The relative merits of these two classes of charge generation theory have been debated extensively by Mason (1976) and Moore (1976). Here we will briefly review the basic concepts and their relationship to the dynamic structure of the cloud.

### *8.14.2.1. Convection Charging Theory*

The convection theory is intimately coupled with the overall dynamics of the cloud system. Advocates of this theory of charge separation include Grenet (1947), Vonnegut (1955, 1963), and Wagner and Telford (1981). According to this theory, a normal fair-weather electric field establishes a net concentration of positive ions in the lower troposphere. As convective updrafts form, they carry

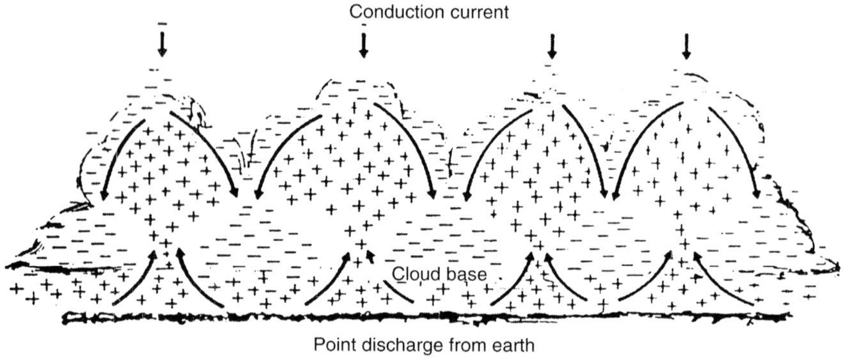


**FIGURE 8.65 Schematic representation of a group of thunderstorm cells illustrating how the electrification process might be maintained by convection.** According to this representation, the negative charge attracted to the top of the cloud is carried to the lower part of the cloud by downdrafts while positive charge created by point discharge at the ground is carried to the upper part of the cloud by updrafts. *(From Vonnegut (1963))*

the positive space charges into the cloud layer, causing the cloud initially to be positively charged. As the cloud penetrates to higher levels in the troposphere, it encounters air in which the mobility of free ions (or conductivity of the air) increases with increasing height. These ions are produced in the ionosphere or at heights above 6 km by cosmic radiation. The rising positively charged cumulus preferentially attracts the negative free ions, causing the cloud top to become negatively charged. Instead of neutralizing the positive space charge in the cloud, Vonnegut hypothesizes that convective downdrafts transport the negative ions to the lower part of the cloud while updrafts carry the positive ions to the upper part of the cloud. It is hypothesized that the resulting buildup of positive space charge near the earth's surface then causes preferential point discharge of positive ions, which are then transported into the cloud by updrafts. The resulting increase of positive charge in the cloud enhances the flow of negative ions to cloud top, leading to an exponentially increasing cloud polarity. Figure 8.65 illustrates the process as hypothesized by Vonnegut.

Chiu and Klett (1976) attempted to simulate the convective charging theory using a simple steady-state, axisymmetric cloud model developed by Gutman (1963, 1967). They found that the theory could not produce sufficient cloud charging to induce a flux of positive space charge near the earth's surface by point discharge. In fact, when their simulated cloud had a base height at observed levels, the cloud exhibited weak, negatively charged cores and relatively weak upper layers of positive charge. Only when the cloud-base height was lowered to a few tens of meters above the ground did a weak charge of opposite polarity develop.

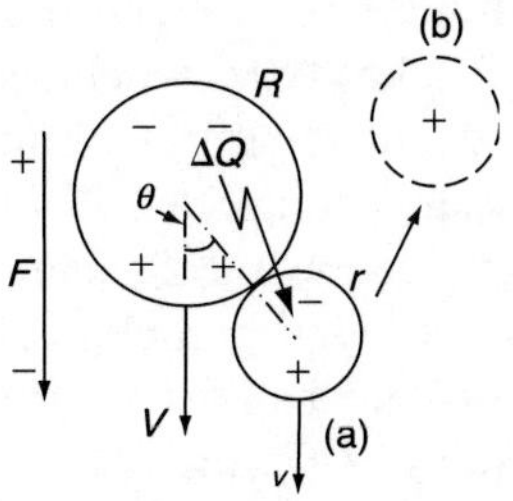


**FIGURE 8.66 Sequence of events during the polarization charging of cloud hydrometeors. (a) During contact; (b) after separation.** *(From Levin and Ziv (1974). Copyright by the American Geophysical Union)*

There have been other adaptations of the convective charging theory such as Wagner and Telford's (1981) application to non-precipitating cumuli but these have not been developed further in the context of simulated cloud dynamics.

### *8.14.2.2. Precipitation Charging Theories*

The precipitation charging theories do not depend directly or solely on the convective motions of the cloud for charge separation. However, they do depend indirectly upon the dynamic structure of the cloud for vertical and horizontal redistribution of precipitation elements. We will now review the induction-precipitation charging theory and noninduction-precipitation charging theories.

### *8.14.2.3. Induction Charging Theory*

The particle-charging mechanism by induction has enjoyed a long list of advocates (Elster and Geitel, 1913; Mullet-Hillebrande, 1954, 1955; Sartor, 1961; Latham and Mason, 1962; Mason, 1968, 1976; Scott and Levin, 1975; Levin, 1976; Colgate et al., 1977; Illingworth and Latham, 1977; Chiu, 1978). The basic concept is that in the presence of a fair-weather field, cloud and precipitation elements become polarized, as shown in Fig. 8.66, with the lower part of the cloud particle being positively charged and its upper part being negatively charged. It is hypothesized that when a precipitation particle and a cloud droplet or small ice crystal collide and rebound, the larger particle becomes negatively charged and the smaller one becomes positively charged. The resultant positively charged small precipitation elements are then swept into the upper portions of the cloud, while the larger, negatively charged particles settle in the lower portions. This process leads to a cloud polarization with a positive space charge in the upper part of the cloud and negative space charge residing on the larger precipitation elements in the lower part of the cloud. Using greatly different models of particle physics and cloud dynamics, both Levin (1976) and Chiu (1978) conclude that particle charging by induction involving collision between liquid droplets can only develop an electric field strength of breakdown potential within a typical cloud lifetime of the order of 1000 s. Illingworth and Latham (1977), however, concluded that liquid-liquid induction charging is not capable of thunderstorm electrification. They concluded that collision between ice crystals and hail pellets is a far more powerful induction-charging mechanism. The induction mechanism needs a high frequency of collision and rebound between cloud particles. This condition is more likely to be satisfied for ice-ice collisions than for water-water or water-ice collisions. However, as noted by Gaskell (1979), the relaxation times to conduct charge through ice is relatively long, possibly too long for significant charge transfer. Also, when Rawlins (1982) examined the ice-ice induction mechanism in a cloud model, he found that if multiple collisions of each ice particle with more than one hail particle were included, a breakdown field was not simulated.

### *8.14.2.4. Noninduction Charging of Graupel and Hail Particles*

There has been considerable circumstantial evidence over the years that the presence of graupel is linked to the separation of charge in thunderstorms. As a consequence, there have been a number of hypotheses and laboratory experiments aimed at explaining its role. Some of these hypotheses, such as the thermoelectric effect (Latham and Mason, 1961), are now thought to have too small a charging rate to be of significance (Marshall et al., 1978).

A number of laboratory studies have exhibited a complex variation in charging rates with temperature and relative humidity. The laboratory experiments by Takahashi (1978), for example, suggested that at temperatures colder than $-10$ °C, the sign of charging during riming depended on temperature and liquid-water content. At temperatures warmer than $-10$ °C, positive charge occurred regardless of the liquid-water content. Figure 8.67 illustrates variations in the sign of charging depending on temperature and cloud water content. Saunders and Peck (1998) also found similar charge reversals but at different temperatures and as a function of simulated graupel critical ice accretion rates. Jayaratne et al. (1983) reported on somewhat different regimes for charge reversal.

An advantage of the non-induction riming hypothesis over the ice-ice induction hypothesis is that charge is rapidly transferred between charged states at the interface between rimed graupel/hail and vapor-grown ice crystals. In contrast to induction charging, where short contact times limit the amount of bulk charge that can be conducted across the interface, charge transfer by this process is quite rapid. It should be noted that this process requires a particular mix of ice-particle types. The process operates most effectively when there is a relatively high concentration of vapor-grown crystals that can collide with large, rapidly falling rimed ice particles such as graupel or hail. The results of laboratory, modeling, and observational studies (MacGorman and Rust, 1998; Helsdon et al., 2002; Kuhlman et al., 2006) suggest that the magnitude, sign of charge, and charging rates can be well described by non-induction charging.

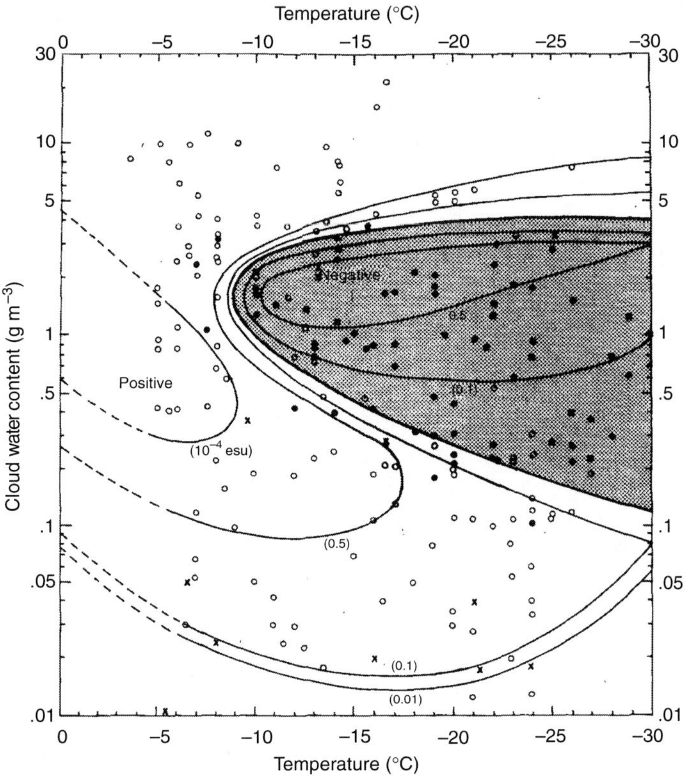


**FIGURE 8.67 Electrification of rime using drop distributions (A)direct condensation of outside moist air introduced in the cold room and (B) water bath in the main chamber.** Open circles show positive charge, solid circles negative charge and crosses represent uncharged cases. The electric charge of rime per ice crystal collision is shown in units of $10^{-4}$ esu. *(From Takahashi (1978))*

### *8.14.2.5. Variation of Cloud Charge Structures and Polarity with Storm Dynamics and Regional Environments*

We conclude this section by noting observations suggesting that different intensities of lightning activity occur with different storm types and different regional environments.

First of all it is generally known that tropical convection over land exhibits much higher lightning flash rates than corresponding convection over the ocean (Williams et al., 2005). This was originally hypothesized to be due to differences in CAPE; with CAPE over land being higher than CAPE over the oceans. As we have seen in Chapter 7, this is not normally the case as CAPE can be similar in both regions (Williams and Renno, 1993; Lucas et al., 1996). The motivation for linking CAPE to lightning flash rate are the model results of Baker et al. (1999), which indicate the lightning flash rate is proportional to the forth power of updraft velocity and, as we have seen, updraft velocity scales with CAPE. But, Lucas et al. (1996) noted that variations in cloud base height might be the factor that controls differences in updraft strengths between marine and continental tropical convection. They argue that lower cloud bases over the oceans would produce smaller-scale updraft elements (thermals) that entrain environmental air, rapidly destroying the buoyancy of the clouds. Whereas clouds forming in similar CAPE environments over land have higher cloud base heights and thereby have larger radius updraft elements, which entrain less and thereby achieve higher vertical velocities. Modeling studies by McCaul and Cohen (2002) support the idea that cloud base height has a controlling influence on the size of updrafts and thus updraft strength. Williams et al. (2005) thus argue that cloud base height is a major determining factor for lightning flash rates.

Williams et al. (2002) also suggest that differences in CCN populations between marine and continental tropical convection may have a controlling influence on lightning flash densities. In a clean marine environment rapid warm cloud precipitation processes would essentially limit the amount of water that could be transported into supercooled levels where non-inductive charging would be efficient. Whereas over land where high CCN concentrations are prevalent, the reduced efficiency of collision and coalescence would favor larger amounts of water being thrust into supercooled levels thus favoring non-inductive cloud charging. Williams et al. (2002) suggested, however, that relatively large CAPE would dominant over aerosol effects.

Several studies have examined electrical activity in tropical cyclones (i.e. Cicil et al., 2008; Samsury and Orville, 1994; Lyons and Keen, 1994). First of all there is a great deal of variability of electrical activity from storm to storm some with relatively low activity and others higher. The differences appear to be linked to the amount of convective activity versus stratiform precipitation, with the latter generally favoring lower electrical activity. The main centers of electrical activity appear to be the eyewall region and the outer rainbands, while the inner rainbands do not exhibit much electrical activity. Lyons and Keen (1994) linked bursts of cloud-to-ground (CG) lightning activity to the formation of supercells or intense hot spots of convective activity in the eyewall region.

As to mid-latitude, continental convection, the focus of research has been on establishing links to the severity of storms and to the sign of CG activity. Overall lightning activity is most prevalent in the US where there is a high frequency of ordinary thunderstorms such as in the southeast US (see Fig. 8.68). As noted by Watson et al. (1994) the locations of preferred lightning activity are regions characterized by high available moisture, high CAPE, and some form of triggering mechanism for convection such as sea-breeze convergence zones or mountains.

The results of studies relating the severity of thunderstorms to electrical activity is mixed. Early studies by Pakiam and Maybank (1975) of multicellular

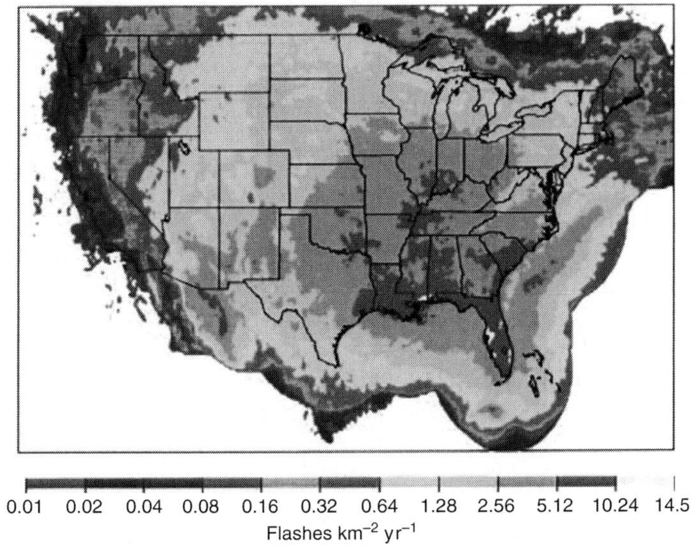


**FIGURE 8.68 Average annual flash density from 1995 to 1999.** *(From Zajac and Rutledge (2001))*

and supercell hail-producing thunderstorms over Alberta, Canada suggested that:

1. If the storm was of an ordinary multicellular type with limited depth (6-7.5 km MSL in Alberta), then rain and small hail occurred with a low frequency of lightning flashes.
2. If the thermodynamic instability was greater, the storms became better organized multicellular storms with high cloud tops (7.5-12 km MSL in Alberta); rain and hail and the lightning frequency increased appreciably.
3. With even greater instability and higher shear, organized multicellular storms with tops well into the stratosphere (12 + km MSL in Alberta) prevailed. In these storms, large hail forms, and the frequency of lightning depends on the number and proximity of thunderstorm cells. A system composed of five cells may produce 35 flashes $min^{-1}$, a number of which are intracloud. An isolated cell produces only 2-3 flashes $min^{-1}$.
4. In one case of a severe, hail-producing supercell storm, the frequency of flashes was only 2-3 $min^{-1}$. They concluded that such a storm is only a single electrical cell resulting in a low frequency of lightning flashes.

Likewise, Perez et al. (1997) examined 42 violent tornado-producing supercell storms and found no correlation with the formation of tornadoes, and the amplitudes of flash rate changes to be highly variable. In fact Knupp et al. (2003) found that storms with high radar reflectivity (greater than 65 dBZ) were inversely correlated with CG frequency. They hypothesized that when large hail was present the reduction of large number concentrations of precipitation-sized particles in the mixed-phase region suppressed CG activity. This is consistent with the modeling results of Kuhlman et al. (2006) which showed that total storm flash rate is well correlated with graupel volume, updraft volume, and updraft mass flux, and poorly correlated with maximum updraft speed.

Goodman and MacGorman's (1986) study of cloud-to-ground lightning activity in mesoscale convective complexes supports the concept that the frequency of lightning discharges is a function of the depth and number of multicellular convective elements. As will be shown in Chapter 9, the MCC represents a very well-organized mesoscale convective system which is composed of numerous cumulonimbus cells. Goodman and MacGorman showed that MCCs produce maximum ground lightning strikes of 54 $\text{min}^{-1}$ averaged over one hour or a sustained lightning frequency in excess of 17 $\text{min}^{-1}$ for nine consecutive hours! Goodman and MacGorman noted that such a high, sustained lightning frequency over the lifetime of an MCC suggests that the passage of a single MCC over a given location can produce 25% of the estimated mean annual strike density for that site. Furthermore they found that the ratio of ground discharges in MCCs to ordinary thunderstorms observed in Florida is 4:1, whereas for severe or multicell storms in the High Plains of the United States the ratio is in excess of 20:1. It is interesting that Goodman and MacGorman (1986) found that the peak flash rate in MCCs occurs at the time that McAnelly and Cotton (1985) found that the storm exhibits its coldest cloud-top temperatures. This corresponds to the time that the MCC achieves its peak rainfall rates. It also corresponds to the time that the MCC is composed of numerous, convectively active cells.

Not only is the frequency of lightning activity of interest but so also is the sign of CG discharge. On average, over 90% of CG lightning strikes transport negative charge to the ground while only 10% transport positive charge. However, there are regions of the US such as the north-central High Plains where positive CGs exceed 20% of total CGs (see Fig. 8.69). The cause of this behavior is still under debate. Figure 8.70 is a cartoon from Williams (2001) illustrating the various theories proposed. They range from the tilted dipole theory originally proposed by Marx Brook (Krehbiel et al., 1983) to unshielded upper positive charge, to inverted dipole, to a tripole charge structure. It appears most of these hypotheses can lead to enhanced positive CGs for some storms or parts of the lifetimes of storms. Lyons et al. (1998) even linked the occurrence of enhanced positive CG activity to aerosols, specifically smoke from forest fires.

In summary, we have seen that the electrical activity of a storm varies with the organization of the storm system, the locations (marine vs. continental) and the aerosols in the storm environment. On the scale of an individual cell, charge separation appears to be linked to local regions of updraft/downdraft shear, to regions of horizontal shear of the horizontal wind, and favored where there is a large volume of graupel, a large updraft volume, and the updraft mass flux is large. The cloud-charging mechanisms are most active when ice-phase precipitation processes are prevalent. However, the centers of charge

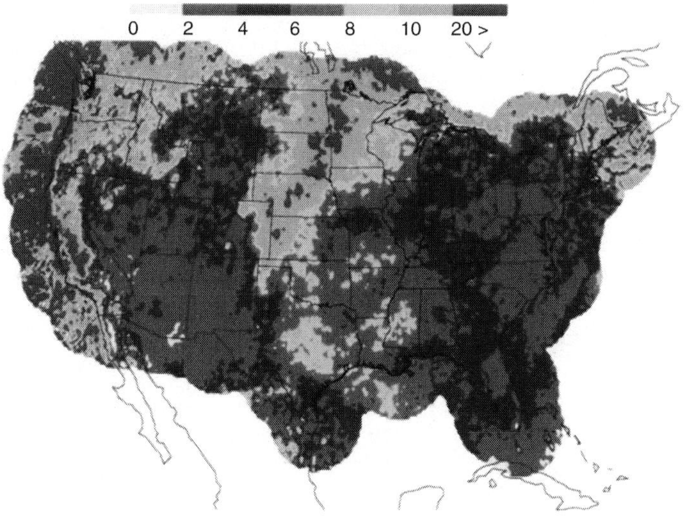


**FIGURE 8.69 The percentage of flashes lowering positive charge to the ground.** *(From Orville and Huffines (2001))*

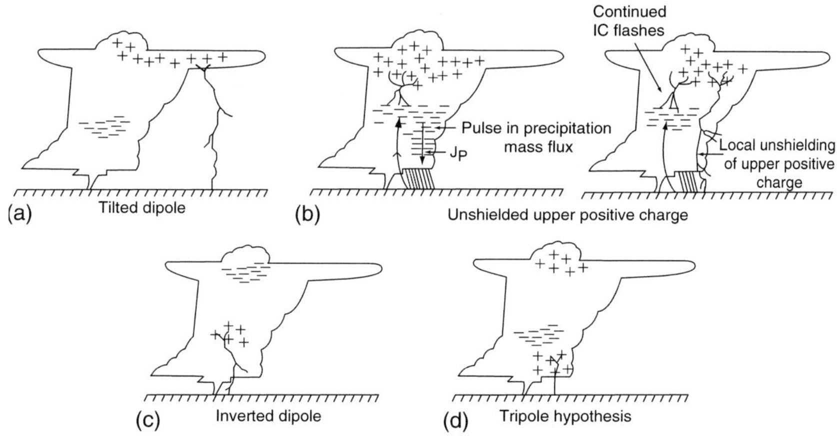


**FIGURE 8.70 Illustration of hypotheses for positive ground flash production in severe storms: (a) tilted dipole, (b) "unshielded" tilted dipole, (c) inverted dipole, and (d) tripole.** *(Adapted from Williams (2001))*

do not coincide with the regions of highest radar reflectivity or precipitation content. The observational and modeling evidence supports the non-inductive charging mechanism as being the dominant cloud charging mechanism. Finally, the most prolific producers of cloud-to-ground lightning are thunderstorm systems organized on the mesoscale that are composed of a number of multicellular convective elements. We will examine the characteristics of mesoscale convective systems more fully in Chapter 9.

## REFERENCES


Achtemeier, G. L. (1975). Doppler velocity and relectivity morphology of a severe left-moving split thunderstorm. In "Radar Meteorol. Conf., 16th, Houston, Tex." pp. 9–98. Am. Meteorol. Soc., Boston, Massachusetts. Preprints.

Ackerman, A. S., Toon, O. B., Taylor, J. P. J., Doug, W., Hobbs, P. V., and Ferek, R. J. (2000). Effects of aerosols on cloud albedo: Evaluation of Twomeys parameterization of cloud susceptibility using measurements of ship Tracks. J. Atmos. Sci. 57, 2684–2695.

Agee, E. M., Snow, J. T., Nickerson, F. S., Clare, P. R., Church, C. R., and Schaal, L. A. (1977). An observational study of the West Lafayette, Indiana, tornado of 20 March 1976. Mon. Weather Rev. 105, 893–907.

Atkins, N. T., Weisman, M. L., and Walker, L. J. (1999). The influence of preexisting boundaries on Supercell evolution. Mon. Weather Rev. 127, 2910–2927.

Atlas, D. (1966). The balance level in convective storms. J. Atmos. Sci. 23, 635–651.

Auer Jr., A. H., and Marwitz, J. D. (1968). Estimates of air and moisture flux into hailstorms on the high plains. J. Appl. Meteorol. 7, 196–198.

Baker, G. L. (1981). Boundary layers in laminar flows. Ph.D. Thesis, Purdue University, West Lafayette, Indiana.

Baker, M. B., Blyth, A. M., Christian, H. J., Latham, J., Miller, K. L., and Gadian, A. M. (1999). Relationships between lightning activity and various thundercloud parameters: satellite and modeling studies. Atmos. Res. 51, 221–236.

Barnes, S. L. (1978a). Oklahoma thunderstorms on 29-30 April 1970. Part I: Morphology of a tornadic storm. Mon. Weather Rev. 106, 673–684.

Barnes, S. L. (1978b). Oklahoma thunderstorms on 29-30 April 1970. Part II: Radar-observed merger of twin hook echoes. Mon. Weather Rev. 106, 685–696.

Barnes, G., and Garstang, M. (1982). Subcloud layer energetics of precipitating convection. Mon. Weather Rev. 110, 102–117.

Battan, L. J. (1964). Some observations of vertical velocities and precipitation sizes in a thunderstorm. J. Appl. Meteorol. 13, 415–420.

Battan, L. J. (1975). Doppler radar observations of a hailstorm. J. Appl. Meteorol. 14, 98–108.

Battan, L. J. (1980). Observations of two Colorado thunderstorms by means of a zenith-pointing Doppler radar. J. Appl. Meteorol. 19, 580–592.

Battan, L. J., and Theiss, J. B. (1970). Measurements of vertical velocities in convective clouds by means of a pulsed Doppler radar. J. Atmos. Sci. 27, 293–298.

Battan, L. J., and Theiss, J. B. (1973). Observations of vertical motion and particle sizes in a thunderstorm. J. Atmos. Sci. 23, 78–87.

Beebe, R. G., and Bates, F. C. (1955). A mechanism for assisting in the release of convective instability. Mon. Weather Rev. 83, 1–10.

Bernardet, L. R., and Cotton, W. R. (1998). Multi-scale evolution of a derecho-producing MCS. Mon. Weather Rev. 126, 2991–3015.

Betts, A. K. (1976). The thermodynamic transformation of the tropical subcloud layer by precipitation and downdrafts. J. Atmos. Sci. 33, 1008–1020.

Betts, A. K. (1984). Boundary layer thermodynamics of a High Plains severe storm. Mon. Weather Rev. 112, 2199–2211.

Blanchard, D. O., and Straka, J. M. (1998). Some possible meachanisms for tornadogenesis failure in a supercell. In "19th Conf. on Severe Local Storms, Minneapolis, MN." pp. 116–119. Amer. Meteor. Soc. Preprints.

Bluestein, H. B., and Jain, M. H. (1985). The formation of mesoscale lines of precipitation: Severe squall lines in Oklahoma during the spring. J. Atmos. Sci. 42, 1711–1732.

Bluestein, H. B., and Sohl, C. J. (1979). Some observations of a splitting severe thunderstorm. Mon. Weather Rev. 107, 861–873.

Bluestein, Howard B., and Parks, Carlton R. (1983). A synoptic and photographic climatology of low-precipitation severe thunderstorms in the southern plains. Mon. Weather Rev. 111, 2034–2046.

Bluestein, Howard B., and Woodall, Gary R. (1990). Doppler-radar analysis of a low-precipitation severe storm. Mon. Weather Rev. 118, 1640–1665.

Bosart, L. F., and Sanders, Frederick (1981). The Johnstown Flood of July 1977: A long-lived convective system. J. Atmos. Sci. 38, 1616–1642.

Boyden, C. J. (1963). A simple instability index for use as a synoptic parameter. Meteorol. Mag. 92, 198–210.

Brady, R. H., and Szoke, E. J. (1989). A case study of nonmesocyclone tornado development in northeast Colorado: Similarities to waterspout formation. Mon. Weather Rev. 117, 843–856.

Braham, R. R. (1952). The water and energy budgets of the thunderstorm and their relation to thunderstorm development. J. Meteorol. 9, 227–242.

Brandes, E. A. (1984). Vertical vorticity generation and mesocyclone sustenance in tornadic thunderstorms: The observational evidence. Mon. Weather Rev. 112, 2253–2269.

Bretherton, C. S., and Smolarkiewicz, P. K. (1989). Gravity waves, compensating subsidence and detrainment around cumulus clouds. J. Atmos. Sci. 46, 740–759.

Brown, J. M., and Knupp, K. R. (1980). The Iowa cyclonic-anticyclonic tornado pair and its parent thunderstorm. Mon. Weather Rev. 108, 1626–1646.

Brown, J. M., Knupp, K. R., and Caracena, R. (1982). Destructive winds from shallow, high-based cumuli. In "Conf. Severe Local Storms, 12th." pp. 272–275. Am. Meteorol. Soc., Boston, Massachusetts. Preprints.

Browning, K. A. (1962). Cellular structure of convective storms. Meteorol. Mag. 91, 341–350.

Browning, K. A. (1963). The growth of large hail within a steady updraft. Q. J. R. Meteorol. Soc. 89, 490–506.

Browning, K. A. (1964). Airflow and precipitation trajectories within severe local storms which travel to the right of the winds. J. Atmos. Sci. 21, 634–639.

Browning, K. A. (1965). Some inferences about the updraft within a severe local storm. J. Atmos. Sci. 22, 669–677.

Browning, K. A. (1968). The organization of severe local storms. Weather 23, 429–434.

Browning, K. A. (1977). The structure and mechanisms of hailstorms. Meteorol. Monogr. 16 (38), 1–43.

Browning, K. A., and Donaldson Jr., R. J. (1963). Airflow and structure of a tornadic storm. J. Atmos. Sci. 20, 533–545.

Browning, K. A., and Foote, G. B. (1976). Airflow and hail growth in supercell storms and some implications for hail suppression. Q. J. R. Meteorol. Soc. 102, 499–533.

Browning, K. A., and Ludlam, F. H. (1960). Radar analysis of a hailstorm. Tech. Note No. 5, Dep. Meteorol. Imperial College, London.

Browning, K. A., and Ludlam, F. H. (1962). Airflow in convectioe storms. Q. J. R. Meteorol. Soc. 88, 117–135.

Browning, K. A., Ludlam, F. H., and Macklin, W. R. (1963). The density and structure of hailstone. Q. J. R. Meteorol. Soc. 89, 75–84.

Browning, K. A., and Wexler, R. (1968). The determination of kinematic properties of a wind field using Doppler radar. J. Appl. Meteorol. 7, 105–113.

Bryan, G. H. (2005). Spurious convective organization in simulated squall lines owing to moist absolutely unstable layers. Mon. Weather Rev. 133, 1968–1997.

Bryan, G. H., and Rotunno, R. (2008). Gravity currents in a deep anelastic atmosphere. J. Atmos. Sci. 65, 536–556.

Burgers, J. M. (1948). A mathematical model illustrating the theory of turbulence. Adv. Appl. Mech. 1, 171–199.

Burgess, D. W., and Davies-Jones, R. P. (1979). Unusual tornadic storms in Eastern Oklahoma on 5 December 1975. Mon. Weather Rev. 107, 451–457.

Byers, H. R., and Braham, R. R. (1949). The Thunderstorm. US Weather Bur., Washington, DC.

Caracena, F., Maddox, R. A., Hoxit, L. R., and Chappell, C. F. (1979). Forecasting likelihood of microbursts along the front range of Colorado-Results of the JAWS project. In "Conf. Severe Local Storms, 13th." pp. 262–264. Am. Meteorol. Soc., Boston, Massachusetts. Preprints.

Charba, J. (1974). Application of gravity current model to analysis of squall line gust front. Mon. Weather Rev. 102, 140–156.

Charney, J. G., and Eliassen, A. (1964). On the growth of the hurricane depression. J. Atmos. Sci. 21, 68–75.

Chisholm, A. J. (1973). Radar case studies and airflow-models. Part I, Alberta hailstorms. Meteorol. Monogr. 36, 1–36.

Chisholm, A. J., and Renick, J. H. (1972). The kinematics of multicell and supercell Alberta hailstorms, Alberta Hail Studies, 1972. Res. Counc. Alberta Hail Stud. Rep. No. 72–2, pp. 24–31.

Chiu, C.-S. (1978). Numerical study of cloud electrification in an axisymmetric, time-dependent cloud model. J. Geophys. Res. 83, 5025–5049.

Chiu, C. S., and Klett, J. D. (1976). Convective electrification of clouds. J. Geophys. Res. 81, 1111–1124.

Church, C. R., Snow, J. T., Baker, G. L., and Agee, E. M. (1979). Characteristics of tornadolike vortices as a function of swirl radio: A laboratory investigation. J. Atmos. Sci. 36, 1755–1776.

Church, D. R., and Snow, J. T. (1993). Laboratory models of tornadoes. In "The Tornado: Its Structure, Dynamics, Prediction, and Hazards" (C. Church, D. Burgess, C. Doswell and R. Davies-Jones, Eds.), In Geophys. Monorgr., vol. 79. pp. 277–295. American Geophysical Union.

Cicil, D. J., Zipser, E. J., and Nesbitt, S. W. (2008). Reflectivity, ice scattering, and lightning characteristics of hurricane eyewalls and rain. Part I: Quantitative description. Mon. Weather Rev. 130, 769–784.

Clark, T. L. (1979). Numerical simulations with a three-dimensional cloud model: Lateral boundary condition experiments and multicellular storm simulations. J. Atmos. Sci. 36, 2191–2215.

Colgate, S. (1975). Comment on "On the relation of electrical activity to tornadoes by R P. Davies-Jones and J. H. Golden". J. Geophys. Res. 80, 4556.

Colgate, S. A., Levin, Z., and Petschek, A. G. (1977). Interpretation of thunderstorm charging by the polarization-induction mechanism. J. Atmos. Sci. 34, 1433–1443.

Cooper, H. J., Garstang, M., and Simpson, J. (1982). The diurnal interaction between convection and peninsular-scale forcing over South Florida. Mon. Weather Rev. 110, 486–503.

Cotton, W. R. (1972). Numerical simulation of precipitation development in supercooled cumuli, Part 1. Mon. Weather Rev. 100, 757–763.

Cotton, W. R., and Anthes, R. A. (1989). "Storm and Cloud Dynamics." In International Geophysics Series, vol. 44. Academic Press, Inc., San Diego, 883 pp.

Cotton, W. R., and Pielke, R. A. (2007). "Human Impacts on Weather and Climate." 2nd Edition, Cambridge University Press.

Cotton, W. R., Gannon, P. T., and Pielke, R. A. (1974). Numerical experiments of the influence of the mesoscale circulation on the cumulus scale. In "Conf. Cloud Phys., Tucson, Ariz." pp. 424–429. Am. Meteorol. Soc., Boston, Massachusetts.

Cotton, W. R., George, R. L., and Knupp, K. R. (1982). An intense, quasi-steady thunderstorm over mountainous terrain. Part I: Evolution of the storm-initiating mesoscale circulation. J. Atmos. Sci. 39, 328–342.

Crow, E. L., Summers, P. W., Long, A. B., Knight, C. A., Foote, G. B., and Dye, J. E. (1976). Experimental Results and Overall Summary. Vol. I. Final Rep., Natl. Hail Res. Exp. Randomized Seeding Exp. 1972–1974. Natl. Cent. Atmos. Res., Boulder, Colorado.

Curtis, L. (2004). Midlevel dry intrusions as a factor in tornado outbreaks associated with landfalling tropical cycles from the Atlantic and Gulf of Mexico. Wea. Forecasting 19, 411–427.

Danielsen, E. F. (1977). Inherent difficulties in hail probability prediction. Meteorol. Monogr. 38, 135–143.

Danielsen, E. F., Bleck, R., and Morris, D. (1972). Hail growth by stochastic collection in a cumulus model. J. Atmos. Sci. 29, 133–155.

Das, P. (1962). Influence of wind shear on the growth of hail. J. Atmos. Sci. 19, 407–414.

Das, P., and Subba Rao, M. C. (1972). The unsaturated downdraft. Indian J. Meteorol. Geophys. 23.

Davies, H. C. (1979). Phase-lagged wave-CISK. Q. J. R. Meteorol. Soc. 105, 325–353.

Davies-Jones, R. P. (1982). Tornado dynamics. In "Thunderstorms: A Social. Scientific, and Technical Documentary, vol. 2" (E. Kessler, Ed.) US Dept. of Commerce, Washington, DC.

Davies-Jones, R. P. (1983). Tornado dynamics. In "Thunderstorm Morphology and Dynamics, vol. 2" (E. Kessler, Ed.), pp. 197–236. University of Oklahoma Press, Norman, OK.

Davies-Jones, R. (1984). Streamwise vorticity: The origin of updraft rotation in supercell storms. J. Atmos. Sci. 41, 2991–3006.

Davies-Jones, R., and Brooks, H. (1993). Mesocyclogenesis from a theoretical perspective. In "The tornado: Its structure, dynamics, prediction, and hazards." Geophys. Monogr. 79, 105–114.

Davies-Jones, R. P., and Golden, J. H. (1975a). On the relation of electrical activity to tornadoes. J. Geophys. Res. 80, 1614–1616.

Davies-Jones, R. P., and Golden, J. H. (1975b). Reply. J. Geophys. Res. 80, 4557–4558.

Davies-Jones, R. P., and Golden, J. H. (1975c). Reply. J. Geophys. Res. 80, 4561–4562.

Davies-Jones, R. P., and Kessler, E. (1974). Tornadoes. In "Weather Climate Modification" (W. N. Hess, Ed.), pp. 552–595. Wiley, New York.

Davies-Jones, R. P., Trapp, R. J., and Bluestein, H. B. (2001). Tornadoes and tornadic storms. In "Severe Convective Storms" (C. A. Doswell III, Ed.), In Meteorological Monographs, vol. 28, No. 50. pp. 167–222.

Davis, M. H. (1961). The forces between conducting spheres in a uniform electric field. RM-2607. Rand Corp., 1700 Main St., Santa Monica, California 90406.

Davis, M. H. (1964a). Two charged spherical conductors in a uniform electric field: Forces and field strength. RM-3860-PR. Rand Corp.

Davis, M. H. (1964b). Two charged spherical conductors in a uniform electric field: Forces and field strength. Q. J. Mech. Appl. Math. 17, 499–511.
Dennis, A. S. (1980). "Weather Modification by Cloud Seeding." Academic Press, New York.
Dennis, A. S., Schock, C. A., and Koscielski, A. (1970). Characteristics of hailstorms of Western South Dakota. J. Appl. Meteorol. 9, 127–135.
Dennis, A. S., Schleusener, R. A., Hirsch, J. H., and Koscielski, A. (1973). Meteorology of the Black Hills Flood of 1972. Rep. No. 73–4, Inst. Atmos. Sci., South Dakota Sch. Mines Technol., Rapid City.
Donaldson Jr., R. J., and Wexler, R. (1969). Flight hazards in thunderstorms determined by Doppler velocity variance. J. Appl. Meteorol. 8, 128–133.
Doswell III, C. A. (1985). The Operational Meteorology of Convective Weather Volume II: Storm Scale Analysis. NOAA Tech. Memo. ERL ESG-15, Boulder, 240 pp.
Doswell III, C. A., and Burgess, D. W. (1993). Tornadoes and tornadic storms: A review of conceptual models. In "The tornado: Its Structure, Dynamics, prediction, and hazards." In Geophys. Monogr., vol. 79. pp. 161–172. Amer. Geophys. Union.
Doswell III, C. A., Moller, A. R., and Przybylinski, R. (1990). A unified set of conceptual models for variations on a supercell theme. In "Proc. 16th Conf on Severe Local Storms, Kananaskis Park." pp. 40–45. Amer. Meteor. Soc., Alberta, Canada.
Doswell, C. A., Brooks, H. E., and Maddox, R. A. (1996). Flash flood forecasting: An ingredients-based methodology. Wea. Forecasting 11, 560–581.
Droegemeier, K. K., and Wilhelmson, R. B. (1985a). Three-dimensional numerical modeling of convection produced by interacting thunderstorm outflows: Part I. Control simulation and low-level moisture variations. J. Atmos. Sci. 42, 2381–2403.
Droegemeier, K. K., and Wilhelmson, R. B. (1985b). Three-dimensional numerical modeling of convection produced by interacting thunderstorm outflows: Part II. Variations in vertical wind shear. J. Atmos. Sci. 42, 2404–2414.
Dutton, J. A. (1976). "The Ceaseless Wind." McGraw-Hill, New York.
Eastin, M. D., Link, M. C., and Anderson, H. B. (2007). Analysis of offshore deep convection within landfalling hurricanes just prior to tornadogenesis. In "7th Conference on Coastal Processes." American Meteorological Society, San Diego, CA. Preprints.
Ellrod, G. P., and Marwitz, J. D. (1976). Structure and interaction in the subcloud region of thunderstorms. J. Appl. Meteorol. 15, 1083–1091.
Elster, J., and Geitel, H. (1913). Zur Influenztheorie der Niederschlagselektrizitat. Phys. Z 14, 1287.
English, M. (1973). Alberta hailstorms. Part II: Growth of large hail in the storm. Meteorol. Monogr. 36, 37–98.

Fankhauser, J. C. (1976). Structure of an evolving hailstorm, Part II: Thermodynamic structure and airflow in the near environment. Mon. Weather Rev. 104, 576–587.

Fankhauser, J. C. (1982). The 22 June 1976 case study: Large-scale influences, radar echo structure and mesoscale circulations. In "Hailstorms of the Central High Plains, vol. 2" (C. A. Knight and P. Squires, Eds.), pp. 1–33. Colorado Assoc. University Press, Boulder.

Fawbush, E. F., and Miller, R. (1953). A method for forecasting hailstone size at the earth's surface. Bull. Am. Meteorol. Soc. 34, 235–244.

Finley, C. A. (1997). Numerical simulation of intense multi-scale vortices generated by supercell thunderstorms. Ph.D. Dissertation, Colorado State University, Dept. of Atmospheric Science, Fort Collins, CO 80523, 297 pp. (Atmospheric Science Paper No. 640).

Finley, C. A., Cotton, W. R., and Pielke, R. A. (2001). Numerical simulation of tornadogenesis in a high-precipitation supercell. Part I: Storm evolution and transition into a bow echo. J. Atmos. Sci. 58, 597-1,629.

Foote, G. B. (1984). A study of hail growth utilizing observed storm condition. J. Clim. Appl. Meteorol. 23, 84–101.

Foote, G. B., and Fankhauser, J. C. (1973). Airflow and moisture budget beneath a Northeast Colorado hailstorm. J. Appl. Meteorol. 12, 1330–1353.

Foote, G. B., and Frank, H. W. (1983). Case study of a hailstorm in Colorado. Part III: Airflow from triple Doppler measurements. J. Atmos. Sci. 40, 686–707.

Foote, G. B., and Knight, C. A. (1979). Results of a randomized hail suppression experiment in Northeast Colorado. Part I: Design and conduct of the experiment. J. Appl. Meteorol. 18, 1526–1537.

Foote, G. B., and Wade, C. G. (1982). Case study of a hailstorm in Colorado. Part I: Radar echo structure and evolution. J. Atmos. Sci. 39, 2828–2846.

Forbes, G. S., and Wakimoto, R. M. (1983). A concentrated outbreak of tornadoes, downbursts and microbursts, and implications regarding vortex classification. Mon. Weather Rev. 110, 220–235.

Foster, D. S., and Bates, F. (1956). A hail size forecasting technique. Bull. Am. Meteorol. Soc. 37, 135–141.

Frank, H. W., and Foote, G. B. (1982). The 22 July 1976 case study: Storm airflow, updraft structure, and mass flux from triple-Doppler measurements. In "Hailstorms of the Central High Plains, vol. 2" (C. A. Knight and P. Squires, Eds.), pp. 131–162. Colorado Assoc. University Press, Boulder.

Frisby, E. M. (1962). Relationship of ground hail damage patterns to features of the synoptic map in the Upper Great Plains of the United States. J. Appl. Meteorol. 1, 348–352.

Frisch, A. S., and Clifford, S. F. (1974). A study of convection capped by a stable layer using Doppler radar and acoustic echo sounders. J. Atmos. Sci. 31, 1622–1628.

Frisch, A. S., and Strauch, R. G. (1976). Doppler-radar measurements of turbulence kinetic energy dissipation rates in a northeastern Colorado convective storm. J. Appl. Meteorol. 15, 1012–1017.

Fritsch, J. M. (1975). Cumulus dynamics: Local compensating subsidence and its implications for cumulus parameterization. Pure Appl. Geophys. 113, 851–867.

Fritsch, J. M., and Rodgers, D. M. (1981). The Fort Collins Hailstorm-An example of the short-term forecast enigma. Bull. Am. Meteorol. Soc. 62, 1560–1569.

Fujita, T. T. (1959). Precipitation and cold air production in mesoscale thunderstorm systems. J. Meteorol. 16, 454–466.

Fujita, T. T. (1960). A detailed analysis of the Fargo tornadoes of June 20, 1957. USWB Res. Pap. No. 42, Chicago, Illinois.

Fujita, T. T. (1970). The Lubbock tornadoes. A study of suction spots. Weatherwise 23, 161–173.

Fujita, T. T. (1974). Overshooting thunderheads observed from ATS and Learjet. Satellite Mesometeorol. Res. Pap. No. 117, Dep. Geophys. Sci., University of Chicago.

Fujita, T. T. (1978). Manual of downburst identification for project NIMROD. Satellite Mesometeorol. Res. Pap. No. 156, Dep. Geophys. Sci., University of Chicago.

Fujita, T. (1981). Tornadoes and downbursts in the context of generalized planetary scales. J. Atmos. Sci. 38, 1511–1534.

Fujita, T. T. (1985). The downburst. SMRP Research Paper No. 210, the University of Chicago, 122 pp. [NTIS PB-148880.].

Fujita, T. T. (1986). The downburst. SMRP Research Paper No. 217, the University of Chicago, 155 pp. [NTIS PB-86-131638.].

Fujita, T. T., and Byers, H. R. (1977). Spearhead echo and downburst in the crash of an airliner. Mon. Weather Rev. 105, 129–146.

Fujita, T. T., and Caracena, F. (1977). An analysis of three weather-related aircraft accidents. Bull. Am. Meteorol. Soc. 58, 1164–1181.

Fujita, T., and Grandoso, H. (1968). Split of a thunderstorm into anticyclonic and cyclonic storms and their motion as determined from numerical model experiments. J. Atmos. Sci. 25, 416–439.

Fujita, T., Hjelmfelt, M. R., and Changnon, S. A. (1977). Mesoanalysis of record Chicago rainstorm using radar, satellite, and rainguage data. In "10th Conf. on Severe Local Storms." pp. 65–72. Amer. Meteor. Soc., Omaha, NE. Preprints.

Fujita, T., and Wakimoto, R. M. (1981). Five scales of airflow associated with a series of downbursts on 16 July 1980. Mon. Weather Rev. 109, 1438–1456.

Fujita, T., Watanabe, K., Tsuchiya, K., and Schimada, M. (1972). Typhoon-associated tornadoes in Japan and new evidence of suction vortices in a tornado near Tokyo. J. Meteorol. Soc. Jpn 50, 431–453.

Gall, R. (1983). A linear analysis of the multiple vortex phenomenon in simulated tornadoes. J. Atmos. Sci. 40, 2010–2024.

Gall, R., and Staley, D. O. (1981). Nonlinear barotropic instability in a tornado vortex. In "3rd Conf. Atmos. Waves Stab." Amer. Meteorol. Soc., Boston, Mass.

Galway, J. G. (1956). The lifted index as a predictor of latent instability. Bull. Am. Meteorol. Soc. 37, 528–529.

Gaskell, W. (1979). Field and laboratory studies of precipitation charge. Ph.D. Thesis, University of Manchester.

Gaudet, B. J., and Cotton, W. R. (2006). Low-level mesocyclonic concentration by non-axisymmetric processes. Part I: Supercell and mesocyclone evolution. J. Atmos. Sci. 63, 1113–1133.

Gaudet, B. J., Cotton, W. R., and Montgomery, M. T. (2006). Low-level mesocyclonic concentration by non-axisymmetric processes. Part II: Vorticity dynamics. J. Atmos. Sci. 63, 1134–1150.

Gentry, R. C. (1983). Genesis of tornadoes associated with hurricanes. Mon. Weather Rev. 115, 1793–1805.

Goff, R. C. (1975). Thunderstorm outflow kinematics and dynamics. NOAA Tech. Memo. ERL NSSL-75, Nad. Severe Storms Lab., Norman, Oklahoma.

Glimore, M. S., and Wicker, L. J. (2002). Influences of the local environment on supercell cloud-to-ground lightning, radar characteristics, and severe weather on 2 June 1995. Mon. Weather Rev. 130, 2349–2372.

Gilmore, M. S., Straka, J. J., and Rasmussen, E. N. (2004). Precipitation uncertainty due to variations in precipitation particle parameters within a simple microphysics scheme. Mon. Weather Rev. 132, 2610–2627.

Goff, R. C. (1976). Vertical structure of thunderstorm outflow. Mon. Weather Rev. 104, 1429–1440.

Golden, J. H., and Purcell, D. (1978). Airflow characteristics around the Union City Tornado. Mon. Weather Rev. 106, 22–28.

Goodman, S. J., and MacGorman, D. R. (1986). Cloud-to-ground lightning activity in mesoscale convective complexes. Mon. Weather Rev. 114, 2320–2328.

Grandia, K. L., and Marwitz, J. D. (1975). Observational investigations of entrainment within the weak echo region. Mon. Weather Rev. 103, 227–234.

Grassl, H (1975). Albedo reduction and radiative heating of clouds by absorbing aerosol particles. Contribution to Atmos. Phys. 48, 199–210. Oxford.

Grasso, L. D. (1992). A numerical simulation of tornadogenesis. M.S. Thesis, Colorado State University, Dept. of Atmospheric Science, Fort Collins, CO 80523, Atmospheric Science Paper No. 495, 102 pp.

Grasso, L. D. (1996). Numerical simulation of the May 15 and April 26, 1991 thunderstorms. Ph.D. dissertation, Colorado State University, Dept. of Atmospheric Science, Fort Collins, CO 80523, Atmospheric Science Paper No. 596, 151 pp.

Grasso, L. D., and Cotton, W. R. (1995). Numerical simulation of a tornado vortex. J. Atmos. Sci. 52, 1192–1203.

Gray, W. M. (1965). Calculation of cumulus vertical draft velocities in hurricanes from aircraft observations. J. Appl. Meteorol. 4, 47–53.

Grenet, G. (1947). Essai d'explication de la charge electrique des nuage d'orages. Ann. Geophys. 3, 306–307.

Grzych, M. L., Lee, B. D., and Finley, C. A. (2007). Thermodynamic analysis of Supercell rear-flank downdrafts from Project ANSWERS. Mon. Weather Rev. 135, 240–246.

Gutman, L. N. (1963). Stationary axially symmetric model of a cumulus cloud. Dokl. Akad. Nauk SSSR 150 (1).

Gutman, L. N. (1967). Calculation of the velocity of ascending currents in a stationary convective cloud. In "Formation of Precipitation and Modification of Hail Processes" (E. K. Fedorov, Ed.), p. 12. Isr. Program Sci. Transl., Jerusalem.

Haagenson, P. L., and Danielsen, E. (1972). Operational steady-state model. NCAR Intern. Rep., Natl. Cent. Atmos. Res., Boulder, Colorado.

Hamilton, R. A., and Archbold, J. W. (1945). Meteorology of Nigeria and adjacent territory. Q. J. R. Meteorol. Soc. 71, 231–265.

Hansen, J., Sato, M., and Ruedy, R. (1997). Radiative forcing and climate response. J. Geophys. Res. 102, 6831–6864.

Helsdon, J. H., Gattaleeradapan, S., Farley, R. D., and Waits, C. C. (2002). An examination of the convective charging hypothesis: Charge structure, electric fields, and Maxwell currents. J. Geophys. Res. 107, 4630. doi:10.1029/2001JD001495.

Heymsfield, A. J., and Musil, D. J. (1982). Case study of a hailstorm in Colorado. Part II: Particle growth processes in mid-levels deduced from in-situ measurements. J. Atmos. Sci. 39, 2847–2866.

Heymsfield, A. J., Jameson, A. R., and Frank, H. W. (1980). Hail growth mechanisms in a Colorado storm. Part II: Hail formation processes. J. Atmos. Sci. 37, 1779–1807.

Hjelmfelt, M. R. (1987). The microbursts of 22 June 1982 in JAWS. J. Atmos. Sci. 44, 1646–1665.

Hobbs, P. V., Matejka, T. J., Herzegh, P. H., Locatelli, J. D., and Houze, R. A. (1980). The mesoscale and microscale structure and organization of clouds and precipitation in midlatitude cycles. I: A case of a cold front. J. Atmos. Sci. 37, 568596.

Hoecker, W. H. (1960a). Windspeed and air flow patterns in the Dallas tornado of April 2, 1967. Mon. Weather Rev. 88, 167–180.

Hoecker, W. H. (1960b). The dimensional and rotational characteristics of the tornadoes and their parent cloud systems. USWB Res. Pap. No. 41, pp. 53–112. Washington, DC.

Holle, R. L., and Maier, M. (1980). Tornado formation from downdraft interaction in the FACE mesonetwork. Mon. Weather Rev. 108, 1010–1028.

Hookings, G. A. (1965). Precipitation maintained downdrafts. J. Appl. Meteorol. 4, 190–195.

Houze Jr., R. A. (1977). Structure and dynamics of a tropical squall-line system. Mon. Weather Rev. 105, 1541–1567.

Hoxit, L. R., Chappell, C. F., and Fritsch, J. M. (1976). Formation of mesolows or pressure troughs in advance of cumulonimbus clouds. Mon. Weather Rev. 104, 1419–1428.

Hoxit, L. R., Maddox, R. A., Chappell, C. F., Zuckerberg, F. L., Mogil, H. M., Jones, I., Greene, D. R., Safe, R. E., and Scofield, R. A. (1978). Meteorological analysis of the Johnstown, Pennsylvania, flash flood, 19–20 July 1977. NOAA Tech. Rep. ERL 401-APCL 43.

Humphreys, W. J. (1928). The uprush of air necessary to sustain the hailstone. Mon. Weather Rev. 56, 314.

Illingworth, A. J., and Latham, J. (1977). Calculations of electric field structure and charge distributions in thunderstorms. Q. J. R. Meteorol. Soc. 103, 281–295.

Iribarne, J. V., and de Pena, R. G. (1962). The influence of particle concentration on the Hailstones. Nubila 5, 7–30.

Jameson, A. R., and Heymsfield, A. J. (1980). Hail growth mechanisms in a Colorado storm. Part I: Dual-wavelength radar observations. J. Atmos. Sci. 37, 1763–1778.

Jayaratne, E. R., Saunders, C. P. R., and Hallett, J. (1983). Laboratory studies of the charging of soft-hail during ice crystal interactions. Q. J. R. Meteorol. Soc. 109, 609–630.

Johns, R. H., and Hirt, W. D. (1983). The derecho-A severe weather producing convective system. In "Conf Severe Local Storms, 13th." pp. 178–181. Am. Meteorol. Soc., Boston, Massachusetts. Preprints.

Johnson, R. H., and Nicholls, M. (1983). A compositive analysis of the boundary layer accompanying a tropical squall line. Mon. Weather Rev. 111, 308–319.

Jorgensen, D. P. (1984). Mesoscale and convective-scale characteristics of mature hurricanes. Ph.D. Thesis, Colorado State University.

Kamburova, P. L., and Ludlam, F. H. (1966). Rainfall evaporation in thunderstorm downdrafts. Q. J. R. Meteorol. Soc. 92, 510–518.

Khain, A. P., Pokrovsky, A., BenMoshe, N., and Rosenfeld, D. (2005). Simulating green-ocean smoky and pyro-clouds observed in the Amazon region during the LBA-SMOCC campaign. J. Atmos. Sci. 61, 2963–2982.

Klemp, J. B. (1987). Dynamics of tornadic thunderstorms. Ann. Rev. Fluid Mech. 19, 369–402.

Klemp, J. B., and Rotunno, R. (1983). A study of the tornadic region within a supercell thunderstorm. J. Atmos. Sci. 40, 359–377.

Klemp, J. B., and Wilhelmson, R. B. (1978). The simulation of three-dimensional convective storm dynamics. J. Atmos. Sci. 35, 1070–1096.

Klemp, J. B., Wilhelmson, R. B., and Ray, P. (1981). Observed and numerically simulated structure of a mature supercell thunderstorm. J Atmos. Sci. 38, 1558–1580.

Klemp, J. B., and Wilhelmson, R. B. (1978a). The simulation of three-dimensional convective storm dynamics. J. Atmos. Sci. 35, 1070–1096.

Klemp, J. B., and Wilhelmson, R. B. (1978b). Simulations of right- and left-moving storms produced through storm splitting. J. Atmos. Sci. 35, 1097–1110.

Kloth, C. M., and Davies-Jones, R. P. (1980). The relationship of the 300-mb jet stream to tornado occurrence. NOAA Tech. Memo. ERL NSSL-88, Natl. Severe Storm Lab., Norman, Oklahoma.

Knight, C. A., Knight, N. C., Dye, J. E., and Toutenhoofd, V. (1974). The mechanism of precipitation formation in Northeastern Colorado cumulus. 1. Observations of the precipitation itself. J. Atmos. Sci. 31, 2142–2147.

Knupp, K. R. (1985). Precipitation convective downdraft structure: A synthesis of observations and modeling. Ph.D. Thesis, Dep. Atmos. Sci., Colorado State University.

Knupp, K. R. (2006). Observational analysis of a gust gront to bore to solitary wave transition within an evolving nocturnal boundary layer. J. Atmos. Sci. 63, 2016–2035.

Knupp, K. R., and Cotton, W. R. (1982a). An intense, quasi-steady thunderstorm over mountainous terrain. Part 11: Doppler radar observations of the storm morphological structure. J. Atmos. Sci. 39, 343–358.

Knupp, K. R., and Cotton, W. R. (1982b). An intense, quasi-steady thunderstorm over mountainous terrain-Part III: Doppler radar observations of the turbulence structure. J. Atmos. Sci. 39, 359–368.

Knupp, K. R., Paech, S., and Goodman, S. (2003). Variations in cloud-to-ground lightning characteristics among three adjacent tornadic Supercell storms over the Tennessee Valley region. Mon. Weather Rev. 131, 172–188.

Koch, S. E., and Clark, W. L. (1999). A nonclassical cold front observed during COPS-91: Frontal structure and the process of severe storm initiation. J. Atmos. Sci. 56, 2862–2890.

Koren, I., Kaufman, Y. J., Rosenfeld, D., Remer, L. A., and Rudich, Y. (2005). Aerosol invigoration and restructuring of Atlantic convective clouds. Geophys. Res. Lett. 32, L14828. doi:10.1029/2005GL023187.

Koscielski, A. (1965). 110 tornado forecasts and reasons why they did or did not verify. Unpublished manuscript, US Weather Bur. (NSSFC, Rm. 1728, Federal Bldg., 601 E. 12th St., Kansas City, Missouri 64106).

Krauss, T. W., and Marwitz, J. D. (1984). Precipitation processes within an Alberta supercell hailstorm. J. Atmos. Sci. 41, 1025–1034.

Krehbiel, P. R., Brook, M., Lhermitte, R. L., and Lennon, C. L. (1983). Lightning charge structure in thunderstorms. In "Proceedings in Atmospheric electricity" (Lothar H. Ruhnke and John Latham, Eds.), pp. 408–410. A. Deepak Publ., Hampton, Virginia.

Kropfli, R. A., and Miller, L. J. (1976). Kinematic structure and flux quantities in a convective storm from dual-Doppler radar observation. J. Atmos. Sci. 33, 520–529.

Krumm, W. R. (1954). On the cause of downdrafts from dry thunderstorms over the plateau area of the United States. Bull. Am. Meteorol. Soc. 35, 122–126.

Kuhlman, K. M., Ziegler, C. L., Mansell, E. R., MacGorman, D. R., and Straka, J. M. (2006). Numerically simulated electrification and lightning of the 29 June 2000 STEPS Supercell storm. Mon. Weather Rev. 134, 2734–2757.

Latham, J. (1969). Experimental studies of the effect of electric fields on the growth of cloud particles. Q. J. R. Meteorol. Soc. 95, 349–361.

Latham, J., and Mason, B. J. (1961). Generation of electric charge associated with the formation of soft hail in thunderclouds. Proc. R. Soc. London, Ser. A 260, 537–549.

Latham, J., and Mason, B. J. (1962). Electrical charging of hail pellets in a polarizing field. Proc. R. Soc. London, Ser. A 266, 387–401.

Leary, C. A., and Houze Jr., R. A. (1979a). The structure and evolution of convection in a tropical cloud cluster. J. Atmos. Sci. 36, 437–457.

Leary, C. A., and Houze Jr., R. A. (1979b). Melting and evaporation of hydrometers in precipitation from the anvil clouds of deep tropical convection. J. Atmos. Sci. 36, 669–679.

Lee, B. D., and Wilhelmson, R. B. (1997). The numerical simulation of nonsupercell tornadogenesis. Part I: Evolution of a family of tornadoes along a weak outflow boundary. J. Atmos. Sci. 54, 2387–2415.

Lemon, L. R. (1976). The flanking line, a severe thunderstorm intensification source. J. Atmos. Sci. 33, 686–694.

Lemon, L. R., and Doswell III, C. A. (1979). Severe thunderstorm evolution and mesocyclone structure as related to tornadogenesis. Mon. Weather Rev. 107, 1184–1197.

Lemon, L. R., Burgess, D. W., and Brown, R. A. (1975) Tornado production and storm sustenance. In "Conf. Severe Local Storms, 9th, Norman, Okla." pp. 100–104. Am. Meteorol. Soc., Boston, Massachusetts. Preprints.

LeMone, M. A., and Zipser, E. J. (1980). Cumulonimbus vertical velocity events in GATE. Part I: Diameter, intensity and mass flux. J. Atmos. Sci. 37, 2444–2457.

Lerach, D. G., Gaudet, B. J., and Cotton, W. R. (2008). Idealized simulations of aerosol influences on tornadogenesis. Geophys. Res. Lett. 35, L23806. doi:10.1029/2008GL035617.

Levin, Z. (1976). A refined charge distribution in a stochastic electrical model of an infinite cloud. J. Atmos. Sci. 33, 1756–1762.

Levin, Z., and Ziv, A. (1974). The electrification of thunderclouds and the rain gush. J. Geophys. Res. 79, 2699–2704.

Levy, G., and Cotton, W. R. (1985). A numerical investigation of mechanisms linking glaciation of the ice-phase to the boundary layer. J. Clim. Appl. Meteorol. 23, 1505–1519.

Lewellen, W. S. (1976). Theoretical models of the tornado vortex. In "Proceedings of Symposium on Tornados. Texas Tech." (R.E. Peterson, Ed.). University, Lubbock, pp. 107–143.
Lewellen, W. S. (1993). Tornado vortex theory. In "The tornado: Its structure, dynamics, prediction, and hazards" (C. Church, D. Burgess, C. Doswell and R. Davies-Jones, Eds.), In Geophysical Monogr., vol. 79. pp. 19–39. American Geophysical Union.
Lewellen, D. C., and Lewellen, W. S. (2007a). Near-surface intensification of tornado vortices. J. Atmos. Sci. 64, 2176–2194.
Lewellen, D. C., and Lewellen, W. S. (2007b). Near-surface intensification through corner flow collapse. J. Atmos. Sci. 64, 2195–2209.
Lewellen, D. C., Lewellen, W. S., and Xia, J. (2000). The influence of a local swirl ratio on tornado intensification near the surface. J. Atmos. Sci. 57, 527–544.
Lilly, D. K. (1986a). The structure, energetics and propagation of rotating convective storms. Part I: Energy exchange with the mean flow. J. Atmos. Sci. 43, 113–125.
Lilly, D. K. (1986b). The structure, energetics and propagation of rotating convective storms. Part II: Helicity and storm stabilization. J. Atmos. Sci. 43, 126–140.
Lindblad, N. R., and Semonin, R. G. (1963). Collision efficiency of cloud droplets in electric fields. J. Geophys. Res. 68, 1051–1057.
Lindzen, R. S. (1974). Wave-CISK in the tropics. J. Atmos. Sci. 31, 156–179.
Lindzen, R. S., and Tung, K. K. (1976). Banded convective activity and ducted gravity waves. Mon. Weather Rev. 104, 1602–1617.
List, R. (1982). Properties and growth of hailstones. In "Thunderstorms: A Social, Scientific, and Technological Documentary, Vol. 2, Thunderstorm Morphology and Dynamics." pp. 409–445. US Dep. Commer., Washington, DC.
Lucas, C., Zipser, E., and LeMone, M. A. (1994). Vertical velocity in oceanic convection off tropical Australia. J. Atmos. Sci. 51, 3183–3193.
Lucas, C., Zipser, E., and LeMone, M. A. (1996). Reply. J. Atmos. Sci. 53, 1212–1214.
Ludlam, F. H. (1963). Severe local storms: A review. In "Severe Local Storms." In Meteorol. Monogr., vol. 27. pp. 1–30.
Lynn, B., Khain, A., Dudhia, J., Rosenfeld, D., Pokrovsky, A., and Seifert, A. (2005). Spectral (bin) microphysics coupled with a mesoscale model (MM5). Part 2: Simulation of a CaPerain event with squall line. Mon. Weather Rev. 133, 59–71.
Lyons, W. A., and Keen, C. S. (1994). Observations of ligtning in convective supercells within tropical storms and hurricanes. Mon. Weather Rev. 122, 1897–1916.
Lyons, W. A., Nelson, T. E., Williams, E. R., Cramer, J. A., and Turner, T. R (1998). Enhanced positive cloud-to-ground lightning in stunderstorms inegesting smoke from fires. Science 282, 77–80.

MacDonald, A. E. (1976). Gusty surface winds and high level thunderstorms. Natl. Weather Serv. West. Region Tech. Attachment No. 76-14.

MacGorman, D. R., and Rust, W. D. (1998). "The Electrical Nature of Storms." Oxford University Press, 422 pp.

Maddox, R. A. (1976). An evaluation of tornado proximity wind and stability data. Mon. Weather Rev. 104, 133–142.

Maddox, R. A. (1980). Mesoscale convective complexes. Bull. Am. Meteorol. Soc. 61, 1374–1387.

Maddox, R. A., Chappell, C. F., Hoxit, L. R., and Caracena, F. (1978). Comparison of meteorological aspects of the Big Thompson and Rapid City flash floods. Mon. Weather Rev. 106, 375–389.

Maddox, R. A., Chappell, C. F., and Hoxit, L. R. (1979). Synoptic and meso-$\alpha$-scale aspects of flash flood events. Bull. Am. Meteorol. Soc. 60, 115–123.

Maddox, R. A., and Doswell, C. A. (1982). An examination of jet stream configurations, 500 mb vorticity advection and low-level thermal advection patterns during extended periods of intense convection. Mon. Weather Rev. 110, 184–197.

Mal, S., and Desai, N. (1938). The mechanism of thundery conditions at Karachi. Q. J. R. Meteorol. Soc. 64, 525–537.

Mapes, B. E. (1993). Gregarious tropical convection. J. Atmos. Sci. 50, 2026–2037.

Market, P., Allen, S., Scofield, R., Kuligowski, R., and Gruber, A. (2003). Precipitation efficiency of warm season midwestern mesoscale convective systems. Wea. Forecasting 18, 12731285.

Markowski, P. M., Rasmussen, E. N., and Straka, J. M. (1998). The occurrence of tornadoes in supercells interaction with boundaries during VORTEX-95. Wea. Forecasting 13, 852–859.

Markowski, P. M., Rasmussen, E. N., and Straka, J. M. (2000). Surface thermodynamic characteristics of RFDs as measured by a mobile mesonet. In "20th Conf. on Severe Local Storms, Orlando Floria." pp. 251–254. Amer. Met. Soc. Preprints.

Markowski, P. M., Straka, J. M., and Rasmussen, E. N. (2002). Direct surface thermodynamic observations within the rear-flank downdrafts of nontornadic and tornadic supercells. Mon. Weather Rev. 130, 1692–1721.

Markowski, P. M., Straka, J. M., and Rasmussen, E. N. (2003). Tornadogenesis resulting from the transport of circulation by a downdraft: Idealized numerical simulations. J. Atmos. Sci. 60, 795–823.

Marshall, B. J. P., Latham, J., and Saunders, C. P. R. (1978). A laboratory study of charge transfer accompanying the collision of ice crystals with a simulated hailstone. Q. J. R. Meteorol. Soc. 104, 163–178.

Marwitz, J. D. (1972a). The structure and motion of severe hailstorms. Part I: Supercell storms. J. Appl. Meteorol. 11, 166–179.

Marwitz, J. D. (1972b). The structure and motion of severe hailstorms. Part II: Multicell storms. J. Appl. Meteorol. 11, 180–188.

Marwitz, J. D. (1972c). The structure and motion of severe hailstorms. Part III: Severely sheared storms. J. Appl. Meteorol. 11, 189–201.
Marwitz, J. D. (1972d). Precipitation efficiency of thunderstorms on the high plains. J. Rech. Atmos. 6, 367–370.
Marwitz, J. D. (1973). Trajectories within the weak echo regions of hailstorms. J. Appl. Meteorol. 12, 1174–1182.
Mason, B. J. (1968). The generation of electric charges and fields in precipitating clouds. In "Proc. Int. Conf. Cloud Phys., Toronto." pp. 657–662. Am. Meteorol. Soc., Boston, Massachusetts.
Mason, B. J. (1971). "The Physics of Clouds." 2nd ed., Oxford University Press, Clarendon, London.
Mason, B. J. (1976). In reply to a critique of precipitation theories of thunderstorm electrification by C. B. Moore. Q. J. R. Meteorol. Soc. 102, 219–225.
Matthews, D. A. (1981). Observations of a cloud are triggered by thunderstorm outflow. Mon. Weather Rev. 109, 2140–2157.
Maxwell, J. B. (1974). Unpublished LMA diagnostic results. Atmos. Environ. Serv., Toronto.
McAnelly, R. A., and Cotton, W. R. (1985). The precipitation lifecycle of mesoscale convective complexes. In "Conf. Hydrometeorol., 6th, Indianapolis, Indiana." pp. 197–204. Am. Meteorol. Soc. Boston, Massachusetts. Preprints.
McCaul Jr., E. W. (1991). Buoyancy and shear characteristics of hurricane-tornado environments. Mon. Weather Rev. 119, 1954–1978.
McCaul Jr., E. W., and Cohen, C. (2002). The impact on simulated storm structure and intensity of variations in the mixed layer and moist layer depths. Mon. Weather Rev. 130, 1722–1748.
McCaul, E. W., and Weisman, M. L. (1996). Simulations of shallow supercells in landfalling hurricane environments. Mon. Weather Rev. 124, 408–429.
McCaul Jr., E. W., Buechler, D. E., Goodman, S. J., and Cammarata, M. (2004). Doppler radar and lightning network observations of a severe outbreak of tropical cyclone tornadoes. Mon. Weather Rev. 132, 1747–1763.
Miller, M. J. (1978). The Hampstead storm: A numerical simulation of a quasi-stationary cumulonimbus system. Q. J. R. Meteorol. Soc. 104, 413–427.
Miller, M. J., and Moncrieff, M. W. (1983). Dynamics and simulation of organized deep convection. In "Proc. NATO Adv. Study Inst. Mesoscale Meteorol.-Theor., Obs. Models, Bonas, Fr., 1982." pp. 451–496. Reidel, Dordrecht, Netherlands.
Miller, M. J., and Pearce, R. P. (1974). A three-dimensional primitive equation model of cumulonimbus convection. Q. J. R. Meteorol. Soc. 100, 133–154.
Miller, L. J., Tuttle, J. D., and Knight, C. K. (1988). Airflow and hail growth in a severe northern High Plains supercell. J. Atmos. Sci. 45, 736–762.
Miller, R. C. (1967). Notes on analysis and severe storm forecasting procedures of the Military Weather Warning Center. Air Weather Serv. Tech. Rep. No. 200, Scott AFB, Illinois.

Miller, R. C. (1972). Notes on analysis and severe storm forecasting procedures of the Air Force Global Weather Central. Air Weather Serv. Tech. Rep. No. 200, Scott AFB, Illinois.

Miller, R. C., and McGinley, J. A. (1977). Response to inherent difficulties in hail probability prediction and forecasting hailfall in Alberta. Meteorol. Monogr. 38, 153–154.

Mitchell, K. E., and Hovermale, J. B. (1977). A numerical investigation of the severe thunderstorm gust front. Mon. Weather Rev. 105, 657–675.

Modahl, A. C. (1969). The influence of vertical wind shear on hailstorm development and structure. Pap. No. 137, Dep. Atmos. Sci., Colorado State University.

Moller, A. R., and Doswell, C. A. (1988). A proposed advanced storm spotters training program. In "15th Conf. on Severe Local Storms, Baltimore, MD." pp. 173–177. Amer. Meteor. Soc. Preprints.

Moller, A. R., Doswell III, C. A., and Przbylinski, R. (1990). High-precipitation supercells: A conceptual model and documentation. In "16th Conf. on Severe Local Storms, Kananaskis Park, Alberta, Canada." pp. 52–57. Amer. Meteor. Soc. Preprints.

Moller, A. R., Doswell III, C. A., Foster, M. P., and Woodall, G. R. (1994). The operational recognition Supercell thunderstorm envioronments and storm structures. Wea. Forecasting 9, 327–347.

Moller, A. R. (2001). Severe local storms forecasting. In "Severe Convective Storms" (C. A. Doswell III, Ed.), In Meteorological Monographs, vol. 28, No. 50. pp. 433–480.

Moore, C. B. (1976). Reply (to B.J. Mason). Q. J. R. Meteorol. Soc. 102, 225–240.

Moore, C. B., Vonnegut, B., Vrablik, E. A., and McCraig, D. A. (1964). Gushes of rain and hail after lightning. J. Atmos. Sci. 21, 646.

Moncrieff, M. W. (1981). A theory of organized steady convection and its transport properties. Q. J. R. Meteorol. Soc. 107, 29–50.

Moncrieff, M. W., and Green, J. S. A. (1972). The propagation and transfer properties of steady convective overturning in shear. Q. J. R. Meteorol. Soc. 98, 336–352.

Moncrieff, M. W., and Miller, M. J. (1976). The dynamics and simulation of tropical cumulonimbus and squall lines. Q. J. R. Meteorol. Soc. 102, 373–394.

Morgan Jr., G. M. (1970). An examination of the wet bulb zero as a hail forecasting parameter in the Po Valley, Italy. J. Appl. Meteorol. 9, 537–540.

Morton, B. R. (1966). Geophysical vortices. Prog. Aeronaut. Sci. 7, 145–193.

Morton, B. R. (1970). The physics of firewhirls. Fire Res. Abstr. Rev. 12, 1–19.

Mueller, C.K., and Hildebrand, P. H. (1983). The structure of a microburst: As observed by ground-based and airborne Doppler radar. In "Conf. Radar Meteorol., 21st." pp. 602–608. Am. Meteorol. Soc., Boston, Massachusetts. Preprints.

Mullet-Hillebrande, D. (1954). Charge generation in thunderstorms by collisions of ice crystals with graupel falling through a vertical field. Tellus 6, 367–381.

Mullet-Hillebrande, D. (1955). Zur Frage des Ursprunges der Gewitterelektrizitat. Ark Geofys. 2, 395.

Musil, D. J., Smith, P. L., Miller, J. R., Killinger, J. H., and Halvorson, J. L. (1977). Characteristics of vertical volocities observed in T-28 penetrations of hailstorms. In "Conf. Weather Modif., 6th." pp. 161–169. Am. Meteorol. Soc., Boston, Massachusetts. Preprints.

National Research Council, (1983). "Low-Altitude Wind Shear and its Hazard to Aviation." Natl. Acad. Press, Washington, DC.

Nelson, S. P. (1977). Rear flank downdraft: A hailstorm intensification mechanism. In "Conf. Severe Local Storms, 10th." pp. 521–525., Boston, Mass. Preprints.

Nelson, S. P. (1983). The influence of storm flow structure on hail growth. J. Atmos. Sci. 40, 1965–1983.

Nelson, S. P. (1987). The hybrid multicellular-supercellular storm an efficient hail producer. Part II: General characteristics and implications for hail growth. J. Atmos. Sci. 44, 2060–2073.

Newton, C. W. (1950). Structure and mechanism of the prefrontal squall line. J. Meteorol. 7, 210–222.

Newton, C. W. (1963). Dynamics of severe convective storms. Meteorol. Monogr. (27), 33–58.

Newton, C. W. (1966). Circulations in large sheared cumulonimbus. Tellus 18, 699–712.

Newton, C. W., and Newton, H. R. (1959). Dynamical interactions between large convective clouds and environment with vertical shear. J. Meteorol. 16, 483–496.

Newton, C. W., Miller, R. C., Fosse, E. R., Booker, D. R., and McManamon, P. (1978). Severe thunderstorms: Their nature and their effects on society. Interdiscip. Sci. Rev. 3, 71–85.

Nicholls, M. (1987). A numerical investigation of tropical squall lines. Ph.D. Thesis, Dep. Atmos. Sci., Colorado State University.

Nicholls, M. E., Pielke, R. A., and Cotton, W. R. (1991). A two-dimensional numerical investigation of the interaction between sea-breezes and deep convection over the Florida Peninsula. Mon. Weather Rev. 119, 298–323.

Normand, C. W. B. (1946). Energy in the atmosphere. Q. J. R. Meteorol. Soc. 72, 145–167.

Novlan, D. J., and Gray, W. M. (1974). Hurricane-spawned tornadoes. Mon. Weather Rev. 102, 476–488.

Ogura, Y., and Liou, M. T. (1980). The structure of a mid-latitude squall line: A case study. J. Atmos. Sci. 37, 553–567.

Ooyama, K. (1964). A dynamical model for the study of tropical cyclone development. Geofis. Int. 4, 187–198.

Orville, H. D. (1978). A review of hailstone-hailstorm numerical simulations. Meteorol. Monogr. (38), 49–61.

Orville, R. E., and Huffines, G. R. (2001). Cloud-to-ground lightning in the USA: NLDN results in the first decade 1989–1998. Mon. Weather Rev. 129, 1179–1193.

Pakiam, J. E., and Maybank, J. (1975). The electrical characteristics of some severe hailstorms in Alberta, Canada. J. Meteorol. Soc. Jpn 53, 363–383.

Paluch, I. R. (1978). Size sorting of hail in an three-dimensional updraft and implications for hail suppression. J. Appl. Meteorol. 17, 763–777.

Perez, A. H., Wicker, L. J., and Orville, R. E. (1997). Characteristics of cloud-to-ground lightning associated with violent tornadoes. Wea. Forecasting 12, 428–437.

Proctor, F. H. (1988). Numerical simulations of an isolated microburst. Part I: Dynamics and structure. J. Atmos. Sci. 45, 3137–3160.

Proctor, F. H. (1989). Numerical simulations of an isolated microburst. Part II: Sensitivity experiments. J. Atmos. Sci. 45, 2143–2165.

Pruppacher, H. R., and Klett, J. D. (1980). "Microphysics of Clouds and Precipitation." Reidel, Dordrecht, Netherlands.

Purdom, J. F. W. (1976). Some uses of high-resolution GOES imagery in the mesoscale forecasting of convection and its behavior. Mon. Weather Rev. 104, 1474–1483.

Purdom, J. F. W. (1979). The development and evolution of deep convection. In "Conf Severe Local Storms, 11th, Kansas City, Mo." pp. 143–150. Am. Meteorol. Soc., Boston, Massachusetts. Preprints.

Purdom, J. F. W. (1982). Subjective interpretation of geostationary satellite data for nowcasting. In "Nowcasting" (Keith Browning, Ed.), pp. 149–156. Academic Press, New York.

Purdom, J. F. W., and Marcus, K. (1982). Thunderstorm triggered mechanisms over the Southeast United States. In "Conf. Severe Local Storms, 12th, San Antonio, Tex." pp. 487–488. Am. Meteorol. Soc., Boston, Massachusetts. Preprints.

Rasmussen, E. N., and Straka, J. M. (1996). Mobile mesonet observations of tornadoes during VORTEX-95. Preprints. In "18th Conf. on Severe Local Storms, San Francisco, CA." pp. 1–5. Amer. Meteor. Soc.

Rasmussen, W. N., and Blanchard, S. O. (1998). A baseline climatology of sound-derived Supercell and tornado forecast parameters. Wea. Forecasting 13, 1148–1164.

Rasmussen, W. N., Richardson, S., Straka, J. M., Markowski, P. M., and Blanchard, D. O. (2000). The association of significant tornadoes with a baroclinic boundary on 2 June 1995. Mon. Weather Rev. 128, 174–191.

Rawlins, F. (1982). A numerical study of thunderstorm electrification using a three-dimensional model incorporating the ice phase. Q. J. R. Meteorol. Soc. 108, 779–800.

Ray, P. S., Johnson, B. C., Johnson, K. W., Bradberry, J. S., Stephens, J. J., Wagner, K. K., Wilhelmson, R. B., and Klemp, J. B. (1981). The morphology of several tornadic storms 20 May 1977. J. Atmos. Sci. 38, 1643–1663.
Raymond, D. J. (1975). A model for predicting the movement of continuously propagating convective storms. J. Atmos. Sci. 32, 1308–1317.
Raymond, D. J. (1976). Wave-CISK and convective mesosystems. J. Atmos. Sci. 33, 2392–2398.
Raymond, D. l. (1983). A wave-CISK in mass flux form. J. Atmos. Sci. 40, 2561–2572.
Renick, J. H. (1971). Radar reflectivity profiles of individual cells in a persistent multicellular Alberta hailstorm. In "Conf. Severe Local Storms, 7th, Kansas City, Mo." pp. 63–70. Am. Meteorol. Soc., Boston, Massachusetts. Preprints.
Renick, J. H., and Maxwell, J. B. (1977). Forecasting hailfall in Alberta. Meteorol. Monogr. 38, 145–151.
Riehl, H., Badner, J., Hovde, J. E., LaSeur, N. E., Means, L. L., Palmer, W. C., Schroeder, M. J., and Snellman, L. W. (1952). Forecasting in the middle latitudes. Meteorol. Monogr. 5.
Rodi, A. R., Elmore, K. L., and Mahoney, W. P. (1983). Aircraft and Doppler air motion comparisons in a JAWS microburst. In "Conf. Radar Meteorol., 21st." pp. 624–629. Am. Meteorol. Soc., Boston, Massachusetts. Preprints.
Rosenfeld, D., Lohmann, U., Raga, G. B., ODowd, C., Kulmala, M., Fuzzi, S., Reissell, A., and Andreae, M. O. (2008). Flood or drought: How do aerosols affect precipitation?. Science 321, 1309–1313. Samsury and Orville, 1994.
Rotunno, R. (1977). Numerical simulation of a laboratory vortex. J. Atmos. Sci. 34, 1942–1956.
Rotunno, R. (1980). Vorticity dynamics of convective swirling boundary layer. J. Fluid Mech. 97, 623–640.
Rotunno, R. (1981). On the evolution of thunderstorm rotation. Mon. Weather Rev. 109, 577–586.
Rotunno, R. (1984). An investigation of a three-dimensional asymmetric vortex. J. Atmos. Sci. 41, 283–298.
Rotunno, R., and Klemp, J. B. (1982). The influence of shear-induced pressure gradients on thunderstorm motion. Mon. Weather Rev. 110, 136–151.
Rotunno, R., and Klemp, J. B. (1985). On the rotation and propagation of simulated supercell thunderstorms. J. Atmos. Sci. 42, 271–292.
Rotunno, R., Klemp, J. B., and Weisman, M. L. (1988). A theory for strong, long–lived squall lines. J. Atmos. Sci. 45, 463–485.
Rotunno, R., and Lilly, D. K. (1981). A numerical model pertaining to the multiple vortex phenomenon. Contract Rep. NUREG/CR-1840, US Nucl. Regul. Comm., Washington, DC.
Sartor, J. D. (1961). Calculations of cloud electrification based on a general charge separation mechanism. J. Geophys. Res. 66, 831–843.
Sartor, J. D., and Miller, J. S. (1965). Relative cloud droplet trajectory computation. In "Proc. Int. Cloud Phys. Conf." pp. 108–112. Meteorol. Soc., Japan, Tokyo/Sapporo.

Samsury, Christopher E., and Orville, Richard E. (1994). Cloud-to-ground lightning in tropical cyclones: a study of hurricanes Hugo (1989) and Jerry (1989). Mon. Weather Rev. 122, 1887–1896.

Saunders, C. P. R., and Peck, S. L. (1998). Laboratory studies of the influence of the rime accretion rate on charge transfer during crystal/graupel collisions. J. Geophys. Res. 103 (13), 949-13,956.

Saunders, C. P. R., and Wahab, N. M. A. (1975). Influence of electric fields on the aggregation of ice crystals. J. Meteorol. Soc. Jpn 53, 121–126.

Schlamp, R. J., Grover, S. N., Pruppacher, H. R., and Hamielec, A. E. (1976). A numerical investigation of the effect of electric charges and vertical external fields on the collision efficiency of cloud drops. J. Atmos. Sci. 33, 1747–1755.

Schlamp, R. J., Grover, S. N., Pruppacher, H. R., and Hamielec, A. E. (1979). A numerical investigation of the effect of electric charges and vertical external electric fields on the collision efficiency of cloud drops: Part 11. J. Atmos. Sci. 36, 339–349.

Schlesinger, R. E. (1975). A three-dimensional numerical model of an isolated deep convective cloud: Preliminary results. J. Atmos. Sci. 32, 934–957.

Schlesinger, R. E. (1978). A three-dimensional numerical model of an isolated thunderstorm. Part 1. Comparative experiments for variable ambient wind shear. J. Atmos. Sci. 35, 690–713.

Schlesinger, R. E. (1980). A three-dimensional numerical model of an isolated thunderstorm. Part 11: Dynamics of updraft splitting and mesovortex couplet evolution. J. Atmos. Sci. 37, 395–420.

Schmidt, J. M. (1991). Numerical and observational investigations of long-lived, MCS-induced, severe surface wind events: The derecho. Ph.D. dissertation, Colorado State University, Dept. of Atmospheric Science, Fort Collins, CO 80523, 196 pp.

Schmidt, J. M., and Cotton, W. R. (1990). Interactions between upper and lower tropospheric gravity waves on squall line structure and maintenance. J. Atmos. Sci. 47, 1205–1222.

Schneider, D., and Sharp, S. (2007). Radar signatures of tropical cyclone tornadoes in central North Carolina. Wea. Forecasting 22, 479–501.

Schonland, B. F. J. (1950). "The Flight of Thunderbolts." pp. 150–151. Oxford University Press, New York.

Schleusener, R. A. (1962). On the relation of the latitude and strength of the 500 millibar west wind along 110 degrees west longitude and the occurrence of hail in the lee of the Rocky Mountains. Atmos. Sci. Tech. Pap. No. 26, Civ. Eng. Sect., Colorado State University.

Scott, W. D., and Levin, Z. (1975). A stochastic electric model of an infinite cloud: Charge generation and precipitation development. J. Atmos. Sci. 32, 1814–1828.

Seifert, A., and Beheng, K. D. (2006). A two-moment cloud microphysics parameterization for mixed-phase clouds. Part II: Maritime vs. continental deep convective storms. Meteorol. and Atmos. Phys. 92, 67–82.

Seitter, K. L. (1987). A numerical study of atmospheric density current motion including the effects of condensation. J. Atmos. Sci. 43, 3068–3076.
Semonin, R. G., and Plumlee, H. R. (1966). Collision efficiency of charged cloud droplets in electric fields. J. Geophys. Res. 71, 4271–4278.
Showalter, A. K. (1953). Synoptic conditions associated with tornadoes. US Weather Bureau Hydrometeorol. Sect. Report.
Silva-Dias, M. F., Betts, A. K., and Stevens, D. E. (1984). A linear spectral model of tropical mesoscale systems: Sensitivity studies. J. Atmos. Sci. 41, 1704–1716.
Simpson, J. E. (1969). A comparison between laboratory and atmospheric density currents. Q. J. R. Meteorol. Soc. 95, 758–765.
Simpson, J. E., and Britter, R. E. (1980). A laboratory model of an atmospheric mesofront. Q. J. R. Meteorol. Soc. 106, 485–500.
Simpson, J., and Wiggert, V. (1969). 1968 Florida seeding experiment: Numerical mode results. Mon. Weather Rev. 97, 471–489.
Simpson, J., Simpson, R. H., Andrews, D. A., and Eaton, M. A. (1965). Experimental cumulus dynamics. Rev. Geophys. 3, 387–431.
Sinclair, P. C. (1973). Severe storm velocity and temperature structure deduced from penetrating aircraft. In "Conf. Severe Local Storms, 8th." pp. 25–31. Am. Meteorol. Soc., Boston, Massachusetts. Preprints.
Sinclair, P. C. (1979). Velocity and temperature structure near and within severe storms. In "Conf. Severe Local Storms, Il th, Colorado State University".
Sinclair, P. C., and Purdom, J. F. W. (1983). The genesis and development of deep convective storms. Final Rep. for NOAA Grant NA80AA-D-00056.
Smith, R. B., Jiang, Q., Fearon, M. G., Tabary, P., Dorninger, M., Doyle, J. D., and Beniot, R. (2003). Orographic precipitation and air mass transformation: An Alpine example. Q. J. R. Meteorol. Soc. 129, 433–454.
Smith, R. K., and Leslie, L. M. (1978). Tornadogenesis. Q. J. R. Meteorol. Soc. 104, 189–199.
Smith, R. K., and Leslie, L. M. (1979). A nuermical study of tornadogenesis in a rotating thunderstorm. Q. J. R. Meteorol. Soc. 105, 107–127.
Snook, N., and Xue, M. (2008). Effects of microphysical drop size distribution on tornadogenesis in supercell thunderstorms. Geophy. Res. Lett. 35, L24803. doi:10.1029/2008GL035866.
Snow, J. T. (1978). On inertial instability as related to the multiple-vortex phenomenon. J. Atmos. Sci. 35, 1660–1677.
Snow, J. T. (1982). A review of recent advances in tornado vortex dynamics. Rev. Geophys. Space Phys. 20, 953–964.
Spratt, S. M., Marks, F. D., Dodge, P. P., and Sharp, D. W. (2000). Examining the pre-landfall environment of mesovortices within a hurricane Bonnie (1998) outer rainband. In "24th Conf. on Hurricanes and Tropical Meteorology, Fort Lauderdale, FL." pp. 300–301. Amer. Meteor. Soc. Preprints.
Steiner, R., and Rhyne, R. H. (1962). Some measured characteristics of severe storm turbulence. US Weather Bur., Nad. Severe Storms Proj. Rep. No. 10.

Stevens, D. E., and Lindzen, R. S. (1978). Tropical wave-CISK with a moisture budget and cumulus friction. J. Atmos. Sci. 35, 940–961.

Stout, G. E., and Changnon, S. A. Jr. (1968). Climatology of hail in the central United States. CHIAA Res. Rep. No. 38, prepared for Crop Hail Insurance Actuarial Association.

Strauch, R. G., and Merrem, F. H. (1976). Structure of an evolving hailstorm. Part III: Internal structure from Doppler radar. Mon. Weather Rev. 104, 588–595.

Sulakvelidze, G. K., Bibilashvili, N. S., and Lapcheva, V. F. (1967). Formation of Precipitation and Modification of Hail Processes. Isr. Program Sci. Transl., Jerusalem.

Summers, P. W. (1972). Project Hailstop: A review of accomplishments to date. In "Alberta Hail Studies 1972." Hail Stud. Rep. 72-2, pp. 47–53, Res. Counc. Alberta.

Suzuki, O., Niino, H., Ohno, H., and Nirasawa, H. (2000). Tornado-producing mini supercells associated with Typhoon 9019. Mon. Weather Rev. 128, 1868–1882.

Takahashi, T. (1978). Riming electrification as a charge generation mechanism in thunderstorms. J. Atmos. Sci. 35, 1536–1548.

Takeda, T. (1971). Numerical simulation of a precipitating convective cloud: The formation of a long–lasting cloud. J. Atmos. Sci. 28, 350–376.

Tepper, M. (1950). A proposed mechanism of squall lines: the pressure jump line. J. Meteorol. 7, 21–29.

Thorpe, A. J., and Miller, M. J. (1978). Numerical simulations showing the role of downdraft in cumulonimbus motion and splitting. Q. J. R. Meteorol. Soc. 104, 873–893.

Thorpe, A. J., Miller, M. J., and Moncrieff, M. W. (1980). Dynamical models of two-dimensional downdrafts. Q. J. R. Meteorol. Soc. 106, 463–484.

Thorpe, A. J., Miller, M. J., and Moncrieff, M. W. (1982). Two-dimensional convection in non-constant shear: A model of mid-latitude squall lines. Q. J. R. Meteorol. Soc. 108, 739–762.

Trapp, R. J. (1999). Observations of nontornadic low-level mesocyclones and attendant tornadogenesis failure during VORTEX. Mon. Weather Rev. 127, 1693–1705.

Tripoli, G. J., and Cotton, W. R. (1982). The Colorado State University three-dimensional cloud/mesoscale model-1982. Part I: General theoretical framework and sensitivity experiments. J. Rech. Atmos. 16, 185–220.

Tripoli, G. J., and Cotton, W. R. (1986). An intense quasi-steady thunderstorm over mountainous terrain. Part IV: Three-dimensional numerical simulation. J. Atmos. Sci. 43, 894–912.

Tripoli, G., and Cotton, W. R. (1989a). A numerical study of an observed orogenic mesoscale convective system. Part 1. Simulated genesis and comparison with observations. Mon. Weather Rev. 117, 273–304.

Tripoli, G., and Cotton, W. R. (1989b). A numerical study of an observed orogenic mesoscale convective system. Part 2. Analysis of governing dynamics. Mon. Weather Rev. 117, 305–328.

Uccellini, L. W., and Johnson, D. R. (1979). The coupling of upper and lower tropospheric jet streaks and implications for the development of severe convective storms. Mon. Weather Rev. 107, 682–703.

van den Heever, S. C., Carri, G. G., Cotton, W. R., DeMott, P. J., and Prenni, A. J. (2006). Impacts of nucleating aerosol on Florida storms. Part I: Mesoscale Simulations. J. Atmos. Sci. 63, 1752–1775.

van den Heever, S. C., and Cotton, W. R. (2004). The impact of hail size on simulated supercell storms. J. Atmos. Sci. 61, 1596–1609.

van den Heever, S., and Cotton, W. R. (2007). Urban aerosol impacts on downwind convective storms. J. Appl. Met. 46, 828–850.

Vasiloff, S. V., Brandes, E. A., and Davies-Jones, R. P. (1986). An investigation of the transition from multicell to supercell storms. J. Clim. Appl. Meteorol. 25, 1022–1036.

Vaughan, D. H., and Vonnegut, B. (1974). Luminous electrical phenomena associated with nocturnal tornadoes in Huntsville. Bull. Am. Meteorol. Soc. 57, 1220–1224. Alabama.

Vonnegut, B. (1955). Possible mechanism for the formation of thunderstorm electricity. In "Conference on Atmospheric Electricity" Res. Pap. No. 42. Geophys. Res. Dir., Air Force Cambridge Res. Cent., Bedford, Massachusetts, pp. 169–181.

Vonnegut, B. (1960). Electrical theory of tornadoes. J. Geophys. Res. 65, 203–212.

Vonnegut, B. (1963). Some facts and speculations concerning the origin and role of thunderstorm electricity. Meteorol. Monogr. 5, 224–241.

Vonnegut, B. (1975). Comment on "On the relation of electrical activity to tornadoes by R. P. Davies-Jones and J. H. Golden". J. Geophys. Res. 80, 4559–4560.

Wade, C. G., and Foote, G. B. (1982). The 22 July 1976 case study: Low-level airflow and mesoscale influences. In C. A. Knight, and P. Squires, (Eds.), Hailstorms of the Central High Plains, vol. 2. pp. 115–130. Colorado Assoc. University Press, Boulder.

Wagner, P. B., and Telford, J. W. (1981). Cloud dynamics and an electric charge separation mechanism in convective clouds. J. Rech. Atmos. 15, 97–120.

Wakimoto, R. M. (1982). The life cycle of thunderstorm gust fronts as viewed with Doppler radar and rawinsonde data. Mon. Weather Rev. 110, 1050–1082.

Wakimoto, R. M. (2001). Convectively driven high winds events. In "Severe Convective Storms" (C. A. Doswell III, Ed.), In Meteorological Monographs, vol. 28, No. 50. pp. 255–298.

Wakimoto, R. M., and Bringi, V. N. (1988). Dual-polarization observations of microbursts associated with intense convection: The 20 July storm during the MIST Project. Mon. Weather Rev. 116, 1521–1539.

Wakimoto, R. M., and Cai, H. (2000). Analysis of nontornadic storm during VORTEX 95. Mon. Weather Rev. 128, 565–592.

Wakimoto, R. M., Kessinger, C. J., and Kingsmill, D. E. (1994). Kinematic, thermodynamic, and visual structure of low-reflectivity microbursts. Mon. Weather Rev. 122, 72–92.

Wakimoto, R. M., and Liu, C. (1998). The Garden City, Kansas, storm during VORTEX 95. Part II: The wall cloud and tornado. Mon. Weather Rev. 126, 372–392.

Wakimoto, R. M., and Wilson, J. W. (1989). Non-supercell tornadoes. Mon. Weather Rev. 110, 1060–1082.

Walko, R. L. (1993). Tornado spin-up beneath a convective cell: Required basic structure of the near-field boundary layer winds. In "The Tornado: Its structure, Dynamics, Prediction, and Hazards." In Meteor. Monogr., vol. 79. pp. 89–95. Amer. Meteor. Soc.

Walko, R., and Gall, R. (1984). A two-dimensional linear stability analysis of the multiple vortex phenomenon. J. Atmos. Sci. 41, 3456–3471.

Ward, N. B. (1972). The exploration of certain features of tornado dynamics using a laboratory model. J. Atmos. Sci. 29, 1194–1204.

Watkins, D. C., Cobine, J. D., and Vonnegut, B. (1978). Electric discharges inside tornadoes. Science 199, 171–174.

Watson, A. I., Lopez, R. L., and Holle, R. L. (1994). Diurnal cloud-to-ground lightning patterns in Arizona during the Southwest monsoon. Mon. Weather Rev. 122, 1716–1725.

Watson, A. I., Jamski, M. A., Turnage, T. J., Bowen, J. R., and Kelley, J. C. (2005). The tornado outbreak across North Florida Panhandle in association with Hurricane Ivan. In "32nd Conf. on Radar Meteorol. 22–28 October 2005." Amer. Met. Soc., Paper 10R.3. Preprints.

Weaver, J. F., and Nelson, S. P. (1982). Multiscale aspects of thunderstorm gust fronts and their effects on subsequent storm development. Mon. Weather Rev. 110, 707–718.

Weisman, M., and Klemp, J. (1982). The dependence of numerically simulated convective storms on vertical wind shear and buoyancy. Mon. Weather Rev. 110, 504–520.

Weisman, M. L., and Klemp, J. B. (1984). The structure and classification of numerically simulated convective storms in directionally varying wind shears. Mon. Weather Rev. 112, 2479–2498.

Wetzel, P. J., Cotton, W. R., and McAnelly, R. L. (1982). The dynamic structure of the mesoscale convective complex-Some case studies. In "Conf Severe Local Storms, 12th." pp. 265–268. Am. Meteorol. Soc., Boston, Massachusetts

Wetzel, P. J., Cotton, W. R., and McAnelly, R. L. (1983). A long-lived mesoscale convective complex. Part II: Morphology of the mature complex. Mon. Weather Rev. 111, 1919–1937.

Wicker, L. J., and Wilhelmson, R. B. (1995). Simulation and analysis of tornado development and decay within a three-dimensional Supercell thunderstorm. J. Atmos. Sci. 52, 2675–2703.

Wilczak, J. M., Christian, T. W., Wolfe, D. E., Zamora, R. J., and Stankov, B. B. (1992). Observations of a Colorado tornado. Part I: Mesoscale environment and tornadogenesis. Mon. Weather Rev. 120, 497–520.

Wilhelmson, R. B., and Chen, C. S. (1982). A simulation of the development of successive cells along a cold outflow boundary. J. Atmos. Sci. 39, 1466–1483.

Wilhelmson, R. B., and Klemp, J. (1978). A numerical study of storm splitting that leads to long-lived storms. J. Atmos. Sci. 35, 1974–1986.

Wilhelmson, R. B., and Klemp, J. (1981). A three-dimensional numerical simulation of splitting severe storms on 3 April 1964. J. Atmos. Sci. 38, 1581–1600.

Williams, E. R. (2001). The electrification of severe storms. In "Severe Convective Storms." In Meteor. Monogr., vol. 50. pp. 527–561. Amer. Meteor. Soc.

Williams, E. R., and 32 coauthors, (2002). Contrasting convective regimes over the Amazon: Implications for cloud electrification. J. Geophys. Res. 107, 8082. doi:10.1029/2001JD000380.

Williams, E. R., and Lhermitte, R. (1983). Radar test of the precipitation hypothesis for thunderstorm electrification. J. Geophys. Res. 88 (10), 984–10, 992.

Williams, E. R., Mushtak, V., Rosenfeld, D., Goodman, S., and Boccippio, D. (2005). Thermodynamic conditions favorable to superlative thunderstorm updraft, mixed phase microphysics and lightning flash rate. Atmos. Res. 76, 288–306.

Williams, E. R., and Renno, N. (1993). An analysis of the conditional instability of the tropical atmosphere. Mon. Weather Rev. 121, 21–36.

Wilson, J. W., and Fujita, T. (1979). Vertical cross section through a rotating thunderstorm by Doppler radar. In "Conf. Severe Local Storms, 11th." pp. 447–452. Am. Meteorol. Soc., Boston, Massachusetts. Preprints.

Wilson, T., and Rotunno, R. (1982). Numerical simulation of a laminar vortex flow. In "Proc. Int. Conf. Comput. Methods." pp. 203–215. Springer-verlag, Berlin.

Wilson, J. W., Roberts, R. D., Kessinger, C., and McCarthy, J. (1984). Microburst wind structure and evolution of Doppler radar for airport wind shear detection. J. Clim. Appl. Meteorol. 23, 898–915.

Wilson, T. (1981). Vortex boundary layer dynamics. M.S. Thesis, University of California, Davis.

Young, K. C. (1977). A numerical examination of some hail suppression concepts. In "Hail: A review of hail science and hail suppression" (G. B. Foote and C. A. Knight, Eds.), In Meteorol. Monogr., vol. 16, No. 38. pp. 195–214.

Yuder, S. E., and Houze Jr., R. A. (1995). Three-dimensional kinematic and microphysical evolution of Florida cumulonimbus. Part I: Spatial distribution of updrafts, downdrafts, and precipition. Mon. Weather Rev. 123, 1921–1940.

Xu, J.-L. (1983). Hail growth in a three-dimensional cloud model. J. Atmos. Sci. 40, 185–203.

Zajac, B. A., and Rutledge, S. A. (2001). Cloud-to-ground lightning activity in the contigous United States from 1995 to 1999. Mon. Weather Rev. 129, 999–1019.

Zhang, J., Lohmann, U., and Stier, P. (2005). A microphysical parameterization for convective clouds in the ECHAM5 Climate Model: 1. Single column model results evaluated at the Oklahoma RM site. J. Geophys. Res. 110, D15S07. doi:10.1029/2004JD005128.

Zipser, E. J. (1969). The role of organized unsaturated convective downdrafts in the structure and rapid decay of an equatorial disturbance. J. Appl. Meteorol. 8, 799–814.

Zipser, E. J. (1977). Mesoscale and convective-scale downdrafts as distinct components of squall-line structure. Mon. Weather Rev. 105, 1568–1589.

Zipser, E. J. (2003). Some views on 'hot tower's after 50 years of tropical field programs and two years of TRMM data. Meteorol. Monogr. 29, 49–58.

Zipser, E. J., and LeMone, M. A. (1980). Cumulonimbus vertical velocity events in GATE. Part II: Synthesis and model core structure. J. Atmos. Sci. 37, 2458–2469.

Zrnic, D. S. (1976). Magnetometer data acquired during nearby tornado occurrences. J. Geophys. Res. 81, 5410–5412.

Zverev, A. S., (Ed.) (1972). In "Practical Work in Synoptic Meteorology" pp. 225–252. Hydrometeorol. Publ. House, Leningrad.

Chapter 9

# Mesoscale Convective Systems

## 9.1. INTRODUCTION

In this chapter we discuss the dynamics and characteristics of precipitating mesoscale convective systems (MCSs), focusing on the mesoscale features of the earth's dominant precipitating convective systems: tropical and mid-latitude squall lines, tropical and mid-latitude cloud clusters, including mesoscale convective complexes (MCCs), mesoscale convective vortices (MCVs), genesis of tropical cyclones, and impacts of MCSs.

MCSs are important systems to study because they are important contributors to rainfall (Fritsch et al., 1986; Jirak et al., 2003; Ashley et al., 2003) that is used to support agriculture and provide water resources for human consumption and energy production. On the negative side they can produce loss of human lives and property loss by producing severe weather (e.g. flash floods, severe straight line winds or derechos, hail, lightning, and tornadoes) over large areas (Maddox, 1980; Houze et al., 1990; Jirak et al., 2003)

We first discuss conceptual models of MCS characteristics and then focus on the climatology of MCSs, followed by examination of the fundamental dynamical characteristics of MCSs, followed next by a discussion of the role of MCSs in the genesis of tropical cyclones, and ending with a discussion of the impacts of MCSs such as severe winds and rainfall.

## 9.2. DEFINITION OF MESOSCALE CONVECTIVE SYSTEMS

The term MCS describes a deep convective system that is considerably larger than an individual thunderstorm and that is often marked by an extensive middle to upper tropospheric stratiform-anvil cloud of several hundred kilometers in the horizontal dimension. The cloud systems typically have lifetimes of six to twelve hours and, on some occasions, the stratiform-anvil portion of the system can survive for several days. In some regions of the world, including many portions of the oceanic tropics and subtropics and even parts of the High Plains of the United States, MCSs are the dominant contributor to the annual precipitation. Thunderstorms embedded within MCSs are often the source of flood-producing rainfalls, strong, damaging winds, and, in mid-latitudes, violent storms producing tornadoes and hail. MCSs are also major venters of pollutants

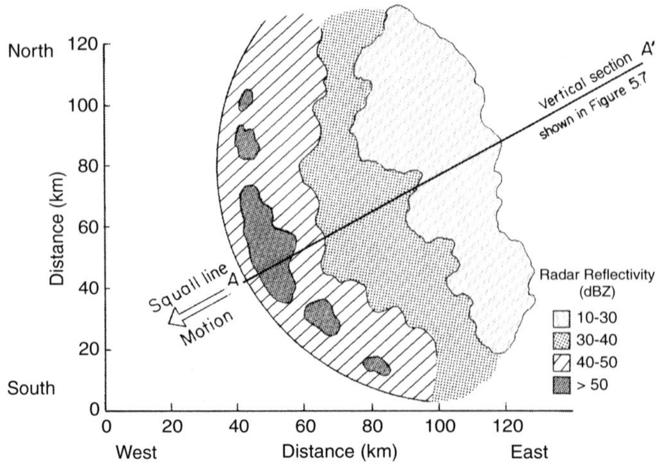


**FIGURE 9.1 Schematic diagram of radar reflectivity at low levels (0.5-1.5 km) for a tropical squall line.** *(Adapted from Chauzy et al. (1985))*

from the boundary layer into the mid and upper troposphere (Lyons et al., 1986; Cotton et al., 1995).

We first investigate tropical mesoscale systems, such as squall and nonsquall clusters, and then their mid-latitude cousins, squall lines, MCSs, and MCCs.

## 9.3. CONCEPTUAL MODELS OF MCSS

### 9.3.1. Tropical Squall Lines

The tropical squall line is identifiable by a line of vigorous convective cells from one hundred to several hundred kilometers along its major axis. At the surface, the passage of the squall line is noted by a distinct roll cloud followed by a sudden wind squall of 12-25 m $s^{-1}$ (Hamilton and Archbold, 1945). Immediately behind the surface squall, a heavy downpour sets in, producing as much as 30 mm of rainfall in 30 min. In the tropics, the heavy downpour is followed by several hours of relatively steady and uniform rainfall from the stratiform anvil of the system. A schematic diagram of a low-level horizontal radar display of an approaching squall line (Fig. 9.1) is earmarked by a convex-shaped line of convective radar echoes that advances at speeds of 15-20 m $s^{-1}$.

A schemematic vertical cross section through the advancing squall line can be seen in Fig. 9.2. The strong updraft at the leading edge of the line is fed by warm boundary layer air. Maximum updraft speeds of 13 m $s^{-1}$ occur at the 2.5 km level. Heavy precipitation forms in this tilted updraft, as evidenced by high radar reflectivity below it. Between 1.0 and 1.5 km, the updraft is continuous along the leading edge of the line. Above 2.0 km, the updraft splits into two parts, with the most intense core in the northern region. Associated

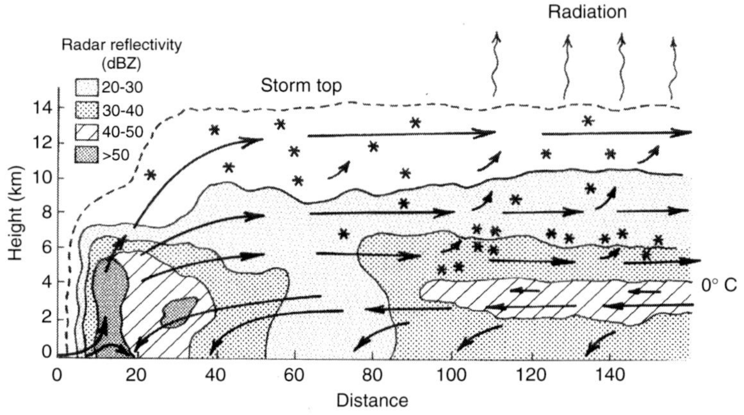


**FIGURE 9.2 Schematic of vertical cross section of radar reflectivity along the A-A′.** Also shown are storm-relative streamlines of flow through the squall line. *(Adapted from Chauzy et al. (1985))*

with the split updrafts is an intrusion of mid-tropospheric rear-to-front flow; a similar feature is common in mid-latitude squall lines.

Behind the leading updraft line is a band of convective-scale downdrafts, most evident at high levels. Maximum downdraft speeds of 4 m $s^{-1}$ are observed and are similar in magnitude to those observed from aircraft by Zipser and LeMone (1980). The horizontal flow is marked by (1) a front-to-rear flow at all levels at the system's leading edge and (2) below 3 km a rear-to-front flow behind the leading squall line.

As the squall line matures the cold-pools spreads out behind the leading line. Aloft, detrained cloudy debris forms a thickening stratiform-anvil cloud. Latent heating from vapor deposition growth of ice particles and freezing of supercooled water drops either on contact with ice particles during riming growth or heterogeneous freezing, leads to warming and ascent in the stratiform-anvil cloud. As will be discussed in the MCS dynamics section, the response to the elevated heating is the generation of gravity waves which favor the further upscale growth of the MCS, and the development of balanced circulations which leads to widespread ascent and stratiform precipitation in the stratiform-anvil cloud. Feeding the stratiform-anvil cloud is detrainment of moist air and cloudy debris in the leading convective line and a slantwise front-to-rear ascending flow branch. Earlier it was thought that this front-to-rear flow had its origins solely in the detrainment from convective cells, but more recent studies (Cotton et al., 1995; Houze, 2004) suggest that, at least in the mature stages of the MCS, the ascending slantwise front-to-rear branch has its origins in a layer of air above the boundary layer roughly 3-5 km above the surface. Moistening of this above boundary layer air is critical to the formation of a deep, precipitating stratiform-anvil cloud. As will be discussed more later, this slantwise ascending front-to-rear flow branch is likely driven by balanced dynamics of the MCS.

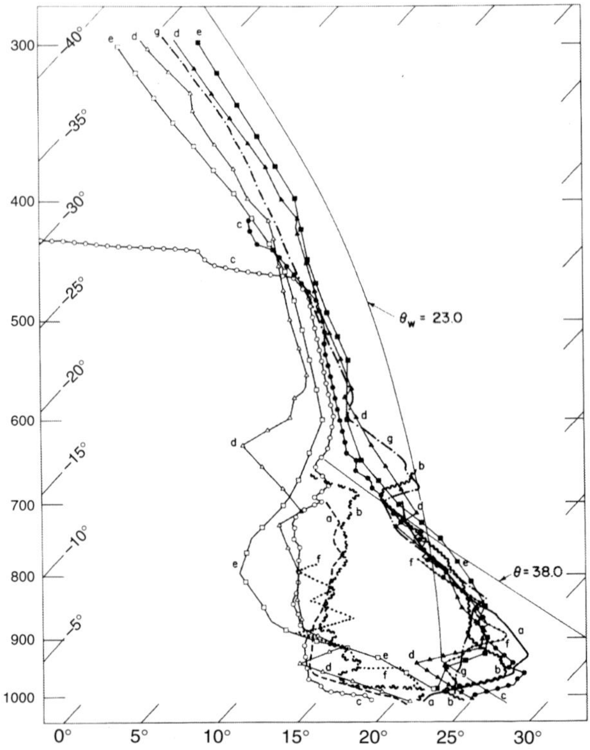


**FIGURE 9.3 Characteristic soundings in postsquall regions.** (a) 5 September 1974 by NCAR *Queen Air*, mostly in rain; (b) 1630 GMT, 12 September 1974, for the *Fay* (GATE position 28A, about 200 km behind leading edge and 50 km behind trailing precipitation); (c) 1804 GMT, 12 September 1974, for the *Oceanographer* (GATE position 4, about 250 km behind leading edge and 100 km behind trailing precipitation); (d) 1130 GMT, 12 September 1974, for the *Poryv* (GATE position 10, about 350 km behind leading edge and 174 km behind trailing precipitation); (e) 200 AST, 28 August 1969, Anaco, Venezuela (about 50 km behind trailing precipitation); (f) aircraft sounding; (g) Barbados sounding (no $T_d$ available). *(From Zipser (1977))*

Also in the stratiform-anvil cloud, either blocking of propagating gravity waves by environmental flow aloft, and/or evaporation and melting of precipitation, forms mesoscale downdrafts and descending rear-to-front flow. The cold-pool associated with the diverging, descending rear-to-front flow may contribute to the low-level layer ascend in the front-to-rear flow. The low-level subsaturated descent beneath the stratiform anvil cloud results in warming and drying which produced the so called "onion" sounding illustrated in Fig. 9.3.

## 9.3.2. Conceptual Model of Tropical Non-squall Clusters

Although the tropical squall cluster may be the most spectacular form of the MCS as far as rain intensity and wind gusts are concerned, the nonsquall cluster

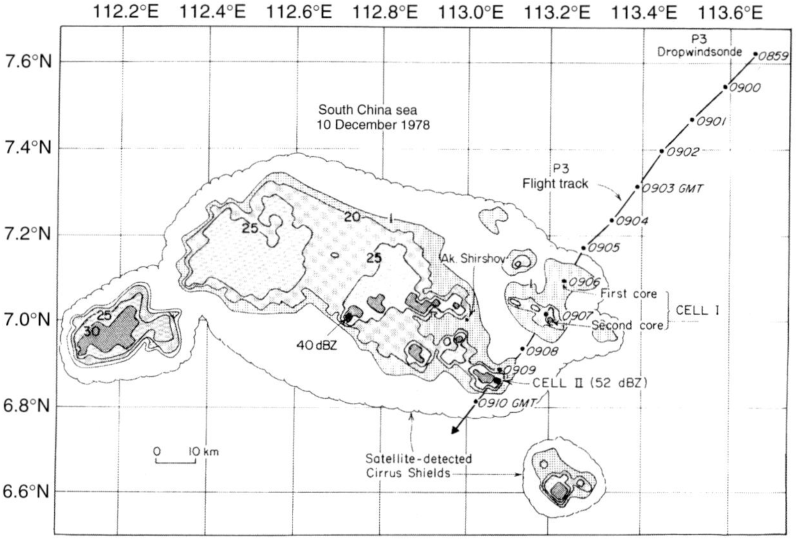


**FIGURE 9.4 Radar and satellite depiction of cluster A.** The scalloped outline indicates the approximate boundary of the cloud shield. The contours indicate precipitation at 1, 20, 25, 30, and 40 dBZ detected by the P3's lower fuselage radar at 0908 GMT. The flight track of the P3 is annotated with time (GMT) to indicate the aircraft position. The two convective cells penetrated by the P3 are labeled I and II. The flight level was 7.8 km. *(From Churchill and Houze (1984a))*

is the most frequent form and probably plays a more important role in the overall energetics, moisture transports, and rainfall climatology of the tropics.

In contrast to the squall cluster, which exhibits a distinct mesoscale precipitation structure, the nonsquall cluster exhibits a variety of mesoscale precipitation features. Figure 9.4 illustrates a satellite and radar depiction of a nonsquall tropical cluster.

In some cases, a squall cluster is a substructural feature of a larger scale nonsquall cluster exhibiting a variety of convective organizations (Leary and Houze, 1979). The only precipitation feature common to all of these cloud clusters is the large stratiform-anvil cloud and precipitation region surrounding the convective precipitation features. In fact, the stratiform region of nonsquall clusters differs little from that observed in squall clusters.

Tropical clusters can become so large that they are described as "superclusters" (Nakazawa, 1988; Mapes and Houze, 1993). Mapes and Houze (1993) defined a supercluster as an MCS with a cold cloud area (at −65 °C) viewed by IR satellite imagery greater than 5000 km$^2$ and having a duration of more than 48 hours. Some superclusters have been observed to have durations greater than 100 hours.

The life cycle of nonsquall clusters is characterized by a formative stage in which convective cells prevail, a mature stage in which convective cells and stratified precipitation features coexist and contribute nearly equally to observed

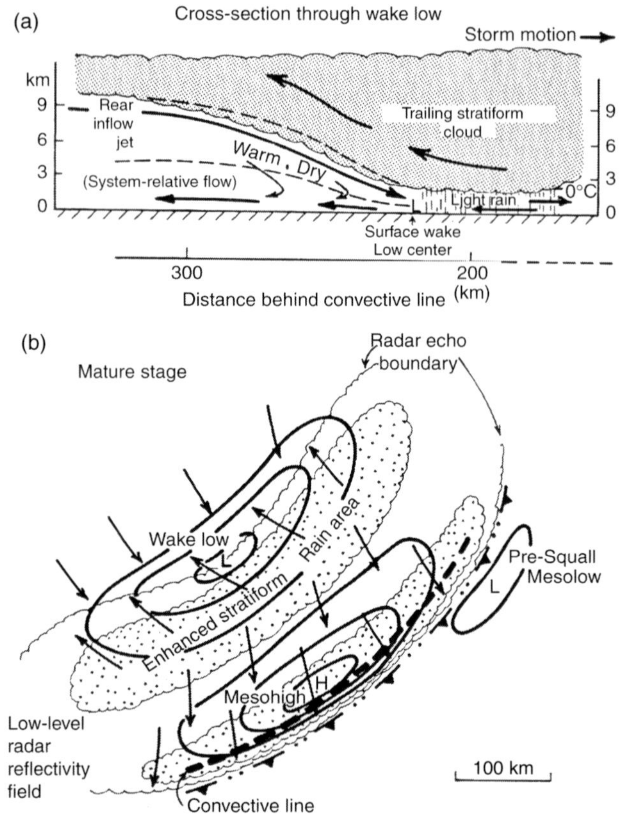


**FIGURE 9.5 Schematic cross-section through wake low (a) and surface pressure and wind fields and precipitation distribution during squall line mature stage (b).** Dashed line in (a) denotes zero relative wind. Arrows indicate streamlines, not trajectories. *(Adapted from Johnson and Hamilton (1988))*

rainfall, and a dissipating stage in which the stratiform precipitation prevails and slowly dissipates.

## 9.3.3. Conceptual Models of Mid-Latitude Squall Lines

The middle latitude squall line shown in Fig. 9.5 often exhibits a leading convective line followed by a trailing stratiform-anvil cloud. As in tropical squall lines, the trailing stratiform-anvil cloud is characterized by a slantwise ascending front-to-rear flow branch. Air in the front-to-rear flow branch ascends at a few tenths of a meter per second and originates in a layer of air above the boundary layer roughly 3-5 km above the surface (Cotton et al., 1995; Houze, 2004). This should be contrasted with the leading convective line which has updrafts of 10 to 20 m s$^{-1}$. Beneath the slantwise ascending branch is a slantwise descending rear-to-front flow branch. This descending flow branch is driven by downward directed pressure gradient forces associated with blocking of the diverging outflow in the ascending front-to-rear branch aloft, and by low-level divergence of air caused by evaporative cooling of precipitation beneath the trailing convective downdrafts and stratiform precipitation region. Melting and evaporation of precipitation in the stratiform region enhances the descending flow. At the surface the pressure field shown in Fig. 9.5(b) is characterized by low pressure immediately ahead of the squall line. This is probably caused by slow sinking motion in response to the rapid ascending air in the leading convective line. The slow sinking motion warms the air adiabatically thus contributing to a hydrostatic lowering of pressure. Behind the convective line is a pronounced region of high pressure. This is caused by the evaporation and melting of heavy precipitation in that region. Beneath the trailing stratiform-anvil cloud is a region of low pressure driven by subsaturated descent of air associated with the rear-to-front flow.

A feature that distinguishes the middle latitude, continental squall lines from their tropical oceanic cousins is the strength of the convective-scale updrafts and downdrafts. In the tropics, where numerous thunderstorms and mesoscale convective systems drive the stability of the atmosphere close to the wet adiabatic rate, clouds are not very buoyant and updrafts are typically 7 to 10 m s$^{-1}$. Also in the moist marine air, cloud-base heights are very low, being of the order of 500 to 600 meters above ground level. As a result thunderstorm downdrafts are shallow and do not obtain a great deal of negative buoyancy from evaporation of raindrops. Typically, downdraft strengths are only 2 to 3 m s$^{-1}$compared to 7 to 10 m s$^{-1}$in extra-tropical continental storms. Thunderstorms over the High Plains of the United States, for example, often have cloud bases 2 to 3 km above a relatively dry sub-cloud layer. Downdrafts are therefore stronger and extend over greater depths. Also, because of the stronger surface heating over the interior continental regions and stronger wind shears in middle-latitudes, convective updrafts in squall-line storms can be very intense, being as high as 25 to 40 m s$^{-1}$. This means that the mass transports and resultant flow structures of middle-latitude, continental squall lines are more strongly influenced by thunderstorm-scale drafts and flow structures than their tropical cousins.

For example, consider the conceptual model of the ordinary squall line thunderstorm system shown in Fig. 9.6. Like most ordinary squall lines, this squall line observed near Miles City, Montana, exhibited a bow-shaped leading squall front with thunderstorm cells aligned parallel to the squall front. Behind the leading squall line is a stratiform anvil region that extends several hundred kilometers. A two-dimensional cross-section perpendicular to the leading line through the cell labeled G1 is shown at the top of Fig. 9.6. This cross section shows the characteristic ascending front-to-rear flow in the trailing anvil region and a descending rear-to-front flow. The three-dimensional flow field illustrated

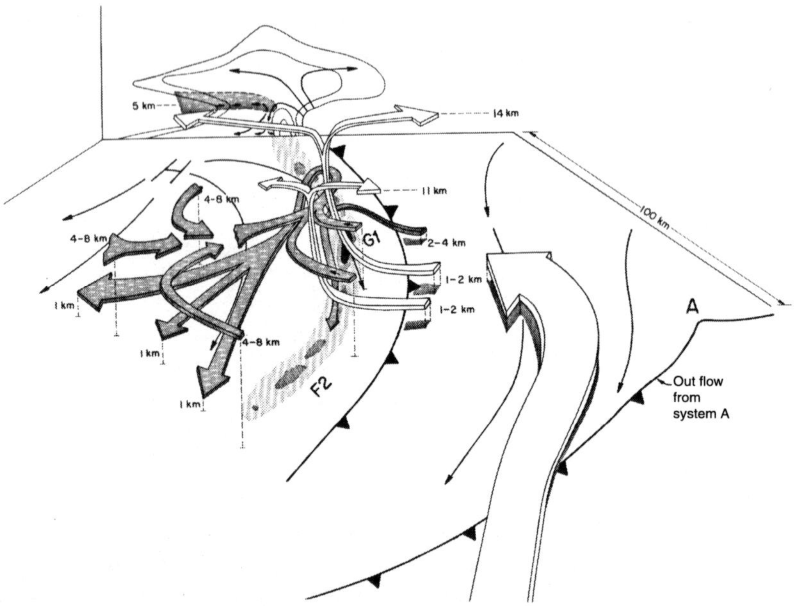


**FIGURE 9.6 Schematic depiction summarizing the 2-D and 3-D flow features for the 2 August 1981 CCOPE squall line showing mesoscale outflow boundaries, surface streamlines (thin arrows), convective reflectivity structure (stippled), overriding flow (bold arrow), and storm relative flow (thin ribbons).** The vertical cross-section corresponds to a representative depiction of the storm core G1 and shows reflectivity (thin solid lines), schematic storm relative flow (thin arrows), and location of the midlevel upshear inflow (shaded). The bold A refers to an isolated supercell system, G1 and F2 represent the cell groups along the squall line and the bold H represents the location of the surface mesohigh. Labeling of the storm relative flow ribbons refers to height AGL. *(From Schmidt and Cotton (1989))*

in Fig. 9.6 reveals a more complicated structure with middle level air in front of the line flowing about the cell G1 and descending in the rear-to-front flow branch. This middle level flow entering the rear of the storm resembles what Weisman and Davis (1998) calls "bookend vortices". Moreover, the air feeding the squall line is not a surface-based flow. Instead it glides over very stable-cool air left in the wake of an earlier supercell thunderstorm.

Clearly, the flow field is much more complicated than the two-dimensional depictions of ordinary squall lines shown previously. Part of the complicated flow structure arises from the fact that this storm is not isolated, but is overtaking a more slowly moving supercell. Another reason for its more complicated flow structure is that the storm formed in an environment with strong vertical shear of the horizontal wind through a large depth of the troposphere. This strong vertical shear was responsible for the formation of the supercell storm that preceded the squall line. Furthermore, the strong shear was responsible for creating the supercell-like storm in the squall line labeled G1. Cell G1 was a steady cell

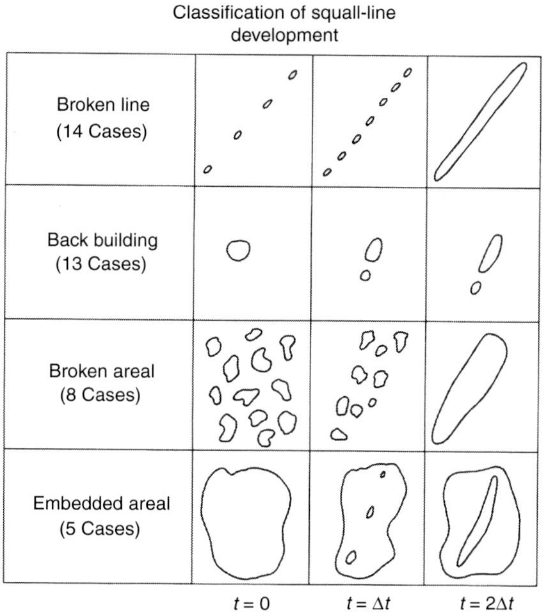


**FIGURE 9.7 Idealized depiction of squall-line formation.** *(From Bluestein and Jain (1985))*

that lasted several hours, contained a rotating updraft structure, and exhibited a bounded weak echo region (BWER). Other cells along the line labeled F2 were ordinary thunderstorm cells having lifetimes of 35 min to one hour. The supercell-like storm G1 so dominated the squall line flow fields that much of the complicated flow structure in Fig. 9.6 can be attributed to its presence.

This middle-latitude squall line was also noteworthy because it produced a nearly continuous swath of severe surface winds covering a four-state area from Montana to Minnesota. Such a severe, straight-line, wind-producing storm system is called a "derecho". Derechoes are noted for producing swaths of severe wind damage 100 km or more in width and extending 500-1000 km in length (see Section 9.6).

Bluestein and Jain (1985) identified four different classes of severe squall-line development: broken lines, back-building lines, broken areal lines, and embedded area lines, as illustrated in Fig. 9.7. They defined the Richardson number $R_\mathrm{i}$ as

$$R_\mathrm{i} = \frac{\mathrm{CAPE}}{\frac{1}{2}[(\overline{U}_6 - \overline{U}_{0.5})^2 + (\overline{V}_6 - \overline{V}_{0.5})^2]}, \tag{9.1}$$

where $\overline{U}_6$ and $\overline{U}_{0.5}$ represent pressure-weighted means over the lowest 6 and 0.5 km. They found that the average $R_\mathrm{i}$ for broken-line storms is 111 which, as found by Weisman and Klemp (1982), favors multicell storms. The average $R_\mathrm{i}$

for back-building lines is 32, similar to the value Weisman and Klemp found for supercell storms. Bluestein and Jain argue that the low values of $R_i$ for back-building lines are a result of strong vertical shear in their environment. Thus, wind shear influences the particular mode of organization of squall lines and their speed of propagation. Because isolated supercells and back-building lines exist in the same $R_i$ regimes, $R_i$ cannot distinguish between environments supporting one of these storm types. Bluestein et al. (1987) extended Bluestein and Jain's analysis to nonsevere squall lines. The principal difference between the environments of severe and nonsevere squall lines is that the convective available potential energy in the nonsevere squall-line environment is about 60% of that found in severe squall-line environments. Vertical shear is less in the nonsevere squall-line environments and, therefore, the bulk Richardson numbers are larger.

The largest and most violent squall lines are the pre-frontal lines that form in middle latitudes. Typically, they form along, or ahead of, a cold front associated with a vigorous, mid-latitude cyclonic storm. Often, the line forms in the warm sector of the cyclone, where the low-level jet brings warm, moist air into the region ahead of the cold front. This region of the cyclonic storm is ideal for manufacturing severe storms; it contains plenty of fuel to sustain intense thunderstorms and strong vertical shear of the horizontal wind, which helps organize the thunderstorms into efficient machines. Figure 9.8 illustrates a pre-frontal squall line observed on 20 May 1949. The squall line is ahead of the surface front and extends from the warm sector of the cyclone on its southern flank to north of the surface warm front. The squall line is depicted relative to the middle troposphere pressure chart (500 mb) in Fig. 9.8(a) and the lower tropospheric pressure chart (850 mb) in Fig. 9.8(b). An example of a very long, pre-frontal squall line is shown in infrared satellite imagery at two different times in Fig. 9.9. This particular squall line extends from southern Texas northeastward to the Illinois, Great Lakes region!

A schematic cross section perpendicular to a severe pre-frontal squall line is shown in Fig. 9.10. The flow structure is similar to the ordinary squall line shown in Fig. 9.5, except that the storm is deeper and the updraft and downdraft strengths are much greater. Due to the strong shear in the pre-frontal air mass, the pre-frontal squall line exhibits a more pronounced leading stratiform-anvil than the ordinary squall line. At low levels, a layer of warm, moist air feeds the thunderstorm updrafts. Intense updrafts extend through much of the troposphere and even penetrate to great heights in the lower stratosphere. Some of the updraft air encounters large quantities of precipitation where water loading, melting, and mixing with dry air entering the rear of the storm contribute to negative buoyancy. This creates a vigorous convective-scale "up-down" downdraft component and the updraft air is forced downward before it reaches the middle troposphere.

At middle levels, dry air enters the storm from the rear, where it mixes with cloudy air. Raindrops then settle into the dry air causing evaporative cooling and

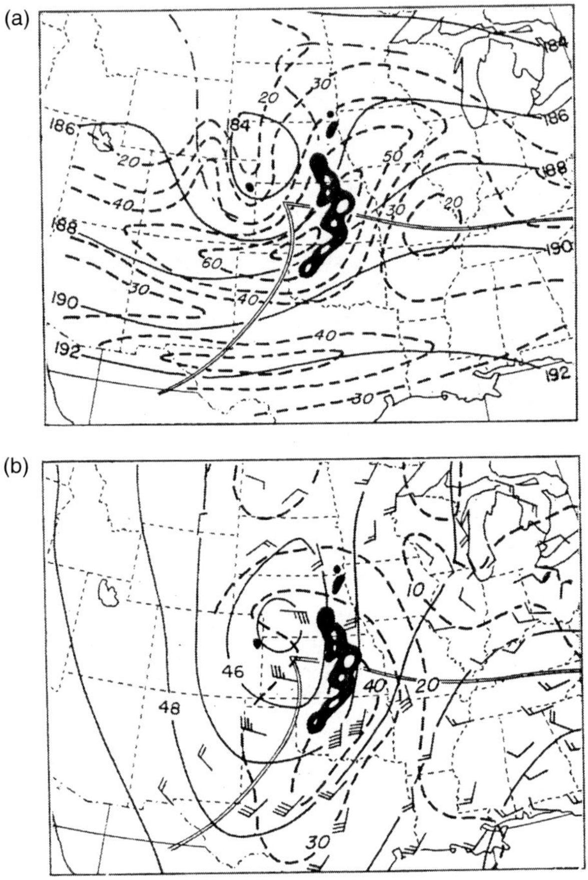


**FIGURE 9.8 (a) 500-mb (middle troposphere) and (b) 850-mb (lower troposphere) charts 2100 CST 20 May 1949.** Solid lines, contours (hundreds of ft); dashed lines, isotachs (knots); thin double lines, surface fronts. Blacked-in areas, rainfall in excess of 0.20 in per hour; inner light areas, 0.50 in or more. *(From Newton and Newton (1959))*

general sinking motion. A feature of such intense, squall-line thunderstorms is the notable tilt of the updrafts upshear (in this case upwind), which leads to a more efficient storm system. This is because precipitation settling out of the updrafts falls into the downdraft regions, thus increasing their strength rather than settling primarily into the updrafts and weakening them.

Frequently, the thunderstorm cells along the pre-frontal squall line are of the supercell variety, capable of producing tornadoes, grapefruit-size hailstones and wind damage. Because of the great length of the squall line, pre-frontal squall lines are responsible for the major outbreaks of severe weather. Fortunately, the conditions favorable for generating the severe, pre-frontal squall line occur rather rarely.

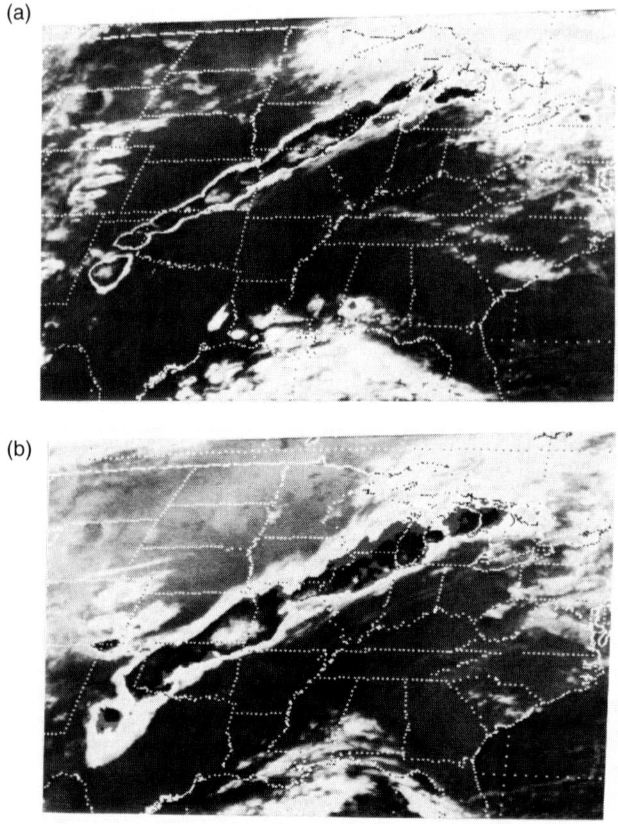


**FIGURE 9.9 (a) Satellite infrared image showing the squall line at 2000 CDT fairly early in life. (b) As in (a) but at 2330 CDT showing increase in width of cloud shield.** *(From Srivastava et al. (1986))*

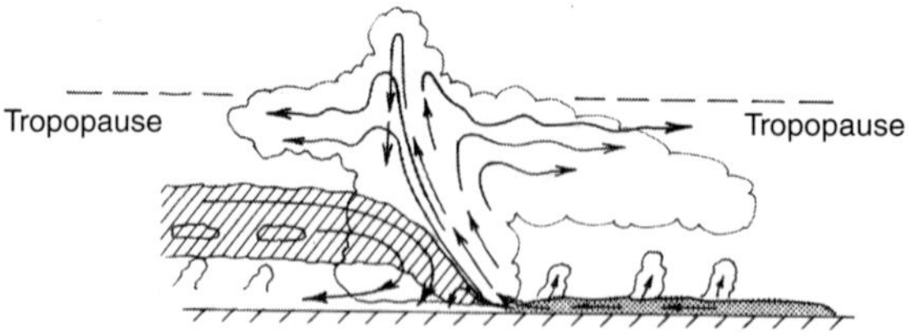


**FIGURE 9.10 Schematic cross-section perpendicular to a severe, pre-frontal squall line.** Slant-left shading represents depth of warm, moist air feeding the thunderstorms. Slant-right shading indicates layer of dry middle level air that becomes cold and negatively buoyant as cloud droplets and raindrops evaporate in it. *(Adapted from Newton (1966))*

**FIGURE 9.11 Enhanced infrared (IR) image of the United States at 0300 MDT 4 August 1977, from the GOES satellite at 70 °W longitude.** The stepped gray shades of medium gray, light gray, dark gray and black are thresholds for areas with apparent blackbody temperatures colder than −32, −42, −53 and −59 °C, respectively. Temperatures progressively lower than −63 °C appear as a gradual convective development of 4 August over the Colorado mountains. The intense meso-$\alpha$-scale convective complex (MCC) centered over eastern Kansas originated in the eastern Rockies and western plains the previous evening. *(From McAnelly and Cotton (1986))*

## 9.3.4. Conceptual Models of Mid-Latitude Non-squall Line MCSs and MCCs

Mesoscale convective systems which are not organized in a squall line structure are quite common in middle latitudes. They range in size from the scale of slightly larger than a multicellular thunderstorm, having three or four cells to the scale of a *mesoscale convective complex* or MCC. The MCC shown in Fig. 9.11 was defined by Maddox (1980) in terms of the scale, duration and eccentricity of the MCS as identified from infrared satellite imagery. Figure 9.11 illustrates a satellite image of an MCC at its mature stage. As seen in Table 9.1 (Maddox, 1980), an MCC is defined as a mesoscale convective system which exhibits infrared cloud top temperatures colder than −32 °C over an area exceeding 100,000 km$^2$, an interior cloud top temperature colder than −52 °C with an area greater than 50,000 km$^2$, for a period of six hours or more. The MCC also has an *eccentricity* (minor axis/major axis) greater than 0.7 at the time the cloud shield exhibits its maximum extent. The eccentricity criterion eliminates the large, squall-line, mesoscale convective systems such as the prefrontal squall line as being MCC's. Smaller squall lines are often embedded beneath the nearly circular cloud shield that is characteristic of an MCC. In some cases, the organization of the thunderstorm cells is a consequence of the early genesis stages of the MCC. Figure 9.12 illustrates the evolution of an MCC whose roots can be traced to the Rocky Mountains. In this case, a line of cells formed along the length of the Colorado Rocky Mountains. This line of cells subsequently moved over the High Plains, where it encountered low-level moisture that is usually brought into the High Plains by a low-level jet that reaches maximum

**TABLE 9.1** Mesoscale Convective Complex

| | |
|---|---|
| Size: | (A) Cloud shield with IR temperature $\leq -32$ °C must have an area $\geq$ 100,000 km$^2$<br>(B) Interior cold cloud region with temperature $\leq -52$ °C, must have an area $\geq$ 50,000 km$^2$ |
| Initiate: | Size definitions A and B are first satisfied |
| Duration: | Size definitions A and B must be met for a period $\geq$ 6 h |
| Maximum extent | Continuous cold cloud shired (IR temperature $\leq -32$ °C) reaches maximum size |
| Shape: | Eccentricity (minor axis/major axis) $\geq$ 0.7 at time of maximum extent |
| Terminate | Size definitions A and B no longer satisfied |

Source: From Maddox (1980)

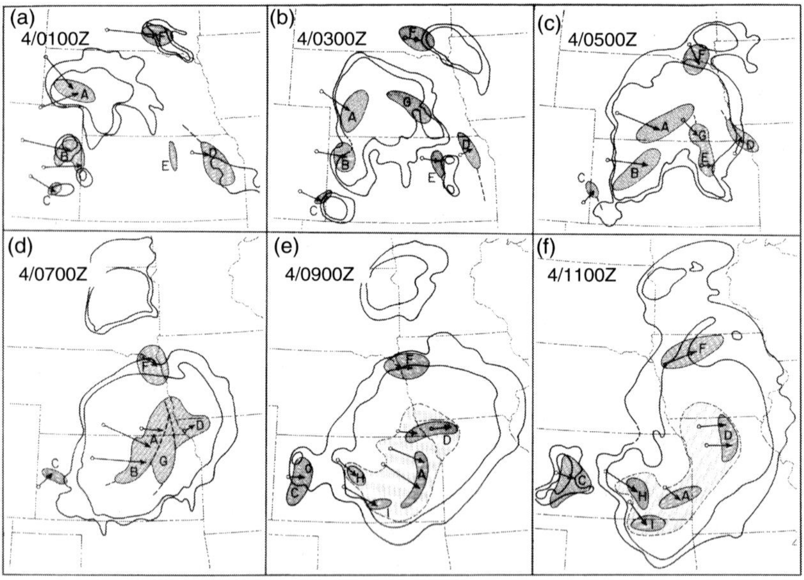


**FIGURE 9.12 Schematic infrared (IR) satellite and radar analysis at 2-h intervals, from 01 to 11 UTC 4 Aug 1977, for the western MCC #1.** The anvil cloud shields are indicated by the $-32$ and $-53$ °C IR contours (outer and inner solid lines, respectively), re-mapped from satellite images at the labeled times. Darkly shaded regions (identified by letters) denote significant radar-observed, mesoscale convective features at about 25 min after the indicated whole hour, with the vectors showing their previous 2-h movements. The dashed line segments extending from the mesoscale convective features indicate flanking axes of weaker convection. In the more developed MCC stages, in (e) and (f), the light-shaded area within the dashed envelope indicates weaker, more uniform and widespread echo. *(From McAnelly and Cotton (1986))*

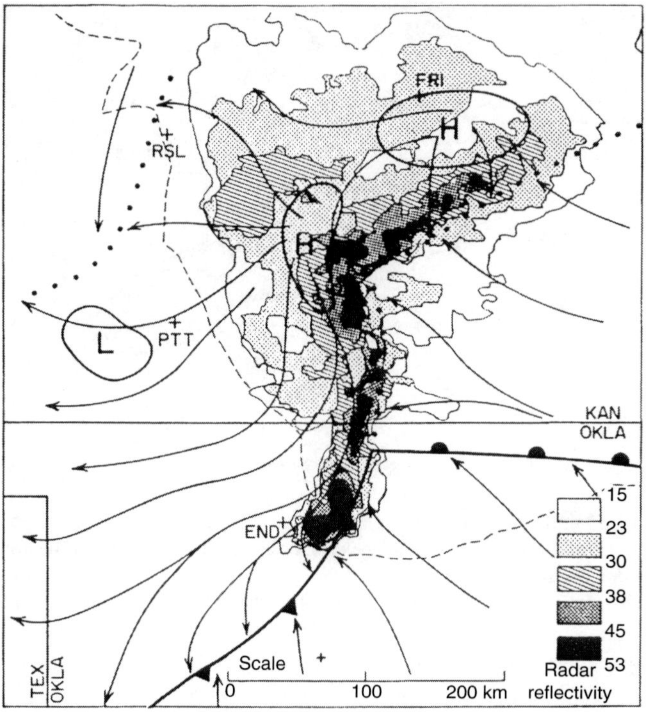


**FIGURE 9.13 Mesoscale surface analysis of the mature system B at 0100 UTC.** Pressure centers, streamlines, convergence lines (dotted), and fronts superimposed on reflectivity pattern (shaded) from Wichita radar, and the −54 °C contour of the margin of the cloud shield (thin dashed). *(From Fortune (1989))*

strength in the nighttime hours. At the same time, the mountain line of cells overtook a cluster of cells (labeled G, E, and D) that formed over the Plains. Thus, the system at its mature stage (see Fig. 9.12) can be seen to be composed of cells that formed initially over the mountains and cells that formed over the plains.

In some cases, MCC's develop a thunderstorm cell organization that resembles a miniature extratropical cyclone. Figure 9.13 illustrates a MCC, in which a north/south oriented squall line is joined by a nearly east/west oriented line of cells that forms a wave-cyclone pattern. Evidence of the wave-cyclone pattern can also be seen in the surface pressure field and in the upper-level wind fields.

An idealized schematic of the flow structure of an MCC resembling a small wave-cyclone is shown in Fig. 9.14. You are viewing the storm from the rear, southwest flank. The airflow structure is composed of three dominant airstreams which we call, conveyor belts, in analogy to extratropical cyclones. Ahead of the storm, the low-level jet serves as a *warm conveyor belt* bringing warm, moist

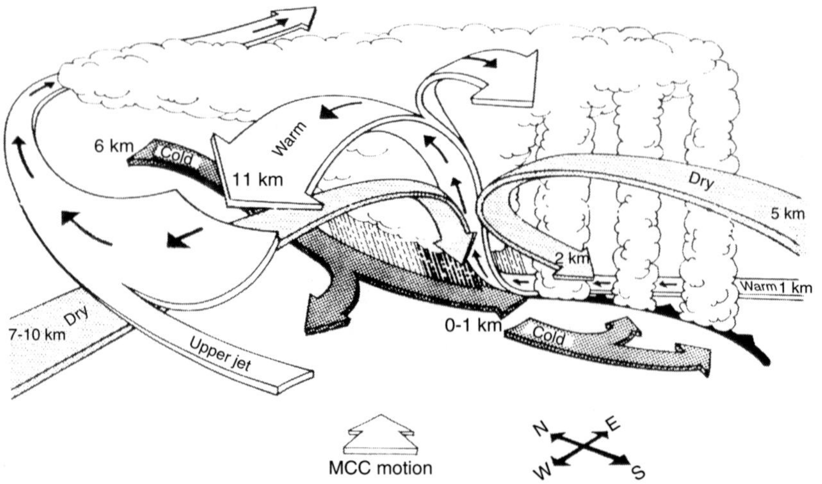


**FIGURE 9.14 Perspective drawing of an idealized, weakly rotating mesoscale convective complex as viewed from the rear.** The apex region is located where the heavier rain is drawn. The cumulonimbus towers at the right form on the mesoscale cold front. The mesoscale warm front and the forward part of the stratiform region are not shown in this view. *(From Fortune (1989))*

air from the south into the region of active convection. The warm, moist air rapidly ascends in cumulonimbus updrafts along the north-south oriented cold front. The warm conveyor belt also glides over the east-west oriented large-scale cold front, and some of it ascends rapidly in convective updrafts, while some of the air gently ascends in slantwise ascent, similar to the slantwise front-to-rear slantwise ascent of squall lines.

Behind the east/west-oriented cold front is another low-level airstream that we call the *cold, conveyor belt.* This airstream is confined vertically between the ground and the gently ascending warm conveyor belt. The warm conveyor belt creates a strong inversion in temperature, thus serving as a lid to convective overturning in the cold conveyor belt airstream. The cool conveyor belt air is frequently quite moist as steady, stratiform rainfall settles into the layer and moistens it as it evaporates.

Figure 9.14 illustrates a third airstream which originates to the northwest of the storm at the 7-10 km level. This airstream, which we call the *dry airstream* descends as it enters the storm. It is the counterpart of the middle-level rear-to-front flow that we have seen is common in squall line systems.

An MCC that resembles a small extratropical cyclone has also been called an "occluded MCS" described by Blanchard (1990). Figure 9.15(d), (e), (f) illustrate the evolution of an MCS into an occluded structure.

Like squall lines, an MCC is composed of both convective updrafts and slowly ascending motion. Instead of primarily trailing the convective line as in squall lines, the MCC experiences slow ascent in the middle and upper troposphere in a broad region surrounding the main convective cores. The life

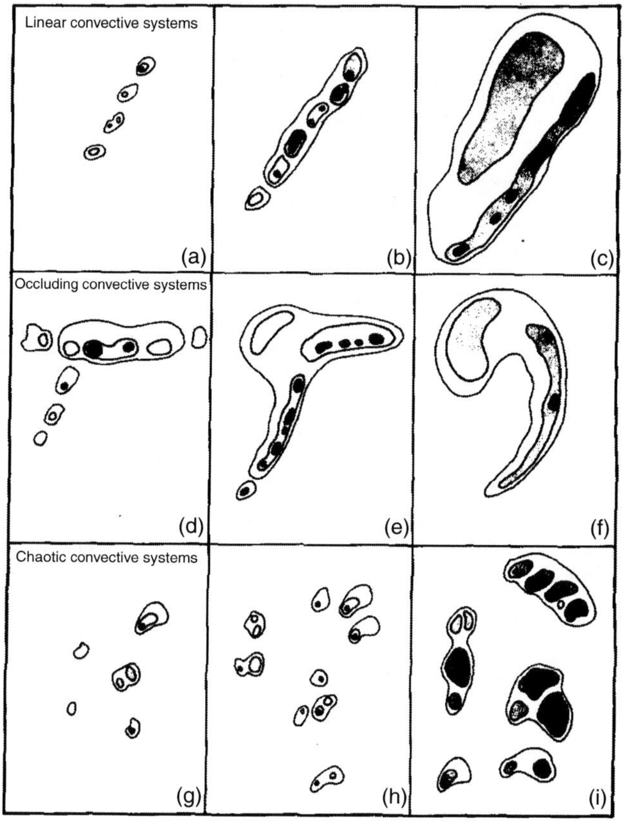


**FIGURE 9.15 Schematic of the evolution of the three convective modes observed during the PRE-STORM program: (a-c) linear convective systems, (d-f) occluding squall-line convective system, and (g-i) chaotic convective systems.** Contours and shading indicate increasing reflectivity roughly corresponding to 20, 40, and 50 dBZ. *(From Blanchard (1990))*

cycle of MCC's is such that early in their development, thunderstorm-scale motions prevail. The rainfall is quite showery at this point and there is a great likelihood that the system will produce severe weather, such as tornadoes, hail and flash floods. As the system reaches maturity and produces a stratiform-anvil cloud of maximum horizontal extent, the rainfall becomes light and steady, or more stratiform in character. At this point, the chance that the storm will produce tornadoes, hail, or very intense rainfall diminishes, although the storm produces the maximum volume of rainfall at its mature stage (see Fig. 9.24). The storm system slowly decays, producing lighter and lighter rainfall.

While much of the literature on MCSs has focused on MCCs that fit Maddox's (1980) definition or classical squall lines, it must be recognized that there exists a broad spectrum of MCS sizes, and organizations. Table 9.2 shows a much broader range of MCS classifications. Besides the classic MCC, it

**TABLE 9.2** MCS Definitions Based Upon Analysis of IR Satellite Data

| MCS category | Size | Duration | Shape |
|---|---|---|---|
| MCC | Cold cloud region $\leq -52$ °C with area $\geq 50{,}000$ km$^2$ | Size definition met for $\geq 6$ h | Eccentricity $\geq 0.7$ at time of maximum extent |
| PECS | | | $0.2 \leq$ Eccentricity $<0.7$ at time of maximum extent |
| M$\beta$MCC | Cold cloud region $\leq -52$ °C with area $\geq 30{,}000$ km$^2$ & maximum size must be $\geq 50{,}000$ km$^2$ | Size definition met for $\geq 3$ h | Eccentricity $\geq 0.7$ at time of maximum extent |
| M$\beta$PECS | | | $0.2 \leq$ Eccentricity $<0.7$ at time of maximum extent |

Source: From Jirak et al. (2003).

includes PECS which Anderson and Arritt (1998) identified as a more elongated version of MCCs, having eccentricities between 0.2 and 0.7, whereas MCCs have eccentricities greater than 0.7. Likewise, Jirak et al. (2003) identified smaller scale MCSs as meso-$\beta$ systems in keeping with Orlanski's (1975) definition of meso-$\beta$ scale weather systems. They identified meso*beta* circular systems (meso-$\beta$ CCSs) and meso-$\beta$ elongated systems (M-$\beta$ ECSs), as smaller MCSs having a satellite identified area greater than 30,000 km$^2$. This size criterion was based on the fact that the most coherent MCSs maintained a size of at least that area.

These definitions will be used in the next section when discussing the climatological characteristics of MCSs.

## 9.4. CLIMATOLOGY OF MCSS

### 9.4.1. Climatology of Squall Lines

Squall lines are ubiquitous phenomena that are observed from the deep tropics to very high latitudes such as Siberia, Alaska, and northern Canada.

Common to all squall lines is the presence of conditional instability in the lower troposphere and some form of forcing mechanism, such as easterly wave troughs in the tropics and frontal forcing and upper level short waves in middle latitudes. In addition, Frank (1978) noted that strong vertical shear of the low-level winds is critical to the organization of convection in squall lines. In their analysis of mid-latitude squall lines, Bluestein and Jain (1985) concluded that wind shear influences the particular mode of convection organization as well as their propagation speeds (see Fig. 9.7).

The large-scale environment of prefrontal squall lines has the features of the warm sector of a mid-latitude cyclonic storm and a surface cold front (see Fig. 9.8). A cyclonic storm with its low-level and upper-level jets provides the

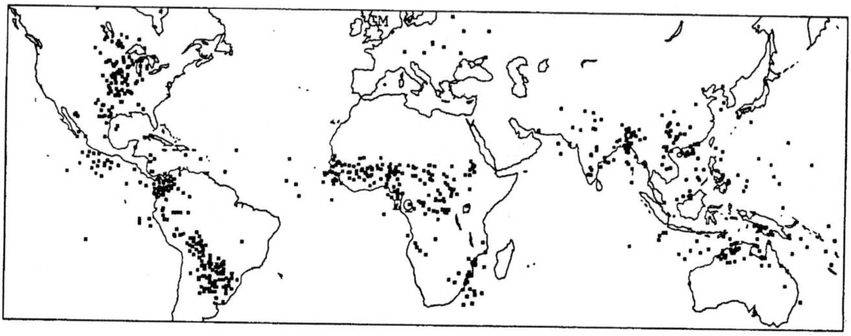

**FIGURE 9.16 Locations of MCCs based on 1-3-yr regional samples of satellite imagery (from Laing and Fritsch, 1997).** Locations are shown for the time of maximum extent of the cold-cloud shield. *(From Fritsch and Forbes (2001))*

necessary vertical shear of the horizontal wind for squall-line maintenance. Differential thermal advection caused by the warm, low-level jet and by the upper-level jet are destabilizing influences in the warm sector of the storm. Moreover, the moist air advected by the low-level jet is a source of conditional instability. In a developing springtime cyclone to the lee of the Rocky Mountains in the United States, lower tropospheric to mid-tropospheric air frequently originates over the warm, dry Mexican plateau. As it is swept northeastward by the evolving cyclonic storm, it encounters warm moist air originating over the Gulf of Mexico. The horizontal contrast in moisture forms the "dry line", a well-known source region for severe weather. As the warm dry air migrating from the Mexican plateau moves over the slightly cooler, Gulf of Mexico air, a pronounced capping inversion is created which inhibits the consumption of moist static energy by small convective elements (Carlson et al., 1980). Combination of all these factors creates an environment capable of sustaining long-lived squall-line convective systems.

Miller (1959) emphasized the significance of the dry-air intrusions between 850 and 700 mbar, not only in sustaining severe convection, but also as an important trigger for initiating those systems.

## 9.4.2. Climatology of Non-Squall Line MCSs and MCCs

We begin with a discussion of the climatology of the largest form of MCS which is called an MCC. Table 9.1 shows the criteria Maddox (1980) identified based on visible satellite imagery.

Climatological studies by Maddox (1980), Velasco and Fritsch (1987), Miller and Fritsch (1991), Laing and Fritsch (1993a,b, 1997, 2000), reveal that MCSs are quite ubiquitous worldwide and generally form over land, are nocturnal, develop in zonally-elongated belts in easterlies or westerlies, and form within 1500 km downstream of north-south oriented mountain ranges. Figure 9.16 illustrates the global distribution of MCCs. Figure 9.17 illustrates

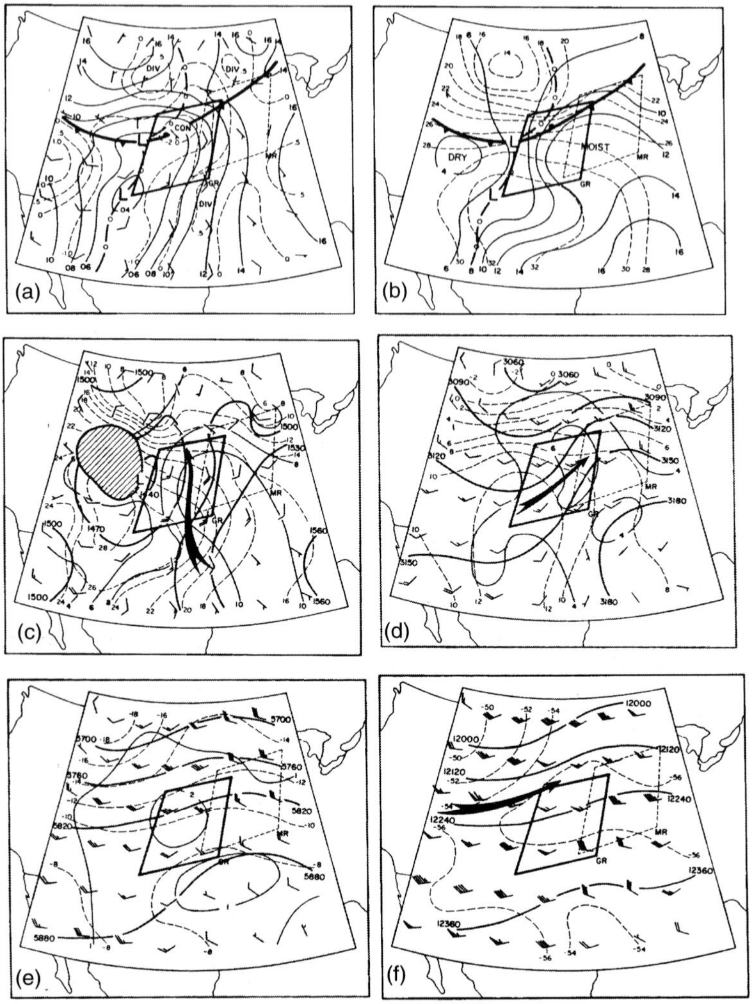


**FIGURE 9.17** **(a) Analysis of surface features prior to MCC development.** Surface winds are plotted at every other grid point (full barb = 5 m $s^{-1}$). Isobars of pressure reduced to sea level are shown as solid lines, with surface divergence ($\times 10^{-5}$ $s^{-1}$) shown as dashed lines. **(b) Analysis of surface features prior to MCC development.** Isopleths of mixing ratio (g $kg^{-1}$) are solid lines and isotherms (°C) are dashed. **(c) Analysis of the 850-mbar level prior to MCC development.** Heights (meters) are heavy solid contours, isotherms ($\delta eC$) are dashed, and mixing ratio (g $kg^{-1}$) is indicated by light solid contours. Winds (full barb = 5 m $s^{-1}$) are plotted at every other grid point and dark arrow shows axis of maximum winds. The cross-hatched region indicates terrain elevations above the 850-mbar level. **(d) As in (c), but for the 700-mbar level prior to MCC development. (e) As in (c), but for the 500-mbar level prior to MCC development. (f) As in (c), but for the 200-mbar level prior to MCC development (note that wind flat = 25 m $s^{-1}$).** *(From Maddox (1983))*

the environment of the MCC genesis region as determined by Maddox (1983). At low levels, MCCs are associated with a nearly stationary surface front oriented in a general east-west direction. The low-level flow (south of the front) is marked by warm advection. Fritsch and Forbes (2001) note that some MCCs form in more barotropic environments in which frontal forcing is weak or absent, and development occurs as a result of the cold pools generated by a cluster of convection. They refer to these MCSs as Type-2 MCCs. At 850 mbar, a pronounced low-level jet (LLJ) feeds warm, moist air into the MCC precursor environment. The preference for MCC formation to the lee of major mountain ranges is in large measure associated with the preferred formation of LLJs to the lee of major mountain ranges (Velasco and Fritsch, 1987; Miller and Fritsch, 1991; Augustine and Caracena, 1994). The systems reach maturity at the time that the diurnally varying low-level jet reaches its maximum intensity over the High Plains of the United States. At 700 mbar, a weak short wave is evident with warm-air advection over the generating region. In several case studies, McAnelly and Cotton (1986) found that individual meso-$\beta$-scale thunderstorm elements followed the contours of the 700-mbar height fields and merged into a meso-$\alpha$-scale MCC system in the confluence zone of the height field. Maddox noted a weak shortwave trough at 500 mbar west of the generating zone of the systems. This shortwave trough progresses eastward throughout the storm's life cycle and is evidenced by larger amplitude perturbations in the 500-mbar height and temperature fields as the storm system matures.

Laing and Fritsch (2000) further emphasized that MCCs around the world typically form in regions of strong low-level wind shear (see Fig. 9.18).

The early stages of convective organization often take place in the form of meso-$\beta$-scale thunderstorm systems outside the genesis region defined by Maddox (1983). These systems move into the region characterized in Fig. 9.17. Figure 9.12 illustrates such a sequence of events as seen by satellite and radar. The stratiform region reaches its fullest areal extent by 0900 GMT (0300 CST). MCCs typically reach full maturity between 0100 and 0500 local standard time and are characterized as nocturnal thunderstorm systems.

Maddox (1983) described the mature stage of an MCC by compositing standard rawinsonde soundings taken at 1200 GMT, which is close to the average time of maximum areal extent of MCCs over the central United States. Figure 9.19 illustrates Maddox's composite analysis for a mature MCC from the surface to 200 mbar and should be compared with the pre-storm conditions in Fig. 9.17. At 1200 GMT, the surface front has moved southward, and the surface divergence pattern suggests a weak mesohigh and outflow boundary on the southern side of the front. The surface air mass has been moistened considerably from moist downdrafts and rainfall evaporation. At 850 mbar the low-level jet has increased slightly in speed and veered to a more southwesterly direction. Associated with the low-level jet is a region of higher moisture content.

The flow at 700 mbar has strengthened (>5 m $s^{-1}$ increase) to a west-southwesterly jet over the mature MCC region. The 700-mbar shortwave trough

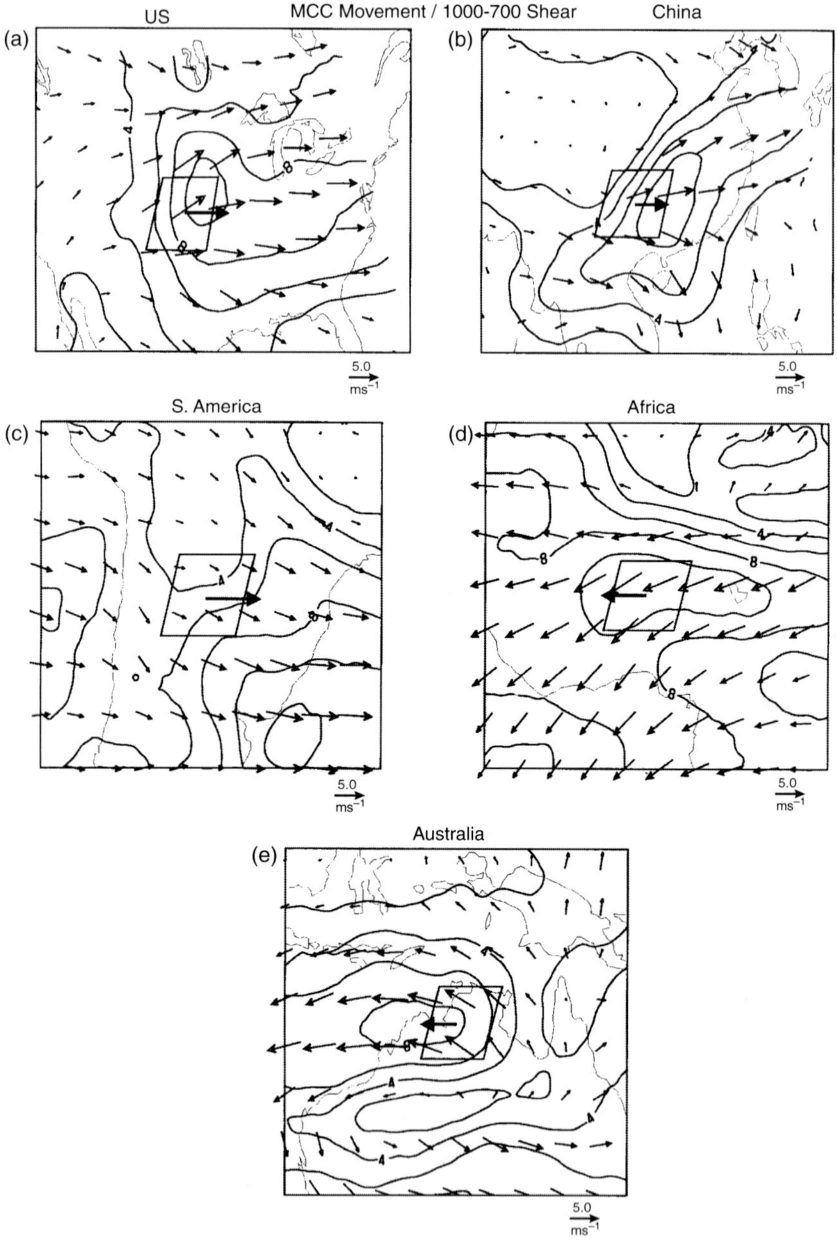


**FIGURE 9.18 Shear vectors (arrows) and shear magnitude (solid lines, m s$^{-1}$) in the 1000-700-mb layer.** Bold arrows in the genesis regions (quadrilaterals) indicate the speed (m s$^{-1}$) and direction (in the rotated coordinate system) of the convective systems. *(From Laing and Fritsch (2000))*

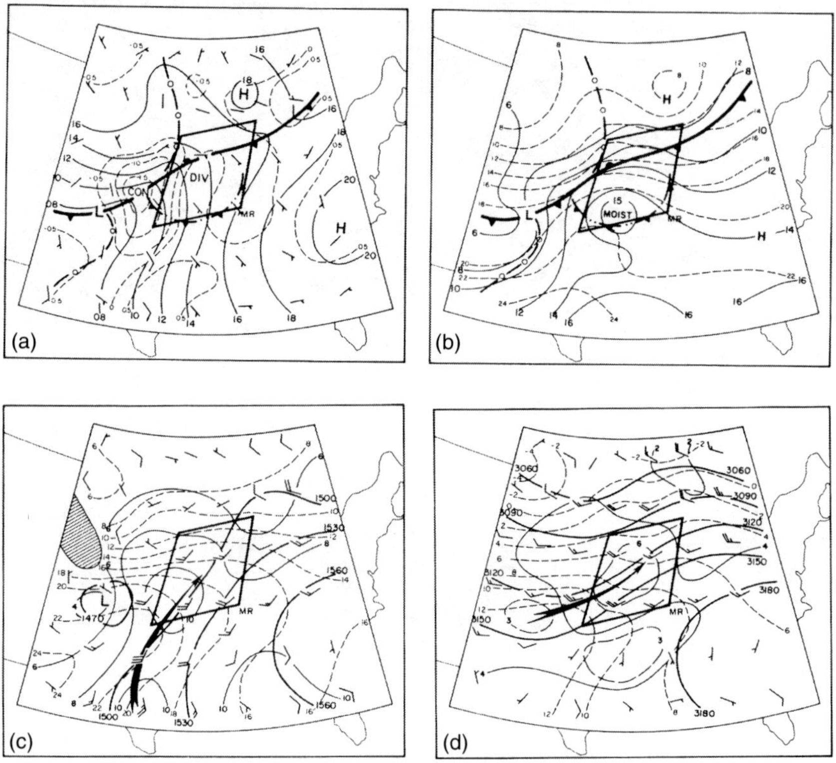


**FIGURE 9.19 Same as Fig. 9.17, except at the time of maximum MCC.** *(From Maddox (1983))*

and ridge pattern has moved eastward with the speed of the system. The shortwave trough is also evident at 500 mbar. Maddox emphasized the fact that the entire composite MCC life cycle is linked to the eastward progression of a meso-$\alpha$-scale shortwave trough having a wavelength of nearly twice the MCC diameter (1500 km).

The flow at 200 mbar is characterized by a pronounced jet streak to the north of the MCC. The anticyclonically curved jet streak exhibits speeds $>15$ m s$^{-1}$, greater than any observed wind at 200 mbar at 0000 GMT. Lin (1986) and Cotton et al. (1989) extended Maddox's 10-case composite analysis by considering over 128 MCC cases. Lin also attempted to define better the MCC life cycle by stratifying MCC life cycles into a total of seven time bins, each of a 2-h duration, centered at the time of MCC maximum maturity, as identified by infrared satellite imagery. MCCs show considerable variability, so at the time the operational 1200 GMT soundings are taken, systems may be anywhere in their life cycle from maturity to 2 to 4 h before or after maturity. Likewise, some systems may be in the early stages of growth, or within several hours of maturity at the time the 0000 GMT soundings are taken, and some may be in the final stages of decay when the 1200 GMT soundings are taken. Thus, a

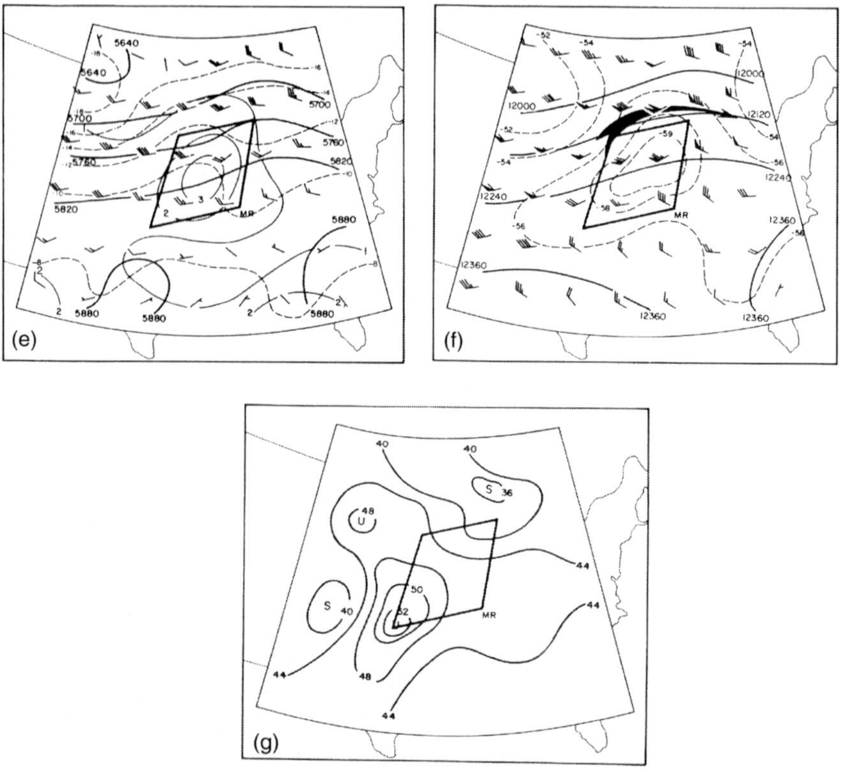


FIGURE 9.19 (*continued*)

composite model of an MCC life cycle can be obtained, even though high-time resolution soundings may not be available.

Figure 9.20 shows the divergence profile associated with Lin's (1986) composite data set. Early in its life cycle, the MCC is characterized by a layer of convergence up to 750 mbar and divergence aloft. By the time of MCC maturity (6 h later), the layer of convergence has deepened to 400 mbar, while the layer of divergence aloft has strengthened with its center at 200 mbar. A low-level region of divergence is evident below 900 mbar, presumably because of the development of a mesohigh as a result of evaporation of precipitation. At the time of MCC dissipation (6 h beyond MCC maturity), the layer of mid-tropospheric convergence has weakened and so has the low-level convergence zone. The region of upper tropospheric divergence reaches its maximum intensity very close to the time of system decay.

The corresponding composite vertical motion in Fig. 9.21 shows that the strongest upward motion is centered near 650 mbar early in the system's life cycle, similar to tropical clusters. Later in the life cycle, the upward motion

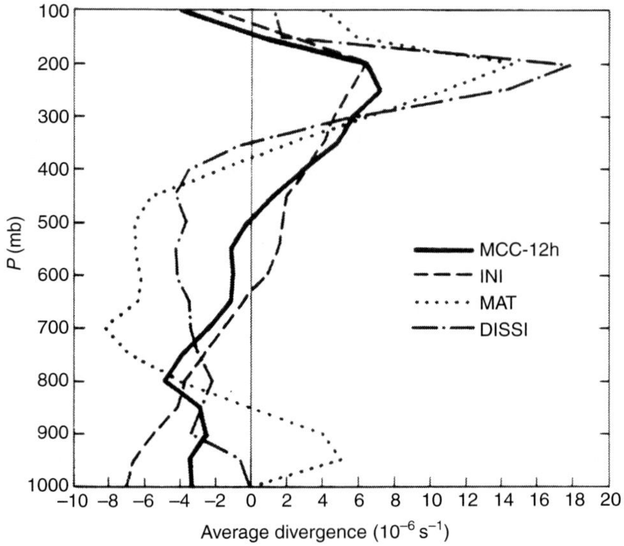


**FIGURE 9.20 Vertical profiles of horizontal divergence (horizontally averaged over the 3 x 3 central grid points at 50-mbar intervals) at the MCC 12-h, initial, mature, and dissipation stages.** Units: $10^{-6}$ $s^{-1}$. *(From Cotton et al. (1989))*

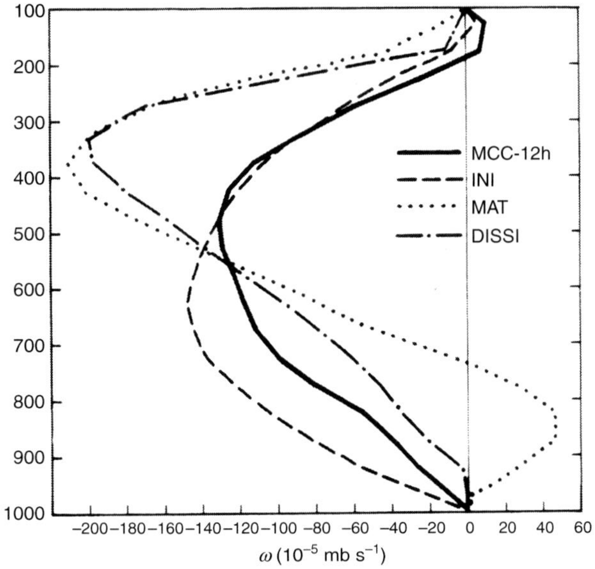


**FIGURE 9.21 Vertical profiles of vertical velocity ($\omega$) at the MCC 12-h, initial, mature, and dissipation stages, calculated by integrating the corresponding divergence profiles in Fig. 10.41. They represent average $\omega$ over the 3 x 3 central grid-point region (4.4 x $10^5$ $km^2$).** Units: $10^{-5}$ mbar $s^{-1}$. *(From Cotton et al. (1989))*

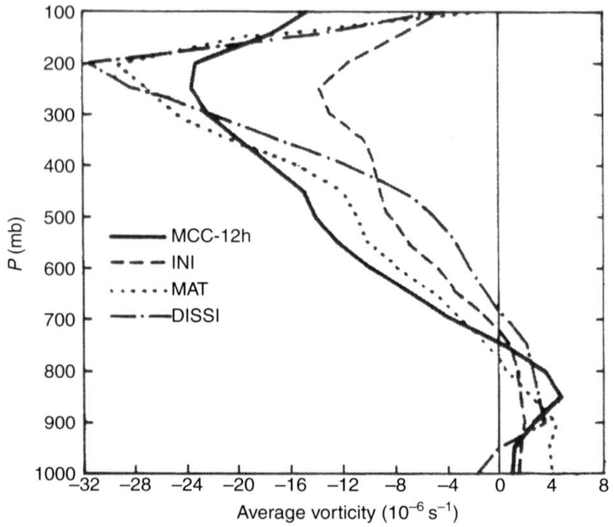


**FIGURE 9.22 Vertical profiles of relative vorticity (horizontally averaged over the 3 x 3 central grid points at 50-mbar intervals) at the MCC 12-h, initial, mature, and dissipation stages.** Units: $10^{-6}$ $s^{-1}$. *(From Cotton et al. (1989))*

maximum shifts upward to 350 mbar. In Lin's composite analysis, the maximum vertical motion at 350 mbar remains strong through the dissipating stage of the system.

The composite vorticity profile is shown in Fig. 9.22. The environment of the MCC is characterized by weak cyclonic vorticity below 700 mbar throughout the MCC life cycle, with a maximum at 900 mbar at maturity. Likewise, above 700 mbar, the environment is characterized by anticyclonic flow, with a maximum at 200 mbar toward the end of the system's life cycle. A region of cyclonic vorticity increase between 450 and 600 mbar is evident in the latter stages of the system's life cycle. Cyclonic vorticity increases at 800 to 900 mbar, while anticyclonic vorticity increases near 200 mbar. The maximum in anticyclonic vorticity at 200 mbar occurs at the decay stage of the composite system. This corresponds to the time at which vertical motion is also a maximum.

The midlevel cyclonic vorticity maxima in the latter stages of the MCC are probably associated with the convergence that reaches a maximum at middle levels at the time of system maturity. A similar midlevel positive vorticity center was analyzed by Bosart and Sanders (1981) in their investigation of the Johnstown Flood MCC. They speculated that the positive vorticity center resulted from interactions between a low-level cold-pool, which intensified a preexisting lower tropospheric jet. In the analysis of one MCC, Leary and Rappaport (1987) found that the stratiform precipitation region was organized in a set of curved bands that were aligned with cyclonic inflow at 500 mbar. These

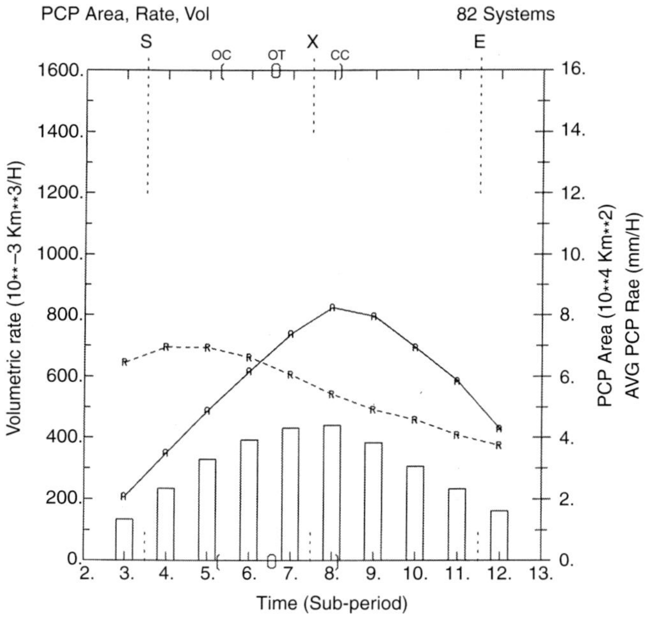


**FIGURE 9.23 Time evolution of MCC precipitation characteristics, averaged over 82 systems.** The time scale represents a normalized life cycle (based on satellite-observed cloud shield size), where the MCC is defined to start, maximize, and end at times 3.5, 7.5 and 11.5, respectively. Solid curve shows area of active precipitation, and dashed curve shows average rainfall rate within that area. Their product gives the volumetric rainfall rate of the MCC, depicted by the bars. *(From McAnelly and Cotton (1986))*

curved bands are evidence of a mesoscale convective vortex (MCV) which will be discussed more fully in a later section.

Analysis of hourly surface precipitation associated with MCCs reveals a well-defined precipitation life cycle that is consistent with the composite kinematic fields and satellite appearance. McAnelly and Cotton (1986) determined the composite precipitation life cycle for an episode of MCCs observed in 1977, and, more recently, the composite precipitation life cycle for MCCs using the same set of cases that Lin used in his composite study. Figure 9.23 shows that the average precipitation rates increase in intensity through the developmental stages of the MCC and reach their peak at the time the MCC exhibits its maximum cloud shield. During this same period, the rainfall rates are intense and convective in scale. The volumetric rainfall reaches a maximum, however, during the mature stage of the system when its meso-$\alpha$-scale circulation is best organized. Later in the MCC life cycle, the rain area continues to expand, but the precipitation rate becomes lighter. This is consistent with the weakening mesoscale circulation in Lin's composite kinematic fields. McAnelly and Cotton (1986) found that the precipitation was largely confined to the meso-$\beta$-scale convective features embedded in the systems.

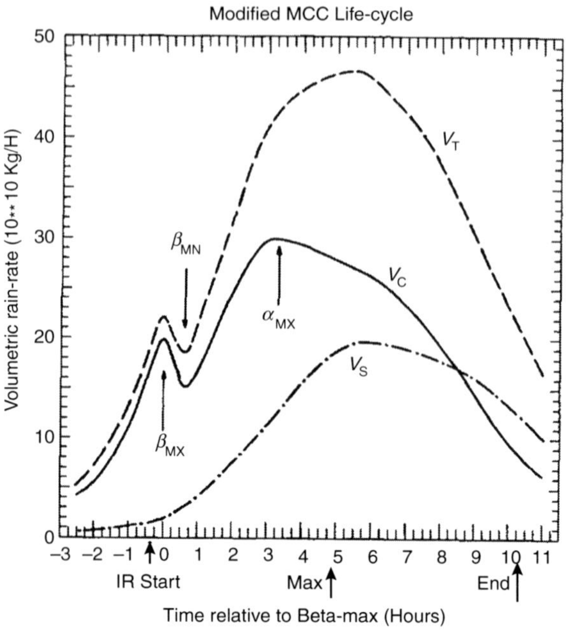


**FIGURE 9.24 Modified MCC precipitation life cycle, in terms of volumetric rain rate due to convective, stratiform, and total echo.** The growth stage through the $\alpha$-scale maximum in $V_C$. *(From McAnelly and Cotton (1992))*

Figure 9.23 is a bit deceptive in that it implies that the precipitation in MCCs undergoes a smooth transition from the early convective stages to the mature and dissipating stages. However, McAnelly and Cotton (1992) analyzed the rainfall in MCCs in more detail, and found a characteristic transition from the stages dominated by deep convection to that dominated by more stratiform rainfall. Figure 9.24 illustrates that the volumetric rainfall rain increases rapidly but it is then followed by a period of weakening of rainfall and a rebound to higher rain volumes during the mature stages of the storm. McAnelly et al. (1997) attributed this behavior to the response of the storm dynamics to gravity wave behavior as the system undergoes a transition from a heating profile characteristic of deep convection to a heating profile that is characterized by more slantwise, ascent and stratiform rainfall.

In summary, the MCC represents a large mesoscale convective system that exhibits a lifetime in excess of 6 h. Convection within the MCC takes on a variety of forms and, therefore, is not a distinguishing property of these systems. The presence of a stratiform-anvil cloud of considerable areal extent, however, is a feature of all MCCs. The upper levels of MCCs are characterized by divergent anticyclonic flow while the flow at midlevels is convergent and cyclonic. At the surface, the MCC is dominated by a divergent mesohigh of large areal extent. As the system reaches maturity, deep convection becomes a less dominant feature of the MCC, and, instead, slantwise ascent and descent with associated stratiform rainfall dominates.

In some cases, MCCs form in episodes on several consecutive days (Wetzel et al., 1983; Tuttle and Carbone, 2004; Tuttle and Davis, 2006). Wetzel et al. (1983) analyzed an 8-day episode in early August 1977 in which one or more MCCs formed. In analyzing the large-scale conditions that prevailed throughout the period as well as shortly before and after the episode, Culverwell (1982) showed that two large-scale circulation features combine to establish a deep, moist, conditionally unstable atmosphere over the High Plains of the United States. At low levels over the High Plains, the Atlantic subtropical high establishes itself far enough westward to drive warm, moist air northward over the High Plains. This circulation favors the development of a vigorous nocturnal low-level jet laden with moist air. The generally southerly flow associated with the Atlantic subtropical high is a factor in retarding the southward penetration of weak synoptic-scale surface fronts and aids the establishment of an east-west-oriented quasi-stationary front across the central High Plains.

At the same time, Culverwell ascertained that the midlevel moistening of the air mass was a consequence of the development of the southwest monsoon (Rasmussen, 1967; Hales, 1972; Brenner, 1974). This circulation pattern, which usually reaches maximum intensity between 15 July and 15 August, transports moisture-laden air from the Gulf of California and the southern Pacific Ocean northeastward over the central Colorado Mountains. Deep convection over the mountains aids the vertical transport of moist air into mid-tropospheric levels as it passes over the Rocky Mountain barrier. These two streams of moist air become vertically aligned over the High Plains and provide a favorable environment for sustaining MCC-scale convection. That both sources of moistening were needed to sustain MCCs was evidenced by the fact that the episode ended with the weakening of the southwest monsoon circulation which cut off the supply of middle-level moisture. Only a few small thunderstorm clusters could be sustained over Texas in the days that followed.

Repeated episodes of MCCs are so common in the climatological record that they contribute to the well known nocturnal maximum in warm season precipitation first identified by Wallace (1975). Carbone et al. (2002) analyzed warm season precipitation episodes using WSR-88D radar data for the period 1997 to 2000. Using Hövmöller diagrams (see Fig. 9.25) they found that coherent rainfall events having a zonal width of about 1000 km and one day duration occurred repeatedly; nearly once per day. Many of the events were episodic much like the one described above. The phase speed of the rainfall anomalies generally exceeded the zonal winds in the low to middle troposphere. Tuttle and Davis (2006) showed that such coherent rainfall features were largely composed of a succession of MCSs occurring in a corridor, that follow similar paths that last several days. They found that the dominant large-scale feature associated with these episodes was their positioning relative to the northern terminus of the LLJ.

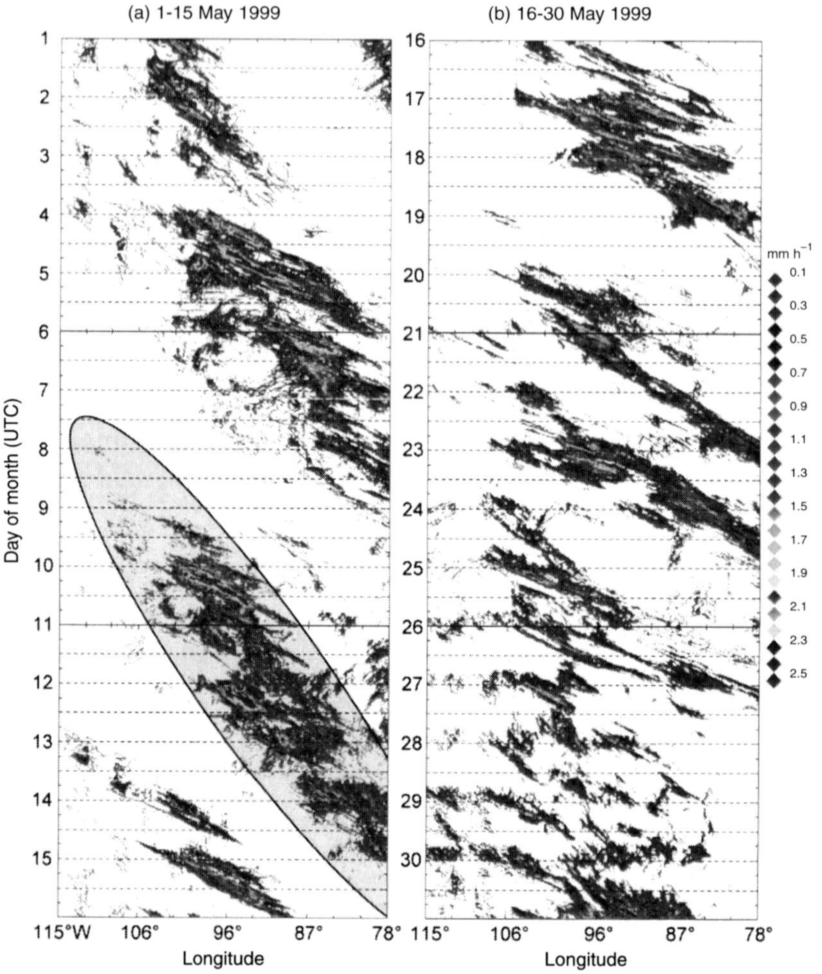


**FIGURE 9.25 Radar-derived rain-rate. Hövmöller diagram for (a) 1-15 May 1999, and (b) 16-30 May 1999.** Note the slow eastward propagation of precipitation envelopes in (a), within which there are faster propagating rain streaks. The shaded, elliptical area denotes one such envelope. In (b), there are mixed regimes including a more obvious component of diurnal modulation late in the period. *(From Carbone et al. (2002))*

Much of the climatology of MCSs has focused on the nearly circular, large systems called MCCs. Jirak et al. (2003) however, examined the climatological characteristics of several hundred MCSs that fit the more general classification defined in Table 9.2. This study included more elongated systems called PECs and smaller, meso-$\beta$-scale systems that are either circular or elongated. That study revealed that PECs were the most dominant form of MCS in terms of their size, frequency of occurrence, likelihood of severe weather production,

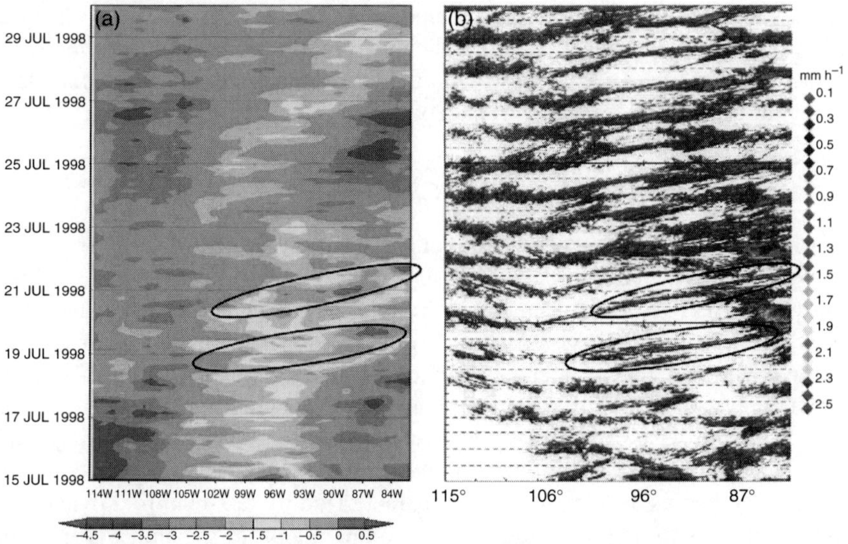


**FIGURE 9.26 Hövmöller diagrams for 1-29 Jul 1998 (in UTC) of (a) the MCS index and (b) the radar-derived rain rate (mm h$^{-1}$) (modified from Carbone et al. (2002)).** The data are averaged from 30° to 48 °N. The elliptical areas encompass examples of precipitation episodes and the corresponding values of the MCS index. *(From Jirak and Cotton (2007))*

and total amounts of rainfall produced. The more circular MCCs came next, while the meso-$\beta$-scale MCSs formed in somewhat more stable environments and thus were overshadowed by their larger cousins.

Using this more general classification of MCSs Jirak and Cotton (2007, 2008), re-examined the conditions that favored the upscale growth of convective systems into MCSs. Low level warm advection, low-level vertical wind shear, and convective instability were found to be the dominant factors in determining the genesis of MCSs from concentrated areas of convection. They developed a MCS index to aid forecasters in predicting the genesis of MCSs. Equation (9.2) defines the MCS index to be composed of three terms, a stability parameter or lifted index (LI), 0-3 km wind shear, and 700-hPa temperature advection (TA). Conditions are unfavorable for MCS genesis if the MCS index is less than −1.5, marginally favorable if between −1.5 and 0, favorable if between 0 and 3, and very favorable if greater than 3.

$$\text{MCS Index Op} = \frac{-(LI + 4.4)}{3.3} + \frac{(0-3\text{ km shear} - 11.5\text{ m s}^{-1})}{4.1\text{ m s}^{-1}} + \frac{(700\text{ mb TA} - 4.5\times10^{-5}\text{ K s}^{-1})}{1.6\times10^{-4}\text{ K s}^{-1}}. \tag{9.2}$$

It is interesting that when the MCS index is presented in the form of a Hövmöller index as shown in Fig. 9.26 for the period 15-29 July 1998, it reveals a structure similar to the precipitation episodes described by Carbone et al. (2002). The implication is that the most important factor in creating such episodes is the time and spatial variation of the dominant large-scale parameters that control MCS genesis and propagation.

### 9.4.3. Dynamical Aspects of MCSs

In our study of cumulus clouds and cumulonimbus clouds in Chapters 7 and 8, we have seen that their dynamics is largely governed by buoyancy-driven upright convective updrafts and downdrafts, cold pools, and gravity waves. We refer to these processes as "fast-manifold" dynamics that are fundamentally nonhydrostatic in nature. As noted previously, an MCS is not only characterized by such fast-manifold dynamics but also by "slow-manifold" dynamics or "potential vorticity" (PV) thinking associated with slantwise ascending and descending flows in the stratiform-anvil region of an MCS. We shall see that an MCS often begins as a storm system characterized by fast-manifold dynamics, and transitions to a system dominated by slow manifold dynamics. We begin in this section by first discussing fast-manifold dynamical aspects of MCSs, focusing on cold-pool dynamics and gravity wave dynamics. We will then discuss the evidence for the slow-manifold dynamical character of MCSs.

#### *9.4.3.1. Cold-Pool Dynamics*

We have seen in Chapter 7 that even weakly precipitating tower cumuli or trade wind cumuli produce cold pools which then spread out leading to the propagation of the cloud systems and even a shift in scale of convective elements to larger scale cumuli. Likewise, we have seen in Chapter 8 that cumulonimbus clouds propagate by generating strong evaporatively-driven downdrafts and cold-pools. The ordinary thunderstorm conceptual model proposed by Byers and Braham (1949) is distinguished by the formation of downdrafts and cold pools that favor the sustained supply of moist, unstable air into the updrafts during the mature and dissipating stages of the storm.

In Section 8.8 we reviewed the theories for autopropagation of cumulonimbi and focused on cold-pool propagation. In particular, we summarized the RKW (Rotunno et al., 1988) theory in which they argued that the optimum situation for maintaining steady, upshear-tilted squall line convection is one in which the negative vorticity associated with a cold-pool and the positive vorticity associated with low-level shear are approximately balanced. Figure 8.27 illustrates the concept.

Fritsch and Forbes (2001) identified MCSs that form in warm sector environments without the benefit of synoptic-scale frontal forcing as Type-2 events (Kane et al., 1987; Johnson et al., 1989; Geldmeier and Barnes, 1997). They noted that type-2 events typically develop from moist downdraft generated cold-pool formed by evaporating and melting precipitation from cumulonimbi rooted in the moist boundary layer. They noted that the slantwise ascending front-to-rear flow does not begin until a mature, large cold-pool forms.

Thus we see that mechanical lifting on the flanks of large cold-pools is an important feature of MCS genesis and propagation.

### 9.4.3.2. Gravity Wave Dynamics

We have also seen in Chapter 8 that propagation by gravity waves is an important property of cumulonimbus clouds. The behavior (speed and amplitude) of gravity waves in a stably stratified environment is dependent upon the degree of stability and wind shear (see Section 8.8.2 for details). They are also dependent on the particular character of the vertical profiles of heating associated with deep convection and MCSs. Figure 8.30 illustrates that as the heating profile evolves from one characterized by isolated cumulonimbi to an MCS with stratiform anvil clouds, vertical motion associated with propagating waves around the maturing MCSs evolves to one with subsidence aloft and upward motion at low levels. This transformation leads to the formation of what Mapes (1993) refers to as buoyancy bores, that favor low-level lifting in the MCS environment and the upscale growth of MCSs or what he calls "gregarious" convection. Mapes argues that gravity wave dynamics is responsible for the formation of super-clusters in the deep tropics.

We have seen that the volumetric rainfall in an MCS does not monotonically increase from the convective stages to the mature stages. Instead it increases rapidly and is then followed by a period of weakening and a rebound to higher rain volumes during the mature stages of the storm (see Fig. 9.24). McAnelly et al. (1997) applied a linear gravity wave model similar to that used by Mapes (1993) and Nicholls et al. (1991) to show that this behavior can be attributed to the response of the storm dynamics to gravity wave behavior as the system undergoes a transition from a heating profile characteristic of deep convection to a heating profile that is characterized by more slantwise, ascent and stratiform rainfall (see Fig. 9.27). They found that the initial response to increased diabatic heating in a rapidly intensifying cumulonimbus ensemble is deep compensation subsidence in near environment. The subsidence reduces CAPE and acts to weaken convection. After passage of the subsidence wave ascent rebuilds. Low-level precipitation evaporation and elevated heating in the stratiform anvil produces ascent and develops sinking at low levels. Thus the convection weakens and rainfall volume diminishes. After $\beta$-burst, destabilized low-level air with increased integrated CAPE is made available to the ensemble of storms. This creates a positive feedback for convective re-intensification similar to Mapes' arguments.

Gravity waves can also be responsible for the propagation of MCSs, particularly squall lines that are decoupled from the boundary layer. We have seen in Chapter 8 that trapping of gravity wave energy can occur whenever the scorer parameter, $\ell^2$ is less than $k^2$, where $k$ is the horizontal wave number of a wave (see Eq. (8.12)). Figure 9.28 illustrates a sounding taken in the vicinity of

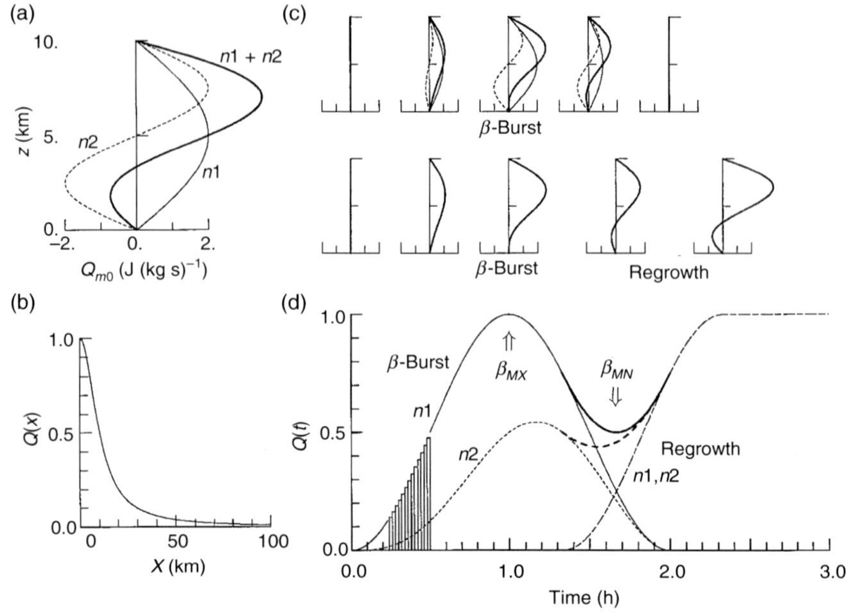


**FIGURE 9.27 Thermal forcing specifications for the linearized model.** (a) Vertical distribution of maximum heating rate magnitude $Q_{m0}$ for $n1$ mode (thin solid line), $n2$ mode (thin dashed line), and $n1+n2$ modes (heavy solid line). (b) Horizontal dependence factor $Q(x)$ on heating profiles in (a). (c) Time-dependent heating profiles at $x=0$ for several times along time axis in (d). Limits of axes are as in (a). Top row: profiles for meso-$\beta$ convective burst with $n1$ mode (thin solid line) and $n1+1n2$ modes (heavy solid line); contribution of $n2$ mode is shown (thin dashed line). Bottom row: profiles for meso-$\beta$ burst followed by regrowth, due to $n1+1n2$ modes. (d) Time dependence factor $Q(t)$ on heating profiles in (a). For the meso-$\beta$ burst (02 h), factors for $n1$ mode (thin solid line) and, when included, $n2$ mode (short dashed line) are shown. For the meso$\beta$ burst followed by regrowth (beginning at 1.33 h), the $n1+n2$ burst is used, and regrowth factors are identical for $n1$ and $n2$ modes (thin long dashed line). Overlapping period from 1.33 to 2.0 h shows added factors from the two heating phases for $n1$ (heavy solid line) and $n2$ (heavy dashed line) modes. Combined $n1$ curve defines meso-$\beta$ maximum and minimum. Thin bars following $n1$ curve from 0.25 to 0.5 h illustrate a series of short-duration, constant-amplitude heat pulses used to build the time-dependent heating. *(From McAnelly et al. (1997))*

a derecho-producing MCS discussed by Schmidt and Cotton (1989). The MCS formed over a cold-pool produced by the outflow from a previous supercell storm. As a result, the low level cold pool was quite stable, and was capped by a nearly adiabatic layer, with a stable upper troposphere. In addition the wind profile exhibited strong vertical wind shear through the depth of the troposphere. These are ideal conditions for the trapping of gravity wave energy. The two dimensional simulations reported by Schmidt and Cotton (1990) revealed that near the surface, gravity waves are trapped by the nearly adiabatic layer above and the ground below, as $N^2$ is quite small in the adiabatic layer. In the upper troposphere, gravity waves are trapped by the adiabatic layer below, and by

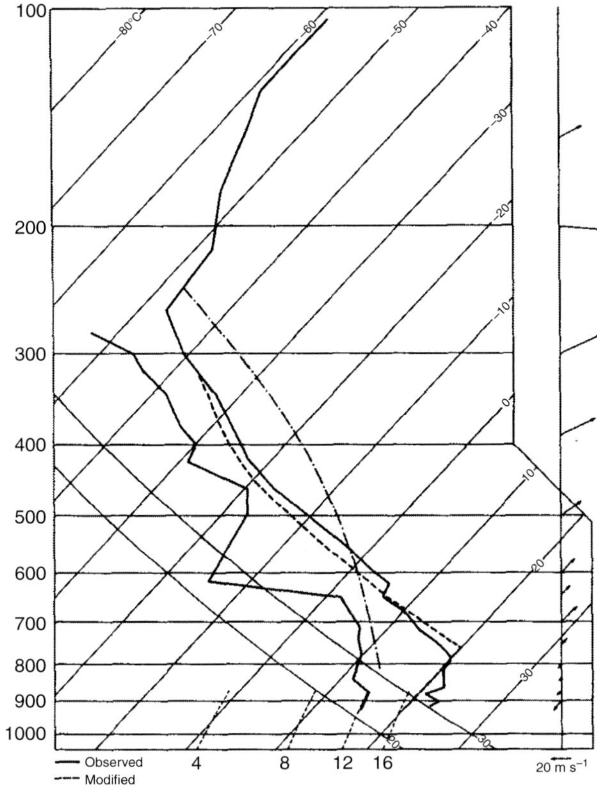


**FIGURE 9.28 Environmental sounding of temperature and dewpoint temperature of the pre-squall environment from Knowlton, Montana on 2 August 1981 (solid lines) used to initialize the control experiment.** The sounding was modified below 60 kPa with aircraft soundings taken immediately ahead of the squall line. The thin dashed line represents the simplification made to the observed temperature profile used to initialize experiments S0-S3 and FS2. Parcel ascent based on the observed environment is shown as the dot-dash line. *(From Schmidt and Cotton (1990))*

strong shear near the tropopause. Schmidt and Cottons simulations revealed that low levels were so stable that penetrative downdrafts could not reach the surface. As a result, the squall line propagated by the upward lifting of air above the stable layer by the trapped low level gravity waves. Figure 9.29 is a conceptual model illustrating the response of the simulated squall line to varying shear for the thermodynamic profile shown in Fig. 9.28. When large scale flow is absent (Fig. 9.29a), waves generated in the upper troposphere move relative to lower tropospheric waves, resulting in decoupling of the vertical motions forced by them,and the convective system rapidly dies. As illustrated in Fig. 9.29b, c, as shear is increased in amplitude, the waves are Doppler-shifted in phase speed such that the waves in the upper troposphere become coupled with the near surface waves and a more upright, long-lived squall line results. In contrast to

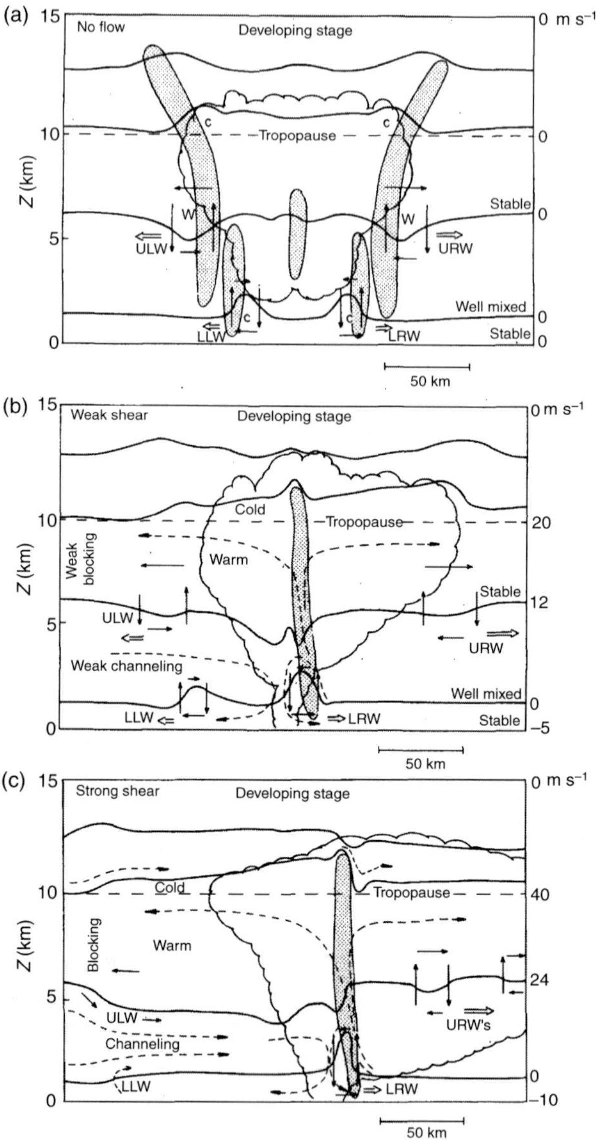


**FIGURE 9.29 Conceptual diagram summarizing: (a) experiment S0 (no shear), (b) experiment S1 (weak shear), and (c) experiment S2 (strong shear).** The solid lines represent $\theta$ surfaces associated with primary waves discussed in Schmidt and Cotton (1990). The labels C/W in (a) represent, respectively, regions of cold/warm perturbations in $\theta$. Double arrows represent wave movement. Thin arrows denote perturbations in $w$ and $u$. The primary updraft zones are shaded. The numerical values on the right figure boundary refer to the ambient horizontal wind speed in meters per second. Schematic streamlines in (b) and (c) are meant to suggest the flow resulting from the primary wave modes. *(From Schmidt and Cotton (1990))*

the RKW theory for squall line propagation in a convectively unstable boundary layer, deep tropospheric shear is important in determining the longevity of the system over a stable boundary layer. Moreover, with deep tropospheric shear with a stable boundary layer, stronger shear favors the formation of a high amplitude wave with stagnation at low levels that lifts low level air to the level of free convection.

Schmidt and Cotton also concluded that shear in the upper troposphere led to the vertical displacement of isentropes on the upshear flank of the storm which led to partial blocking of the storm-relative flow aloft and channeling of the flow below. The channeling of the storm-relative flow was responsible for the excitation of the rear-inflow jet.

It is interesting to note that modeling studies by Bernardet and Cotton (1998) suggest that long lived squall lines can transition from propagation in a daytime unstable boundary layer driven by cold-pool dynamics much like the RKW concepts, to gravity wave propagation in a nocturnal boundary layer. There is observational evidence that the reverse can happen in which an MCS is driven by gravity wave propagation over a low level stable layer and then transition to one driven by cold-pool dynamics (Marsham, 2009; Personal communication). The question is what is the glue that keeps an MCS intact during such transitions? We hypothesize that it is the balanced flow dynamics that are discussed in the next section.

However, Parker (2008) performed idealized simulations in which the transition from afternoon convection to nocturnal, elevated convection was mimicked by slow introduction of low level cooling in a high resolution cloud resolving model. No changes in winds such as the formation of LLJs were simulated. The simulations revealed that a squall line underwent a transition from a cold-pool dominated density-current driven system, to a density-current like bore, and then to a gravity-wave like bore that provided continued lift as an elevated convective system much as in the simulations of Schmidt and Cotton (1990) and Bernardet and Cotton (1998).

Pandya and Durran (1996) carried out a series of idealized two dimensional squall line simulations, including the explicit treatment of cloud and precipitation processes,and then identical simulations in which all microphysics was turned off and either a constant or a time-varying heating profile was applied to mimic convective heating and cooling. They showed that much of the structure of the simulated squall line with explicit microphysics could be reproduced with applied heating including the slantwise, ascending front-to-rear circulation, the slantwise descending rear-to-front flow, upper level rear-to-front flow ahead of the squall line, and an upper-level cold anomaly to the rear of the thermal forcing. Their results imply that the circulations in the trailing stratiform-anvil region are a result of gravity waves forced primarily by the low-frequency components of latent heating and cooling in the leading convective line. Even though steady thermal forcing was applied, the cellular character of updrafts and downdrafts was represented by the dry simulation, albeit crudely.

Another example of the role of gravity waves in MCSs is in the observed discrete propagation of squall lines. Cram et al. (1992a,b) described observations and modeling studies of a pre-frontal squall line that moved much faster than a front. Squall line propagation was discrete in that new cells formed ahead of the gust front. In fact, at one point it was observed that the squall line convection dissipated and reformed 100 km ahead! Numerical simulations of the observed squall line revealed that the squall line propagated as a gravity wave that was trapped between the tropopause and a low level layer having a small Scorer parameter.

In summary, there is strong theoretical and observational evidence that MCSs can propagate by both cold pool dynamics and gravity wave propagation and that much of the flow structure of MCSs is at least triggered if not determined by gravity wave dynamics. As we shall see, another important aspect of MCS dynamics is the balanced character of mature MCSs.

### *9.4.3.3. Viewing MCS Dynamics from a Balanced Dynamical Perspective or PV Thinking*

We have seen that the presence of a stratiform-anvil cloud of considerable areal extent is a feature of all MCCs. Moreover, observations and modeling studies show that once a system develops a large stratiform-anvil cloud, slantwise ascending and descending flow branches become a characteristic of MCSs (Ogura and Liou, 1980; Smull and Houze, 1985; Cotton et al., 1995; Fritsch and Forbes, 2001; Mecham et al., 2002; Houze, 2004; Browning et al., 2008). These flow branches are referred to as slantwise ascending front-to-rear and descending rear-to-front branches in squall lines. In some cases the rear-to-front flow branches may consist of multi-level branches (Fortune et al., 1992; Browning et al., 2008). It is also often thought that the slantwise ascending flow branches have their origins in the boundary layer and feed the upright convective updrafts. But the trajectory analyses in simulated tropical and mid-latitude MCSs reported by Cotton et al. (1995) and Mecham et al. (2002) revealed that the slantwise ascending front-to-rear flow branches can have their origin 3 to 5 km above the ground and above the convectively unstable boundary layer. Moreover, the slantwise ascending air was found to directly feed the stratiform-anvil cloud circulations rather than feed upright convective updrafts which detrain into the stratiform anvil cloud. Bryan and Fritsch (2000) showed that slantwise layer lifting in MCSs can lead to moist absolutely unstable layers (MAULs) wherein $\partial\theta_e/\partial Z < 0.0$. Whenever MAULs are present, rather than smooth slantwise ascent, one would expect considerable small-scale turbulent overturning to be present. As Mecham et al. (2002) have shown, a moist mid-level layer is essential to the formation of above-boundary layer slantwise ascent. They suggested that preexisting moistening by detraining cumulus and small cumulonimbi (Johnson et al., 1999) or evaporation from pre-existing stratiform rain may moisten the layer. Another mechanism discussed later is the presence of an elevated, moist LLJ that is frequently present in the MCS environment.

The idealized two-dimensional modeling studies by Pandya and Durran (1996) and Schmidt and Cotton (1989) suggest that these slantwise flow branches associated with MCSs are a result of "fast manifold" gravity wave dynamics. However, we shall show that similar features can be represented by balanced circulations much like those associated with mid-latitude cyclones.

Other features of MCSs that are consistent with a balanced dynamical view is that for the larger systems like MCCs, their lifetimes are in excess of 6 h. Convection within the MCC takes on a variety of forms and, therefore, is not a distinguishing property of these systems. The upper levels of MCCs are characterized by divergent anticyclonic flow while the flow at midlevels is convergent and cyclonic. At the surface, the MCC is dominated by a divergent mesohigh of large areal extent. As the system reaches maturity, deep convection becomes a less dominant feature of the MCC, and, instead, stratiform cloud and rainfall dominates. Moreover, one of the more dramatic features associated with MCSs is the formation of a mesoscale convective vortex (MCV) later in the life cycle of MCSs, particularly MCCs (Leary and Rappaport, 1987; Zhang and Fritsch, 1988a,b,c; Menard and Fritsch, 1989; Brandes, 1990; Johnson and Bartels, 1992; Fritsch et al., 1994). These features motivated Cotton et al. (1989) and later modified by Olsson and Cotton (1997a,b) to conclude that "a mature MCC represents an inertially stable MCS that is in a nearly balanced dynamical state and whose horizontal scale is comparable to or greater than a locally-defined Rossby radius of deformation".

The Rossby radius (see Eq. (9.3)) crudely identifies the scale at which the inertial stability of a system becomes important.

$$\lambda_{\mathrm{R}} = \frac{NH}{(\zeta + f)^{1/2}(2VR^{-1} + f)^{1/2}}. \tag{9.3}$$

The parameter $N$ is the Brunt-Väisälä frequency, $H$ the scale height of the circulation, $\zeta$ the vertical component of relative vorticity, $f$ the Coriolis parameter, and $V$ the tangential component of the wind at the radius of curvature $R$.

The significance of $\lambda_{\mathrm{R}}$ is that it identifies the scale at which rotational influences or the inertial stability of a system become important. As a tropical cyclone develops, for example, the relative vorticity and $2\,V/R$ become larger than $f$, which increases the inertial stability of the system. Thus if the scale of a disturbance exceeds $\lambda_{\mathrm{R}}$, the system is nearly balanced such that the circulations evolve slowly, and vertical motions are largely controlled by those primary circulations. As shown by Schubert et al. (1980), a barotropic model with a single phase speed predicts that the mass field adjusts to the wind field when the scale of an initial disturbance is small compared to $\lambda_{\mathrm{R}}$, and the wind field adjusts to the mass field when it is large compared to $\lambda_{\mathrm{R}}$. This is illustrated in Fig. 9.30, which is adapted from Ooyama (1982) and Frank (1983). Thus in region III, the

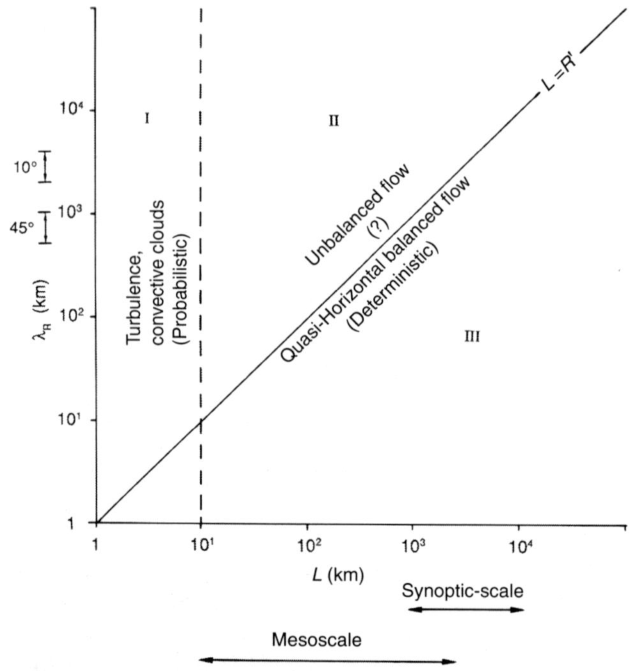


**FIGURE 9.30 Schematic of relationships between horizontal scale ($L$), the Rossby radius of deformation ($\lambda_R$), and modes of circulation. [After Ooyama (1982).] Typical values of $\lambda_R$ for systems at latitudes 10° and 45° are indicated.** *(Adapted from Frank (1983))*

circulations are approximately in geostrophic balance and two-dimensional in nature. Convective systems smaller than $\lambda_R$ represent unbalanced circulations in which convective heating excites gravity waves that propagate vertically and horizontally away from the convective disturbance. Thus, instead of sustained local subsidence occurring just outside the region of convection, transient gravity waves are produced that can distribute the compensating motions well away from the region of convection. Region I includes the scale of convective clouds, at which deep convective overturning takes place. Vertical motions in region I are typically of the order of 10 m $s^{-1}$. In region II, by contrast, vertical motions are of the order of 1 m $s^{-1}$ or less and are characterized by vertical displacements of a few kilometers.

An important feature illustrated in Fig. 9.30 is that the geometric scale of a disturbance does not completely describe a system as being dynamically large or small. A disturbance having a radius of 700 km in the tropics may be dynamically small, while a similar-sized system in mid-latitudes would be classified as dynamically large. Moreover, as a system develops relative vorticity, it can transform from a transient dynamically small system to a quasi-balanced, dynamically large system without any significant change in geometric scale. Owing to the larger values of $f$ in mid-latitudes, however, small changes in relative vorticity or the rotational component of the wind results in little change in the magnitude of $\lambda_R$.

The concept that MCSs, MCCs and MCVs, in particular, are nearly balanced mesoscale storms has led a number of researchers to apply PV thinking to develop models of MCSs. Examples include Raymond and Jiang (1990), Jiang and Raymond (1995), Davis and Weisman (1994), Schubert et al. (1989), Hertenstein and Schubert (1991), and Olsson and Cotton (1997b). Except for Davis and Weisman (1994), these studies have involved rather idealized course resolution simulations, in which heating profiles are specified, or a convective parameterization scheme is deployed to represent the heating profiles.

Consider Eq. (9.4) which is an approximate form of the Ertel potential Vorticity (EPV) introduced by Ertel (1942)

$$Q = \frac{1}{\rho}(\vec{\zeta} \cdot \nabla\theta), \tag{9.4}$$

where $\zeta$ is the vector absolute vorticity. Raymond (1992) expressed PV in terms of diabatic heating as

$$\frac{dq}{dt} = -\rho^{-1}(\nabla \cdot \vec{Y}) \tag{9.5}$$

where $\rho$ is density, and $Y$ is the non-advective flux defined as

$$\vec{Y} = -H\vec{\zeta} + \nabla\theta \times \vec{F} \tag{9.6}$$

where $H$ is the diabatic heating rate and $F$ is friction. When motion is frictionless and adiabatic ($H = 0$), $q$ is conserved.

By use of the hydrostatic approximation and introduction of a balance condition, which relates the thermal and wind fields, PV can be determined by specification of only one property, such as pressure or geopotential, while other quantities can be diagnosed. Combining the balanced condition with the PV equation produces a so-called "invertibility principle" (IP). In essence, the IP permits a unique combination of vorticity and $\theta$ fields from an infinite combination of those fields which satisfy the definition of PV. It is important to recognize that the solution of the system using PV inversion is a global rather than a local problem and that purely local information at one point is insufficient.

For illustrative purposes consider a simplification of Eq. (9.5) for the case of horizontally homogeneous PV, in frictionless flow in which absolute vorticity is mostly vertical, such that

$$\frac{\partial q}{\partial t} \approx -w\frac{\partial q}{\partial z} + \rho^{-1}\eta\frac{\partial H}{\partial z} \tag{9.7}$$

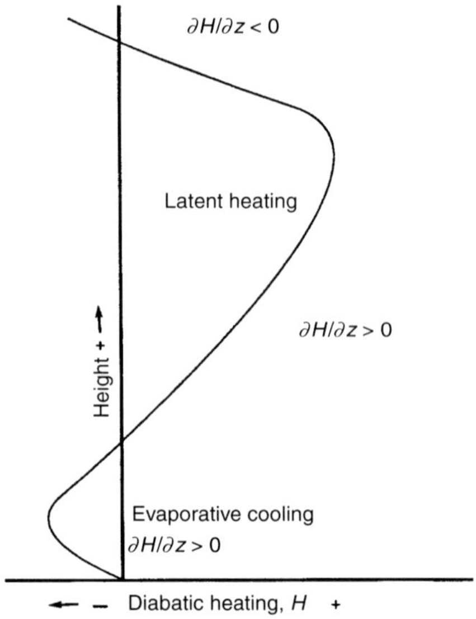


**FIGURE 9.31 Vertical profile of the diabatic heating rate typical of that measured in a mesoscale convective system.** The cooling near the surface is due to evaporation. *(From Olsson (1994))*

where $w$ is vertical velocity, and $\eta$ is the vertical component of vorticity which is positive in sign (inertially stable).

Figure 9.31 illustrates a conceptual diagram of heating associated with a mature MCS with stratiform-anvil cloud. This is based on observations reported by Yanai et al. (1973); Gamache and Houze (1982, 1985); Gallus and Johnson (1991). Figure 9.32 is a conceptual diagram that illustrates the expected vertical structure of the PV anomaly field proposed by Raymond and Jiang (1990). It shows that associated with the elevated heating and low-level cooling of the stratiform-anvil cloud of an MCS, a lower troposphere +PV anomaly and upper troposphere PV anomaly would occur. These features are consistent with the circulation features discussed earlier and that of MCVs discussed in the next section. Raymond and Jiang (1990) also presented a conceptual diagram of the expected behavior of a PV anomaly caused by, say, heating in a stratiform anvil cloud, in the presence of shear such as that associated with a LLJ, shown in Fig. 9.33. This shows that shear such as that associated with a LLJ will cause ascent beneath the low-level +PV anomaly (e.g. the stratiform-anvil cloud). They also show that in the presence of westerly shear, a +PV anomaly will force air up the east side of the anomaly and sinking on the west side. These features are consistent with the observed slantwise ascending or front-to-rear flow, and a descending rear-to-front flow found in MCSs with a pronounced stratiform-anvil cloud. Raymond and Jiang noted that the induced PV anomalies will have a vertical radius of influence of only a few kilometers under normal atmospheric

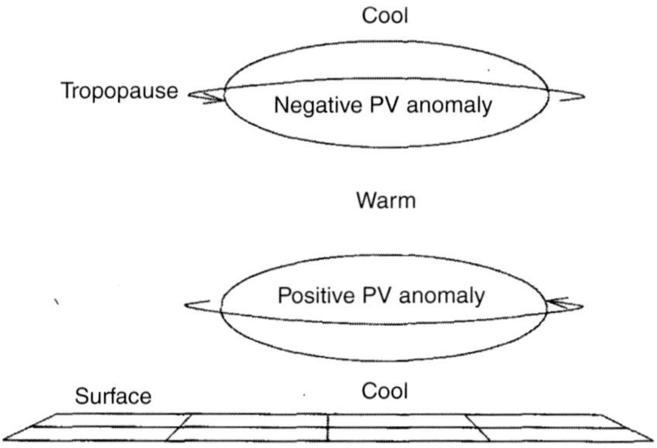


**FIGURE 9.32 Postulated structure of potential vorticity anomalies produced by a region of convection and the associated changes in temperature and wind structure.** The circulation is cyclonic around the lower, positive anomaly, and anticyclonic around the upper anomaly, as shown by the arrows. *(From Raymond and Jiang (1990))*

conditions. Thus a LLJ may not only provide the low-level shear favorable for the ascending motions illustrated in Fig. 9.33, but also provides the moisture and energy supply to sustain the heating profile shown in Fig. 9.31. In other words, a LLJ is an important agent in the transformation of an MCS from a system dominated by upright convection to one dominated by slantwise ascending and descending motions. We speculate that gravity wave dynamics as discussed by Schmidt and Cotton (1989) and Pandya and Durran (1996) are like "nucleation mechanisms" that transform the shape of the motion fields in an MCS to an environment favorable for sustained quasi-balanced flow dynamics.

Olsson and Cotton (1997a,b) performed mesoscale numerical simulations with a hydrostatic primitive equation version of RAMS (PE model) of several observed MCSs which compared favorably with observations. They then applied a nonlinear balance diagnostic model to the simulated MCS to access the degree of balance in the simulated MCSs. They found that the balanced model replicated many features found in the NWP representation of the storms. In particular, the PE model nondivergent components of the winds were in a nearly balanced state from the initial state to the dissipation of the simulated MCC. This almost instantaneous adjustment of the nondivergent winds to a balanced state was also diagnosed by Davis and Weisman (1994) in their diagnosis of balance flow in their nonhydrostatic simulation of an idealized squall line and associated MCV. Olsson and Cotton were surprised to find that a significant part of the PE model divergent winds was captured by the balanced model, though less than the nondivergent flow. The divergent balanced flow included features such as the slantwise descending rear-to-front flow. Unfortunately, specific focus on the existence of a slantwise non-convective ascending flow was not made in their analysis, so we can not say for certain that it is a component

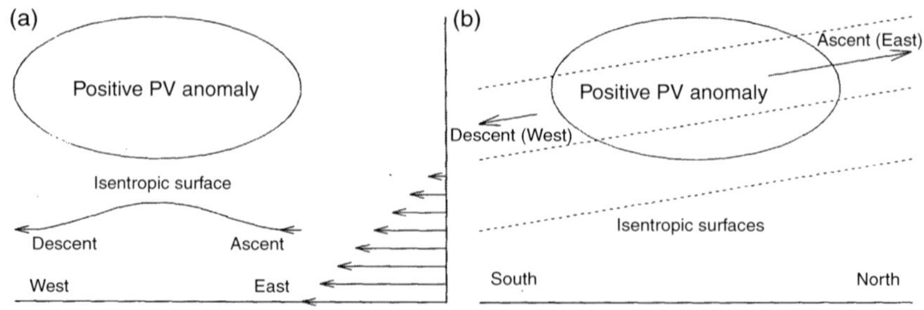


**FIGURE 9.33 Sketches of mechanisms by which lifting might occur in the presence of a potential vorticity anomaly in shear.** The ambient shear is confined to the east-west plane. (a) Ambient shear is limited to below the potential vorticity anomaly here for clarity. In a frame in which the anomaly is stationary, the relative environmental wind causes flow on the perturbation isentropic surface caused by the potential vorticity anomaly. Ascent and descent occurs as illustrated. (b) View of the potential vorticity anomaly from the east. The tilted isentropic surfaces (dashed lines) are associated with uniform ambient westerly shear through the depth of the illustration. The cyclonic circulation around the anomaly causes ascent in the northward-moving air on the east side and descent in the southward-moving air on the west side as illustrated. *(From Raymond and Jiang (1990))*

of balanced flow dynamics. The balanced model provided clear signatures of ascending motions associated with stratiform anvil heating, but the link to slantwise ascending flow originating between 3 to 6 km above the ground was not made. The balanced flow dynamics, however, presents a clear signature of the middle troposphere MCV. Clearly absent in the diagnosed balanced flow were downward motions associated with compensating subsidence, as we have seen these motions are represented by fast manifold gravity wave dynamics.

To illustrate the interplay of fast and slow manifold dynamics during the initiation and evolution of MCSs, consider the conceptual model that Tripoli and Cotton (1989a,b) derived from two-dimensional simulations of orographically-generated MCSs. Figures 9.34–9.38 illustrate five of the six stages. Stage 1, not shown, begins in the morning hours as the nocturnal downslope drainage flow is replaced by an upslope flow driven by heating of the elevated terrain. About 60 km east of the ridge top, a convergence zone forms as a result of the interaction between the mountains' and plains' thermally-driven upslope circulations and the ambient flow over the mountain. The mountain lee wave (gravity wave) flow inflects upward at the convergence zone. The meridional flow is sheared anticyclonically slightly above ridge top early in the day; this residual effect of the nocturnal regime accelerates the development of upslope flows. Upright cumulus convection at this stage can be characterized as cumulus and towering cumulus. During this and subsequent stages, the mountains' and plains' circulations exert a strong stabilizing influence over the eastern plains. The shallow upslope flow cools the low-level air, while sinking motion above the upslope flow causes warming and strengthens the formation of a strong capping inversion.

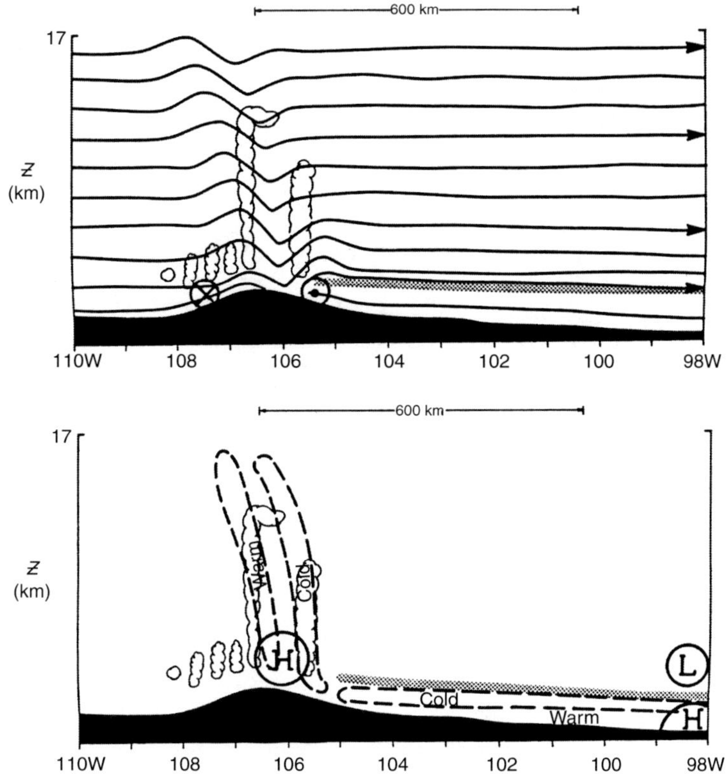


**FIGURE 9.34 Conceptual model showing flow field and position of convective elements at the time deep convection forms (stage 2).** Axes and topography are as described. The fields are derived from 86-km running averages of actual model-predicted variables. The surface topography is depicted by the black shading. The vertical axis is height in kilometers above mean sea level and the horizontal axis is west longitude. The stippled line represents the position of the plains inversion. Regions of cloud are indicated. Top: The flow field with ground-relative streamlines. Circles depict flow perturbation normal to plane. Bottom: The pressure and temperature response. Pressure centers are depicted by solid closed contours and temperature by dashed contours. The length scale of 600 km ($2L_R$) is indicated. *(From Tripoli and Cotton (1989b))*

The commencement of stage 2 (shown in Fig. 9.34) is indicated by deep convection that first forms over the mountain ridge top as the prevailing westerly flow, augmented by the slope circulation, advects moisture into the higher elevations, where it is lifted by the terrain to trigger cumulonimbi.

As the deep convection moves eastward with the prevailing westerly flow, low-level western slope moisture is carried along with it. When the eastward-migrating cumulonimbus system approaches the convergence zone between the eastern upslope flow and the mountain gravity wave, all elements favoring explosive development of intense convection come together, including moisture, surface convergence, and preexisting cumulus convection. As explosive

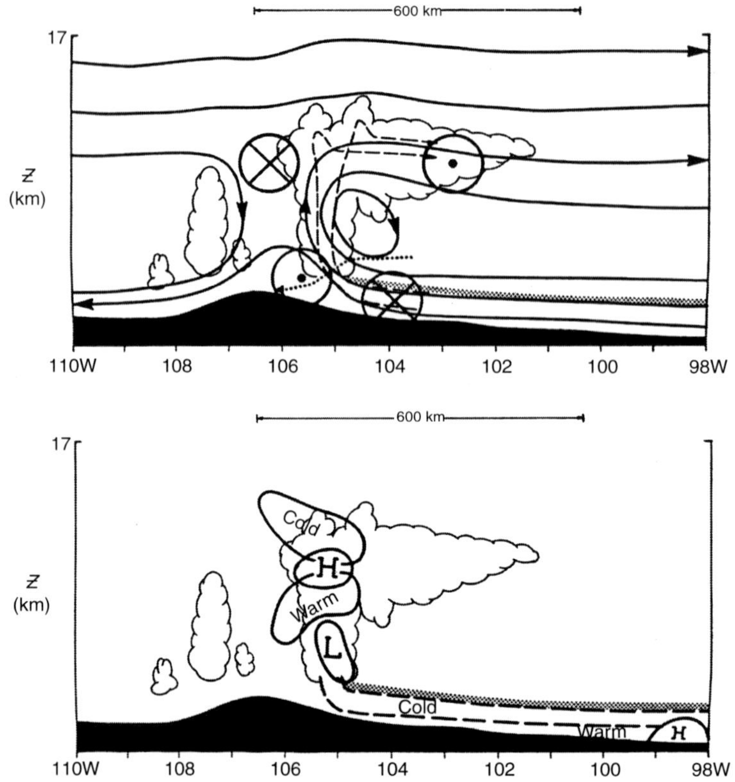


**FIGURE 9.35 Same as Fig. 9.34, except for stage 3. Also, flow is storm-relative flow.** Individual parcel paths are given by dashed (updraft) and dotted (downdraft) lines. *(From Tripoli and Cotton (1989b))*

development of an ensemble of mesoscale convection ensues, moisture advection from the eastern plains is enhanced, further invigorating convection.

Stage 3 (Fig. 9.35) thus begins when an organized convective line forms along the eastern slopes. Until this time, the mountain and plains solenoidal circulation was confined to below 5 km MSL. Under dry conditions, when deep convection does not occur, the solenoid circulation remains below 5 km MSL and slowly migrates eastward in the presence of prevailing westerly flow. Under sufficiently moist conditions, deep convection organized in a mesoscale line develops and the solenoidal circulation deepens to 12 km or the depth of the troposphere. In association with the deepened solenoidal cell, a pronounced thunderstorm outflow forms in the 10- to 12-km layer. Geostrophic adjustment to the persistent circulation forms anticyclonic shear of meridional flow aloft and cyclonic shear at low levels. This represents the first symptoms of the development of a balanced component to the storm dynamics.

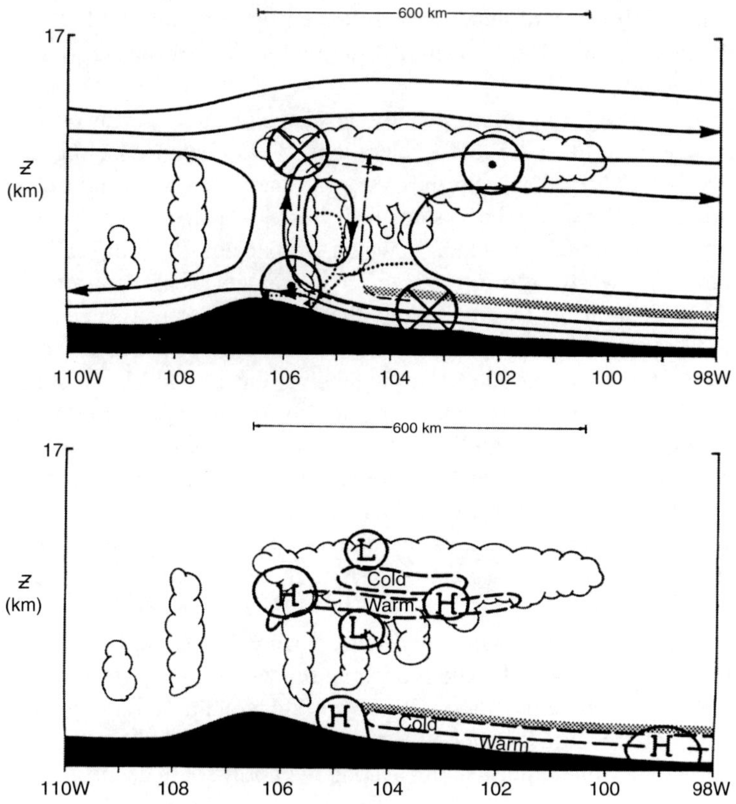


**FIGURE 9.36** **Same as Fig. 9.34, except for stage 4.** *(From Tripoli and Cotton (1989b))*

Stage 4 (Fig. 9.36) begins when the convective system moves from the mountains over the eastern plains, where a zone of suppression is encountered; the zone of suppressed convection is a consequence of several topographically related phenomena. At the juncture between the steep mountain slope and the more gentle sloping plains, two solenoidal circulations interact. Associated with the heated steep mountain slope, an intense deep solenoidal circulation develops with its upward branch focused over the mountain peaks and its downward branch centered over the transition between the mountain slope and the plains slope. Associated with the gentle sloping plains is another solenoidal circulation with its ascending branch just east of the sinking branch of the mountain solenoid and its subsidence branch located much farther east. [Figure 9.39 illustrates these cells, obtained by Dirks (1969) with a simple numerical model.] As the embryonic MCS moves onto the plains, it encounters greater subsidence as well as surface divergence as the low-level air is forced up the mountain slope or away from the system core. Because of the greater subsidence in the system core, upward motion within the system core collapses. This feature appears to be similar to the growth and decay, followed by rebound in MCS life cycle that McAnelly et al. (1997) attributed to gravity wave response to convective heating.

The upper-level anticyclone and low-level cyclone persist, however, because they slowly adjust to Coriolis accelerations. These balanced circulations act like a large flywheel that keep the system in tack even as upright convection decays. The collapse of the system also initiates a deep gravity wave 150-200 km in horizontal wavelength which propagates laterally eastward and westward. The westward-propagating mode triggers weak convection, where it encounters moisture resupplied to the mountain convergence zone. The eastward-propagating mode fails to trigger deep convection, however, because the plains inversion prevents it from tapping the low-level moisture. More importantly, the collapsing core, augmented by precipitation loading and evaporation, overshoots equilibrium, and after condensate is exhausted, it adiabatically warms. The warmed core causes the upward motion to rebound, while at the same time the core has moved eastward out of the zone of suppression, where it encounters the influx of moisture-rich air from beneath the plains inversion. The convective system, therefore, intensifies and matures to MCS proportions.

At this juncture, the system enters stage 5 (Fig. 9.37). Until now, the system evolution is very site specific and is a strong function of the particular properties of the Rocky Mountains' thermally-driven slope/plains circulations and the availability of moisture unique to the synoptic-scale circulations of that region. This is not to say that other locations along the north-south extent of the Rocky Mountain barrier, spanning Alberta, Canada, to the New Mexico mountains, may not also be favorable for such a development scenario. Similar "orogenic" forcing of MCSs may occur along other major mountain ranges (see, e.g. Velasco and Fritsch, 1987; Miller and Fritsch, 1991; Augustine and Caracena, 1994; Laing and Fritsch, 2000). Upon entering stage 5, the convective system enters a stage of its life cycle that is more generic in its MCS characteristics. The simulated MCS is not steady; it undergoes repeated cycles of growth, overdevelopment, and weakening, although the system does not collapse as it did at the end of stage 4. Each breakdown cycle initiates oppositely propagating transient gravity waves that expand radially from the system core. This also resembles McAnelly et al.'s (1997) attribution of the growth, decay, rebound cycle of MCSs to gravity wave dynamics. As the system moves eastward, the mean-core circulation gradually strengthens as the kinetic energy generated by latent heating is partially retained while the system undergoes geostrophic adjustment. That only a minor fraction of the kinetic energy generated by latent heat is retained as a balanced circulation is not surprising, since this is characteristic of systems smaller than $\lambda_R$. Instead, as demonstrated by Schubert et al. (1980), the vast majority of energy is radiated vertically and horizontally as gravity-wave energy.

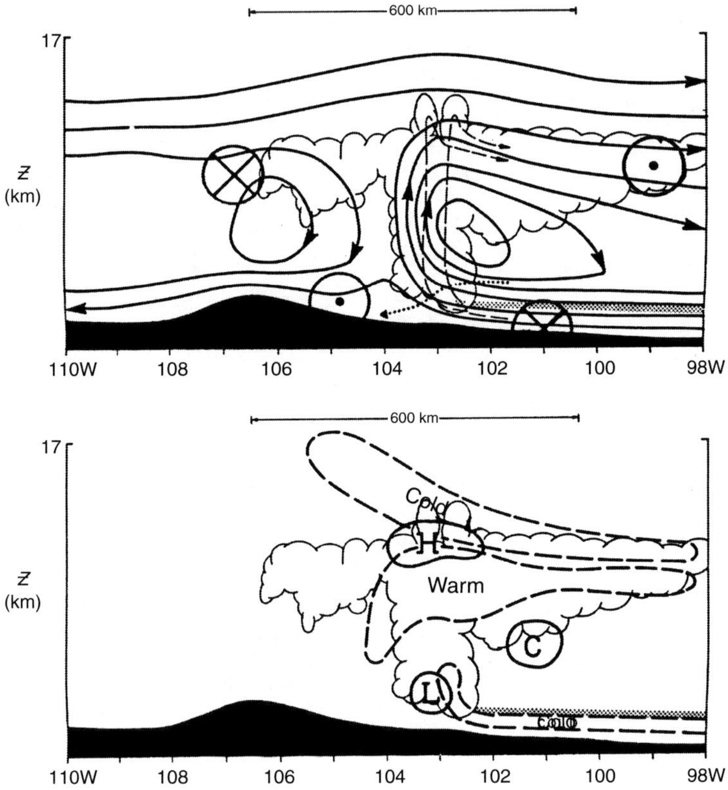


**FIGURE 9.37** **Same as Fig. 9.34, except for stage 5.** *(From Tripoli and Cotton (1989b))*

Throughout stage 5, the convective core of the MCS remains localized at the western edge of the plains inversion. Short-lived convective cells residing over the plains inversion are triggered by gravity waves emitted from the system core, but they do not intensify because they cannot tap the moist air below it. The western boundary of the plains inversion is continually eroded by the MCS core circulation. Here, adiabatic cooling associated with the mesoscale ascent, and turbulent mixing by the convection itself, destroy the inversion interface. Precipitation is advected eastward where its evaporation cools the air above the inversion and destabilizes the air on the westernmost edge of the inversion. Overall, the system moves at about 10 m $s^{-1}$, which corresponds to the speed of the upper tropospheric wind. The solenoidal circulation of the mature MCS resembles the observationally-derived squall-line model of Ogura and Liou (1980), as well as the more detailed flow features depicted by Smull and Houze (1985, 1987a,b). A rear midlevel inflow jet is absent in this simulation, however. This may be a consequence of the moist air mass at middle levels which would inhibit evaporative cooling in mesoscale downdrafts.

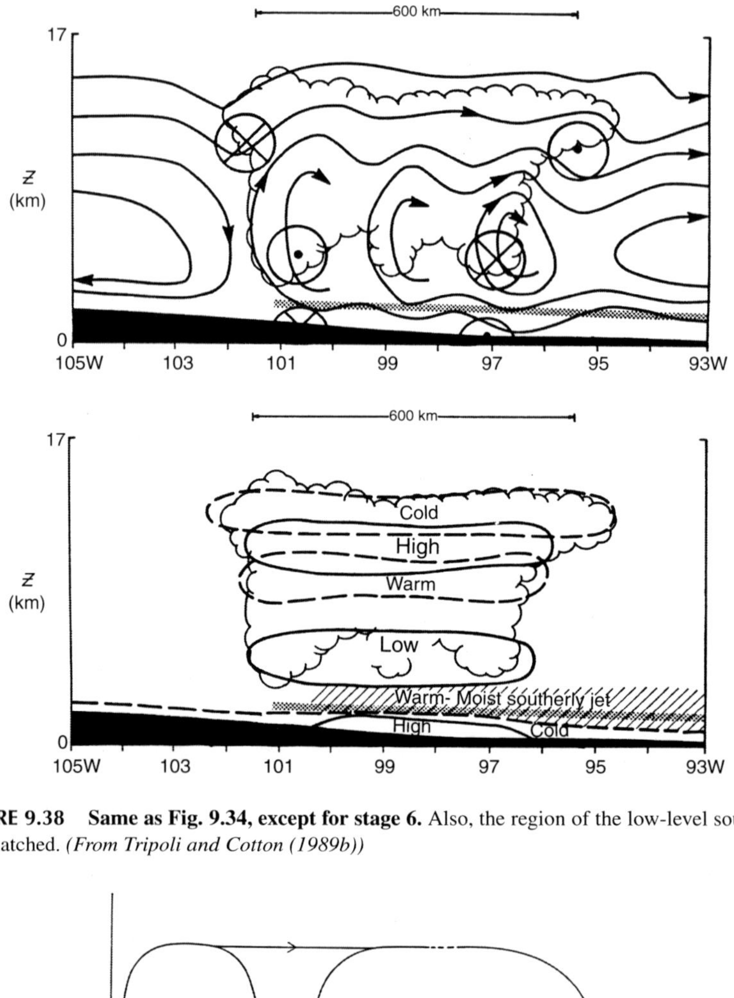


**FIGURE 9.38 Same as Fig. 9.34, except for stage 6.** Also, the region of the low-level southerly jet is hatched. *(From Tripoli and Cotton (1989b))*

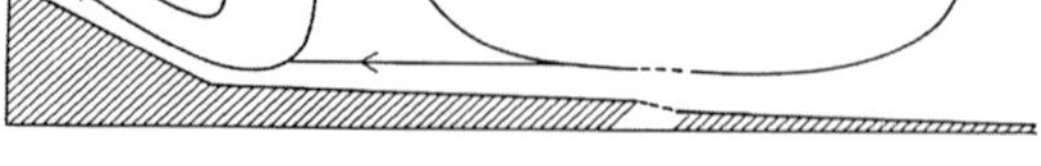

**FIGURE 9.39 Schematic illustration of the idealized Rocky Mountain-Great Plains daytime circulation.** *(From Dirks (1969))*

Similar to the tropical and mid-latitude MCSs we have examined, the system forms a deep stratiform cloud aloft. A large part of the stratiform cloud extends in advance of the main convective core. The radiative effects of the anvil become more important as darkness approaches. Prior to this period, heating at cloud top from absorption of solar radiation largely offsets longwave radiation cooling at the cloud top. As night approaches, longwave radiative cooling at cloud top and heating at the base of the stratiform layer lead to further destabilization of the stratiform layer. The intensity of the MCS, therefore, increases with peak vertical velocity being simulated at 2000 MST. Another consequence of the greater dominance of longwave radiative flux divergence to the radiative budget of the MCS is that the Brunt-Väisälä frequency is lowered in magnitude in the stratiform layer, particularly near cloud top. As a result, the phase speed of propagation of gravity waves is reduced, causing a partial refraction of the vertically-propagating gravity wave modes. A greater fraction of the vertically-propagating gravity wave energy becomes trapped beneath the stratiform cloud top. As the stratiform cloud gains greater prominence, so does the intensity of the anticyclonic shear of the meridional winds aloft and the cyclonic shear at low levels.

Stage 6 of the system (Fig. 9.38) represents the transformation from an unsteady MCS having a scale less than the Rossby radius of deformation, to the more balanced MCC system greater in scale than $\lambda_R$. Of the stages of the conceptual model derived from Tripoli and Cotton's numerical simulation, this is the most speculative because the observed systems are strongly influenced by the diurnally varying low-level jet (LLJ) over the High Plains of the United States. The two-dimensional simulations were unable to simulate properly the MCS response to the LLJ for the following reasons: (1) The eastern boundary of the model domain was near the climatological centroid of the southerly LLJ over the High Plains. (2) The LLJ is not only a function of regional sloping terrain and diurnally varying boundary layer effects, but it is also dependent upon the position and strength of the subtropical high. (3) The horizontal advection of heat and moisture by the southerly LLJ cannot be easily simulated in an east-west-oriented two-dimensional model domain.

Nonetheless, Tripoli and Cotton's simulations suggest that some significant transformations take place in the genesis of an MCC from an ordinary MCS as nighttime approaches. The surface begins to cool, and low-level convective available potential energy is reduced. Furthermore, as a low-level nocturnal inversion forms, convective updrafts no longer draw on surface air and, instead, begin to draw on air residing above the nocturnal inversion. At this point, the nocturnal LLJ intensifies and fuels the upright convective updrafts and slantwise ascending flow into the stratiform anvil clouds of observed systems. Without the thermal and moisture advection associated with the LLJ in Tripoli and Cotton's simulation, the simulated MCS begins to weaken. Second, as a consequence of the greater trapping of internal gravity-wave energy, gravity waves propagating away from the system core have greater amplitude. With increased moisture available to them above the inversion, the higher amplitude gravity waves can excite new convective elements away from the main convective core. Consistent with the observations, convection becomes more widespread, though weaker in intensity, beneath the stratiform cloud. The resultant more widespread convection favors the projection of a greater fraction of the kinetic energy generated by latent-heat release on to a scale comparable to or greater than the Rossby radius of deformation. At the same time, the system is developing greater anticyclonic shear aloft and cyclonic shear at low levels. Thus, the system becomes increasingly dominated by balanced flow dynamics.

Tripoli and Cotton's simulations illustrate dramatically that the genesis of an MCS and, subsequently, the more balanced MCC, evolves through a complex series of steps. Thunderstorms are initiated in an organized low-level convergence field. If the supply of CAPE is great enough, the anvils emitted by the neighboring cumulonimbi may merge to form a stratiform-anvil cloud layer extending from the melting level to the tropopause. This will result in the elevation of the level of heating to the upper troposphere. At the same time, downdraft outflows and associated cold-pools from the neighboring storms merge to form a mesohigh at low levels. In a sense, the embryonic MCS creates its own baroclinicity. Both processes favor the upscale growth of a convective system to mesoscale proportions. As the system organizes on the mesoscale and the vertical heating profile changes from upright convection dominated to stratiform-anvil contributions, anticyclonic shear develops aloft and cyclonic shear forms at low levels, and a greater fraction of the kinetic energy generated by latent-heat release is projected onto balanced flow. Greater trapping of gravity-wave energy by a radiatively destabilized stratiform-anvil cloud in the nocturnal regime further contributes to the projection of kinetic energy generated by latent heating on to the more balanced meso-$\alpha$-scale.

Of course, more often than not, one or more of these processes do not take place and the upscale growth of a convective system into MCS or MCC proportions is curtailed. Furthermore it is not essential that an MCS evolve step-by-step in the way we have just outlined. Instead, the large-scale environment can provide the vertical motion field, cyclonic and anticyclonic shears, conditional instability and LLJs, which can immediately organize convection on the meso-$\beta$-scale and completely circumvent the first stages of MCS genesis. It is entirely possible that some large-scale environments will support the immediate organization of a convective system on to the more balanced meso-$\alpha$-scale, favoring the nearly spontaneous development of an MCC and MCV.

## 9.5. MCVS AND TROPICAL CYCLONE GENESIS

Mesoscale convective vortices (MCVs) were first identified with visible satellite imagery, which exhibited an elevated cloud banded structure, shown in Fig. 9.40, during the decay stages of an MCS. They were first observed with decaying tropical MCSs (Leary and Thompson, 1976; Houze, 1977; Leary, 1979; Fortune, 1980; Gamache and Houze, 1982, 1985; Houze and Rappaport, 1984; Wei and Houze, 1987; Chong et al., 1987; Kuo and Chen, 1990; Keenan

**FIGURE 9.40 Visible satellite photograph (1530 GMT, 14 May 1984) of decaying MCC showing circulation remaining long after active convection has ceased.**

and Rutledge, 1993; Jou and Yu, 1993; Harr and Elsberry, 1996; Yu et al., 1999; Chong and Bousquet, 1999; Bousquet and Chong, 2000; May et al., 2008). They have also been observed with middle-latitude MCSs and MCCs (Ogura and Liou, 1980; Johnston, 1982; Smull and Houze, 1985; Menard et al., 1986; Leary and Rappaport, 1987; Menard and Fritsch, 1989; Stirling and Wakimoto, 1989; Johnson et al., 1989; Smith and Gall, 1989; Murphy and Fritsch, 1989; Tollerud et al., 1989; Brandes, 1990; Verlinde and Cotton, 1990; Hales, 1990; Biggerstaff and Houze, 1991; Bartels and Maddox, 1991; Johnson and Bartels, 1992; Wang et al., 1993; Fritsch et al., 1994; Scott and Rutledge, 1995; Bartels et al., 1997; Trier and Davis, 2002; Knievel and Johnson, 2002; Galarneau and Bosart, 2005; Davis and Trier, 2007; Trier and Davis, 2007; Schumacher and Johnson, 2008; James, 2009).

The synoptic scale environment in which MCVs are generally observed is not too different from that found for MCSs in general, and those MCVs associated with a parent MCS, including mid-level convergence and ascent within the MCS stratiform region, upper level divergence and an ageostrophic jet streak, a pronounced trough in the height field, a warm core and a positive PV anomaly at middle levels, and a LLJ (James, 2009). Fritsch et al. (1994) likewise found that a strong low level jet or low-level flow at 850-900 mb, provided a strongly sheared low level storm environment. According to the composite study of James (2009), the upper level jet streak and associated divergence, as well as middle-level convergence and ascent within the stratiform region of an MCS is strongest during the early stages of MCV formation.

James (2009) identified five classes of MCVs in his 45 member composite study. They are "collapsing stratiform region MCV", "rear inflow jet MCV", "surface-penetrating MCV", "cold-pool dominated MCV", and "remnant

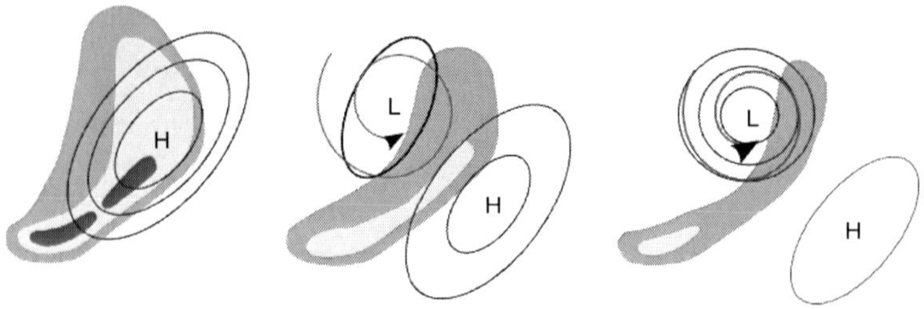


**FIGURE 9.41 Conceptual model of the evolution of collapsing stratiform region MCVs.** Time runs from left to right. Colors represent composite radar reflectivity. Green indicates 30-40 dBZ, yellow indicates 40-50 dBZ, and red indicates >50 dbZ. Thin black lines represent surface isobars (contour interval 0.5 hPa). *H* indicates center of mesohigh, and *L* indicates center of mesolow. Red spiral indicates developing midlevel vortex. The evolution is with respect to a system moving towards the right. (For interpretation of the references to color in this figure legend, the reader is referred to the web version of this book.) *(From James (2009))*

circulation MCV". The collaping stratiform region MCV illustrated in Fig. 9.41, display a characteristic organization of precipitation consisting of a weak convective line that is on average oriented ~45 degrees relative to the storm direction of motion. The surface mesohigh is generally centered near the leading edge of precipitation, between the stratiform and convective regions. The middle-level MCV first becomes detectable during the middle stage of the system when convection has weakened and the stratiform-anvil region begins to dissipate. During the dissipating stages a developing mesolow is detectable towards the rear of the stratiform region as it dissipates. The collapsing stratiform region MCVs are generally smaller and weaker than the other MCV types.

The rear inflow jet MCVs illustrated in Fig. 9.42 exhibit strong surface pressure perturbations associated with the formation of a wake low (Johnson and Hamilton, 1988). These MCVs typically exhibit a leading line trailing stratiform MCS organization (Parker and Johnson, 2000), and evolve into a asymmetric type of squall line (Houze et al., 1990) with an embedded MCV in the northern part of the system.

Of the 45 member composite study of James (2009) only one was a surface-penetrating MCV. This exhibited a closed surface circulation with a well-defined low pressure center, a deep vertical column of high PV which extended to the middle and upper troposphere, and a very moist environment. The MCV did not exhibit any particular pattern of organized precipitation. A shallow layer near the surface is also quite evident and James (2009) speculated that it is probably responsible for the lack of organized surface-based convection near the center of the circulation. Similar features were noted by Fritsch et al. (1994). Such surface-penetrating MCS may provide insights into the genesis of tropical cyclones.

The cold pool dominated MCVs exhibited deep and extensive surface cold pools. These MCVs did not have significant surface pressure perturbations

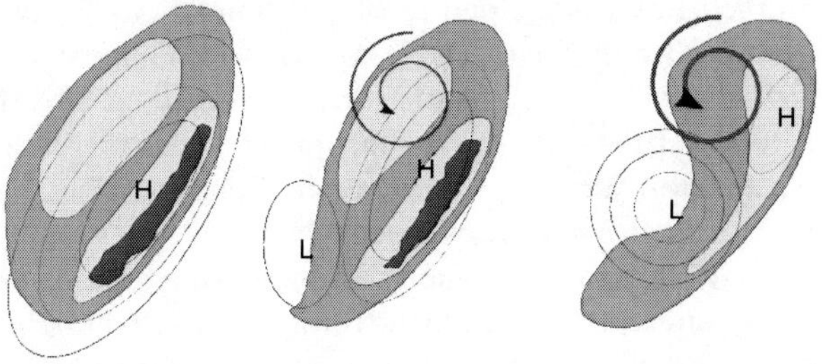


**FIGURE 9.42 As in Fig. 9.41, but of the evolution of rear inflow jet MCVs.** *(From James (2009))*

associated with the wake low. Local high pressure anomalies associated with the cold-pools likely masked the effects of the elevated low pressure. Likewise, the remnant circulation MCVs did not exhibit features of a surface wake low. These MCVs did not precipitate significantly. James (2009) speculated that since melting and evaporation of precipitation is hypothesized to be a major driving force for subsidence and wake low development, its absence may account for the lack of a identifiable surface wake low.

Numerical modeling and theoretical studies of MCVs can be broken down into those using parameterized convection, and those using explicit, cloud-resolving models. Likewise, diagnostic studies of observed and simulated MCVs can be identified as those employing PV thinking and those using vorticity diagnostics, such as are used in explicit supercell simulations discussed in Chapter 8, or a combination of the two. From a PV-thinking perspective the middle-level MCV vortex is initiated and driven by the heating profile in MCSs such as the idealized profile illustrated in Fig. 9.31. This heating profile develops a PV field that includes the middle level MCV vortex and an upper tropospheric anticyclonic circulation. Interactions of the middle tropospheric +PV anomaly with shear, often associated with an LLJ, drives ascending motion that can supply moisture and energy to the stratiform-anvil cloud and destabilize the lower atmosphere to excite deep-upright convection. Likewise, the cooling associated with melting and evaporation drives lower tropospheric subsidence which creates a wake low that sometimes can be found at the surface. The PV diagnostic studies of Raymond and Jiang (1990), Jiang and Raymond (1995), and Olsson and Cotton (1997b) are examples of such studies. From the PV thinking perspective, an MCV is nothing more than a more visible, and perhaps stronger vortex extension of a nearly balanced MCC.

The other perspective to MCV formation is from a vorticity dynamics or vorticity budget analysis. Such analyses have been performed on simulated model output data using coarse resolution mesoscale models with parameterized convection (e.g. Zhang and Fritsch, 1988a,b,c; Zhang, 1992; Kirk, 2003, 2007). They have also been done from idealized higher resolution explicitly resolved convection models (Skamarock et al., 1994; Davis and Weisman,

1994) and more recently using cloud-resolving simulations for actual observed case studies (Davis and Trier, 2002; Davis and Galarneau, 2009). The vorticity budget analysis is based on the vorticity equations (Eqs (8.14)–(8.16)) that we used to describe vorticity production in supercell thunderstorms or mesocyclones, and tornado vortex formation. Equation (8.15) for example, describes the horizontal and vertical components of vorticity resulting from convergence and tilting, and baroclinicity. In some papers the convergence term is referred to as the stretching term and the tilting term as the tipping term. Many of the analyses, especially those using data from coarser resolution mesoscale simulations using parameterized convection, focus almost exclusively on the budget for the vertical component of vorticity. As noted in Chapter 8, one must be cautious about only interpreting one component of vorticity, such as the vertical, because a vortex tube strength can increase due contributions from the two horizontal components. In other words, it is most appropriate to view vorticity changes as one moves along the pathway of a meandering vortex tube in three dimensional space.

From a vorticity dynamics perspective, an MCV can be viewed as arising from convergence of ambient or synoptic scale vorticity, dominated by the earth's rotation (Bartels and Maddox, 1991; Skamarock et al., 1994). Shortwave troughs in the pre-MCS environment can also provide a significant source of vorticity that can be converged by MCS heating and dynamics. Idealized cloud-resolving simulations without the earth's rotation (no Coriolis force) develop two identical counter rotating vortex pairs that are essentially book-end vortices (Weisman and Davis, 1998). However, when the earth's rotation is taken into account, a single dominant middle-level vortex forms. Trapp and Weisman (2003) conclude that when low level environmental shear is above a certain threshold, tilting in downdrafts of initially crosswise horizontal baroclinic vorticity, forms two counter rotating low-level vortices. Under the influence of the earth's rotation, a single, dominant cyclonic vortex emerges and convergence then leads to a single dominant middle-level MCV.

Kirk (2007) presented evidence of multiple pathways to the formation of MCVs. In some cases, heating in the middle troposphere can lead to MCV formation, whereas when heating is less dominant, tilting can be instrumental to MCV formation. We emphasize if there is pre-existing vorticity in the environment, such as that supplied by shortwave troughs, then convergence associated with MCS heating can rapidly form an MCV.

We now turn to the even more controversial topic of how an MCV can lead to surface vortices and tropical cyclone genesis. It has been known for some time that MCVs are instrumental in the genesis of tropical cyclones. The large scale conditions that favor the formation of tropical cyclones (TCs) are also conducive to forming middle tropospheric MCVs, including lower tropospheric cyclonic vorticity and upper tropospheric anticyclonic vorticity, and weak vertical wind shear (Gray, 1968; McBride, 1981; Zehr, 1992; Kurihara and Tuleya, 1981). TC genesis is generally associated with monsoon troughs (Briegel and Frank, 1997;

Ritchie and Holland, 1997) or easterly waves (Frank and Clark, 1980; Carlson, 1969; Landsea, 1993; Thornkroft and Hodges, 2001; Frank and Roundy, 2006). Easterly waves are thought to be instrumental in MCV formation because they provide a favorable background vorticity and relatively weak wind shear. In fact, Dunkerton et al. (2008) argue that a closed circulation in the middle troposphere of easterly waves provides a "protective pouch or marsupial pouch" that nurtures the formation of a TC by allowing MCVs to intensify to TCs by providing low shear and, through repeated deep convection cycles, moistening of the lower troposphere.

Briegel and Frank (1997) summarized the general characteristics of the environment in which TCs form:

- Sea surface temperatures of 26.5-27.0 °C coupled with a deep ocean mixed layer (e.g. large ocean heat content).
- A deep surface-based layer of conditional instability.
- Cyclonic low-level absolute vorticity.
- Organized deep convection in an area of large scale ascending motion and high middle level humidity.
- Weak to moderate (preferably easterly) vertical wind shear.

There are numerous examples of middle latitude MCVs transforming into TCs (Bosart and Sanders, 1981; Velasco and Fritsch, 1987; Miller and Fritsch, 1991). But MCVs generally remain as middle tropospheric vorticities with little if any direct connection to the surface. Examples of where this was not true are the case studies simulated by Davis and Galarneau (2009). They found that the MCV extended downwards to low-levels as the MCS cold-pool boundary formed a line-end vortex that was distinct from the MCV aloft. Owing to appropriate vertical wind shear, the line-end vortex moved in a storm-relative sense rearward beneath the middle tropospheric MCV. Once they became vertically-aligned, a deep vertical column of vorticity ensued. As the MCS matured moistening of the lower troposphere and changes in the divergence profile further favored vorticity generation near the surface. As a result a deep surface-based vortex developed that grew at the expense of the vorticity and PV of the MCV aloft. They concluded that moistening in a stabilized lower tropospheric environment was crucial to the formation of a surface-based MCV. Widespread surface-based instability was not needed.

In order for an MCV to transform into a self-sustaining TC, strong sustained surface winds must develop over a wide area which can provide a sustained flux of heat and moisture into the MCV, a process now referred to as WISHE (Ooyama, 1969; Rotunno and Emanuel, 1987). How then can an MCV transform into a TC? The concepts can be broken down into "top-down" vs. "bottom-up" theories similar to the theories for tornado genesis discussed in Chapter 8. There we saw that early theories for tornado genesis in supercell storms involved a top-down concept in which the mesocyclone aloft built downward to the surface through the action of a pressure-deficit tube.

However, recent evidence suggests a bottom-up mechanism of tornado genesis can also operate, in which vortices form along the rear flank and forward flank downdraft gust fronts, and merge to form larger, more intense surface-based vortices which then merge with the mesocyclone aloft. Crucial to this process is the vertical coupling between the surface-based vortices and mesocyclone aloft. One of the conditions that favor such a coupling is relatively warm surface cold-pools arising from very moist low level air masses and reduced evaporation in downdrafts due to lower precipitation rates and larger raindrops and hailstones, which favors less evaporative and melting cooling. We will see that a similar evolution of top-down vs. bottom-up concepts for TC genesis from MCVs is emerging.

### 9.5.1. Top-Down vs. Bottom-Up Theories for TC Genesis

Much discussion has occurred in recent years regarding whether TC genesis is a top-down vs. a bottom-up process. Emanuel (1993) and Bister and Emanuel (1997) argue that a TC can form from the descent of a single MCV to the surface. They argue that sustained stratiform precipitation from the stratiform-anvil of an MCS or MCV will gradually saturate the lower troposphere. At the same time the slantwise descending flow in the MCS will advect cyclonic vorticity associated with the MCV to the surface. It is argued that the moistening and cooling by the stratiform precipitation destabilizes the boundary layer to support deep convection and also weakens convective downdrafts. They argue that convergence of lower tropospheric mean vorticity in a region of convection where downdrafts are weak, favors the spin-up of the low-level vortex. We suggest that in analog to tornado genesis, the weakened cold-pools as a result of moistening of the lower troposphere may aid the genesis of a TC by favoring the sustained vertical coupling between the MCV aloft and an evolving low-level vortex.

Another example of a top-down theory of TC genesis is that proposed by Ritchie and Holland (1997) and Simpson et al. (1997). Using modeling studies and supporting observations, they argue that the merger of middle level MCVs on scales of 100-200 km will result in larger, more intense MCVs whose circulations have greater penetration depths thus favoring the spin-up of a low-level vortex.

Nolan (2007) describes the results of a series of idealized simulations which are interpreted as the top-down genesis of a TC. As in the above studies, the TC genesis process is intimately associated with an MCV and associated stratiform precipitation. He notes that after several days of amplification of the MCV, when it contracts to a radius of maximum winds of 60 km and a maximum wind speed of 12 m $s^{-1}$, a single strong updraft or cluster of updrafts form near the center of the MCV aloft. In response to the strong updrafts, a single, dominant low-level vortex forms that becomes the central core of the developing TC. He notes that this single, dominant low-level vortex does not form until low-levels become nearly saturated (by evaporation of stratiform precipitation and convective showers) and the MCV becomes inertially stable (well balanced).

The bottom-up theories proposed by Montgomery and Enagonio (1998), Enagonio and Montgomery (2001), Hendricks et al. (2004) focus more on the MCS convective region where low-level convergence enhances vorticity rather than the descent of the MCV itself. Hendricks et al. (2004) introduced the concept of vortical hot towers (VHTs) wherein strong convective updrafts generate strong cyclonic and anticyclonic eddies, principally by the tilting of horizontally-oriented vortex tubes as in the theories for formation of mesocyclones in supercells discussed in Chapter 8. Idealized modeling studies (Montgomery et al., 2006) suggest that the VHTs merge to form stronger and larger low-level vortices, which then can serve as the embryos for genesis of the low-level TC vortex. Important to the bottom-up theory is that initially off-center low-level vortices become vertically-aligned with the MCV aloft to produce a deep tropospheric vortex. As in bottom-up theories of tornadogenesis, moistening of the lower troposphere by evaporating precipitation would result in relatively less cold, cold-pools which would favor a reduction in storm-relative motion of the low-level vortex relative to the MCV thus favoring coupling through the depth of the troposphere. As noted by Davis and Galarneau (2009) optimum low-level vertical wind shear, can favor the movement of a low-level vortex in a storm-relative sense rearward beneath the middle tropospheric MCV. There is observational evidence (Reasor et al., 2005; Sippel et al., 2006) that VHTs and their merger occur in developing TCs. Thus both top-down and bottom-up theories have observational support.

It must be kept in mind that preexisting upper tropospheric troughs, and upper tropospheric disturbances can also play an important role in TC genesis (e.g. Davis and Bosart, 2002, 2006).

## 9.6. IMPACTS OF MCSS

We have just seen that MCSs and MCVs, in particular, are major contributors to the genesis of tropical cyclones. But MCSs are best known for the rain they produce. Fritsch et al. (1986) found that MCSs account for 30%-70% of the warm season (growing season) rainfall in the region between the Rocky Mountains and the Mississippi River. Similar contributions to rainfall can be found in most areas frequented by MCSs throughout the world. We have seen that MCSs often occur in one week long or two week long episodes when storms form daily and track across the US for long distances. If the storms track through the same regions repeatedly, extreme rainfall and flooding can result. Tollerud and Collander (1993a) showed that MCCs account for 20%-40% of the hourly rainfall rates in the central US during May through August even though they contributed only 7% of the total number of observations of measurable rainfall. Studies by Kane et al. (1987), McAnelly and Cotton (1989), Collander (1993) show that the heaviest rainfall rates occur on the equatorward side of the cold cloud shield.

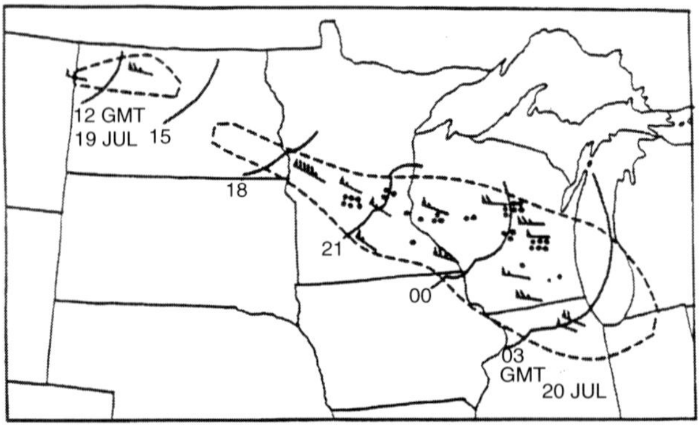


**FIGURE 9.43 Area affected by widespread downburst activity during derecho occurrence of 19-20 July 1983 (bounded by dashed line).** Surface wind gusts indicated by wind flag = 50 kt, full barb = 10 kt, half barb = 5 kt. Dots represent personal injuries. Three-hourly squall front positions indicated in Greenwich Mean Time (GMT). *(From Johns and Hirt (1985))*

MCSs are also associated with severe weather, such as tornadoes, hail, and straight-line winds. The preponderance of severe events occur during the early growth phase of the MCS (Maddox, 1980; Houze et al., 1990; Tollerud and Collander, 1993b). Some of the more unique forms of severe weather events associated with MCSs are straight-line severe wind events called "derechos". The characteristics of derechos as defined by Johns and Hirt (1987) are:

- Surface wind speeds greater than 24 m $s^{-1}$.
- Average duration of 9 hours.
- Damage swaths exceeding 400 km in length and over 100 km wide.
- 90% of derechos form on cold side of stationary or slowly-moving synoptic-scale frontal boundaries.

Figure 9.43 illustrates damage reports associated with such storms. Derechos are associated with bow-echo shaped squall line systems but are on a scale larger than a typical downburst event (e.g. Lee et al., 1992a,b). It is generally thought that the severe surface winds arise from the surfacing of the slantwise descending rear-to-front flow that produces the bow-shaped structure of the squall line. But back-trajectory analysis of simulated derecho producing MCSs by Schmidt (1991) and Bernardet and Cotton (1998) suggested that the air producing the strongest downdrafts and surface winds originated in "up-down" downdraft components. Stable air near the ground was lifted by strong upward directed pressure gradient forces associated with mesocyclone-like circulations at the apex of the bow line. Once this lifted stable air encountered air cooled by precipitation evaporation and melting, it descended rapidly to the surface producing strong outflows. How often derecho winds are produced by such phenomena has still to be determined.

MCSs are also major venters of low-level pollutants into the upper troposphere. Lyons et al. (1986) presented analysis of a case with a number of MCSs that vented huge amounts of pollutants over a large region of the southeastern US. They estimated that the total amount of sulfate aerosol removed by all MCSs in the region on a particular day amounted to about 16 to $32 \times 10^6$ kg. Cotton et al. (1995) used cloud-resolving simulations of MCSs and satellite cloud climatology to estimate venting of boundary layer air by a wide variety of cloud types ranging from ordinary cumuli, to ordinary cumulonimbi, to MCSs to tropical and extratropical cyclones. They found that MCSs ranked second only to extratropical cyclones in venting boundary layer air to the upper troposphere on a global annual basis.

The analysis by Cotton et al. (1995) suggests that MCSs are a dominant contributor to the tropical global-annual heat budget.

## REFERENCES

Anderson, C. J., and Arritt, R. W. (1998). Mesoscale convective complexes and persistent elongated convective systems over the United States during 1992 and 1993. Mon. Weather Rev. 126, 578–599.

Ashley, W. S., Mote, T. L., Dixon, P. G., Trotter, S. L., Powell, E. J., Durkee, J. D., and Grundstein, A. M. (2003). Distribution of mesoscale convective complex rainfall in the United States. Mon. Weather Rev. 131, 3003–3017.

Augustine, J. A., and Caracena, F. (1994). Lower-tropospheric precursors to nocturnal MCS development over the central United States. Wea. Forecasting 9, 116–135.

Bartels, D. L., and Maddox, R. A. (1991). Midlevel cyclonic vortices generated by mesoscale convective systems. Mon. Weather Rev. 119, 104–118.

Bartels, D. L., Brown, J. M., and Tollerud, E. I. (1997). Structure of a midtropospheric vortex induced by a mesoscale convective system. Mon. Weather Rev. 125, 193–211.

Bernardet, L. R., and Cotton, W. R. (1998). Multi-scale evolution of aderecho-producing MCS. Mon. Weather Rev. 126, 2991–3015.

Biggerstaff, M. I., and Houze Jr., R. A. (1991). Midlevel vorticity structure of the 10-11 June 1985 squall line. Mon. Weather Rev. 119, 3066–3079.

Bister, M., and Emanuel, K. A. (1997). The genesis of Hurricane Guillermo: TEXMEX analyses and a modeling study. Mon. Weather Rev. 125, 2662–2682.

Blanchard, D. O. (1990). Mesoscale convective patterns of the Southern High Plains. Bull. Amer. Met. Soc. 71, 994–1005.

Bluestein, H. B., and Jain, M. H. (1985). The formation of mesoscale lines of precipitation: Severe squall lines in Oklahoma during the spring. J. Atmos. Sci. 42, 1711–1732.

Bluestein, H. B., Jain, M. H., and Marx, G. T. (1987). Formation of mesoscale lines of precipitation: Non-severe squall lines in Oklahoma during the spring. In "Conf Mesoscale Processes, 3rd, Vancouver, BC." pp. 198–199. Am. Meteorol. Soc, Boston, MA.

Bosart, L. F., and Sanders, F. (1981). The Johnstown Flood of July 1977: A long-lived convective system. J. Atmos. Sci. 38, 1616–1642.
Bousquet, O., and Chong, M. (2000). The oceanic mesoscale convective system and associated mesovortex observed 12 December 1992 during TOGA-COARE. Q. J. R. Meteorol. Soc. 126, 189–211.
Brandes, E. A. (1990). Evolution and structure of the 6-7 May 1985 mesoscale convective system and associated vortex. Mon. Weather Rev. 118, 109–127.
Brenner, I. S. (1974). A surge of maritime tropical air—Gulf of California to the southwestern United States. Mon. Weather Rev. 102, 375–389.
Briegel, L. M., and Frank, W. M. (1997). Large-scale influences on tropical cyclogenesis in the Western North Pacific. Mon. Weather Rev. 125, 1397–1413.
Browning, K. A., Marsham, J. H., Blyth, A. M., Mobbs, S. D., Cl Nicol, J., Perry, F. M., and White, B. A. (2008). Observations of dual slantwise circulations above a cool undercurrent in a mesoscale convective system. Q. J. R. Meteorol. Soc. 00, 1–25.
Bryan, G. H., and Fritsch, J. M. (2000). Moist absolute instability: The sixth static stability state. Bull. Amer. Meteorol. Soc. 81, 1207–1230.
Byers, H. R., and Braham, R. R. (1949). "The Thunderstorm." US Weather Bur., Washington, DC.
Carbone, R. E., Tuttle, J. D., Ahijevich, D. A., and Trier, S. G. (2002). Inferences of predictability associated with warm seaon precipitation episodes. J. Atmos. Sci. 59, 2033–2056.
Carlson, T. N. (1969). Synoptic histories of three African disturbances that developed into Atlantic hurricens. Mon. Weather Rev. 97, 256–276.
Carlson, T. N., Anthes, R. A., Schwartz, M., Benjamin, S. G., and Baldwin, D. G. (1980). Analysis and prediction of severe storms environment. Bull. Am. Meteorol. Soc. 61, 1018–1032.
Chauzy, S., Chong, M., Delannoy, A., and Despiau, S. (1985). The June 22 tropical squall line observed during COPT 81 experiment: Electrical signature associated with dynamical structure and precipitation. J. Geophys. Res. 90, 6091–6098.
Chong, M., and Bousquet, O. (1999). A mesovortex within a near-equatorial mesoscale convective system during TOGA COARE. Mon. Weather Rev. 127, 1145–1156.
Chong, M., Amayenc, P., Scialom, G., and Testud, J. (1987). A tropical squall line observed during the COPT 81 experiment in West Africa. Part I: Kinematic structure inferred from dual-doppler radar data. Mon. Weather Rev. 115, 670–694.
Churchill, D. D., and Houze Jr., R. A. (1984a). Development and structure of winter monsoon cloud clusters on 10 December 1978. J. Atmos. Sci. 41, 933–960.
Collander, R. S. (1993). A ten-year summary of severe weather in mesoscale convective compleses, Part 2: Heavy rainfall. In "17th conf. on Severe Local Storms, St. Louis, MO." pp. 638–641. Amer. Met. Soc. Preprints.

Cotton, W. R., Lin, M. S., McAnelly, R. L., and Tremback, C. J. (1989). A composite model of mesoscale convective complexes. Mon. Weather Rev. 117, 765–783.

Cotton, W. R., Weaver, J. F., and Beitler, B. A. (1995). Anunusual summertime downslope wind event in Fort Collins, Colorado on 3July 1993. J. Weather Forecasting 10, 786–797.

Cram, J. M., Pielke, R. A., and Cotton, W. R. (1992a). Numerical simulation andanalysis of a prefrontal squall line. Part I: Observations and basicsimulation results. J. Atmos. Sci. 49, 189–208.

Cram, J. M., Pielke, R. A., and Cotton, W. R. (1992b). Numerical simulation andanalysis of a prefrontal squall line. Part II: Propagation of the squall lineas an internal gravity wave. J. Atmos. Sci. 49, 209–225.

Culverwell, A. H. (1982). An analysis of moisture sources and circulation fields associated with an MCC episode. M.S. Thesis, Dep. Atmos. Sci., Colorado State University.

Davis, C., and Bosart, L. F. (2002). Numerical simulations of the genesis or Hurricane Diana (1984). Part I: Sensitivity of track and intensity prediction. Mon. Weather Rev. 130, 1100–1124.

Davis, C., and Bosart, L. F. (2006). The formation of hurricane Humberto (2001): The importance of extra-tropical precursors. Q. J. R. Meteorol. Soc. 132, 2055–2085.

Davis, C. A., and Galarneau Jr., T. J. (2009). The vertical structure of mesoscale convective vortices. J. Atmos. Sci. 66, 686–704.

Davis, C. A., and Trier, S. B. (2002). Cloud-resolving simulations of mesoscale vortex intensification and its effect on a serial mesoscale convective system. Mon. Weather Rev. 130, 2839–2858.

Davis, C. A., and Trier, S. B. (2007). Mesoscale convective vortices observed during BAMEX. Part I: Kinematic and thermodynamic structure. Mon. Weather Rev. 135, 2029–2049.

Davis, C. A., and Weisman, M. L. (1994). Balanced dynamics of mesoscale vortices produced in simulated convective systems. J. Atmos. Sci. 51, 2005–2030.

Dirks, R. (1969). A theoretical investigation of convective patterns in the lee of the Colorado Rockies. Atmos. Sci. Pap. No. 154, Dep. Atmos. Sci., Colorado State University.

Dunkerton, T. J., Montgomery, M. T., and Wang, Z. (2008). Tropical cyclogenesis in a tropical wave critical layer. Easterly waves. Atmos. Chem. Phys. Disc. 8, 11149–11292.

Emanuel, K. A. (1993). The physics of tropical cyclogenesis over the eastern Pacific. In “Tropical cyclone disasters. Proceedings of ICSU/WMO International Symposium on Tropical Cyclone Disasters” (J. Lighthill, Z. Zhemin, G. Holland and K. Emanuel, Eds.), pp. 136–142. Peking University Press.

Enagonio, J., and Montgomery, M. T. (2001). Tropical cyclogenesis via convectively forced vortex Rossby waves in a shallow water primitive equation model. J. Atmos. Sci. 58, 685–706.

Ertel, H. (1942). Ein neuer hydrodynamischer Wirbelsatz. Meteor. Z. 59, 277–281.

Fortune, M. A. (1980). Properties of African squall lines inferred from time-lapse satellite imagery. Mon. Weather Rev. 108, 153–168.

Fortune, M. A. (1989). The evolution of vortical patterns and vortices in mesoscale convective complexes. Ph.D. Dissertation, Colorado State University, Dept. of Atmospheric Science, Fort Collins, CO 80523.

Fortune, M. A., Cotton, W. R., and McAnelly, R. L. (1992). Frontal-wave-like evolution in some mesoscale convective complexes. Mon. Weather Rev. 120, 1279–1300.

Frank, N. L., and Clark, G. (1980). Atlantic tropical systems of 1979. Mon. Weather Rev. 108, 966–972.

Frank, W. M. (1978). The life cycles of GATE convective systems. J. Atmos. Sci. 35, 1256–1264.

Frank, W. M. (1983). The cumulus parameterization problem. Mon. Weather Rev. 111, 1859–1871.

Frank, W. M., and Roundy, P. E. (2006). The role of tropical waves in tropical cyclogenesis. Mon. Weather Rev. 134, 2397–2417.

Fritsch, J. M., and Forbes, G. S. (2001). Mesoscale convective systems. In "Severe Convective Storms." In Meteor. Monogr., Vol. 28, No. 50. pp. 323–357. Amer. Meteor. Soc.

Fritsch, J. M., Kane, R. J., and Chelius, C. R. (1986). The contribution of mesoscale convective weather system to the warm-season precipitation in the United States. J. Climate Appl. Meteor. 25, 1333–1345.

Fritsch, J. M., Murphey, J. D., and Kain, J. S. (1994). Warm core vortex amplification over land. J. Atmos. Sci. 51, 1780–1807.

Galarneau, T. J., and Bosart, L. F. (2005). An examination of the long-lived MCV of 10-13 June 2003. In "11th Conf. on Mesoscale Processes, Albuquerque, NM." pp. 5–7. Amer. Meteor. Soc. Preprints.

Gallus Jr., W. A., and Johnson, R. H. (1991). Heat and moisture budgets of an intense midlatitude squall line. Mon. Weather Rev. 48, 122–146.

Gamache, J. F., and Houze Jr., R. A. (1982). Mesoscale air motions associated with a tropical squall line. Mon. Weather Rev. 110, 118–135.

Gamache, J. F., and Houze, R. A. (1985). Further analysis of the composite wind and thermodynamic structure of the 12 September GATE squall line. Mon. Weather Rev. 113, 1241–1259.

Geldmeier, M. F., and Barnes, G. M. (1997). The "footprint" under a decaying tropical mesoscale convective system. Mon. Weather Rev. 125, 2879–2895.

Gray, W. M. (1968). Global view of the origins of tropical disturbances and storms. Mon. Weather Rev. 96, 669–700.

Hales Jr., J. E. (1972). Surges of maritime tropical air northward over the Gulf of California. Mon. Weather Rev. 100, 298–306.

Hales, J. E. (1990). An examination of the development and role of a mesoscale vorticity center in the Council Bluffs tornado on 15 July 1988. In "16th Conf. on Severe Local Storms, Kananaskis Park, AB, Canada." pp. 446–449. Amer. Meteor. Soc. Preprints.

Hamilton, R. A., and Archbold, J. W. (1945). Meteorology of Nigeria and adjacent territory. Q. J. R. Meteorol. Soc. 71, 231–265.

Harr, P. A., and Elsberry, R. L. (1996). Structure of a mesoscale convective system embedded in Typhoon Robyn during TCM-93. Mon. Weather Rev. 124, 634–652.

Hendricks, E. A., Montgomery, M. T., and Davis, C. A. (2004). On the role of "vortical" hot towers in hurricane forms. J. Atmos. Sci. 61, 1209–1232.

Hertenstein, R. F. A., and Schubert, W. H. (1991). Potential vorticity anomalies associated with squall lines. Mon. Weather Rev. 119, 1663–1672.

Houze, R. A. (2004). Mesoscale convective systems. Rev. Geophys. 42, RG4003, doi:10.1029/2004RG000150.

Houze, R. A. (1977). Structure and dynamics of a tropical squall-line system. Mon. Weather Rev. 105, 1540–1567.

Houze Jr., R. A., and Rappaport, E. N. (1984). Air motions and precipitation structure of an early summer squall line over the eastern tropical Atlantic. J. Atmos. Sci. 41, 553–574.

Houze Jr., R. A., Smull, B. F., and Dodge, P. (1990). Mesoscale organization of springtime rainstorms in Oklahoma. Mon. Weather Rev. 118, 613–654.

James, E. (2009). An observational climatology of midlatitude mesoscale convective vortices. M.S. thesis, Colorado State University, Dept. of Atmospheric Science, Fort Collins, CO 80523, 180 pp.

Jiang, H., and Raymond, D. J. (1995). Simulation of a mature mesoscale convective system using a nonlinear balance model. J. Atmos. Sci. 52, 161–175.

Jirak, I. L., and Cotton, W. R. (2007). Observational analysis of the predictability of mesoscale convective systems. Wea. Forecasting 22, 813–838.

Jirak, I. L., and Cotton, W. R. (2008). Reply. Wea. Forecasting 24, 356–360.

Jirak, I. L., Cotton, W. R., and McAnelly, R. L. (2003). Satellite and radar survey of mesoscale convective system development. Mon. Weather Rev. 131, 2428–2449.

Johns, R. H., and Hirt, W. D. (1985). The derecho of 19-20 July 1983 A case study. National Weather Digest 10, 17–32.

Johns, R. H., and Hirt, W. D. (1987). Derchos: Widespread convectively induced windstorms. Wea. Forecasting 2, 32–49.

Johnson, R. H., and Bartels, D. L. (1992). Circulations associated with a mature-to-decaying midlatitude mesoscale convective system. Part II: Upperlevel features. Mon. Weather Rev. 120, 1301–1320.

Johnson, R. H., and Hamilton, P. J. (1988). The relationship of surface pressure features to the precipitation and air flow structure of an intense midlatitude squall line. Mon. Weather Rev. 116, 1444–1472.

Johnson, R. H., Chen, S., and Toth, J. J. (1989). Circulations associated with a mature-to-decaying midlatitude mesoscale convective system. Part I: Surface features heat bursts and mesolow development. Mon. Weather Rev. 117, 942–959.

Johnson, R. H., Rickenbach, T. M., Rutledge, S. A., Ciesielski, P. E., and Schubert, W. H. (1999). Trimodal characteristics of tropical convection. J. Climate 12, 2397–2418.

Johnston, E. C. (1982). Mesoscale vorticity centers induced by mesoscale convective complexes. In "Ninth Conf. On Weather Analysis and Forecasting, Seattle, WA." pp. 196–200. Amer. Meteor. Soc. Preprints.

Jou, B. J. D., and Yu, C. K. (1993). Structure of a tropical convective system and associated vortex: A TAMEX case study. In "Proc. 26th Int. Radar Conf." pp. 65–67. Amer. Meteor. Soc., Norman, OK.

Kane, R. J., Chelius, C. R., and Fritsch, J. M. (1987). The precipitation characteristics of mesoscale convective weather systems. J. Climate Appl. Meteorol. 26, 1323–1335.

Keenan, T. D., and Rutledge, S. A. (1993). Mesoscale characteristics of monsoonal convection and associated stratiform precipitation. Mon. Weather Rev. 121, 352–374.

Kirk, J. R. (2003). Comparing the dynamical development of two mesoscale convective vortices. Mon. Weather Rev. 131, 862–890.

Kirk, J. R. (2007). A phaseplot method for diagnosing vorticity concentration mechanisms in mesoscale convective vortices. Mon. Weather Rev. 135, 801–820.

Knievel, J. C., and Johnson, R. H. (2002). The kinematics of a midlatitude, continental mesoscale convective system and its mesoscale vortex. Mon. Weather Rev. 130, 1749–1770.

Kuo, Y. H., and Chen, G. T. J. (1990). The Taiwan Area Mesoscale Experiment (TAMEX): An overview. Bull. Amer. Meteorol. Soc. 71, 488–503.

Kurihara, Y., and Tuleya, R. E. (1981). A numerical simulation study on the genesis of a tropical storm. Mon. Weather Rev. 109, 1629–1653.

Laing, A. G., and Fritsch, J. M. (1993a). Mesoscale convective complexes over the Indian monsoon region. J. Climate 6, 911–919.

Laing, A. G., and Fritsch, J. M. (1993b). Mesoscale convective complexes in Africa. Mon. Weather Rev. 121, 2254–2263.

Laing, A. G., and Fritsch, J. M. (1997). The global population of mesoscale convective complexes. Q. J. R. Meteorol. Soc. 123, 389–405.

Laing, A. G., and Fritsch, J. M. (2000). The large scale environments of the global populations of mesoscale convective complexes. Mon. Weather Rev. 128, 2756–2776.

Landsea, C. W. (1993). A climatology of intense (or major) Atlantic hurricanes. Mon. Weather Rev. 121, 1703–1713.

Leary, C. A. (1979). Behavior of the wind field in the vicinity of a cloud cluster in the Intertropical Convergence Zone. J. Atmos. Sci. 36, 631–639.

Leary, C. A., and Houze Jr., R. A. (1979). The structure and evolution of convection in a tropical cloud cluster. J. Atmos. Sci. 36, 437–457.

Leary, C. A., and Rappaport, E. N. (1987). The life cycle and internal structure of a mesoscale convective complex. Mon. Weather Rev. 115, 1503–1527.

Leary, C. A., and Thompson, R. O. (1976). A warm-core disturbance in the western Atlantic during BOMEX. Mon. Weather Rev. 104, 443–452.

Lee, W.-C., Wakimoto, R. M., and Carbone, R. E. (1992a). The evolution and structure of a bow-echo-microburst event. Part I: The microburst. Mon. Weather Rev. 120, 2188–2210.

Lee, W.-C., Wakimoto, R. M., and Carbone, R. E. (1992b). The evolution and structure of a bow-echo-microburst event. Part II: The bow echo. Mon. Weather Rev. 120, 2211–2225.

Lin, M.-S. (1986). The evolution and structure of composite meso-a-scale convective com plexes. Ph.D. Thesis, Colorado State University.

Lyons, W. A., Calby, R. H., and Keen, C. S. (1986). The impact of mesoscale convective systems on regional visibility and oxidant distributions during persistent elevated pollution episodes. J. Clim. Appl. Met. 25, 1518–1531.

Maddox, R. A. (1980). Mesoscale convective complexes. Bull. Am. Meteorol. Soc. 61, 1374–1387.

Maddox, R. A. (1983). Large-scale meteorological conditions associated with midlatitude, mesoscale convective complexes. Mon. Weather. Rev. 111, 1475–1493.

Mapes, B. (1993). Gregarious tropical convection. J. Atmos. Sci. 50, 2026–2037.

Mapes, B. E., and Houze Jr., R. A. (1993). An integrated view of the Australian monsoon and its mesoscale convectives systems. Part II: Vertical structure. Q. J. R. Meteorol. Soc. 118, 927–963.

May, P. T., Mather, J. H., Vaughan, G., Jakob, C., McFarquhar, G. M., Bower, K. N., and Mace, G. G. (2008). The tropical warm pool international cloud experiment. Bull. Amer. Meteorol. Soc. 89, 629–645.

McAnelly, R. L., and Cotton, W. R. (1986). Meso-$\beta$-scale characteristics of an episode of meso-$\alpha$-scale convective complexes. Mon. Weather Rev. 114, 1740–1770.

McAnelly, R. L., and Cotton, W. R. (1989). The precipitation life cycle ofmesoscale convective complexes over the central United States. Mon. Weather Rev. 117, 784–808.

McAnelly, R. L., and Cotton, W. R. (1992). Early growth of mesoscale convective complexes: A meso-$\beta$-scale cycle ofconvective precipitation? Mon. Weather Rev. 120, 1851–1877.

McAnelly, R. L., Nachamkin, J. E., Cotton, W. R., and Nicholls, M. E. (1997). Upscale evolution of MCSs: Doppler radar analysis and analyticalinvestigation. Mon. Weather Rev. 125, 1083–1110.
McBride, J. L. (1981). Observational analysis of tropical cyclone formation. Part II: comparison on non-develping verus developing systems. J. Atmos. Sci. 38, 1132–1151.
Mecham, D. B., Houze Jr., R. A., and Chen, S. S. (2002). Layer inflow into precipitating convection over the western tropical Pacific. Q. J. R. Meteorol. Soc. 128, 1997–2030.
Menard, R. D., and Fritsch, J. M. (1989). A mesoscale convective complex-generated inertially stable warm core vortex. Mon. Weather Rev. 117, 1237–1261.
Menard, R. D., Merritt, J. H., Fritsch, J. M., and Hirschberg, P. A. (1986). Mesoanalysis of a convectively generated, inertially stable mesovortex. In "11th Conf. on Weather Forecasting and Analysis, Kansas City, MO." pp. 194–199. Amer. Meteor. Soc. Preprints.
Miller, R. C. (1959). Tornado-producing synoptic patterns. Bull. Amer. Meteorol. Soc. 40, 465–472.
Miller, D., and Fritsch, J. M. (1991). Mesoscale convective complexes in the western pacific region. Mon. Weather Rev. 11, 2978–2992.
Montgomery, M. T., and Enagonio, J. (1998). Tropical cyclogenesis via convectively forced vortex Rossy waves in a three-dimensional quasigeostrophic model. J. Atmos. Sci. 55, 3176–3207.
Montgomery, M. T., Nicholls, M. E., Cram, T. A., and Saunders, A. (2006). A vortical hot tower route to tropical cyclogenesis. J. Atmos. Sci. 63, 355–386.
Murphy, J. D., and Fritsch, J. M. (1989). Multiple production of mesoscale convective systems by a convectively-0generated mesoscale vortex. In "12th Conf. on Weather Analysis and Forecasting, Monterey, CA." pp. 68–73. Amer. Meteor. Soc. Preprints.
Nakazawa, T. (1988). Tropical superclusters within interseasonal variations over the western Pacific. J. Meteor. Soc. Japan 66, 823–839.
Newton, C. W. (1966). Circulations in large sheared cumulonimbus. Tellus 18, 699–712.
Newton, C. W., and Newton, H. R. (1959). Dynamical interactions between large convective clouds and environment with vertical shear. J. Meteorol. 16, 483–496.
Nicholls, M. E., Pielke, R. A., and Cotton, W. R. (1991). A two-dimensionalnumerical investigation of the interaction between sea-breezes and deepconvection over the Florida Peninsula. Mon. Weather Rev. 119, 298–323.
Nolan, D. S. (2007). What is the trigger for tropical cyclogenesis? Aus. Met. Mag. 56, 241–266.
Ogura, Y., and Liou, M.-T. (1980). The structure of a midlatitude squall line: A case study. J. Atmos. Sci. 37, 553–567.

Olsson, P. Q. (1994). Evolution of balanced flow in a simulated mesoscale convective complex. Ph.D. dissertation, Colorado State University, Dept. of Atmospheric Science, Fort Collins, CO 80523, 177 pp.

Olsson, P. Q., and Cotton, W. R. (1997a). Balanced and unbalanced circulations in a primitive equation simulation of a midlatitude MCC. Part I: The numerical simulation. J. Atmos. Sci. 54, 457–478.

Olsson, P. Q., and Cotton, W. R. (1997b). Balanced and unbalanced circulations in a primitive equation simulation of a midlatitude MCC: Part II: Analysis of balance. J. Atmos. Sci. 54, 479–497.

Ooyama, K. V. (1982). Conceptual evolution of the theory and modeling of the tropical cyclone. J. Meteor. Soc. Japan 60, 369–379.

Ooyama, K. V. (1969). Numerical simulation of the life cycle of tropical cyclones. J. Atmos. Sci. 26, 3–40.

Orlanski, I. (1975). A rational subdivision of scales for atmospheric processes. Bull. Amer. Meteorol. Soc. 56, 527–530.

Pandya, R. E., and Durran, D. R. (1996). The influence of convectively generated thermal forcing on the mesoscale circulation around squall lines. J. Atmos. Sci. 53, 2924–2951.

Parker, M. D. (2008). Response of simulated squall lines to low-level cooling. J. Atmos. Sci. 64, 1323–1341.

Parker, M. D., and Johnson, R. H. (2000). Organizational modes of midlatitude mesoscale convective systems. Mon. Weather Rev. 128, 3413–3436.

Rasmussen, E. M. (1967). Atmospheric water vapor transport and the water balance of North America: Part 1. Characteristics of the water vapor field. Mon. Weather Rev. 95, 403–426.

Raymond, D. J. (1992). Nonlinear balance and potential vorticity thinking at large Rossy number. Q. J. R. Meteorol. Soc. 119, 987–1015.

Raymond, D. J., and Jiang, H. (1990). A theory for long-lived mesoscale convective systems. J. Atmos. Sci. 47, 3067–3077.

Reasor, P. D., Montgomery, M. T., and Bosart, L. F. (2005). Mesoscale observations of the genesis of Hurricane Dolly (1996). J. Atmos. Sci. 62, 3151–3171.

Ritchie, E. A., and Holland, G. J. (1997). Scale interactions during the formation of Typhoon Irving. Mon. Weather Rev. 125, 1377–1396.

Rotunno, R., and Emanuel, K. A. (1987). An air-sea interaction theory for tropical cyclones. Part II: Evolutinonary study using a nonhydrostatic axisymmetric numerical model. J. Atmos. Sci. 44, 542–561.

Rotunno, R., Klemp, J., and Weisman, M. (1988). A theory for strong, long-lived squall lines. J. Atmos. Sci. 45, 463–485.

Schmidt, J. M. (1991). Numerical and observational investigations of long-lived, MCS-induced, severe surface wind events: The derecho. Ph.D. dissertation, Colorado State University, Dept. of Atmospheric Science, Fort Collins, CO 80523, 196 pp.

Schmidt, J. M., and Cotton, W. R. (1989). A high plains squall line associated with a derecho. J. Atmos. Sci. 46, 281–302.

Schmidt, J. M., and Cotton, W. R. (1990). Interactions between upper and lowertropospheric gravity waves on squall line structure and maintenance. J. Atmos. Sci. 47, 1205–1222.

Schubert, W. H., Fulton, S. R., and Hertenstein, R. F. (1989). Balanced atmospheric response to squall lines. J. Atmos. Sci. 46, 2478–2483.

Schubert, W. H., Hack, J. J., Silva-Dias, P. L., and Fulton, S. R. (1980). Geostrophic adjustment in an axisymmetric vortex. J. Atmos. Sci. 37, 1464–1484.

Schumacher, R. S., and Johnson, R. H. (2008). Mesoscale processes contributing to extreme rainfall in a midlatitude warm-season flash flood. Mon. Weather Rev. 136, 3964–3986.

Scott, J. D., and Rutledge, S. A. (1995). Doppler radar observations of an asymmetric mesoscale convective system and associated vortex couplet. Mon. Weather Rev. 123, 3437–3457.

Simpson, J., Ritchie, E. A., Holland, G. J., Halverson, J., and Stewart, S. (1997). Mesoscale interactions in tropical cyclone genesis. Mon. Weather Rev. 125, 2643–2661.

Sippel, J. A., Nielsen-Gammon, M. W., and Allen, S. E. (2006). The multiple-vortex nature of tropical cyclogenesis. Mon. Weather Rev. 134, 1796–1814.

Skamarock, W. C., Weisman, M. L., and Klemp, J. B. (1994). Three-dimensional evolution of simulated long-lived squall lines. J. Atmos. Sci. 51, 2563–2584.

Smith, W. P., and Gall, R. L. (1989). Tropical squall lines of the Arizona monsoon. Mon. Weather Rev. 117, 1553–1569.

Smull, B. F., and Houze Jr., R. A. (1985). A midlatitude squall line with a trailing region of stratiform rain: Radar and satellite observations. Mon. Weather Rev. 113, 117–133.

Smull, B. F., and Houze, R. A. (1987a). Dual-Doppler radar analysis of a mid-latitude squall line with a trailing region of stratiform rain. J. Atmos. Sci. 44, 2128–2148.

Smull, B. F., and Houze, R. A. (1987b). Rear inflow in squall lines with trailing stratiform precipitation. Man. Weather Rev. 115, 2869–2889.

Srivastava, R. C., Matejka, T. J., and Lorello, T. J. (1986). Doppler radar study of the trailing anvil region associated with a squall line. J. Atmos. Sci. 43, 356–377.

Stirling, J., and Wakimoto, R. M. (1989). Mesoscale vortices in the stratiform region of a decaying mid-latitude squall line. Mon. Weather Rev. 117, 452–458.

Thornkroft, C., and Hodges, K. (2001). African easterly wave variability and its relationship to Atlantic tropical cyclone activity. J. Clim. 14, 1166–1179.

Tollerud, E. I., Brown, J. M., and Bartels, D. L. (1989). Structure of an MCS-induced mesoscale vortex as revealed by VHF profiler, Doppler radar, and satellite observations. In "12th Conf. on Weather Analysis and Forecasting, Monterey, CA." pp. 81–86. Amer. Meteor. Soc. Preprints.

Tollerud, E. I., and Collander, R. S. (1993a). A ten-year summary of severe weather in mesoscale convective complexes. Part I: High wind, tornadoes, and hail. In "17th Conf. on Severe Local Storms, St. Louis, MO." pp. 533–537. Amer. Meteor. Soc. Preprints.

Tollerud, E. I., and Collander, R. S. (1993b). A ten-year synopsis of record station rainfall produced in mesoscale convective complexes. In "13th Conf. on Weather Analysis and Forecasting, Vienna, VA." pp. 430–433. Amer. Meteor. Soc. Preprints.

Trapp, R. J., and Weisman, M. L. (2003). Low-level mesovortices within squall lines and bow echoes. Part II: Their genesis and implications. Mon. Weather Rev. 131, 2804–2823.

Trier, S. B., and Davis, C. A. (2002). Influence of balanced motions on heavy precipitation within a long-lived convectively generated vortex. Mon. Weather Rev. 130, 877–899.

Trier, S. B., and Davis, C. A. (2007). Mesoscale convective vortices observed during BAMEX. Part II: Influences on secondary deep convection. Mon. Weather Rev. 135, 2051–2075.

Tripoli, G., and Cotton, W. R. (1989a). A numerical study of an observedorogenic mesoscale convective system. Part 1. Simulated genesis andcomparison with observations. Mon. Weather Rev. 117, 273–304.

Tripoli, G., and Cotton, W. R. (1989b). A numerical study of an observedorogenic mesoscale convective system. Part 2. Analysis of governingdynamics. Mon. Weather Rev. 117, 305–328.

Tuttle, J. D., and Carbone, R. E. (2004). Coherent regeneration and the role of water vapor and shear in a long-lived convective episode. Mon. Weather Rev. 132, 192–208.

Tuttle, J. D., and Davis, C. A. (2006). Corridors of warm season precipitation in the central United States. Mon. Weather Rev. 134, 2297–2317.

Velasco, I., and Fritsch, J. M. (1987). Mesoscale convective complexes in the Americas. J. Geophys. Res. 92, 9591–9613.

Verlinde, J., and Cotton, W. R. (1990). A mesoscale vortex couplet observed in the trailing anvil of a multicellular convective complex. Mon. Weather Rev. 118, 993–1010.

Wallace, J. M. (1975). Diurnal variations in precipitation and thunderstorm frequency over the conterminous United States. Mon. Weather Rev. 103, 406–419.

Wang, W., Kuo, Y.-H., and Warner, T. T. (1993). A diabatically driven mesoscale vortex in the lee of the Tibetan plateau. Mon. Weather Rev. 121, 2542–2561.

Wei, T., and Houze Jr., R. A. (1987). The GATE squall line of 9-10 August 1974. Advances in Atmospheric Sciences 4, 85–92.

Weisman, M. L., and Davis, C. A. (1998). Mechanisms for the generation of mesoscale vortices within quasi-linear convective systems. J. Atmos. Sci. 55, 2603–2622.

Weisman, M., and Klemp, J. (1982). The dependence of numerically simulated convective storms on vertical wind shear and buoyancy. Mon. Weather Rev. 110, 504–520.

Wetzel, P. J., Cotton, W. R., and McAnelly, R. L. (1983). A long-lived mesoscale convective complex. Part II: Evolution and structure of the mature complex. Mon. Weather Rev. 111, 1919–1937.

Yanai, M., Esbensen, S., and Chu, J. (1973). Determination of bulk properties of tropical cloud clusters from large-scale heat and moisture budgets. J. Atmos. Sci. 36, 53–72.

Yu, C.-K., Jou, B. J.-D., and Smull, B. F. (1999). Formative stage of a long-lived mesoscale vortex observed by airborne Doppler radar. Mon. Weather Rev. 127, 838–857.

Zehr, R. (1992). Tropical cyclogenesis in the western North Pacific. NOAA Tech. Rep. NESDIS 61, 181 pp.

Zhang, D.-L. (1992). The formation of a cooling-induced mesovortex in the trailing stratiform region of a midlatitude squall line. Mon. Weather Rev. 120, 2763–2785.

Zhang, D.-L., and Fritsch, J. M. (1988a). Numerical sensitivity experiments of varying model physics on the structure, evolution and dynamics of two mesoscale convective systems. J. Atmos. Sci. 45, 261–293.

Zhang, D.-L., and Fritsch, J. M. (1988b). Numerical simulation of the meso-scale structure and evolution of the 1977 Johnstown flood. Part III: Internal gravity waves and the squall line. J. Atmos. Sci. 45, 1252–1268.

Zhang, D.-L., and Fritsch, J. M. (1988c). A numerical investigation of a convectively generated, inertially stable, extratropical warm-core mesovortex over land. Part I: Structure and evolution. Mon. Weather Rev. 116, 2660–2687.

Zipser, E. J. (1977). Mesoscale and convective-scale downdrafts as distinct components of squall-line structure. Mon. Weather Rev. 105, 1568–1589.

Zipser, E. J., and LeMone, M. A. (1980). Cumulonimbus vertical velocity events in GATE. Part II: Synthesis and model core structure. J. Atmos. Sci. 37, 2458–2469.

Chapter 10

# The Mesoscale Structure of Extratropical Cyclones and Middle and High Clouds

## 10.1. INTRODUCTION

Extratropical cyclones have been found to account for about half of the warm season precipitation over the United States (Heideman and Fritsch, 1988), and the warm conveyor belts associated with these cyclones produce approximately half of the wintertime precipitation in middle and high latitudes (Eckhardt et al., 2004). Thus these storms, also referred to as mid-latitude cyclones, play an important role in the hydrological cycle of the mid-latitudes. In Chapter 9 we considered how the dynamical behavior of an atmospheric circulation system can be characterized by the scaling parameter $\lambda_R$, the Rossby radius of deformation [Eq. (9.3)]. A substitution of typical values of the parameters in Eq. (9.3) for tropical and extratropical cyclones (averaged over the entire cyclone) shows that for the largest scale of these cyclonic disturbances, the atmosphere behaves as a two-dimensional, quasi-balanced fluid (Table 10.1). However an enormous spatial variation of $\lambda_R$ exists within extratropical cyclones and thus a variety of clouds and mesoscale precipitating phenomena are embedded within these systems. As a result, extratropical cyclones also participate in the global energy balance. Finally, these storms may also produce a wide range of severe weather. In an analysis of what has been referred to as *The Storm of the Century*, Kocin et al. (1995) described the widespread heavy snowfall, coastal flooding, squall lines, thunderstorms and tornadoes associated with an extratropical cyclone that developed and progressed along the southern and eastern coasts of the United States between the 12 and 14 March, 1993. Debate continues as to whether the mesoscale phenomena associated with mid-latitude cyclones are simply a consequence of the large-scale motions of the cyclone, or whether they may modulate the cyclone through upscale forcing. Zhang and Harvey (1995), for example, document a case in which a squall line enhanced the large-scale baroclinic environment, making it more favorable for subsequent surface cyclogenesis. The focus in this chapter is on the processes and properties of various mesoscale features generated by extratropical cyclones.

**TABLE 10.1** Horizontal scale and Rossby radius of deformation ($\lambda_R$) for the large-scale structure of extratropical and tropical cyclones

| | $L$ (km) | $H$ (km) | $N$ ($s^{-1}$) | $f$ ($s^{-1}$) | $\zeta$ ($s^{-1}$) | $2V/R$ ($s^{-1}$) | $\lambda_R$ (km) |
|---|---|---|---|---|---|---|---|
| Extratropical cyclone | 10,000 | 8 | 0.01 | $10^{-4}$ | $f$ | $f$ | 400 |
| Tropical cyclone | 5,000 | 8 | 0.01 | $0.5 \times 10^{-4}$ | $2f$ | $2f$ | 500 |

Synoptic-scale forcing and processes are only discussed when relevant to the mesoscale features of interest.

## 10.2. LARGE-SCALE PROCESSES THAT DETERMINE MESOSCALE FEATURES

Mesoscale cloud and precipitation features are initiated by two mechanisms—forcing on the mesoscale by inhomogeneities in the surface (such as terrain features) and instabilities in the larger scale environment. Terrain-forced features are discussed in Chapter 11. Here we discuss the large-scale processes that produce an environment that is stable or unstable with respect to mesoscale cloud and precipitation systems. We have seen that the type of cloud and precipitation of a system is determined by six factors: (1) water vapor content of the air (both relative and absolute humidity), (2) temperature, (3) aerosol types and amounts, (4) static stability, (5) vertical motion, and (6) vertical shear of the horizontal wind. As these atmospheric properties vary greatly throughout extratropical cyclones, these storms contain a rich variety of clouds and mesoscale precipitating systems. After a brief description of the physical processes determining these parameters on large scales of motion (100 km and greater), we discuss their variation in extratropical cyclones, which in turn affects the variation of clouds and precipitation in these systems.

### 10.2.1. Water Vapor Content

Chapters 2 and 4 describe in detail the equations that determine the temporal variation of water vapor as well as liquid water and ice. For convenience, we repeat the continuity equation for the water vapor mixing ratio appropriate for large-scale models,

$$\frac{\partial r_v}{\partial t} = -\mathbf{V} \cdot \nabla r_v - w\frac{\partial r_v}{\partial z} + E - C + F_{r_v}, \tag{10.1}$$

where $E$ represents evaporation (from the surface or from precipitation), $C$ represents condensation (including sublimation), $F_{r_v}$ represents unresolvable (subgrid-scale) transports, $V$ is the horizontal velocity vector, and $\nabla$ is the horizontal gradient operator. Over short periods of time and for the large scales of motion considered here, horizontal and vertical transports are the major processes contributing to local changes in water vapor. For longer time periods (greater than 12 h), evaporation from the surface is important over water and over land surfaces that are moister and warmer than the air immediately above the surface. Condensation is a key removal mechanism in precipitating systems, whereas evaporation of cloud water and precipitation can be locally significant over short time periods. Subgrid-scale (turbulent) transports are greatest in the unstable planetary boundary layer where water vapor evaporated from the surface is transported upward.

### 10.2.2. Temperature

Detailed forms of the thermodynamic equation are presented in Chapter 2. A simplified form appropriate for interpreting the temporal variation of the large-scale temperature is

$$\frac{\partial \theta}{\partial t} = -\mathrm{V} \cdot \nabla \theta - w\frac{\partial \theta}{\partial z} + \frac{\theta}{c_{\mathrm{p}} T} Q + F_{\theta}, \tag{10.2}$$

where $\theta$ is potential temperature and $Q$ is the net grid scale averaged diabatic heating. All of the other variables have been defined previously. The first term of the right-hand side represents horizontal advection of potential temperature, while the second term is the vertical advection of potential temperature, which in a statically stable atmosphere produces cooling with upward motion and warming with subsiding motion. The diabatic heating term $Q$ includes latent heating and cooling effects associated with condensation or evaporation, as well as radiation. In precipitation systems, there is a close balance between adiabatic cooling associated with vertical advection and diabatic heating. The last term $F_{\theta}$ is the subgrid-scale transport of heat and includes sensible heating from the surface and the upward transport in the PBL. On the large scale, above the PBL and in the absence of precipitation, horizontal and vertical advection are the largest terms. During the daytime in the heated PBL, the last term, dominates.

### 10.2.3. Aerosol Types and Amounts

Condensation does not usually occur at relative humidities of exactly 100% (Chapter 4). The presence and type of cloud depend on the amount and distribution of aerosols in the atmosphere, in particular, cloud condensation nuclei and ice nuclei. The effects of these aerosols are discussed in Chapter 4 and are mentioned here for completeness. They are not presently explicitly represented in most large-scale models, however some cloud resolving models now include explicit aerosol schemes (e.g. Saleeby and Cotton, 2004; Ekman et al., 2004).

### 10.2.4. Static Stability

The evolution of cloud and precipitation systems depends on the mean vertical motion and the atmospheric static stability. Static stability is described in detail in Chapter 2. As described in this chapter, conditional instability refers to the state of a parcel of air as it is lifted through its environment. Under such conditions a parcel of air at the environmental temperature is stable to all vertical displacements if is unsaturated, but is unstable to upward vertical displacements if it is saturated, and is unstable to downward vertical displacements if it is saturated and contains cloud water. Convective instability, in contrast, refers to the state of instability when lifting an entire layer of air. If such a layer is lifted until it becomes completely saturated, then the layer will become unstable. The condition for convective instability, which is sometimes referred to as *potential instability*, is that equivalent potential temperature decreases throughout the layer,

$$\partial \theta_e / \partial z < 0. \tag{10.3}$$

It is possible for a layer of air to be convectively unstable but conditionally stable. Convective instability is often associated with the large-scale environment prior to the development of severe convective storms and tornadoes, when warm, dry air overlies warm, moist air, with an associated rapid decrease of $\theta_e$ with height. If the layer is lifted, it rapidly becomes unstable and favorable for severe thunderstorm development. When the mean vertical motion is near zero and the atmosphere is conditionally stable, fogs or layered clouds occur (Chapter 6). When the mean vertical velocity is upward (typically a few centimeters per second) and the atmosphere is conditionally stable, deep layers of nonconvective (stratiform) clouds are produced. Under conditionally unstable conditions, convective clouds and precipitation can occur, even with near-zero mean (large-scale) vertical velocities or weak subsidence.

If we define a static stability parameter $\gamma_\theta \equiv \partial\theta/\partial z$, a simple equation describing the temporal variation of the large-scale static stability can be derived from Eq. (10.2):

$$\frac{\partial \gamma_\theta}{\partial t} = \frac{-\partial \mathrm{V} \cdot \nabla\theta}{\partial z} - \frac{\partial w \gamma_\theta}{\partial z} + \frac{\partial}{\partial z}\left(\frac{\theta}{c_\mathrm{p} T} Q\right) + \frac{\partial F_\theta}{\partial z}. \tag{10.4}$$

The first term on the right-hand side of Eq. (10.4) represents the variation of the horizontal advection of potential temperature with height; for example, cold advection overlying warm advection contributes to destabilization. The second term represents the effect of vertical stretching of a column, i.e. if $\gamma_\theta$ is constant in the vertical, an increase of upward motion with height (stretching) represents destabilization. This process is effective in producing or destroying temperature inversions. The third term represents the effect of differential heating in the vertical. For example, a decrease in diabatic heating with height, as occurs above the region of maximum latent heating associated with cumulus convection, destabilizes the environment. Another example of this process is radiative cooling near the tops of layered clouds which destabilizes this region. The final term represents the vertical variation of turbulent heat fluxes and is largest in the heated PBL.

Conditional and convective instabilities do not consider the effects of rotation, but rotation affects the stability of fluid motions. Bennetts and Hoskins (1979) and Emanuel (1979, 1982, 1983a,b) discuss the combined influence of rotation and static stability in a theory of *conditional symmetric instability* (CSI) in which the atmosphere is convectively stable and inertially stable, and yet is unstable to slantwise ascent. We start by examining dry symmetric instability. As just stated, the condition for dry absolute instability is that the potential temperature decreases with height ($\partial\theta/\partial z < 0$). The condition for inviscid, inertial instability in the Northern Hemisphere is $\partial M_g/\partial x < 0$, where $M_g = v_g + fx$ is the geostrophic absolute momentum of a geostrophically balanced mean state, $v_g$ is the geostrophic wind in the direction perpendicular to the temperature gradient (and is in thermal wind balance), $x$ is the distance in the direction of the temperature gradient (perpendicular to the thermal wind) with $x$ increasing toward the warmer air, and $f$ is the Coriolis parameter. In the absence of friction, $M_g$ is approximately conserved. A parcel may be absolutely stable to vertical displacement ($\partial\theta/\partial z > 0$), and inertially stable to horizontal displacements ($\partial M_g/\partial x > 0$), but unstable with respect to slantwise displacements by dry symmetric instability. The condition for dry symmetric instability is that the $M_g$ surfaces slope less steeply than the isentropic surfaces. Dry symmetric instability can therefore be viewed either as dry absolute instability on an $M_g$ surface, or inertial instability on an isentropic surface. Any slantwise displacement that occurs at an angle between the slopes of the $M_g$ and isentropic surfaces will therefore release the symmetric instability.

For moist slantwise convection, CSI occurs at each height where the environmental lapse rate along an $M_g$ surface lies between the dry- and saturated-adiabatic lapse rates, and hence is conditionally unstable along an $M_g$ surface. The CSI condition is then one in which the saturation equivalent potential temperature decreases with height along an $M_g$ surface. Thus if a parcel moves upward along a surface of constant $M_g$ and becomes warmer than its environment due to the release of latent heat, the atmosphere is in a state of conditional symmetric instability. A conditionally stable atmosphere may possess conditional symmetric instability. Convection arising from the release of CSI is called moist slantwise convection. The release of CSI is thought to be an important process in producing rainbands in extratropical cyclones. In environments near the frontal zones of extratropical cyclones a variety of instability types may be found, ranging from convectively unstable air in the warm sector, to CSI north of the frontal boundary, and weak symmetric stability to the north of that.

Finally, as described by Schultz and Schumacher (1999) in a review article on the use and misuse of CSI as a diagnostic tool, equivalent potential temperature instead of saturation equivalent potential temperature is often, although incorrectly, used to assess CSI. When using equivalent potential temperature, it is convective or potential symmetric instability (PSI), rather than conditional symmetric instability, that is being assessed. As there can be significant differences between equivalent potential temperature and saturated equivalent potential temperature, there could be a significant difference between conditional and convective symmetric instabilities. Thus the term CSI should only be employed when using saturated equivalent potential temperature, while PSI should be utilized when using equivalent potential temperature. However, Moore et al. (2005) make the point that the instability is not realized unless the atmosphere becomes saturated in the presence of large-scale lifting, at which time the equivalent potential temperature and the saturated equivalent potential temperature are equal. It has been found to be more convenient operationally to use the equivalent potential temperature, and require that the relative humidity be at least 80%, in order to diagnose regions that are susceptible to CSI (Market and Cissell, 2002).

## 10.2.5. Vertical Motion

Along with moisture content and static stability, vertical motion is one of the most important properties of the large-scale environment that determines the presence and type of clouds and precipitation systems. Large-scale upward motion favors cloud and precipitation development because it both cools and destabilizes the air. In addition, the low-level convergence associated with rising motion in the middle troposphere is associated with moisture convergence in all but the driest air masses. By contrast, sinking air becomes more stable, relative humidity decreases, and low-level moisture divergence usually results.

A useful diagnostic equation for isolating the large-scale physical processes associated with vertical motion is the quasi-geostrophic omega equation, derived from the vorticity equation and the first law of thermodynamics using the assumption that the vorticity and the large-scale wind are in quasi-geostrophic balance (see Holton (1979) for a derivation). A convenient form of the omega equation with pressure as the vertical coordinate can be written as

$$\nabla^2\omega + \frac{f_0^2}{\sigma}\frac{\partial^2\omega}{\partial p^2} = \frac{f}{\sigma}\frac{\partial}{\partial p}[\mathrm{V_g}\cdot\nabla(\zeta + f)] + \frac{R}{p\sigma}\nabla^2\mathrm{V_g}\cdot\nabla T - \frac{R}{c_p p\sigma}\nabla^2 Q - \frac{f_0}{\sigma}\frac{\partial F_\zeta}{\partial p}, \qquad (10.5)$$

where the static stability parameter $\sigma$ is given by

$$\sigma = -(1/\rho\theta)(\partial\theta/\partial p),$$

and $F_\zeta$ represents the contribution of subgrid-scale effects (friction) to the temporal rate of change of the vertical component of the relative vorticity ($\zeta$).

To interpret Eq. (10.5), wave forms may be assumed for the horizontal and vertical variation of $\omega$ (Holton, 1979),

$$\omega \approx \sin[\pi(p/p_0)]\sin(kx)\sin(ly), \tag{10.6}$$

where $k$ and $l$ are horizontal wave numbers and $p_0$ is a reference pressure. Using the assumptions made in Eq. (10.6), the left-hand side of Eq. (10.5) can be written as

$$\left(\nabla^2 + \frac{f_0^2}{\sigma}\frac{\partial^2}{\partial p^2}\right)\omega \approx -\left[(k^2 + l^2) + \frac{1}{\sigma}\left(\frac{f_0\pi}{p_0}\right)^2\right]\omega. \tag{10.7}$$

Equation (10.7) demonstrates that the left-hand side of Eq. (10.5) is proportional to $-\omega$, and hence that the left-hand side of Eq. (10.5) is proportional to $w$. Thus, the Omega equation, Eq. (10.5), can be interpreted as follows. The first term on the right-hand hand side represents the vertical derivative of the horizontal vorticity advection. In most situations in the atmosphere, the vorticity advection is much smaller in the lower troposphere than in the middle to upper troposphere, and the sign of this term is thus generally determined by the horizontal vorticity advection. Therefore, rising (sinking) vertical motion is generally proportional to positive (negative) vorticity advection in the Northern Hemisphere. The second term on the right-hand side represents the Laplacian of the horizontal temperature advection, and thus $-\omega \propto V_g \cdot \nabla T$. In regions of warm air (cold air) advection, this term is positive (negative) and contributes to rising (sinking) motion. The third term on the right-hand side is the Laplacian of diabatic heating and thus $-\omega \propto Q$ (as the sign on this term is negative). Regions of maximum diabatic heating are therefore associated with upward motion.

The last term in Eq. (10.5) represents the effect of subgrid-scale motions (turbulence). In the PBL, its effect may be estimated by assuming a quadratic stress law for the frictional terms in the equations of motion,

$$\frac{\partial u}{\partial t} = \cdots - C_D|\mathbf{V}|u/h,$$
$$\frac{\partial v}{\partial t} = \cdots - C_D|\mathbf{V}|v/h, \tag{10.8}$$

where $C_D$ is the drag coefficient, $u$ and $v$ are the mean horizontal wind components in the PBL, and $h$ is the depth of the PBL. With the frictional terms represented by Eq. (10.11), the linearized form of $F_\zeta$ is:

$$F_\zeta(p_s) \approx -K\zeta(p_s), \tag{10.9}$$

where $K$ is a mean value of $C_D|\mathbf{V}|/h$. Thus, using Eqs (10.5), (10.7) and (10.9), the vertical velocity near the top of the PBL is approximately

$$\omega \propto \frac{\partial F_s}{\partial p} \propto F_\zeta(p_S), \tag{10.10}$$

where we have used the fact that $F_\zeta$ vanishes near the top of the PBL. From (10.9) and (10.10) we see that surface friction induces upward motion in cyclonic systems and downward motion in anticyclonic systems. This effect is sometimes called *Ekman pumping* or *Ekman suction* depending on its sign.

### 10.2.6. Vertical Shear of the Horizontal Wind

A fifth property of the environment important in the development of some types of clouds and convective systems is the vertical shear of the horizontal wind. In the PBL, the wind shear organizes fair-weather cumulus clouds into bands, rolls, rings, and streets (Chapter 7). In addition, strong vertical wind shear is a major factor in determining the organization and structure of cumulonimbus clouds. Wind shear affects the entrainment rate, the strength, movement, precipitation efficiency, and lifetime of convective clouds and storms (Chapter 8). Wind shear is also a factor in the splitting of severe thunderstorms and the development of rotating storms and hail and tornado-producing thunderstorms.

The development of wind shear in the large-scale environment is closely tied to the development of baroclinicity, since the thermal wind balance is approximately satisfied for these scales of motion. Figure 10.1 shows a horizontal cross section of a strong baroclinic zone and the associated wind shear (Shapiro et al., 1984). Baroclinicity on these scales is produced primarily by two mechanisms - frontogenetic processes and differential heating associated with latent heat release. Frontogenetic processes are important in extratropical cyclone systems. Confluence and deformation in the large-scale environment and an ageostrophic response of the atmosphere to thermal-wind imbalances produced by the changing baroclinicity are key elements of the frontogenesis process. A local increase in the horizontal temperature gradient through deformation of the wind, disrupts thermal wind balance as the horizontal temperature gradient becomes too large for the associated vertical wind shear. The atmosphere produces a thermally-direct ageostrophic circulation transverse to the baroclinic zone, thereby re-establishing thermal wind balance (Koch, 1984; Sanders and Bosart, 1985; Keyser and Shapiro, 1986). Frontogenesis therefore enhances vertical motion through this thermally-direct ageostrophic circulation. Figure 10.2 shows the transverse ageostrophic circulation associated with frontogenesis in a numerical model. This vertical circulation, in addition

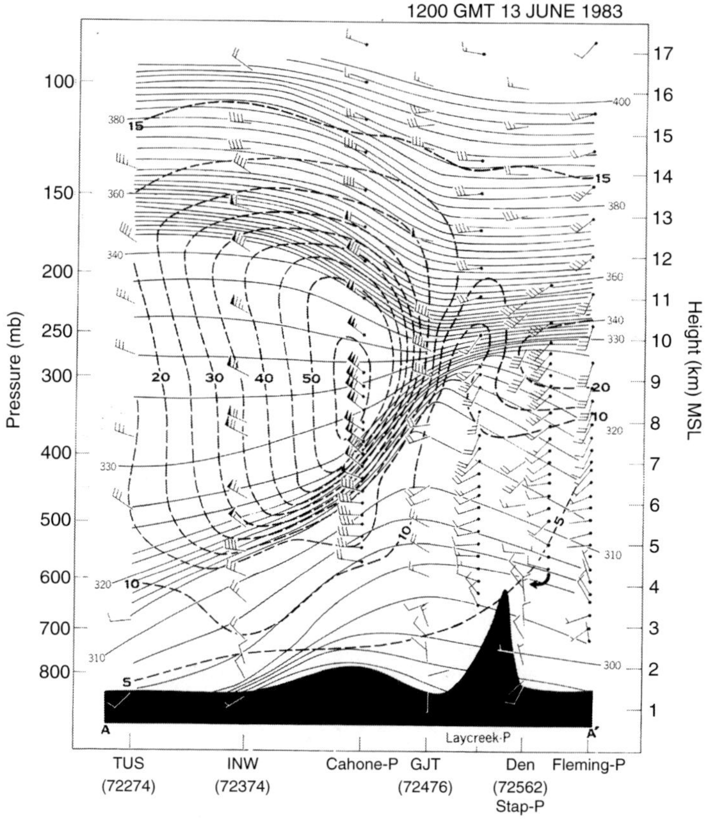


**FIGURE 10.1 Cross-sectional analysis of wind speed (m s$^{-1}$, dashed lines) and potential temperature (K, solid lines) at 1200 GMT, 13 June 1983 along a SW-NE line from Tucson, Arizona to Fleming, Colorado.** Analysis is a composite of rawinsonde winds and radar wind profiles. Profile soundings are designated by the letter P at the horizontal axis. Flag = 25 m s$^{-1}$; full barb = 5 m s$^{-1}$; half barb = 2.5 m s$^{-1}$. *(From Shapiro et al. (1984))*

to playing an essential role in the frontogenesis process and the development of wind shear, destabilizes the environment on the warm, moist side of fronts and triggers clouds and precipitation systems. Baroclinicity can also be produced by differential heating associated with the release of latent heat.

Another mechanism for producing low-level wind shear is surface friction. The general decrease of frictional effects with height in the lower troposphere results in vertical wind shear. For example, the diurnal variation in the depth of the PBL and the intensity of turbulent mixing of momentum can lead to the development of low-level jets (Blackadar, 1957; Bonner, 1966, 1968). The horizontal convergence and wind shear associated with these jets can significantly affect the development of convective storms.

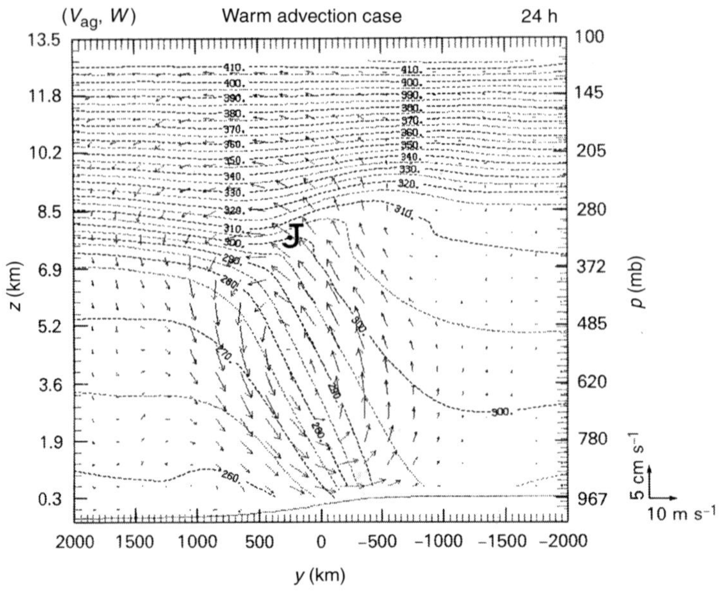


**FIGURE 10.2 Cross section of transverse ageostrophic circulation ($v_{ag}$, $w$) and potential temperature (dashed lines, contour interval 5 K) after a 24-h integration of a two-dimensional primitive-equation model of frontogenesis (Keyser and Pecnick, 1985a,b) due to confluence in the presence of advection.** Location of upper-level jet in along-front-velocity component is indicated by J; magnitudes of components of transverse ageostrophic circulation are represented by vector scales on lower right margins of figure. *(From Keyser and Shapiro (1986))*

### 10.2.7. Ocean versus Land Extratropical Cyclones

The conceptual model of extratropical cyclone development discussed below is modified somewhat for cyclones over the ocean. Two major differences over the ocean are the greater availability of moisture and the smoother, more homogeneous surface. The latter allows more rapid development and ultimately more intense cyclones. Rapid development can be pronounced when the sea surface is relatively warm, thus destabilizing lower levels and leading to explosively deepening cyclones, often referred to as "bombs" (Sanders and Gyakum, 1980). In addition, the more homogeneous surface favors the development of organized mesoscale convective clouds in the PBL (Chapter 7).

## 10.3. MESOSCALE STRUCTURE OF EXTRATROPICAL CYCLONES

### 10.3.1. Introduction

Driven inexorably by differential radiative heating between high and low latitudes, the middle-latitude atmosphere is characterized by large-scale horizontal temperature gradients and, through the thermal wind relationship,

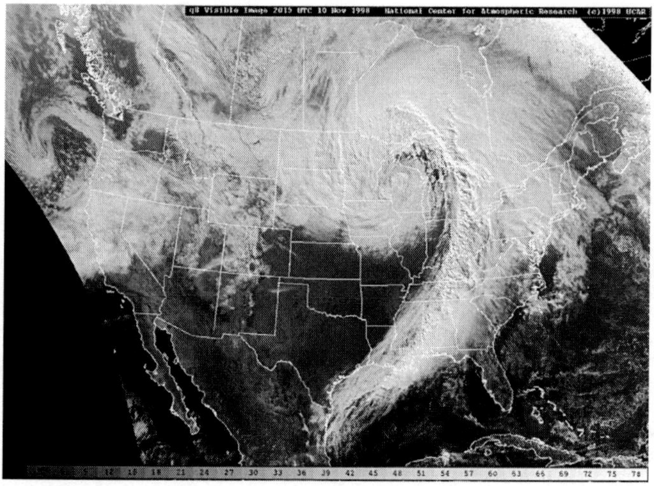

**FIGURE 10.3 Visible satellite photograph of occluding cyclone over North America at 2200 GMT, 23 February 1977.** *(Copyright © 1998 by University Corporation for Atmospheric Research. Reprinted with permission.)*

westerly winds that normally increase with height throughout the troposphere. The poleward decrease of temperature is rarely uniform, and, instead, is usually concentrated in relatively narrow baroclinic zones or fronts. These baroclinic zones become unstable with respect to wavelike perturbations (Charney, 1947; Eady, 1949), and the result is the development of cyclones in the baroclinic zone. The wavelength of maximum instability depends on the static stability and horizontal temperature gradient (Staley and Gall, 1977). On the average it is around 3000 km.

As cyclones develop, cold air is carried southward to the rear of the cyclone while warm air is carried northward ahead of the cyclone. For convenience, we refer to the Northern Hemisphere in these discussions. Confluence and deformation associated with the developing circulation produce increasing horizontal temperature gradients in narrow bands, i.e. warm and cold fronts. Temperature changes associated with horizontal advection and vertical motion destroy thermal wind balance, and the resulting ageostrophic motions produce organized regions of divergence, convergence, and associated vertical motion. Extratropical cyclones dominate the large-scale variability of the weather of middle latitudes. Figure 10.3 shows visible satellite imagery of a mature cyclone over North America. The circulation of this cyclone covers a significant portion of North America and adjacent waters east of the Rocky Mountains.

Many aspects of the mesoscale structure of extratropical cyclones, discussed in detail in this text, include fogs and stratocumulus clouds (Chapter 6), cumulus clouds (Chapter 7), severe thunderstorms and tornadoes (Chapter 8), squall lines and other mesoscale convective systems (Chapter 9), and middle- and high-level

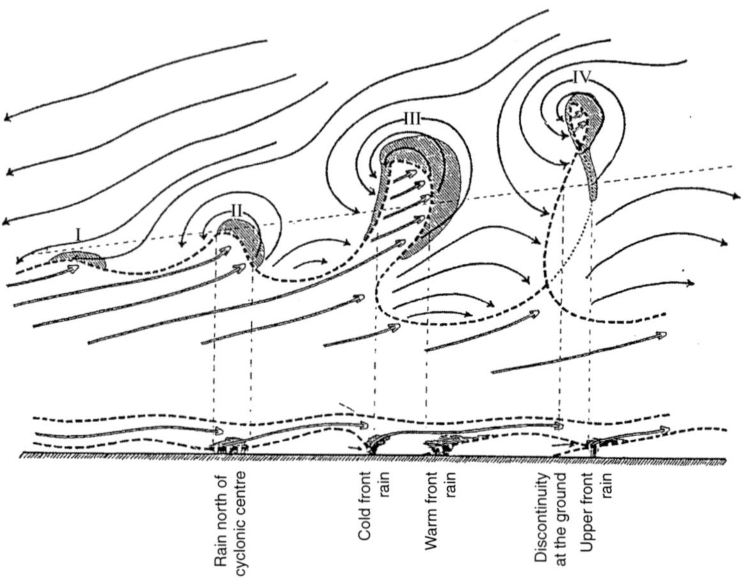


**FIGURE 10.4 The Norwegian frontal cyclone model.** *(From Bjerknes and Solberg (1922); After Shapiro and Keyser (1990))*

cloud systems (later in this chapter). In this section we interpret the formation of these features in terms of the physical processes and characteristics of the environment discussed in the preceding section.

## 10.3.2. Conceptual Models of Extratropical Cyclones

While the focus of this chapter is on the mesoscale aspects of extratropical cyclones, it is useful to briefly describe the synoptic-scale organization of these systems, as the synoptic-scale features not only influence the mesoscale aspects, but are also influenced by the mesoscale features of these storms. Since its development during the early part of the 20th century, the Norwegian Cyclone model (Bjerknes, 1919; Bjerknes and Solberg, 1921, 1922) has formed the basis of our understanding of the life cycle of extratropical cyclones, an amazing achievement given that this conceptual model was developed during a time period of highly limited upper air data. The Norwegian Cyclone model is shown in Fig. 10.4. The development of the frontal wave from its incipient phase, through the mature cyclogenesis phase, and finally terminating in a frontal occlusion phase are all demonstrated. At the northern tip of the occluded front, a seclusion of warm air from the warm sector occurs as a result of being trapped during the occlusion process. The Norwegian Cyclone model also includes a description of the vertical structure of the warm, cold and occluded fronts, and the precipitation and cloud development associated with these fronts (bottom portion of Fig. 10.4), and Fig. 10.5 illustrates the typical cloud types in various

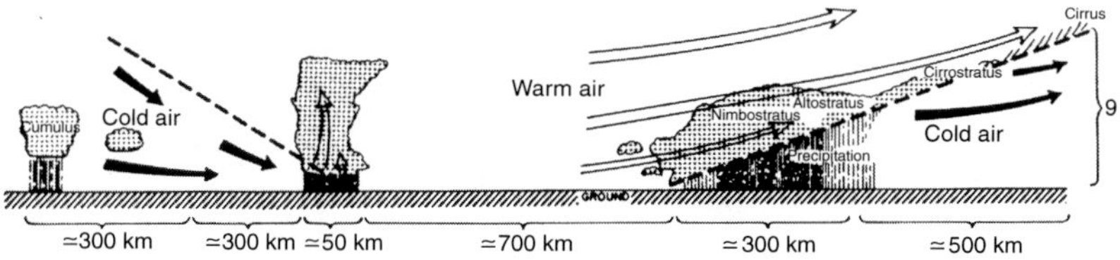


**FIGURE 10.5 Idealized vertical cross section through a mid-latitude cyclone, according to the Norwegian Cyclone model.** (Vertical scale is stretched by a factor of about 30 compared to horizontal scale.). *(From Houze and Hobbs (1982))*

regions of the cyclone. This model, sometimes referred to as the ideal or classical cyclone model, was developed over 90 years ago and remains a useful, although somewhat simplified conceptual model.

Research conducted after World War II began to demonstrate that there may be more to the dynamic and microphysical aspects of extratropical cyclones than was suggested by the Norwegian Cyclone model. Observational studies indicated that the cyclone and frontal structures did not always extend continuously from the surface to the upper atmosphere, and that different dynamics were responsible for upper and lower-level fronts; early baroclinic wave simulations (e.g. Hoskins, 1976) produced frontal characteristics that had not as yet been observed. Based on previous modeling and observational studies, as well as observations from three field experiments investigating marine cyclone development, Shapiro and Keyser (1990) proposed several significant modifications to the Norwegian Cyclone model.

In the Shapiro-Keyser model (Fig. 10.6) the frontal cyclone develops on a continuous and broad frontal zone (~400 km) (Fig. 10.6 I). As the cyclone develops, the previously continuous cold front separates or "fractures" from the warm front and advances into the warm sector air, and the temperature gradients of the warm and cold fronts contract inwards (~100 km) (Fig. 10.6 II). As the cold front moves eastward into the narrowing warm sector, the warm front develops westward resulting in cyclogenesis now occurring in the northerly flow to the west (rear) of the storm within the cold air advancing behind the cold front. The extension of the warm front starts to wrap around in the cold, northerly flow to the rear of the cyclone forming the *bent-back* front. This bent-back front has the structure of a warm front, and not that of an occluded front as in the classifical model. The *T-bone* term is used as the cold front is orientated perpendicular to the bent-back extension of the warm front (Fig. 10.6 III). Finally, in the warm core frontal seclusion phase (Fig. 10.6 IV), the phase of maximum intensity, the cold front moves further east of the cyclone center (~500 km). The bent-back warm front and the cold polar air encircling the low, partially or totally enclose a pocket of relatively warm air at its center, thus forming a warm-core *seclusion*. This seclusion forms within the polar air, and unlike the Norwegian Cyclone model, does not include air originating from the warm sector.

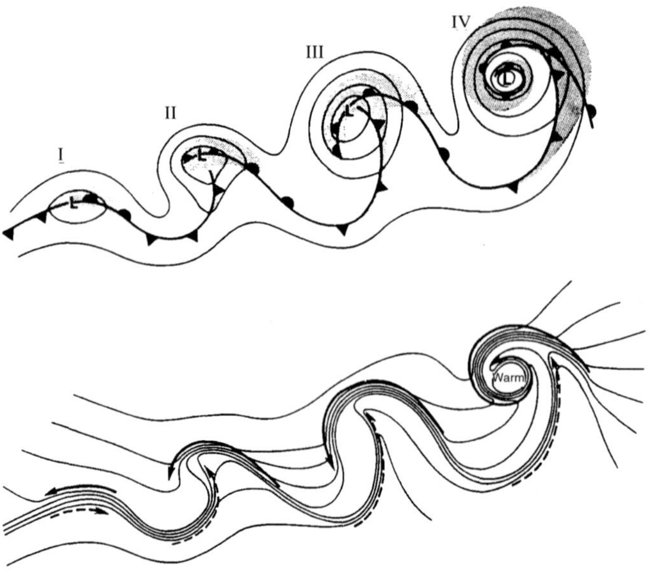


**FIGURE 10.6 The life cycle of the marine extratropical cyclone: (I) incipient frontal cyclone; (II) frontal fracture; (III) bent-back warm front and frontal T-bone; (IV) warm-core frontal seclusion.** Upper: sea-level pressure, solid lines; front, bold lines; and cloud signature, shaded. Lower: temperature, solid lines; cold and warm air currents, solid and dashed arrows, respectively. *(After Shapiro and Keyser (1990))*

While some observational studies over ocean (e.g. Wakimoto et al., 1992; Neiman and Shapiro, 1993; Neiman et al., 1993; Blier and Wakimoto, 1995) and land (Martin, 1998a,b) have demonstrated the features described by the Shapiro-Keyser model, others have found the Norwegian cyclone model more applicable, or that different elements of both models may be applicable to the same storm (e.g. Mass and Schultz, 1993). Several idealized (e.g. Hoskins, 1976; Hoskins and West, 1979; Polavarapu and Peltier, 1990) and case study simulations (e.g. Anthes et al., 1983; Kuo et al., 1991) have also exhibited evolutions and structures that resemble the Shapiro and Keyser model, while other model simulations, in particular, those conducted over land, or those using surface friction values representative of land, have found the resultant structure more representative of the Norwegian cyclone model (Kuo et al., 1992; Schultz and Mass, 1993). Neiman and Shapiro (1993) emphasize that the frontal evolution described by the Shapiro-Keyser model is not characteristic of all maritime cyclogenesis, and that this model should be considered a companion to, rather than a replacement of, the classical Norwegian cyclone model. Mass and Schultz (1993) speculated that the differences they observed between the storm and the Keyser-Shapiro model may be attributed to the differences in surface properties of land and ocean. Martin (1998a,b) also suggested that the evolution of the bent back front may be different over land from that over the ocean, although the structure is more robust above the friction layer, implying that there may be greater similarities between the frontal structure of marine and continental cyclones than was previously thought. It thus appears that both the Norwegian cyclone model and the Shapiro-Keyser model should be considered when examining the development of extratropical cyclones.

Recently CloudSat (Stephens et al., 2002) transects through a cold front, warm front and occluded front of three different maritime extratropical cyclones were used to examine the internal structure of frontal clouds and precipitation (Posselt et al., 2008). These transects are shown in Fig. 10.7. The transect through the cold front case (Fig. 10.7a) shows a classic cold frontal structure. Shallow convection is evident in the cold and relatively unstable air to the north of the front, and deep convection can be seen associated with the narrow region of instability at the front's leading edge. The general cloud distribution is in keeping with the Norwegian cyclone model, however, the CloudSat reflectivity data shows further details of the internal cloud structure not included in the model. For example, multiple precipitating low-level convective showers are evident in the cold air behind (to the north of) the front; the cirrus cloud does not decrease uniformly ahead of the front, but rather shows pockets of high reflectivity, suggesting the presence of multi-cellular convection; and most of the shallow convective clouds in the weakly stratified cold air behind the front appear to be producing precipitation at the surface.

The cross-section through the warm front case (Fig. 10.7b) demonstrates a broad area of cloud and precipitation along and ahead of the warm front itself. The classic sloping distribution of clouds associated with the warm front is evident, and the general cloud distribution is similar to that of the Norwegian cyclone model. As with the cold front, the CloudSat data once again reveal greater details of the warm front cloud and precipitation structures, including that instead of sloping upward with height as in the Norwegian cyclone model, the observed cloud tops remain relatively constant. Large variability in the reflectivity, and hence cloud water content, may be observed. Also, the enhanced region of reflectivity near the base of the cloud with increasing distance from the warm front indicates that large ice particles are settling into the lower regions of the cloud. Finally the extensive region of cloud to the south of the surface warm front produces less precipitation than the clouds directly on top of the warm front, despite similar reflectivity values. This is part of the warm conveyor belt airstream (discussed below), which was not included in the Norwegian cyclone model.

While the cold and warm front transects show characteristics similar to those described in the Norwegian cyclone model, the cross section through the occlusion (Fig. 10.7c) demonstrates a cloud distribution that is significantly different from the classical model, and more in keeping with the Shapiro-Keyser

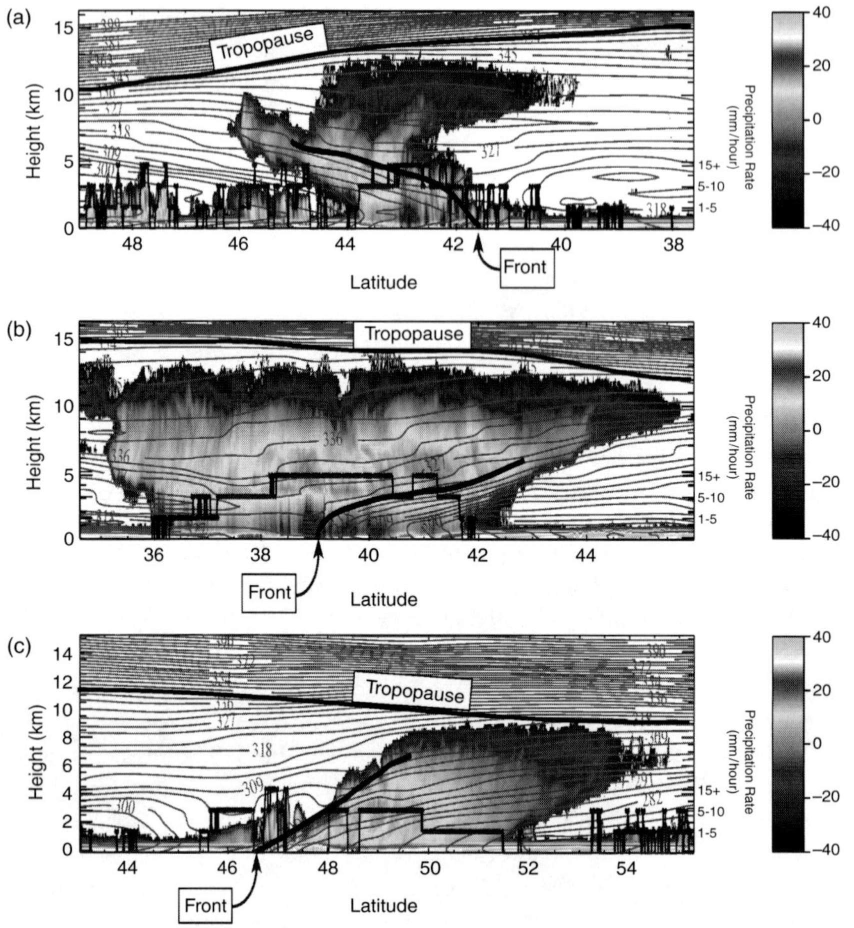


**FIGURE 10.7 CloudSat observed radar reflectivity (dBZ, color shading) overlaid with ECMWF-analyzed equivalent potential temperature (K, solid red lines) for transects through a (a) cold front (0448 UTC 22 November 2006), (b) warm front (1730 UTC 22 November 2006), and (c) the occluded sector (0553 UTC 5 December 2006) of three different mid-latitude cyclones.** Note that the ground clutter tends to obscure the radar signal from the cloud below 1 km above the surface. The positions of the front and tropopause are marked with a heavy black line. CloudSat-estimated precipitation rates are depicted at the base of the plot. (For interpretation of the references to color in this figure legend, the reader is referred to the web version of this article.) *(Adapted after Posselt et al. (2008))*

model. The cloud and precipitation fields are similar to those of the warm front, although the precipitation rates are lower. These observations demonstrate, as stated above, that both the classic and the Shapiro-Keyser models should be considered when examining the structure and life cycles of extratropical cyclones.

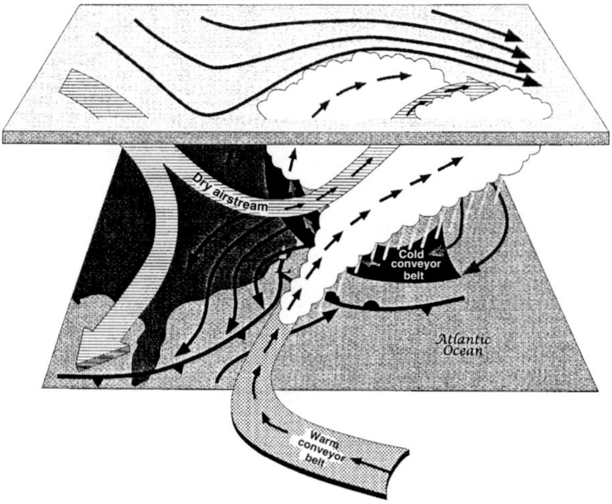


**FIGURE 10.8 The conveyor-belt model of airflow through a northeast US snowstorm.** *(Adapted from Kocin and Uccellini 1990, Fig. 26, based on the Carlson (1980, Figs 9 and 10) conceptual model [after Schultz (2001)])*

## 10.3.3. Conveyor Belt Concepts

Browning and his collaborators observed similarities in the mesoscale cloud and precipitation features of extratropical storms over the United Kingdom and began to relate these features to commonly observed airstreams or *conveyor belts* in these storms (Browning and Harrold, 1969; Browning, 1971; Browning et al., 1973; Harrold, 1973). They focused specifically on the warm, ascending airstream that flowed over the warm front, and a drier mid-tropospheric flow that capped the warm, ascending air, thereby generating potential instability as the two airstreams rose over the warm front. They also identified a dry, cool airstream that descended from the downstream anticyclone and flowed under the warm front. Palmén and Newton (1969) had also depicted three airstreams in their study of extratropical cyclones. Schultz (2001) offers an excellent review of the development of these early airstream models. Carlson (1980) extended these airstream ideas to develop a conveyor belt model for extratropical cyclones. His model, shown in Fig. 10.8, consists of three streams of air: (a) the warm conveyor belt, (b) the cold conveyor belt and (c) the dry airstream.

### *10.3.3.1. The Warm Conveyor Belt*

The warm conveyor belt is a stream of relatively warm moist air that originates in the low levels of the southeast quadrant of the storm and flows northward and westward toward the center of the cyclone (in the Northern Hemisphere). As isentropic surfaces slope upward to the north, this air rises as it flows through the warm sector and above the surface warm front. Interpreted in terms of the quasi-geostrophic omega equation [Eq. (10.5)] on constant-pressure surfaces this air, which is ahead of an advancing upper-level trough, is usually associated with positive vorticity advection and warm air advection.

If the air in the warm conveyor belt is convectively unstable, the lifting will destabilize the air and lead to convective clouds and precipitation systems, including rain-bands, squall lines, and severe thunderstorms. These systems are often most intense along and ahead of the cold front where ageostrophic vertical circulation and frictional convergence at the surface cold front produces maximum updraft speeds. If the air in the warm conveyor belt is absolutely stable, then lifting will produce extensive layers of stratiform clouds, including nimbostratus. Often parts of the warm conveyor belt will be convectively unstable, such as in the low levels in the warm sector, while other parts will be absolutely stable, such as in the upper levels of the cool sector north of the surface warm front. Thus, convective clouds and precipitation systems are most likely in the warm sector, while stratiform clouds and precipitation are predominant in the cool sector. Diabatic heating is important in modifying the air in the warm conveyor belt. In the low levels, surface heating destabilizes the air when sufficient solar radiation is available, and fair-weather cumulus clouds may populate this region during the daytime. When either convective or nonconvective precipitation occurs in the warm conveyor belt, latent heat release reduces the adiabatic cooling of the rising air and represents a significant energy source to the cyclone.

The air in the warm conveyor belt eventually reaches the upper troposphere and, according to Carlson (1980), turns anticyclonically toward the northeast, representing the southwesterly flow ahead of the upper-level trough on a constant-pressure surface. More recently it has been argued that, during the early stages of cyclogenesis, if the upper-level flow is an open wave, then the warm conveyor belt turns anticyclonically at jet level, but that later if the upper-level flow is characterized by closed flow, then some of the airflow turns cyclonically around the low center (Bader et al., 1995; Browning and Roberts, 1996; Wernli, 1997). This flow is the *trowal airstream* discussed in further detail below. Mass and Schultz (1993) also observed that rising trajectories within the warm sector (corresponding roughly to the warm conveyor belt) appear to fan out as they ascend, with some turning cyclonically to the west and others anticyclonically to the east. Such a cyclonic flow regime allows for the contribution of the warm conveyor belt to the comma-shaped cloud mass. By the time the air reaches the upper troposphere, it is usually absolutely stable, so the clouds are extensive sheets of altostratus and cirrostratus. Downstream of the cyclone, the leading edge of the warm conveyor belt is evident by thin cirrus or cirrostratus clouds (Fig. 10.5).

The left (west) edge of the warm conveyor belt originates farthest south and most closely approaches the cold front. Because of its origin, this air is the warmest, moistest, and most unstable air in the cyclone system. As it approaches the cold front, it also experiences the greatest rate of lifting. Due to the strong baroclinicity in the frontal region, strong wind shear is present as well. The combination of the moist convectively unstable air, rapid lifting, and strong wind shear makes this region favorable for the severe mesoscale convective phenomena discussed in Chapters 8 and 9. Recently, Eckhardt et al. (2004) presented the first climatology of warm conveyor belts. They found from their analysis that the mean specific humidity at the starting points of warm conveyor belts in various regions varies from 7 to 12 g $kg^{-1}$, and that most of this available moisture is precipitated out, leading to an increase of potential temperature of between 15 and 22 K along warm conveyor belt trajectories. Over a time period of three days, a warm conveyor belt trajectory produces on average about 4 (6) times as much precipitation as a global (extratropical) average trajectory starting from 500m above the ground does.

#### *10.3.3.2. The Dry Conveyor Belt*

Due to the important role of the dry conveyor belt in convective destabilization and cyclogenesis it has received significant attention (e.g. Carr and Millard, 1985; Young et al., 1987; Browning and Golding, 1995; Browning, 1997). The dry conveyor belt originates in the upper troposphere or lower stratosphere west of the upper-level trough and descends into the middle and lower troposphere. As the air descends, it warms and dries. The northern portion of the dry air stream separates from the descending anticyclonic flow, crosses the trough axis, and flows northeastward parallel to the left edge of the warm conveyor belt, where it then ascends over the warm/occluded front (Fig. 10.8). Although this air stream ascends, it is dry and hence normally cloud free and is sometimes referred to as the *dry tongue* or *dry slot*. The boundary between the dry and warm conveyor belts can be extremely sharp, resulting in a surface feature frequently observed over the High Plains of the United States called a *dryline*. The satellite image in Fig. 10.3 shows a dry tongue and the sharp boundary between the dry and warm conveyor belts. The southern portion of the dry conveyor belt continues to descend, warm and dry as it approaches the surface. Strong surface winds associated with the lower portions of this sinking air stream can produce significant blowing dust in the arid regions of the southwestern United States, south of the cyclone.

#### *10.3.3.3. The Cold Conveyor Belt*

The stream of air called the cold conveyor belt originates to the northeast of the cyclone. This air flows westward (relative to the cyclone) beneath the warm conveyor belt and north of the warm front. Thus the warm front separates the cold and warm conveyor belts. Over flat terrain, the most pronounced lifting of the cold conveyor belt occurs just north of the cyclone center and under the southwesterly jet aloft. According to Carlson (1980), this usually stable air rises, turns anticyclonically, and flows parallel to the upper regions of the warm conveyor belt, although studies prior to 1980 showed the cold conveyor belt turning cyclonically around the low center, and then remaining in the lower troposphere behind the cold front. A recent study by Schultz (2001) has shown that the cold conveyor belt has both cyclonic and anticyclonic paths, with the anticyclonic path representing a transition between the warm conveyor belt and the cyclonic path of the cold conveyor belt.

As any precipitation produced within the warm conveyor belt falls through the cold conveyor belt, the temperature and the humidity of the cold conveyor belt can significantly influence the type and amount of precipitation reaching the surface. For example, if the cold conveyor belt is relatively dry, evaporation and/or sublimation of hydrometeors will reduce the amount of precipitation reaching the surface, whereas a nearly saturated cold conveyor belt could result in substantial surface precipitation. Also, if the temperatures of the cold conveyor belt are below freezing, ice pellets or freezing rain may be expected at the surface. Saturation is enhanced by evaporation of precipitation falling from the warm conveyor belt aloft, and hence thick fog and stratus clouds may occur in the cold conveyor belt, north of the surface warm front. Along the eastern slopes of the Rocky Mountains in the United States, the cold conveyor belt experiences significant lifting as it follows the rising terrain, and these "upslope" conditions are responsible for almost all of the nonconvective precipitation in this region (see Chapter 11).

Carlson (1980) concluded that the air in the comma-shaped cloud heads of extratropical cyclones originates in the anticyclonic path of the cold conveyor belt, and Liu (1997) showed that as much as 70% of the moisture transported within extratropical cyclones could be linked to the cold conveyor belt. Others, however, have emphasized the role of the warm conveyor belt in generating heavy precipitation (e.g. Martin, 1999; Schultz, 2001). Schultz (2001) argues that while shallow clouds could be produced within the relatively stable air of both the anticyclonic or cyclonic branches of the cold conveyor belt, heavy precipitation associated with strong vertical uplift and deep clouds is not likely to be produced within the stable environment of the cold conveyor belt. The latter is more likely to be generated at mid-levels by the cyclonic path of the warm conveyor belt. The cool cyclonically flowing low-level air to the rear of the extratropical cyclone is however frequently moist enough that the frictionally-induced rising motion may produce an extensive layer of stratocumulus or fair-weather cumulus clouds at the top of the PBL.

The conveyor-belt model has been criticized in the past for a number of reasons, a detailed overview of which is provided by Schultz (2001). Some studies have found no evidence of a well-defined cold conveyor belt (e.g. Reed et al., 1994; Browning et al., 1995), sharp boundaries between the warm and cold conveyor belts (e.g. Kuo et al., 1992; Browning and Roberts, 1994), or any anticyclonic turning of the cold conveyor belt (Mass and Schultz, 1993). Concerns have been expressed in the literature as to whether the conveyor belts represent streamlines, trajectories, streak lines or some combination of these.

It has been suggested that the airstreams cannot be represented as steady-state, flat belt-like structures or three-dimensional tubes with well-defined boundaries due to their complex geometries, and should rather be thought of as flexible tubes that evolve over time (Kuo et al., 1992; Reed et al., 1994; Wernli, 1997). It has also been argued that the three-belt model is an oversimplification of the airflow, with more than three air streams being needed to show the various parcel trajectories (Mass and Schultz, 1993; Reed et al., 1994; Bader et al., 1995). In spite of these criticisms, the conveyor-belt model does provide some useful guidance as to the basic airflow through extratropical cyclones and continues to be utilized.

## 10.3.4. Frontal Occlusion and the Trowal Airstream

The concept of *occluded fronts*, in which a cold front overtakes a warm front thus lifting the warm sector air above the frontal intersection, was first introduced within the Norwegian cyclone model. In an ideal warm (cold) occlusion, the air preceding the warm front is colder (warmer) than the air behind the cold front so that the cold (warm) front rides aloft over the warm (cold) front. More recently, Schultz and Mass (1993) defined the occlusion process as one in which the low center becomes progressively separated from the warm sector of the cyclone resulting in a pressure drop, and an associated tongue of intermediate temperature air that extends from the low center to the warm sector. Debate continues to exist in the literature regarding the occlusion process. Some have suggested that the important feature of an occlusion is the trough of warm air that is lifted aloft ahead of the cold front rather than the surface occluded front (e.g. Crocker et al., 1947; Godson, 1951; Penner, 1955; Galloway, 1958, 1960). Others have shown that an occluded structure can occur without the classical occlusion process occurring (e.g. Reed, 1979; Locatelli et al., 1989), or that there is little evidence that supports the occlusion process as described in the Norwegian cyclone model (e.g. Shapiro and Keyser, 1990). Still others have found that while the surface-based cold front does catch up to the warm front it does not ride over the warm front, but that the upper-level frontal zone appears to ascend over the surface-based warm front (Schultz and Mass, 1993). Posselt and Martin (2004) described a case in which the occlusion first occurred at mid-tropospheric levels and then at the surface. Finally, Schultz and Mass (1993) suggest that warm-type occlusions, rather than the cold-type occlusions, are the normal structural form.

As just stated, one of the main results of the warm occlusion is the production of a wedge of warm air aloft, which is displaced poleward of the surface warm and occluded fronts, as is shown schematically in Fig. 10.9a. This trough of warm air, referred to as the *trowal* (trough of warm air aloft), was observed to correspond more closely to the cloud and precipitation characteristics of the occluded cyclone than to the surface occluded front (Crocker et al., 1947; Godson, 1951; Penner, 1955; Galloway, 1958, 1960).

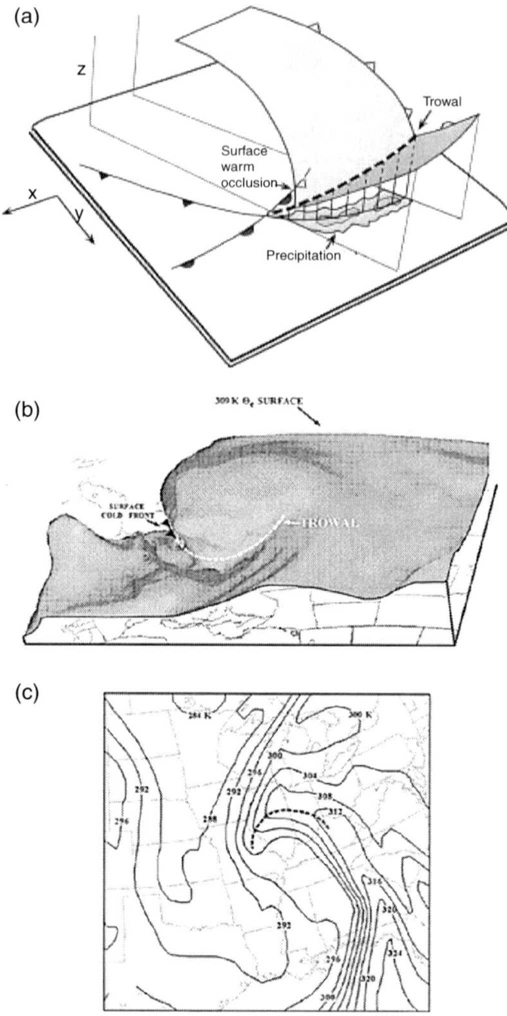


**FIGURE 10.9** **(a) Schematic illustration of the trowal conceptual model.** The lightly shaded surface represents the warm edge of the cold frontal baroclinic zone. The darker shaded surface represents the warm edge of the warm frontal baroclinic zone. The thick dashed line (marked "TROWAL") represents the three-dimensional sloping intersection between the cold and warm frontal zones characteristic of warm occlusions. Schematic precipitation band is indicated as are the positions of the surface warm, cold and occluded fronts (after Martin, 1999). **(b) Elevated view from the north of the 309 K $\theta_e$ surface from an 18-h forecast of the UW-NMS valid at 1800 UTC 19 January 1995.** The dashed line indicates the trowal position, which slopes upward from near the surface to 8 km. Position of the surface cold front is also indicated. **(c) Equivalent potential temperature at 2 km from an 18-h forecast of the UW-NMS valid at 1800 UTC 19 January 1995.** Solid lines are $\theta_e$ labeled in K and contoured every 4 K. Thick dashed line is the axis of maximum $\theta_e$ characteristic of a warm occluded cyclone and used to approximately locate the trowal. *(Adapted after Martin (1999))*

While the trowal was initially indicated on a surface map showing the intersection of the cold and warm front aloft (Crocker et al., 1947; Godson, 1951; Penner, 1955), the trowal concept has evolved over time to represent the warm anomaly aloft (Martin, 1998a,b, 1999; Market and Cissell, 2002; Posselt and Martin, 2004; Moore et al., 2005). The trowal may be identified either as a ridge of high (equivalent) potential temperature on a horizontal cross-section, or as a three-dimensional sloping canyon on an isosurface of (equivalent) potential temperature, as shown in Fig. 10.9 (b, c). Martin (1998a) defined the trowal as marking the "3-D sloping intersection of the upper cold-frontal portion of the warm occlusion with the warm-frontal zone". The upward-sloping tongue of warm air wraps cyclonically around the cyclone.

In satellite imagery of maturing continental cyclones it is evident that the northwestern boundary of the dry slot typically coincides with the southeastward boundary of the cloud pattern produced in association with the trowal airstream (Grim et al., 2007). A number of numerical simulations of occluded cyclones (Schultz and Mass, 1993; Reed et al., 1994; Martin, 1998a,b) have exhibited the thermal characteristics of the occluded structure just described. They have also shown a cloud- and precipitation-producing airstream that originates in the warm sector boundary layer and that ascends cyclonically in the occluded quadrant of the cyclone, which Martin (1998b) called the *trowal airstream.* Martin (1999) found that convergence of the along-isentrope component of the Q-vector simultaneously creates the thermal ridge and provides most of the Q-G forcing for upward motion within the occluded quadrant and of the trowal airstream. The trowal airstream is associated with the wrap-around cloud and precipitation that is commonly observed with occluded extratropical cyclones, and has been linked to heavy snowfall produced by intense snowbands, as will be seen below.

## 10.3.5. Cloud and Precipitation Characteristics

### *10.3.5.1. CSI and Precipitation Bands*

As will be seen in the next section, clouds and precipitation associated with extratropical cyclones tend to be organized into bands. The roles of frontogenesis and CSI in producing mesoscale precipitation bands, and the fact that these processes may operate together to generate such bands have received significant attention in the literature (e.g. Bennetts and Hoskins, 1979; Emanuel, 1979, 1983a,b, 1985; Sanders and Bosart, 1985; Sanders, 1986; Xu, 1989; Shields et al., 1991; Nicosia and Grumm, 1999). Emanuel (1979, 1983a,b, 1985) showed that rising motion in association with frontogenesis was contracted and enhanced when CSI or even weak symmetric stability were present, thus suggesting a relationship between frontogenesis and symmetric stability in producing precipitation bands. Xu (1989) demonstrated that frontogenesis occurring in the presence of CSI can, in theory, produce long-lived precipitation bands. Moore and Lambert (1993) developed a two-dimensional form of equivalent potential vorticity (EPV), later extended to a more general, three-dimensional form by McCann (1995), that could be used to diagnose regions of CSI. They found that CSI could be diagnosed in a region of negative EPV, a region which tends to be saturated and characterized by strong vertical wind shear and weak convective stability.

The relationship between frontogenesis, CSI and EPV has been discussed in detail by others such as Bluestein (1993) and Nicosia and Grumm (1999), and is summarized here. As just stated, CSI is present when EPV is negative; CI may also be present in conjunction with CSI when EPV is negative. For a layer characterized by negative EPV, CSI and/or CI can only be released if the layer becomes saturated in association with rising motion. Also, when CSI and CI are both present, it is expected that CI-associated upright convection will dominate over the CSI-associated slantwise convection. It can be shown for frictionless, adiabatic flow that EPV will be reduced in a region where the moisture gradient lies in the same direction as the thermal wind vector. Such a scenario regularly occurs for extratropical cyclones located within a baroclinic zone with the cold air to the north. In this scenario, moist air is found to the east, while the drier air tends to be located to the west. When frontogenesis occurs in a region of negative EPV and CSI, stronger vertical motion develops and becomes constricted to a smaller scale (Emanuel, 1979, 1983a,b, 1985). Stronger vertical motion leads to stronger convergence in the lower levels which enhances the temperature gradient and hence frontogenesis. This in turn increases the temperature gradient, which further reduces the EPV. The continual decrease in EPV results in the development of CSI, and CI to a lesser extent, in regions of frontogenesis. As the rising air reaches saturation, CSI and CI are released, thereby enhancing the vertical motion within the region of frontogenesis and subsequent band development. In this way a positive feedback is established.

As Nicosia and Grumm (1999) pointed out, the fact that frontogenesis occurs in or near a region of CSI is not just a coincidence. In regions of frontogenesis, the potential temperature gradient increases, which requires an increase in the geostrophic wind shear through thermal wind arguments. Such geostrophic wind shear produces differential moisture advection and a subsequent increase in the slope of the $\theta_e$ isentropes. The weakening of the convective stability and the strengthening of the vertical wind shear results in negative EPV and CSI. The reduction in EPV can take place at quite a distance from the surface cyclonic circulation (Moore et al., 2005). It is the unique juxtaposition of the three conveyor belts within an extratropical cyclone that results in a region of negative EPV and associated CSI within a region of midlevel frontogenesis (Nicosia and Grumm, 1999). EPV is significantly reduced on the warm side of the midlevel frontogenesis due to the passage of the dry conveyor belt over the cold conveyor belt. The continual reduction in the EPV produces a deep layer of negative EPV located within a region of warm advection and uplift associated with the warm conveyor belt.

**FIGURE 10.10 Visible satellite photograph of occluding extratropical cyclone over North Pacific.** Note cellular cumulus convection south of cyclone center and small cloud system southeast of center behind cold-frontal band of clouds. This system is an incipient mesoscale cyclone. Time is 1845 GMT, 27 February 1980.

### *10.3.5.2. Rainbands*

The mesoscale organization of clouds and precipitation within extratropical cyclones has been extensively studied using both in situ and remotely-sense observations, as well as numerical simulations (e.g. Browning and Harrold, 1969; Browning, 1971; Browning et al., 1973; Houze et al., 1976; Hobbs and Locatelli, 1978; Herzegh and Hobbs, 1980; Matejka et al., 1980; Houze and Hobbs, 1982; Hobbs and Persson, 1982; Houze, 1993; Braun et al., 1997; Jorgensen et al., 2003). A summary of the early history of the research leading to our knowledge of the mesoscale precipitation structure of extratropical storms is provided by Atkinson (1981). These studies have shown that clouds and precipitation tend to be organized in band-like structures referred to as rainbands (discussed here) and/or snowbands (discussed in the next section). An example of such bands may be seen in Fig. 10.10, which shows an occluding extratropical cyclone over the North Pacific. As the upper-level trough deepens and becomes a closed cyclone, the low- and middle-level clouds are wrapped around the cyclone center. A region of organized, cellular cumulus convection is present in the unstable PBL south of the cyclone center. Bands of convective clouds occur along the cold front and in the convectively unstable warm sector ahead of the front. Rainbands are typically orientated parallel to one of the fronts, contain smaller regions of more intense precipitation, are found throughout extratropical

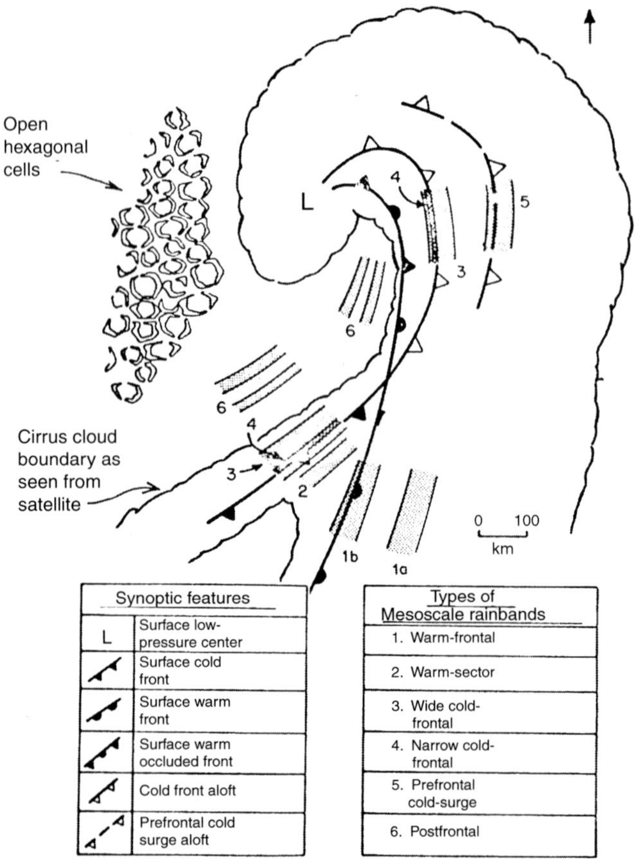


**FIGURE 10.11** **Schematic depiction of types of rainbands observed in extratropical cyclones.** *(From Houze and Hobbs (1982); adapted from Hobbs (1978). Copyright by the American Geophysical Union)*

cyclones, and are classified as wide (50-75 km) or narrow (5-25 km) according to their width.

Six different types of rainbands typically associated with Pacific storms have been identified in a classification scheme developed by Hobbs (1978), Matejka et al. (1980) and Houze and Hobbs (1982). The scheme, which we will refer to that the HHM scheme, was based on the location and the morphology of the bands, the location of which is shown in Fig. 10.11. The six types of rainband include the following:

1. **Warm frontal rainbands:** Warm frontal bands, depicted in cross-sectional form in Fig. 10.12, are oriented parallel to the surface warm front where a deep layer of ascending warm, moist air exists. They are typically of the order of 50 kilometers wide and several hundred kilometers long. Ice particles may form above the warm front, often within shallow convective cells, and may

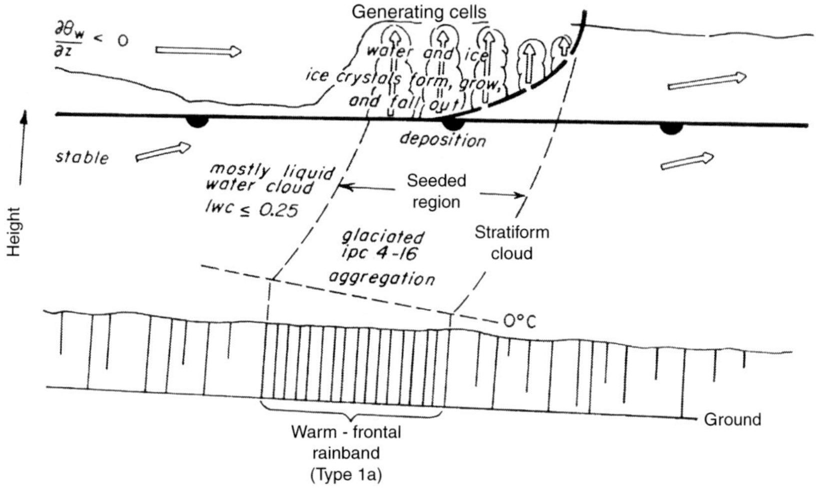


**FIGURE 10.12 Model of a warm-frontal rainband in vertical cross section.** Structure of clouds and predominant mechanisms for precipitation growth are indicated. Vertical hatching below cloud bases represents precipitation: density of the hatching corresponds qualitatively to precipitation rate. Heavy broken line branching out from front is a warm-frontal zone with convective ascent in generating cells. Ice-particle concentrations (ipc) are given in numbers per liter; cloud liquid-water contents (lwc) are in g m$^{-3}$. Motion of the rainband in the figure is from left to right. *(From Houze and Hobbs (1982); adapted from Hobbs (1978) and Matejka et al. (1978). Copyright by the American Geophysical Union)*

grow through seeder-feeder processes (see below) as they fall through the stable air mass below.

2. **Warm sector rainbands:** Warm sector bands (~50 km wide) are located ahead of and parallel to the cold front, and often resemble squall lines. Younger convective bands in this region tend to occur ahead of older, less convective bands, and contain both liquid water and ice compared with their nearly glaciated, older counterparts.
3. **Narrow cold frontal rainbands:** A narrow (~5 km) cold frontal rainband (NCFR) is typically observed at the leading edge of the surface cold front and is probably caused by intense frictionally induced convergence at the front (Keyser and Anthes, 1982). NCFRs are characterized by vigorous, narrow convective updrafts and narrow downdrafts, are often associated with heavy convective rainfall ( 10-50 mm hr$^{-1}$) and may produce severe weather. In spite of the strength of the updrafts associated with the NCFRs (Carbone (1982) observed updrafts up to 18 m s$^{-1}$), which tend to be nearly vertical, the rainbands typically only extend to heights of between 3.5 and 5 km. They appear to behave like density currents in which the vertical circulation is characterized by an approximate balance between the baroclinicly generated vorticity of the cold front and the ambient vertical wind shear (Rotunno et al., 1988; Parsons, 1992; Jorgensen et al., 2003). Sometimes the NCFR

occurs near the leading edge of a larger area of stratiform rainfall induced by post-frontal, slantwise convection, although it appears that they are more typically embedded within the large stratiform cloud region; occasionally they may be neither (Koch and Kocin, 1991). It should be noted here that a front having the local character of a gravity current is called a front with a gravity current-like structure (Smith and Reeder, 1988). We will use the term density current to refer to the density difference across a cold front, and the term gravity current-like structure to refer to a cold front that locally has the characteristics of a gravity current. Gravity currents can propagate ahead of cold fronts, where they may be associated with warm sector rainbands.

4. **Wide cold frontal rainbands:** Wide (~50-75 km) cold frontal rainbands (WCFRs, 50 km in width) are oriented parallel to the cold front and either straddle the front or occur behind the front. In occlusions, they can be associated with the cold front aloft. WCFRs tend to produce regions of enhanced stratiform precipitation. The upward motion of these bands is characterized by mean ascent aloft, although they can also include shallow convective cells that contribute to precipitation through seeder-feeder processes (see below). The WCFR generally forms behind the NCFR but tends to move faster than the NCFR and thus may be co-located with the NCFR or positioned ahead of the low-level cold front. WCFRs may also occur without NCFRs (Evans et al., 2005; Bond et al., 2005). Parsons and Hobbs (1983) describe several cases in which a WCFR moved across a surface cold front. This usually led to a decrease in the frontal convergence and a subsequent reduction in the organization of the NCFR, even occasionally the dissipation of the band. Finally, sub-bands have also been observed within WCFRs (Evans et al., 2005).
5. **Prefrontal cold-surge bands:** Prefrontal cold-surge bands are associated with surges of cold air above the warm front and ahead of the cold front in occluded cyclones.
6. **Postfrontal rainbands:** Postfrontal bands are lines of convective clouds that form in the unstable cold air behind the cold front.

The relatively intense precipitation associated with wide rainbands appears to be generated by a seeder-feeder process, in which the seeding of supercooled water by ice crystals falling from above enhances precipitation (Fig. 10.12). The region where the ice crystals are produced is called the seeder zone. The remaining precipitation is produced in the zone below, termed the feeder zone because of the strong advective supply of moisture by the cyclone-scale circulation. The seeder-feeder process may also enhance the upward motion and precipitation of the other types of bands depicted in Fig. 10.11. The seeder-feeder process has been discussed by Browning and Harrold (1969), Herzegh and Hobbs (1980), and Houze et al. (1981) and modeled by Rutledge and Hobbs (1983), and is described more fully in Chapter 11 with respect to its role in orographic precipitation.

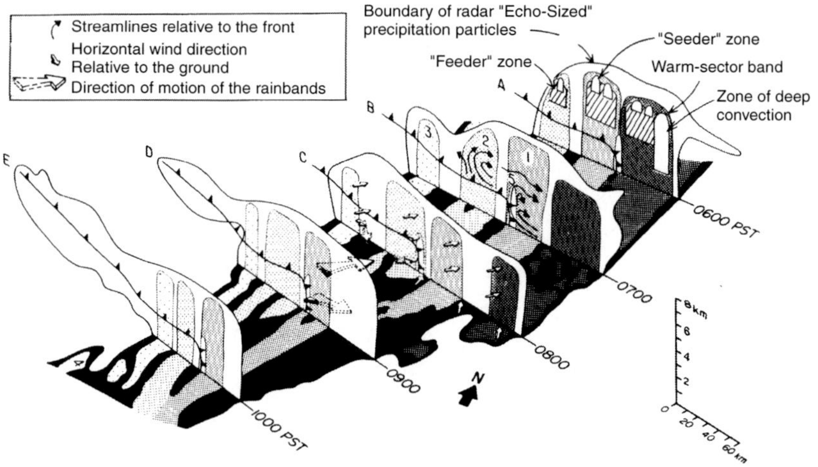


**FIGURE 10.13 Schematic summary of structure and organization of clouds and precipitation and airflow associated with a typical cold front.** Outline of the radar-echo pattern in a vertical plane normal to the front is shown for five different times. Various rainbands (1-5 and the warm-frontal band) are indicated by different shadings. Various features of the structure are highlighted in each vertical section. **(A) Locations of "feeder-seeder" zones and regions of deeper convection; (B) streamlines of airflow relative to the front; (C) horizontal winds; and (D) motions of wide and narrow cold-frontal rainbands.** *(From Hobbs et al. (1980))*

A schematic model developed by Hobbs et al. (1980) depicts the band structure of a typical cold front (Fig. 10.13). In this model, the heaviest precipitation is located in mesoscale rainbands oriented parallel to the cold front. These include a warm-sector rainband consisting of a series of mesoscale convective sub-bands, a NCFR, and four WCFRs. Also illustrated are generating zones of ice crystals and the feeder zones which supply moist, cloudy air from which seeder crystals grow by vapor deposition, aggregation, and riming of cloud droplets. Hobbs et al. (1980) estimated that about 20% of the precipitation in the wide cold-frontal rainbands originates in the seeder zones and ~80% originates in the feeder zones.

As reported by Ryan (1996), there are many studies confirming the HHM rainband classification scheme for numerous parts of the world including the United States, the United Kingdom, Russia, Finland, China, Israel and Australia (see Ryan, 1996 for related references), and rainband structures have been observed in Spain, Morocco, Japan and South Africa. The survey conducted by Ryan, although limited, does indicate that the scheme appears valid for a wide range of environmental conditions. However, while the scheme is relatively well established, the mechanisms responsible for the variety of bands in extratropical cyclones are still under debate. This is partly due to the limitations of theoretical models, the variation in the mechanisms from storm to storm and even within the same storm, the range of scales (synoptic, mesoscale, convective, microscale) involved that make observing and numerically simulating all the processes involved difficult, and the fact that a combination of mechanisms may be operating at once.

A number of studies have focused on the role of different instability mechanisms in generating rainbands, particularly within the warm sector. For example, in more severe cyclonic storms, the warm sector can be conditionally unstable, leading to the formation of prefrontal squall lines, where SI (Ogura et al., 1982) may be responsible for triggering the linear structure of the squall lines. Bennetts and Ryder (1984) compared the predictions of CSI and symmetric wave-CISK (Emanuel, 1982) to the observed banded structure in the warm sector of an extratropical cyclone. The primary energy source for SI is the kinetic energy of the basic flow, while CSI receives additional energy from latent heat release. Symmetric wave-CISK in contrast, derives its energy principally from latent heat release, while additional energy is supplied by the mean flow (e.g. vertical wind shear). Both theories predicted banded rolls, but the CSI theory predicted that the rolls move with the mean wind while the wave-CISK theory predicted that the rolls exhibit a propagation velocity relative to the mean flow of 6 m $s^{-1}$. The observations suggest that CSI theory was more applicable to this case than the wave-CISK theory. As noted in Chapter 8, however, a number of variations in the parameterizations used in wave-CISK theory can account for the faster propagation of Emanuel's model. Raymond's (1983) advective wave-CISK model, for example, predicts a propagation velocity comparable to the mean wind speed. Thus, in the warm sector of many extratropical cyclones, both CSI and wave-CISK may occur, the differences being so slight that variations in the details of the parameterizations used in the models are greater than the fundamental differences between the two theories.

In another extratropical cyclone, Parsons and Hobbs (1983) found that CSI appeared to be the stronger candidate for explaining the observed warm-sector rainbands, and also noted that the CSI theory was consistent with the observed structure of the wide cold-frontal rainbands in the same case. Lemaitre et al. (1989) noticed the role of CSI, as well as other mechanisms, in their analysis of a WCFR over southwestern France. They found that the band was associated with synoptic-scale ascent of the WCB, slantwise convective along the tilted frontal zone and upright convection in the first 10 km at the leading edge of the system. Frontogenetic forcing and CSI appeared to be important to band organization on the mesoscale, while the gravity current-like behavior of the low level cold air produced new cells ahead of the system. Schultz and Knox (2007) examined the development of several mesoscale rainbands that formed over Montana and the Dakotas during July 2005. The bands were located to the north or poleward of a region of frontogenesis. The formation and organization of the rainbands in this case appears to have been dependent on the release of dry symmetric instability with possible contributions from inertial instability within a zone of frontogenesis. However, once the bands began to grow and develop they were able to tap conditionally unstable air.

In Carbone's (1982) Doppler radar study of the kinematic and thermodynamic structure of a NCFR over California, he found that SI, as well as vertical shear, which occurred in a direction nearly parallel to the band, were likely important factors in the development of this band. This band produced a variety of severe mesoscale weather conditions, including heavy precipitation, strong winds, electrical activity, and tornadoes. The Doppler radar observations showed a nearly two-dimensional updraft of magnitude 15-20 m $s^{-1}$ that was associated with a gravity current that propagated ahead of the cold front toward a strong prefrontal low-level jet. Diabatic cooling associated with the melting of ice was important in maintaining the density contrast across the gravity current. Hobbs and Persson (1982) and Parsons et al. (1987) also proposed that the dynamics of NCFRs are similar to that of a gravity current and that diabatic heating associated with the microphysics of the rainband can influence the density across the gravity current. Condensation in the air rising immediately ahead of the line convection warms the air through the release of latent heat, thereby offsetting the adiabatic cooling associated with lifting. This heating, together with the cooling due to melting and evaporation of precipitation behind the line convection, work in unison to enhance the density contrast across the gravity current. Further warming ahead of the gravity current can occur due to warm air advection by the pre-frontal, low-level jet.

Moncrieff (1989) showed that NCFR dynamics can be represented by steady-state models, and that once a gravity current balance has been generated by precipitation processes, a squall line can be maintained in the presence of negligible potential instability. Shapiro et al. (1985) suggested that non-precipitating fronts may also develop gravity-current-like features, however, the observed circulations were not typical of those that occur in association with classical atmospheric gravity currents. A similar concern was raised by Smith and Reeder (1988) regarding the precipitating studies of Browning and Harrold (1970), Testud et al. (1980) and Carbone (1982). In the idealized study of Knight and Hobbs (1988) a NCFR was generated by frictional convergence in the PBL that was free of precipitation, and that showed no similarities to a gravity current.

Others have also examined the diabatic effects of condensation and evaporation on rainband formation. Hsie et al. (1984) used an explicit model for clouds and precipitation to study the diabatic effects of condensation and evaporation on an idealized model of a cold front. Without moisture, frontogenesis occurred in the two-dimensional model as a result of geostrophic shearing deformation. Upward motion occurred just ahead of the surface cold front (SCF) (Fig. 10.14) as a result of deformational and frictional processes. A northerly jet developed at the top of the PBL behind the SCF, while a low-level southerly jet developed in the warm sector ahead of the SCF.

The moist simulation contained some distinctive features not present in the dry simulation. The vertical velocity showed a banded structure in the warm sector (Fig. 10.15). The first band was associated with the frictional convergence

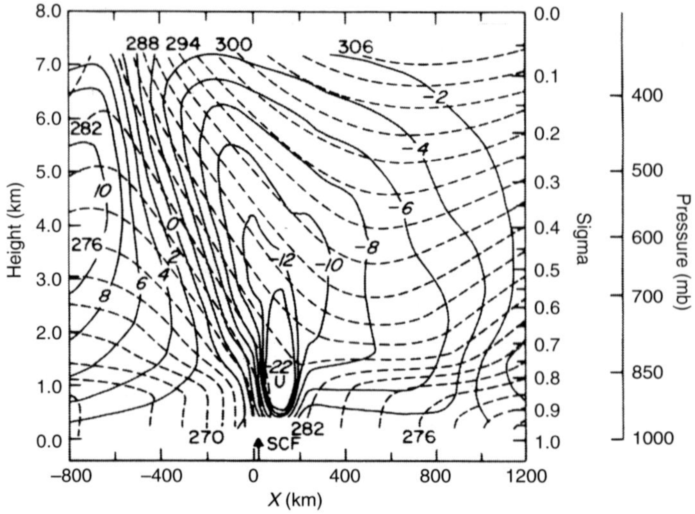


**FIGURE 10.14 Cross section in $x$ and $z$ plane through cold front in dry simulation.** Dashed lines are isentropes (contour interval 2 K). Solid lines are contours of vertical velocity $\omega$ in millibars per hour. *(From Hsie et al. (1984))*

around the pressure minimum at the SCF, and was stronger than the updraft in the dry case (Fig. 10.14). The second and third bands were associated with moist convection; these bands formed in a region of CSI. The horizontal wavelength of 200-300 km of the convective bands is similar to the shorter wavelength mode predicted by Emanuel (1982) for CSI. Initially, the formation of the cloud was from the large-scale upgliding component of the frontal circulation. Later, this large-scale motion broke down into the banded structure. The narrow band that formed in the low levels close to the surface cold front (band 1 in Fig. 10.15) is similar to the NCFR and moved with the same speed as the SCF (Hobbs et al., 1980). The wider band straddling the cold front at higher levels (centered at $x = 180$ km and $z = 4.5$ km in Fig. 10.15) is similar to the WCFR. Two other bands resembled observed warm sector bands. Both the warm-sector and WCFRs moved at a speed faster than the SCF. The behavior of the bands in the moist simulation agrees qualitatively with the observations of Hobbs et al. (1980).

In addition to producing the rainbands and a stronger low-level jet, the effects of latent heat on the cold frontal circulations are summarized as follows:

1. Latent heating produces a stronger horizontal potential temperature gradient across the front, especially in the middle and upper levels. The gradient in the low levels is relatively unaffected.
2. Convection results in a stronger static stability across the front and a weaker static stability in the convective region.

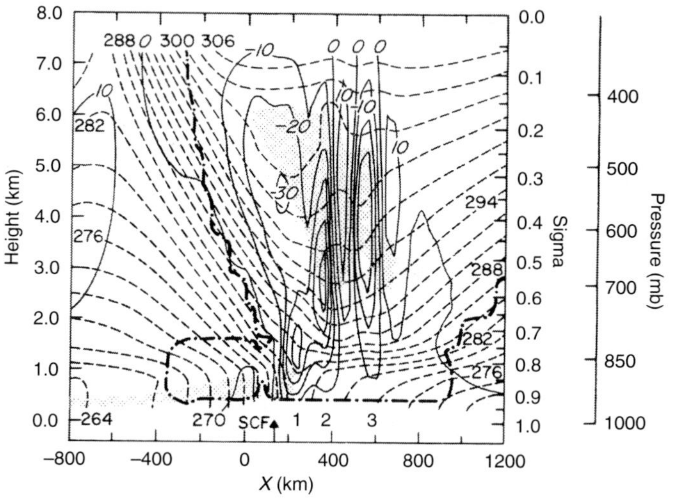


**FIGURE 10.15 As in Fig. 10.14, but for moist simulation.** Cloud boundary is indicated by a thick dash-dot line; position of surface cold front (SCF) is indicated by vertical arrow at about 160 km. Shaded areas denote regions of negative equivalent potential vorticity. *(From Hsie et al. (1984))*

3. Both the upper- and lower-level jets are stronger when latent heating is present. The horizontal wind shear (or relative vorticity) is stronger in the moist simulation, especially in the low levels.
4. Convection intensifies the ageostrophic circulation around the frontal zone. The circulations in the warm sector appear to be dominated by convection.
5. Moist convection increases the speed of the SCF and reduces the slope of the front.

Rutledge (1989) found from his 3-D cloud simulation of a NCFR that diabatic processes were important in the maintenance of the cross-frontal temperature gradient, and that a positive feedback exists between cloud microphysical processes of the rainband and the maintenance of the density current. The diabatic processes of evaporation, melting and condensation associated with the rainband enhance the density current as described above. The enhanced density current, in turn, drives a stronger vertical circulation, thereby maintaining the rainband. Rutledge suggested that cooling associated with the melting of ice may also contribute to the preservation of the density current, as the stable layer generated by melting could limit the dissipative effects of shear-instabilities and wave-breaking behind the head of the gravity current.

Barth and Parsons (1996) investigated the microphysical processes within a NCFR through the use of a 2-D nonhydrostatic cloud model with ice microphysics. Their results showed that intense but shallow updrafts along the NCFR produced significant amounts of cloud water which led to rain and graupel, and, to a lesser extent, snow, within the region of heavy precipitation of the NCFR. The ice phase was needed in order to represent the stratiform precipitation produced by this band. Like Rutledge (1989), they also found positive feedbacks between the microphysics and the dynamics of the NCFR, with melting and sublimation enhancing the cooling within the cold air mass, thus intensifying the circulations that support rainband development. However, their simulations indicated that the prime role of these ice phase processes was to enhance cooling and hence baroclinicity across the front, as opposed to decreasing the mixing across the frontal air mass through the melting-induced stabilization.

A NCFR associated with a cold front with a gravity current-like structure was examined by Koch and Kocin (1991) and Chen et al. (1997) using observational and numerical model output, respectively. This case was characterized by frontal merging and strong frontal contraction that generated severe line convection, and that had a gravity current-like structure along the leading edge of the surface cold front. The NCFR developed in an environment devoid of potential instability, thus requiring significant forced lifting. Koch and Kocin suggested that, while frontal merger processes and hydraulic jumps generated by the flow over the Appalachian mountains were potential mechanisms for forcing this uplift, they had insufficient data to address this. Their analysis did however suggest that the gravity current-like structure of the cold front that developed following the formation of the NCFR may have arisen from the evaporation and melting of hydrometeors. Strong lifting of the air may thus have been possible due to a near balance between the solenoidal circulation associated with the negative vorticity of the gravity current and the circulation associated with the positive vorticity of the vertical shear of the low-level jet (Rotunno et al., 1988).

Chen et al. (1997) simulated this case in order to determine whether the gravity-current-like structure was generated by precipitation melting and evaporation. Their simulations revealed that neglecting the moist processes in the model did not have a significant effect on the gravity-current-like structure of the front, and that planetary boundary layer frictional processes appear to play a much greater role. These frictional processes generate cross-frontal low-level wind shear, which produces low-level convergence and steepens the isentropes along the leading edge of the front. Thus the isolines of $\theta_e$ may become steeper than the absolute momentum surfaces, resulting in negative PV in the planetary boundary layer. A negative PV anomaly would inhibit the geostrophic adjustment process, thereby maintaining a front with gravity-current-like structure. Therefore, the observed, intense NCFR was primarily related to frictionally-induced PBL processes.

Others have investigated the role that PBL processes can play in the formation of rainbands. Keyser and Anthes (1982) found that PBL processes can be important in altering ageostrophic circulations with fronts, thus producing a narrow intense updraft at the warm edge of the front through frictionally induced convergence. Reeder (1986) noted that differential heating in the PBL resulted in the development of an intense prefrontal updraft, and that a gravity-current-like structure developed at the leading edge of the front. Knight and Hobbs (1988) observed from their modeling study, in which precipitation processes were included, that the formation of NCFRs was aided by frictional convergence in the PBL. PBL processes were also found to be important in the generation of NCFRs by Benard et al. (1992) and Koch et al. (1995).

Four types of rainbands, two narrow and two wide, were identified by Benard et al. (1992) in their 2-D nonhydrostatic simulation of moist frontogenesis in an idealized baroclinic wave. A NCFR, positioned at the surface cold front, produced heavy precipitation and consisted predominantly of a line of shallow convection that was triggered by frictionally induced instability. Narrow free-atmosphere rainbands driven by a series of updrafts and downdrafts were located above the convective line of the NCFR. Gravity wave theory, rather than CSI or conditional convective instability, appears to explain the characteristics of these bands. WCFRs were found to repeat periodically in the frontal zone, with a lifetime limited to six to nine hours, and appeared to be associated with slantwise convection. A single warm-sector rainband was identified at 300 to 400 kilometers ahead of the surface cold front and produced widespread precipitation.

Finally, Szeto et al. (1999) described the role of the surface in their simulations of a severe ice storm associated with the passage of a warm front over the east coast of Canada. Their goal was to understand the cloud and mesoscale processes that affected the development of freezing rain in this system. One of the mechanisms for the formation of freezing rain is the presence of an above-freezing inversion layer (AFIL) at low levels with a layer below the AFIL of subzero temperatures. Such low-level inversions are often associated with surface fronts, however, the dynamics driving the AFIL are not well understood. This is further complicated by the fact that cooling produced by melting within the warm inversion layer, and warming associated with refreezing in the lower levels, can alter this vertical temperature structure. Analysis of their model results showed that as the surface warm front approached Newfoundland, the change in surface characteristics from ocean to land disturbed the quasi-thermal wind balance, the result of which was accelerated warm frontogenesis through the intensification of the ageostrophic cross-frontal circulation. This, in turn, produced an extensive AFIL, which may be associated with ice pellets or freezing rain depending on the depth of the subzero layer. Potential instability and CSI appeared to be the mechanisms that produced the banded precipitation features within the model.

### *10.3.5.3. Snowbands*

Winter cyclones, whether explosively deepening (e.g. Schneider, 1990; Powers and Reed, 1993; Marwitz and Toth, 1993; Mass and Schultz, 1993; Pokrandt et al., 1996), or less intense (e.g. Hakim and Uccellini, 1992; Shea and

Przybylinski, 1993), are often accompanied by a combination of strong winds, freezing rain, subfreezing temperatures, heavy snow and blizzard conditions. Modest intensity extratropical cyclones make up the majority of all cyclone events (Roebber, 1984). These cyclones can produce significant amounts of precipitation through a variety of forcing mechanisms that influence the precipitation processes but that may not all simultaneously contribute to the cyclone development (Martin, 1998a,b). This situation occurs frequently in the central United States where the synoptic-scale forcing does not generate a strong surface disturbance, and yet large amounts of low-level moisture and a weakly stratified lower troposphere can result in a significant response to frontal forcing. Such a case occurred over the central United States in January 1995 when moist, relatively warm low-air from the Gulf of Mexico, and warm frontogenesis processes, produced heavy snow in association with snowbands that developed parallel to the warm front (Martin, 1998a,b).

Snowbands are intense, narrow (5-40 km) bands of heavy snow that are typically located within larger (100-500 km wide) regions of light to moderate snow. Numerous studies have been conducted on the factors affecting the development of snowbands (e.g. Sanders, 1986; Martin, 1998a,b; Nicosia and Grumm, 1999; Schultz and Schumacher, 1999; Jurewicz and Evans, 2004). A number of different processes have been identified that may enhance snowband formation including jet streak interactions, frontogenesis in the presence of weak moist symmetric stability or convective instability, and local orographic forcing. Research by Novak et al. (2004) and Martin (1998a,b) indicated that snowbands producing heavy snowfall generally occur in regions where frontogenesis and instability are co-located, typically to the northwest of surface cyclones in large-scale flow that is highly amplified. However, several more recent studies demonstrate that less intense, but still significant snowbands can occur in other locations relative to a surface cyclone, and in other types of large-scale environments such as to the northeast of the surface cyclone (e.g. Banacos, 2003; Schumacher, 2003) or in situations of no surface cyclone (e.g. Skerritt et al., 2002).

An explosively developing extratropical cyclone occurred over the United States during December 1987. This record-breaking snowstorm, which moved from New Mexico to the Great Lakes region between 13 and 16 December producing heavy snowfall along its path, has been the subject of numerous studies (e.g. Schneider, 1990; Powers and Reed, 1993; Pokrandt et al., 1996). Over a foot of snow over Oklahoma was produced by a northeast-southwest orientated warm-frontal snowband. Marwitz and Toth (1993) studied the mechanisms producing this snowband and found that frontogenesis was occurring in the warm frontal region, and that a direct circulation occurred around the warm front. Above the warm front, ageostrophic winds forced the conditionally unstable air to rise, thereby releasing its instability. Thus both frontogenetic forcing and convective buoyancy were responsible for forcing the snowband.

Martin (1998b) examined a long, narrow snowband (~1100 km) that developed in association with a moderate extratropical cyclone on 19 January 1995. The snowband produced thunder, lightning, heavy winds and record-breaking snowfall. Similarly to Marwitz and Toth (1993), Martin concluded that the snowband was forced by the direct vertical circulation associated with warm frontogenesis. The vertical circulation was supplied with high-$\theta_e$ air from the Gulf of Mexico, and the release of convective instability within the ascending branch of the vertical circulation resulted in the convective characteristics of the band. Lifting of warm, moist air within the trowal also contributed to the heavy snowfall, with frontogenesis along the warm-front portion of the occluded structure being the mechanism by which the air was lifted into and through the trowal. While the precipitation bands were orientated parallel to the warm front, thus suggesting that the release of CSI may have played a role in the band formation, saturated regions of CSI could not be found in the region of the warm front within the model, although the model resolution was relatively coarse.

As discussed above, CSI, and at times CI, are present when EPV is negative in a baroclinic atmosphere. When low EPV occurs in a region of frontogenesis, stronger vertical motion develops and becomes constricted to a smaller scale (Emanuel, 1985). Stronger vertical motion leads to stronger convergence in the lower levels which enhances frontogenesis and the temperature gradient. The increased temperature gradient is associated with further reductions in the EPV. The continual decrease in EPV results in the development of CSI, and CI to a lesser extent, in regions of frontogenesis. As the rising air reaches saturation, CSI and CI are released, thereby enhancing the vertical motion and the associated snowband development. The enhanced vertical motion within the frontogenetic region enhances low-level convergence and a positive feedback is established.

Nicosia and Grumm (1999) made use of both model output and radar data to examine intense mesoscale band formation in three northeastern snow storms associated with extratropical cyclones. The heaviest total snowfall rates in each of these storms was co-located with the location of these mesoscale snowbands. Rates reached as high as 6 in $h^{-1}$. Their analysis showed that the mesoscale snowbands formed in a region to the north of the developing cyclone at midlevels, which was characterized by intense midlevel frontogenesis and a deep layer of negative EPV. The location of the dry conveyor belt over the moisture-rich cold conveyor belt to the north of the low-level cyclone resulted in the significant reduction of the EPV. As air parcels ascended north over the warm front CSI, and CI to a lesser extent were released upon saturation, resulting in enhanced vertical motion and mesoscale band formation. As both CSI and CI were present, it is likely that upright convection associated with the release of CI dominated over the slantwise convection associated with the release of CSI, thereby generating heavier snowfall rates. Therefore the mesoscale snowbands in these cases formed in association with intense midlevel frontogenesis and a deep layer of negative EPV.

Jurewicz and Evans (2004) compared two banded snowstorms that occurred over the northern mid-Atlantic region during January 2002 under very different synoptic-scale forcing. They found that in both cases the snowbands were co-located with mid-tropospheric frontogenesis and reduced stability. In the first case, a layer of PSI occurred just above a deep sloping frontogenetic zone with near-saturated conditions, while in the second case, a layer of PI occurred just above a shallow sloping frontogenetic zone in associated with decreasing humidity with height. Jurewicz and Evans concluded that the difference in the snowband characteristics arose from differences in the synoptic-scale forcing, the frontogenetical forcing, the amount of moisture available, the degree of instability and the thermal profiles.

Moore et al. (2005) examined the processes associated with a long (1000 km), narrow snowband that produced heavy snowfall from the Texas panhandle to northwest Missouri on 4-5 December 1999. Thundersnow was also reported. The snowband was located along a northeast-southwest frontal boundary several hundred kilometers to the northwest of a weak, surface cyclone. A region of negative EPV and associated CSI was located to the north of the surface low near where the dry conveyor belt overlay the warm, moist air of the trowal airstream. Mid-tropospheric frontogenesis occurred to the northwest of the negative EPV region as the trowal airstream became confluent with the cold air to the north of the cyclone. The snowband formed to the north of the negative EPV zone but to the south of the frontogenesis zone. Moore et al. noted that a gradual southeast-to-northwest transition existed in the stability of the atmosphere from convective instability near the surface, to elevated convective instability, to CSI, to weak symmetric stability. Unlike Nicosia and Grumm (1999) who found that the CCB played an important role in generating a deep, moist layer, Moore et al. (2005) found that the WCB provided the deep moisture along the trowal airstream. Nevertheless the interaction between the conveyor belts, whether for cyclones in the northeastern (Nicosia and Grumm, 1999) or central (Moore et al., 2005) United States, provides a mesoscale region of moisture, lift and instability that is conducive to the formation of snowbands. Others have found similar processes to be important in the development of heavy snowfall-producing bands (e.g. Novak and Horwood, 2002; Novak et al., 2006).

Grim et al. (2007) used high resolution observations to compare the trowal and warm frontal structures of two winter cyclones that produced heavy snow swaths across Illinois, Wisconsin and Michigan. The cyclones had different origins, with one originating over the Colorado Rockies and the other over the Gulf of Mexico. The trowal structure in both of these storm systems was different from the classical trowal structure in which the trowal axis is located at the intersection of the warm front and cold front, the cold front having overtaken and ascended the warm front. In both cases, the movement of the DCB over the warm front resulted in wedging the trowal air mass between the dry air and the warm front. The majority of the heavy snowfall produced by these storms was associated with the precipitation band coincident with the trowal and which stretched around the north and northwest portions of the cyclone. While both cyclones included a CCB, the CCB did not appear to play a significant role in generating a deep moist layer or in the precipitation production. Rather, the trowal air mass appeared to be the most influential factor in precipitation production, similar to the findings of Martin (1998a,b) and Schultz (2001). While the trowal was bounded by the warm front to the north and an upper-level front to the south, the upper-level front was an upper-level humidity front in the one case and a cold front aloft (see below) in the other.

Finally, Novak et al. (2008) used both observational and model data to investigate the formation of a mesoscale snowband that occurred over the northeastern United States during December 2002. The formation of the band occurred within a region of increasing midlevel frontogenesis due to the sharpening of a midlevel trough, and was characterized by conditional and inertial instability. The mature stage of the band was associated with a significant increase in frontogenetic forcing and an increase in conditional stability in association with the release of the conditional instability. During dissipation the conditional stability continued to strengthen while the frontogenetic forcing decreased. The changes in moisture appeared to play a smaller role than changes in rising motion in the demise of the band. This case differs from a number of the previous snowband studies in that the maximum rising motion was found close to the frontogenesis maximum, as opposed to being located 50 to 200 km away on the warm side of the front. CI was also evident at least 1.5 hours before the band, and the release of this CI through frontogenetical forcing resulted in the formation of the band and increased conditional stability. Traditionally band formation has been associated with decreasing conditional and symmetric stability. Even more recently, simulations conducted by Novak et al. (2009) showed that latent heat released by the band itself was highly important to both the formation and the maintenance of the band.

#### *10.3.5.4. Mesoscale Features of the Shapiro-Keyser Model*

The original rainband classification studies were conducted for extratropical cyclones more typical of the Norwegian cyclone model. The distribution of precipitation associated with the features of the Shapiro-Keyser cyclone model will be discussed in this section. Some of the earlier observations of these features came from the examination of extratropical cyclone development during the Experiment on Rapidly Intensifying Cyclones over the Atlantic (ERICA; Hadlock and Kreitzberg, 1988; Wakimoto et al., 1992; Neiman and Shapiro, 1993; Neiman et al., 1993). Much of the analysis focused on the record intensity cyclone that occurred on 4-5 January 1989 over the Atlantic Ocean east of North Carolina shown in Fig. 10.16. As described in detail by Neiman and Shapiro (1993) and summarized here, a baroclinic leaf cloud initially observed at 0000 UTC evolved into a comma-shaped cloud system by 0600 UTC (Fig. 10.16a). Cold cloud tops associated with deep (8-10 km)

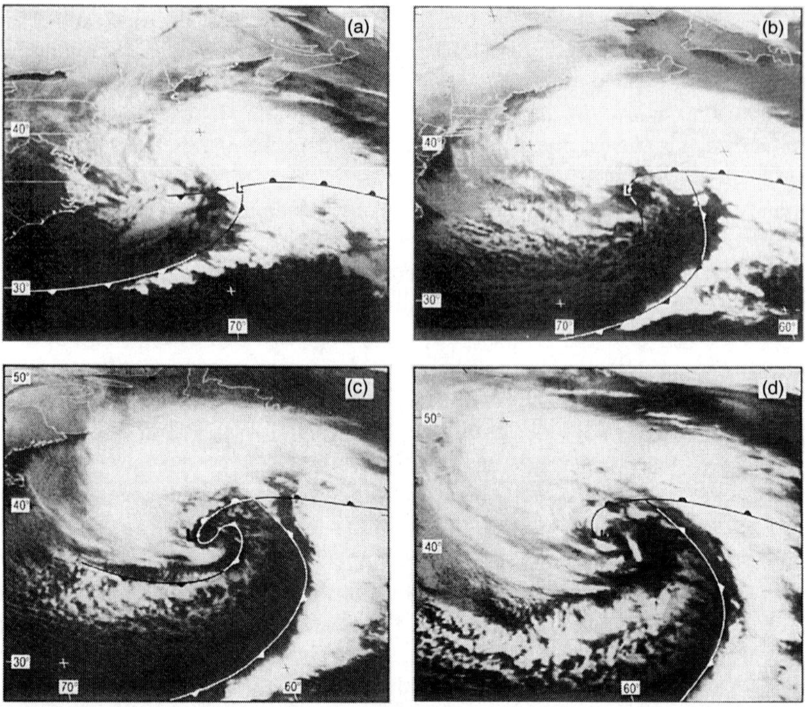


**FIGURE 10.16 NOAA GOES-East infrared satellite images at (a) 0600 UTC, (b) 1200 UTC, (c) 1800 UTC 4 January 1989, and (d) 0000 UTC 5 January 1989.** Surface frontal positions are shown. *(After Neiman and Shapiro (1993))*

cumulus convection were located parallel and to the east of the cold front, and extended northward into the region of the warm front. The comma head cirrus clouds had grown in extent, and descent of dry stratospheric air to the northwest of the cold front is evident. Also evident at this time was the development of the bent-back front as it extended westward into the polar airstream.

By 1200 UTC (Fig. 10.16b) the T-bone phase became evident and surface cyclogenesis was focused on the bent-back front. The cloud band associated with the cold front to the east formed the tail of the comma head and narrowed in cross-frontal scale. Stratocumulus cloud streaks were evident to the west of the cyclone, and orientated parallel to the north-northwesterly flow of cold, continental air over the cool (~5 °C) ocean surface. These clouds were initially limited to the shallow boundary layer (~1 km deep) but developed into deeper cellular cumulus bands (~3 km) further downstream as this airmass moved over the warmer waters (~20 °C) of the Gulf Stream. By 1800 UTC (Fig. 10.18c) the bent-back front and the comma-head began to wrap around the cyclone center in association with the development of the warm core seclusion phase, and a mesoscale cloud-free "eye" was evident. The cold-topped clouds of the comma head tail were situated $\sim$100 km ahead of the surface cold front. The only clouds located at the leading edge of the cold front were located within $\sim$250 km of the triple point, and were relatively deep (>8 km) convective clouds. While the cold front was collocated with a line of intense convection, the causes of the ascent under the cold cloud tops could not be ascertained. Cloud streaks associated with the offshore flow surrounded the western, southern and eastern regions of the cyclone, and vertically-enhanced cloud streaks were located along the secondary cold front where air-sea temperature and moisture differences were greatest. By 0000 UTC (Fig. 10.16d) the low clouds associated with the bent-back front, surrounding the "eye" were still obvious. The eastward displacement of the comma head tail relative to the cold front was also still evident, although the deep convection located along the cold front to the south of the triple point was no longer obvious.

Wakimoto et al. (1992) and Neiman et al. (1993) both made use of airborne radar data to examine the mesoscale cloud and precipitation characteristics of the case just described. Intense mesoconvective precipitation bands extending to between 8 and 10 km were observed along each front, as well as the region of frontal fracture. While deep convection had previously been observed along the cold front (Hobbs and Biswas, 1979; James and Browning, 1979; Hobbs and Persson, 1982; Parsons and Hobbs, 1983), such convective organization had not been observed along the warm front. In addition to the deep convective elements, shallower (<4 km) widespread precipitation was also produced along the warm front, being found in the regions of frontal upglide north of the front, including its bent-back extension. The heaviest precipitation along the cold front was produced by narrow rain bands spaced $\sim$25-40 km apart and orientated at $\sim 30°$ to the front.

Along the warm front, the southerly ascending airstream of the warm conveyor belt produced shallow (<4 km) upglide precipitation and vertical motions of the order of 2-4 m $s^{-1}$. The front in these regions was approximately 20 km wide, and had a slope of $\sim$1:9. However, in the mesoconvective elements the warm conveyor belt ascended to $\sim$10 km over a horizontal distance of $\sim$10 km. The reflectivities were of the order of 45-50 dBZ, the cross-frontal convergence collapsed to storm-scale ($\sim$3 km) and the frontal slope was $\sim$1:1, making it much steeper than previous observations of warm fronts (e.g. Locatelli and Hobbs, 1987). The implied ascent along the axis of the conveyor belt in these elements (> 25 m $s^{-1}$) is comparable with the vertical motions within mid-latitude continental storms. The ascent of the warm, southerly air over the warm front therefore alternated between the "escalator" of the slantwise frontal upglide in regions of shallow precipitation and the "elevator" within the mesoconvective updrafts, where the convective cloud scale processes influenced the warm front circulations. A schematic showing this escalator-elevator concept is shown in Fig. 10.17.

The convection associated with the bent-back front was of moderate intensity with cloud tops extending to $\sim$8 km. In contrast to the intense, nearly

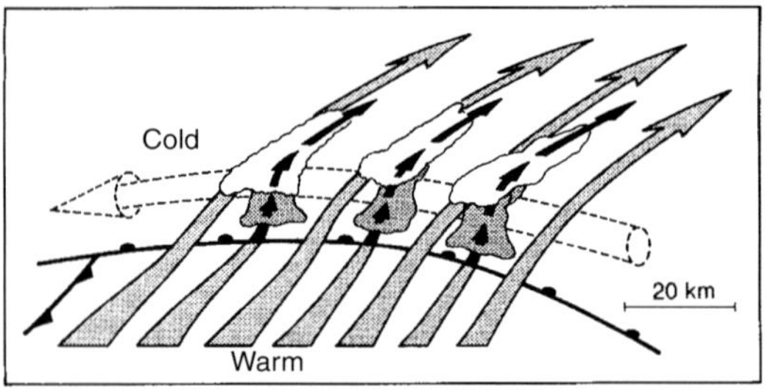


**FIGURE 10.17 Schematic of the "escalator-elevator" perspective of warm-frontal ascent, as the warm conveyor belt (flat, lightly stippled arrows) rises over the cold conveyor belt (tubular dashed arrow).** Mesoconvective ascent (the elevator, solid arrows) and convective clouds (stippled with white anvils) are shown at regular intervals between regions of upglide ascent (the escalator). *(After Neiman et al. (1993))*

continuous lines of convection associated with the cold and warm fronts, the convection was more scattered in nature, occurring along and to the north of the bent-back warm front. During the initial stages of the warm core seclusion phase, the convective elements encircled the western and southern regions of the developing warm core seclusion. In the mature warm seclusion phase, the bent-back front completely encircled the cyclone center, thus secluding a mesoscale region (125 km) of air with high $\theta_e$ (20-30 K higher than on the cold side of the bent-back front). The equivalent potential temperature of this air increased by 8-10 K between 1800 and 0000 UTC, even though it was secluded from the warm sector. This occurred due to an increase in boundary layer moisture resulting from upward latent heat fluxes acting on air parcels spiraling into the warm core seclusion.

In an examination of an extratropical cyclone that developed during IOP5 (19-20 January) of ERICA, Blier and Wakimoto (1995) noted that the strongest radar echoes occurred along the primary cold front and the bent-back to the northeast of the low, and that the widths of the precipitation bands associated with the cold, warm and bent-back fronts were comparable. Also, while precipitation increased in intensity along the bent-back front in the direction toward the low center, there was no convective activity or convection along the secondary cold front that extended beyond the low. Both the bent-back front and the warm front had narrow cross-frontal scales.

Until Martin's (1998a,b) observational and modeling study of extratropical cyclone development over the central United States during January 1995, the bent-back front had not been identified in continental cyclones. The cyclone was characterized by a deep warm front, a cold front and a bent-back front that showed a similar structure to the bent-back front of the maritime cyclones previously described, although the evolution of this feature may have been different to that observed for the maritime cyclones. In this case, the bent-back front was stronger above the friction layer, suggesting that there may be stronger similarities between continental and maritime cyclones than was previously thought. Martin also found that this cyclone produced the classical occluded structure, first at upper levels and then at the surface, and concluded that a warm-occluded structure may exist at mid-levels while a fractured frontal structure occurs at the surface, and that at least some cyclones that exhibit warm-occluded structures develop those structures as a result of the seclusion mechanism.

Finally, it should be noted, that these observational studies have demonstrated that the scale of significant frontal and circulation features within the Shapiro-Keyser model, such as the bent-back front and the fracture of the cold front from the warm front, are mesoscale processes. Such considerations should be borne in mind when numerically simulating these systems if the cloud and precipitation processes associated with these features are to be accurately represented.

### *10.3.5.5. Orographic Influences on Precipitation Bands*

When an extratropical cyclone moves over a mountainous region, the horizontal distribution of precipitation typically associated with fronts and rainbands is influenced by orographic forcing, thus making rainband classification in the various cyclone sectors difficult. The precipitation processes of extratropical cyclones tracking over the mountains of the Pacific Northwest have been researched for several decades (e.g. Hobbs et al., 1975; Houze et al., 1976; Marwitz, 1987). Several recent studies have focused specifically on the role of orography in rainband initiation and formation. Yu and Smull (2000) examined a cold frontal system as it made landfall along the mountainous coast of Oregon, and found that upstream blocking by the coastal terrain led to the rapid genesis of a NCFR, and the enhancement of two prefrontal rainbands. The blocking of the low-level prefrontal flow occurred as a result of the acute orientation of the front to the coast. This resulted in significantly intensifying the prefrontal, low-level along-barrier flow and changing the cross-frontal vertical wind shear. Finally, as the front moved southward the along-barrier flow decreased and the NCFR dissipated. Colle et al. (2002) found from their modeling studies of the same case study that rapid development of the NCFR occurred, even in the absence of topography. However, the coastal topography helped to strengthen the thermal gradients through enhanced deformation frontogenesis driven by the terrain-enhanced prefrontal flow, thereby strengthening the vertical circulation and the NCFR.

Braun et al. (1997) investigated the changes in the mesoscale characteristics of an intense frontal system that moved toward the Oregon shoreline on 8 December 1993. As the frontal system approached the shore the low-level prefrontal flow increased while the wind perpendicular to the front decreased, thereby affecting the evolution of the frontal rainband. The offshore NCFR was characterized by deep convection which, together with the lightning frequency, decreased as the frontal system approached the coast, thus suggesting a stabilization of the lower troposphere in this region. Even though the radar echo depths and maximum updrafts decreased as the NCFR moved toward the coast, the maximum reflectivity at lower levels didn't change significantly, implying a change in the microphysical characteristics of the frontal bands as the system approached the coast, an observation supported by changes in lightning frequency. The stronger mid-to-upper levels updrafts in the offshore bands appear to have increased the production and upward transport of ice particles to form the cloud and stratiform precipitation observed ahead of the NCFR.

Kingsmill et al. (2006) analyzed ten extratropical cyclones that underwent landfall during the California Landfalling Jets (CALJET) experiment using data from two sites, one in the mountains and one in the Central Valley in northern California. In their 10-case composite they noted a distinct change in the slope of the radar reflectivity about 2.5 km above the brightband that represents a change in the hydrometeor growth rate. This structure, which they call a "shoulder", remained at approximately 2.5 km above the brightband in the cold sector, warm front, warm sector, cold front and cool sector regimes and was evident at both sites indicating that this feature occurs independently of orographic effects. About one third of the examined profiles did not show the brightband. These non-brightband profiles exhibited an increase in reflectivity with decreasing altitude in the lower levels, suggesting that growth was occurring by collision-coalescence processes. The relationship between the surface rainfall rate and the low-level radar reflectivity implied that a larger number of small drops occurred in the mountainous profiles than would be obtained using a Marshal-Palmer drop size distribution, a trend especially obvious for the non-brightband profiles. The drop size distributions of the valley profiles, on the other hand, more closely followed the Marshall-Palmer size distribution, therefore indicating larger droplets in these profiles than at the mountainous site, and that non-brightband rainfall occurs less frequently in the valley. Finally, the rainfall rates were greatest in the cold-frontal regime, and the non-brightband rainfall was most frequent during the warm-frontal, warm-sector and cool-sector regions.

A synergistic interaction between frontal and orographic processes was suggested by Woods et al. (2005) in their analysis of the passage of a strong extratropical system over the Cascades. In this context, the term synergistic means that the combination of the two processes produces a greater amount of precipitation than if either process were acting alone. The storm was characterized by a lower-tropospheric frontal zone that tipped forward with height, and was overlaid by a backward tilted upper cold front. A broad rainband was located ahead of the surface cold front. The structure was similar to a warm occlusion, however a distinct warm front could not be found and hence their use of the term tipped-forward cold front. Tipped-forward cold fronts have also been identified in other studies (e.g. Locatelli et al., 2002a, 2005). Woods et al. suggested four possible combinations of cross-barrier flow (easterly and westerly) and lower-tropospheric frontal (tipped-forward or tipped-backward) structures (Fig. 10.18). In each case, a mid-to-upper level frontal cloud produces

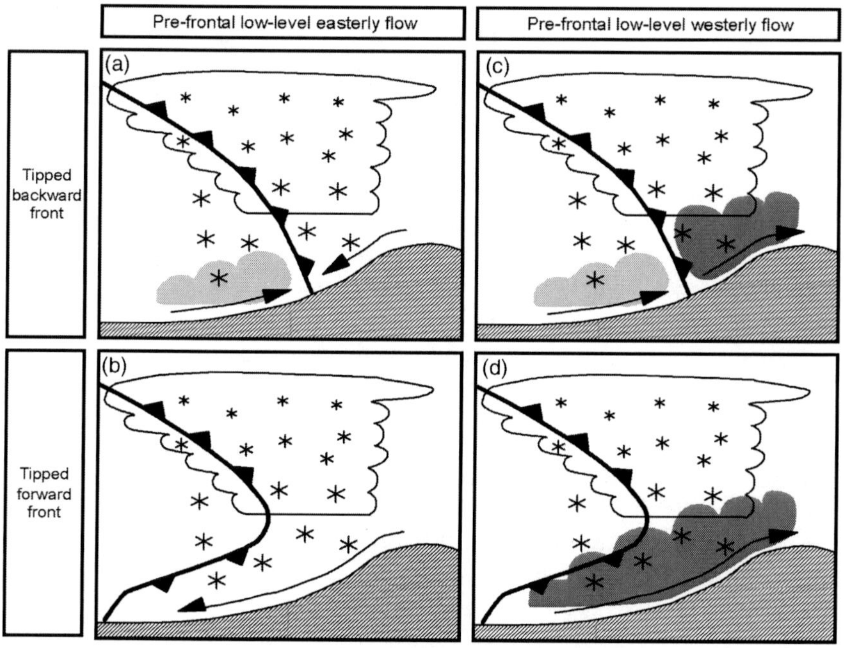


**FIGURE 10.18 Schematic representation of potential frontal structure and cloud liquid-water production above a mountain barrier in response to low-level flow patterns.** Darkly shaded regions indicate clouds with the highest cloud liquid-water. **(a) For a tipped-backward front and prefrontal easterly flow. (b) As in (a) but for a tipped-forward front. (c) For a tipped-backward front and prefrontal westerly flow. (d) As in (c) but for a tipped-forward front.** *(From Woods et al. (2005))*

ice crystals which fall to the lower troposphere. In Fig. 10.18a, the low-level cross-barrier flow is easterly and the frontal structure is backward-tilted. Under this scenario, precipitation rates are not enhanced due to the lack of orographically-generated cloud liquid associated with the easterly downslope flow. The westerly flow behind the front may produce some lower level orographic cloud which in turn may enhance the growth rates of the ice particles falling from the rear of the upper-level frontal cloudband. Following the frontal passage, the precipitation is purely orographic in association with the postfrontal westerlies. The setup in Fig. 10.18b is similar to that in Fig. 10.18a, except that the lower-tropospheric front is now tipped forward. In this case, there is no low-level upslope flow underneath the upper-level precipitation as the prefrontal, downslope easterlies extend all the way to the surface front. Thus any form of enhancement through orographic related processes cannot occur. This situation has not been described in the literature, although tipped-forward fronts with prefrontal easterlies are common in the Pacific Northwest and were observed in the Improvement of Microphysical Parameterization through Observational Verification Experiment (IMPROVE-1) (e.g. Locatelli et al., 2005; Evans et al., 2005).

In Fig. 10.18c, the setup differs from Fig. 10.18a in that the prefrontal flow is now a warm, moist southwesterly flow which produces significant amounts of cloud liquid-water through orographic forcing. However, only some of the upper-level frontal precipitation falls into this cloud. Most of the upper-level precipitation falls behind the surface front into the more shallow, postfrontal orographic clouds. Such a scenario was observed in the 8-9 December 2001 case of IMPROVE-2 (Bond et al., 2005). Finally, the southwesterly, moist cross-barrier flow and the tipped-forward frontal structure observed in Wood et al. 's case study are shown in Fig. 10.18d. This scenario represents the situation that has the greatest potential for producing the most precipitation. The low-level flow produces significant amounts of cloud liquid-water through orographic forcing, and the tipped-forward front results in the entire upper-level frontal precipitation band being located over the lower orographic cloud, thus providing the optimal situation for ice particles from above to seed the lower orographic cloud, thereby removing cloud water that otherwise may not have been removed. Woods et al. (2005) estimated that such liquid-water scavenging clouds result in precipitation enhancements of 0.4 to 1.0 mm $hr^{-1}$. Finally, if the upward motion associated with the orographic forcing is sufficiently deep, it can enhance snow production in the upper-level cloud band (Colle, 2004).

Medina et al. (2007) analyzed 16 extratropical cyclones that crossed the Oregon Cascade Mountains during IMPROVE-2 (Stoelinga, 2003) in November 2001. They examined the mountain-induced variations in precipitation of a storm by focusing on the vertical structures observed at a particular site. They defined three basic sectors of extratropical maritime cyclones: (1) the early sector which is the first to pass over a surface site, is identified with the warm advection region of the cyclone and which often contains a warm front that slopes gradually down toward storm center; (2) the middle sector in which the warm advection transitions to cold advection, the warm front often reaches the surface, and cold and occluded frontal structures may occur; and (3) the late sector which is the cold, unstable zone behind the cold and/or occluded front and is often called the postfrontal period. The radar echo structures associated with each of these stages are shown in Fig. 10.19.

The early sector of the storm's passage over the windward slopes is characterized by a leading edge echo (LEE) in which a deep layer of precipitation appears aloft initially, and then descends toward the surface until a deep stratiform echo extends from the surface to ~6 to 7 km (Fig. 10.19a). Updraft cells may be embedded in the LEE at upper levels. The LEE period echoes do not appear to be qualitatively influenced by the orography. The precipitation over the windward slope within the middle sector of the storm extends in a vertically continuous column from the mountainside up to a height of about 5-6 km (Fig. 10.19b). The vertical structure of the echo is one of a double maximum echo (DME) in which the lower echo is the melting-generated brightband, and the second region of high reflectivity is located 1 to 2.5 km above the brightband. The latter is not evident when the middle sector of the

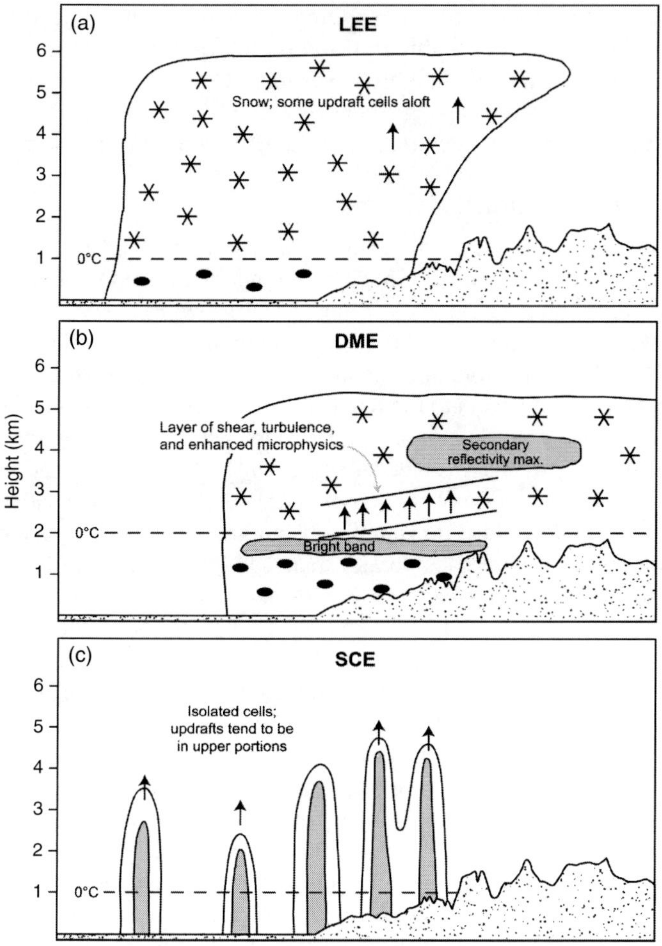


**FIGURE 10.19 Schematic illustration of the typical reflectivity structures observed in the (a) LEE, (b) DME and (c) SCE periods of mid-latitude Pacific cyclones as they progress toward the terrain of the Oregon Cascade Range.** The solid contours enclose areas of moderate reflectivity, while the shading indicates areas of increased reflectivity. The stars indicate snow and the ellipses indicate rain. The speckled area shows the orography. The arrows represent updrafts. *(After Medina et al. (2007))*

storm passes over the regions upwind of the Cascades but becomes obvious over the windward slope, apparently as a result of the dynamic interaction of the storm with the terrain. A layer of turbulent overturning exists between the two reflectivity maxima with small-scale updraft cells ($w > 0.5$ m s$^{-1}$) that may play an important role in the precipitation enhancement on the windward slope. The late sector is characterized by generally isolated shallow convection echoes (SCEs) (Fig. 10.19c). These cells that are observed in the late sector of extratropical cyclones offshore of Oregon and Washington become broader on the windward slope, probably in response to orographic uplift. The shallow convective echoes have top heights lower than those observed during the LEE and DME periods, and disappear more rapidly on the lee side of the Cascades compared to the lee side precipitation of the LEE and DME periods.

### *10.3.5.6. Precipitation Cores and Gaps*

Radar reflectivity has shown that the rainfall within NCFRs is typically not uniform but rather is organized into alternating cores of precipitation and gap regions (Hobbs and Biswas, 1979; James and Browning, 1979; Locatelli et al., 2005; Wakimoto and Bosart, 2000). The precipitation cores tend to be regularly spaced ellipsoidal cells orientated at an angle to the surface cold front, although their shapes and orientation to the front are not always uniform, even within the same front. While precipitation cores have also been observed within WCFRs, they tend to be irregularly shaped without any distinct organization to them (Moore, 1985). Interactions between the cores depend on the relative strength of each core and how close they are to one another. The gap regions in NCFRs tend to vary in size, with gaps greater than 10-12 km occurring less than 20% of the time and being referred to as "large" gaps by Locatelli et al. (1995). The large gaps are associated with weak surface wind discontinuities, lack strong updrafts and appear to be dynamically different from smaller gaps. Gaps have been found to influence the evolution of the cores, and may move along the front at speeds slower than that of the cores.

Several mechanisms have been suggested regarding the formation and control of the precipitation core and gap structure of the NCFR, including horizontal shear instability along the cold front (Hobbs and Persson, 1982; Moore, 1985; Wakimoto and Bosart, 2000; Jorgensen et al., 2003), trapped gravity waves (Brown et al., 1999), and differential advection of precipitation (Locatelli et al., 1995; Brown et al., 1999). To further investigate the horizontal shear hypothesis Moore (1985) performed a linear stability analysis on an idealized frontal zone that consisted of a line of convection coincident with a line of cyclonic shear. The air within the rainband was assumed to be unstable. Moore found that the presence of the horizontal wind shear resulted in a mode with a short wave cutoff, that the coupling between the convective processes and shear instability in this mode was strong, and that its most unstable wave has characteristics similar to the precipitation cores in NCFRs. Moore also noted that as the shear increased, the cells become elongated and narrow.

Locatelli et al. (1995) observed that spacing between the precipitation cores, the height of the head of the density current associated with the cold front, and the strength of the cold front were all positively correlated to the precipitation strength, thus suggesting that the precipitation cores are strongly influenced by precipitation. Based on their observations they developed a new conceptual model. In their model, while the formation of precipitation cores may initially be controlled by processes other than precipitation processes, the latter soon become dominant. As the precipitation cores develop, diabatic frontogenetic processes associated with precipitation evaporation result in an increase in the height of the hydraulic head in the regions of the cores. The hydraulic head moves at a speed that is predicted well by density current theory. The greater the height of the head, and hence the stronger the frontal zone, the greater the speed of the hydraulic head and the associated cores. This results in a positive feedback in that, the greater the speed of the hydraulic head, the stronger the updraft and the associated precipitation. Enhanced precipitation results in increased hydraulic head heights and local enhancements of frontal strength. This in turn is associated with local increases in vorticity (through greater convergence and vertical velocity) within the cores, and hence a perturbation in the frontal boundary. This model has several implications including, that stronger cores move faster than weaker cores, that stronger cores can overwhelm weaker cores, and that fronts will be strongest where the cores are located.

Browning and Roberts (1996) have observed that classical cold fronts and split cold fronts (see section below) can change from one to the other either in space or in time, and that such a change may influence the precipitation structure along the frontal boundary. A common scenario is one in which the cold front is a split front just south of the center of a developing extratropical cyclone, but transitions to a classical cold front further away from the cyclone, such as occurred over the United Kingdom on 8 December 1994. On this day, a dry intrusion moved over a tongue of relatively moist air and advanced ahead of the surface cold front near the center of the cyclone, thus forming a split front structure, while to the south of this, along the trailing portion of the cold front, the dry air undercut the moist air ahead of the front in a classical cold front structure. While line convection occurred along the entire NCFR, the almost continuous line convection in the region of the classical cold front structure transitioned to distinct, broken convective elements that were orientated at approximately 30° or more to the surface cold front in the split front region, as shown in Fig. 10.20. Individual convective elements developed at the northern end of the classical part of the front and advected to the split front part of the cold front, where they were tracked for over 4 hours (Fig. 10.20a). The transition to the broken line appears to have been primarily associated with a change in the orientation of the surface cold front with respect to the preferred orientation of the convective line elements, and only partly due to a change in the orientation of the line elements themselves (Fig. 10.20b). This mechanism therefore appears to be different from shearing mechanisms suggested previously in which the distinct convective elements result from the reorientation of sections of the convective line as a result of shearing instabilities.

The output of a 3-D numerical simulation of an observed NCFR conducted by Brown et al. (1999) demonstrated precipitation cores approximately 40 km in length and rotated counterclockwise from the cold front, results that were

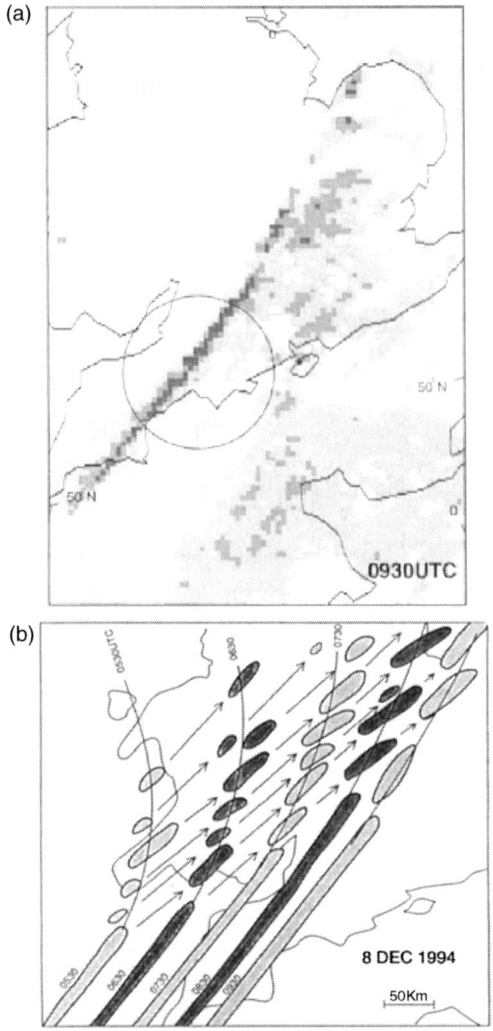


**FIGURE 10.20** **(a) Part of a radar-network picture at 0930 UTC 8 December 1994, showing the precipitation distribution at 5 km resolution.** Grey shades represent rainfall rates exceeding approximately 0.5, 2 and 4 mm $hr^{-1}$ averaged over 5 km squares. (b) Analysis of the precipitation cores along the NCFR at 0530, 0630, 0730, 0830, and 0930 UTC 8 December 1994. *(Adapted after Browning and Roberts (1996))*

consistent with observations. Brown et al. found that convection associated with each precipitation core generated a gravity wave, similar to the scenario in which air flows over an isolated mountain peak. These gravity waves are V-shaped, with the apex of the V orientated slightly counterclockwise to the cold front. The eastward branch of the V, or that branch closest to the cold front was found to extend over the cold front. The motion of the first wave oscillation is downward in direction, strongest in magnitude, and results in warming aloft thereby producing a gap region. The next oscillation of the gravity wave is weaker, upward in direction and induces convection further north. In this manner, trapped gravity waves produce a series of precipitation cores and gaps. Brown et al. also noticed that the elliptical shapes and orientations of the precipitation cores appear to be due to the advection of precipitation by the core-relative winds.

Wakimoto and Bosart (2000) conducted a detailed analysis of airborne Doppler radar data of an oceanic cold front exhibiting precipitation core and gap regions. They found from their observations that the precipitation cores were due to both horizontal shearing instability and the advection of hydrometeors by the precipitation core-relative winds. Also, unlike previous work, a strong discontinuity did not exist along the whole length of the precipitation core. While the southern regions of the precipitation cores may be associated with a discontinuity and lighter precipitation, the variables changed more slowly in the northern regions of the cores and the precipitation was more intense. The cold front in the regions of the precipitation cores appeared to propagate as a density front, although density current theory did not appear applicable to the overall motion of the cold front.

Many of the features previously observed to occur with NCFRs were evident in a study by Jorgensen et al. (2003) of an intense NCFR observed using airborne Doppler radar during the Pacific Coastal Jet Experiment, and simulated using MM5. These include gaps in the narrow precipitation zone, a strong frontal updraft at the leading edge of the cold air, strong shear in the low levels of the inflowing air, a low-level jet parallel to the NCFR with a maximum near 1.5 km AGL, and a rear-infow jet of westerly air peaking at ~1 km AGL behind the front. Like Wakimoto and Bosart (2000), Jorgensen et al. (2003) also noticed differences between the northern and southern regions of the precipitation cores within the NCFR. In particular they found that differences in the updraft tilt exist consistently on either side of the NCFR gap regions (Fig. 10.21). Within the precipitation core the updrafts tend to be erect (Fig. 10.21 A-A′) with narrow rainfall regions, representing the optimal vertical shear balance (Rotunno et al., 1988). Near the south end of the precipitation core (Fig. 10.21 B-B′) the environmental shear is weak relative to the cold air circulation, thus causing the updraft to tilt upshear. This region is associated with a broader zone of precipitation. Finally, at the northern end of the precipitation core, the environmental low-level shear is stronger and the updraft tilts downshear with height (Fig. 10.21 C-C′). Rain is this region falls ahead of the cold air. It is thus the interaction of the density current and the ambient cross-frontal vertical wind shear, rather than the density current dynamics alone, that explains the updraft and precipitation characteristics. The prefrontal cross line flow shows the greatest variation along the NCFR, while the depth and strength of the cold air varies little across the gaps. The gaps will not be self-sustained, as

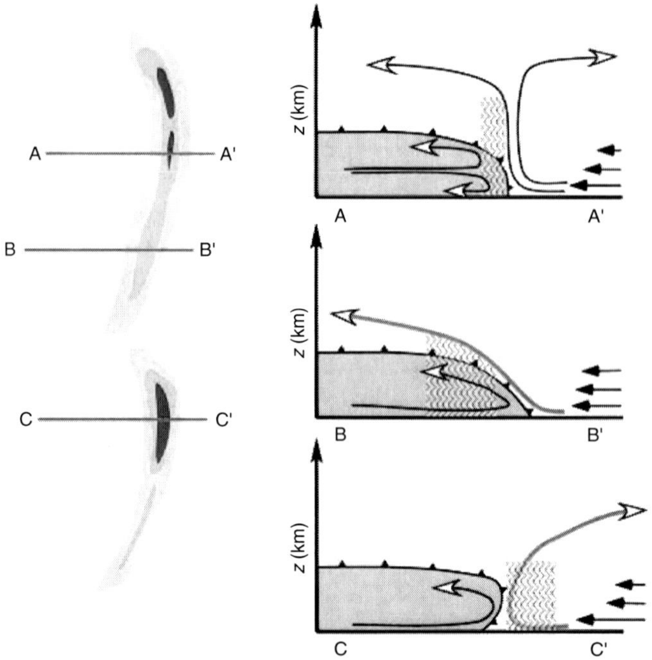


**FIGURE 10.21 Schematic depiction of the relationship between rainfall location, updraft location and tilt, and low-level vertical shear of the ambient flow at various locations along the NCFR.** The horizontal reflectivity depiction of the NCFR showing a predominant break is shown at the left. The locations of three vertical cross sections are indicated by the lines A-A′, B-B′ and C-C′. Schematic vertical circulation and structure is shown for the three cross sections at the right. The low-level shear of the ambient environmental flow is indicated by the arrows at the right in the vertical cross sections. Locations of rainfall maxima are indicated by the shaded regions on the cross sections. *(After Jorgensen et al. (2003))*

convection to the south of the gap will weaken with time compared with its northern counterpart, due to evaporative cooling as the inflow air passes through the rainshaft. The observations showed that the gaps only lasted for several hours.

### 10.3.5.7. Split Fronts and Cold Fronts Aloft

The limitations of the Norwegian cyclone model in explaining all of the observed precipitation distributions of extratropical cyclones, particularly within the warm sector (such as squall lines that form at least 200 km ahead of surface cold fronts), led to the identification of split fronts (e.g. Harrold, 1973; Browning and Monk, 1982). More recently, the recognition of the influence of topography on frontal structures in the central United States resulted in investigations into cold fronts aloft (CFA) (e.g. Hobbs et al., 1990; Locatelli et al., 1995, 1998, 2002b; Stoelinga et al., 2000), and the subsequent development of a more general conceptual model for cyclones in this region referred to as the Structurally Transformed by Orography Model (STORM) (Hobbs et al., 1996). As Locatelli et al. (2002b) pointed out, cold fronts aloft (where by this we are generically referring to baroclinic zones above the surface that are significantly ahead of those at the surface) and their associated wide cold frontal rainbands form an important part of the classical model's warm occlusion (Bjerknes and Solberg, 1922), the trowal model (Galloway, 1958), the split front model (Browning and Monk, 1982) and the STORM model (Hobbs et al., 1996). The common feature of the cold front aloft in all four of these models, even though called by different names (upper cold front in the split front model and cold front aloft in the warm occlusion and STORM model) is that it is a baroclinic zone associated with a thermally-direct circulation. These cold fronts aloft differ from those upper level fronts previously described (e.g. Keyser and Shapiro, 1986), which are more closely associated with dynamical processes near the tropopause. The role and forcing specifically associated with the orography are addressed in Chapter 11, whereas the mesoscale and precipitation characteristics of cold fronts aloft are addressed here.

Browning and Monk (1982) studied nine cold fronts associated with cyclones crossing the United Kingdom in order to develop a simple model of split fronts. In this split front model the "upper cold front" is located up to several hundred kilometers ahead of the surface cold front as a result, either of overrunning the surface cold front, or of the modification of the low-level air (Fig. 10.22a). The warm conveyor belt extends ahead of the surface and upper cold fronts and above the surface-based warm front. Low-$\theta_e$ air within the dry conveyor belt overruns the warm conveyor belt (Fig. 10.22a), thereby creating potential instability. When this potential instability is released in association with the upper cold front, a convective rainband is produced ahead of the surface cold front. As the classical cyclone model does not distinguish between upper and surface cold fronts, it cannot describe the rainbands associated with these split fronts. The band itself is often observed to extend from within the warm sector, where the precipitation associated with the upper cold front tends to be more intense and cellular in character, to beyond the warm front where the precipitation is produced by upgliding over the warm front and hence is more frontal in nature.

Behind the upper cold front a sharp drop in cloud cover is typically observed (although bands may be evident), followed by a shallow moist zone made up of warm conveyor belt air (2-3 km deep) located between the upper cold front and the surface cold front (Fig. 10.22b). Uplift of the warm conveyor belt air by weak convection may produce patches of light rain and drizzle in this region, although precipitation is frequently suppressed due to the deep layer of dry air aloft that limits the depth of the moist layer. Widespread stratus may also develop after the rainfall produced by the upper cold front has passed, the base of which tends to lower until lifted by the passage of the surface cold

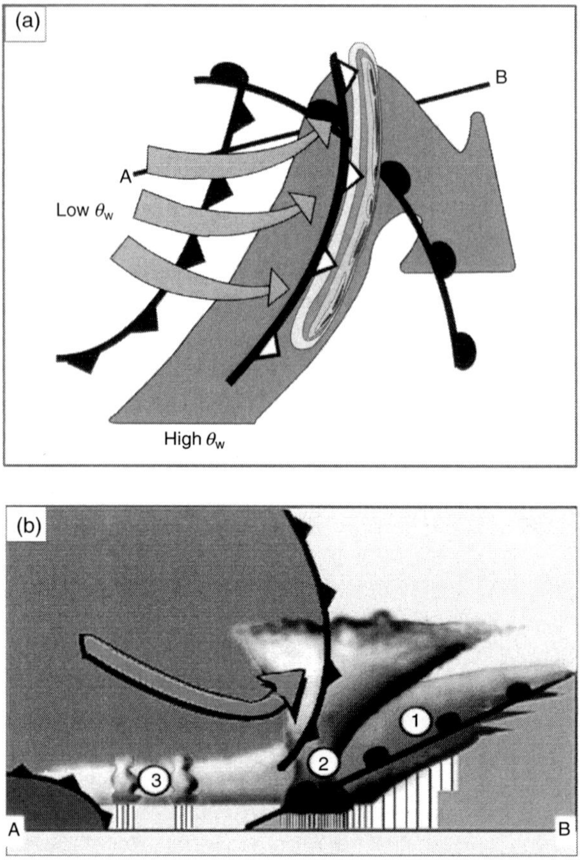


**FIGURE 10.22 (a) Schematic plan view of a split front (from Koch, 2001, after Browning and Monk, 1982).** The broad arrow depicts the warm conveyor belt (high-$\theta_W$ air) which gently rises in an isentropic coordinate systems relative to the moving cyclone. The narrow arrows represent the flow of low-$\theta_W$ air within the dry conveyor belt, which descends while overrunning the warm conveyor belt. The leading edge of the low-$\theta_W$ air marks the position of the split cold front (open symbols) just ahead of which appears a rainband. Surface fronts are shown by standard symbols. **(b) Vertical cross section along A-B showing (1) regions of light precipitation along the warm front, (2) intense precipitation produced by the split-front rainband, and (3) shallow, light precipitation occurring in the warm sector ahead of the surface cold front.** The dry conveyor belt is depicted by the broad arrow behind the split front. *(After Koch (2001))*

front. The surface cold front tends to produce further light rain and drizzle, although occasionally a narrow band of convection develops. While changes in temperature across the upper and surface fronts tend to be small due to the subsidence of air behind the front, the drop in humidity is significant. Finally, split fronts can be distinguished from trowels by the presence of unoccluded warm air between the upper and lower fronts.

Hobbs et al. (1990) proposed the cold front aloft model in order to explain frontal features often observed to the east of the Rocky Mountains. In this model there is a warm or Arctic front at the surface, a CFA and its associated short wave aloft, and a surface trough located 200-300 km behind the leading edge of the CFA. The surface trough may take the form of a cold front, a warm occluded front, a lee trough, a lee trough/Arctic trough combination, or a dry trough (a lee trough that has the characteristics of a dryline) as described in the STORM conceptual model (Hobbs et al., 1996). When the surface trough is a cold front, the CFA model is equivalent to the split cold front model of Browning and Monk (1982). In the CFA model, the air to the east of surface trough is characterized by high-$\theta_e$ values. As dry descending air from over the mountains overlays this warm moist air to the east of the trough, potentially unstable conditions develop. The upward motion then produced by the passage of the CFA can lift this potentially unstable air thereby producing a CFA rainband. This rainband can be composed of several squall lines and may be associated with tornadoes, hail, and flash flooding over several states. Hobbs et al. (1990) proposed adding squall lines that form in association with CFA to the two previously identified squall line types (prefrontal and ordinary). As the precipitation produced by these events is associated with the leading edge of the CFA, the surface precipitation falls well to the east of the surface trough, as it does in the split front model.

The main difference between the split front and the CFA is that the split front is an upper cold front located over the warm sector, ahead of the surface cold front (Fig. 10.22), whereas the CFA is ahead of a surface trough. Thus, as stated by Locatelli et al. (1995), in the CFA model "the region analogous to the classic warm sector of a warm occlusion is located west of the surface position of the lee trough, not east of it as required by the split-front model." Also, the CFA forms as a result of orographic blocking of the lower portion of an advancing cold front and subsequent adiabatic warming by the descent of the flow over the lee slope of the mountains, both of which act to destroy the cold front near the surface. In the split front model, which was developed based on case studies observed over the United Kingdom, no such topographical influences were necessary.

Following their original CFA study (Hobbs et al., 1990), and based on a number of subsequent studies by Locatelli et al. (1989, 1995), Martin et al. (1990, 1995) and Wang et al. (1995), and Hobbs et al. (1996) went on to develop a more generalized conceptual model of cold fronts aloft. The model incorporates a number of features including a dry trough, an Arctic front, a low-level jet, and two rainbands, the CFA rainband (Locatelli et al., 1995, 1998) and the pre-dry-trough rainband (Martin et al., 1995). Both of these rainbands can produce heavy precipitation and severe weather ahead of the dry trough. The aspects of this conceptual model are described in detail in Hobbs et al. (1996) and only briefly summarized here. As a shortwave trough moves eastward, westerly downslope flow over the Rockies increases, producing adiabatic warming and a lee trough. Confluence of warm, dry air off the Rockies and warm, moist air from the Gulf of Mexico produces a west-east moisture gradient, and a trough with the characteristics of a dryline may develop called the dry trough. This dry trough has warm front-like circulations that can lift potentially unstable air. The moist air flowing northward from the Gulf rises and turns toward the northeast as it approaches the dry trough. The dry, downslope air off the Rockies reaches its lowest point over the dry trough and then rises above the warm, moist Gulf air. In late winter, well-mixed heated air off the Mexican plateau may flow eastward above these two airstreams.

As the dry tough develops, the strong southerly flow ahead of the dry trough enhances the formation of a southerly LLJ jet which in turn enhances the transport of warm, moist air northward. The airflow associated with the dry trough results in dry warm (low-$\theta_e$) air being superimposed over warm moist (high-$\theta_e$) air, thereby producing a potentially unstable environment ($\theta_e$ is decreasing with height), which when lifted becomes unstable and produces convective clouds and precipitation (Fig. 10.23a). The resultant convective rainband is referred to as the pre-dry-trough rainband (Martin et al., 1995) (Fig. 10.23b). The pre-dry-trough rainband typically moves northward and eastward away from the dry trough, may be composed of several sub-bands, is often associated with severe weather and may extend over many states. The southward movement of the Arctic front and the lifting of warm moist air around the associated low pressure center, produces precipitation that is often in the form of freezing rain or snow (Stoelinga et al., 2000), and which can at times be heavy should the flow be suitably located to be upslope flow. Thus precipitation that is both stratiform (Arctic front) and convective (pre-dry-trough) may occur simultaneously.

The most important feature of the STORM model is the rainband that forms in association with the CFA, called the CFA rainband (Fig. 10.23c). A surface front may accompany a CFA as it moves over the Rockies, although the surface front tends to become eroded with the adiabatic warming associated with the downslope motion of the front. As the CFA moves over the dry trough it occludes with the high-$\theta_e$ air ahead of the dry trough in a structure typical of warm occlusions. Upward motion produced by the CFA can then lift the potentially unstable air to the east (ahead) of the dry trough, producing what is called the CFA rainband. This rainband may consist of several squall lines, and can produce severe weather including tornadoes, large hailstones and flash floods, from the Rockies to the east coast of the United States (e.g. Locatelli et al., 1989, 1995; Sienkiewicz et al., 1989; Hobbs et al., 1990; Martin et al., 1995; Stoelinga et al., 2000). Finally, Hobbs et al. do point out that while they are not proposing that all of the cyclones that develop to the lee of the Rockies fit their model, many of the cyclones that do develop in winter, spring and even early summer appear to be better described by this model than the classic cyclone model. The use of the classic model would result in these rainbands and their associated severe weather not being forecast.

The impacts of cold fronts aloft (in the generic sense) on various mesoscale systems and processes continue to receive attention. The precipitation produced

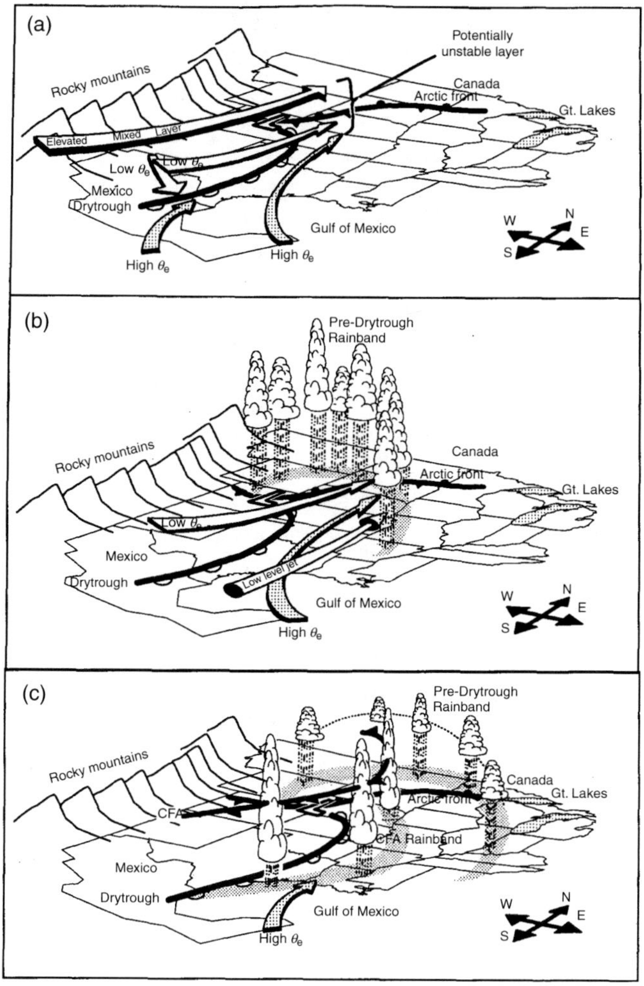


**FIGURE 10.23 Schematic showing the formation of a CFA rainband containing several squall lines, east of the Rockies. (a) The flow of warm, dry low-$\theta_e$ air from the Rockies and warm, moist high-$\theta_e$ air from the Gulf of Mexico form an upward-sloping potentially unstable region east of the dry trough. (b) The regions of upward air motion associated with the CFA. (c) The location of the CFA rainband with respect to the CFA and the dry trough.** *(After Locatelli and Hobbs (1995))*

in association with cold fronts in the upper air can be nonconvective, weakly convective or severely convective, being dependent on the stability of the atmosphere being lifted by the cold front. The structure of cold fronts aloft is ideal for supporting squall lines in that the cool, dry air behind the nose of the front supplies the rear inflow air that is required to evaporate precipitation, thereby maintaining the squall line (Locatelli et al., 1995). A positive feedback mechanism can even exist between the CFA and the squall line, in that the evaporative cooling behind the CFA and the latent heat release ahead of it, associated with the squall line development, strengthens the CFA, which in turn strengthens the squall line and so on. Locatelli et al. (1998) observed a squall line embedded within a rainband associated with a CFA, although the structure was quite different from the typical leading line/trailing stratiform squall line model. They concluded that the CFA, rather than the lower-level cold-pool, was responsible for the squall line generation and maintenance. Geerts and Hobbs (1991) noticed in their observations of a rainband associated with a CFA that the CFA interacted with the boundary layer not only through evaporative processes but also by a shallow downdraft to the rear of the rainband's rainshaft. Locatelli and Hobbs (1995) reanalyzed a record rainfall-producing storm that occurred over Holt, Missouri on 22 June 1947. This storm produced approximately 1 foot of rain within 42 minutes. It was previously analyzed by Lott (1954) who concluded that the storm was the result of local intensification of a warm sector convective storm a short distance ahead of the surface cold front. Locatelli and Hobbs demonstrated that the STORM conceptual model more accurately described this storm.

More recently, Brennan et al. (2003) made use of a numerical model to examine the impacts of a split front rainband on cold air damming (CAD) to the east of the Appalachian mountains. They found from their simulations that, as the rainband moved over the cold dome, the vertically-integrated latent heat released by the rainband resulted in low-level pressure falls, isallobaric convergence within the cold dome and the resultant inland surge of the coastal front, which is synonymous with the retreat of the CAD cold dome. It should be noted that the cold dome was nearly saturated before the split front passed over the region, thereby reducing any evaporative cooling effects. Thus, should a saturated dome begin to erode before the passage of a split front rainband, the presence of the rainband would not prevent this erosion, and may in fact accelerate the process. This offers an alternative view to the idea that precipitation associated with the rainband would act to strengthen the CAD. Businger and Baik (1991) also examined the impacts of a CFA on CAD and found that the significant source of instability associated with the CFA allowed for the development of severe convection above the cold and highly stable CAD air mass.

More recently, Han et al. (2009) noted a synergistic interaction within an intense cold-frontal rainband between a low- and upper-level front that led to variations in the updraft structure and ice generation along the front. The low-level front was associated with a NCFR that produced heavy precipitation, while the upper-level front was associated with a WCFR that was located either behind or in conjunction with the NCFR along its central and northern regions, and that generated stratiform rain. In the central regions of the front, the link between lifting along the NCFR and lifting associated with the upper-level front resulted in a zone of deep upward motion that produced snow and graupel, while those regions to the north and south produced relatively little snow and graupel as ascent associated with the upper-level front did not occur in these regions.

### *10.3.5.8. Gravity Wave Influences*

Mesoscale gravity wave disturbances typically have durations greater than 4 hours, periods between 0.5 and 4 hours, wavelengths of 30 to 400 km, and 2 mb pressure perturbations, although pressure jumps of up to 11 mb in 15 minutes (Schneider, 1990) and wind gusts up to 68 kts have been observed. They have been associated with many mesoscale weather phenomena including squall lines, heavy snowstorms, precipitation bands and thunderstorm initiation (Uccellini and Koch, 1987; Bosart and Seimon, 1988; Feretti et al., 1988; Koch and Golus, 1988; Schneider, 1990). These wave disturbances have been observed as both wave packets (Bosart and Sanders, 1986) and singular waves (Ramamurthy et al., 1990), and various mechanisms have been proposed to explain the formation of the gravity waves including geostrophic adjustment (e.g. Koch and Dorian, 1988; Bosart and Seimon, 1988; Ramamurthy et al., 1993), convection (e.g. Bosart and Cussen, 1973; Uccellini, 1975), vertical wind shearing instability (Einaudi and Lalas, 1973; Stobie et al., 1983), and triggering by the leading edge of downslope flow (e.g. Karyampudi et al., 1995; Rauber et al., 2001). It has also been argued that gravity waves initiate convection, rather than convection initiating gravity waves, or that gravity waves and convection are mutually dependent (e.g. Koch and Golus, 1988; Powers and Reed, 1993). The occurrence of mesoscale gravity waves with extratropical cyclones has been observed to enhance the conditions produced by the cyclone, strengthen the cyclone itself, and contribute to the formation of precipitation bands within the storm system. Due to their small scale, clear weather and severe blizzard conditions may be separated by only several kilometers.

An excellent case of gravity wave development in association with a rapidly developing extratropical cyclone occurred over the Midwest and lower Great Lakes on 15 December 1987. This case has been extensively examined using both observational data and numerical output. The waves lasted over 10 hours, produced pressure falls of up to 11 mb in 15 minutes, had horizontal wavelengths between 100 and 200 km, propagation speeds of 30 m $s^{-1}$, and produced cloud-to-ground lightning and localized heavy snow (Schneider, 1990). While the heavy snowfall produced by the cyclone was correctly forecast, blizzard conditions associated with the wave disturbances were not well predicted. The wave disturbances significantly influenced the onset time of the heavy snowfall, and the impacts of the waves on the surface pressure and wind fields made identification of the surface cyclone low difficult, thereby further complicating the forecast. In addition to their remarkable magnitude, these waves appeared to interact with the extratropical cyclone itself (Schneider, 1990). The largest amplitude wave propagated through the cyclone center during its rapid intensification phase, and contributed to a 7 mb pressure fall within one hour. The wave at this stage had a surface pressure minimum lower than that of the cyclone itself.

Schneider (1990) examined previously suggested mechanisms for mesoscale wave formation in this case, including geostrophic adjustment, convection and vertical wind shear. He found that the atmosphere in which the waves propagated was suited to atmospheric wave ducting, which reduces the vertical propagation of gravity waves, and hence enhances long-lived waves (Lindzen and Tung, 1976). To maintain such long-lived gravity waves requires a deep, stable layer in the lower levels, overlain by a conditionally unstable layer above, that will reflect the waves, hence providing a wave duct. A dynamically unstable critical level characterized by vertical wind shear that is capable of generating and/or maintaining gravity wave energy was also present within the conditionally unstable layer. While the growth and maintenance of such waves seemed to be supported by wave ducting, the mechanisms responsible for forming these waves were less apparent. Schneider suggested that, either a mass-momentum adjustment (Uccellini and Koch, 1987) associated with a shortwave in the region, or convectively induced subsidence zones, may have played a role in the wave generation and/or amplification. Gravity wave generation by vertical shear did not appear to be important, as even though vertical shear was present in the conditionally unstable layer, the region covered by the wave duct was significantly more extensive than the localized gravity wave response.

Powers and Reed (1993) were the first to conduct numerical simulations of the 15 December 1987 case in order to examine the dynamics of this system, in particular the roles played by wave ducting, wave-CISK, convection, shearing instability and geostrophic adjustment in the maintenance of these wave disturbances. While the location and time of occurrence of the simulated waves in their control run were similar to those observed, the speed of the model waves was significantly greater than that of the observed waves. A sensitivity test in which the latent heat of condensation was turned off did not produce strong mesoscale gravity waves. Based on observational and modeling data, Powers and Reed suggested that the model waves were generated by convection on the mesoscale and that they were maintained and enhanced by wave-CISK processes, while the observed waves were maintained by both wave-CISK processes and an environment supportive of ducting processes. Calculations of the duct efficiencies showed that the duct was not perfect, and hence some of the wave energy would propagate vertically out of the duct. A further energy source would therefore be required to maintain the wave for several hours. Wave-CISK processes could supply this energy, a suggestion supported by accounts of thunder as the wave disturbance passed by. Powers and Reed also suggested while convection was primarily responsible for the generation of the observed waves and the lack of latent heating was the limiting factor in wave development, that wind shear (e.g. Stobie et al., 1983) may also have played a relevant role. Geostrophic adjustment (Uccellini and Koch, 1987; Koch and Dorian, 1988) did not however, appear to be important.

Pokrandt et al. (1996) analyzed several weaknesses with the previous studies of the 15 December 1987 case and based on observational and modeling evidence suggested a mechanism different from convection for the formation of these gravity waves. They pointed out that the meso-$\beta$ scale of the waves was not addressed, and that convection in the atmosphere usually forms on the meso-$\gamma$ scale, unless other factors organize it on the meso-$\beta$ scale. Infrared satellite imagery and radar data from the genesis region just before the waves formed suggest that convection was not the primary feature at this time. Instead, a meso-$\beta$ scale comma-shaped cloud present in this region appeared to develop into the wave disturbances. The comma-shaped cloud appeared to originally be associated with the left exit region of an approaching subtropical jet. Through the use of both observational and model data, Pokrandt et al. developed a new hypothesis for the explanation of the observed gravity waves. The rising motion of the transverse circulation of the jet streak produced a band of enhanced cloudiness and precipitation that moved at the same speed as the jet streak. The transverse circulation also transported potential vorticity from the stable, cold low-level air to the mid-levels, where it formed thin, PV anomalies on the meso-$\beta$ scale. This circulation resulted in frontogenesis at midlevels. A wave duct was also present. With the spinup of the large-scale circulation as it moved into the duct region, the midlevel PV anomaly was rotated and stretched, thereby generating small-scale vertical motion perturbations within the duct. The waves then became stronger through wave-CISK processes.

A long-lived gravity wave event that occurred on 14-15 February 1992 during the Storm-scale Operational and Research Meteorology-Fronts Experiment Systems Test (STORM-FEST) has been studied by numerous researchers (e.g. Jin and Koch, 1998; Jewett et al., 1999; Trexler and Koch, 2000). Using observations, Rauber et al. (2001) observed that the gravity wave originated at the leading edge of a dry air mass that was associated with downslope flow east of the Rocky Mountains. The gravity wave and a weak rainband were generated simultaneously just behind the leading edge of this dry air mass as the dry air mass ascended over a warm front to the east of a lee cyclone. In the second paper in this series, Yang et al. (2001) noted that as the dry air mass moved over the denser, cold air below the warm front it caused a wave perturbation in the dense fluid, thus forming the mesoscale gravity wave. The air above the warm-frontal inversion was stable to parcel ascent, but close to neutral for lifting of the layer. As the dry air ascended over the frontal inversion, sufficient lifting occurred to generate a weak convective updraft and rainband. Rain produced by the rainband fell into the dry environment behind the leading edge of the dry air mass. Evaporative cooling resulted in descending air and a depression of the inversion height. Using the numerical experiments with and without evaporation processes, Jewett et al. (2003) demonstrated that evaporatively-generated downdrafts produced by the rainband appeared to be

FIGURE 10.24 **A GOES visible satellite image of small comma cloud (see arrow) located at the west of a synoptic-scale frontal cyclone.** The length of the arrow is approximately 400 km. The West Coast of the United States is visible on the right-hand side of the image. *(After Businger and Reed, 1989a; from Reed, 1979)*

important to wave genesis. The evaporative processes resulted in the depression of the inversion across the warm front. This in turn produced surface pressure falls and the initiation of the gravity wave.

## 10.3.6. Polar Lows

Meoscale features other than cloud bands may also be associated with extratropical cyclones, such as that shown in Fig. 10.24 (Businger and Reed, 1989a). This system is an example of a mesoscale cyclonic system that frequently occurs in polar airstreams behind or poleward of cold fronts in the north Pacific, North Sea, north Atlantic (particularly south of Iceland), and other locations where cold air flows over relatively warm water. These subsynoptic scale cyclones that develop poleward of the polar front and jet stream are called polar lows. Although often only a trough of low pressure occurs at the surface, some polar lows do develop closed cyclonic circulations.

Polar lows appear to be associated either with a comma-shaped or a spiral-shaped cloud structure (Rasmussen, 1983). The spiral cloud pattern of a polar low developing over the Bering Sea is shown in Fig. 10.25. Polar lows associated with the spiral-shaped cloud pattern (e.g. Ernst and Matson, 1983; Rasmussen and Zick, 1987) typically form in the cold air away from the polar front and its associated baroclinic forcing (Rasmussen, 1979, 1983). They appear to develop when cold air in the form of a trough or vortex moves over relatively warm water, thereby generating deep convection through CISK. These

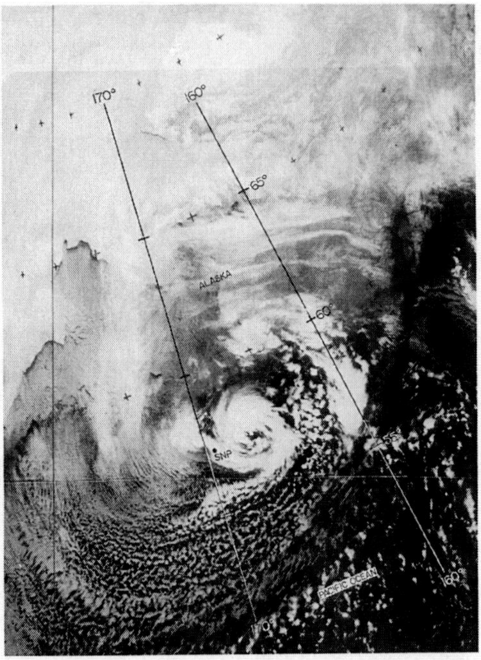


**FIGURE 10.25 A NOAA-5 infrared satellite photograph of a polar low and cloud streets over the Bering Sea at 2100 UTC 8 March 1977.** SNP indicated the location of the rawinsonde station at St. Paul Island. *(After Businger and Reed, 1989a)*

lows are referred to as polar lows, Arctic lows, Arctic instability lows, or spiral form lows. They are most frequently observed in regions near the edges of ice sheets or ice covered surfaces such as over the Gulf of Alaska and the Bering Sea (e.g. Businger, 1987), the coast of Norway (e.g. Lystad, 1986), and off the coasts of Antarctica (e.g. Carleton, 1979; Carleton and Carpenter, 1989, 1990).

The polar lows associated with a comma-shaped cloud pattern (Fig. 10.24) tend to be larger than their spiral form counterparts, and develop closer to the polar front, often just poleward of a pre-existing frontal boundary, in regions of positive vorticity advection at mid-tropospheric levels (Reed, 1979; Mullen, 1983; Reed and Blier, 1986a,b; Businger and Walter, 1988). The term comma-shaped cloud pattern is often abbreviated to comma cloud, which should not be confused with the comma cloud that develops in association with extratropical cyclones. Frontal structures appear to accompany these comma-shaped polar lows, although on a smaller scale than their extratropical cyclone counterparts. Baroclinicity seems to play a more important role than CISK in these systems. These comma-shaped polar lows may occur in the North Atlantic (e.g. Carleton, 1985), the North Pacific (e.g. Reed and Blier, 1986a,b; Businger and Hobbs, 1987), and the Southern Oceans (Carleton, 1979).

Polar lows are warm core vortices, which resemble tropical cyclones in their structure including a clear "eye" surrounded by deep convection, and can be accompanied by light up to hurricane force winds (Rasmussen, 1979; Reed, 1979; Locatelli et al., 1982; Rasmussen, 1981, 1983; Mayengon, 1984; Forbes and Lottes, 1985; Shapiro et al., 1987). They range in size from several hundred kilometers to approximately 1000 km in diameter, their size distinguishing them from extratropical cyclones. These systems have been observed to produce severe storms including tornadoes (e.g. Reed and Blier, 1986b), and heavy rain and snowfall (Harrold and Browning, 1969). The most favorable environments within which polar lows tend to develop are characterized by cyclonic flow or shear (Reed, 1979; Mullen, 1979), strong heat and moisture fluxes from the surface, typically the ocean (e.g. Reed and Blier, 1986b; Shapiro et al., 1987), and neutral or unstable conditions in the boundary layer, with conditionally unstable conditions up to the middle or upper troposphere (e.g. Rasmussen, 1977; Mullen, 1979; Businger, 1987). Thus they are most frequently found in areas with large temperature contrasts such as in the region of the polar front or at the interface of relatively warm water and cold ice sheets. In developing a seven year winter polar low climatology of the North Pacific Ocean, Yarnal and Henderson (1989) found that early season polar lows were located further north in association with the land or ice edge, while midseason polar lows were found to occur further south away from the land as the polar vortex expands. The polar low frequency decreases late in the season as the polar vortex breaks down.

Polar lows form on a small synoptic or subsynoptic scale and often intensify rapidly (e.g. Rasmussen, 1985; Shapiro et al., 1987; Rabbe, 1987). The temporal and spatial scale of these systems, and the fact that they form in data-sparse regions, makes them difficult to forecast, although recent advancements in satellite technology and increased forecast model resolutions have helped. Research into polar lows has been conducted since the 1960s and 1970s (e.g. Lyall, 1972; Harrold and Browning, 1969; Mansfield, 1974; Rasmussen, 1977, 1979, 1981). Businger and Reed (1989b) provide an excellent in-depth overview of the polar low research conducted until this time. Even though several field campaigns, such as the Arctic Cyclone Experiment (ACE, Shapiro and Fedor, 1986) and the Lofotes cyclone experiment (LOFZY 2005) have been conducted, in situ data of these systems are rare due to their time and spatial scales, with studies by Shapiro et al. (1987) and Douglas et al. (1995) being the exception rather than the rule.

A number of more recent investigations have made use of satellite data in their investigations of polar lows. Lieder and Heinemann (1999) used AVHRR data and ERS and SSM/I retrievals over the Antarctic region of the Southern Pacific and observed that the polar low of interest developed through baroclinic forcing associated with an upper-level trough. Moore and Vonder Haar (2003) used data from the Advanced Microwave Sounding Unit (AMSU) to examine a polar low that occurred over the Labrador Sea on 17-18 March, 2000 and found that the warm core structure of the polar low was clearly identifiable, being

2-3 K warmer than the background environment. This could be used to track storm motion. Blechschmidt (2008) utilized thermal infrared satellite imagery and satellite derived wind speeds to develop a two-year polar low climatology of 90 polar low events for the Nordic seas.

Numerous modeling studies of polar lows have been conducted over a variety of regions including Hudson Bay (e.g. Albright and Reed, 1995), the Labrador sea (e.g. Pagowski and Moore, 2001), the Denmark Strait region (e.g. Sardie and Warner, 1985), the north coast of Norway (e.g. Nordeng and Rasmussen, 1992), the Bering Sea (e.g. Bresch et al., 1997) and the Japan Sea (e.g. Yanase et al., 2004). Numerical simulations have enhanced our understanding of these systems primarily by isolating the dominant mechanisms of formation within model sensitivity runs, but also through representing processes that are typically under sampled in the data-sparse regions of polar low formation.

While polar lows appear to have many attributes in common, there is no widely accepted classification scheme for these systems (Rasmussen and Lystad, 1987). Several different schemes have been suggested based on distinctive synoptic setups (Lystad et al., 1986; Rasmussen and Lystad, 1987; Reed, 1987; Businger and Reed, 1989a), and on cloud patterns and shapes observed within satellite imagery (Carleton, 1985; Forbes and Lottes, 1985). Rasmussen (1983) differentiated between the spiral-shaped polar lows and comma-shaped polar lows, and referred to the spiral-shaped case as real or true polar lows. Businger and Reed (1989a) distinguished three types of polar lows based on distinctive synoptic patterns, which provided a distinct degree and distribution of baroclinicity, static stability, and surface latent and sensible heat fluxes. Their three types included (1) the short-wave/jet-streak type (comma clouds) which are characterized by deep, moderate baroclinicity, modest surface fluxes, and a secondary vorticity maximum and PVA aloft; (2) the Arctic-front type which are characterized by shallow baroclinicity, strong surface fluxes and ice boundaries; and (3) the cold-low type which are characterized by weak baroclinicity, strong surface fluxes and deep convection. A combination of these types can exist too. Businger and Reed (1989b) also noted that the comma shape arises as a result of the location of a positive vorticity center within a moderate background wind flow; when the latter is weaker, a spiral pattern tends to develop.

Rasmussen and Turner (2003) tried to improve on the Businger and Reed (1989a) classification scheme by developing a scheme that was based partly on the synoptic setting, and partly on the mechanisms driving polar low formation. Their scheme consisted of seven different types of polar lows observed to occur over the Norwegian Sea (trough systems, cold lows, orographic polar lows, comma clouds, boundary layer fronts, reverse shear systems and baroclinic wave-forward shear types). More recently, a classification scheme for polar low events over the Nordic seas was developed using satellite observations and NCEP re-analysis data that divides polar lows into four types (western polar lows, eastern polar lows, Greenland lee polar lows and storm track polar lows) (Blechschmidt et al., 2009). These types are distinct in sea level pressure, upper level geopotential height, and the ocean and upper level temperature difference. Pronounced upper level cold troughs or lows of the circumpolar vortex, and large differences between the upper level and sea surface temperatures indicative of strong vertical instability, were common to all of the types.

Debate continues as to whether there are a number of different types of polar lows that are generated by different forcing mechanisms, or whether a range of polar low morphology develops from the same forcing mechanism or combination of forcing mechanisms. A number of mechanisms for the development and intensification of polar lows have been suggested, including baroclinic instability, CISK, air-sea interaction instability/wind-induced surface heat exchange instability, barotropic instability, and the interaction between mobile upper-level PV anomalies and lower-level PV anomalies. While barotropic instability has been observed in association with polar low development (e.g. Mullen, 1979), the contributions from this instability to rapid polar low development appear to be minor (Reed, 1979; Sardie and Warner, 1985) and will not be discussed further here.

Significant baroclinicity has been observed through deep layers of the atmosphere in association with polar lows (e.g. Harrold and Browning, 1969; Mullen, 1979; Reed, 1979; Locatelli et al., 1982). Harrold and Browning (1969) demonstrated that precipitation associated with a polar low passing over England occurred primarily as a result of slantwise ascent, typical of baroclinic disturbances, and Nordeng (1987) found, from their simulations of two polar lows, that the inclusion of a slantwise convection parameterization scheme resulted in the development of stronger low-level winds in both cases. Reed (1979) analyzed two polar low cases in detail and found that they formed poleward of the jet stream where the atmosphere was conditionally unstable and characterized by weak to moderate baroclinicity. A composite analysis of 22 polar lows by Mullen (1979) demonstrated that polar lows developed within deep baroclinic zones on the low-pressure side of well-developed jet streams in regions of strong cyclonic shear. The lower troposphere was conditionally unstable and strongly heated by the ocean in the early stages of development, which was 2-6 K warmer than the air. These conditions were also present in a case study by Rasmussen (1985). Reed and Mullen both concluded that the formation of polar lows is probably a result of baroclinic instability in the presence of low static stability in the lower troposphere as cold air flowing around the large-scale cyclone is heated by the warmer ocean surface. The weak static stability explains the mesoscale size of the system (Staley and Gall, 1977). Mansfield (1974) and Duncan (1977) both showed that polar low development could be supported over realistically short spatial and temporal scales by an environment characterized by a shallow baroclinic layer and low static stability.

Locatelli et al. (1982) found that a number of comma clouds have wind, temperature and precipitation patterns similar to their larger extratropical cousins that form along the polar front. Relatively intense comma clouds have occasionally been observed to develop over land in the absence of significant surface sensible and latent heat fluxes, thus indicating that baroclinic instability alone may be sufficient to drive these systems (Reed, 1979). Businger (1985) analyzed a composite of 42 cases of well-developed polar lows over the Norwegian and Barents seas and found large-scale conditions similar to those observed previously (Reed, 1979; Mullen, 1979; Rasmussen, 1983). In particular, the lows developed in a strong baroclinic region of very low static stability under a region of positive vorticity advection. His study also showed an outbreak of deep convection at the time of rapid deepening, thus suggesting that latent heating may play a role in the deepening process. The conditionally unstable air mass in which polar lows occur favors the development of convective clouds. A series of modeling studies (e.g. Mansfield, 1974; Duncan, 1977; Staley and Gall, 1977; Orlanski, 1986) demonstrated the importance of reduced static stability and moist processes within a region of baroclinicity in enhancing polar low development. Mudrick (1987) found that the initial stages of polar low development could be explained by dry baroclinic instability. However, the simulated polar lows in the dry baroclinic simulations moved at greater speeds than the observed systems (Reed and Duncan, 1987), thus indicating the potentially important role that latent heat release may have on the structure and growth of these systems.

Shapiro et al. (1987) were the first to make use of research aircraft data to study polar lows. They suggested that the initiation of polar lows was due to baroclinic forcing on a synoptic scale. It was not possible from their observations, however, to determine the relative roles of mesoscale baroclinicity or convection. Bond and Shapiro (1991) examined two polar lows that developed over the Gulf of Alaska during the OCEAN STORMS field experiment. The polar lows developed near the center of an occluded, synoptic-scale extratropical cyclone, within a zone of low-level mesoscale baroclinic forcing. Satellite and radar imagery demonstrated that convection was insignificant during the growth phase of the polar lows, and they concluded that the polar low development in this case was primarily due to moist, baroclinic processes.

More recently, Yanase and Niino (2005, 2007) performed idealized numerical experiments to examine the impacts of baroclinicity, stratification and average temperature on the development of polar lows. In their first study they found that a polar low develops within a simplified atmosphere due only to the effects of baroclinic instability and diabatic effects. They also observed that a polar low with spiral cloud bands, an "eye" and a warm core develops when the baroclinicity of the basic state is low, whereas a larger-scale polar low with a comma shape cloud developed when the basic state baroclinicity is high. In their second study they showed that baroclinicity is the dominant factor controlling polar low dynamics. When the baroclinicity was weak, positive interactions between the vortex flow and cumulus convection resulted in a small, weak, almost axisymmetric vortex. Surface friction was found to play an important role in the organization of cumulus convection within the vortex through its impacts on the transport of heat and moisture to the vortex. When the baroclinicity was strong, a larger vortex developed that contained a comma shape cloud pattern. In this case, condensational heating was found to be important, as was the generation of eddy kinetic energy from both the eddy available potential energy and from the mean kinetic energy through the vertical shear. The former case was sensitive to the initial perturbation while the latter was not. Finally, the characteristics of the polar low changed smoothly without any significant regime shifts as the baroclinicity was changed.

Some researchers have found that CISK or other heating mechanisms are important in polar low formation. CISK (Charney and Eliassen, 1964) processes represent a positive feedback between cumulus convection and the low-level convergence involved. In the context of a polar low, convergence associated with the rotating vortex enhances the moisture available for convection, which in turn supplies latent heating that intensifies the circulation, which then provides more moisture to the convection. Numerical simulations suggest that this process appears to play an important role in the development of polar lows (e.g. Økland, 1977, 1987; Rasmussen, 1977, 1979; Bratseth, 1985). Both Bratseth (1985) and Økland (1987) showed that convective heating needs to reach a maximum at lower levels for CISK processes to be significant. Locatelli et al. (1982) analyzed the mesoscale structure of three polar lows and found convective rainbands and cells similar to those that occur with large-scale Pacific extratropical cyclones. The latent heating in the convection may have been the primary mechanism for system deepening as discussed by Rasmussen (1979, 1981).

An observational (Fu et al., 2004) and modeling study (Yanase et al., 2004) was conducted of a 200 km wide polar low that developed over the Japan Sea. The polar low initially developed in association with an E-W orientated cloud band and produced a spiral-shaped cloud pattern with an "eye" structure. A number of model sensitivity tests demonstrated that condensational heating was the primary factor causing the rapid development of this polar low, and that surface fluxes were important for maintaining an environment that supports the development of the low-level vortex. Vortex development was strongly suppressed in the absence of surface fluxes due to the resultant stabilization of the boundary layer.

Miner et al. (2000) recently documented the development of a polar low, similar to the cold-low class described by Businger and Reed (1989a), except that this polar low developed during early autumn over the interior of North America, and subsequently intensified as it moved over the Great Lakes. This cyclone began as a cold-core cyclone but evolved into a warm-core system, and developed an eye and spiral bands. The heat and moisture fluxes from the Great Lakes appear to have played a significant role in the development of the system based on the fact that the cyclone deepened rapidly in the presence of weak baroclinicity, the surface heat and moisture fluxes were comparable in magnitude to those of other polar low and Category 1 hurricane cases, the low strengthened more at lower levels than at upper levels, and the cyclone evolved from a cold-core structure into a warm-core structure.

Two polar lows developed, one behind the other, on 7 March 2005 over the warm Norwegian current. This area is a region of frequent polar low formation (e.g. Mokhov et al., 2007; Bracegirdle and Gray, 2008). Such serial polar low developments have been observed elsewhere (e.g. Reed and Duncan, 1987; Hewson et al., 2000). Brümmer et al. (2009) analyzed the properties of these polar lows which developed within the left exit region of a jet streak. They found that both polar lows had a radius of 100-130 km and extended to about 2.5 km in height. The systems were warm-core systems with temperature anomalies of 1-2 K compared with the ambient environment. In situ mass, water and sensible heat budgets showed that about twice as much of the moisture supply in the subcloud layer came from evaporation compared with convergence, and that, while almost all of the condensed water within these systems was converted into precipitation, only about half of the precipitation at cloud base actually reached the surface. Interactions between the two polar lows observed previously (e.g. Renfrew et al., 1997) could not be detected here.

While some studies of polar low formation have emphasized the role of either baroclinic effects or CISK (or other heating mechanisms), other research has suggested that the interaction of both these mechanisms may be important. Simulations conducted by Sardie and Warner (1983) suggested that both CISK and moist baroclinicinty were important in polar low formation, but that moist baroclinic processes alone were insufficient to generate comma clouds. Forbes and Lottes (1985) also observed the importance of both mechanisms from their climatology of Atlantic polar low cases. Sardie and Warner's (1985) simulation of a Denmark Strait polar low demonstrated that low-level baroclinicity was sufficient to initiate development of the low, but that sensible heating from the surface and latent heating associated with convective and nonconvective heating were essential to sustain the development as the low moved away from the baroclinic zone. In their simulation of a Pacific low, baroclinicity and latent heating were also important, but sensible heating from the surface had little effect on the time scale of the development. They concluded from their modeling studies that the development of polar lows depends on both CISK and moist baroclinicity, but that the relative importance of these processes varied from environment to environment. Moist baroclincity dominates the formation of the comma-shaped polar lows in the North Pacific, while CISK is the dominant contributor to the spiralform polar low formation in the North Atlantic. The former formed over water, near the polar front and away from ice edges, whereas the latter developed along the marginal ice zone. Businger (1987) showed that ice edge processes were important to the formation of polar lows over the North Pacific. Craig and Cho (1988) also examined the roles of CISK and baroclinic instability by combining the Eady model of baroclinic instability with wave-CISK. They found that the role of heating was to reduce the static stability, which resulted in faster growth and shorter wavelengths. For small heating rates, perturbations about a baroclinic base state resemble a baroclinic wave. As the heating was increased, the instability transitioned and took on the characteristics of a CISK disturbance.

As only a few polar lows have been observed to occur over land, researchers generally agree that air-sea interaction processes are important to their development. Emanuel (1986) proposed that an air-sea interaction instability (ASII), more recently renamed as wind-induced surface heat exchange instability (WISHE) (Emanuel et al., 1994), plays a significant role in the formation of tropical cyclones. Craig and Gray (1996) provide a detailed comparison of CISK and WISHE. Strong surface winds and decreasing pressure associated with tropical cyclones lead to anomalous sea-surface fluxes of sensible and latent heat, which in turn lead to enhanced temperature anomalies, and hence to decreases in central pressure and increases in surface winds, and so on. Given the large latent and sensible heat fluxes associated with polar lows (e.g. Shapiro et al., 1987), values that are comparable in some cases with those observed in tropical cyclones, this instability may be important in polar low formation. Rasmussen and Lystad (1987) do note, however, that CAPE is large in the cold air outbreaks associated with polar lows and hence the reasoning applied to tropical cyclones may not be applicable to polar lows. Emanuel and Rotunno (1989) used simulations to argue for ASII and against CISK as the dominant energy source in polar low development. They did, however, note that a triggering mechanism is needed before the ASII occurs.

Craig and Gray (1996) used a cloud-resolving axisymmetric simulation to examine the relative contributions of CISK and WISHE to polar low intensification. As they point out, for both WISHE and CISK, the rate of growth is controlled by boundary layer processes. In the WISHE theory, the rate limiting process is the rate at which latent and sensible heat is fluxed from the ocean surface. For the CISK mechanism, rapidly acting surface fluxes are assumed to maintain CAPE, and hence the rate limiting process in this theory is the fractional mass convergence. Thus, the rate of growth is controlled by heat and moisture fluxes in WISHE, but by frictional convergence in CISK. Craig and Gray's modeling results demonstrate that the rate at which their simulated cyclone systems intensified, increased with increasing values of the heat and moisture transfer coefficients, and hence with increasing surface heat and moisture fluxes. Frictional convergence was found to be of secondary importance. Hence they conclude from their results that the intensification of modeled polar lows is due to WISHE, rather than CISK.

More recently, the role of upper-level PV anomalies in the initiation and development of polar lows have been recognized (e.g. Mullen, 1983; Businger, 1987; Nordeng, 1990; Businger and Baik, 1991; Shapiro et al., 1987; Montgomery and Farrell, 1991, 1992; Nordeng and Rasmussen, 1992). Nordeng (1990) suggested that polar low development occurs in two phases in which baroclinic interaction first occurs between an upper-level trough and a surface disturbance, followed by an ASII. The results of simulations by Nordeng and Rasmussen (1992) indicated that upper-tropospheric forcing plays an important role in the organization of the ascent that leads to the spinup of the lower tropospheric vortex. Through the use of a two-dimensional semigeostrophic Eady model Montgomery and Farrell (1991) found that an initial disturbance with interior PV exhibited strong baroclinic coupling between the upper and lower disturbances compared with those situations of uniform interior PV. Their simulations also demonstrated a two phase development of polar lows, beginning with an initial baroclinic growth phase, and being followed by a long, slow intensification due to diabatic effects. Montgomery and Farrell (1992) then used their results from a three-dimensional nonlinear geostrophic momentum model to further develop their two stage conceptual model of polar low development. During the first stage of development, a stage they refer to as induced self-development, an interaction occurs between upper-level and lower-level PV anomalies in a nearly moist neutral baroclinic atmosphere. Ascending motion ahead of the advancing trough results in rapid spinup in the lower levels and the production of PV anomalies at low and midlevels, which enhances the baroclinic interaction between the upper-level and lower-level systems. In the second stage, referred to as diabatic destabilization, a secondary intensification occurs as a result of the production of low-level PV in ascending regions through diabatic processes. The secondary phase appears to progress more slowly than the primary phase. Diabatic destabilization thus provides a mechanism through which polar lows can maintain or enhance their intensity until they reach land, as is often observed. While CISK and/or ASII may contribute to system enhancement during the later stage of development, neither appeared to be necessary.

Douglas et al. (1995) compared the characteristics of a 300 km diameter polar low that developed over the northern Gulf of Alaska during the Alaska Storms Program (Douglas et al., 1991) and a 400 km diameter polar low that formed along the ice edge of east Greenland during the 1989 CEAREX experiment. Frontal zones were observed to occur with the CEAREX polar low, and the cloud field was more similar to that of an extratropical cyclone than the Alaskan polar low. The vortex of the Alaskan low was warmer than its surroundings at lower levels, and suppressed cloudiness characterized the vortex center, possibly occurring as a result of warm air seclusion similar to in the Shapiro-Keyser model of extratropical cyclone development discussed previously in this chapter. Unlike the Alaskan low, the CEAREX low did not have a clearly defined warm inner core. Convection was deeper in the Alaskan case. In spite of the fact the CEAREX polar low occurred over open water, significant intensification did not appear to occur, suggesting that surface heat and moisture fluxes were not necessary for further development, although it is possible that the conditions were not suitable for ASII processes to occur. Also, as the convection was shallow, this case suggests that the upper-tropospheric short-wave was not coupled to the lower-level vortex via convection. Instead, it appears that the formation and intensification of the polar low occurred through the interaction between an upper-level mobile PV anomaly and a lower-level PV anomaly (Montgomery and Farrell, 1992).

Finally, Bresch et al. (1997) simulated the development of a polar low that occurred over the western Bering Sea in order to examine the processes important to its development. Observations of this case study showed that the polar low formed in a region of moderate low-level baroclinicity near the ice edge when a region of high PV associated with an upper-level trough advected into the region. Various sensitivity tests were conducted including, turning off the surface fluxes which failed to develop a polar low, switching on the surface fluxes after 24 hours in which only a weak low developed, and increasing the distance from the ice edge which had small positive effects. Experiments which included surface fluxes but no latent heating, and then latent heating but no surface fluxes, produced polar lows of similar intensity, demonstrating that these two processes are equally effective in enhancing development. They concluded from their experiments that the development of polar lows of this type are similar to that of marine extratropical cyclones, and that polar lows require the interaction between a mobile upper-level PV anomaly and a low-level PV anomaly generated either through thermal advection or diabatic heating.

From the discussion above it seems that neither a single mechanism nor a single structure or shape appears to govern the initiation and development of polar lows, and that the mechanisms are not discrete but rather continuous. Perhaps as Bresch et al. (1997) suggest, polar lows should be viewed as part of a wide spectrum of maritime cyclones that varies based on the strength of the upper-level forcing, tropospheric stability, degree of baroclinicity, deep moist convection, the amount of latent heat release and the magnitude of the surface heat and moisture fluxes, rather than as a discrete type of oceanic cyclone.

### 10.3.7. Lake-Effect Storms

During the fall and winter months, when cold Arctic air associated with the passage of extratropical cyclones sweeps across the relatively warm waters of the Great Lakes, or other large lakes, local, often heavy snowstorms occur along the lee shores. Most of the heavy snowfall does not occur during the passage of cold fronts, but several hours afterward. In some of the more severe storms, snowfall accumulations of more than 75 cm per day are not uncommon (Wiggin, 1950) and snowfall rates of 30 cm hr$^{-1}$ have been reported (E.S.S.A., 1966). A single storm event produced 175 cm (68.9 in) of snow in Ohio (Schmidlin and Kosarik, 1999), and totals of 150-250 cm over several days have been observed (Niziol, 1989). An important feature of these storms is their persistence for several days over limited and sharply defined regions. These events often also pose a forecasting hazard as one location may receive more than 100 cm of snow over several days while locations only 20 km may receive just a trace.

Lake effect events have previously been defined as those events where heat and moisture fluxes from a lake surface result in the development of an internal convective boundary layer and associated clouds (e.g. Lenschow, 1973; Chang and Braham, 1991; Kristovich and Laird, 1998). An excellent overview of the climatology, characteristics and factors influencing lake effect snow is provided by Niziol et al. (1995), a summary of which is included here. Lake effect precipitation tends to be greatest earlier in the cold season, as the decline in the lake temperatures and the air-lake temperature difference, and the increase in lake ice, all act to reduce fluxes of heat and moisture from the surface of the lake as winter progresses (Jiusto et al., 1970). In general, lake effect processes are most effective when the prevailing wind blows across the greatest fetch of water. Orography on the lee side of the lake may further enhance lake effect snow. Muller (1966) observed an increase of 12-20 cm in the annual snowfall for every 330 m increase in elevation to the lee of the lakes. The transfer of sensible and latent heat from the lake to the atmosphere can trigger convection that is typically organized into long, linear features known as cloud streets or cloud bands, the orientation of which is determined by the wind profile between the lake surface and the subsidence inversion associated with the Arctic air mass. The bands tend to be aligned parallel to the steering wind, the exception being when thermally-induced circulations dominate in conditions of weak winds. While lake effect snow is generally thought to occur on the eastern sides of the Great Lakes in association with typical westerly flow over the lakes, cyclones that track to the south of the Great Lakes can produce lake effect snow on the western lake shore due to the easterly flow over the lakes. Lake effect processes and frontal processes can also interact, in that the mid- to upper-level frontal clouds can seed the lower lake-induced clouds (Kristovich et al., 2000). Lake effect snowstorms appear to be produced by complex interactions between processes on a wide range of scales from the microphysical through to meso-$\alpha$ scale, and it is this complexity and range of spatial and temporal scales of the processes that make lake effect snow difficult to forecast. Several recent studies have also shown that lake effect snowfall has increased over the twentieth century, a statistic attributed to warmer lake waters and decreased ice cover (Norton and Bolsenga, 1993; Burnett et al., 2003; Kunkel et al., 2009).

Arctic air masses are often accompanied by a strong subsidence inversion, which limits the depth to which convection can grow. The fluxes of heat and moisture from the lake surface can lift and/or erode the inversion (Lavoie, 1972). Thus the low-level instability (air-lake temperature difference) and the depth of the unstable layer (height and intensity of the inversion) are important factors controlling the intensity of lake effect snow. Also, differences in the temperature and surface roughness between the lake and the shore generate low-level thermal gradients and frictional convergence, which result in organized vertical motion that can aid in lifting the capping inversion, thereby enhancing convective development and precipitation (Hjelmfelt, 1990; Niziol et al., 1995). A conceptual model of a major snowstorm over Lake Erie developed by

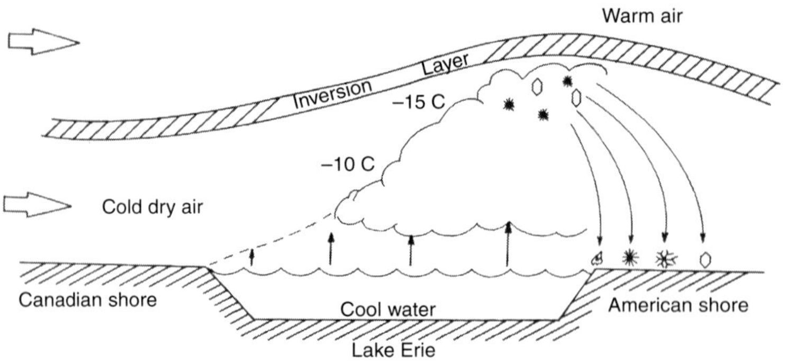


**FIGURE 10.26 Schematic cross section of lake effect snow.** *(After Davis et al. (1968))*

Davis et al. (1968) is shown in Fig. 10.26. Illustrated is a stratocumulus cloud that deepens over the lake as it nears the lee shore. Corresponding to the deepening cloud layer is an upward displacement of the capping inversion. The faster falling, heavily rimed snow crystals such as graupel particles precipitate just onshore. This is also the location of the heaviest snowfall amounts. Lightly rimed dendrites and plates are carried further inland. Aggregates of snow crystals having terminal velocities between graupel and unrimed single crystals settle somewhere in between.

Four general types of lake effect storms have been identified over Lake Michigan (Braham and Kelly, 1982; Forbes and Merritt, 1984; Hjelmfelt, 1988). Hjelmfelt (1990) was able to simulate all four types simply by varying the lake-land temperature difference, the wind direction and strength, and the static stability. Schematics of these four storm types are shown in Fig. 10.27, and include the following:

(1) Widespread coverage/wind parallel bands (e.g. Kelly, 1982, 1984; Braham, 1986; Kristovich, 1993; Steve, 1996): the winds in association with these bands show no reversal on the lee shore; the bands may become organized into wind parallel bands (Kelly, 1982, 1984), open convective cells (Braham, 1986) or a combination of these two types (e.g. Kristovich et al., 1999; Cooper et al., 2000); they occur during strong westerly winds, strong static stability, and strong lake-land temperature differences. With band widths of 2-4 km and spacings of 8-20 km, these bands tend to produce widespread snowfall of low intensity. Wind parallel bands may join to form less regular cloud morphologies.

(2) Shoreline parallel bands (e.g. Braham, 1983; Hjelmfelt and Braham, 1983; Schoenberger, 1986): these bands occur with a well-developed land breeze on the lee shore (Braham, 1983); they occur during moderate westerly winds, weaker static stability and a strong lake-land temperature difference.

(3) Midlake bands: these bands are associated with low-level convergence centered over the lake (e.g. Passarelli and Braham, 1981); they occur when

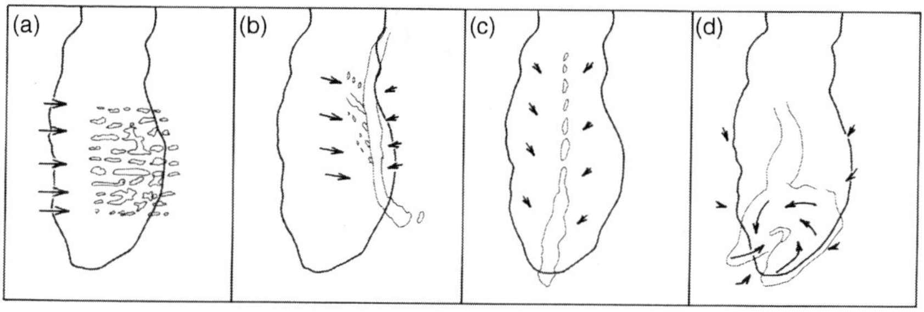


**FIGURE 10.27 Schematic of four morphological types of lake effect snow storms over Lake Michigan as defined from observations: (a) broad area coverage, (b) shoreline band, (c) midlake band, and (d) mesoscale vortex.** *(From Hjelmfelt (1988))*

westerly and easterly winds from both sides of the lake converge, and in association with moderate lake-land temperature differences and weak static stability.

(4) Mesoscale vortices: these vortices are associated with a well-developed cyclonic flow in the boundary layer, usually occurring with weak surface-pressure gradients and a ridge of high pressure centered over the lake or to the west of the region (e.g. Peace and Sykes, 1966; Forbes and Merritt, 1984; Pease et al., 1988); they occur under conditions of strong lake-land temperature differences, weak northerly winds and weak stability.

A mix of these morphologies may also occur (Shoenberger, 1986). It should be noted that the emphasis in the modeling study by Hjelmfelt (1990) was on the east coast of Lake Michigan, and hence the conditions listed above are somewhat specific to this region. These morphologies could however be expected to exist under other flow regimes over different lakes. Five snow band types have been identified for the eastern Great Lakes (Niziol et al., 1995). In the classification above the wind parallel bands were not separated into those that form when the winds blow parallel to the long axis of the lake and those that form when they blow perpendicular to the long axis of the lake, an effect that turns out to being important for the eastern Great Lakes.

A number of typical lake effect morphologies occurring simultaneously are evident in the GOES visible satellite image shown in Fig. 10.28 (Laird et al., 2003a). A widespread coverage event, consisting of both wind parallel bands and cellular convection is evident over and downwind of Lake Superior, being associated with unidirectional winds across the lake, and producing light to moderate snowfall over and downwind of the lake. A shoreline band is evident over Lake Michigan. This band develops in association with a thermally-driven land breeze circulation and/or prevailing winds nearly parallel to the major axis of the lake, and typically produces heavy snow and the strongest updrafts of all four band types. Finally, a vortex is evident over Lake Huron. These events

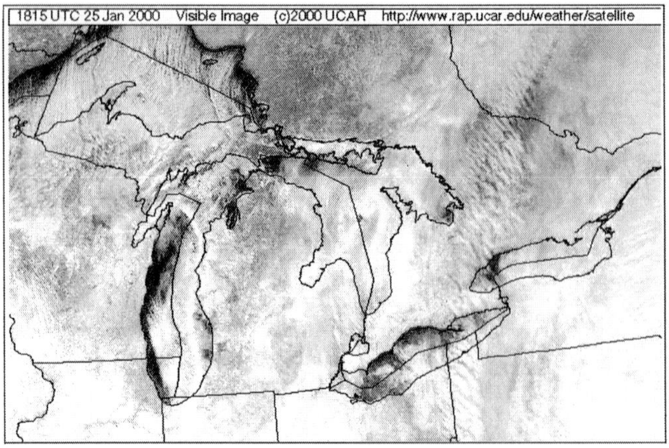


**FIGURE 10.28 Typical mesoscale lake effect structures.** Geostationary Operational Environment Satellite (GOES) visible image at 1815 UTC 25 January 2000 of a mesoscale vortex, shoreline band and widespread coverage event over Lakes Huron, Michigan and Superior, respectively. *(From Laird et al. (2003a))*

are typically associated with light snow and weak land breezes, and may have a cloud-free region in the center.

The most common convective pattern over most lakes appears to be the widespread type, which accounts for nearly 60% of cloudiness during December and January (Kelly, 1986; Kristovich and Steve, III, 1995). Also, the widespread wind-parallel bands of convection associated with boundary layer rolls are generally associated with strong cross-lake winds. Convective bands that develop parallel to the long axis of a Great Lake tend to occur in conditions of high air-lake temperature differences and weak contributions from the cross-lake wind component. The midlake bands and the shoreline bands tend to be wider than the wind parallel bands (Braham and Dungey, 1995), and can produce very heavy snowfall over and near the lake shore. Midlake bands and shoreline bands both form from a combination of dynamic and diabatic forcing. Midlake bands may develop when strong winds blow along the lake major axis resulting in enhanced surface heat fluxes, the generation of a mesoscale low pressure over the lake, the development of a land-breeze circulation, and the subsequent development of a midlake convergence zone and associated midlake snowband (e.g. Braham and Kelly, 1982; Niziol et al., 1995). Shoreline parallel bands can develop when the winds are calm, and the strong horizontal temperature gradients associated with large lake-land temperature difference, results in the development of a mesoscale overlake low, the formation of a land breeze, and the development of the shoreline parallel bands (e.g. Passarelli and Braham, 1981). The shoreline bands may even occur in the absence of a land breeze, providing strong convergence develops on the lee side of the lake, although these systems will tend to be weaker and shorter-lived than those associated with a well-developed land breeze. Given the similarities in their formation, midlake and shoreline bands are occasionally grouped together into the same category (e.g. Laird et al., 2003a,b). Lake vortices appear to be a weak version of the polar lows described above. They have been observed to occur with shoreline parallel or midlake snowbands (e.g. Pease et al., 1988; Grim et al., 2004), or with other lake vortices (Laird, 1999). These systems range in size from 10 km (Schoenberger, 1986) to over 100 km (Forbes and Merritt, 1984). In the vortex studies of Schoenberger (1986) and Grim et al. (2004), a land breeze front propagated westward into an area of horizontal convective rolls. Upward tilting of the rolls by the land breeze front appeared to be responsible for the vortex generation, although the observations in the Grim et al. (2004) could not confirm this.

The most intense snowstorms are the shoreline parallel bands which occur over the Great Lakes following the passage of a cold front when the following conditions are met (Sheridan, 1941; Wiggin, 1950; Rothrock, 1969):

(1) the difference between the lake surface temperature and the 850 mb temperature exceeds 13 °C corresponding approximately to a dry adiabatic lapse rate between the surface and 850 mb;
(2) an onshore wind with an overwater fetch of more than 100 km is present;
(3) over water, low-level wind speeds are moderate to strong; and
(4) the height of the boundary layer capping inversion exceeds 1000 m.

Ideas regarding lake effect snow were developed as early as the 1950s using observations. Wiggin (1950) noticed that the width of the snowbands was proportional to the depth of the cold air, while Peace and Sykes (1966) suggested that pressure troughs and narrow zones of confluence were important to snowband formation. A number of numerical simulations were also conducted of lake effect storms in the 1970s and 1980s. Using a simple mixed-layer model, Lavoie (1972) demonstrated the importance of wind speed and direction to the strength of the mesoscale disturbance, with the longest fetch of air across the lake surface producing the strongest disturbance. Both Lavoie (1972) and Ellenton and Danard (1979) showed that the heat and moisture fluxes from the lake were the principal causes of precipitation, while shoreline convergence and orographic lift played secondary roles. Likewise, Hsu (1987) showed that the coupling between surface heat fluxes and winds were responsible for the formation of patterns of low-level convergence that generated precipitation. Hjelmfelt and Braham (1983) simulated the lee-shore parallel band over Lake Michigan, including the formation of a surface mesolow beneath the major cloud band. Latent heat release was found to be important in strengthening the convection, although the land-breeze circulation that was associated with the band formed even without latent heat release. Ballentine et al. (1992), like Hjelmfelt (1990), made use of sensitivity tests to examine the significance of water temperature, land and air temperature, wind speed and direction, and the humidity on a lake effect snowband.

Recent advances in both satellite and radar technology, as well as computer modeling, have enhanced our understanding of lake effect processes. Agee and Gilbert (1989) made use of data from the Lake-Effect Snow Studies (LESS) field campaign and identified a penetrative convective layer located between the top of the fully mixed layer and the part of the inversion layer that has not been modified. This penetrative convective layer creates a large amount of turbulent exchange between the mixing layer and the capping inversion. Chang and Braham (1991) examined the development of the convective thermal internal boundary layer over Lake Michigan and found that the average slope of the boundary layer was 1% over a fetch of ~125 km. Byrd et al. (1991) used sounding data to investigate lake effect bands and suggested that the depth of the mixed layer may be more important than the strength of the instability precipitation intensity, while Burrows (1991) identified low-level mass divergence as the primary parameter controlling the amount of snow produced. Recent observations from the Lake-Induced Convection Experiment (Lake-ICE) demonstrated the importance of air-lake fluxes and the vertical structure of the atmosphere in the development of lake-effect storms (Scott and Sousounis, 2001).

The presence of ice on a lake surface reduces the heat and moisture fluxes from the surface, and thus will influence lake effect storms (e.g. Niziol et al., 1995; Laird and Kristovich, 2004). Using observations, Gerbush et al. (2008) estimated that surface sensible heat fluxes increased to open water values as the ice concentration over Lake Erie decreased from 100% to 70%, while numerical simulations by Zulauf and Krueger (2003) demonstrated that increases in the ice thickness from 0 to 10 cm (10 to 20 cm) were accompanied by decreases in the sensible heat flux of ~50% (25%). The significance of thinner ice on the enhancement of snowbands has also been observed recently over Lake Erie (Cordeira and Laird, 2008). The lake temperature also appears to influence the lightning characteristics of lake effect storms, with lightning occurring predominantly between September and December due to the warmer lake temperatures and the greater convective cloud depths (Moore and Orville, 1990). Lake effect storms with lightning also have significantly higher temperatures and dewpoint temperatures in the lower troposphere, lower lifted indices, higher lake-induced equilibrium levels and CAPE, lower wind shear, and an increase in the mean height of the −10 °C level compared with non-electrified storms (Schultz, 1999; Steiger et al., 2009). Lightning producing storms were also strong, single-band storms.

Several studies of lake effect events have been more climatological in nature. Ellis and Leathers (1996) showed for the eastern Great Lakes that, while the large-scale synoptic situation was similar for each lake effect type, variations in the sea level pressure patterns, 850 mb temperatures and heights, 500 mb heights, seasonality, fetch and the strength of the flow produced significant differences in the location, magnitude and frequency of the lake effect snowfall. In their study of autumnal (Sep-Nov) lake effect precipitation downwind of Lake

Erie, Miner and Fritsch (1997) found that lake effect precipitation occurs on approximately one out of every five days, with a diurnal peak in precipitation intensity during the afternoon and evening. They also noted that the greatest number of lake effect days occurred in October, that lake enhanced precipitation actually begins in late summer (when the lake temperature already starts to exceed that of the surrounding air), and that the precipitation transitions from rain to snow during November. Kristovich and Spinar (2005) examined the diurnal characteristics of lake effect events over Lakes Superior and Michigan and found a morning maximum and an afternoon/evening minimum in the frequency of lake effect precipitation.

The impacts of the intensity, track and duration of synoptic-scale systems were found to have a significant effect on lake effect storms (Ballentine et al., 1998; Schmidlin and Kosarik, 1999). Lake effect processes that are enhanced during the passage of a synoptic low pressure system have been referred to as lake-enhanced snowfall (Eichenlaub, 1979). More recently, Lackmann (2001) observed from a composite of 32 lake effect events over Rochester, New York that all were accompanied either by a mobile upper level trough or a closed low at the 500 mb level. An examination of a particularly intense event suggested that the upper trough results in an increase in the inversion altitude and relative humidity in the lower troposphere. Liu and Moore (2004) developed a climatology of lake effect snowstorms over southern Canada for the years 1992-1999. They observed that a low pressure and cold-temperature anomaly over Hudson Bay, north of the Great Lakes, provides an environment conducive to lake effect storms over southern Ontario, and that the track of the low can have a significant impact on the development of these storms.

Studies of lake effect snowstorms have also been performed for smaller lakes including the Great Salt Lake (e.g. Carpenter, 1993; Slemmer, 1998; Steenburgh et al., 2000; Steenburgh and Onton, 2001; Onton and Steenburgh, 2001), Lake Champlain (e.g. Tardy, 2000; Payer et al., 2007; Laird et al., 2009a), and the New York State Finger Lakes (e.g. Cosgrove et al., 1996; Watson et al., 1998; Sobash et al., 2005; Laird et al., 2009b), as well as other small lakes in the United States (e.g. Wilken, 1997; Cairns et al., 2001; Schultz et al., 2004). Some of the lake effect events have produced significant amounts of snow in spite of the lake size and the localized nature of the storms. Steenburgh and Onton (2001) examined the development lake effect snowstorms over the Great Salt Lake and found that the primary snowband formed along a land-breeze front. They concluded from their study that even though the Great Salt Lake is relatively small in comparison with the Great Lakes, that thermally-driven circulations and banded precipitation structures similar to those over the Great Lakes can still occur. In a follow-on paper, Onton and Steenburgh (2001) conducted a modeling study of snowband development over the Great Salt Lake. They found that moisture fluxes from the lake were necessary for the development of the snowband. However, the saline composition of the lake resulted in a reduction in the moisture fluxes compared to a freshwater lake, causing a 17% decrease in snowfall.

Widespread lake effect events are often composed of boundary layer rolls. Roll vortices have been observed to occur when the convective pattern is organized into wind-parallel bands of heavier precipitation. These rolls have been shown to influence the surface layer fluxes of heat and moisture (e.g. LeMone, 1976). Numerous factors have been suggested for their formation including high wind shear, curvature of wind speed profiles parallel to the roll axes, inflection points in the wind profiles, boundary layer mean shear, and gravity waves (Kristovich et al., 1999 and the references therein). However, Kristovich (1993) found that none of these criteria were met in every one of the cases of roll convection that he studied, and suggested that low-level shear was the primary factor in roll development. Transitions between nonroll and roll convection have been observed. Such transitions may be associated with large changes in the snow spatial coverage and mean snowfall rate (Atkinson and Zhang, 1996). The two most important factors for differentiating between roll and cellular convection according to Atkinson and Zhang (1996) are atmospheric dynamic forcing and thermal instability.

Kristovich et al. (1999) examined the mechanisms that result in the transitions between boundary layer rolls and the more cellular convective structures observed in association with lake effect snow. They found that roll formation occurred following increases in the low-level wind speed and speed shear below $\sim 0.3zi$ where $zi$ is the boundary layer depth. Mass overturning rates were greatest at midlevels in the boundary layer when rolls were dominant and decreased when cellular-type convection was dominant. Mean snowfall rates demonstrated little change with the transitions from one form to the other, but the heaviest snow was more concentrated in the updraft regions when the rolls were dominant. Cooper et al. (2000) found from their simulations that the variation in wind speed shear below 200 m played a major role in the degree of linearity of the convection; directional shear was not a requirement for roll convection. Tripoli's (2005) idealized simulations showed that shoreline geometry effects were sufficient to generate a near-surface streamwise vorticity, which served as the seed for roll development at the most efficient mode of roll convection.

Combined heating and moistening from all of the Great Lakes has been found to influence the atmospheric conditions near the individual lakes, which in turn affects the lake effect storms that may develop. This combined influence of lakes is referred to as the lake-aggregate effect (e.g. Sousounis and Mann, 2000) or collective lake disturbances (Weiss and Sousounis, 1999). Lake-aggregate circulations are on the meso-$\alpha$ scale (200-2000 km) compared with the meso-$\beta$ scale (20-200 km) of the individual lake circulations. An upstream lake can impact the distribution of local lake effect snow (Byrd et al., 1995), and the aggregate effect has been found to enhance lake effect precipitation in northern lower Michigan and in southern Ontario, but reduce it in regions south and east of Lakes Erie and Ontario (Sousounis and Mann, 2000). A factor separation analysis (Stein and Alpert, 1993) of model output to assess the contributions from adjacent lakes (both upstream and downstream) to lake-effect storm evolution showed that interactions amongst lake-scale processes contributed to the development of the regional-scale disturbance and that the relative influence of the adjacent lakes increased as the collective lake disturbance matured (Mann et al., 2002). During the development of the collective lake disturbance, contributions from the lake-lake interactions tended to offset the individual lake contributions, however, as the regional-scale disturbance matured, the lake-lake interactions then enhanced the individual lake contributions. For example, the eastern Great Lakes (downstream lake) were found to reduce the precipitation amounts over southern Lake Michigan by as much as 25% through reduced convective instability effects, and the impacts of Lake Superior (upstream lake) on Lake Michigan caused a delay in the maximum snowfall occurrence. Finally, as the collective lake disturbance matured, the Lake Superior processes were found to have a significant effect on the snowband morphology over Lake Michigan.

Laird et al. (2003a) conducted numerous simulations to examine the factors controlling the meso-$\beta$ scale circulations of lake effect storms. They found that the morphological regimes could be predicted using the ratio of wind speed to maximum fetch distance (U/L). Environmental conditions with low values of U/L ($\sim< 0.02$ m s$^{-1}$ km$^{-1}$) were associated with vortex circulations, those with moderate values of U/L (between 0.02 and 0.09 m s$^{-1}$ km$^{-1}$) with shoreline bands, and those with higher U/L values ($>\sim$ 0.9 m s$^{-1}$ km$^{-1}$) with widespread events. They also noted that transitions from one regime to the next were continuous, and that within transitional regions, more than one regime may exist. Laird et al. (2003b) noticed a change in morphology from vortices to bands to widespread coverage as U/L increased, and that the conditions supporting multiple morphologies were more favorable for elliptical lakes than circular lakes. Laird and Kristovich (2004) then assessed the U/L criteria using observational data and found that, even though the U/L provides useful information regarding the band morphology, the criterion only provides limited use when used for actual forecasting of the band morphology.

Many of the more recent studies of lake effect bands and precipitation have been associated with the Lake-Induced Convection Experiment (Lake-ICE) and the Snowband Dynamics Project (SNOWBAND) conducted during the winter of 1997/98 (Kristovich et al., 2000). Kristovich et al. (2003) examined the microphysical and thermodynamic characteristics of the boundary layer and convective organization of a lake effect event that occurred over Lake Michigan during this campaign and found that, while the horizontal scale of the convective structures grew across the lake, it did so less rapidly than the depth of the convective boundary layer, thus the aspect ratio decreased, which is contrary to previous findings. Miles and Verlinde (2005a,b) observed transient linear organization of the convection for a case during Lake-ICE, even though the conditions should only have supported cellular organization. Mode switching has been previously observed (e.g. Braham, 1986; Kristovich et al., 1999). Miles and Verlinde (2005a) found that the transition between the linear and cellular modes showed no correlation with the mean or low-level shear, surface buoyancy fluxes or stability parameters, thus suggesting that those factors that normally control the linear organization of convection did not affect the transition. In the second paper in this series, Miles and Verlinde (2005b) investigated the possible role of non-linear interactions between different scales of motion in the transient linear organization and found that the net nonlinear interactions between the roll and turbulence scales were significant in magnitude. Nonlinear interactions may thus help to explain the observed transitional linear organization.

Finally, interactions may occur between the synoptic-scale circulations of extratropical cyclones and mesoscale circulations of lake effect storms through boundary layer growth rates and seeding effects (e.g. Lenschow, 1973; Agee and Gilbert, 1989; Chang and Braham, 1991). Schroeder et al. (2006) examined the interaction of a synoptic cyclone with the convective boundary layer in a lake effect event over Lake Michigan. They found that both the precipitation rates and the growth of the convective boundary layer were enhanced through seeding from the clouds above, and the impact of the cyclone on the thermodynamic characteristics of the air over the lake. Not only was the convective boundary layer growth enhanced due to the interaction of the convection with a layer of reduced stability above, but the convective boundary layer was also deeper in the seeded regions compared with the non-seeded regions and appears to be related either to enhanced latent heat release or to mesoscale updrafts. The snowfall rates were similar to those produced in previous lake effect events when the surface heat fluxes were much larger, but the system was not interacting with an extratropical cyclone.

In summary, three major streams of air are involved in the three-dimensional circulation of extratropical cyclones. Because of their different origins and the variety of dynamical and physical processes that they experience in their passage through the cyclone system, the characteristics of the environment vary greatly with time and space. These environments, with different temperatures, moisture content, vertical motion, static stability, and wind shear, host the enormous variety of mesoscale phenomena and clouds present in extratropical cyclones.

## 10.4. MIDDLE- AND HIGH-LEVEL CLOUDS

### 10.4.1. Introduction

Middle- and high-level clouds play an important role in both the radiative and water budgets of the earth, an effect made all the more important given their extensive coverage. On an annual average, clouds cover between 55% and 60% of the earth (Matveev, 1984), and much of this cloud cover

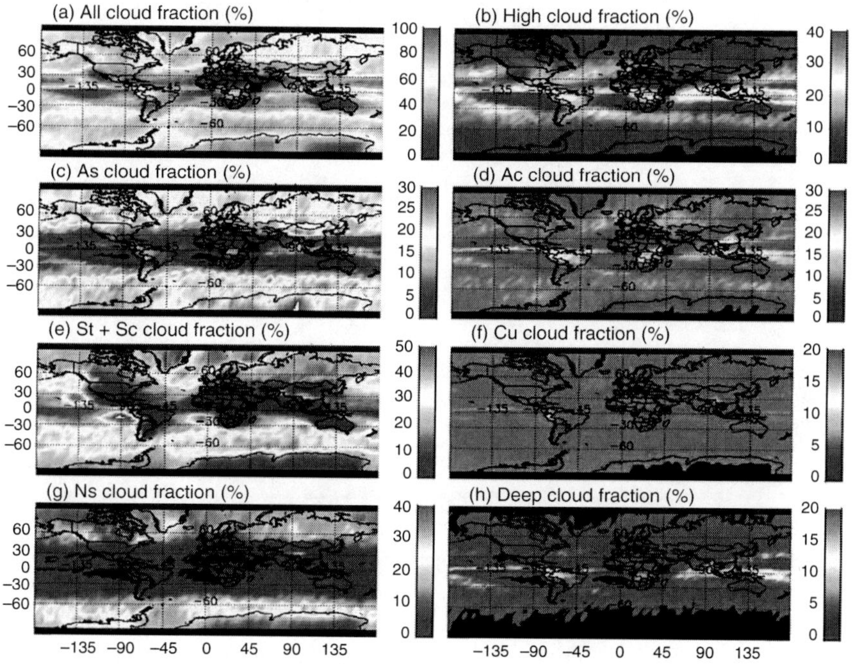


**FIGURE 10.29** **2.5 x 2.5° average of cloud type distributions based on CloudSat measurements over the initial 1-year period.** *(From Sassen and Wang, 2008)*

consists of vast sheets of middle (altostratus and altocumulus) and high (cirrus, cirrostratus, and cirrocumulus) clouds. Global distributions of cloud type generated using CloudSat data (Sassen and Wang, 2008) are shown in Fig. 10.29, and are compared with those from the International Satellite Cloud Climatology Program (ISCCP) (Rossow and Schiffer, 1999) and surface observation reports (Hahn and Warren, 1999) in Table 10.2. As is evident from the "all cloud" fraction panel in Fig. 10.29a, cloud fractions over the mid-latitude storm tracks are high in both the northern and southern hemispheres, reaching a value of ~80% in the southern storm track. High cloud distributions (Fig. 10.29b) are greatest over the ITCZ, coinciding with deep convective systems (Fig. 10.29h; Mace et al., 2006a). The uneven distribution of high clouds over the mid-latitudes is due to the variety of cirrus-forming mechanisms, including those associated with the synoptic jet stream, deep convection and orographic uplift. High frequencies of middle clouds are found poleward of 30° (Fig. 10.29c,d). The distribution of altostratus corresponds to the extratropical cyclone storm tracks, whereas altocumulus clouds are more closely connected with deep convection. Middle and high clouds can produce significant precipitation in association with organized tropical and extratropical cyclonic storm systems.

**TABLE 10.2** Comparison of 1-year CloudSat global cloud type frequency averages over land and ocean with annual means of extended surface observer reports (Hahn and Warren, 1999) and ISSCP annual means from 1986-1993 (Rossow and Schiffer, 1999).

| | CloudSat | | Surface | | ISCCP | |
|---|---|---|---|---|---|---|
| Type | Land | Ocean | Land | Ocean | Land | Ocean |
| High | 9.6 | 10.9 | 23.1 | 14.0 | 19.3 | 15.6 |
| As | 12.7 | 12.0 | 4.8 | 6.5 | 8.7 | 9.7 |
| Ac | 6.8 | 6.7 | 17.2 | 17.1 | 8.6 | 10.2 |
| St + Sc | 13.5 | 22.5 | 18.9 | 39.4 | 10.7 | 18.3 |
| Cu | 1.7 | 1.7 | 4.2 | 9.8 | 7.7 | 12.7 |
| Ns | 8.6 | 8.3 | 6.3 | 7.9 | 3.2 | 3.0 |
| Deep | 1.8 | 1.9 | 3.2 | 5.3 | 2.5 | 2.4 |

**Source:** From Sassen and Wang (2008)

Middle-level clouds refer to altocumulus and altostratus, which may be composed entirely of liquid water or a mix of liquid water and ice. Nimbostratus may be classified either as middle or low clouds and are not considered here. While the elevation of middle-level clouds may vary considerably with season and latitude, a typical elevation in middle latitudes is ~3 km, or about 700 mb. High-level clouds refer to cirrus (including cirrostratus and cirrocumulus) clouds which are composed primarily of ice, although cirrocumulus may be mixed-phase clouds. The elevation of cirrus clouds may also vary considerably; a typical height is 10 km, or 250 mb. Cirrus trap longwave radiation and reflect shortwave radiation, the net effect of which is dependent on cloud thickness; precipitating ice crystals from cirrus clouds can trigger glaciation of warm clouds such as altocumulus clouds, thereby changing the precipitation efficiency of these clouds; the scavenging of aerosols and soluble trace gases and the subsequent ice crystal sedimentation can lead to their redistribution throughout the troposphere; and optically thin cirrus in the tropical tropopause layer (TTL) can dehydrate air entering the tropical tropopause layer (e.g. Holton and Gettleman, 2001). Thus these clouds play an important role in the water budget and the radiation balance in the upper troposphere, and hence in the global climate.

The physical properties of middle-and high-level clouds and their association with extratropical weather systems are discussed in this section, as are the factors influencing their development, and aspects that need to be considered when representing such clouds in GCMs. Given the location of extratropical weather systems, the focus of this chapter will be on high- and middle-level clouds in the mid-latitudes, although comparisons with their tropical counterparts will be made where applicable. Arctic clouds will not be considered here.

### 10.4.2. Characteristics and Formation of Middle- and High-Level Clouds

Middle- and high-level clouds form when the lifting of moist air, often on large scales as in association with extratropical cyclones, adiabatically cools the air to its dew-point temperature. Even at temperatures well below freezing, the clouds almost always form first as water droplets, with the initial condensation occurring on the typically abundant CCN. Direct transformation from water vapor to ice crystals (sublimation) is possible, but rarely occurs due to the scarcity of sublimation nuclei in the atmosphere. The evolution of the cloud after the formation of the first cloud drops depends on two factors - the temperature of the cloud and the vertical motion. At temperatures below freezing, ice processes are likely to be important because of the relative abundance of freezing nuclei in the atmosphere. These nuclei trigger the freezing of supercooled water to ice. Once temperatures drop below $\sim -33$ °C, homogeneous freezing of supercooled droplets becomes likely, and at temperatures below $-40$ °C and relative humidities great than $\sim 80\%$, homogeneous freezing of haze droplets becomes likely. Once ice appears, the difference in saturation vapor pressure over ice and water favors the rapid growth of ice crystals at the expense of the water drops, referred to as the Bergeron-Findeisen process.

If the upward vertical velocities are sufficiently strong (10 cm $s^{-1}$ or greater) and persist for enough time to cause further cooling of the layer, the water drops or ice crystals will grow to sizes wherein fall velocities relative to the air motion become significant and precipitation will occur. The physical processes that produce this growth are collision and coalescence in all water clouds. The Bergeron-Findeisen process, riming of supercooled cloud droplets and aggregation among ice crystals are important in clouds containing both ice and water. For weaker upward velocities (1-2 cm $s^{-1}$), the temperature change resulting from adiabatic expansion is reduced and the temperature changes from radiative effects become more significant. Cooling at cloud top due to longwave radiative flux divergence and warming at cloud base as a result of the longwave radiative flux convergence destabilize the cloud layer and lead to internal convective circulations in the cloud layer. This destabilization is aided by the release of latent heat near cloud base and evaporative cooling near cloud top.

Middle- and high-level clouds frequently occur in layers or sheets of great horizontal dimensions (hundreds or even thousands of kilometers). Because their horizontal dimensions are much greater than their vertical extent, which is typically a kilometer, they are called stratiform or layered clouds. Although this terminology suggests a statically stable lapse rate, thin layers of conditional or convective instability are often present, and these instabilities are conducive to the formation of small-scale convection embedded in the cloud layer. The type of middle or high cloud is strongly affected by the static stability within the cloud layer. Altostratus clouds are produced by the lifting of a layer of air in which the lapse rate is less than the saturated adiabatic lapse rate. For layers in which the lapse rate exceeds the saturated adiabatic lapse rate, vertical convection results in altocumulus clouds. For high clouds, absolutely stable layers are associated with cirrostratus clouds, while cirrus and cirrocumulus clouds occur when the lapse rate exceeds the saturated adiabatic lapse rate. Alternating regions of upward- and downward-moving air associated with gravity waves in a stable cloud layer can result in a banded cloud structure.

### *10.4.2.1. Middle-Level Clouds*

Altocumulus and altostratus clouds together cover about 22% of the earth's surface (Warren, 1988). They often extend through the freezing level and thus may either be liquid-phase or mixed-phase clouds. The relative proportions of liquid water and ice have important implications for microphysical processes, aircraft icing, radiative transfer, and remote sensing. However, thin midlevel clouds are not well represented within GCMs. In a recent study of GCMs, all of the models considered, significantly under-predicted thin, midlevel clouds such as altocumuli, and over-predicted thick midlevel clouds such as nimbostratus (Zhang et al., 2005). Others have also shown that thicker midlevel clouds are also not well represented in large-scale models (e.g. Ryan et al., 2000; Xu, 2005). Compared with other cloud systems, relatively few studies have been conducted on the structure and characteristics of mid-level, mixed-phase clouds. The majority of these studies (e.g. Gedzelman, 1988; Heymsfield et al., 1991; Sassen, 1991; Pinto, 1998; Field, 1999; Cober et al., 2001; Larson et al., 2001; Lawson et al., 2001; Sassen et al., 2001; Fleishauer et al., 2002; Hogan et al., 2002; Korolev et al., 2003; Wang et al., 2004; Carey et al., 2008 and others) are observational studies, with far fewer modeling studies (e.g. Starr and Cox, 1985a,b; Liu and Krueger, 1998; Larson et al., 2006, 2007; Larson and Smith, 2009).

Altocumulus clouds tend to form in distinct layers that may be less than 100 m thick. The formation of these clouds is due to a number of different mechanisms including mountain waves, spreading and decaying convective updrafts and anvils, radiational cooling of moist layers, and the lifting of humid air away from its source in sheared environments such as in frontal situations. The existence of liquid water within middle-level clouds results from the balance between the generation of liquid water through vertical motion and its loss due to the Bergeron-Findeisen mechanism. Riming of liquid drops also represents a loss of liquid water. For conditions of saturation with respect to water, the supersaturation with respect to ice and the Bergeron-Findeisen mechanism both increase with decreasing temperature. The concentration, size, and shape of the ice crystals represent the dominant controls on the Bergeron-Findeisen process.

Heymsfield et al. (1991) sampled two altocumulus clouds and found that the clouds were similar to stratocumulus clouds in that there was extensive cloud top entrainment, a capping inversion and a dry layer above these clouds. Once formed the altocumulus behaved much like stratocumulus with radiatively-cooled air from near cloud top descending into the cloud layer, although the radiative forcing is enhanced in altocumulus clouds as there is more longwave heating at cloud base. Fleishauer et al. (2002) examined six midlevel clouds that occurred over the Great Plains during the Complex Layered Cloud Experiments (CLEX): a single-layer cloud composed primarily of liquid water, three mixed-phase, single-layer clouds, and two multi-layer mixed-phase clouds. These clouds were found between 2400 and 7000 m, had depths of 500-600 m, and in-cloud temperatures from just below 0 °C to − 31 °C. In contrast to Heymsfield et al. (1991), significant temperature inversions or wind shears were not observed in association with any of these clouds, possible reasons for which include: that the clouds did not live long enough to form them, that there were no pre-existing inversions, and as the clouds moved with the wind rather than being anchored to a location, that the wind shear was weak. The absence of inversions and shear-generated turbulence has important implications for dry air entrainment rates.

A common feature of midlevel, single-layer, mixed-phase clouds is the dominant presence of supercooled liquid water in the upper portion of the cloud with ice in the lower regions. Supercooled altocumulus clouds overlaying a deep layer of ice virga are a typical example of such clouds (Hogan et al., 2002; Wang et al., 2004). Temperature inversions at cloud top prevent further vertical development, and a region of ice-saturated relative humidity supports the growth of ice crystals precipitating from the cloud. Two thirds of thirty vertical profiles of mid-latitude, midlevel mixed-phase altocumulus clouds taken over western Nebraska were supercooled, liquid-topped altocumulus clouds with mixed-phase conditions extending to cloud base (Carey et al., 2008). The other ten were glaciated clouds with little evidence of supercooled liquid water. For the mixed-phase clouds, peak liquid water contents occurred in the upper regions of the cloud, typically within tens of meters of cloud top, and ice virga were evident well below cloud base.

### *10.4.2.2. High-Level Clouds*

The global coverage of cirrus is of the order of 40% as is evident in Table 10.2 (the CloudSat estimates are lower; see Sassen and Wang (2008) for a discussion on the differences). The greatest coverage occurs in tropical regions, although mid-latitude cirrus are also extensive (Fig. 10.29). Cirrus are by definition semitransparent (Sassen, 2002), and can be difficult to measure observationally. Lidar can detect cirrus with optical depths as low as $10^{-4}$ for wavelengths in the visible part of the spectrum, while satellite sensors need optical depths of ~0.1 in order for their detection. As described by Kärcher and Spichtinger (2009), cirrus clouds are somewhat unique for several reasons. Firstly, cirrus develop in relatively stable thermodynamic environments. Secondly, they are typically associated with relatively high ice nucleation thresholds (supersaturations with respect to ice of tens of percent) and long growth and sublimation time periods due to the low temperatures of the environment in which they form. This results in regions of long-lived saturations both within and outside of the cloud, as well as difficulties in distinguishing between primary (vapor diffusion processes) and secondary (aggregation and precipitation) ice processes. Thirdly, the transition between clear and cloudy air in cirrus regions is relatively continuous as ice crystals can survive for long periods of time in subsaturated conditions. Ice crystal sedimentation is important in cirrus development and longevity. Finally, rather than originating due to a single mechanism, cirrus clouds can develop from several different mechanisms or sources, and it is these mechanisms that we now examine.

Cirrostratus may form in association with airflow over the warm front in extratropical cyclones. Air flowing over mountains may generate orographic wave clouds of water and/or ice that extend to great heights in the troposphere. Cirrus clouds sometimes form in the vicinity of the jet stream in association with small-scale vertical circulations that develop around the jet. Cirrus also develop in association with closed upper-level lows. Cumulus convection produces significant cirrus clouds through detrainment upon encountering a stable layer, most often the tropopause. TTL cirrus cover large regions near the cold tropical tropopause (Gettleman et al., 2002) and are difficult to detect. Tropical cirrus that occur primarily between 10 and 15 km are optically thicker, contain more condensate, show greater temporal and spatial variability, and tend to be more frequently associated with deep convection at some point in their development than TTL cirrus (Mace et al., 2006b). Finally, condensation trails (contrails) from jet aircraft can produce cirrus clouds.

Cirrus clouds assume a variety of forms, depending on the vertical velocity, wind shear, relative humidity, and static stability. The formation of cirrus clouds is primarily dynamically driven by variations in the vertical wind field on the mesoscale (gravity waves, turbulence) that occur within regions supersaturated with respect to ice. Ordinary cirrus tend to be generated through in situ ice nucleation, whereas ice in anvil and frontal cirrus is typically formed within mixed-phase clouds and then transported aloft. In situ nucleation may then occur following the sedimentation of the previously transported ice. Initial ice crystal concentrations within cirrus are a strong function of vertical velocity and the associated cooling rates. In regions of relatively low cooling rates, heterogeneous ice nucleation may influence cirrus formation. Few numerical modeling studies have been conducted to date that assess the dynamic and aerosol controls of cirrus formation and development.

Anvil cirrus may spread horizontally outward and cover areas many times the size of their generating convective updrafts (Fig. 10.30). The anvils from neighboring cumulonimbus clouds can merge into an extensive stratiform layer in the middle and upper troposphere (Chapter 9). Anvil cirrus occur frequently in the tropics, and are also a significant source of ice in the upper troposphere of the mid-latitudes (Tiedtke, 1993). The mass and number concentration of ice within anvil cirrus are primarily determined by the drop size distributions within the convective updraft. Given the range of convective updrafts that produce

**FIGURE 10.30 Satellite photograph of cumulus convection over Florida peninsula and adjacent waters.** A thunderstorm in south Florida is producing a massive shield of cirrus clouds.

cirrus anvils, the anvils are characterized by a range of sizes, longevities and optical properties. Thin, TTL cirrus occur most frequently in regions of deep convection (Dessler et al., 2006). These clouds develop in association with the strong updrafts of overshooting tops, which allow for the freezing of liquid particles and the formation of thin cirrus above the level of the main convective detrainment, and hence above typical anvil cirrus.

Contrails are perhaps the only single cloud type that are purely anthropogenically generated (Fig. 10.31). Unlike adiabatic cooling processes that produce middle and high clouds, contrails are generated through the mixing of warm, moist exhaust air with colder, drier environmental air. This mechanism, made possible by the exponential increase of saturation vapor pressure with temperature, is the same mechanism that produces steam fogs when cold, dry air flows over warmer water (Chapter 6). Aspects such as the ambient temperature, pressure and relative humidity with respect to ice (RHI), as well as the specific heat of jet fuel, and the aircraft propulsion efficiency all influence the maximum temperature and minimum relative humidity at which contrails form (Schumann, 1996). In corridors of high aircraft density, individual linear contrails can merge to form a thin blanket of contrail cirrus clouds, a process influenced by the ambient conditions and dynamical processes controlling ice supersaturation. Under suitable conditions contrail cirrus may cover extensive regions (up to 100,000 km$^2$; Duda et al., 2001). It has been suggested that contrails are the reason for the sudden increase in high-level

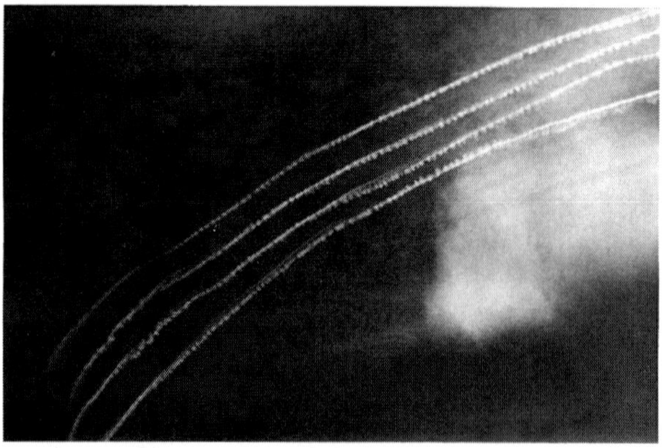

FIGURE 10.31 **Photograph of condensation trails. (Photo by R. Anthes.).**

cloud cover observed over the United States at the start of the jet era (Liou et al., 1990). The radiative forcing of contrails and contrail cirrus is a strong function of their optical depth, which typically varies from 0.1 to 0.5 (Minnis et al., 2005) but can be much smaller or larger.

Dense layers of cirrostratus occur under conditions of gentle, uniform upward motion, saturated air, and high static stability. Under reduced stability and weaker mean upward motion, convection may form cirrus uncinus (mare's tails), dense patches of cirrus, which produce ice crystals large enough to acquire appreciable terminal velocities. In the presence of wind shear these ice crystal fallstreaks may form trails of considerable length (Fig. 10.32). Because of the high RHI compared to that of water, the crystals may survive falls of more than 5 km through clear air (Hall and Pruppacher, 1976). Sometimes these ice crystals "seed" middle-level supercooled water clouds, thus stimulating the growth of precipitation at the lower levels.

Heymsfield (1975a,b) studied the dynamics and microphysics of cirrus uncinus clouds using aircraft observations. A conceptual model for cases of positive wind shear (west wind increasing with height) throughout the cloud system is presented in Fig. 10.33a. According to this model, clouds are initiated in an updraft that develops in a layer with a nearly dry adiabatic lapse rate. The updraft velocity in this convective head is 1.0-1.5 m $s^{-1}$, considerably larger than the average updraft velocities in cirrus clouds. As the ice crystals rise, they grow and move downshear relative to their point of origin. When their size and terminal velocity increase and/or they move out of the convective updraft, they fall and form a trail extending upshear of the generating point. When the head forms in a region of negative wind shear, the head curves the opposite way (Fig. 10.33b). The particles grow and are carried downshear from the generating region until they become large enough to fall out of the updraft. In a relative sense, they then move back toward their point of origin until they reach the region of positive shear, when they again move upshear away from

**FIGURE 10.32** **Photograph of cirrus uncinus and fall streaks. (NCAR photograph.)**

the point of origin. The resulting cloud resembles a reversed question mark. An almost infinite variety of trails can be produced depending on the wind shear. Evaporation of the fallstreaks in the lower levels forms part of the transferral of moisture from the upper to lower levels via these cirrus systems.

Atlas et al. (2006) examined the transition of contrails to cirrus uncinus. They noted that downward "pendants" created by the wake dynamics of the aircraft generate the convective turrets that produce the cirrus uncinus. The turrets may grow to 1-2 km in horizontal size. The pendants are characterized by high concentrations of small ice crystals and high IWCs, and grow as a result of heating by longwave radiation from the ground. Mesoscale uncinus complexes have also been observed (Wang and Sassen, 2008).

Lidar is sensitive to optically thin cirrus layers. By comparing visual and lidar detections, two classes of ice clouds were distinguished by Sassen et al. (1989): visible and subvisible clouds. It is not yet clear whether these two cloud classes are due to different formation processes. Thin or subvisible cirrus have been observed in 50% of the lidar observations taken over Taiwan between 1993 and 1995 (Nee et al., 1998), and thin cirrus layers above 15 km occurred in 29% of lidar observations during the Central Equatorial Pacific Experiment (CEPEX) and the Coupled Ocean-Atmosphere Response Experiment of the Tropical Ocean and Global Atmosphere Programme (TOGA COARE). Haladay and Stephens (2009) studied tropical cirrus between 20°N and 20°S using CloudSat radar data and Cloud-Aerosol Lidar and Infrared Pathfinder Satellite

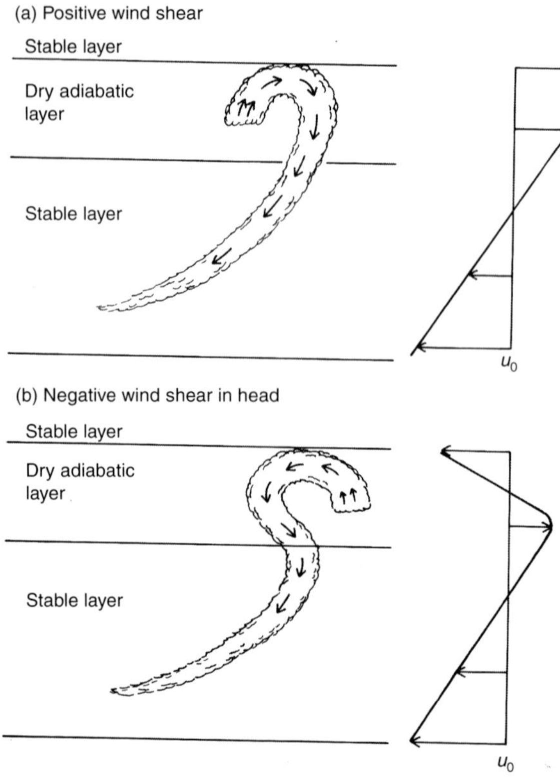


**FIGURE 10.33 Schematic west-east vertical cross sections illustrating development of cirrus uncinus with fallstreaks in two types of wind shear. (a) Positive wind shear (west wind increasing with height) and (b) negative wind shear in generating region (head) of cloud.** *(After Heymsfield, 1975a,b)*

Observations (CALPISO) lidar data and found that optically thin cirrus covered about 30% of the region, about one third of which occurred as single cloud layers without any clouds below them, and that they demonstrated seasonal variations in association with the annual cycle of convection.

Apart from the identification of visible and subvisible classes, a number of other cirrus classification schemes have also been developed. Keckhut et al. (2006) identified four cirrus classes based on geometric height and thickness, three of which contributed similar proportions (~30%) to the total cirrus detected. The first class consisted of thin clouds occurring slightly above the local tropopause (~11.5 km). These clouds may be related to anvil clouds advected to the mid-latitudes or to moisture transport from the tropical upper troposphere to the stratosphere (Garrett et al., 2004). The second class was also made up of thin cirrus, however they occurred at lower altitudes (~8.6 km). In the third class the clouds were thick (~3 km) and were located between the other two classes (~9.8 km), just below the local tropopause height. Other cirrus classification schemes have been based on clustering using Meteosat imagery (Desbois et al., 1982) and on shape ratios (Noel et al., 2002).

Two classes of tropical cirrus clouds were identified by Pfister (2001). The first class is linked to the detrainment of particles and water vapor from cumulonimbus clouds, although much of the cirrus at this level may be generated and maintained through processes other than detrainment. The second class occurs in the regions of the TTL, regions characterized by stable conditions. While these cirrus may be associated with deep convection, they do not appear to be directly related to mass detrainment. Pfister (2001) suggested that they are due to isentropic uplift, while Boehm and Verlinde (2000) linked them to vertically propagating Kelvin and gravity waves.

Wang and Sassen (2002) found that most of the cirrus clouds over the ARM SGP site between 1996 and 2000 were typically optically thin (mean of 0.58) with a low IWP (mean of 12.19 g $m^{-2}$), and a general effective radius between 30 and 50 microns. The IWC, general effective radius and extinction coefficients were found to depend strongly on temperature, although these properties did vary significantly at a given temperature thus demonstrating that factors other than temperature are important in controlling cirrus properties. Massie et al. (2002) examined the distribution of tropical cirrus and deep convection and found that approximately one half of their cirrus observations occurred in association with deep convection, while the other half developed due to in situ processes.

Mace et al. (2006b) used a year of observations to compare the properties of tropical cirrus (excluding the TTL type) that occurred over the ARM sites, Manus and Nauru islands, in the tropical West Pacific. Manus is located within the center of the Pacific warm pool, a region of frequent and widespread deep convection, whereas Nauru is located on the eastern edge of the warm pool where deep convection is less frequent. Cirrus above 7km over Manus occurred 48% of the time compared with only 23% of the time over Nauru, and were thicker and warmer on average. Less than half of the cirrus observed at both islands could be traced to deep convection within the past 12 hours indicating that cirrus in the Tropics must evolve into structures that are maintained by other dynamical processes that support their longevity. A clear temporal evolution in cirrus properties was evident, with both the radar Doppler moments and the IWP decreasing with increasing cloud age. The cirrus properties also appeared to be sensitive to the properties of the deep convective source. Over Manus during the boreal winter, the cirrus had higher water contents and higher small particle concentrations compared to the summer cirrus, being due to the fact that the trajectories associated with cirrus during the winter originate to the south and east of the island where deep convection occurs frequently in association with the winter monsoon, whereas during the boreal summer the trajectories originate in the maritime regions to the north and east of the island where convection is less frequent.

Properties of cirrus are closely tied to the distance of the cirrus from the parent convection, with larger crystals being found nearer cloud base and closer to the connective core (McFarquhar and Heymsfield, 1996). Hoyle et al. (2005) noticed regions of high number concentrations of ice particles (as high as 50 $cc^{-1}$) embedded within more widespread regions of lower concentrations (0.1-50 $cc^{-1}$) from measurements made of in situ-formed cirrus (without convective or orographic influences). They attributed these observations to small-scale temperature fluctuations associated with gravity waves, mechanical turbulence or other small-scale air motions. The waves were found to have frequencies of 10 $hr^{-1}$ and peak-to-peak amplitudes of 1-2 K, with instantaneous cooling rates of up to 60 K $hr^{-1}$. Hoyle et al. suggested that the properties of young, in situ forming cirrus clouds are primarily determined by homogeneous freezing forced by a range of small-scale fluctuations (on the period of several minutes) with moderate to high cooling rates (1-100 K $hr^{-1}$).

Garrett et al. (2005) examined a case study of deep convection and its associated anvil cirrus over Florida during the Cirrus Regional Study of Tropical Anvils and Cirrus Layers—Florida Area Cirrus Experiment (CRYSTAL-FACE) (Jensen et al., 2004). They found that the detrained cloud mass was separated into two vertical layers, a cirrus anvil with cloud top temperatures of $\sim -45$ °C and a thin tropopause cirrus (TTC) layer with temperatures of $\sim -70$ °C which had similar dimensions to the anvil layer, and which lay ~1.5 km above the anvil layer. The anvil layer was optically thick, while the second layer only had a visible optical depth of 0.3 and IWCs of ~ 1 mg $m^{-3}$ in spite of being 1.5 km thick. The unimodal size distribution of the TTC was characterized by small ice particles (effective radius of 4-10 microns), and the microphysical properties remained relatively uniform over time. This may have been due to the fact that the lower anvil shielded the TTC from surface infrared radiation thereby cooling it (Hartmann et al., 2001). Garrett et al. suggested that these TTCs might evolve into widespread subvisible cirrus over low-latitude regions.

In the anvil, the ice crystals larger than 50 microns were found to aggregate and precipitate out, thus resulting in reduced IWCs and an ice crystal population dominated by smaller sizes. The anvil cloud thinned from ~2.5 to 0.5 km during this time. The top-of-atmosphere (TOA) radiative forcing ranged from strong cooling near the leading edge of the anvil and tended toward zero as the anvil thinned downstream of the convective core, being influenced both by a decrease in IWP and an increase in effective radius. The TTC on the other hand made little contribution to the TOA except when the anvil became transparent. The anvil cirrus appeared to spread laterally due to the radiatively-driven pressure gradients at the cloud boundaries rather than convection through the entire layer. Finally, while the TTC layer also spread out, it did not dissipate, possibly due to the shielding provided by the anvil cirrus.

Cirrus cloud structures play important roles in satellite retrievals, radiative calculations and cloud dynamic forcing (e.g. Quante and Starr, 2002; Hogan and Kew, 2005). These structures are difficult to analyze as they are typically non-stationary and comprise a number of different length scales. For example, cirrus uncinus cells (~1 km) may be made up of groups of smaller updrafts (~100 m) and can also occur in larger uncinus complexes (~10-100 km) (Sassen et al., 1989). Various dynamic phenomena may also be observed in cirrus clouds, such as small convective cells, gravity waves (wavelengths of 2-9 km), quasi-two-dimensional waves (wavelengths of 10-20 km) and larger two-dimensional mesoscale waves (wavelengths of ~100 km) (Gultepe and Starr, 1995). While most cirrus layers are horizontally inhomogeneous, locally periodic structures are common in cirrus and include cirrus uncinus cells at cloud top, cirrus mammata at cloud base, and KH waves embedded within the cirrus (Sassen et al., 2007). Even though these structures have different formation mechanisms, they have similar average length scales (1-10 km), can coexist, and tend to occur with greatest frequency in the lower regions of the cloud layer.

### 10.4.3. Environmental Properties of Middle- and High-Level Clouds

The upper troposphere is often supersaturated with respect to ice, the evidence of which is the large spatial scale of anvil cirrus and the development of contrails in the absence of other upper level clouds. Starr and Cox (1980) examined temperature, moisture, and wind soundings associated with more than 3600 cloud cases to determine the characteristics of the environment of middle and high clouds. Knowledge of these characteristics is useful to develop methods of parameterizing clouds in large-scale models as a function of the large-scale thermodynamic and dynamic variables predicted by the model. The parameters examined by Starr and Cox included static stability, defined by

$$\sigma \equiv \partial\theta/\partial z, \tag{10.11}$$

vertical wind shear,

$$S \equiv |\partial V/\partial z| \tag{10.12}$$

and the Richardson number,

$$R_i = (g/\theta)(\sigma/S^2), \tag{10.13}$$

in addition to the mean temperature and relative humidity in the cloud layer.

A common property of most cloud observations was that the lapse rate was rarely saturated adiabatic. Other properties of the environment showed considerable variation from case to case, indicating the difficulty of parameterizing the cloud effects in large-scale models. There were statistical differences between the characteristics of thick cloud layers (defined as extending through a depth of greater than 50 mb) and thin clouds (50 mb

**TABLE 10.3** Characteristics of thick and thin middle- and high-level clouds as deduced from radiosonde ascents

| | |
|---|---|
| A. Thick clouds (depth greater than 50 mb) | |
| Static stability: | Decreases with height from subcloud layer to above cloud layer, with a typical range of 5 K $km^{-1}$ in the subcloud layer to 3 $km^{-1}$ near the cloud top. |
| Vertical wind shear: | Usually positive (increasing wind speed with height): magnitude ranging from about 5 m $s^{-1}$ $km^{-1}$ in summer to 6.5 m $s^{-1}$ $km^{-1}$ in winter. |
| Richardson number: | Average values of $R_i$ throughout cloud layer 12-18; very small percentage (about 15%) of soundings showed $R_i$ less than 1.0. |
| B. Thin clouds (depth less than or equal to 50 mb) | |
| Static stability: | Quite variable, but on average greatest stability (5.5 K $km^{-1}$) above cloud layer and least stability (3.5 K $km^{-1}$) in and below cloud layer; stable layer on top may correspond to tropopause. |
| Wind shear: | No significant difference from thick clouds. |
| Richardson number: | Average values slightly greater (18-24) than for thick clouds. |

**Source:** Adapted from Starr and Cox (1980).

thickness or less). In general, thick clouds were associated with frontal circulations and cyclonic storms and thin clouds were not. The general characteristics of thick and thin clouds are summarized in Table 10.3.

Two conclusions may be drawn from the summary of average cloud conditions in Table 10.3. First, the layers are generally statically stable in an absolute sense, with lapse rates less than the saturation-adiabatic rates with respect to either water or ice. The fact that thick clouds are generally above stable layers indicates that they are associated with upper-level fronts. By contrast, thin clouds are usually below stable layers, which indicates that they are located below the tropopause. A second conclusion from Table 10.3 is that shear-induced turbulence (Kelvin-Helmholtz instability) is not likely in the vicinity of most middle- and high-level cloud systems. A necessary condition for Kelvin-Helmholtz instability is that the Richardson number be less than 1/4. As shown in Table 10.3, the average values of $R_i$ are much greater than this value. The coarse vertical resolution used to evaluate $R_i$ in this study may bias the estimates of $R_i$ toward higher values; it is possible that considerably smaller values of $R_i$ exist locally in thin layers where strong destabilization associated with radiative effects is important. An exception to this conclusion occurs in shallow layers in the vicinity of jet streams, where mesoscale vortices produce turbulence and extensive bands of cirrus.

Cirrus clouds develop primarily in ice-supersaturated regions which, in the upper troposphere at mid-latitudes, are typically ~150 km in extent, although they may be larger (~1000 km), and are ~0.5-1.5 km deep, although they may be as deep as 3-5 km. Three mechanisms appear to play a role in enhancing the RHI including (1) adiabatic vertical motion on a range of spatial and temporal scales; (2) turbulent mixing of air parcels due to wind shear or dynamic instabilities; and (3) diabatic effects associated with shortwave and longwave radiation. Turbulent mixing and diabatic effects typically occur on timescales too long to initiate cloud formation, but appear to influence the cloud life cycle once formed (Kärcher and Spichtinger, 2009). While variations in water vapor affect the formation of cirrus, this is of secondary importance to changes in temperature and the associated changes in RHI. Cirrus clouds thus differ from lower tropospheric clouds where water vapor variations are more important. The properties of cirrus appear to be only weakly dependent on rising motion on the synoptic scale, thus suggesting that vertical velocity on the small scale may play a more significant role in determining their properties (Mace et al., 2001; Quante and Starr, 2002). Kärcher and Ström (2003) demonstrated that the properties of cirrus cannot be simulated when only making use of synoptic cooling rates, and that mesoscale variability in vertical velocity and the associated temperature variations are the primary controllers of the microphysical characteristics of cirrus. Such mesoscale variations in vertical velocity and temperature are associated with gravity waves produced by factors such as orographic forcing, convection, baroclinic instability and geostrophic adjustment.

As cirrus characteristics and properties vary on a seasonal basis, this demonstrates their dependency on synoptic patterns (Mace et al., 2006a). Sassen and Campbell (2001) identified three major synoptic patterns of importance: a zonal jet stream flow, split-jet flow and strong amplitude ridges. Cirrus in the tropics appear to be associated with deep convection at ~50% of the time, but the origin of the other 50% is not known. In the mid-latitudes, cirrus characteristics appear to be a stronger function of the large-scale vertical motion within cloud systems (Mace et al., 2006a). Deng and Mace (2008) compared the properties of mid-latitude and tropical cirrus clouds and found that the cirrus clouds in the tropics had larger particle sizes, greater ice masses, were more likely to be associated with ascending air motions, and were colder and higher in altitude. The mid-latitude cirrus showed strong seasonal variations whereas interannual variations (such as variations due to ENSO) were stronger than seasonal variations for tropical cirrus. The seasonal variations in the tropics and subtropics are associated with convective cycles and the movement of the ITCZ (Stubenrauch et al., 2006). Cirrus clouds were found to be more frequent and thicker during the summer than in winter in the mid-latitudes. Diurnal variations in cirrus were observed both in the tropics and the mid-latitudes, although tropical cirrus showed a greater variation. Tropical cirrus occurred higher in the atmosphere and were thicker in the morning and then decreased in the afternoon. In the mid-latitudes there was no obvious diurnal variation in winter, however, during the summer, cirrus were found at higher altitudes and were thicker during the afternoon, which is indicative of the development of deep convection and the associated spreading of anvils. In the western tropical Atlantic ITCZ, dual maxima (morning and late afternoon) have been found, the causes of which are not well understood.

Mace et al. (2006a) observed a seasonal variation in the macroscale properties of cirrus over the ARM SGP site. A minima in frequency occurred in late summer, while the maxima existed in late winter and early spring when synoptic-scale systems and strong jet streams are associated with deep convection. Cirrus occurred over lower-level clouds 1/3 of the time, and were more frequent in winter. Cirrus tended to occur within regions of upper-troposphere humidity maxima and just downstream of a peak in upper-troposphere vertical motion. Those cirrus that formed within large-scale ascent regions upstream of mid-tropospheric ridges had higher water content than those forming in regions of large-scale subsidence downwind of the ridge axis. In summer, while cirrus still appeared to be associated with large-scale ascent the anomalies were less distinct, suggesting that warm season cirrus tend to be more frequently associated with detrainment from deep convection.

The condensate detrainment associated with deep convection first forms cirrostratus and then thin cirrus. It has been estimated that it takes ~6-12 hours for the development of cirrostratus, followed by ~1 day for the transition to thin cirrus. The cirrus amount has an e-folding time of ~5 days, and the mean lifetime of convectively driven cirrus is ~10 days (Luo and Rossow, 2004). Contrails can form and persist in environments that are saturated with respect to ice, whereas natural cirrus require high ice supersaturations to form. Thus contrails can occur in environments that may otherwise be cloud free. The life cycle of contrail cirrus is poorly understood.

## 10.4.4. Microphysical Properties of Middle- and High-Level Clouds

### *10.4.4.1. Middle-Level Clouds*

Measurements of the microphysical properties of middle clouds are relatively sparse. In situ measurements of droplets made by Borokvikov et al. (1963) showed that droplets averaged about 10 microns in size. A more detailed investigation into the microphysical characteristics of a 300 m deep altocumulus sampled between $-9°$ and $-12$ °C showed that the mean droplet size ranged from 7 to 10 microns, droplet concentrations averaged ~ 300 $cm^{-3}$, and the LWC ranged from 0.03 to 0.09 g $m^{-3}$(Herman and Curry, 1984). Retrieved LWPs and effective radii for the supercooled liquid regions of altocumulus of 15 g $m^{-2}$ and 6 microns, respectively, have also been observed (Wang et al., 2004).

Using observations and simple model calculations of droplet growth in altocumulus clouds Heymsfield et al. (1991) observed typical droplet growth patterns beginning with a rapid increase in droplet concentrations and mean diameters immediately above cloud base, followed by nearly constant peak drop concentrations and diameters that increase in size much more slowly, and finally, a rapid decrease in mean diameters and cloud droplet concentrations near cloud top due to mixing with the dry, overlying air. Peak supersaturations are much higher than typically calculated for cumulus clouds due to low droplet concentrations and comparatively low droplet growth rates relative to the vapor supply rate. Few, if any, ice crystals were observed in the altocumulus clouds they examined, which they attribute to the absence of ice nuclei.

Observations of the evolution of the size spectra of ice crystals larger than 800 microns within an altostratus cloud associated with a cold front showed variations to occur on horizontal scales of ~5 km (Field, 1999). Between −40 °C and −20 °C, particle growth was dominated by diffusional growth, although aggregation was also evident. Between −20 °C and −10 °C growth was dominated by aggregation. Rauber and Grant (1986) observed that supercooled water can be produced in regions in the cloud where the condensational rate exceeds the diffusional rate of ice crystals.

In the CLEX study of single and multiple layer mixed-phase midlevel clouds, Fleishauer et al. (2002) noted in the thin, mixed-phase single-layer clouds that the LWC increase with altitude, whereas the IWC maximizes in the mid- to lower regions of the cloud. This is in contrast to the multilayered systems in which ice was more evenly distributed throughout the cloud layer. In the thin, single-layer clouds, drop sizes were found to increase with increasing altitude. Peak LWCs were 0.35 g $m^{-3}$ with mean values ranging from 0.01 to 0.15 g $m^{-3}$. The seeder-feeder mechanism was examined as a potential mechanism causing the differences in vertical ice distribution between the single layer and multilayer cases. While some of the ice crystals had habits originating from colder regions than where they were observed, and ice crystals were detected in the clear air between the cloud layers, the number of crystals between layers was small, thus suggesting that this mechanism was not significant. Finally, their observations showed a poor correlation between IWC and temperature.

Larson et al. (2001) examined the causes of the dissipation of one of the altocumulus clouds observed by Fleishauer et al. (2002) during CLEX. An analysis of the liquid water budget of this cloud showed that the cloud did not dissipate due to precipitation fallout. Rather, the largest contribution to the liquid water decay was subsidence drying. The net effect of radiative transfer on this cloud was unclear. In a subsequent modeling study of the same case, Larson et al. (2006) made use of three-dimensional, LES simulations in order to assess the role played by radiative heating or cooling, large-scale subsidence, diffusional growth of ice and turbulent mixing of dry air into the cloud in the dissipation time of the midlevel clouds. They found that subsidence and ice diffusional growth had non-positive contributions at all altitudes, thus leading to a strong decrease in liquid water. They also estimated from their sensitivity tests that increasing the solar zenith angle by 1 (the maximum possible change) reduces the cloud lifetime by the same amount as increasing the subsidence velocity by 1.21 cm $s^{-1}$, increasing the ice number concentration by 781 $m^{-3}$ or decreasing the total water mixing ratio above cloud by 0.597 g $kg^{-1}$ for this particular case.

Hobbs and Ragno (1985) observed two types of altocumulus (castellanus and floccus) and found that the LWC maxima occurred at cloud top, together with a small number of ice particles, while large concentrations of ice particles occurred in the lower regions of the cloud, together with little or no LWC. Carey et al. (2008) also found peak LWCs at or near cloud top and peak IWCs in the lower half of the cloud in their altocumulus study. They also observed that as the long-lasting (9-10 hours) cloud field dissipated, the altocumulus cloud appeared to "deglaciate" in that while the LWC remained the same, the IWC decreased dramatically as the cloud progressed from the early through to the dissipation stage. A similar "deglaciation" was also noted in another mixed-phase study by Kankiewicz et al. (2000). This deglaciation process was observed to occur even while there was still significant supercooled water present, and may have resulted due to decreases in the ice nuclei supply. Ice nuclei measurements were not however made during the field campaign. Larson and Smith (2009) conducted LES simulations of three thin, midlevel layer clouds in order to investigate the sensitivity of the glaciation rate of altocumulus clouds to habit type. They found that the relationships were complex, but that, for the cases they examined, dendrites tend to glaciate altocumulus clouds more rapidly than plates.

### *10.4.4.2. High-Level Clouds*

The size, shape and concentration of ice crystals in cirrus clouds strongly influence the earth's radiation budget. Using a climate model, Kristjánsson et al. (2000) showed that variations in ice crystal habit and size may have a significant influence on the simulated climate change. Stephens et al. (1990) concluded that our understanding of the impacts of cirrus on climate is limited by our knowledge of the relationship between ice crystal size and shape, and the broad radiative properties of cirrus. Measurements of ice crystal sizes and shapes are also important to obtain accurate remote sensing products. One of the primary reasons for our lack of understanding of the processes involved in ice crystal formation within cirrus is the difficulty in observing the formation of cirrus in situ. Cirrus clouds are cold clouds that form high in the troposphere, especially in the Tropics. Also, unlike the formation of ice particles in cumulus clouds and wave clouds, cirrus clouds do not have an obvious initial stage, and thus it is difficult to place aircraft in the right locations at the right time to observe ice crystal formation.

Aircraft observations of middle- and high-level clouds have revealed much about their structure and microphysical characteristics. As summarized in Table 10.4, cirrus clouds contain ice crystals typically 0.5 mm in length and concentrations that vary widely from $10^4$ to $10^6$ $m^{-3}$. The terminal velocity of the ice crystals is about 0.5 m $s^{-1}$, which is generally of the same order as

**TABLE 10.4** Microphysical characteristics of Middle and High Cirrus Clouds

| Property or variable | Value | Reference |
|---|---|---|
| Concentration of ice crystals ($m^{-3}$) | $10^5$-$10^6$ | Braham and Spyers-Duran (1967) |
| | $2 \times 10^5$-$5 \times 10^5$ | Heymsfield (1975a) |
| | $1.0 \times 10^4$-$2.5 \times 10^4$ | Heymsfield and Knollenberg (1972) |
| | $6.0 \times 10^5$-$38.0 \times 10^5$ | Ryan et al. (1972) |
| | $2.0 \times 10^4$-$8.0 \times 10^4$ | Houze et al. (1981) |
| | $1.0 \times 10^4$-$5.5 \times 10^5$ | Churchill and Houze (1984) |
| Length of crystals | Up to 0.17 mm | Braham and Spyers-Duran (1967) |
| | 0.6-1.0 mm | Heymsfield and Knollenberg (1972) |
| | 0.35-0.9 mm | Heymsfield (1975a) |
| Terminal velocity of crystals | Typically 50 cm $s^{-1}$; max 120 cm $s^{-1}$; | Heymsfield (1975a) |
| Ice-water content | 0.15-0.25 g $m^{-3}$ | Heymsfield and Knollenberg (1972) |
| | 0.15-0.30 g $m^{-3}$ | Heymsfield (1975a) |
| | 0.10-0.50 g $m^{-3}$ | Rosinski et al. (1970) |
| Precipitation rate | 0.5-0.7 mm $h^{-1}$ | Heymsfield and Knollenberg (1972) |
| Updraft velocity | 1.0-1.5 m $s^{-1}$ (in cirrus uncinus) | Heymsfield (1975a) |
| | 2.0-10.0 cm $s^{-1}$ (warm front overrunning) | Heymsfield (1977) |
| | 25-50 cm $s^{-1}$ (closed low aloft) | Heymsfield (1975a) |
| Lifetime individual cloud | 15-25 min | Heymsfield (1975a) |

the updraft speed. The water content of cirrus clouds is typically 0.2 g $m^{-3}$. Observations have shown that IWC is strongly dependent on the vertical velocity and the temperature. Figure 10.34 shows plots of observed IWCs versus temperature for vertical velocities ranging from 1 to 50 cm $s^{-1}$. The observations were made under various synoptic situations, including warm-frontal overrunning, warm-frontal occlusions, closed lows aloft, and jet stream cloudiness (Heymsfield, 1977). The data shown in Fig. 10.34 indicate, for a

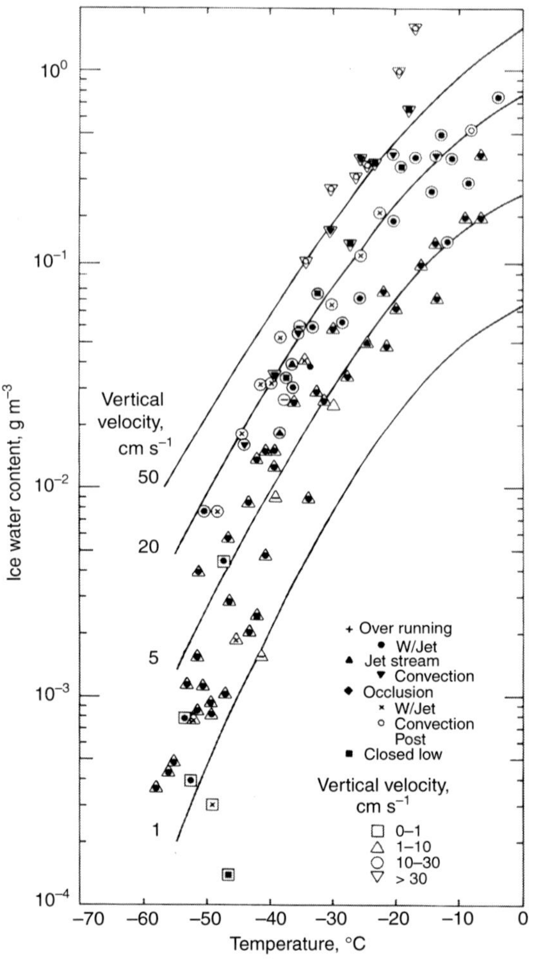


**FIGURE 10.34 Ice content (g m$^{-3}$) plotted against temperature and parameterized in terms of vertical air velocity: small, solid inner symbols, synoptic type; larger, open outer symbols, vertical velocity range.** *(From Heymsfield (1977))*

given updraft speed (W), a nearly exponential increase of water content with increasing temperature, from low values of less than $10^{-3}$ g m$^{-3}$ at temperatures below $-50$ °C to maximum values around 0.3 g m$^{-3}$ at temperatures in the range of 0 to $-10$ °C. At a given temperature (e.g. $-30$ °C), the water content increases from $10^{-2}$ g m$^{-3}$ for an updraft of 0.01 m s$^{-1}$ to $2.0 \times 10^{-1}$ g m$^{-3}$ for an updraft speed of 0.5 m s$^{-1}$. The empirical equation fitting the data in Fig. 10.34 is

$$\mathrm{IWC} = 0.072W^{0.78}\exp[-0.01W^{0.186}(-T)^{1.59W^{-0.04}}]. \tag{10.14}$$

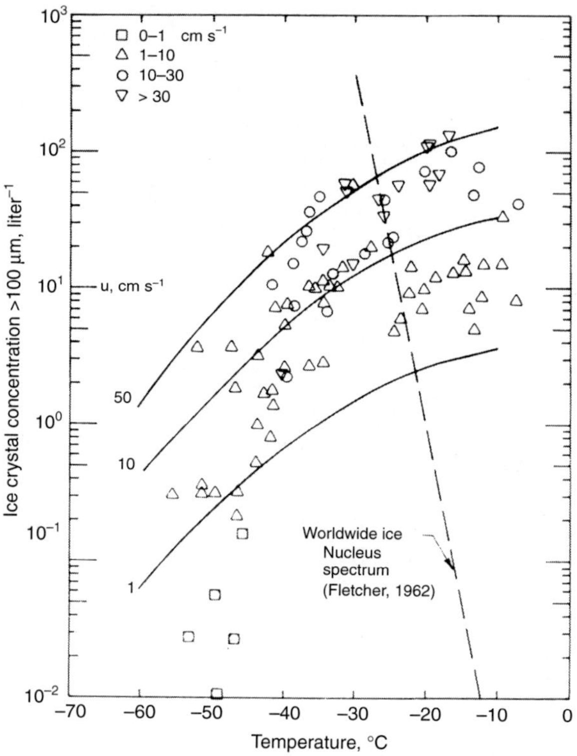


**FIGURE 10.35 Total ice-crystal concentration (>100 μm) plotted against temperature and parameterized in terms of air velocity.** Worldwide ice-nucleus spectrum is also plotted. *(From Heymsfield (1977))*

The total ice-crystal concentration also shows a strong dependence on temperature and vertical velocity. As shown in Fig. 10.35, the concentration varies from less than $10^{-1}$ liter$^{-1}$ (100 m$^{-3}$) for weak (0.01 m s$^{-1}$) updrafts and temperatures below $-50$ °C to 100 liter$^{-1}$ ($10^5$ m$^{-3}$) for temperatures around $-10$ °C and updraft speeds of 0.5 m s$^{-1}$. Observations also indicate a relationship between crystal size and IWC. Figure 10.36 is a plot of observed mean and maximum crystal lengths as a function of IWC. The maximum lengths in particular, show an increase with increasing IWC, from about 0.5 mm at $10^{-4}$ g m$^{-3}$ to 5 mm at 1 g m$^{-3}$.

A property of precipitating cloud systems having practical consequences is the close relationship between the precipitation rate R and IWC. Figure 10.37 shows this strong positive correlation, which is well represented by

$$R = 3.6(\mathrm{IWC})^{1.17}, \tag{10.15}$$

where IWC is expressed in g m$^{-3}$ and R is in mm h$^{-1}$. A very similar expression

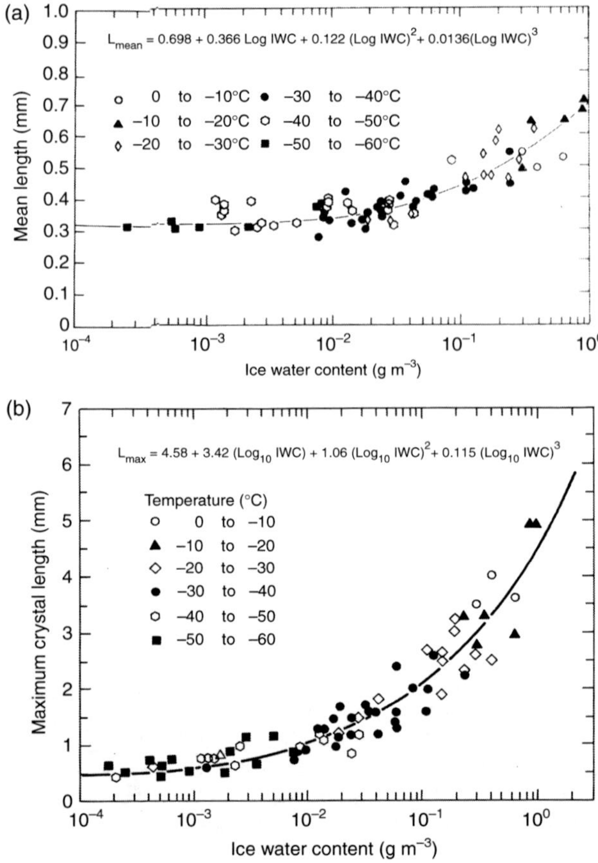


**FIGURE 10.36 Mean (a) and maximum (b) crystal lengths plotted against ice-water content.** Maximum crystal length corresponds to particles in concentrations of 1 $m^{-3}$ per millimeter size class. Sampling temperature is indicated. *(From Heymsfield (1977))*

was derived independently by Sekhon and Srivastava (1970),

$$R = 5.0(\mathrm{IWC})^{1.16}. \tag{10.16}$$

This empirical relationship is useful because radar reflectivity Z is a measure of IWC (Fig. 10.38),

$$Z = 750(\mathrm{IWC})^{1.98}, \tag{10.17}$$

where $Z$ is expressed in units of $mm^6\ m^{-3}$. Equations (10.16) and (10.17), or other similar empirical expressions, can be used with radar measurements to estimate precipitation rates from cirrus clouds.

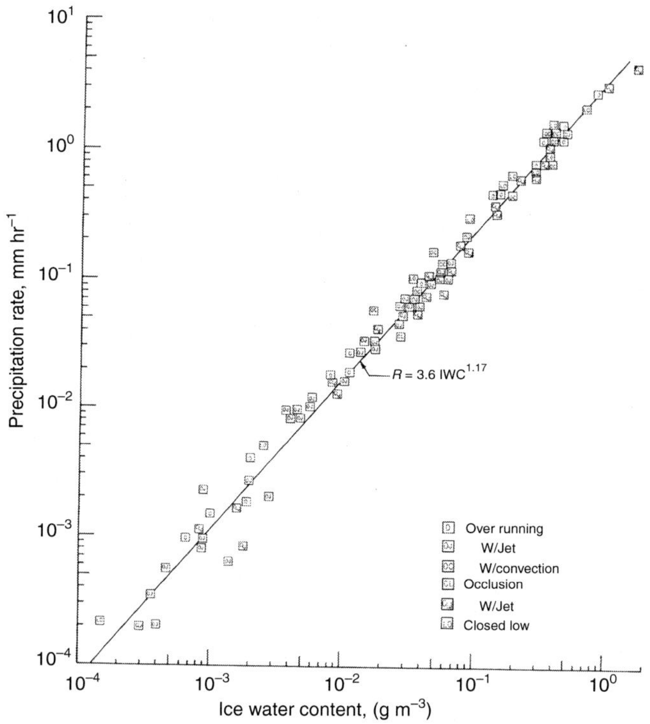


**FIGURE 10.37 Calculated precipitation rate plotted against ice content in synoptic systems sampled.** Best-fit line through data points is indicated. *(From Heymsfield (1977))*

Aircraft observations of low-latitude anvil cirrus (e.g. Griffith et al., 1980; Knollenberg et al., 1993; McFarquhar and Heymsfield, 1997; Heymsfield et al., 1998) show that the IWC varies horizontally, tends to decrease with height and can range from $\sim$0.01 g m$^{-3}$ to more than $\sim$0.1 g m$^{-3}$. Observations from field campaigns demonstrate that the size and shape of ice particles in cirrus clouds vary significantly, both spatially and temporally. Processes such as aggregation and sedimentation result in large ice crystals (1000 microns or larger) near the cloud base. Aggregation does not always occur however, and appears to be less frequent than in other clouds. It is sometimes difficult to distinguish aggregation from the complex crystals arising out of diffusional growth processes. Near cloud top, concentrations of ice crystals smaller than 100 microns may be greater than 100 cc$^{-1}$. Measurements in cirrus that formed independent of convective or orographic influences show number concentrations as high as 50 cc$^{-1}$ within regions of concentrations of 0.1-50 cc$^{-1}$. Others have shown number concentrations of the order of 0.1 cc$^{-1}$ with values as high as 10 cc$^{-1}$, especially in younger cirrus (Mace et al., 2001; Kärcher and Ström, 2003). Optically thin cirrus may have concentrations of the order of 0.1 cc$^{-1}$(Mace et al., 2001).

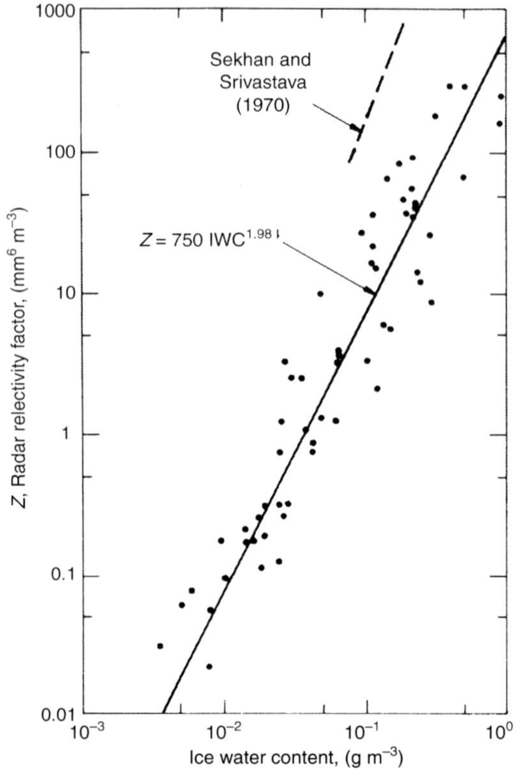


**FIGURE 10.38 Calculated radar reflectivity factor plotted against ice content.** Best-fit line and Sekhon and Srivastava (1970) curve for aggregate snow are shown for comparison. *(From Heymsfield (1977))*

Microphysical properties of cirrus clouds appear to be weakly correlated with ambient environmental conditions (Korolev et al., 2001), although particles appear to become more regularly shaped as the temperature and supply of water vapor decreases (Magano and Lee, 1966). The size and structure of cirrus ice crystals determines their sedimentation velocity, which in turn influences cloud lifetime and the humidity of the upper troposphere. If the total available water is spread over a smaller number of large ice crystals then the cloud dehydrates more rapidly due to greater sedimentation rates than if the same total water was spread over a larger number of smaller ice crystals. The microphysical properties of contrail cirrus are likely to differ from those of natural cirrus, at least during the initial stages of development, due to the nucleating properties of aircraft particles.

The crystal habit and size has important radiative implications. Heymsfield and Platt (1984) observed that the predominant crystal types for temperatures between −20 °C and −40 °C were polycrystalline forms, with some columns, plates and bullets. Between −40 °C and −50 °C, hollow columns were predominantly observed, and at temperatures below −50 °C, hollow and solid columns, together with some hexagonal plates and thick plates were primarily seen. Bullet rosettes were also sometimes found, although they tended to be infrequent. More recent observations of ice crystals within cirrus have shown that mid-latitude and high-latitude cirrus may contain higher concentrations of rosette-shaped crystals than previously recorded.

Lawson et al. (2006) analyzed data from 22 mid-latitude cirrus clouds. The cloud temperatures of these cirrus clouds ranged from −28 °C to −61 °C. They showed that the ice particle size distributions of cirrus are primarily bimodal, with a maximum in mass, area and number concentration near 30 microns, and the second mode, a smaller maximum, occurring between 200-300 microns. Rosette shapes were found to be the predominant particle shape. They also observed that crystal habit categories are a function of particle size. Spheroids dominate the particle size distribution for sizes less than 30 microns; between 30 and 150 microns, small irregular crystals contribute to most of the mass with smaller contributions from columns in this size range; budding rosettes contribute ~25%-40% of the mass for sizes between 75 and 300 microns, and rosette shapes contribute more than 80% of the mass for crystal sizes larger than 200 microns. The percentage contribution of rosette shapes increases with increasing temperature, from ~40% at −50 °C, to 55% at −40 °C to 80% at −30 °C. The trend in surface areas is similar. Rosette shapes made up 50% of the surface area and mass of ice particles greater than 50 microns in size, with irregular shapes making up 40% of the remaining 50%, columns and spheroidal shapes constituting several percent, and plates accounting for 1% of the total mass. Particles that are less than 50 microns in size contribute ~99% of the total number concentration, ~40% of the mass and ~70% of the shortwave extinction in these mid-latitude cirrus. Also, the average particle number concentrations was found to be of the order of 1 $cc^{-1}$, although they were greater than 5 $cc^{-1}$ occasionally. Table 10.5 summarizes various ice crystal characteristics for three different temperature zones. Finally, the microphysical properties of these cirrus clouds were found to be similar to wave clouds thus suggesting that ice particles in cirrus first convert to liquid water and/or solution drops before freezing.

The results of Lawson et al. (2006) are similar to those of Heymsfield and Miloshevich (1995) regarding bimodal size distributions and particle shapes, however, the number concentrations observed by the latter are one to two orders of magnitude lower. Also, Heymsfield and Miloshevich found that deep cirrus clouds have a distinct structure with an ice crystal generation region near cloud top that contains the highest concentrations of small particles, an ice-saturated region below this that supports the growth of ice particles, and finally a lower region in which ice particles sublimate. However, Lawson et al. suggest that cirrus clouds are much more complex in that there is a lack of consistency in particle habits as a function of position in cloud, particles that are nucleating,

**TABLE 10.5** Means, standard deviations and extreme values of microphysical properties from flights through 22 cirrus clouds. The measurements are separated into three temperature zones and $N$ is the number of legs flown in each temperature zone

| | Mean | Std dev | Max | Min |
|---|---|---|---|---|
| | −50 °C to −63 °C ($N = 36$) | | | |
| IWC (g m$^{-3}$) | 0.005 | 0.007 | 0.024 | 0.001 |
| Concentration (L$^{-1}$) | 846 | 853 | 3679 | 36 |
| Extinction (km$^{-1}$) | 0.464 | 0.585 | 3.00 | 0.023 |
| $R_{eff}$ (μm) | 15.4 | 4.7 | 26.0 | 8.1 |
| Radar reflectivity (dBZ) | −27.5 | −21.6 | −13.8 | −51.6 |
| | −40 °C to −49 °C ($N = 49$) | | | |
| IWC (g m$^{-3}$) | 0.013 | 0.017 | 0.077 | 0.001 |
| Concentration (L$^{-1}$) | 968 | 951 | 3817 | 30 |
| Extinction (km$^{-1}$) | 0.920 | 1.060 | 4.779 | 0.017 |
| $R_{eff}$ (μm) | 21.3 | 7.10 | 41.46 | 9.99 |
| Radar reflectivity (dBZ) | −21.3 | −18.59 | −11 | −53.5 |
| | −30 °C to −39 °C ($N = 30$–) | | | |
| IWC (g m$^{-3}$) | 0.037 | 0.038 | 0.125 | 0.002 |
| Concentration (L$^{-1}$) | 2170 | 2159 | 8267 | 81 |
| Extinction (km$^{-1}$) | 2.404 | 2.663 | 11.094 | 0.140 |
| $R_{eff}$ (μm) | 26.5 | 7.5 | 43.1 | 13.0 |
| Radar reflectivity (dBZ) | −15.5 | −14.8 | −8.8 | −39.1 |

**Source:** After Lawson et al. (2006)

growing and sublimating can be found throughout the cloud, and almost any of the crystal shapes can be found either at cloud top or cloud base.

Wu et al. (2000) conducted numerical simulations in order to investigate radiative effects on the diffusional growth of ice particles in cirrus clouds. They found such effects to be both significant and complex. Even in a radiatively-cooled environment, ice particles may undergo radiative warming as the net radiation received by an ice particle depends on the emission from the particle, and the local upwelling and downwelling radiative fluxes. Radiative warming of an ice particle restricts the diffusional growth of the particle. For sufficiently large ice crystals, the associated high radiative warming produces surface ice saturation vapor pressures that are so large as to produce sublimation of the crystals, while smaller particles may grow by vapor deposition under such conditions. Radiative cooling of ice particles on the other hand enhances ice production. For the cirrus case study that Wu et al. (2000) simulated they found that the overall influence of radiative feedbacks was to significantly reduce the total ice mass of the cloud, especially the production of large ice crystals. In subsequent LES simulations, Cheng et al. (2001) simulated both a thin and a thick cirrus event. They found that, while latent heat release within the thin cirrus clouds was insufficient to generate positive buoyancy, positively buoyant cells could be produced within thick cirrus layers. The pressure perturbations induced by the associated updrafts subsequently affected the cloud evolution. Cheng et al. (2001) also found that radiative cooling and latent heating were of similar magnitudes in well-developed, deep cirrus layers and that latent heating thus needs to be taken into account when developing cirrus models. Also, while probability density functions (PDFs) of the vertical velocity in the more radiatively-driven, thin cirrus were approximately Gaussian in distribution, those for the deep, more convectively unstable cirrus were multimodal and broad. These variations in cirrus characteristics make the parameterization of such clouds within numerical models a very difficult task.

### 10.4.5. Radiative Properties of Middle- and High-Levels Clouds

Clouds affect both solar (shortwave) and infrared (longwave) radiation. Shortwave cloud radiative forcing depends on the LWP and IWP, as well as on cloud particle size and habit, and is usually negative at the TOA as clouds reflect sunlight. Longwave cloud radiative forcing at the TOA depends on the cloud top temperature. This forcing is usually positive, and largest for clouds with high cold cloud tops. The net cloud radiative forcing at the TOA is then dependent on the difference between the longwave and shortwave radiative forcing. Therefore high clouds in the Tropics can have either strong warming or cooling effects on the TOA radiation budget depending on their optical depth, whereas low-topped clouds always have a net radiative cooling effect.

Many previous studies of cloud radiative forcing have focused on the effects of clouds at the TOA however, as pointed out by Stephens (2005), important feedbacks may also exist in association with the separate effects of clouds on the atmospheric and surface radiation budgets. Relative to clear skies, clouds at low latitudes warm the *atmosphere* due to increased IR absorption and emission at colder temperatures, and cool the *surface* due to the reflection of solar radiation back to space. These two effects are often largely reciprocal as observed at the TOA. Clouds at high latitudes tend to affect the radiation balance in a reverse manner to those at low latitudes. The effect of clouds on the atmospheric radiation balance is dependent on the vertical location of the clouds. High cold clouds tend to warm the atmospheric column (relative to clear skies) especially at low latitudes, whereas low clouds enhance the cooling of the atmosphere, especially at high latitudes (Slingo and Slingo, 1988).

Several energy balance studies have been conducted linking ISCCP cloud properties to radiative fluxes at the TOA, the surface and within the atmosphere (e.g. Okhert-Bell and Hartmann, 1992; Chen et al., 2000). Okhert-Bell and Hartmann correlated ISCCP and Earth Radiation Budget Experiment (ERBE) data in order to assess what types of clouds contribute most significantly

**TABLE 10.6** The contributions to the longwave, shortwave and net cloud forcings by the five different cloud classes identified by Okhert-Bell and Hartmann (1992) The cloud amount ($N$) for each cloud type is also given

| | Type 1 high, thin | | Type 2 high, thick | | Type 3 mid, thin | | Type 4 mid, thick | | Type 5 low | | |
|---|---|---|---|---|---|---|---|---|---|---|---|
| | JJA | DJF | JJA | DJF | JJA | DJF | JJA | DJF | JJA | DJF | Sum Avg |
| N | 10.2 | 10.0 | 8.5 | 8.8 | 10.7 | 10.7 | 6.5 | 8.2 | 27.2 | 25.9 | 63.3 |
| OLR | 6.5 | 6.3 | 8.4 | 8.8 | 4.8 | 4.9 | 2.4 | 2.4 | 3.5 | 3.5 | 25.8 |
| Albedo | 1.2 | 1.1 | 4.1 | 4.2 | 1.1 | 1.0 | 2.7 | 3.0 | 5.8 | 5.6 | 14.9 |
| Net | 2.4 | 2.3 | −6.4 | −7.5 | 1.4 | 0.8 | −6.6 | −8.5 | −15.1 | −18.2 | −27.6 |

**Source:** From Hartmann et al. (1992). After Stephens (2005)

to longwave and shortwave forcing at the TOA, a study extended to the atmospheric and surface radiation budgets by Chen et al. (2000). They used a simplified group of clouds (5 instead of 9) from the ISSCP database. The results of their study are shown in Table 10.6. Their analysis demonstrated the dominance of high clouds on the outgoing longwave radiation (OLR), at least in low latitudes, as well as the important role of thicker, higher clouds in the OLR of the Tropics and low clouds on the shortwave fluxes in the mid- and high-latitudes.

Cirrus clouds exert varying but significant effects on the radiation budget of the earth. When the optical depth of cirrus is sufficiently low, the absorption and re-emission of IR dominates the solar albedo effect and cirrus clouds then exert a warming effect. As the optical depth increases, the albedo effect increases which leads to a net cooling. In terms of the surface and atmospheric radiation budgets, thin cirrus cool the surface, but exert a net warming within and at the top of the atmosphere. Optically thick cirrus, on the other hand, still warm the atmosphere on the whole, but cool the surface and the upper atmosphere (Chen et al., 2000). In spite of their importance in the radiation budget, the understanding of the global net radiative responses to the various cirrus cloud types is still not well known.

As stated above, the altitude and vertical extent of clouds are important parameters in the radiation balance of the earth. A cirrus cloud at high altitudes will exert a more significant impact on the IR flux than the same cirrus cloud at lower altitudes. A cirrus cloud at lower altitudes may exert less of an impact on the IR radiative flux, and hence the albedo effect may become dominant. For a given cloud, the cloud may transition from cooling the surface to warming the surface relative to clear-sky conditions if the surface albedo is increased. The radiative effects of cirrus depend strongly on their microphysical properties in contrast to those of low and middle clouds. For example, for the same optical thickness or IWP, the forcing associated with cirrus clouds can switch sign depending on the shape and size of the ice crystals (Zhang et al., 1999). Unfortunately, as we saw above, the microphysical properties of cirrus are still not well known.

Other factors that influence the radiative effect of cirrus clouds include cloud fraction, variations in horizontal homogeneity, the solar zenith angle, the ice crystal size distribution, the cloud geometrical thickness, IWC and the presence of other clouds or water vapor in the column. Schlimme et al. (2005) noted from their sensitivity study that the order of importance of cirrus cloud properties in solar broadband radiative transfer is optical thickness, ice crystal shape, ice particle size and spatial structure. Cirrus clouds also affect the radiation budget indirectly through their influence on water vapor distribution in the Tropics. In regions of subsidence, away from deep convective outflow, cirrus clouds tend to moisten the upper troposphere, which in turn influences the heating rate.

The optical depths of in situ cirrus are not often greater than 5, while the optical depths of anvil cirrus can range from 10-50 and those for contrails are of the order of 0.1-0.5. Solar radiative effects appear to be more important for optical depths in the range of 1-2, whereas IR effects are more important for thicker layers in which the optical depth is greater than 2. For optical depths of less than 1, radiative effects appear minimal. Subvisible cirrus layers can have heating rates of up to 1 K $day^{-1}$ and radiative forcing of 1.2 W $m^{-2}$, which is equivalent to $\sim$ 0.7 °C change in surface temperature globally (McFarquhar et al., 2000). Haladay and Stephens (2009) estimated that the effects of their thin ice cloud category was $\sim$ 4 W $m^{-2}$ on the average IR heating of the tropics. Radiative forcing can increase both the IWP and cirrus lifetime through dynamic responses, although this is dependent on the cloud layer stability. Anvil cirrus or in situ mid-latitude cirrus can be enhanced through turbulence effects forced by radiative heating in the lower regions of the cloud. Radiative cooling at cloud top may also produce supersaturations that are sufficiently high to initiate ice nucleation.

The effects of radiative heating on convection have been demonstrated to regulate high level cloudiness, such as in the global study by Fowler and Randall (1994). More recently, Stephens et al. (2008) examined the impacts of radiative heating on convection using a cloud-resolving model in radiative convective equilibrium. It was evident from their model sensitivity experiments, one in which the radiative forcing was held constant irrespective of the cloud amounts, and the other in which the optical properties of clouds were turned off, that the lack of upper-tropospheric radiative heating associated with these anvil clouds influences the stability of the atmosphere in such a manner that convection was strengthened, thus producing more high clouds relative to the control experiment in which the radiative forcing and optical properties were not altered. The high clouds were found to increase by a factor of 6-8 over the amount of high clouds produced in the control simulation. When the radiative heating of these high clouds was included, as in the control simulation, the high cloud fraction was significantly reduced, even though the actual fractional area of convection was not significantly altered. Thus the high cloud radiative heating appears to provide a feedback on the convection that regulates the high clouds produced by the convection.

### 10.4.6. Aerosol Effects on Middle- and High-Level Clouds

The relatively high number concentrations of small ice crystals ($\sim$0.1-10 cc$^{-1}$) regularly observed in in situ cirrus clouds in both the tropics and mid-latitudes appear to be associated with the mesoscale temperature variations generated by the variations in vertical velocity within cirrus clouds (Kärcher and Ström, 2003; Jensen and Pfister, 2004; Hoyle et al., 2005). Observations of the tropical and mid-latitude upper troposphere suggest a relatively consistent distribution of mesoscale temperature fluctuations, leading to typical cooling rates of the order of 10 K hr$^{-1}$. The origin of these fluctuations is not well known although they appear to be associated with gravity waves. In such a dynamic environment, the effect of an IN population on cirrus properties depends primarily on the IN number concentration, the ice nucleation threshold (determined by size, chemical composition etc.) and the local cooling rate. As described in greater detail below (references included there), it has been hypothesized that the likely impact of enhanced IN concentrations, keeping all other factors constant, is to reduce homogeneous freezing and decrease the total number of ice crystals formed, thus producing slightly larger effective ice crystal sizes and less bright cirrus. Absorption of thermal emission by large ice crystals may induce internal convective instability (especially in less stable thermodynamic cloud environments), thereby prolonging cirrus lifetimes by additional cooling and possible triggering of turbulence-induced ice nucleation. Unfortunately the response of these processes to enhanced IN concentrations have not as yet been observed in the field. Also, our current inability to accurately measure RHI in the upper troposphere limits our ability to discriminate between the various ice nucleation processes associated with different IN types.

While chemical composition and size are important considerations for a given ice nucleus, the general activation of ice crystals appears to be less sensitive to the size distribution of freezing aerosol and more sensitive to the updraft speeds and associated cooling rates. Homogeneous and heterogeneous nucleation both contribute to the formation of ice in middle and high-level clouds. Secondary ice processes also play a role in middle-level clouds but appear to be less important in cirrus clouds (Cantrell and Heymsfield, 2005). For cirrus clouds, the predominant heterogeneous modes are immersion freezing and deposition nucleation (DeMott, 2002). While immersion nuclei are likely to be observed in the upper troposphere, deposition nuclei often also act as CCN and thus may be removed before reaching the upper troposphere. IN may also be pre-activated in that following sublimation, the IN may later generate ice at lower RHI compared to previously unactivated IN. Twohy et al. (2009) demonstrated that Saharan dust particles can serve both as CCN and

IN. This has important implications for mixed-phase clouds and anvil cirrus as Saharan and Asian dusts activate more efficiently as condensation freezing nuclei than as deposition nuclei at temperatures warmer than ~240 K (Field et al., 2006). The contributions of the homogeneous freezing of liquid particles and heterogeneous ice nucleation involving mixed-phase or insoluble particles vary for cirrus that form in situ. While direct ice nucleation from aerosol particles may contribute during the life cycle of anvil cirrus and contrail cirrus, this does not appear to be the primary factor in in situ cirrus. Heymsfield et al. (1991) note that, as IN concentrations are low in cirrus clouds and yet ice crystals are evident in these clouds, ice crystal generation below −40 °C is due primarily to the homogeneous freezing of water droplets. Within jet exhaust, high concentrations of aerosols are activated ($10^4$-$10^5$ cc$^{-1}$) into water droplets that freeze rapidly. The actual nucleation process appears to be less important in contrails compared to the high number of available aerosols and the rapid cooling rates within these systems. Ice processes are discussed in more detail in Chapter 4.

Aerosol indirect effects associated with IN on middle- and high-level clouds are still not well understood. Changes in the number concentrations and sizes of liquid water particles can produce changes in the total concentrations of ice crystals of the order of a factor of 2 (Kärcher and Lohmann, 2002). When efficient IN and liquid particles compete during the formation of cirrus clouds, aerosol indirect effects may become more significant (DeMott et al., 1997). For a given cooling rate in a rising air parcel, the activation and deposition of water vapor on IN-generated ice crystals at low supersaturations reduces any increases in RHI that would otherwise occur, thus producing fewer homogeneously frozen particles. Adding IN to a population of liquid particles can therefore lead to a reduction in the number of nucleated ice crystals (DeMott et al., 1994). This process has been called the *negative Twomey effect* (Kärcher and Lohmann, 2003) in that while the Twomey effect refers to an increase in the number of activated cloud droplets with increasing CCN concentrations, this effect instead reduces the number of activated ice crystals with enhanced IN concentrations (DeMott et al., 1994). The negative Twomey effect can reduce total ice crystal concentrations by a factor of 10. Model simulations have demonstrated that it is associated with increased effective radii and reduced IWCs, and hence reduced cirrus albedos, as well as changes in optical extinction, subvisible cloud fraction and the occurrence of cirrus (Haag and Kärcher, 2004).

The magnitude of IN aerosol indirect effects depends on the ratio between the cooling rate and the growth rate. If sufficient IN are available for activation and the cooling rates are relatively slow so that only a fraction of these IN are activated, more ice crystals could then form than in the case when no IN are available. However, this effect tends to be weaker than the negative Twomey effect (Kärcher et al., 2006). In situ measurements of heterogeneous nucleation on IN have shown that the presence of IN did not prevent homogeneous freezing from occurring (DeMott et al., 2003), and while the negative Twomey effect has been inferred from satellite data, lidar observations of the same case study were inconclusive (Seifert et al., 2007). It is thus very difficult to directly attribute differences in cirrus properties to variations in IN concentrations given the difficulties in separating the dynamic effects from the aerosol forcing. More recently the negative Twomey effect has been studied in large-eddy simulations on the cloud scale (Spichtinger and Gierens, 2008). These simulations show that under some conditions the macroscale evolution of cirrus may be strongly affected by IN, while overall radiative properties are still dominated by liquid particle freezing. It is thus clear that significant advances are required in our understanding of aerosol indirect effects on cirrus.

While convective storms potentially transport IN from the surface to the upper troposphere it is uncertain as to what fractions of the IN actually reach the anvil cirrus and as to how significant their impacts may be. Very few observational studies have been conducted of the impacts of aerosols on deep convection and their associated anvils. Most of what is understood in this regard has been obtained via numerical modeling. Two recent modeling studies investigated the impacts of CCN and IN on the characteristics of deep convection and their associated anvils observed during the CRYSTAL-FACE field campaign (van den Heever et al., 2006; Carrió et al., 2007). Using a cloud resolving model, van den Heever et al. (2006) demonstrated that, in the presence of enhanced CCN and IN concentrations associated with a Saharan dust intrusion, the anvils covered a smaller area but were better organized and had larger condensate mixing ratios. Higher IN concentrations produced ice at warmer temperatures (and hence lower in the atmosphere) and the anvils were found to be thicker. Based on the results of their modeling study of storms that developed during the same field campaign, Fridlind et al. (2004) advanced the hypothesis that aerosols between 6 and 10 km had the greatest impact on the cirrus anvil microphysics. However, the simulations presented by van den Heever et al. demonstrated that many characteristics of both the convective and anvil stages of storm development were more sensitive to changes in the aerosol concentrations below 4 km. Other modeling studies have shown greater number concentrations of smaller crystals in response to enhanced aerosol concentrations (Khain et al., 2004) while others have observed the opposite trend, which they have attributed to differences in the strength of convection (Cui et al., 2006). Still other modeling studies have emphasized the role of aerosols on the release of latent heat and the feedbacks to subsequent storms dynamics (e.g. Khain et al., 2005; van den Heever et al., 2006).

In a subsequent study, Carrió et al. (2007) made use of an LES model forced by the output data from the mesoscale simulations of van den Heever et al. (2006), to investigate the anvil processes and characteristics in detail. The results from the LES simulations showed that variations in CCN and IN have significant effects on the optical properties and lifetime of anvil cirrus clouds. The anvils were both physically and optically thicker under enhanced aerosol concentrations. The largest differences in optical depth ($\sim$40%), cloud-averaged effective diameter (>25%) and IWC (~75%) occurred during the first hours of the simulation due to a higher frequency of large particles (aggregates). These differences then decreased due to the sedimentation of the aggregates. Toward the end of the simulations, the cloud-averaged effective diameter showed little change compared with the clean case. However, given that the differences in ice particle number concentrations remained large (>30%) throughout the simulation, the impacts of nucleating aerosol on optical depth continued to be significant. All of the microphysical and optical properties demonstrated a monotonic response to enhanced nucleating aerosol concentrations. Whether increasing those aerosols that serve as both CCN and IN, or just those serving as CCN, the mass fraction of particles involved in collisions (aggregates) and the total ice number concentrations increased monotonically, while those ice crystals that grew only by vapor diffusion (pristine ice) showed no significant changes. On the other hand, increasing the concentrations of those aerosols that serve as IN had a significant effect on the mass fraction of pristine ice but little effect on aggregates. The largest differences in the size distributions is associated with those simulations in which aerosols may serve as both CCN and IN. Enhancing CCN had two different effects on the simulated particle size distributions: (1) a large number of smaller, supercooled cloud droplets reached the top of the convective cloud, thereby significantly increasing the number concentration of ice particles with diameters less than 20 microns; and (2) significantly larger numbers of cloud droplets at high levels of the convective cloud increased the probability of binary interactions between particles and therefore the production of aggregates. It appears that the influence of enhanced concentrations available to act as CCN exerted the dominant effect on the particle size distributions of anvil cirrus. Finally, the narrower ice particle size distributions and the lower turbulent kinetic energy associated with enhanced aerosol number concentrations resulted in enhanced anvil-cirrus lifetimes.

## REFERENCES

Agee, E. M., and Gilbert, S. R. (1989). An aircraft investigation of mesoscale convection over Lake Michigan during the 10 January 1984 cold air outbreak. J. Atmos. Sci. 46, 1877–1897.

Albright, M. D., and Reed, R. J. (1995). Origin and structure of a numerically simulated polar low over Hudson Bay. Tellus 47A, 834–848.

Anthes, R. A., Kuo, Y.-H., and Guyakum, J. R. (1983). Numerical simulations of a case of explosive marine cyclogenesis. Mon. Weather Rev. 111, 1174–1188.

Atkinson, B. W. (1981). "Meso-scale Atmospheric Circulations." Academic Press, New York.

Atkinson, B. W., and Zhang, J. W. (1996). Mesoscale shallow convection in the atmosphere. Rev. Geophys. 34, 403–431.

Atlas, D., Wang, Z., and Duda, D. P. (2006). Contrails to cirrus–morphology, microphysics, and radiative properties. J. Appl. Met. Climatol. 45, 5–19.

Bader, M. J., Forbes, G. S., Grant, J. R., Lilley, R. B. E., and Waters, A. J. (1995). "Images in Weather Forecasting: A Practical Guide for Interpreting Satellite and Radar Imagery." Cambrige University Press, 499 pp.

Ballentine, R. J., Chermack, E. C., Stamm, A., Frank, D., Thomas, M., and Beck, G. (1992). Preliminary numerical simulations of the 31 January 1991 lake-effect snowstorm. In "Proc. 49th Annual Eastern Snow Conf., Oswego, NY, CRREL", pp. 115-123 [Available from US Army, CRREL, 72 Lyme Rd., Hanover, NH 03755].

Ballentine, R. J., Stamm, A. J., Chermack, E. E., Byrd, G. P., and Schleede, D. (1998). Mesoscale model simulation of the 4-5 January 1995 lake-effect snowstorm. Wea. Forecasting 13, 893–920.

Banacos, P. C. (2003). Short-range prediction of banded precipitation associated with deformation and frontogenetic forcing. In "10th Conf. on Mesoscale Processes", Portland, OR, Amer. Meteorol. Soc., CD-ROM, P1.7. Preprints.

Barth, M. C., and Parsons, D. B. (1996). Microphysical processes associated with intense rainbands and the effect of evaporation and melting on frontal dynamics. J. Atmos. Sci. 53, 1569–1585.

Bennetts, D. A., and Hoskins, B. J. (1979). Conditional symmetric instability-A possible explanation for frontal rainbands. Q. J. R. Meteorol. Soc. 105, 945–962.

Bennetts, D. A., and Ryder, P. (1984). A study of mesoscale convective bands behind cold fronts. Part I: Mesoscale organization. Q. J. R. Meteorol. Soc. 110, 121–145.

Benard, P., LaFore, J.-P., and Redelsperger, J.-L. (1992). Nonhydrostatic simulation of frontogenesis in a moist atmosphere. Part I: General description and narrow rainds. J. Atmos. Sci. 49, 2200–2217.

Bjerknes, J. (1919). On the structure of moving cyclones. Geophys. Publ. 1, 1–8.

Bjerknes, J., and Solberg, H. (1921). Meteorological conditions for the formationi of rain. Geofys. Publikasjoner 2, 1–60.

Bjerknes, J., and Solberg, H. (1922). Life cycle of cyclones and the polar front theory of atmospheric circulation. Geofys. Publ. 3, 1–18; Republished in English in Mon. Weather Rev. 50, 468–473.

Blackadar, A. K. (1957). Boundary layer wind maxima and their significance for the growth of nocturnal inversions. Bull. Am. Meteorol. Soc. 38, 283–290.

Blechschmidt, A.-M. (2008). A 2-year climatology of polar low events over the Nordic seas from satellite remote sensing. Geophys. Res. Lett. 35, L09815; doi:10.1029/2008GL033706.

Blechschmidt, A.-M., Bakan, S., and Graßl, H. (2009). Large-scale atmospheric circulation patterns during polar low events over the Nordic seas. J. Geophys. Res. 114, D06115. doi:10.1029/2008JD010865.

Blier, W., and Wakimoto, R. M. (1995). Observations of the early evolution of an explosive oceanic cyclone during ERICA IOP 5. Part I: Synoptic overview and mesoscale frontal structure. Mon. Weather Rev. 123, 1288–1310.

Bluestein, H. B. (1993). "Synoptic-Dynamic Meteorology in Midlatitudes. Vol. II, Observations and Theory of Weather Systems." Oxford University Press, 594 pp.

Boehm, M. T., and Verlinde, J. (2000). Stratospheric influence on upper tropospheric tropical cirrus. Geophys. Res. Lett. 27, 3209–3212.

Bond, N. A., and Shapiro, M. A. (1991). Polar lows over the Gulf of Alaska in conditions of reverse shear. Mon. Weather Rev. 119, 551–572.

Bond, N. A., Smull, B. F., Sooelinga, M. T., Woods, C. P., and Haase, A. (2005). Evolution of a cold front encountering steep quasi-2D terrain: Coordinated aircraft observations on 8-9 December 2001 during IMPROVE-2. J. Atmos. Sci. 62, 3559–3579.

Bonner, W. D. (1966). Case study of thunderstorm activity in relation to the low-level jet. Mon. Weather Rev. 94, 167–178.

Bonner, W. D. (1968). Climatology of the low level jet. Mon. Weather Rev. 96, 833–850.

Borokvikov, A. M., Gaivoronskii, IL. I., Zak, E. G., Kostarev, V. V., Mazin, I. P., Minervin, V. E., Khrgian, A. Kh., and Shmeter, S. M. (1963). "Cloud Physics." Israel Program for Scientific Translations, 392 pp.

Bosart, L. F., and Cussen Jr., J. P. (1973). Gravity wave phenomena accompanying East Coast cyclogenesis. Mon. Weather Rev. 101, 446–454.

Bosart, L. F., and Sanders, F. (1986). Mesoscale structure in the megalopolitan snowstorm of 11-12 February 1983. Part III: A large-amplitude gravity wave. J. Atmos. Sci. 43, 924–939.

Bosart, L. F., and Seimon, A. (1988). A case study of an unusually intense gravity wave. Mon. Weather Rev. 116, 1857–1886.

Bracegirdle, T., and Gray, S. (2008). An objective climatology of the dynamical forcing of polar lows in the Nordic Seas. Int. J. Climatol. 28, 1903–1919. doi:10/1002/joc.1686.

Braham Jr., R. R. (1983). The midwest snow storm of 8–11 December 1977. Mon. Weather Rev. 111, 253–272.

Braham, R. R. Jr. (1986). Cloud and motion fields in open-cell convection over Lake Michigan. In "Joint Sessions-23rd Conf. on Radar Meteorology and Conf. on Cloud Physics, Snowmass, CO." Amer. Meteorol. Soc., JP202-JP205. Preprints.

Braham Jr., R. R., and Dungey, M. J. (1995). Lake-effect snowfall over Lake Michigan. J. Appl. Met. 34, 1009–1019.

Braham Jr., R. R., and Kelly, R. D. (1982). Lake-effect snow storms on Lake Michigan, USA. In "Cloud Dynamics" (E. M. Agee and T. Asai, Eds.), pp. 87–101. Reidel, Dordrecht, Netherlands.

Braham Jr., R. R., and Spyers-Duran, P. (1967). Survival of cirrus crystals in clear air. J. Appl. Meteorol. 6, 1053–1061.

Bratseth, A. (1985). A note on CISK in polar air masses. Tellus 37A, 403–406.

Braun, S. A., Houze Jr., R. A., and Smull, B. F. (1997). Airborne Dual-Doppler Observations of an Intense Frontal System Approaching the Pacific Northwest Coast. Mon. Weather Rev. 125, 3131–3156.

Brennan, M. J., Lackmann, G. M., and Koch, S. E. (2003). An anlysis of the impact of a split-front rainband on Appalachian cold-air damming. Wea. Forecasting 18, 712–731.

Bresch, J. F., Ree, R. J., and Albright, M. D. (1997). A Polar-low development over the Bearing Sea: Analysis, numerical simulation, and sensitivity experiments. Mon. Weather Rev. 125, 3109–3130.

Brown, S. A., Locatelli, J. D., Stoelinga, M. T., and Hobbs, P. V. (1999). Numerical modeling of precipitation cores on cold fronts. J. Atmos. Sci. 56, 1175–1196.

Browning, K. A. (1971). Radar measurements of air motion near fronts. Part II. Weather 26, 320–340.

Browning, K. A. (1997). The dry intrusion perspective of extra-tropical cyclone development. Meteorol. Appl. 4, 317–324.

Browning, K. A., and Golding, B. W. (1995). Mesoscale aspects of a dry intrusion within a vigorous cyclone. Q. J. R. Meteorol. Soc. 121, 463–493.

Browning, K. A., and Harrold, T. W. (1969). Air motion and precipitation growth in a wave depression. Q. J. R. Meteorol. Soc. 95, 288–309.

Browning, K. A., and Harrold, T. W. (1970). Air motion and precipitation growth at a cold front. Q. J. R. Meteorol. Soc. 96, 369–389.

Browning, K. A., and Monk, G. A. (1982). A simple model for the synoptic analysis of cold fronts. Q. J. R. Meteorol. Soc. 108, 435–452.

Browning, K. A., and Roberts, N. M. (1994). Structure of a frontal cyclone. Q. J. R. Meteorol. Soc. 120, 1535–1557.

Browning, K. A., and Roberts, N. M. (1996). Variation of frontal and precipitation structure along a cold front. Q. J. R. Meteorol. Soc. 122, 1845–1872.

Browning, K. A., Hardman, M. E., Harrold, T. W., and Pardoe, C. W. (1973). The structure of rainbands within a mid-latitude depression. J. R. Met. Soc. 99, 215–231.

Browning, K. A., Clough, S. A., Davitt, D. S. A., Roberts, N. M., Hewson, T. D., and Healey, P. G. W. (1995). Observations of the mesoscale substructure in the cold air of a developing front cyclone. Q. J. R. Meteorol. Soc. 121, 1229–1254.

Brümmer, B., Müller, G., and Noer, G. (2009). A polar low pair over the Norwegian Sea. Mon. Weather Rev. 137, 2559–2575.

Burnett, A. W., Kirby, M. E., Mullins, H. T., and Patterson, W. P. (2003). Increasing Great Lake-effect snowfall during the Twentieth Century: A regional response to global warming. J. Climate 16, 3535–3542.

Burrows, W. R. (1991). Objective guidance for 0-24-hour and 24-48-hour mesoscale forecasts of lake-effect snow using CART. Wea. Forecasting 6, 357–378.

Businger, S. (1985). The synoptic climatology of polar low outbreaks. Tellus 37A, 419–432.

Businger, S. (1987). The synoptic climatology of polar low outbreaks over the Gulf of Alaska and the Bering Sea. Tellus 39A, 307–325.

Businger, S., and Baik, J.-J. (1991). An Arctic hurricane over the Bearing Sea. Mon. Weather Rev. 119, 2293–2322.

Businger, S., and Hobbs, P. V. (1987). Mesoscale structure of two comma cloud systems over the Pacific Ocean. Mon. Weather Rev. 115, 1908–1928.

Businger, S., and Reed, R. J. (1989a). Cyclogenesis in cold air masses. Wea. Forecasting 4, 133–156.

Businger, S., and Reed, R. J. (1989b). Polar lows. "Polar and Arctic Lows (P. F. Twitchell, E. A. Rasmussen and K. L. Davidson, Eds), pp. 3-45, A. Deepak, Hampton, VA.

Businger, S., and Walter, B. (1988). Comma cloud development and associated rapid cyclogenesis over the Gulf of Alaska. A case study using aircraft and operational data. Mon. Weather Rev. 116, 1103–1123.

Byrd, G. P., Anstett, R. A., Heim, J. E., and Usinski, D. M. (1991). Mobile sounding observations of lake-effect snowbands in western and central New York. Mon. Weather Rev. 119, 2323–2332.

Byrd, G. P., Bikos, D. E., Schleede, D. L., and Ballentine, R. J. (1995). The influence of upwind lakes on snowfall to the lee of Lake Ontario. In "14th Conf. on Weather Analysis and Forecasting, Dallas, TX." pp. 204–207. Amer. Meteorol. Soc. Preprints.

Cairns, M. M., Collins, R., Cylke, T., Deutschendorf, M., and Mercer, D. (2001). A lake effect snowfall in Western Nevada–Part I: Synoptic setting and observations. In "18th Conf. on Weather Analysis and Forecasting/14th Conf. on Numerical Weather Prediction, Fort Lauderdale, FL." pp. 329–332. Amer. Met. Soc. Preprints.

Cantrell, W., and Heymsfield, A. J. (2005). Production of ice in tropospheric clouds: A review. Bull. Amer. Meteorol. Soc. 86, 795–807.

Carbone, R. E. (1982). A severe frontal rainband. Part I: Stormwide hydrodynamic structure. J. Atmos. Sci. 39, 258–279.

Carey, L. D., Niu, J., Yang, P., Kankiewicz, I. A., Larson, V. E., and Vonder Haar, T. H. (2008). The vertical profile of liquid and ice water content in midlatitude mixed-phase altocumulus clouds. J. Appl. Met. Climatol. 47, 2487–2495.

Carlson, T. N. (1980). Airflow through midlatitude cyclones and the common cloud pattern. Mon. Weather Rev. 108, 1498–1509.

Carleton, A. M. (1979). A synoptic climatology of satellite-observed extratropical cyclone activity for the Southern Hemisphere winter. Arch. Meteorol., Geophys. Bioklim. 27B, 265–279.

Carleton, A. M. (1985). Satellite climatological aspects of the "polar low" and "instant occlusion". Tellus 37, 433–450.

Carleton, A. M., and Carpenter, D. A. (1989). Intermediate-scale sea ice-atmosphere interactions over high southern latitudes in winter. Geojournal 18, 87–101.

Carleton, A. M., and Carpenter, D. A. (1990). Satellite climatology of "polar low" and broadscale climatic associations for the Southern Hemisphere. Int. J. Climatol. 10, 219–246. doi:10.1002/joc.3370100302.
Carpenter, D. (1993). The lake-effect of the Great Salt Lake: Overview and forecast problems. Wea. Forecasting 8, 181–193.
Carr, F. H., and Millard, J. P. (1985). A composite study of comma clouds and their association with severe weather over the Great Plains. Mon. Weather Rev. 113, 370–387.
Carrió, G. G., van den Heever, S. C., and Cotton, W. R. (2007). Impacts of nucleating aerosol on anvil-cirrus clouds: A modeling study. Atmos. Res. 84, 111–131.
Chang, S. S., and Braham, R. R. (1991). Observational study of a convective internal boundary layer over Lake Michigan. J. Atmos. Sci. 48, 2265–2279.
Charney, J. G. (1947). The dynamics of long waves in a baroclinig westerly current. J. Meteorol. 4, 135–163.
Charney, J. G., and Eliassen, A. (1964). On the growth of hurricane depression. J. Atmos. Sci. 21, 255–268.
Chen, C., Bishop, C. H., Lai, G. S., and Tao, W.-K. (1997). Numerical simulations of an observed narrow cold-frontal rainband. Mon. Weather Rev. 125, 1027–1045.
Chen, T., Rossow, W. B., and Zhang, Y. (2000). Radiative effects of cloud-type variations. J. Climate 13, 264–286.
Cheng, W. Y. Y., Wu, T., and Cotton, W. R. (2001). Large-eddy simulations of the 26 November 1991 FIRE II cirrus case. J. Atmos. Sci. 58, 1017–1034.
Churchill, D. D., and Houze, R. A. (1984). Development and structure of winter monsoon cloud clusters on 10 December 1978. J. Almos. Sci. 41, 933–960.
Cober, S. G., Isaac, G. A., and Korolev, A. V. (2001). Assessing the Rosemount icing detector with in situ measurements. J. Atmos. Oceanic Technol. 18, 515–528.
Colle, B. A. (2004). Sensitivity of orographic precipitation to changing ambient conditions and terrain geometries: An idealized modeling perspective. J. Atmos. Sci. 61, 588–606.
Colle, B. A., Smull, B. F., and Yang, M.-J. (2002). Numerical simulations of a landfalling cold front observed during COAST: Rapid evolution and responsible mechanisms. Mon. Weather Rev. 130, 1945–1966.
Cooper, K. A., Hjelmfelt, M. R., Derickson, R. G., Kristovich, D. A. R., and Laird, N. F. (2000). Numerical simulation of transitions in boundary layer convective structures in a lake-effect snow event. Mon. Weather Rev. 128, 3283–3295.
Cordeira, J. M., and Laird, N. F. (2008). The influence of ice cover on two lake-effect snow events over Lake Erie. Mon. Weather Rev. 139, 2747–2763.
Cosgrove, B. A., Colucci, S. J., Ballentine, R. J., and Waldstreicher, J. S. (1996). Lake effect snow in the Finger Lakes region. In "15th Conf. on Weather Analysis and Forecasting, Norfolk, VA." pp. 573–576. Amer. Met. Soc. Preprints.

Craig, G., and Cho, H.-R. (1988). Cumulus heating and CISK in the extratropical atmosphere. Part I: Polar lows and comma clouds. J. Atmos. Sci. 45, 2622–2640.

Craig, G. C., and Gray, Z. L. (1996). CISK or WISHE as the mechanism for tropical cyclone intensification. J. Atmos. Sci. 53, 3528–3529.

Crocker, A. M., Godson, W. L., and Penner, C. M. (1947). Frontal contour charts. J. Meteorol. 4, 95–99.

Cui, Z., Carslaw, K. S., Yin, Y., and Davies, S. (2006). A numerical study of aerosol effects on the dynamics and microphysics of a deep convective cloud in a continental environment. J. Geophys. Res. 111, D05201.

Davis, L. G., Lavoie, R. L., Kelley, J. I., and Hosler, C. L. (1968). Lake-effect studies. Final Rep. to Environ. Sci. Servo Adm., Contract No. E22-80-67(N) Dep. Meteorol., Pennsylvania State University.

DeMott, P. J. (2002). Laboratory studies of cirrus cloud processes. In "Cirrus" (D. K. Lynch, K. Sassen, D. O. C. Starr and G. Stephens, Eds.), Oxford University Press, New York, Chapter 5.

DeMott, P. J., Meyers, M. P., and Cotton, W. R. (1994). Parameterization and impact of ice initiation processes relevant to numerical model simulations of cirrus clouds. J. Atmos. Sci. 51, 77–90.

DeMott, P. J., Rogers, D. C., and Kreidenweis, S. M. (1997). The susceptibility of ice formation in upper troposphere clouds to insoluble aerosol components. J. Geophys. Res. 102, 19575–19584.

DeMott, P. J., Cziczo, D. J., and Prenni, A. J. et al. (2003). Measurements of the concentration and composition of nuclei for cirrus formation. PNAS 100, 14655–14660.

Deng, M., and Mace, G. G. (2008). Cirrus microphysical properties and air motion statistics using cloud radar Doppler moments. Part II: Climatology. J. Appl. Met. Climatol. 47, 3221–3235.

Desbois, M., Seze, G., and Szejwach, G. (1982). Automatic classification of clouds on METEOSAT imagery: Application to high-level clouds. J. Appl. Meteorol. 21, 401–412.

Dessler, A. E., Palm, S. P., Hart, W. D., and Spinhirne, J. D. (2006). Tropopause-level thin cirrus coverage revealed by ICE-Sat/Geoscience Laser Altimeter System. J. Geophys. Res. 111, D08203. doi:10.1029/2005JD005686.

Douglas, M. W., Fedor, L. S., and Shapiro, M. A. (1991). Polar low structure after the northern Gulf of Alaska based on research aircraft observations. Mon. Weather Rev. 119, 32–54.

Douglas, M. W., Shapiro, M. A., Fedor, L. S., and Saukkonen, L. (1995). Research aircraft observations of a polar low at the East Greenland ice edge. Mon. Weather Rev. 123, 5–15.

Duda, D. P., Minnis, P., and Nguyen, L. (2001). Estimates of cloud radiative forcing in contrail clusters using GOES imagery. J. Geophys. Res. 106, 4927–4937.

Duncan, C. N. (1977). A numerical investigation of polar lows. Q. J. R. Meteorol. Soc. 103, 225–267.

Eady, E. T. (1949). Long waves and cyclone waves. Tellus 1, 33–52.

Eckhardt, S., Stohl, A., Wernli, H., James, P., Forster, C., and Spichtinger, N. (2004). A 15-year climatology of warm conveyor belts. J. Climate 17, 218–237.

Ekman, A. M. L., Wang, C., Wilson, J., and Ström, J. (2004). Explicit simulations of aerosol physics in a cloud-resolving model: a sensitivity study based on an observed convective cloud. Atmos. Chem. Phys. 4, 773–791.

Eichenlaub, V. L. (1979). Lake effect snowfall to the lee of the Great Lakes: Its role in Michigan. Bull. Am. Meteorol. Soc. 51, 403–412.

Einaudi, F., and Lalas, D. P. (1973). On the growth rate of an unstable disturbance in a gravitationally stratified shear flow. J. Atmos. Sci. 30, 1707–1710.

Ellenton, G. E., and Danard, M. B. (1979). Inclusion of sensible heating in convective parameterization applied to lake-effect snow. Mon. Weather Rev. 107, 551–565.

Ellis, A. W., and Leathers, D. J. (1996). A synoptic climatological approach to the analysis of Lake-effect snowfall: Potential forecasting applications. Wea. Forecasting 11, 216–229.

Emanuel, K. A. (1979). Inertial instability and mesoscale convective systems. Part I: Linear theory of inertial instability in rotating viscous fluids. J. Almos. Sci. 36, 2425–2449.

Emanuel, K. A. (1982). Inertial instability and mesoscale convective systems. Part II: Symmetric CISK in a baroclinic flow. J. Almos. Sci. 39, 1080–1097.

Emanuel, K. A. (1983a). Symmetric instability. In "Mesoscale Meteorology-Theories, Observations and Models" (D. K. Lilly and T. Gal-Chen, Eds.), pp. 217–229. Reidel, Dordrecht, Netherlands.

Emanuel, K. A. (1983b). Conditional symmetric instability: A theory for rainbands within extratropical cyclones. In "Mesoscale Meteorology-Theories, Observations and Models" (D. K. Lilly and T. Gal-Chen, Eds.), pp. 231–245. Reidel, Dordrecht, Netherlands.

Emanuel, K. A. (1985). Frontal circulations in the presence of small moist symmetric instability. J. Atmos. Sci. 42, 1062–1071.

Emanuel, K. A. (1986). An air-sea interaction for tropical cyclones. Part I: Steady-state maintenance. J. Atmos. Sci. 43, 585–604.

Emanuel, K. A., Neelin, J. D., and Bretherton, C. S. (1994). On large-scale circulations in convecting atmospheres. Q. J. R. Meteorol. Soc. 120, 1111–1144.

Emanuel, K. A., and Rotunno, R. (1989). Polar lows as arctic hurricanes. Tellus, Ser. A. 41, 1–17.

Ernst, J. A., and Matson, M. (1983). A Mediterranean tropical storm? Weather 38, 332–337.

E.S.S.A. (1966). "Storm Data," Monthly publication, December. US Gov. Print. Off., Washington, DC.

Evans, A. G., Locatelli, J. D., Stoelinga, M. T., and Hobbs, P. V. (2005). The IMPROVE-1 Storm of 1-2 February 2001. Part II: Cloud structures and the growth of precipitation. J. Atmos. Sci. 62, 3456–3473.

Feretti, R., Einaudi, F., and Uccelini, L. W. (1988). Wave disturbances associated with the Red River valley severe weather outbreak of 10-11 April 1979. Meteorol. Atmos. Phys. 39, 132–168.

Field, P. R. (1999). Aircraft observations of ice crystal evolution in an altostratus clouds. J. Atmos. Sci. 56, 1925–1941.

Field, P. R., Heymsfiled, A. J., and Bansemer, A. (2006). Shattering and particle interarrival times measured by optical array probes in ice clouds. J. Atmos. Oceani Technol. 23, 1357–1371.

Fleishauer, R. P., Larson, V. E., and Vonder Haar, T. H. (2002). Observed microphysical structure of midlevel, mixed-phase clouds. J. Atmos. Sci. 59, 1779–1804.

Forbes, G. S., and Lottes, W. D. (1985). Classification of mesoscale vortices in polar air streams and the influence of the large-scale environment on their evolutions. Tellus 37A, 132–155.

Forbes, G. S., and Merritt, J. H. (1984). Mesoscale vortices over the Great Lakes in wintertime. Mon. Weather Rev. 112, 377–381.

Fowler, L. D., and Randall, D. A. (1994). A global radiative-convective feedback. Geophys. Res. Lett. 21, 2035–2038.

Fridlind, A. M., Ackerman, A. S., and Jensen, E. J. et al. (2004). Evidence for the predominance of mid-tropospheric aerosols as subtropical anvil cloud nuclei. Science 304, 718–722.

Fu, G., Niino, H., Kimura, R., and Kato, T. (2004). A polar low over the Japan Sea on 21 January 1997. Part I: Observational analysis. Mon. Weather Rev. 132, 1537–1551.

Galloway, J. L. (1958). The three-front model: its philosophy, nature, construction and use. Weather 13, 3–10.

Galloway, J. L. (1960). The three front model, the developing depression and the occluding process. Weather 15, 293–301.

Garrett, T. J., Navarro, B. C., Twohy, C. H., Jensen, E. J., Baumgardner, D. G., Bui, P. T., Gerber, H., Herman, R. L., Heymsfield, A. J., Lawson, P., Minnis, P., Nguyen, L., Poellot, M., Pope, S. K., Valero, K. P. J., and Weinstock, E. M. (2005). Evolution of a Florida cirrus anvil. J. Atmos. Sci. 62, 2352–2372.

Garrett, T. J., Heymsfield, A. J., McGill, M. J., Ridley, B. A., Baumgardner, D. G., Bui, P. T., and Webster, C. R. (2004). Convective generation of cirrus near the tropopause. J. Geophys. Res. 109, D21203. doi:10.1029/2004JD004952.

Gedzelman, S. D. (1988). In praise of altocumulus. Weatherwise 41, 143–149.

Geerts, B., and Hobbs, P. V. (1991). Organization and structure of clouds and precipitation on the Mid-Atlantic Coast of the United States. Part IV: Retrieval of the thermodynamic and cloud microphysical structures of a frontal rainband from Doppler radar data. J. Atmos. Sci. 48, 1287–1305.

Gerbush, M. R., Kristovich, D. A. R., and Laird, N. F. (2008). Mesoscale boundary layer and heat fluxes variations over pack ice-covered Lake Erie. J. Appl. Meteorol. Climatol. 47, 668–682.

Gettleman, A. W., Randal, J., Wu, F., and Massie, S. T. (2002). Transport of water vapor in the tropical tropopause layer. Geophys. Res. Lett. 29, 1009. doi:10.1029/2001GL013818.

Godson, W. L. (1951). Synoptic properties of frontal surfaces. Q. J. R. Meteorol. Soc. 77, 633–653.

Griffith, K. T., Cox, S. K., and Knollenberg, R. G. (1980). Infrared radiative properties of tropical cirrus clouds inferred from aircraft measurements. J. Atmos. Sci. 37, 1077–1087.

Grim, J. A., Laird, N. F., and Kristovich, D. A. R. (2004). Mesoscale vortices embedded within a lake-effect shoreline band. Mon. Weather Rev. 132, 2269–2274.

Grim, J. A., Rauber, R. M., Ramamurth, M. K., Jewett, B. F., and Han, M. (2007). High-resolution observations of the trowal–warm-frontal region of two continental winter cyclones. Mon. Weather Rev. 135, 1629–1646.

Gultepe, I., and Starr, D. O. (1995). Dynamical structure and turbulence in cirrus clouds: Aircraft observations during the FIRE. J. Atmos. Sci. 52, 4159–4182.

Haag, W., and Kärcher, B. (2004). The impact of aerosols and gravity waves on cirrus clouds at midlatitudes. J. Geophys. Res. 109, D12202.

Hadlock, R., and Kreitzberg, C. W. (1988). The eexperiment of rapidly intensifying cyclones over the Atlantic (ERICA) field study. Objectives and plans. Bull. Am. Meteorol. Soc. 69, 1309–1320.

Hahn, C. J., and Warren, S. G. (1999). Extended edited cloud reports from ships and lang stations over the globe, 1952-1996. Numer. Data Package NDP-026C, 79 pp., Carbon Dioxide Inf. Anal. Cent., Dep. of Energy, Oak Ridge, Tenn.

Hakim, G. J., and Uccellini, L. W. (1992). Dianosing coupled jet-steak circulations for a northern plains snowband from the operational nested grid model. Wea. Forecasting 7, 26–48.

Haladay, T., and Stephens, G. (2009). Characteristics of tropical thin cirrus clouds deduced from joint CloudSat and CALIPSO observations. J. Geophys. Res. 114, D00A25. doi:10.1029/2008JD010675.

Hall, W. D., and Pruppacher, H. R. (1976). The survival of ice particles falling from cirrus clouds in subsaturated air. J. Atmos. Sci. 33, 1995–2006.

Han, M., Braun, S. A., Persoon, P. O. G., and Bao, J.-W. (2009). A longfront variability of precipitation associated with a midlatitude frontal zone: TRMM observations and MM5 simulation. Mon. Weather Rev. 137, 1008–1028.

Harrold, T. W. (1973). Mechanisms influencing the distribution of precipitation within baro clinic disturbances. Q. J. R. Meteorol. Soc. 99, 232–251.

Harrold, T. W., and Browning, K. A. (1969). The polar low as a baroclinic disturbance. Q. J. R. Meteorol. Soc. 95, 710–723.

Hartmann, D. L., Okhert-Bell, M. E., and Michelson, M. L. (1992). The effect of cloud type on Earth's energy balance. Global analysis. J. Climate 5, 1281–1304.

Hartmann, D. L., Holton, J. R., and Fu, Q. (2001). The heat balance of the tropical tropopause, cirrus, and stratospheric dehydration. Geophys. Res. Lett. 28, 1969–1972.

Heideman, K. F., and Fritsch, J. M. (1988). Forcing mechanisms and other characteristics of significant summertime precipitation. Wea. Forecasting 3, 115–130.

Herman, G. F., and Curry, J. A. (1984). Observational and theoretical studies of solar radiation in arctic stratus clouds. J. Climate Appl. Met. 23, 5–24.

Herzegh, P. H., and Hobbs, P. V. (1980). The mesoscale and microscale structure and organization of clouds and precipitation in midlatitude cyclones. II. Warm-frontal clouds. J. Almos. Sci. 37, 597–611.

Hewson, T. D., Craig, G. C., and Claud, C. (2000). Evolution and mesoscale structure of a polar low outbreak. Q. J. R. Meteorol. Soc. 126, 1031–1063.

Heymsfield, A. J. (1975a). Cirrus uncinus generating cells and the evolution of cirriform clouds. Part I: Aircraft observations of the growth of the ice phase. J. Atmos. Sci. 32, 799–808.

Heymsfield, A. J. (1975b). Cirrus uncinus generating cells and the evolution of cirriform clouds. Part II: The structure and circulations of the cirrus uncinus generating head. J. Atmos. Sci. 32, 809–819.

Heymsfield, A. J. (1977). Precipitation development in stratiform ice clouds: A microphysical and dynamical study. J. Atmos. Sci. 34, 367–381.

Heymsfield, A. J., and Knollenberg, R. G. (1972). Properties of cirrus generating cells. J. Atmos. Sci. 29, 1358–1366.

Heymsfield, A. J., and Miloshevich, L. M. (1995). Relative humidity and temperature influences on cirrus formation and evolution: Observations from wave clouds and FIRE II. J. Atmos. Sci. 52, 4302–4326.

Heymsfield, A. J., and Platt, C. M. R. (1984). A parameterization of the particle size spectrum of ice clouds in term sof the ambient temperature and the ice water content. J. Atmos. Sci. 41, 846–855.

Heymsfield, A. J., Miloshevich, L. M., Slingo, A., Sassen, K., and Starr, D. O'C. (1991). An observational and theoretical study of highly supercooled altocumulus. J. Atmos. Sci. 48, 923–945.

Heymsfield, A. J., Lawson, R. P., and Sachs, G. W. (1998). Growth of ice crystals in a precipitating contrail. Geophys. Res. Lett. 25, 1335–1338.

Hjelmfelt, M. (1988). Numerical sensitivity study of the morphology of lake-effect snow storms over Lake Michigan. In "Int. Cloud Phys. Conf, 10th, Int. Assoc. Meteorol. Atmos. Phys., Bad Homburg, F.R.G." pp. 413–415. Preprints.

Hjelmfelt, M. R. (1990). Numerical study of the influence of environmental conditions on lake effect snowstorms over Lake Michigan. Mon. Weather Rev. 118, 138–150.

Hjelmfelt, M. R., and Braham Jr., R. R. (1983). Numerical simulation of the airflow over Lake Michigan for a major lake-effect snow event. Mon. Weather Rev. 111, 205–219.

Hobbs, P. V. (1978). Organization and structure of clouds and precipitation on the mesoscale and microscale in cyclonic storms. Rev. Geophys. Space Phys. 16, 741–755.

Hobbs, P. V., and Biswas, K. R. (1979). The cellular structure of the narrow cold-frontal ranbands. Q. J. R. Meteorol. Soc. 105, 723–727.

Hobbs, P. V., and Locatelli, J. D. (1978). Rainbands, precipitation cores and generating cells in a cyclonic storm. J. Atmos. Sci. 35, 230–241.

Hobbs, P. V., and Persson, P. O. G. (1982). The mesoscale and microscale structure and organization of clouds and precipitation in mid-latitude cyclones. Part V: The substructure of narrow cold-frontal rainbands. J. Atmos. Sci. 39, 280–295.

Hobbs, P. V., Houze, R. A., and Matejka, T. J. (1975). The dynamical and microphysical structure of an occulded frontal system and its modification by orography. J. Atmos. Sci. 32, 1542–1562.

Hobbs, P. V., Matejka, T. J., Herzegh, P. H., Locatelli, P. H., and Houze Jr., R. A. (1980). The mesoscale and microscale structure and organization of clouds and precipitation in midlatitude cyclones. I: A case study of a cold front. J. Atmos. Sci. 37, 568–596.

Hobbs, P. V., Locatelli, J. D., and Martin, J. E. (1990). Cold fronts aloft and the forecasting of precipitation and several weather east of the Rocky Mountains. Wea. Forecasting 5, 613–626.

Hobbs, P. V., Locatelli, J. D., and Martin, J. E. (1996). A new conceptual model for cyclones generated in the lee of the Rocky Mountains. Bull. Am. Meteorol. Soc. 77, 1169–1178.

Hobbs, P. V., and Ragno, A. L. (1985). Ice particle concentrations in clouds. J. Atmos. Sci. 42, 2523–2549.

Hogan, R. J., and Kew, S. F. (2005). A 3D stochastic cloud model for investigating the radiative properties of inhomogeneous cirrus clouds. Q. J. R. Meteorol. Soc. 131, 2585–2608.

Hogan, R. J., Francis, P. N., Flentje, H., Illingworth, A. J., Quante, M., and Pelon, J. (2002). Characteristics of mixed-phase clouds. Part I: Lidar, radar and aircraft observations from CLARE'98. Q. J. R. Meteorol. Soc. 128, 1–28.

Holton, J. R. (1979). "An Introduction to Dynamic Meteorology." 2nd Ed., Academic Press, New York.

Holton, J. R., and Gettleman, A. (2001). Horizontal transport and the dehydration of the stratosphere. Geophys. Res. Lett. 28, 2799–2802.

Hoskins, B. J. (1976). Baroclinic waves and frontogenesis. Part I: Introduction and Eady waves. Q. J. R. Meteorol. Soc. 102, 103–122.

Hoskins, B. J., and West, N. V. (1979). Baroclinic waves and frontogenesis. Part II: Uniform potential vorticity jet flows–cold and warm fronts. J. Atmos. Sci. 36, 1663–1680.

Houze Jr., R. A. (1993). "Cloud Dynamics." Academic Press, 573 pp.
Houze Jr., R. A., and Hobbs, P. V. (1982). Organization and structure of precipitating cloud systems. Adv. Geophys. 24, 225–315.
Houze, R. A., Hobbs, P. V., Biswas, K. R., and Davis, W. M. (1976). Mesoscale rainbands in extra-tropical cyclones. Mon. Weather Rev. 104, 868–878.
Houze Jr., R. A., Rutledge, S. A., Matejka, T. J., and Hobbs, P. V. (1981). The mesoscale and microscale structure and organization of clouds and precipitation in midlatitude cyclones. III. Air motion and precipitation growth in a warm-frontal rainband. J. Atmos. Sci. 38, 639–649.
Hoyle, C. R., Luo, B. P., and Peter, T. (2005). The origin of high ice crystal number densities in cirrus clouds. J. Atmos. Sci. 62, 2568–2579.
Hsie, E.-Y., Anthes, R. A., and Keyser, D. (1984). Numerical simulation of frontogenesis in a moist atmosphere. J. Atmos. Sci. 41, 2581–2594.
Hsu, H.-M. (1987). Mesoscale lake-effect snowstorms in the vicinity of Lake Michigan: Linear theory and numerical simulations. J. Atmos. Sci. 44, 1019–1040.
James, P. K., and Browning, K. A. (1979). Mesoscale structure of line convection at surface cold fronts. Q. J. R. Meteorol. Soc. 105, 371–382.
Jensen, E. J., and Pfister, L. (2004). Transport and freeze-drying in the tropical tropopause layer. J. Geophys. Res. 109, D02207. doi:10.1029/2003JD004022.
Jensen, E. J., Starr, D., and Toon, B. (2004). Mission investigates tropical cirrus clouds. EOS, Trans. Amer. Geophys. Union 84, 45–50.
Jewett, B. F., Ramamurthy, M. K., and Rauber, R. M. (1999). Initiation and modeled evolution of a STORM-FEST gravity wave. In "Eighth Conf. on Mesoscale Processes, Boulder, CO." pp. 136–139. Amer. Meteorol. Soc. Preprints.
Jewett, B. F., Ramamurthy, M. K., and Rauber, R. M. (2003). Origin, Evolution, and Finescale Structure of the St. Valentine's Day Mesoscale Gravity Wave Observed during STORM-FEST. Part III: Gravity Wave Genesis and the Role of Evaporation. Mon. Weather Rev. 131, 617–633.
Jin, Y., and Koch, S. E. (1998). Model predictability of the gravity wave event of 14 February 1992 during STORM-FEST. In "16th Conf. on Weather Analysis and Forecasting, Phoenix, AZ." pp. 466-468. Amer. Meteorol. Soc. Preprints.
Jiusto, J. E., Paine, D. A., and Kaplan, M. L. (1970). Great Lakes snowstorms. Part 2. Synoptic and climatological aspects. State University of New York at Albany Rep. ESSA E22-13-60, 58 pp.
Jorgensen, D. P., Pu, Z., Persson, P. O. G., and Tao, W.-K. (2003). Variations associated with cores and gaps of a Pacific narrow cold frontal rainband. Mon. Weather Rev. 131, 2705–2729.
Jurewicz, M. L., and Evans, M. S. (2004). A comparison of two banded heavy snowstorms with very different synoptic settings. Wea. Forecasting 19, 1011–1028.

Kankiewicz, J. A., Fleishauer, R. P., Larson, V. E., Reinke, D. L., Davis, J. M., Vonder Haar, T. H., and Cox, S. K. (2000). In-situ and satellite-based observations of mixed phase non-precipitating clouds and their environments. In "13th Int. Conf. on Clouds and Precipitation, Reno, NV, International Commission on Clouds and Precipitation." pp. 697–700. Preprints.

Kärcher, B., and Lohmann, U. (2002). A parameterization of cirrus cloud formation: Homogeneous freezing including effects of aerosol size. J. Geophys. Res. 107, 4698.

Kärcher, B., and Lohmann, U. (2003). A parameterization of cirrus cloud formation: Heteorogeneous freezing. J. Geophys. Res. 108, 4402. doi:10.1029/2002JD003220.

Kärcher, B., and Spichtinger, P. (2009). Cloud controlling factors of cirrus. In "Clouds in the Perturbed Climate System: Their relationship to energy balance, atmospheric dynamics, and precipitation" (J. Heintzenberg and R. J. Charlson, Eds.), pp. 235–267. The MIT Press, Cambridge, MA.

Kärcher, B., and Ström, J. (2003). The roles of dynamical variability and serosols in cirrus cloud formation. Atmos. Chem. Phys. 3, 823–838.

Kärcher, B., Hendricks, J., and Lohmann, U. (2006). Physically based parameterization of cirrus cloud formation for use in global atmospheric models. J. Geophys. Res. 111, D01205.

Karyampudi, V. M., Koch, S. E., Chen, C., Rottman, J. W., and Kaplan, M. L. (1995). The influence of the Rocky Mountains on the 13-14 April 1986 severe weather outbreak. Part II: Evolution of a prefrontal bore and its role in triggering a squall line. Mon. Weather Rev. 123, 1423–1446.

Keckhut, P., Borshi, F., Bekki, S., Hauchecorne, A., and SiLaouina, M. (2006). Cirrus classification at midlatitude from systematic lidar observations. J. Appl. Met. Climatol. 45, 249–258.

Kelly, R. D. (1982). A single Doppler radar study of horizontal-roll convection in a lake-effect snow storm. J. Atmos. Sci. 39, 1521–1531.

Kelly, R. D. (1984). Horizontal rolls and boundary-layer interrelationships observed over Lake Michigan. J. Atmos. Sci. 41, 1816–1826.

Kelly, R. D. (1986). Mesoscale frequencies and seasonal snowfall for different types of Lake Michigan snow storms. J. Clim. Appl. Meteorol. 25, 308–312.

Keyser, D., and Anthes, R. A. (1982). The influence of planetary boundary layer physics on frontal structure in the Hoskins-Bretherton horizontal shear model. J. Atmos. Sci. 39, 1783–1802.

Keyser, D., and Pecnick, M. J. (1985a). A two-dimensional primitive equation model of frontogenesis forced by confluence and horizontal shear. J. Atmos. Sci. 42, 1259–1282.

Keyser, D., and Pecnick, M. J. (1985b). Diagnosis of ageostrophic circulations in a twodimensional primitive equation model of frontogenesis. J. Atmos. Sci. 42, 1283–1305.

Keyser, D., and Shapiro, M. A. (1986). A review of the structure and dynamics of upper-level frontal zones. Mon. Weather Rev. 114, 452–499.

Khain, A., Pokrovsky, A., Pinsky, M., Seifert, A., and Phillips, V. T. J. (2004). Simulation of effects of atmospheric aerosols on deep turbulent convective clouds by using a spectral microphysics mixed-phase cumulus cloud model. Part I: Model description and possible applications. J. Atmos. Sci. 61, 2963–2982.

Khain, A., Rosenfeld, D., and Pokrovsky, A. (2005). Aerosol impact on the dynamics and microphysics of deep convective clouds. Q. J. R. Meteorol. Soc. 131, 2639–2663.

Kingsmill, D. E., Nieman, P. J., and Ralph, F. M. (2006). Synoptic and topographic variability of Northern California precipitation characteristics in landfalling winter storms observed during CALJET. Mon. Weather Rev. 134, 2072–2093.

Knight, D. J., and Hobbs, P. V. (1988). The mesoscale and microscale structure and organization of clouds and precipitation I midlatitude cyclones. Part XV: A numerical modeling study of frontogenesis and cold-frontal rainbands. J. Atmos. Sci. 45, 915–930.

Knollenberg, R. G., Kelley, K., and Wilson, J. C. (1993). Measurements of high number densities of ice crystals in the tops of tropical cumulonimbus. J. Geophys. Res. 98, 8639–8664.

Koch, S. E. (1984). The role of an apparent mesoscale frontogenetical circulation in squall line initiation. Mon. Weather Rev. 112, 2090–2111.

Koch, S. E. (2001). Real-time detection of split fronts using mesoscale models and WSR-88D radar products. Wea. Forecasting 16, 35–55.

Koch, S. E., and Dorian, P. B. (1988). A mesoscale gravity wave event observed during CCOPE. Part III: Wave environment and probable source mechanisms. Mon. Weather Rev. 116, 2570–2592.

Koch, S. E., and Golus, R. E. (1988). A mesoscale gravity wave event observed during CCOPE. Part I: Multiscale statistical analysis of wave characteristics. Mon. Weather Rev. 116, 2527–2544.

Koch, S. E., and Kocin, P. J. (1991). Frontal contraction processes leading to the formation of an intense narrow rainband. Meteorol. Atmos. Phys. 46, 123–154.

Koch, S. E., McQueen, J. T., and Karyampudi, V. M. (1995). A numerical study of the effects of differential cloud cover on cold frontal structure and dynamics. J. Atmos. Sci. 52, 937–964.

Kocin, P. J., Schumacher, P. N., Morales Jr., R. F., and Uccellini, L. W. (1995). Overview of the 12-14 March 1993 superstorm. Bull. Am. Meteorol. Soc. 76, 165–182.

Korolev, A. V., Issac, G. A., Mazin, I. P., and Barker, H. W. (2001). Microphysical properties of continental clouds from in situ measurements. Q. J. R. Meteorol. Soc. 127, 2117–2151.

Korolev, A., Isaac, G. A., Cober, S. G., Strapp, J. W., and Hallett, J. (2003). Microphysical characterization of mixed-phase clouds. Q. J. R. Meteorol. Soc. 129, 39–65.

Kristjánsson, J. E., Edwards, J. M., and Mitchell, D. L. (2000). Impact of a new scheme for optical properties of ice crystals on climates of two GCM's. J. Geophys. Res. 105, 10063–10079.

Kristovich, D. A. R. (1993). Mean circulations of boundary-layer rolls in lake-effect snow storms. Bound.-Layer Meteorol. 63, 293–315.

Kristovich, D. A. R., and Laird, N. F. (1998). Observations of widespread lake-effect cloudiness: Influences of lake surface temperature and upwind conditions. Wea. Forecasting 13, 811–821.

Kristovich, D. A. R., and Spinar, M. L. (2005). Diurnal variations in lake-effect precipitation near the western Great Lakes. J. Hydrometeor. 6, 210–218.

Kristovich, D. A. R., and Steve, III, R. A. (1995). A satellite study of cloud-band frequencies over the Great Lakes. J. Appl. Meteorol. 34, 2083–2090.

Kristovich, D. A. R., Laird, N. F., Hjelmfelt, M. R., Derickson, R. G., and Cooper, K. A. (1999). Transitions in boundary layer meso-$\gamma$ convective structures: An observational case study. Mon. Weather Rev. 127, 2895–2909.

Kristovich, D. A. R., and coauthors, (2000). The lake-induced convection experiment (Lake-ICE) and the snowband dynamics project. Bull. Am. Meteorol. Soc. 81, 519–542.

Kristovich, D. A. R., Laird, N. F., and Hjelmfelt, M. R. (2003). Convective evolution across Lake Michigan during a widespread Lake-effect snow event. Mon. Weather Rev. 131, 643–655.

Kunkel, K. E., Palecki, M., Ensor, L., Hubbard, K. G., Robinson, D., Redmond, K., and Easterling, D. (2009). Trends in twentieth-century US Snowfall using a quality-controlled dataset. J. Atmos. Oceanic Tech. 26, 33–44.

Kuo, Y.-H., Shapiro, M. A., and Donall, E. G. (1991). The interaction between baroclinic and diabatic processes in a numerical simulation of a rapidly intensifying extratropical cyclone. Mon. Weather Rev. 119, 368–384.

Kuo, Y.-H., Reed, R. J., and Low-Nam, S. (1992). Thermal structure and airflow in a model simulation of an occluded marine cyclone. Mon. Weather Rev. 120, 2280–2297.

Lackmann, G. M. (2001). Analysis of a surprise western New York snowstorm. Wea. Forecasting 16, 99–116.

Laird, N. F. (1999). Observation of coexisting mesoscale lake-effect vortices over the Western Great Lakes. Mon. Weather Rev. 127, 1137–1141.

Laird, N. F., and Kristovich, D. A. R. (2004). Comparison of observations with idealized model results for a method to resolve winter lake-effect mesoscale morphology. Mon. Weather Rev. 132, 1093–1103.

Laird, N. F., Kristovich, D. A. R., and Walsh, J. E. (2003a). Idealized model simulations examining the mesoscale structure of winter lake-effect circulations. Mon. Weather Rev. 131, 206–221.

Laird, N. F., Walsh, J. E., and Kristovich, D. A. R. (2003b). Model simulations examining the relationship of lake-effect morphology to lake shape, wind direction, and wind speed. Mon. Weather Rev. 131, 2102–2111.

Laird, N. F., Desrochers, J., and Payer, M. (2009a). Climatology of lake-effect precipitation events over Lake Champlain. J. Appl. Met. Clim. 48, 232–250.

Laird, N. F., Sobash, R., and Hodas, N. (2009b). The frequency and characteristics of lake-effect precipitation events associate with the New York State Finger Lakes. J. Appl. Met. Clim. 48, 873–886.

Larson, V. E., and Smith, A. J. (2009). An analytical scaling law for the depositional growth of snow in thin mixed-phase layer clouds. J. Atmos. Sci. 66, 2620–2639.

Larson, V. E., Fleishauer, R. P., Kankiewicz, J. A., Reinke, D. L., and Vonder Haar, T. H. (2001). The death of an altocumulus cloud. Geophys. Res. Lett. 28, 2609–2612.

Larson, V. E., Smith, A. J., Falk, M. J., Kotenberg, K. E., and Golaz, J.-C. (2006). What determines altocumulus dissipation time? J. Geophys. Res. 111, D19207. doi:10.1029/2005JD007002.

Larson, V. E., Kotenberg, K. E., and Wood, N. B. (2007). An analytic longwave radiation formula for liquid layer clouds. Mon. Weather Rev. 135, 689–699.

Lavoie, R. L. (1972). A mesoscale numerical model of lake-effect storms. J. Atmos. Sci. 29, 1025–1040.

Lawson, R. P., Baker, B., Schmitt, C. G., and Jensen, T. L. (2001). An overview of microphysical properties of Arctic clouds observed in May and July during FIRE ACE. J. Geophys. Res. 106, 14989–15014.

Lawson, R. P., Baker, B., Pilson, B., and Mo, Q. (2006). In situ observations of the microphysical properties of wave, cirrus, and anvil clouds. Part II: Cirrus clouds. J. Atmos. Sci. 63, 3186–3203.

LeMone, M. A. (1976). Modulation of turbulence energy by longitudinal rolls in an unstable planetary boundary layer. J. Atmos. Sci. 33, 1308–1320.

Lemaitre, Y., Scialom, G., and Amayenc, P. (1989). A cold frontal rainband observed during the LANDES-FRONTS 84 experiment: Mesoscale and small-scale structure inferred from dual-Doppler radar analysis. J. Atmos. Sci. 46, 2215–2235.

Lenschow, D. H. (1973). Two examples of planetary boundary layer modification over the Great Lakes. J. Atmos. Sci. 30, 568–581.

Lieder, M., and Heinemann, G. (1999). A summertime Antarctic mesocyclone event over the Southern Pacific during FROST SOP-3: A mesoscale analysis using ANHRR, SSM/I, ERS, and numerical model data. Wea. Forecasting 14, 893–908.

Lindzen, R. S., and Tung, K. K. (1976). Banded convective activity and ducted gravity waves. Mon. Weather Rev. 104, 1602–1617.

Liou, K. N., Ou, S. C., and Koenig, G. (1990). An investigation of the climatic effect of contrail cirrus. In "Air Traffic and the Environment: Background, Tendencies, and Potential Global Atmospheric Effects" (U. Schumann, Ed.), In Lecture Notes in Engineering, pp. 154–169. Springer, Berlin.

Liu, A. Q., and Moore, G. W. K. (2004). Lake-effect snowstorms over southern Ontario, Canada, and their associated synoptic-scale environment. Mon. Weather Rev. 132, 2595–2609.

Liu, G. (1997). Case studies of water supply in oceanic extratropical cyclones using an Eulerian-Lagrangian method. Ph.D. dissertation, Drexel University, 212 pp.

Liu, S., and Krueger, S. K. (1998). Numerical simulations of altocumulus using a cloud resolving model and a mixed layer model. Atmos. Res. 47–48, 461–474.

Locatelli, J. D., and Hobbs, P. V. (1987). The mesoscale and microsacle structure and organization of clouds and precipitation in midlatitude cyclones. Part XIII: Structure of a warm front. J. Atmos. Sci. 44, 2290–2309.

Locatelli, J. D., and Hobbs, P. V. (1995). A world record rainfall rate at Holt, Missouri: Was it due to cold frontogenesis aloft? Wea. Forecasting 10, 779–785.

Locatelli, J. D., Hobbs, P. V., and Werth, J. A. (1982). Mesoscale structures of vortices in polar air streams. Mon. Weather Rev. 110, 1417–1433.

Locatelli, J. D., Martin, J. E., Castle, A., and Hobbs, P. V. (1995). Structure and evolution of winter cyclones in the Central United States and their effects on the distribution of precipitation. Part III: The development of a squall line associated with weak cold frontogenesis aloft. Mon. Weather Rev. 123, 2641–2662.

Locatelli, J. D., Sienkiewicz, J. M., and Hobbs, P. V. (1989). Organization and structure of clouds and precipition on the mid-Atlantic coast of the United States. Part I: Synoptic evolution of a frontal system from the Rockies to the Atlantic. J. Atmos. Sci. 46, 1327–1348.

Locatelli, J. D., Stoelinga, M. T., and Hobbs, P. V. (1998). Structure and evolution of winter cyclones in the Central United States and their effects on the distribution of precipitation. Part V: Thermodynamic and dual-Doppler radar analysis of a squall line associated with a cold front aloft. Mon. Weather Rev. 126, 860–875.

Locatelli, J. D., Sotelinga, T. T., and Hobbs, P. V. (2002a). Organization and structure of clouds and precipitation on the mid-Atlantic coast of the United States. Part VII: Diagnosis of a nonconvective rainband associated with a cold front aloft. Mon. Weather Rev. 130, 278–297.

Locatelli, J. D., Schwartz, R. D., Stoelinga, M. T., and Hobbs, P. V. (2002b). Norwegian-type and cold front aloft-type cyclones east of the Rocky Mountains. Wea. Forecasting 17, 66–82.

Locatelli, J. D., Sotelinga, T. T., Garvert, M. G., and Hobbs, P. V. (2005). The IMPROVE-1 storm of 1-2 February 2001. Part I: Development of a forward-titled cold front and a warm occlusion. J. Atmos. Sci. 62, 3431–3455.

Lott, G. A. (1954). The world-record 42-min Holt, Missouri rainstorm. Mon. Weather Rev. 82, 50–59.

Luo, Z. Z., and Rossow, W. B. (2004). Characterizing tropical cirrus life cycle, evolution, and interaction with upper-tropospheric water vapor using Lagrangian trajectory analysis of satellite observations. J. Climate 17, 4541–4563.

Lyall, I. T. (1972). The polar low over Britain. Weather 27, 378–390.

Lystad, M., (Ed.) (1986). Polar lows in the Norwegian Greenland air Barents Eas. Final Rep., Polar Lows Project. The Norwegian Meteorological Institute, Oslo, 196 pp.

Lystad, M., Hoem, V., and Wilhelmsen, K. (1986). Climatological studies. "Polar Lows Project Final Report" (M. Lystad, Ed). [Norwegian Meteorological Institute (DNMI), P.O. Box 320-Bildern, 0314 Oslo 3, Norway].

Mace, G. G., Clothiaux, E. E., and Ackerman, T. A. (2001). The composite characteristics of cirrus clouds: Bulk properties revealed by one year of continuous cloud radar data. J. Climate 14, 2185–2203.

Mace, G. G., Deng, M., Soden, B., and Zipser, E. (2006a). Association of tropical cirrus in the 10-15-km layer with deep convective sources: An observational study combining millimeter radar data and satellite-derived trajectories. J. Atmos. Sci. 63, 480–503.

Mace, G. G., Benson, S., and Vernon, E. (2006b). Cirrus clouds and the large-scale atmospheric state: Relationships revealed by six years of ground-based data. J. Climate 19, 3257–3278.

Magano, C., and Lee, C. W. (1966). Meteorological classification of natural snow crystals. J. Fac. Sci. Hokkaido University, Series VII 2, 321–335.

Mann, G. E., Wagenmaker, R. B., and Sousounis, P. J. (2002). The influence of multiple lake interactions upon lake-effect storms. Mon. Weather Rev. 130, 1510–1530.

Mansfield, D. A. (1974). Polar lows. The development of baroclinic disturbances in cold air outbreaks. Q. J. R. Meteorol. Soc. 100, 541–554.

Market, P. S., and Cissell, D. (2002). Formation of a sharp snow gradient in a midwestern heavy snow event. Wea. Forecasting 17, 723–738.

Martin, J. E. (1998a). The structure and evolution of a continental winter cyclone. Part I: Frontal structure and the classical occlusion process. Mon. Weather Rev. 126, 303–328.

Martin, J. E. (1998b). The structure and evolution of a continental winter cyclone. Part II: Frontal forcing of an extreme snow event. Mon. Weather Rev. 126, 329–347.

Martin, J. E. (1999). Quasigeostrphic forcing of ascent in the occluded sector of cyclones and Trowal airstream. Mon. Weather Rev. 127, 70–88.

Martin, J. E., Locatelli, J. D., and Hobbs, P. V. (1990). Organization and structure of clouds and precipitation on the mid-Atlantic coast of the United States. Part III: The evolution of a middle-tropospheric cold front. Mon. Weather Rev. 118, 195–217.

Martin, J. E., Locatelli, J. D., Wang, P.-Y., and Castle, J. (1995). Sturcture and evolution of winter cyclones in the Central United States and their effects on the distribution of precipitation. Part I: A synoptic-scale rainband associated with a dryline and lee trough. Mon. Weather Rev. 123, 241–264.

Marwitz, J. D. (1987). Deep orographic storms over the Sierra Nevada. Part II: The precipitation process. J. Atmos. Sci. 44, 174–185.

Marwitz, J. D., and Toth, J. (1993). A case study of heavy snowfall in Oklahoma. Mon. Weather Rev. 121, 648–660.

Mass, C. F., and Schultz, D. M. (1993). The structure and evolution of a simulated midlatitude cyclone over land. Mon. Weather Rev. 121, 889–917.

Massie, S., Gettelman, A., Randel, W., and Baumgardner, D. (2002). Distribution of tropical cirrus in relation to convection. J. Geophys. Res. 107, 4591. doi:10.1029/2001JD001293.

Matejka, T. J., Hobbs, P. V., and Houze, R. A. (1978). Microphysical and dynamical structure of mesoscale cloud features in extratropical cyclones. In "Conf. Cloud Phys. Atmos. Electr., Issaquah, Wash." pp. 292–299. Am. Meteorol. Soc., Boston, Massachusetts. Preprints.

Matejka, T. J., Houze Jr., R. A., and Hobbs, P. V. (1980). Microphysics and dynamics of clouds associated with mesoscale rainbands in extratopical cyclones. Q. J. R. Meteorol. Soc. 106, 29–56.

Matveev, L. T. (1984). "Cloud Dynamics." Reidel, Dordrecht, Netherlands.

Mayengon, R. (1984). Warm core cyclones in the Mediterranean. Mar. Wea. Log 28, 6–9.

McCann, D. W. (1995). Three-dimensional computations of equivalent potential vorticity. Wea. Forecasting 10, 798–802.

McFarquhar, G. M., and Heymsfield, A. J. (1996). Microphysical characteristics of three anvils sampled during the Central Equatorial Pacific Experiment (CEPEX). J. Atmos. Sci. 53, 2401–2423.

McFarquhar, G. M., and Heymsfield, A. J. (1997). Parameterization of tropical cirrus ice crystal size distributions and implications for radiative transfer: Results from CEPEX. J. Atmos. Sci. 54, 2187–2190.

McFarquhar, G. M., Heymsfield, A. J., Spinhirne, J., and Hart, B. (2000). Thin and subvisual tropopause tropical cirrus: Observations and radiative impacts. J. Atmos. Sci. 57, 1841–1853.

Medina, S., Sukovich, E., and Houze Jr., R. A. (2007). Vertical structures of precipitation in cyclones crossing the Oregon Cascades. Mon. Weather Rev. 135, 3565–3586.

Miles, N. L., and Verlinde, J. (2005a). Observations of transient linear organization and nonlinear scale interactions in lake-effect clouds. Part I: Transient linear organization. Mon. Weather Rev. 133, 677–691.

Miles, N. L., and Verlinde, J. (2005b). Observations of transient linear organization and nonlinear scale interactions in lake-effect clouds. Part Ii: Non-linear scale interactions. Mon. Weather Rev. 133, 692–706.

Miner, T. J., and Fritsch, J. M. (1997). Lake-effect rain events. Mon. Weather Rev. 125, 3231–3248.

Miner, T., Sousounis, P. J., Wallman, J., and Mann, G. (2000). Hurrican Huron. Bull. Am. Meteorol. Soc. 81, 223–236.

Minnis, P., Palikonda, R., Walter, B. J., Ayers, J. K., and Mannstein, H. (2005). Contrail properties over the eastern North Pacific from AVHRR data. Meteorol. Z. 14, 515–523.

Mokhov, I. I., Akeperov, M. G., Lagun, V. E., and Lutsenko, E. I. (2007). Intense Arctic mesocyclones. Izv. Atmos. Ocean Phys. 43, 259–265.

Moncrieff, M. W. (1989). Analytical models of narrow cold-frontal rainbands and related phenomena. J. Atmos. Sci. 46, 150–162.

Montgomery, M. T., and Farrell, B. F. (1991). Moist surface frontogenesis associated with interior potential vorticity anomalies in a semigeostrophic model. J. Atmos. Sci. 48, 343–367.

Montgomery, M. T., and Farrell, B. F. (1992). Polar low dynamics. J. Atmos. Sci. 49, 2484–2505.

Moore, G. W. K. (1985). The organization of convection in narrow cold-frontal rainbands. J. Atmos. Sci. 42, 1777–1791.

Moore, J. T., and Lambert, T. E. (1993). The use of equivalent potential vorticity to diagnose regions of conditional symmetric instability. Wea. Forecasting 8, 301–308.

Moore, P. K., and Orville, R. E. (1990). Lightning characteristics in lake-effect thunderstorms. Mon. Weather Rev. 118, 1767–1782.

Moore, R. W., and Vonder Haar, T. H. (2003). Diagnosis of a polar low warm core utilizing the Advanced Microwave Sounding Unit. Wea. Forecasting 18, 700–711.

Moore, J. T., Graves, C. E., Ng, S., and Smith, J. L. (2005). A process-oriented methodology toward understanding the organization of an extensive mesoscale snowband: A diagnostic case study of 4-5 December 1999. Wea. Forecasting 20, 35–50.

Mudrick, S. E. (1987). Numerical simulation of polar lows and comma clouds using simple dry models. Mon. Weather Rev. 115, 2890–2903.

Mullen, S. L. (1979). An investigation of small synoptic scale cyclones in polar air streams. Mon. Weather Rev. 107, 1636–1647.

Mullen, S. L. (1983). Explosive cyclogensis associated with cyclones in polar air streams. Mon. Weather Rev. 111, 1537–1553.

Muller, R. A. (1966). Snowbelts of the Great Lakes. Weatherwise 19, 248–255.

Nee, J. B., Len, C. N., Chen, W. N., and Lin, C. I. (1998). Lidar observation of the cirrus cloud in the tropopause at chung-Li (25°N, 121°E). J. Atmos. Sci. 55, 2249–2257.

Neiman, P. J., and Shapiro, M. A. (1993). The life cyclone of an extratropical marine cyclone. Part I: Frontal-cyclone evolution and thermodynamic air-sea interaction. Mon. Weather Rev. 121, 2153–2176.

Neiman, P. J., Shapiro, M. A., and Fedor, L. S. (1993). The life cycle of an extratropical marine cyclone. Part II: Mesoscale structure and diagnostics. Mon. Weather Rev. 121, 2177–2199.

Nicosia, D. J., and Grumm, R. H. (1999). Mesoscale band formation in three major northeastern United States snowstorms. Wea. Forecasting 14, 346–368.

Niziol, T. A. (1989). Some synoptic and mesoscale interactions in a lake effect snowstorm. In "Second National Winter Weather Workshop, Raleigh, NC, NOAA." pp. 260-269. Preprints.

Niziol, T. A., Snyder, W. R., and Waldstreicher, J. S. (1995). Winter weather forecasting throughout the eastern United States. Part IV: Lake effect snow. Wea. Forecasting 10, 61–77.

Noel, V., Chepfer, H., Ledanois, G., Delaval, A., and Flamant, P. H. (2002). Classification of particle shape ratios in cirus clouds based on the lidar depolarization ratio. Appl. Opt. 41, 4245–4257.

Nordeng, T. E. (1987). The effect of vertical and slantwise convection on the simulation of polar lows. Tellus 39A, 354–375.

Nordeng, T. E. (1990). A model-based diagnostic study of the development and maintenance mechanism of two polar lows. Tellus 42A, 92–108.

Nordeng, T. E., and Rasmussen, E. A. (1992). A most beautiful polar low. A case study of a polar low development in the Bear Island region. Tellus 44A, 81–99.

Norton, D. C., and Bolsenga, S. J. (1993). Spatiotemporal trends in lake effect and continental snowfall in the Laurentian Great Lakes. J. Climate 6, 1943–1956.

Novak, D. R., and Horwood, R. S. W. (2002). Analysis of mesoscale banded features in the 5-6 February 2001 New England snowstorm. In "19th Conf. on Weather and Forecasting, San Antonio, TX." Amer. Meteorol. Soc., J103-J106. Preprints.

Novak, D. R., Bosart, L. F., Keyser, D., and Waldstreicher, J. S. (2004). An observational study of cold season-banded precipitation in northeast US cyclones. Wea. Forecasting 19, 993–1010.

Novak, D. R., Waldstreicher, J. S., Keyser, D., and Bosart, L. F. (2006). A forecast strategy for anticipating cold season mesoscale band formation within eastern US cyclones. Wea. Forecasting 21, 3–23.

Novak, D. R., Colle, B. A., and Yuter, S. E. (2008). High-resolution observations and model simulations of the life cycle of an intense mesoscale snowband over the northeastern United States. Mon. Weather Rev. 136, 1433–1456.

Novak, D. R., Colle, B. A., and McTaggart-Cowan, R. (2009). The role of moist processes in the formation and evolution of mesoscale snowbands within the comma head of northeast US cyclones. Mon. Weather Rev. 137, 2662–2686.

Ogura, Y., Juang, H.-M., Zhang, K.-S., and Soone, S.-T. (1982). Possible triggering mechanisms for severe storms in SESAME-AVE IV (9-10 May 1979). Bull. Am. Meteorol. Soc. 63, 503–515.

Okhert-Bell, M. E., and Hartmann, D. L. (1992). The effect of cloud type on earth's energy balance: Results for selected regions. J. climate 5, 1157–1171.

Økland, H. (1977). On the intensification of small-scale cyclones formed in very cold air masses heated over the ocean. Inst. Rep. Ser. No. 26, Institutt for Geofysikk, Universitet I Oslo, 25 pp.

Økland, H. (1987). Heating by organized convection as a source of polar low intensification. Tellus 31A, 397–407.

Onton, D. J., and Steenburgh, W. J. (2001). Diagnostic and sensitivity studies of the 7 December 1998 Great Salt Lake-effect snowstorm. Mon. Weather Rev. 129, 1318–1338.

Orlanski, I. (1986). Localized barcolinicity: a source for meso-$\alpha$ cyclones. J. Atmos. Sci. 43, 2857–2885.

Pagowski, M., and Moore, G. W. K. (2001). A numerical study of an extreme cold-air outbreak over the Labrador Sea: Sea ice, air-sea interaction and development of polar lows. Mon. Weather Rev. 129, 47–72.

Palmén, E., and Newton, C. W. (1969). "Atmospheric Circulation Systems: Their Sturcture and Physical Interpretation." Academic Press, 603 pp.

Parsons, D. B. (1992). An explanation for intense frontal updrafts and narrow cold-frontal rainbands. J. Atmos. Sci. 49, 1810–1825.

Parsons, D. B., and Hobbs, P. V. (1983). The mesoscale and microscale structure and organization of clouds and precipitation in midlatitude cyclones. VII: Formation, development, interaction and dissipation of rainbands. J. Atmos. Sci. 40, 559–579.

Parsons, D. B., Mohr, C. G., and Gal-Chen, T. (1987). A severe frontal rainband. Part III: Derived thermodynamic structure. J. Atmos. Sci. 44, 1615–1631.

Passarelli Jr., R. E., and Braham Jr., R. R. (1981). The role of the winter land breeze in the formation of Great Lake snow storms. Bull. Am. Meteorol. Soc. 62, 482–491.

Payer, M., Desrochers, J., and Laird, N. F. (2007). A lake-effect snowband over Lake Champlain. Mon. Weather Rev. 135, 3895–3900.

Peace Jr., R. L., and Sykes Jr., R. B. (1966). Mesoscale study of a lake effect snow storm. Mon. Weather Rev. 94, 495–507.

Pease, S. R., Lyons, W. A., Keen, C. S., and Hjelmfelt, M. (1988). Mesoscale spiral vortex embedded within a Lake Michigan snow squall band: High resolution satellite observations and numerical model simulations. Mon. Weather Rev. 116, 1374–1380.

Penner, C. M. (1955). A three-front model for synoptic analyses. Q. J. R. Meteorol. Soc. 81, 89–91.

Pfister, L., and coauthors, (2001). Aircraft observations of thin cirrus clouds near the tropical tropopause. J. Geophys. Res. bf 106, 9765–9786.

Pinto, J. O. (1998). Autumnal mixed-phase cloud boundary layers in the Arctic. J. Atmos. Sci. 55, 2016–2038.

Pokrandt, P. J., Tripoli, G. J., and Houghton, D. D. (1996). Process leading to the formation of mesoscale waves in the Midwest cyclone of 15 December 1987. Mon. Weather Rev. 124, 2726–2752.

Polavarapu, S. M., and Peltier, W. R. (1990). The structure and evolution of synoptic-scale cyclones: Life cycle simulation with a cloud-scale model. J. Atmos. Sci. 47, 2645–2672.

Posselt, D. J., and Martin, J. E. (2004). The role of latent heat release in the formation of a warm occluded thermal structure. Mon. Weather Rev. 132, 578–599.

Posselt, D. J., Stephens, G. L., and Miller, M. (2008). CLOUDSAT adding a new dimension to a classical view of extratropical cyclones. Bull. Am. Meteorol. Soc. 89, 599–609.

Powers, J. G., and Reed, R. J. (1993). Numerical simulation of the large-amplitude mesoscale gravity-wave event of 15 December 1987 in the Central United States. Mon. Weather Rev. 121, 2285–2308.

Quante, M., and Starr, D. O'C. (2002). Dynamic processes in cirrus clouds: A review of observational results. In "Cirrus" (D. K. Lynch et al., Eds.), pp. 346–374. Oxford University Press.

Rabbe, A. (1987). A polar low over the Norwegian Sea, 27 February-1 March 1984. Tellus 39, 326–333.

Ramamurthy, M. K., Rauber, R. M., Collins, B. P., and Kennedy, P. C. (1990). Dramatic evidence of atmospheric solitary waves. Nature 348, 314–317.

Ramamurthy, M., Rauber, R. M., Collins, B. P., and Malhotra, N. K. (1993). A comparative study of large-amplitude gravity-wave events. Mon. Weather Rev. 121, 2951–2974.

Rasmussen, E. (1977).The polar low as a CISK-phenomenon. Rep. No. 6, Inst. Teoret. Meteorol., Lobenhavns Universitet, Copenhagen, 77 pp.

Rasmussen, E. (1979). The polar low as an extratropical CISK-disturbance. Q. J. R. Meteorol. Soc. 105, 531–549.

Rasmussen, E. (1981). An investigation of a polar low with a spiral cloud structure. J. Atmos. Sci. 38, 1785–1792.

Rasmussen, E. A. (1983). A review of meso-scale disturbances in cold air masses. In "Mesoscale Mteorology–Theories, Observations and Models" (D. K. Lilly and T. Gal-Chen, Eds.), pp. 247–283. D. Reidel, Boston, MA.

Rasmussen, E. (1985). A case study of a polar low development. Tellus 37A, 407–418.

Rasmussen, E., and Lystad, M. (1987). The Norwegian polar lows project: a summary of the international conference on polar lows. Bull. Amer. Meteorol. Soc. 68, 801–816.

Rasmussen, E. A., and Turner, J. (2003). "Polar Lows: Mesoscale Weather Systems in the Polar Regions." Cambridge University Press, Cambridge, UK.

Rasmussen, E., and Zick, C. (1987). A subsynoptic vortex over the Mediterranean Sea with some resemblance to polar lows. Tellus 39, 408–425.

Rauber, R. M., and Grant, L. O. (1986). The characteristics and distribution of cloud water over the mountains of northern Colorado during wintertime storms. Part II: Spatial distribution and microphysical characteristics. J. Climate Appl. Meteorol. 25, 489–504.

Rauber, R. M., Yang, M., Ramamurthy, M. K., and Jewett, B. F. (2001). Origin, evolution, and finescale structure of the St. Valentine's Day mesoscale gravity wave observed during STORM-FEST. Part I: Origin and evolution. Mon. Weather Rev. 129, 198–217.

Raymond, D. (1983). Wave-CISK in mass flux form. J. Atmos. Sci. 40, 2561–2574.

Reed, R. J. (1979). Cyclogenesis in polar airstreams. Mon. Weather Rev. 107, 38–52.

Reed, R. J. (1987). Polar lows. "Seminar on the Nature of Prediction of Extratropical Weather Systems." pp. 213–236 [Available from ECMWF, Shinfield Park, Reading, RG2 9AX, UK].

Reed, R. J., Kuo, Y.-H., and Low-Nam, S. (1994). An adiabatic simulation of the ERICA IOP 4 storm: An example of quasi-ideal frontal cyclone development. Mon. Weather Rev. 122, 2688–2708.

Reed, R. J., and Blier, W. (1986a). A case study of comma cloud development in the Eastern Pacific. Mon. Weather Rev. 114, 1681–1695.

Reed, R. J., and Blier, W. (1986b). A further case study of comma cloud development in the Eastern Pacific. Mon. Weather Rev. 114, 1696–1708.

Reed, R. J., and Duncan, C. N. (1987). Baroclinic instability as a mechanism for the serial development of polar lows: A case study. Tellus, Ser. A 39, 376–384.

Reeder, M. J. (1986). The interaction of a surface cold front with a prefrontal thermodynamically well-mixed boundary layer. Austral. Meteorol. Mag. 34, 137–148.

Renfrew, I. A., Moore, G. W., and Clark, A. A. (1997). Binary interactions between polar lows. Tellus 49A, 577–594.

Roebber, P. J. (1984). Statistical analysis and updated climatology of explosive cyclones. Mon. Weather Rev. 112, 1577–1589.

Rosinski, J., Nagamoto, C. T., Langer, G., and Parungo, E. P. (1970). Cirrus clouds as collectors of aerosol particles. J. Geophys. Res. 75, 2961–2973.

Rossow, W. B., and Schiffer, R. A. (1999). Advances in understanding clouds from ISCCP. Bull. Am. Meteorol. Soc. 80, 2261–2287.

Rothrock, H. J. (1969). An aid in forecasting significant lake snows. Tech. Memo. WBTM CR-30, US Dep. Commer., E.S.S.A.

Rotunno, R., Klemp, J. B., and Weisman, M. L. (1988). A theory for strong, long-lived squall lines. J. Atmos. Sci. 45, 463–485.

Rutledge, S. A. (1989). A severe frontal rainband. Part IV: Precipitation mechanisms. J. Atmos. Sci. 46, 3570–3594.

Rutledge, S. A., and Hobbs, P. V. (1983). The mesoscale and microscale structure and organization of clouds and precipitation in midlatitude cyclones. VIII: A model for the "Seeder Feeder" process in warm frontal rainbands. J. Atmos. Sci. 40, 1185–1206.

Ryan, B. F. (1996). On the global variation of precipitating layer clouds. Bull. Am. Meteorol. Soc. 77, 53–70.

Ryan, B. F., Katzfey, J. J., Abbs, D. J., Jakob, C., Lohmann, U., Rockel, B., Rotstayn, L. D., Stewart, R. E., Szeto, K. K., Tselioudis, G., and Yau, M. K. (2000). Simulations of a cold front by cloud-resolving, limited-area and large-scale models, and a model evaluation using in situ and satellite observations. Mon. Weather Rev. 128, 3218–3235.

Ryan, R. T., Blan Jr., H. H., van Thuna, P. C., and Cohen, M. L. (1972). Cloud microstructure as determined by an optical cloud particle spectrometer. J. Appl. Meteorol. 11, 149–156.

Saleeby, S. M., and Cotton, W. R. (2004). A large droplet mode and prognostic number concentration of cloud droplets in the RAMS@CSU model. Part I: Module descriptions and supercell test simulations. J. Appl. Meteorol. 43, 182–195.

Sanders, F. (1986). Frontogenesis and symmetric stability in a major New England snowstorm. Mon. Weather Rev. 114, 1847–1862.

Sanders, F., and Gyakum, J. R. (1980). Synoptic-dynamic climatology of the "bomb". Mon. Weather Rev. 108, 1589–1606.

Sanders, F., and Bosart, L. F. (1985). Mesoscale structure in the megalopolitan snowstorm of 11-2 February 1983. Part I: Frontogeneticla forcing and symmetric instability. J. Atmos. Sci. 42, 1050–1061.

Sardie, J. M., and Warner, T. T. (1983). On the mechanism for the development of polar lows. J. Atmos. Sci. 40, 869–881.

Sardie, J. M., and Warner, T. T. (1985). A numerical study of the development mechanisms of polar lows. Tellus 37A, 460–477.

Sassen, K. (1991). Aircraft-produced ice particles in a highly supercooled altocumulus cloud. J. Appl. Met. 30, 765–775.

Sassen, K. (2002). Cirrus clouds: A modern perspective. In "Cirrus" (D. K. Lynch et al., Eds.), pp. 11–40. Oxford University Press.

Sassen, K., and Campbell, J. R. (2001). A midlatitude cirrus cloud climatology from the Facility for Atmospheric Remote Sensing. Part I: Macrophysical and synoptic properties. J. Atmos. Sci. 58, 481–496.

Sassen, K., and Wang, Z. (2008). Classifying clouds around the globe with the CloudSat radar: 1-year of results. Geophys. Res. Lett. 35, L04805. doi:10.1029/2007GL032591.

Sassen, K., Starr, D. O'C., and Uttal, T. (1989). Mesoscale and microscale structure of cirrus clouds: Three case studies. J. Atmos. Sci. 46, 371–396.

Sassen, K., Comstock, J. M., wang, z., and Mace, G. G. (2001). Cloud and aerosol research capabilities at FARS: The Facility for Atmospheric Remote Sensing. Bull. Am. Meteorol. Soc. 82, 1119–1138.

Sassen, K., Wang, L., Starr, D. O'C., Comstock, J. M., and Quante, M. (2007). A midlatitude cirrus cloud climatology from the Facility for Atmospheric Remote Sensing. Part V: cloud structural properties. J. Atmos. Sci. 64, 2483–2501.

Schlimme, J., Mace, A., and Reichardt, J. (2005). The impact of ice crystal shapes, size districution, and spatial structure of cirrus clouds on solar radiative fluxes. J. Atmos. Sci. 62, 2274–2283.

Schmidlin, T. W., and Kosarik, J. (1999). A record Ohio snowfall during 9-14 November 1996. Bull. Am. Meteorol. Soc. 80, 1107–1116.
Schneider, R. S. (1990). Large-amplitude mesoscale wave disturbances within the intense midwest extratropical cyclone of 15 December 1987. Wea. Forecasting 5, 533–558.
Schoenberger, L. M. (1986). Mesoscale features of the Michigan land breeze using PAM II temperature data. Wea. Forecasting 1, 127–135.
Schroeder, J. J., Kristovich, D. A. R., and Hjelmfelt, M. R. (2006). Boundary layer and microphysical influences of natural cloud seeding on a lake-effect snowstorm. Mon. Weather Rev. 134, 1842–1858.
Schultz, D. M. (1999). Lake effect snowstorms in northern Utah and western New York with and without lightning. Wea. Forecasting 14, 1023–1031.
Schultz, D. M. (2001). Reexamining the cold conveyor belt. Mon. Weather Rev. 129, 2205–2225.
Schultz, D. M., and Knox, J. A. (2007). Banded convection caused by frontogenesis in a conditionally, symmetrically, and inertially unstable environment. Mon. Weather Rev. 135, 2095–2110.
Schultz, D. M., Arndt, D. S., Stensrud, D. J., and Hanna, J. W. (2004). Snowbands during the cold-air outbreak of 23 January 2003. Mon. Weather Rev. 132, 827–842.
Schultz, D. M., and Mass, C. F. (1993). The occlusion process in a midlatitude cyclone over land. Mon. Weather Rev. 121, 918–940.
Schultz, D. M., and Schumacher, P. N. (1999). The use and misuse of conditional symmetric instability. Mon. Weather Rev. 127, 2709–2732.
Schumacher, P. N. (2003). An example of forecasting mesoscale bands in an operational environment. In "10th Conf. on Mesoscale Processes, Portland, OR." Amer. Meteorol. Soc., CD-ROM, P 1.11. Preprints.
Schumann, U. (1996). On conditions for contrail formation from aircraft exhausts. Meteorol. Z. 5, 4–23.
Scott, C. P. J., and Sousounis, P. J. (2001). The utility of additional soundings for forecasting lake-effect snow in the Great Lakes region. Wea. Forecasting 16, 448–462.
Seifert, P., Ansmann, A., and Müller, D. et al. (2007). Cirrus optical properties observed with lidar, radiosonde, and satellite over the tropical Idian Ocean during the aerosol-polluted northeast and clean maritime southwest monsoon. J. Geophys. Res. 112, D17205.
Sekhon, R. S., and Srivastava, R. C. (1970). Snow size spectra and radar reflectivity. J. Atmos. Sci. 27, 299–307.
Shapiro, M. A., and Fedor, L. S. (1986). The Arctic cyclone expedition, 1984: Research and aircraft observations of fronts and polar lows over the Norwegian and Barents Sea, Part I. Polar Lows Project, Technical Rep. No. 20, The Norwegian Meteorological Institute, Oslo, 56 pp.
Shapiro, M. A., and Keyser, D. (1990). Fronts, jet streams and the tropopause. In "Extratropical Cyclones, the Erik Palmén Memorial Volume" (C. W. Newton and E. Holopainen, Eds.), pp. 167–191. Amer. Met. Soc.

Shapiro, M. A., Hample, T., and van de Kamp, D. W. (1984). Radar wind profiles observations of fronts and jet streams. Mon. Weather Rev. 112, 1263–1266.

Shapiro, M. A., Hampel, T., Rotzoll, D., and Mosher, F. (1985). The frontal hydraulic head: A micro-$\alpha$ scale (1 km) triggering mechanism for the mesoconvective weather systems. Mon. Weather Rev. 113, 1166–1183.

Shapiro, M. A., Fedor, L. S., and Hampel, T. (1987). Research aircraft measurements of a polar low over the Norwegian Sea. Tellus 39A, 272–306.

Shea, T. J., and Przybylinski, R. W. (1993). Assessing vertical motion fields in a winter storm using PCGRIDDS. In "13th Conf. on Weather Analysis and Forecasting, Vienna, VA." pp. 10–14. Amer. Met. Soc. Preprints.

Sheridan, L. W. (1941). The influence of Lake Erie on local snows in western New York. Bull. Am. Meteorol. Soc. 22, 393–395.

Shields, M. T., Rauber, R. M., and Ramamurthy, M. K. (1991). Dynamical forcing and mesoscale organization of precipitation bands in a midwest winter cyclonic storm. Mon. Weather Rev. 119, 936–964.

Shoenberger, M. J. (1986). Mesoscale features of a midlake snowband. In "23rd Conf. Radar Meteorology, Snowmass." Amer. Meteorol. Soc., JP206-JP209. Preprints.

Sienkiewicz, J. M., Locatelli, J. D., Hobbs, P. V., and Geerts, B. (1989). Organization and structure of clouds and precipitation on the mid-Atlantic Coast of the United States. Part II: The mesoscale and microscale structures of some frontal rainbands. J. Atmos. Sci. 46, 1349–1364.

Skerritt, D. A., Przybylinski, R. W., and Wolf, R. A. (2002). A study on the 6 December 1995 Midwest snow event: Synoptic and mesoscale aspects. Natl. Wea. Dig 26, 52–62.

Slemmer, J. W. (1998). Characteristics of winter snowstorms near Salt Lake City as deduced from surface and radar observations. M. S. Thesis, Department of Meteorology, Nuniversity of Utah, 138 pp.

Slingo, A., and Slingo, J. M. (1988). Response of a general circulation model to lcoud long-wave radiative forcing. Part I: Introduction and initial experiments. Q. J. R. Meteorol. Soc. 114, 1027–1062.

Smith, R. K., and Reeder, M. J. (1988). On the movement and low-level structure of cold fronts. Mon. Weather Rev. 116, 1927–1944.

Sobash, R. H., Carr, N. F., and Laird, (2005). An investigation of New York State Finger Lakes snow band events. In "11th Conf. on Mesoscale Processes, Albuquerque, NM." Amer. Met. Soc., P3M.3. Preprints.

Sousounis, P. J., and Mann, G. E. (2000). Lake-aggregate mesoscale disturbances. Part V: Impacts on lake-effect precipitation. Mon. Weather Rev. 128, 728–745.

Spichtinger, P., and Gierens, K. (2008). Modelling of cirrus clouds. Part 1: Model description and validation. Atmos. Chem. Phys. Discuss. 8, 601–686.

Staley, D. O., and Gall, R. L. (1977). On the wavelength of maximum baroclinic instability. J. Atmos. Sci. 34, 1679–1688.

Starr, D. O'C., and Cox, S. K. (1980). Characteristics of middle and upper tropospheric clouds as deduced from rawinsonde data. Atmos. Sci. Pap. No. 327, Colorado State University (US ISSN 0 067–0340).

Starr, D. O'C., and Cox, S. K. (1985a). Cirrus clouds. Part I: A cirrus cloud model. J. Atmos. Sci. 42, 2663–2681.

Starr, D. O'C., and Cox, S. K. (1985b). Cirrus clouds. Part II: Numerical experiment on the formation and maintenance of cirrus. J. Atmos. Sci. 42, 2682–2694.

Steenburgh, W. J., and Onton, D. J. (2001). Multiscale analysis of the 7 December 1998 Great Salt Lake-Effect snowstorm. Mon. Weather Rev. 129, 1296–1317.

Steenburgh, W. J., Havorson, S. F., and Onton, D. J. (2000). Climatology of lake-effect snowstorms of the Great Salt Lake. Mon. Weather Rev. 128, 709–727.

Steiger, S. M., Hamilton, R., Keeler, J., and Orville, R. E. (2009). Lake-effect thunderstorms in the Lower Great Lakes. J. Appl. Met. Clim. 48, 889–902.

Stein, U., and Alpert, P. (1993). Factor separation in numerical simulations. J. Atmos. Sci. 50, 2107–2115.

Stephens, G. L. (2005). Cloud feebacks in the climate system: A critical review. J. Climate 18, 237–273.

Stephens, G. L., Tsay, S-C., Stackhouse Jr., P. W., and Flatu, P. J. (1990). The relevance of the microphysical and radiative properties of cirrus clouds to climate and climate feedback. J. Atmos. Sci. 47, 1742–1753.

Stephens, G. L., Vane, D. G., Boain, R. J., Mace, G. G., Sassen, K., Wang, Z., Illingworth, A. J., O'Connor, E. J., Rossow, W. B., Durden, S. L., Miller, S. D., Austin, R. T., Benedetti, A., Mitrescu, C., and CloudSat Science Team, (2002). The CloudSat mission and the A-TRAIN: A new dimension to space-based observations of clouds and precipitation. Bull. Am. Meteorol. Soc. 83, 1771–1790.

Stephens, G. L., van den Heever, S. C., and Pakula, K. (2008). Radiative-convective feedbacks in idealized states of radiative-convective equilibrium. J. Atmos. Sci. 65, 3899–3916.

Steve, R. A. (1996). Evolution of convective elements in lake-effect boundary layers. M. S. Thesis, Dept. of Atmospheric Sciences, University of Illinois at Urbana-Champaign, 67 pp.

Stobie, J. G., Einaudi, F., and Uccelini, L. W. (1983). A case study of gravity waves-convective storms interaction: 9 May 1979. J. Atmos. Sci. 40, 2804–2830.

Stoelinga, M. T., Locatelli, J. D., and Hobbs, P. V. (2000). Structure and evolution of winter cyclones in the central United States and their effects on the distribution of precipitation. Part VI: A mesoscale modeling study of the initiation of convective rainbands. Mon. Weather Rev. 128, 3481–3500.

Stoelinga, M. T., and coauthors, (2003). Improvement of microphysical parameterization through Observational Verification Experiment. Bull. Am. Meteorol. Soc. 84, 1807–1826.

Stubenrauch, C. J., Chedin, A., Radel, G., Scott, N. A., and Serrar, S. (2006). Cloud properties and their seasonal and diurnal variability from TOVS Path B. J. Clim. 19, 5531–5553.

Szeto, K. K., Tremblay, A., Guan, H., Hudak, D. R., Stewart, R. E., and Cao, Z. (1999). The mesoscale dynamics of freezing rain storms over Eastern Canada. J. Atmos. Sci. 56, 1261–1281.

Tardy, A. (2000). Lake-effect and lake-enhanced snow in the Champlain Valley of Vermont. Tech. Memo. 2000-05, NWS Eastern Region, 27 pp.

Testud, J., Breger, G., Amayenc, P., Chong, M., Nutten, B., and Sauvaget, A. (1980). A Doppler radar observation of a cold front: Three dimensional air culations, related precipitation system and associated wavelike motions. J. Atmos. Sci. 37, 78–98.

Tiedtke, M. (1993). Representation of clouds in large-scale models. Mon. Weather Rev. 121, 3040–3060.

Trexler, C. M., and Koch, S. E. (2000). The life cycle of a mesoscale gravity wave as observed by a network of Doppler wind profilers. Mon. Weather Rev. 128, 2423–2446.

Tripoli, G. J. (2005). Numerical study of the 10 January 1998 lake-effects bands observed during Lake-ICE. J. Atmos. Sci. 62, 3232–3249.

Twohy, C. H., Kreidenweis, S. M., Eidhammer, T., Browell, E. V., Heymsfield, A. J., Bansemer, A. R., Anderson, B. E., Chen, G., Ismail, S., DeMott, P. J., and van den Heever, S. C. (2009). Saharan dust particles nucleate droplets in eastern Atlantic clouds. Geophys. Res. Lett. 36, doi:10.1029/2008GL035846.

Uccellini, L. W. (1975). A case study of apparent gravity wave initiation of severe convective storms. Mon. Weather Rev. 103, 497–513.

Uccellini, L. W., and Koch, S. E. (1987). The synoptic setting and possible energy sources for mesoscale wave distrubances. Mon. Weather Rev. 115, 721–729.

van den Heever, S. C., Carrió, G. G., Cotton, W. R., DeMott, P. J., and Prenni, A. J. (2006). Impacts of nucleating aerosol on Florida storms. Part I: Mesoscale Simulations. J. Atmos. Sci. 63, 1752–1775.

Wakimoto, R. M., Blier, W., and Liu, C. (1992). The frontal structure of an explosive oceanic cyclone: Airborne radar observations of ERICA IOP 4. Mon. Weather Rev. 120, 1135–1155.

Wakimoto, R. M., and Bosart, B. L. (2000). Airborner radar observations of a cold front during FASTEX. Mon. Weather Rev. 128, 2447–2470.

Wang, P.-Y., Martin, J. E., Locatelli, J. D., and Hobbs, P. V. (1995). Structure and evolution of winter cyclones in the central United States and their effects on the distribution of precipitation. Part II: Arctic fronts. Mon. Weather Rev. 123, 1328–1344.

Wang, Z., and Sassen, K. (2002). Cirrus cloud microphysical property retrieval using Lidar and radar measurements. Part II: Midlatitude cirrus microphysical and radiative properties. J. Atmos. Sci. 59, 2291–2302.

Wang, Z., and Sassen, K. (2008). Wavelet analysis of cirrus multscale structures from Lidar backscattering: A cirrus uncinus complex case study. J. Appl. Met. Climatol. 47, 2645–2658.

Wang, Z., Sassen, K., Whiteman, D. N., and Demoz, B. B. (2004). Studying altocumulus with ice virga using ground-based active and passive remote sensors. J. Appl. Met. 43, 449–460.

Warren, S. G. (1988). Global distribution of total cloud cover and cloud type amounts over the ocean. NCAR Tech. Note TN-317 STR, 212 pp.

Watson, J. S., Jurewica, L., Ballentine, R. J., Colucci, S. J., and Waldstreicher, J. S. (1998). High-resolution numerical simulations of Finger Lakes snow bands. In "16th Conf. on Weather Analysis and Forecasting, Phoenix, AZ." pp. 308–310. Amer. Met. Soc. Preprints.

Weiss, C. C., and Sousounis, P. J. (1999). A climatology of collective lake disturbances. Mon. Weather Rev. 127, 565–574.

Wernli, H. (1997). A Lagrangian-based analysis of extratropical cyclones. Part II: A detailed case study. Q. J. R. Meteorol. Soc. 123, 1677–1706.

Wiggin, B. L. (1950). Great snows of the Great Lakes. Weatherwise 3, 123–126.

Wilken, G. R. (1997). A Lake-effect snow in Arkansas. NWS/NOAA Tech. Attachment SR/SSD 97-21, 3 pp.

Woods, C. P., Stoelinga, M. T., Locatelli, J. D., and Hobbs, P. V. (2005). Microphysical processes and synergistic interaction between frontal and orographic forcing of precipitation during the 13 December 2001 IMPROVE-2 event over the Oregon Cascades. J. Atmos. Sci. 62, 3493–3519.

Wu, T., Cotton, W. R., and Cheng, W. Y. Y. (2000). Radiative effects on the diffusional growth of ice particles in cirrus clouds. J. Atmos. Sci. 57, 2892–2904.

Xu, K.-M., and coauthors, (2005). Modeling springtime shallow frontal clouds with cloud-resolving and single-column models. J. Geophys. Res. 110, D15S04. doi:10.1029/2004JD005153.

Xu, Q. (1989). Extended Sawyer-Eliassen equation for frontal circulations in the presence of small viscous moist symmetric stability. J. Atmos. Sci. 46, 2671–2683.

Yanase, W., and Niino, H. (2005). Effects of baroclinicity on the cloud pattern and structure of polar lows: A high-resolution numerical experiment. Geophys. Res. Lett. 32, L02806. doi:10.1029/2004GL020469.

Yanase, W., and Niino, H. (2007). Dependence of polar low development on baroclinicity and physical processes: An idealized high-resolution numerical experiment. J. Atmos. Sci. 64, 3044–3067.

Yanase, W., Fu, G., Niino, H., and Kato, T. (2004). A polar low over the Japan Sea on 21 January 1997. Part II: A numerical study. Mon. Weather Rev. 132, 1552–1574.

Yang, M., Rauber, R. M., and Ramamurthy, M. K. (2001). Origin, evolution, and finescale structure of the St. Valentine's Day mesoscale gravity wave observed during STORM-FEST. Part II: Finescale structure. Mon. Weather Rev. 129, 218–236.

Yarnal, B., and Henderson, K. G. (1989). A climatology of polar low cyclogenetic regions over the North Pacific Oean. J. Climate 2, 1476–1491.

Young, M. V., Monk, G. A., and Browning, K. A. (1987). Interpretation of satellite imagery of a rapidly deepening cyclone. Q. J. R. Meteorol. Soc. 113, 1089–1115.

Yu, C.-K., and Smull, B. F. (2000). Airborne observations of a landfalling cold front upstream of steep coastal orography. Mon. Weather Rev. 128, 1577–1603.

Zhang, D.-L., and Harvey, R. (1995). Enhancement of extratropical cyclogenesis by a mesoscaleconvective system. J. Atmos. Sci. 52, 1107–1127.

Zhang, M. H., Lin, W. Y., and Klein, S. A. et al. (2005). Comparing clouds and their seasonal variations in 10 atmospheric general circulation models with satellite measurements. J. Geophys. Res. 110, D15S02.

Zhang, Y., Macke, A., and Albers, F. (1999). Effect of crystal size spectrum and crystal shape on stratiform cirrus radiative forcing. J. Atmos. Sci. 52, 59–75.

Zulauf, M. A., and Krueger, S. K. (2003). Two-dimensional cloud-resolving modeling of the atmospheric effects of Arctic leads based upon midwinter conditions at the Surface Heat Budget of the Arctic Ocean ice camp. J. Geophys. Res. 108, 4312. doi:10.1029/2002JD002643.

Chapter 11

# The Influence of Mountains on Airflow, Clouds, and Precipitation

## 11.1. INTRODUCTION

The emphasis of this chapter is on wintertime clouds and cloud systems that are forced at least in part by orography, but excluding deep convective clouds. We include, however, convective clouds that can be triggered by the release of potential instability during the passage of extratropical cyclones. Our focus is on air motions over mountainous terrain that are conducive to the formation and spatial distribution of precipitation. We also consider the interaction between flow over mountains and larger scale precipitating weather systems such as extratropical cyclones.

We briefly review the theories and models of orographic flows, and then turn our attention to the formation and distribution of precipitation over mountainous terrain.

## 11.2. THEORY OF FLOW OVER HILLS AND MOUNTAINS

The classical theories of flow over small-amplitude hills are based on the linearized equations of motion, the thermodynamic energy equation, and mass-continuity equation. For a more extensive review of this topic, we recommend the excellent review by Smith (1979).

We decompose each variable into a base-state value subscript 0 and a perturbation from the reference state. Thus, for purely two-dimensional flow,

$$\begin{aligned} u(x,z) &= u_0(z) + u'(x,z), \\ w(x,z) &= w'(x,z), \\ \rho(x,z) &= \rho_0(z) + \rho'(x,z), \\ p(x,z) &= p_0(z) + p'(x,z), \\ T(x,z) &= T_0(z) + T'(x,z), \end{aligned} \tag{11.1}$$

where vertical motion in the unperturbed atmosphere is zero $[w_0(z) = 0]$.

Substituting Eq. (11.1) into the governing equations (Chapter 2), assuming the flow field is steady state, and linearizing, we obtain

$$\rho_0[u_o(\partial u'/\partial x) + w'(\partial u_0/\partial z)] = -\partial p'/\partial x, \qquad (11.2)$$

$$\rho_0[u_0(\partial w'/\partial x)] = -\partial p'/\partial z - \rho' g, \qquad (11.3)$$

for the equations of motion. The linearized mass-continuity equation is

$$u_0(\partial \rho'/\partial x) + w'(\partial \rho_0/\partial z) = -\rho_0(\partial u'/\partial x + \partial w'/\partial z). \qquad (11.4)$$

If we now differentiate the equation of state for a dry atmosphere with respect to time,

$$\mathrm{d}p/\mathrm{d}t = RT(\mathrm{d}\rho/\mathrm{d}t) + \rho R(\mathrm{d}T/\mathrm{d}t), \qquad (11.5)$$

and substitute the thermodynamic energy equation for adiabatic motion into Eq. (11.5), we obtain, after some manipulation,

$$\mathrm{d}p/\mathrm{d}t = (c_\mathrm{p}/c_\mathrm{v})RT(\mathrm{d}\rho/\mathrm{d}t). \qquad (11.6)$$

The quantity,

$$(c_\mathrm{p}/c_\mathrm{v})RT = \gamma RT = c^2,$$

where $c^2$ is the speed of sound for dry adiabatic motion. Equation (11.6) thus becomes,

$$\mathrm{d}p/\mathrm{d}t = c^2(\mathrm{d}\rho/\mathrm{d}t). \qquad (11.7)$$

Expansion of the total derivatives in Eq. (11.7) into local and advective contributions and linearization gives

$$u_0(\partial p'/\partial x) + w'(\mathrm{d}p_0/\mathrm{d}z) = c_0^2[u_0(\partial \rho'/\partial x) + w'(\mathrm{d}\rho_0/\mathrm{d}z)]. \qquad (11.8)$$

Equation (11.8) can be rearranged into the form

$$\underset{\text{(a)}}{\frac{u_0}{\rho_0}\frac{\partial \rho'}{\partial x}} = w'\left(-\underset{\text{(b)}}{\frac{1}{\rho_0}\frac{\mathrm{d}\rho_0}{\mathrm{d}z}} + \underset{\text{(c)}}{\frac{g}{c_0^2}}\right) - \underset{\text{(d)}}{\frac{u_0}{\rho_0 c_0^2}\frac{\partial p'}{(\partial x)}}. \qquad (11.9)$$

Equation (11.9) describes the formation of density anomalies which give rise to accelerations in vertical velocity. Term (a) represents the change in density experienced by an observer moving horizontally downstream at a speed $u_0(z)$.

Term (b) represents the change in density due to vertical displacement of the base-state air mass. Term (c) represents the effects of adiabatic expansion or compression of the air parcel during vertical displacement. Term (d) accounts for the pressure deviation during vertical displacement, which is normally small for slow vertical displacements and is only important in the generation of fast acoustic waves.

Employing the definition of potential temperature (Chapter 2) and assuming a hydrostatic base state, we find that terms (b) and (c) can be combined to yield

$$-(1/\rho_0)(\mathrm{d}\rho_0/\mathrm{d}z) + (g/c_0^2) = (1/\theta_0)(\mathrm{d}\theta_0/\mathrm{d}z) = \beta. \tag{11.10}$$

Thus, terms (b) and (c) are a measure of static stability. Ignoring term (d), Eq. (11.9) can be written as

$$(u_0/\rho_0)(\partial\rho'/\partial x) = w'\beta = w'N^2/g \tag{11.11}$$

where $N^2 = g/\theta(\partial\theta/\partial z)$, and $N$ is the Brunt-Väisälä frequency. Substitution of Eqs (11.11) and (11.10) into Eq. (11.4) results in the simplified continuity equation,

$$(\partial u'/\partial x) + (\partial w'/\partial z) = (g/c_0^2)w'. \tag{11.12}$$

Following a straightforward, though lengthy, process of elimination of variables (i.e. pressure in particular), the above equations can be combined to form a single equation for vertical velocity,

$$\frac{\partial^2 w'}{\partial x^2} + \frac{\partial^2 w'}{\partial z^2} - S_0\frac{\partial w'}{\partial z} + \left(\frac{\beta g}{u_0^2} + \frac{S_0}{u_0}\frac{\partial u_0}{\partial z} - \frac{1}{u_0}\frac{\partial^2 u_0}{\partial z^2}\right)w' = 0, \tag{11.13}$$

where

$$S_0 = (\mathrm{d}/\mathrm{d}z)\ln\rho_0(z). \tag{11.14}$$

If we now introduce the new dependent variable,

$$\tilde{w} = [\rho_0(z)/\rho_0(0)]^{1/2}w', \tag{11.15}$$

then (11.4) becomes

$$(\partial^2\tilde{w}/\partial x^2) + (\partial^2\tilde{w}/\partial z^2) + l^2(z)\tilde{w} = 0, \tag{11.16}$$

where

$$l^2(z) = \frac{\beta g}{u_0^2} - \frac{1}{u_0}\frac{\partial^2 u_0}{\partial z^2} + \frac{S_0}{u_0}\frac{\partial u_0}{\partial z} - \frac{1}{4}S_0^2 + \frac{1}{2}\frac{\partial S_0}{\partial z}. \tag{11.17}$$

Equation (11.16) represents the foundation for two-dimensional, steady, linear mountain-wave theory. The parameter $l(z)$, often referred to as the *Scorer parameter*, is dominated by the stability and shear terms [first two terms in Eq. (11.18)]. Neglect of the latter terms involving $S_0$ is equivalent to making the Boussinesq approximation in which density variations are ignored except where multiplied by gravity.

In a Fourier decomposition of the horizontal structure of Eq. (11.16), the vertical velocity $\tilde{w}$ must satisfy the wave solution of the form

$$\tilde{w} = \hat{w}(z)\mathrm{e}^{ikx}. \tag{11.18}$$

The amplitude of the $k$th component $\hat{w}(z)$ is then

$$[\partial^2 \hat{w}(z)/\partial z^2] + [l^2(z) - k^2]\hat{w}(z) = 0. \tag{11.19}$$

Solutions to Eq. (11.19) can be obtained for specified lower and upper boundary conditions. For an upper boundary condition, a radiation condition (e.g. Sommerfeld, 1912, 1948) at $z = \infty$ is imposed to prevent the reflection of upward-propagating waves that could interfere with the waves of interest in the middle and upper troposphere.

A free-slip lower boundary condition is applied such that

$$w(0, x) = u(0, x)[\partial h(x)/\partial x], \tag{11.20}$$

where $h(x)$ is the height of the topography. A variety of orographic shapes have been imposed as lower boundaries. Lyra (1943) used a rectangular orographic profile which a number of investigators have criticized because it could force unrealistic gravity wave modes.

Queney (1947) obtained solutions to the mountain-wave problem for infinite sinusoidal orography. More general isolated orographic shapes can be obtained from the Fourier transform of the mountain shape,

$$\tilde{h}(k) = \frac{1}{\pi}\int_{-\infty}^{\infty} h(x)\mathrm{e}^{-ikx}\,\mathrm{d}x. \tag{11.21}$$

Queney (1947, 1948) used the bell-shaped, so-called "witch of Agnesi" profile,

$$h(x) = z_{\mathrm{m}}a^2/(x^2 + a^2), \tag{11.22}$$

which has the simple Fourier transform,

$$\tilde{h}(k) = z_{\mathrm{m}}a\ \mathrm{e}^{-ka}, \tag{11.23}$$

where $z_{\mathrm{m}}$ is the height of the mountain and $a$ is the mountain half-width.

With the top and bottom boundary conditions specified and with suitable definitions of $l^2(z)$, analytic solutions can be obtained to Eq. (11.19) for small-amplitude hills (i.e. of the order of 10 m or so). The problem is then to find solutions $\hat{w}(z)$ to Eq. (11.19) for all values of $k$ and then evaluate the Fourier integral of the corresponding component solutions yielding the total disturbance in the flow field over a finite mountain barrier. The solutions to Eq. (11.19) are frequently extended to larger amplitude hills by multiplying the results by the relative scaling of the actual mountain. This presumes, of course, that the linear solutions are valid for the larger amplitude mountains.

### 11.2.1. Queney Models

Both Lyra and Queney obtained analytic solutions to the linearized mountain-wave equations for the case of constant stability and wind speed, which implies that $l(z)$ is independent of height [Eq. (11.17)]. Queney obtained several special case solutions to Eq. (11.19). For steady mountain-waves to exist, they must have a phase velocity $C^{\mathrm{p}}$ relative to the fluid, which is equal and opposite to the mean flow $u_0$, and a group velocity directed upward away from the mountain. In this way, the waves produced by the flow against the mountain can remain steady against the prevailing flow. As air parcels flow over a mountain of wavelength $L$, they will experience an orbital frequency $2\pi u_0/L$. The highest natural frequency for gravity waves in the atmosphere is the Brunt-Väisälä frequency $N$. The condition

$$2\pi u_0/L = N \tag{11.24}$$

identifies an important cutoff frequency for pure gravity waves. Thus, for any atmosphere we can identify a cutoff wave number $k_{\mathrm{s}}$, where

$$k_{\mathrm{s}} = N/u_0. \tag{11.25}$$

When the earth's rotation is considered, free oscillations of the geostrophic wind occur. The natural frequency for such oscillations is the Coriolis parameter $f$, with a corresponding cutoff frequency

$$2\pi u_0/L = f \tag{11.26}$$

and a cutoff wave number

$$k_{\mathrm{f}} = f/u_0. \tag{11.27}$$

Inertial-gravity waves, therefore, have frequencies between $k_{\mathrm{f}}$ and $k_{\mathrm{s}}$.

Queney obtained solutions to the linearized wave equation for the three cases in which a finite mountain half-width is (1) $a \ll 1/k_{\mathrm{s}}$, (2) $a \sim 1/k_{\mathrm{s}}$, and (3) $a \sim 1/k_{\mathrm{f}}$. He also obtained several other special solutions (not discussed here).

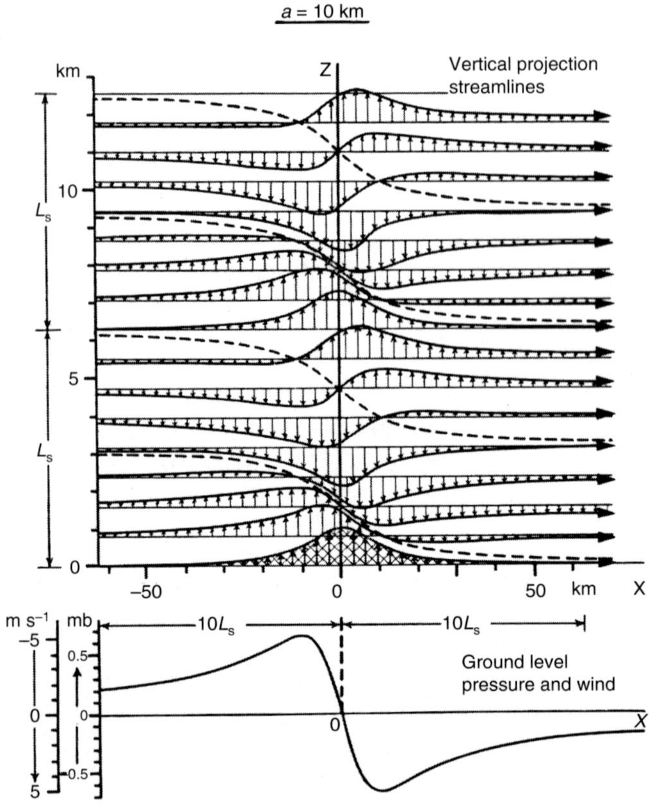


**FIGURE 11.1 Perturbation by a medium-size typical mountain range ($a = L_S/2\pi = 10$ km) in an unlimited uniform stratified current ($u = 10$ m s$^{-1}$).** Upper part: vertical projection of the streamlines; displacement indicated by small arrows. Lower part: perturbation of pressure and longitudinal wind velocity at ground level. The horizontal displacement is negligible. *(From Queney (1948))*

The first two cases can be described in terms of a Froude number Fr (Clark and Peltier, 1977), defined as

$$\mathrm{Fr} = \frac{2\pi/N}{2a/u_0}, \tag{11.28}$$

the ratio of the Brunt-Väisälä period to the mountain-forcing period of the waves. Therefore, case (1) corresponds to Fr $\gg 0(1)$ and case (2) corresponds to Fr $= 0(1)$.

### 11.2.1.1. Case $a \sim 1/k_s$; $Fr = 0(1)$

Case $a \sim 1/k_S$ should apply to moderate-sized mountain ranges (10-20 km wide). Figure 11.1 illustrates the perturbation streamlines for this case. The

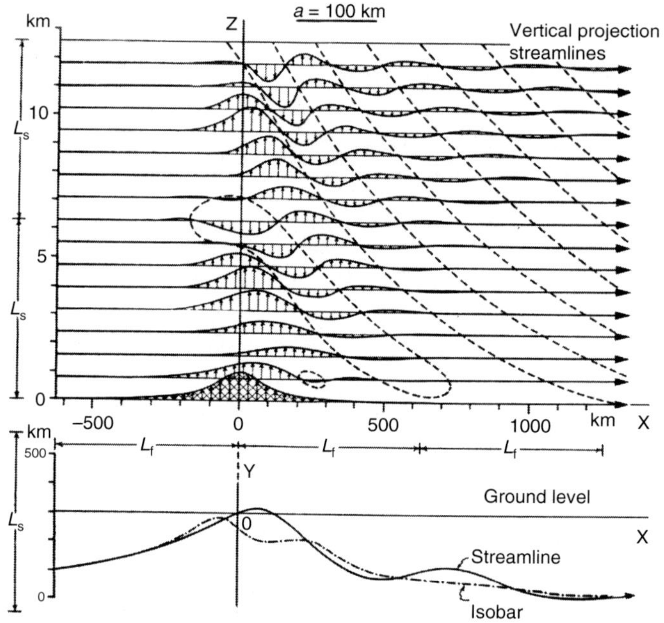


**FIGURE 11.2 Perturbation by a broad typical mountain range ($a = L_f/2\pi = 100$ km) in an unlimited uniform stratified current ($u = 10$ m s$^{-1}$).** Upper part: vertical projection of the streamlines; displacement indicated by small arrows. Lower part: horizontal projection of a streamline and of the asymptotic isobar at ground level. *(From Queney (1948))*

solution is periodic in the vertical, such that the flow exhibits an inverted ridge shape at $z = \pi/L$, which reverts to an upright ridge shape at $z = 2\pi/L$. The flow can be described as a field of nondispersive, vertically-propagating gravity waves. At low levels, the streamlines exhibit a pronounced crest immediately upwind from the mountain crest and a weak trough on the lee side of the mountain. The upwind crest becomes weaker with height and the lee trough becomes more intense. Both features exhibit an upstream tilt of the phase lines associated with downward momentum transport. The wind speed is lowest on the windward slope of the ridge and fastest on the lee slope. There is a corresponding pressure difference across the ridge with high pressure upwind and lower pressure downwind.

At least qualitatively, the solutions exhibit features similar to the observed warm and dry foehn or chinook winds in the lee of major mountain ranges such as the Alps, Sierras, and Rockies. There is no evidence of lee-wave solutions for this case.

### *11.2.1.2. Case $a \sim 1/k_f$*

Queney also obtained solutions (Fig. 11.2) for the case of a broad mountain for which the mountain half-width $a \sim 1/k_f$. The actual dimension of $a$ is 100 km.

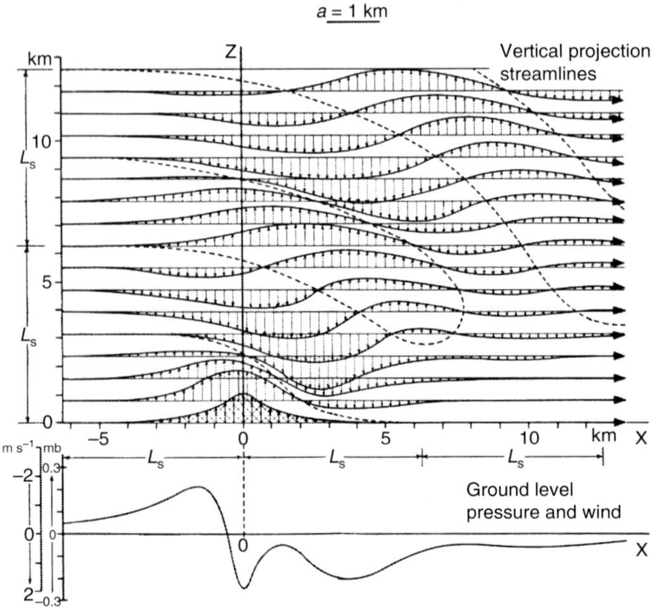


**FIGURE 11.3 Perturbation by a narrow typical mountain range ($a = L_S/2\pi = 1$ km) in an unlimited uniform stratified current ($u = 10$ m s$^{-1}$).** Upper part: vertical projection of the streamlines; displacement indicated by small arrows. Lower part: perturbation of pressure and longitudinal wind velocity at ground level. The horizontal displacement is negligible. *(From Queney (1948))*

The waves that form are long, having a wavelength of approximately 600 km. Because the waves are located exclusively on the lee side of the mountain range, they give the appearance of trapped lee waves. These waves are hydrostatic, and vertical accelerations are small. Trapped lee waves have a different character. The waves in Fig. 11.2, more properly called inertial gravity waves, exhibit a horizontal displacement as the motion attempts to adjust to a geostrophic balance.

### *11.2.1.3.* Case $1/a \gg k_s$; $Fr \gg 0(1)$

Queney obtained solutions for a small mountain for which $a \sim 1$ km and $a \ll 1/k_s$ (Fig. 11.3). The effects of the earth's rotation are negligible for this case. At low levels, the flow resembles Fig. 11.1; at higher levels, and particularly to the lee of the mountain, the flow resembles lee waves. Smith (1979) referred to these as a "dispersive tail" of nonhydrostatic waves with $k$ less than, but not much less than, $l(z)$. As noted by Holmboe and Klieforth (1957), the structure of these lee waves differs markedly from observed lee waves, which have maximum amplitude at low levels and small, if any, tilt in the phase lines with height. Moreover, the amplitude of the observed lee waves is maintained for appreciable distances downwind, while those shown in Fig. 11.3 decay in amplitude rapidly downstream.

### 11.2.2. Linear Theory of Trapped Lee Waves

Queney's and Lyra's solutions were obtained for the case in which $l^2$ in Eq. (11.17) was a constant. Scorer (1949, 1953, 1954) divided the atmosphere into two and three layers, each having a different constant value of $l^2$. By applying the wave equation to each separate layer, he found probably the most important dynamic requisite for periodic lee-wave development, namely, that $l^2$ must be less in an upper layer than in a lower layer of sufficient depth. This can be accomplished either by the wind speed increasing with height, the presence of a very stable layer in the lower atmosphere, or a combination of the two.

A disparity between the Queney and Scorer solutions has led to considerable debate over the years. The discrepancies arise mainly from different upper boundary conditions for long waves. Both Queney and Lyra chose solutions corresponding to upstream sloping phase lines. Queney obtained a unique solution by introducing a weak viscosity similar to Rayleigh (1883), while Lyra (1943) followed a technique pioneered by Kelvin (1886), in which disturbances far upstream of the mountain were removed. Scorer, on the other hand, chose the correct boundary conditions for the waves that traveled downstream, thus allowing him to obtain a trapped lee-wave solution. Unfortunately, Scorer chose the incorrect conditions for the vertically propagating waves: his main waves over the mountain crest exhibit phase lines which tilt downstream, a property inconsistent with a stationary mountain-wave pattern and with observations.

That the two forms of waves are compatible with each other can be seen in the solutions obtained by Sawyer (1960) with a 17-layer numerical model. Figure 11.4 shows trapped lee waves at lower elevations, as well as vertically propagating Queney-type waves at higher elevations. Conceptually, trapped lee waves can be thought of in terms of the movement of wave packets in the atmosphere in which $l^2$ decreases with height. In the stable lower atmosphere, waves with $k^2 < l^2$ propagate upward and downwind of the mountain barrier. When the waves encounter a level where $k^2 > l^2$, the wave cannot propagate so that the wave energy is reflected downward. The wave energy then repeatedly reflects off the ground (or an adiabatic layer near the surface) and the layer where $k^2 > l^2$. The wave energy is said to be trapped, and leads to a standing wave pattern in which the phase lines have no tilt.

Smith (1979) noted several other ways in which trapped lee waves may occur. For example, in an abrupt change in stability at the tropopause level, a vertically propagating wave will be partially reflected downward. The reflected wave will then rebound off the surface of the earth, where it can partially reflect off the tropopause again. Each time, the wave loses a fraction of its energy, because the reflection is only partial; the wave decays downstream.

Scorer's discovery spawned nearly two decades of analytical and observational studies of lee waves and lee-wave clouds. Models were extended

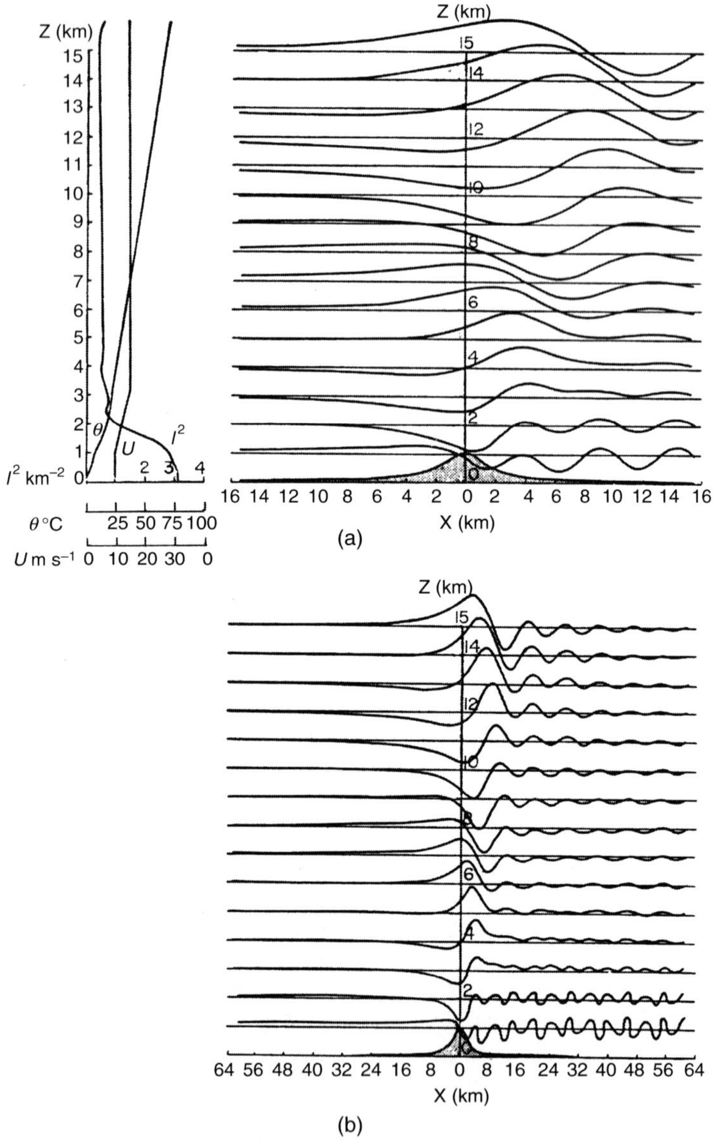


**FIGURE 11.4 Displacement of streamlines computed for an idealized airstream with a layer of high $\ell^2$ values near the ground.** The left-hand section of the diagram shows the assumed potential temperature ($\theta$), wind velocity ($U$), and Scorer's parameter ($\ell^2$) as function of height. The right-hand section of the diagram shows the computed displacements of the streamlines as functions of horizontal distance $x$ at 1-km intervals in the vertical. The vertical displacement is plotted for each level on the same scale as the mountain profile at the bottom of the figure (mountain half-width, $b = 2$ km). *(From Sawyer (1960))*

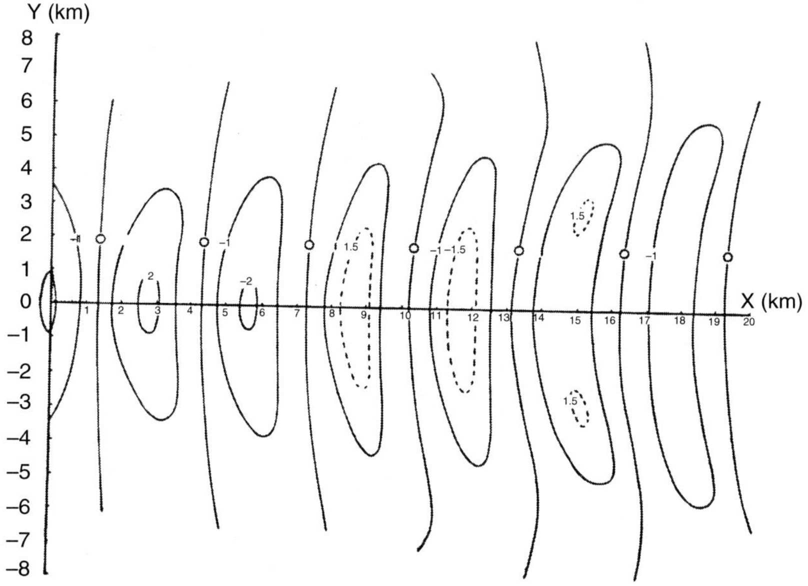


**FIGURE 11.5 Lee-wave component of vertical velocity at 1.5 km for two-layer airstream. Velocities in m s$^{-1}$ for a 100-m hill.** *(From Sawyer (1962))*

to include a number of layers (Wurtele, 1953; Corby and Sawyer, 1958; Sawyer, 1960; Danielsen and Bleck, 1970) and even an infinite number of layers (Palm, 1955). Small-amplitude linear theory has been applied to three-dimensional simulations of trapped lee waves (Scorer and Wilkinson, 1956; Palm, 1958; Sawyer, 1962; Crapper, 1962; Gjevik and Marthinsen, 1977). The three-dimensional wave solutions obtained by Sawyer (1962), shown in Fig. 11.5 for a small isolated hill, resemble surface waves behind a moving ship. The wave pattern is wedge shaped with its apex at the mountain crest.

Clark and Gall (1982) have also investigated some three-dimensional properties of trapped lee waves with a three-dimensional, non-hydrostatic, numerical model. They simulated the flow about Elk Mountain, an isolated peak in southern Wyoming. Although from the ground the mountain appears as an isolated peak, a high ridge starts ~15 km south of the mountain and extends southward. One expects that under prevailing westerly flows or even northeasterly flow, the ridge, being largely downstream, would have little influence on the flow near Elk Mountain. Clark and Gall found that, when the atmosphere supported lee waves, the lee waves formed by the ridges south of Elk Mountain interacted with lee waves generated by Elk Mountain. The resultant constructive and destructive interference among the lee-wave families caused significant changes in the wind field east of Elk Mountain. When the atmosphere supported only freely propagating waves, Elk Mountain remained dynamically isolated from the neighboring ridges. These results show

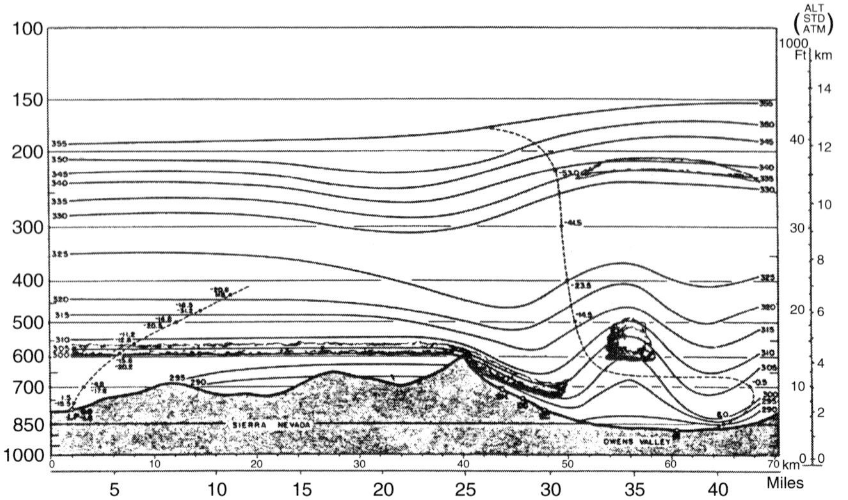


**FIGURE 11.6 West to east vertical cross section of potential temperature across the Sierra Nevada.** Dashed line represents sailplane soundings. Observed Chinook arch or Foehn wall cloud is illustrated over barrier crest, as well as rotor cloud at low levels to the east and lenticular cloud at higher levels. *(From Holmboe and Klieforth (1957))*

that properly formulated and constructed numerical models are useful in the investigation of flow fields and precipitation processes about finite-amplitude mountains.

## 11.2.3. Large-Amplitude Theories

### *11.2.3.1. Analytical Models*

The finite-amplitude linear models reviewed thus far have provided us with a basic understanding of the flow over small hills and ridges and have identified some of the conditions favorable for the formation of trapped lee waves. The solutions obtained by these models cannot explain the formation of orographic clouds in general. First, the flow field is assumed to be steady; numerous observations show that trapped lee waves seldom exhibit steady behavior (e.g. Starr and Browning, 1972). Moreover, as the amplitude of the mountains becomes larger, the flow becomes increasingly nonlinear. The impact on cloud formation is visible in the formation of such cloud forms as "rotor" clouds. Figure 11.6 is a schematic cross section obtained by Holmboe and Klieforth (1957) from sailplane flights across the Sierra Nevada. Noteworthy are the foehn, or wall cloud, that extends some distance upwind and downwind of the mountain crest, the high-level lenticular cloud, and the low-level rotor cloud to the lee of the barrier. Many of these features, especially the rotor cloud, cannot be adequately explained by small-amplitude linear theory.

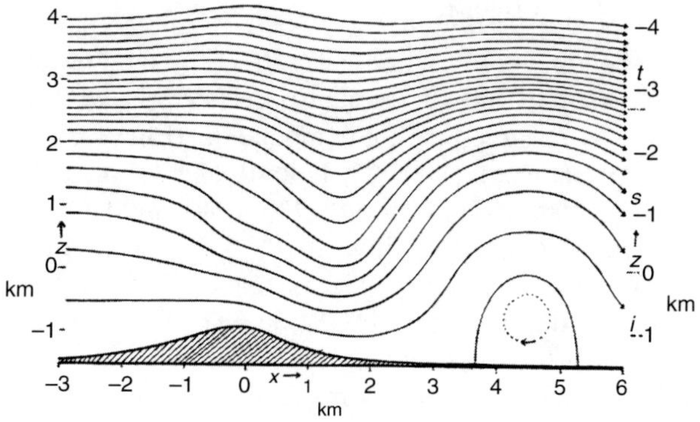


**FIGURE 11.7** **Streamlines for flow over a two-dimensional ridge.** Theoretically, only the first of an infinite train of waves is shown. However, the rotor is necessarily turbulent so that the second lee wave would probably be of smaller amplitude. This flow pattern, though particular, could occur in a restricted class or airstream, but qualitatively it is what might occur in a much wider class. The chief restriction is that the airstream have periodic lee waves of large amplitude. *(From Scorer and Klieforth (1959))*

Long (1953a,b, 1954, 1955, 1972) developed a series of laboratory and nonlinear, steady-state, hydraulic flow models. The flow features produced by the laboratory hydraulic flow models and the analytic models, resemble qualitatively the flow over the Sierra Nevada, particularly the formation of rotor clouds. Kuettner (1958) hypothesized that the flow creating rotor cloud is like a hydraulic jump. Long's models represent special mathematical and physical cases of the governing equations for which the theory becomes exactly linear. Variations in wind field and stability result in different special-case solutions for which analytic solutions are not available. In spite of the apparent qualitative success of Long's laboratory and analytic models, their applicability to atmospheric flow remains questionable, because both laboratory fluid flow and analytic models contain a reflective free surface as an upper boundary. Because the energy reflected off the upper boundary can constructively or destructively interfere with wave energy emitted by the lower topographic surface, the character of the flow may not resemble atmospheric flow. We examine more extensively the application of hydraulic models to orographic flow in our discussion of downslope windstorms and associated cloud formation.

Scorer and Klieforth (1959) obtained solutions for special cases of large-amplitude flow over mountains. They showed that the presence or absence of rotors did not depend on the nonlinearity of the flow. An example of rotor-like flow obtained with linearized equations is shown in Fig. 11.7. The calculated lee waves and rotors, however, actually extend indefinitely downstream of the mountain.

### *11.2.3.2. Numerical Models*

As the mountain height increases, nonlinear effects on the flow are increasingly likely. Beginning in the mid-1960s, numerical models were applied to the simulation of flow over large-amplitude mountains (Eliassen and Palm, 1960; Krishnamurti, 1964; Hovermale, 1965; Foldvik and Wurtele, 1967; Eliassen, 1968; Mahrer and Pielke, 1975; Deaven, 1976; Clark and Peltier, 1977; Anthes and Warner, 1978; Klemp and Lilly, 1978; Nickerson, 1979; Durran and Klemp, 1982b; Tripoli and Cotton, 1982; Cotton et al., 1986; Nickerson et al., 1986). These models differ in the formulation of numerical operators, vertical coordinate systems, top and lateral boundary conditions, the evaluation of pressure (i.e. hydrostatic versus nonhydrostatic), and the inclusion of elasticity (incompressible versus elastic).

The details of the numerical formulation are important in obtaining realistic simulations. For example, the numerical operators should be non-damping, energy-conservative forms that allow a realistic propagation of wave energy through the atmosphere. The top and lateral boundary conditions should be nonreflective to prevent wave energy from reflecting off the top and lateral boundaries and interfering with the wave energy emitted by the mountain. The selection of a vertical coordinate can affect the solutions. To simulate flow over a variety of complex orographic shapes and to reduce truncation error near the surface, a terrain-following coordinate system is desirable. A number of investigators have found an isentropic vertical coordinate to be a natural coordinate for dry isentropic motions in the free atmosphere. Unless a hybrid vertical coordinate is used, however, isentropic coordinates prevent the use of unstable soundings near the earth's surface. If gravity waves break in the free atmosphere, the resultant turbulent mixing can generate locally unstable lapse rates which cannot occur in an isentropic coordinate system.

The technique for evaluating pressure has a substantial bearing on the simulated flow field. As the width of a mountain increases, the direction of the group velocity relative to the mountain becomes increasingly vertical. The hydrostatic limit is thus approached where the group velocity for standing waves is exactly vertical. One consequence of making the hydrostatic assumption is that trapped lee waves are completely filtered out of the solution. In some hydrostatic models (e.g. Hovermale, 1965), a form of lee wave is simulated, but the properties of such a lee-wave solution are probably a computational artifact or a result of the earth's rotation. Hydrostatic models are more suitable for simulating the response of the atmosphere to large mountains of the order of 100 km in width. They can realistically simulate the propagation of inertial-gravity waves emitted by such large mountains (e.g. Eliassen and Rekustad, 1971).

### *11.2.3.3. Severe Downslope Windstorms*

A consequence of nonlinearity of flow over mountains and in orographic cloud formation is in the generation of severe downslope windstorms. Lilly and Zipser (1972) describe one of the best documented cases of severe downslope

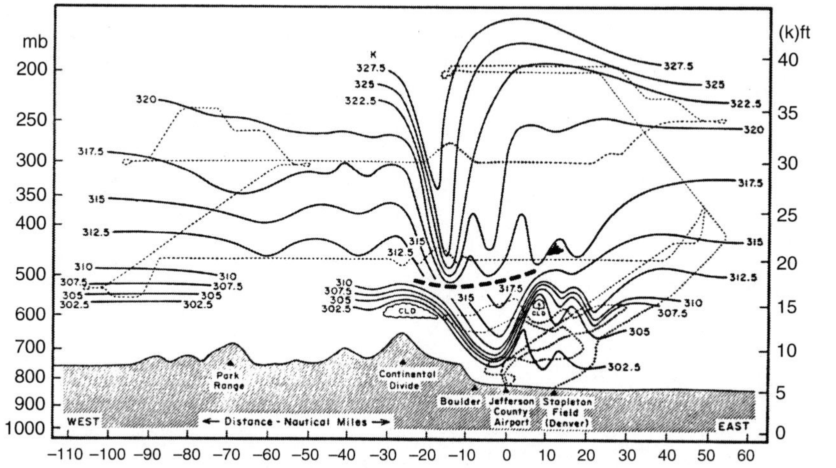


**FIGURE 11.8 Cross section of the potential temperature field in degrees Kelvin over the mountains and foothills as obtained from analysis of NCAR Queen Air and Sabreliner data on 11 January 1972.** Data above 500 mbar are exclusively Sabreliner from 1700-2000 MST. Data below 500 mbar are primarily Queen Air from 1330-1500 MST. Flight tracks are indicated by the dashed lines, except by crosses in turbulent portions. It is not possible to determine if apparent westward displacement with height of the major features is real or related to the time difference between the two flights. Windstorm conditions on the ground extend eastward to where the isentropes rise sharply, a few miles east of the origin at the Jefferson County airport. *(From Lilly and Zipser (1972))*

windstorms that occurred to the lee of the Rocky Mountains. Figure 11.8 illustrates a vertical cross section of potential temperature, clouds, and aircraft profiles obtained on 11 January 1972. Over the mountain crest is a rather typical flow pattern over a broad mountain. A stationary orographic cloud exists over the highest peaks. Directly to the lee of the higher peaks, the flow descends abruptly to the plains elevation. Evidence of trapped lee waves, including a lenticular cloud, can be seen over the plains. At higher levels is a deep trough in which air originating near stratospheric heights descends to below 500 mbar. This very high-amplitude wave is believed to be instrumental in causing surface winds in excess of 50 m $s^{-1}$. A vertical cross section of horizontal wind speeds for the same case (Fig. 11.9) shows that the maximum wind speeds occur along the lee slope of the mountain barrier at low levels.

Klemp and Lilly (1975) first explained the severe downslope wind phenomena with a two-dimensional, linearized, hydrostatic model in isentropic coordinates. Based on these linear results, they concluded that the mechanism leading to strong amplification of the wave is associated with the partial reflection of upward-propagating wave energy by variations in thermal stability. They argued that a strong wave response occurs whenever the mean vertical wavelength is such that an integral number of half-wavelengths can be confined between the ground and the tropopause. Constructive interference can then

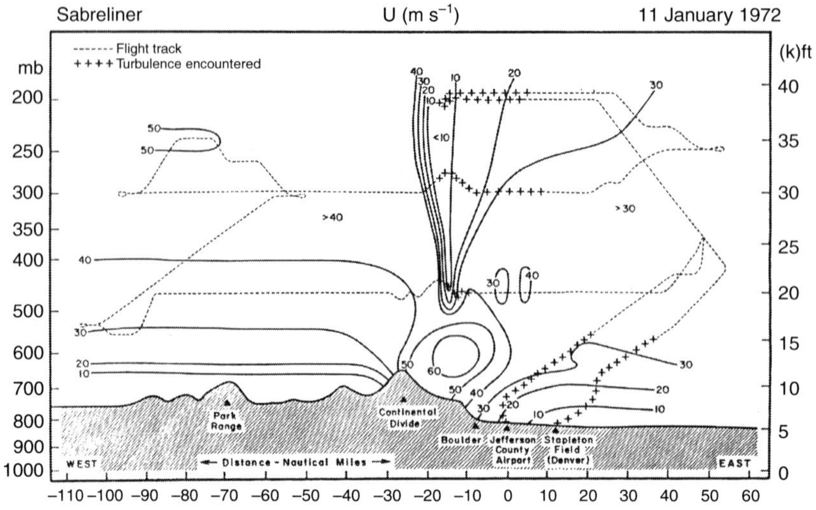


**FIGURE 11.9 Contours of horizontal velocity (m s$^{-1}$) along an east-west line through Boulder, Colorado, as derived from the NCAR Sabreliner data on 11 January 1972.** The analysis below 500 mbar was partially obtained from vertical integration of the continuity equation, assuming two-dimensional steady-state flow. *(From Klemp and Lilly (1975))*

result between wave energy reflected off the surface and wave energy partially reflected from the tropopause temperature inversion.

Subsequently, Klemp and Lilly (1978) refined this concept with a two-dimensional, time-dependent, nonlinear, hydrostatic model also cast in isentropic coordinates. Figure 11.10 illustrates that the simulated flow field exhibits many features in common with the observed flow field (Figs 11.8 and 11.9). Noteworthy is the large-amplitude standing wave over the lee of the mountain. Strong winds are also predicted just above the lee slope. Klemp and Lilly interpreted their linear model results as being generally consistent with the observations. Some evidence of the influence of non-linearity was noted, upstream blocking of the low-level flow by the mountain barrier. One effect of blocking was to alter the vertical wavelength of the gravity waves generated by the mountain, resulting in a linear model's incorrectly predicting the conditions when wave amplification can occur between the tropopause and the mountain. Klemp and Lilly noted other nonlinear effects, especially in the vicinity of critical layers (i.e. layers of zero wind or flow reversal). Linear theory predicts that total absorption of upwelling wave energy will take place in critical layers without reflection (Booker and Bretherton, 1967). Kemp and Lilly's nonlinear model, on the other hand, predicted partial reflections in the vicinity of critical layers. We shall see that the amount of reflection in the vicinity of critical layers is limited by the isentropic vertical coordinate. As the isentropes become vertically oriented, Klemp and Lilly had to activate eddy viscosity to prevent the

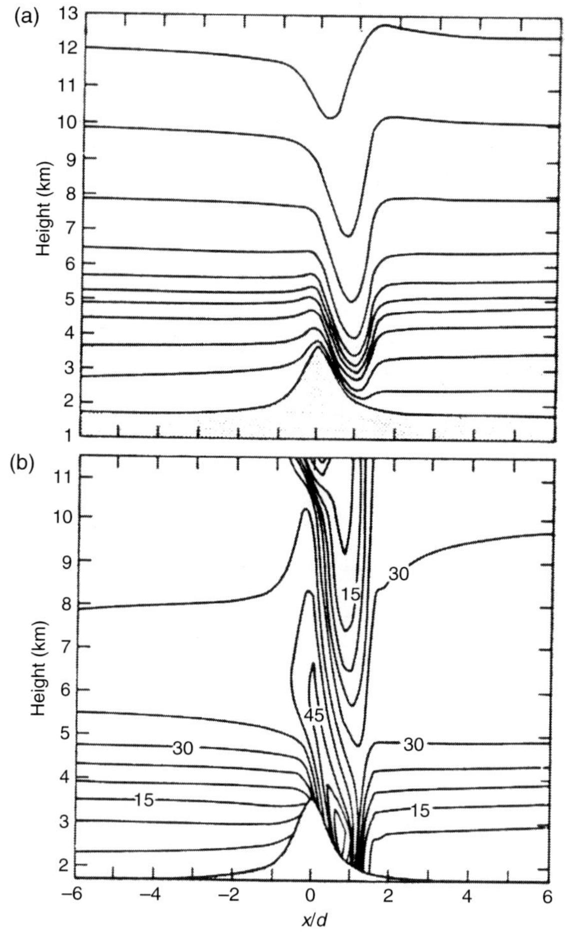


**FIGURE 11.10 Numerical simulation of 11 January 1972 case.** (a) Displacement of potential temperature surfaces; (b) contours of west wind component (m $s^{-1}$). Maximum surface velocity lee of the mountain is 55 m $s^{-1}$. *(From Klemp and Lilly (1978))*

isentropes from folding over. This prevented the formation of turbulent mixed layers that are very reflective to wave energy.

Using a two-dimensional, time-dependent, nonhydrostatic, numerical prediction model, Clark and Peltier (1977) and Peltier and Clark (1979, 1980) examined the consequences of nonlinear behavior in mountain-waves and downslope windstorms. Like Klemp and Lilly (1978), the non-linear model was first tested for its ability to reproduce linear solutions when the small-amplitude conditions are applicable. This served as a foundation of credibility before the models were extended into the unknown territory of nonlinear wave behavior. They obtained several integrations of the nonlinear system, including a simulation of the 11 January 1972 windstorm. Figures 11.11 and 11.12 illustrate the time evolution of the simulated isentropic fields and winds for the 11 January 1972 case. Note the evolution of the high-amplitude standing wave pattern to the lee of the mountain crest. The amplitude of this wave is quite similar to that observed. Peltier and Clark also obtained trapped lee waves in the lower troposphere, similar to those in Fig. 11.8. Trapped lee waves were filtered out of Klemp and Lilly's solutions by the hydrostatic approximation. Peltier and Clark noted that the wavelength of the lee wave predicted with the nonlinear model was consistent with the linear theory predictions. The amplitude of the wave was much greater in the nonlinear solutions, having a maximum vertical velocity of $\sim$8 m s$^{-1}$ compared to $\sim$1 m s$^{-1}$ for the linear solutions.

The peak wind speeds near the surface along the lee slope shown in Fig. 11.12 reach 58 m s$^{-1}$, within a few percent of the observed maximum. Although Peltier and Clark's results are qualitatively similar to those obtained by Klemp and Lilly (except the lee wave), interpretation of the sequence of events leading to the amplification of the large amplitude standing wave differs substantially. Like Klemp and Lilly, Peltier and Clark found a steepening of the isentropes to near-vertical orientation in the lower stratosphere when an integral number of half-wavelengths could be contained between the tropopause and the ground. In contrast to Klemp and Lilly, however, Peltier and Clark found that the wave actually broke, leading to a local wind reversal and a layer of constant potential temperature. As a consequence, wave energy reflected from the earth's surface became trapped between the resultant critical layer and the ground. This reflection cavity produced the large-amplitude streamline deflections that resulted in the strong surface winds. When the stratospheric wind profile was modified to prevent wave breaking, the final phase of wave amplification did not occur, and the results were then similar to linear theory. The surface drag caused by the strong lee-slope winds was over 300% of the value predicted by linear theory, even though the aspect ratio of the mountain was only slightly in excess of the critical value predicted by linear theory.

Clark and Peltier (1984) and Clark and Farley (1984) have elaborated on the role of breaking internal waves in the formation of severe downslope windstorms. Clark and Peltier (1984) showed that whenever a flow reversal or critical level occurred at a height $\frac{3}{4}\lambda_z$ above the level of forcing, where $\lambda_z$ is the vertical hydrostatic wavelength of the internal waves, the direct and reflected waves interfered constructively and intense resonant growth of the low-level wave occurred. Clark and Farley (1984) extended the downslope windstorm simulations to three dimensions. The three-dimensional simulations exhibited strong surface gustiness which they attributed to the development of turbulent eddies in the convectively unstable region of the wave. Downdrafts at the leading edge of the convectively unstable region then transported the turbulence to the surface.

Smith (1985) and Durran (1986) presented further evidence of the transition from strong smooth stratified flow, dominated by a large-amplitude wave, to a

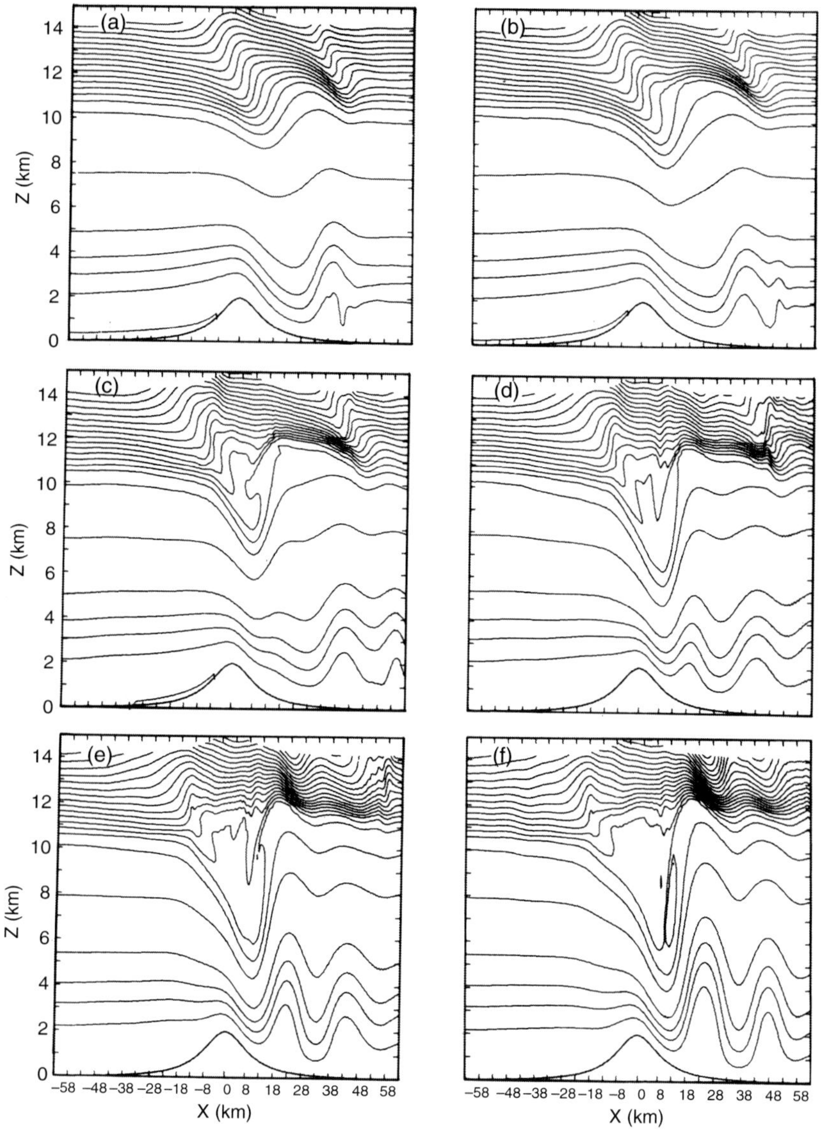


**FIGURE 11.11 Total horizontal velocity field for the Boulder, Colorado, windstorm simulation.** Times are (a) 3200, (b) 4160, (c) 5120, (d) 6020, (e) 7040, and (f) 8000 s. Contour interval is 8 m $s^{-1}$. In (f) the horizontal wind maximum in the lee of the peak is in excess of 60 m $s^{-1}$. *(From Peltier and Clark (1979))*

deep turbulent, high surface-wind state, characterized by the development of highly nonlinear waves. Smith (1985) obtained analytic solutions to Long's equation for flow beneath a breaking wave by assuming that breaking of the high-amplitude wave produces a layer of neutral stability. Figure 11.13 is a

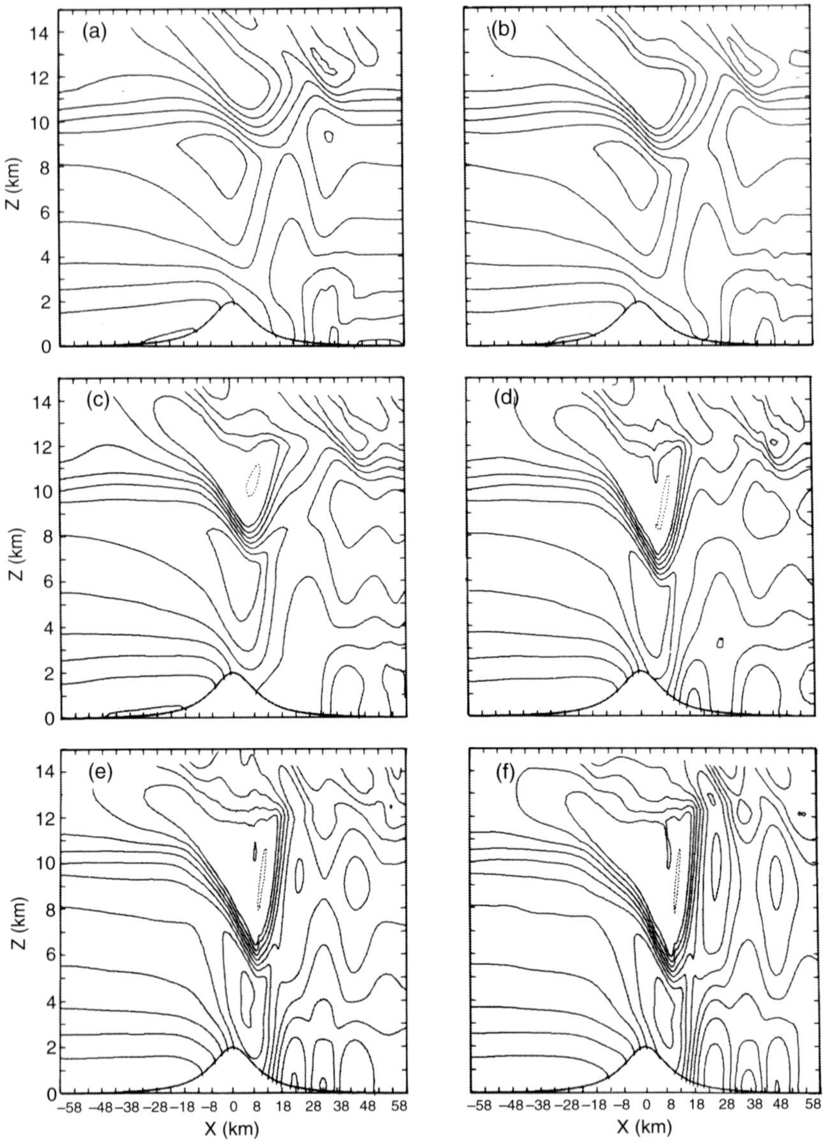


**FIGURE 11.12 Total horizontal velocity field for the Boulder, Colorado, windstorm simulation.** Times are (a) 3200, (b) 4160, (c) 5120, (d) 6020, (e) 7040, and (f) 8000 s. Contour interval is 8 m s$^{-1}$. In (f) the horizontal wind maximum in the lee of the peak is in excess of 60 m s$^{-1}$. *(From Peltier and Clark (1979))*

schematic of the hydraulic flow corresponding to a layer beneath a mixed layer induced by a breaking wave. The height $H_0$ represents the dividing streamline height; above $H_0$, only weak waves are assumed to be present. At $H_0$, the flow splits with the lower branch descending in smooth, steady, nondissipative,

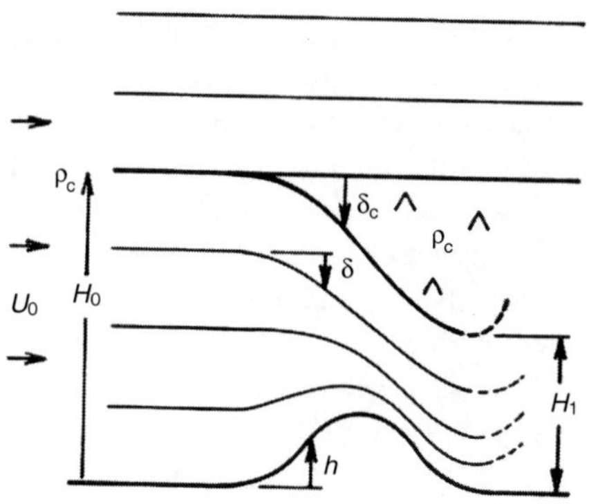


**FIGURE 11.13 Schematic of the idealized high-drag flow configuration, derived from aircraft observations and numerical simulations.** A certain critical streamline divides and encompasses a region of uniform density. The disturbance aloft is small compared to that below. *(From Smith (1985))*

hydrostatic flow. Between the split streamlines on the lee slope, the air is turbulent, well-mixed, and has little mean motion. As in Long (1955), an equation calculating the vertical displacement of a streamline following air moving over and to the lee of the mountain below $H_0$ is obtained.

Smith's theoretical predictions agree qualitatively with Clark and Peltier's numerical results. Some quantitative discrepancy exists but, in general, the hydraulic model is a reasonable representation of severe downslope windstorms in spite of the fact that it does not explicitly contain wave reflection from a critical level. Smith noted, however, "free boundary" resonances appeared in the linear theory as special points in the finite-amplitude hydraulic theory. In the hydraulic theory, the free boundary condition is not applied at a fixed level giving a coherent reflection, but along a strongly deflected streamline of flow. Smith speculated that the linear resonant conditions predicted by linear models such as Klemp and Lilly's may be useful in predicting the conditions suitable for the onset of severe downslope windstorms even though they are not suitable for predicting the final severe-wind state. That is, the linear models may be useful in predicting the level where wave breaking and a midlevel mixed layer may form. This midlevel mixed layer is essential to the application of the hydraulic theory to severe downslope windstorm simulation. It is also consistent with the formation of a resonant cavity as hypothesized by Clark and Peltier. Durran (1986), however, does not believe that linear models are useful in diagnosing conditions where wave breaking will occur.

Smith speculated that a middle-level inversion can trigger a hydraulic jump and the formation of a severe windstorm if the Froude number, based on the height and strength of the inversion $\mathrm{Fr} = u_0/(g'H)^{0.5}$, is less than one, but not too small. Using a nonhydrostatic model, Durran (1986) showed that initial amplification of the surface wind and pressure drag occurred when a low-level inversion was displaced downward along the lee slope producing "supercritical flow." The concept of supercritical flow in hydraulic theory can be explained as follows. As a parcel of air ascends a windward slope of a mountain, it slows and converts kinetic energy to potential energy. After passing the crest, the parcel accelerates as potential energy is converted to kinetic energy. In supercritical flow, nonlinear advection dominates the pressure gradient force and the resultant acceleration is in the same sense as the gravitational force. As a result, kinetic energy is no longer returned to potential energy and the air continues to accelerate as it descends the leeward slope. Durran showed that removal of the low-level inversion prevented the initial amplification of the wave and no windstorm developed. The importance of elevated inversions to the generation of severe downslope windstorms has also been suggested in climatological studies of downslope windstorm phenomena reported by Colson (1954) and Brinkmann (1974).

Durran concludes that Clark and Peltier's critical-layer reflection mechanism and the concept of supercritical flow in hydraulic theory may be alternative frameworks for describing the severe downslope windstorm phenomena. The breaking of high-amplitude waves is analogous to the transition to supercritical flow in hydraulic theory. Durran's numerical experiments suggest, however, that wave breaking is not necessary and sufficient for the transition to supercritical flow and the formation of at least moderately strong windstorms.

## 11.3. EFFECTS OF CLOUDS ON OROGRAPHIC FLOW

Cloud processes can influence mountain-wave flow and thereby feed back on the formation of precipitation in orographic clouds. In deriving the linearized wave equation, Eq. (11.19), we introduced the Scorer parameter $l(z)$. The leading-order term affecting $l(z)$ in Eq. (11.17) is $(\beta g)/(u_0^2)$, which can be equivalently expressed in terms of the Brunt-Väisälä frequency $N$, giving

$$l^2(z) = N^2/u_0^2. \tag{11.29}$$

We have defined $N$ with respect to dry atmospheric motions. A number of investigators have shown that, when condensational heating and evaporational cooling are present, $N$ should be modified accordingly.

Fraser et al. (1973) defined a moist Brunt-Väisälä frequency $N_\mathrm{m}$ as

$$N_\mathrm{m}^2 = (g/T)(\mathrm{d}T/\mathrm{d}z + \gamma_\mathrm{m}), \tag{11.30}$$

where $\gamma_\mathrm{m}$ represents the wet adiabatic lapse rate. They introduced wet processes in a Scorer-type linear model by simply replacing $N$ by $N_\mathrm{m}$ in the wave equation. The inclusion of the effects of clouds in the flow model resulted in a less stable flow because $N_\mathrm{m}$ is less than $N$. Because the buoyancy-restoring force is decreased, the amplitude of the mountain-wave under certain conditions can be significantly weakened. As expected from Eq. (11.30), the importance of the effect of clouds on the simulated flow fields depends on the wind profile as well.

Using a two-dimensional hydrostatic model, Barcilon et al. (1979) demonstrated that cloud processes can significantly modify the momentum drag created by flow over mountains. The effect of cloud processes on lee waves could not be examined because they used the hydrostatic approximation. The effect is quite substantial.

If virtual temperature corrections, including condensate, are considered in the buoyancy term of the vertical equation of motion, Lalas and Einaudi (1974) showed that an additional term should be present in $N_{\mathrm{m}}$,

$$N_{\mathrm{m}}^2 = (g/T)(\mathrm{d}T/\mathrm{d}z + \gamma_{\mathrm{m}})(1 + Lr_{\mathrm{s}}/RT) - g/(1 + r_{\mathrm{w}})(\mathrm{d}r_{\mathrm{w}}/\mathrm{d}z), \quad (11.31)$$

where $r_{\mathrm{s}}$ is the saturation mixing ratio, and $r_{\mathrm{w}} = r_{\mathrm{s}} + r_{\mathrm{l}}$, where $r_{\mathrm{l}}$ is the liquid-water mixing ratio. Durran and Klemp (1982a) defined Eq. (11.32) in terms of the wet conservative variable $\theta_{\mathrm{q}}$ (Paluch, 1979; see also Chapter 7). The saturated Brunt-Väisälä frequency is then

$$N_{\mathrm{m}}^2 = \frac{g}{(1 + r_{\mathrm{w}})}\left[\frac{\gamma_{\mathrm{m}}}{\gamma_{\mathrm{d}}}\left(\frac{\mathrm{d}\ln\theta_{\mathrm{q}}}{\mathrm{d}z}\right) - \frac{\mathrm{d}r_{\mathrm{w}}}{\mathrm{d}z}\right]. \quad (11.32)$$

Durran and Klemp (1983) applied a two-dimensional version of the Klemp and Wilhelmson (1978a,b) convective cloud model to stable orographic flow. They applied the model to the simulation of the 11 January 1972 severe downslope windstorm over Boulder, Colorado. The addition of moisture decreased the downslope wind speed from 45 to 25 m $\mathrm{s}^{-1}$, a result of a weakened mountain-wave amplitude. Durran and Klemp also noted that lee side warming, referred to as the Chinook or Alpine foehn, is often attributed to the release of latent heat on the windward side of the barrier in precipitating clouds and to dry adiabatic descent on the lee side. In a precipitating cloud simulation, they noted that the lee side temperatures were several degrees cooler than those in nonprecipitating flow. This suggests that the most important factor influencing Chinook or foehn wind lee side temperatures is the amplitude of the mountain-wave, which is larger in the dry case.

An even more dramatic illustration of the effects of moist processes on orographic flow was presented by Durran and Klemp (1982b) in their simulation of trapped lee waves. In one case, the Scorer parameter exhibited a sharp decrease with height (Fig. 11.14), which is capable of supporting trapped lee waves in a dry atmosphere. They then added moisture to the atmosphere until the Scorer parameter no longer exhibited a sharp drop-off with height for a saturated atmosphere (Fig. 11.14b). The results of their simulations are shown in Fig. 11.15. In the dry atmosphere, a distinct trapped lee wave is evident in their solutions. The addition of a layer with 90% relative humidity results in the formation of clouds over the mountain crest and in the regions of upward motion of the trapped lee waves. The wave structure is modified somewhat.

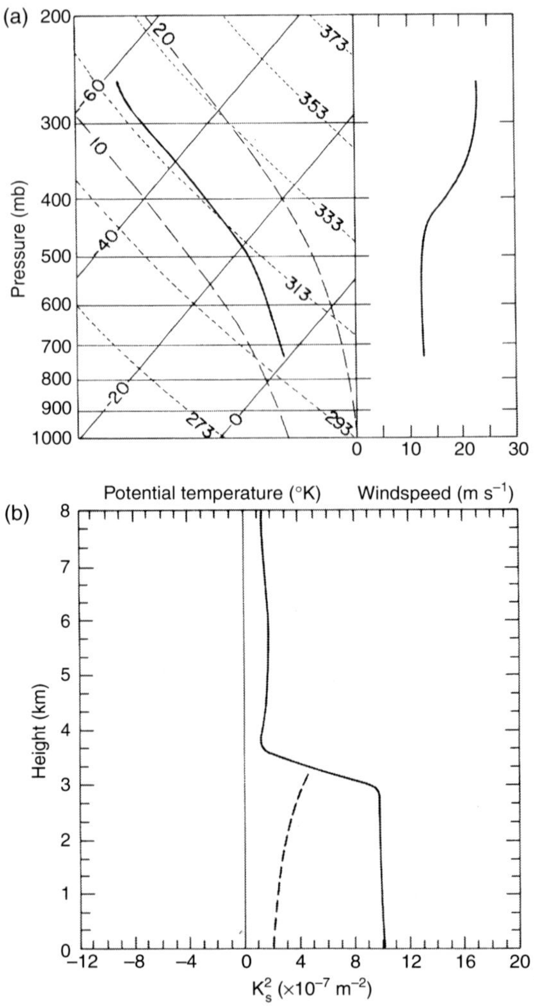


**FIGURE 11.14 Absolutely stable atmosphere favorable for the development of dry lee waves.** (a) Temperature and wind speed profiles; dry adiabats are marked with a short-dash line; moist pseudo-adiabats are marked with a long-dash line. (b) Scorer parameter ($\ell^2$) profiles; the dry $\ell^2$ is marked with a solid line, the equivalent saturated $\ell^2$ is a dashed line. *(From Durran and Klemp (1982b))*

As a result of adding a 100% saturated layer, the flow is modified so that the wavelength of the partially trapped waves is increased significantly. Finally, by adding 0.2 g kg$^{-1}$ of liquid water to the saturated layer, a cloud could be maintained in the wave troughs as well as the wave crests. This so altered the resultant vertical profile of the Scorer parameter that the lee waves became untrapped.

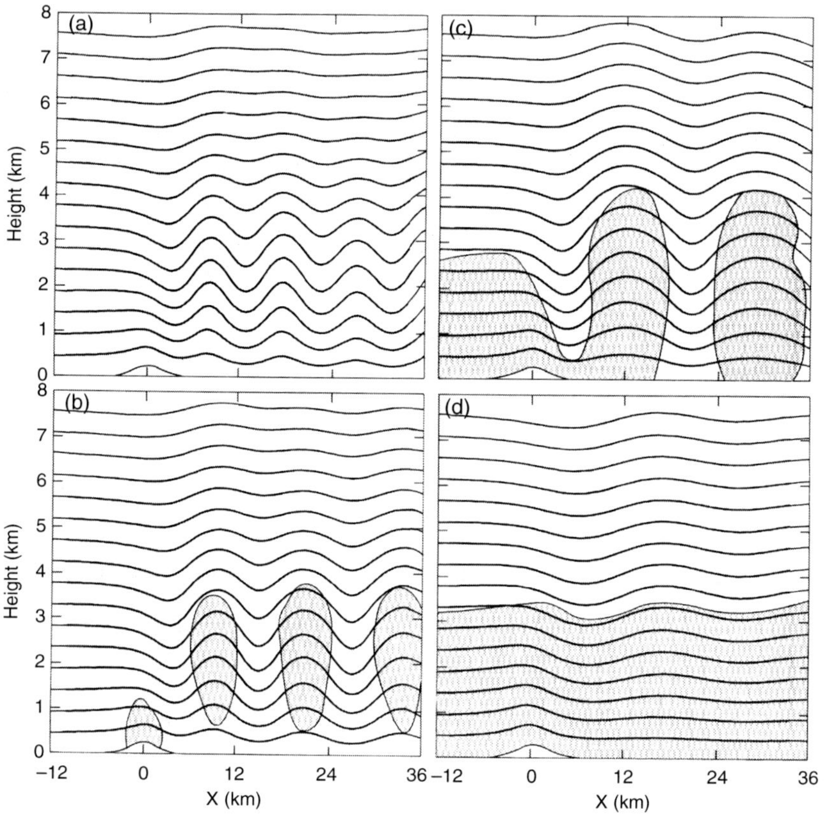


**FIGURE 11.15 Streamlines produced by a 300-m-high mountain in the flow for relative humidity (RH):** (a) RH = 0%, (b) RH = 90%, (c) RH = 100%, and (d) RH = 100% with 0.2 g $kg^{-1}$ of cloud, in the lowest layer upstream. Cloudy regions are shaded. *(From Durran and Klemp (1982b))*

Durran and Klemp also investigated a case in which the Scorer parameter decreased with height, thus supporting dry lee waves. With moisture added, however, the layer became conditionally unstable. In the previous case, latent heat release could not overcome environmental stability. Figure 11.16 illustrates the transformation from an atmosphere supporting steady trapped lee waves to transient convection as moisture is added. The lee wave structure is eventually destroyed by the convection. They also examined the influence of moist layers in the middle troposphere on "detuning" trapped lee waves.

Thus far, we have focused on the influence of nonprecipitating clouds on stable orographic flow. Lilly and Durran (1983) extended the Durran-Klemp calculations to precipitating clouds as well. Using a simple Kessler-type warm rain parameterization (Chapter 4), they investigated the effects of cloud processes, including precipitation, on vertical momentum fluxes over orographic

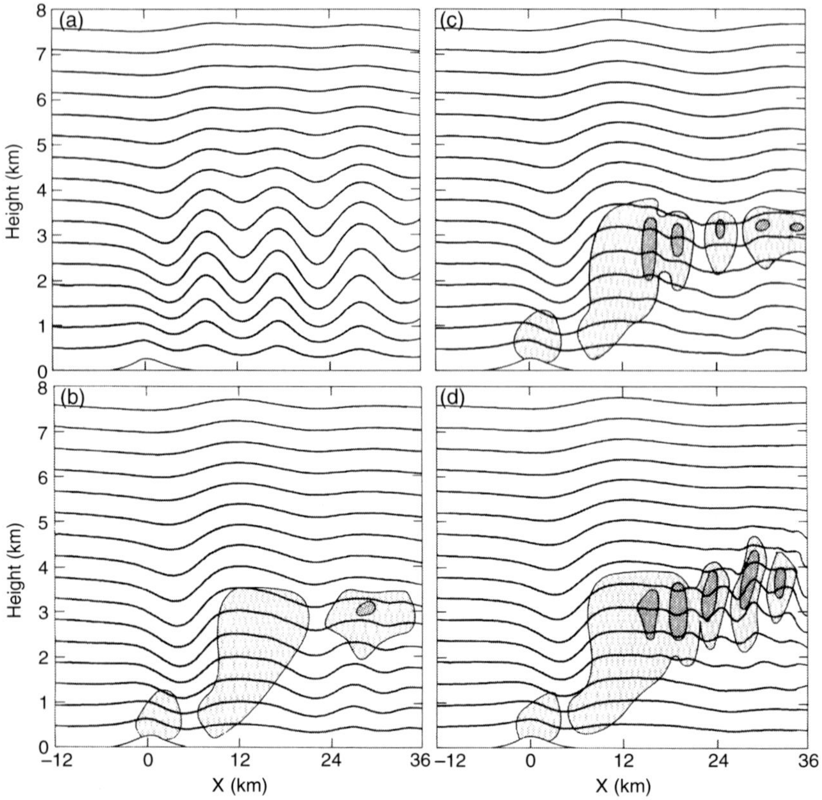


**FIGURE 11.16 Streamlines produced by a 300-m-high mountain in the flow.** (a) Steady solution for RH = 0%. Time-dependent flow for RH = 90% in the lowest upstream layer at (b) $t$ = 8000 s, (c) $t$ = 12,000 s, and (d) $t$ = 16,000 s. Cloud regions are shaded; dark shading indicates cloud densities exceeding 0.3 g $kg^{-1}$. *(From Durran and Klemp (1982b))*

barriers. Figure 11.17 illustrates the calculated vertical momentum fluxes for (a) a case having low clouds and (b) a case saturated everywhere. The fluxes are normalized to fluxes expected for linear mountain-waves ($M_{LC}$). In the dry case, linear theory predicts that the vertical momentum flux remains constant through a vertically homogeneous layer (Eliassen and Palm, 1960). The model-predicted momentum flux profile is relatively constant below 11 km. Above 11 km, a viscous layer is imposed to absorb upward-propagating wave energy and prevent its reflection into the domain of interest. The momentum fluxes are greater than predicted by linear theory (i.e. $M/M_{LC} \sim 1.4$) because the calculations were performed for a 1000-m mountain. Miles and Huppert (1969) and Smith (1977) predicted that nonlinearity forced by large-amplitude mountains should create an amplification of momentum fluxes. In nonprecipitating saturated flow, the momentum flux is reduced by a factor of three or more relative to dry flow, a

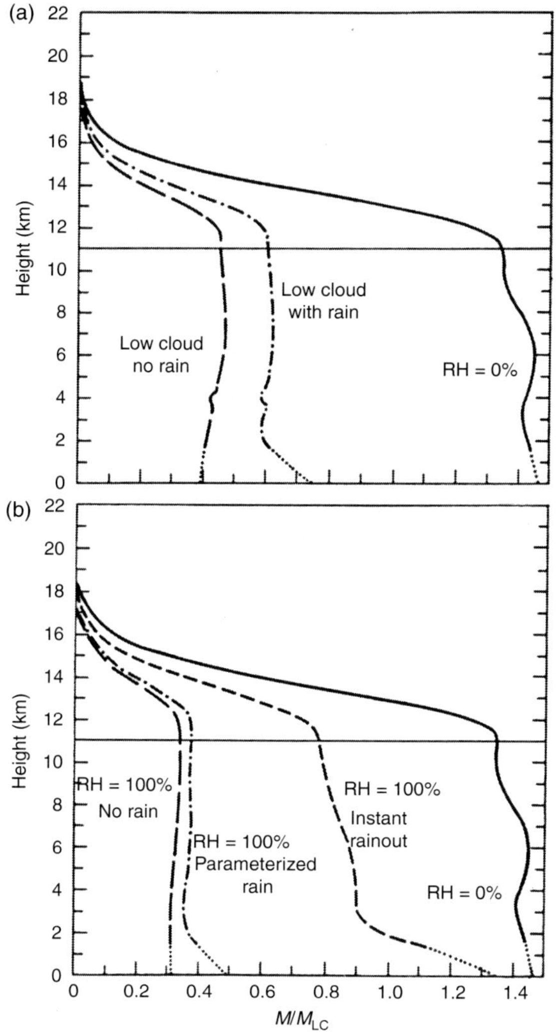


**FIGURE 11.17 The effects of rain on the vertical profiles of momentum flux produced by upstream moisture profiles in which (a) there are low clouds between the heights of 667 and 3000 m, and (b) RH = 100% everywhere.** The fluxes are normalized by $M_{LC}$, the flux associated with linear mountain-waves. *(From Lilly and Durran (1983))*

result of the effective reduction in stability of the atmosphere by condensational heating and evaporational cooling. Precipitation increases the wave momentum flux amplitudes relative to that of the nonprecipitating saturated flow. The momentum fluxes remain well below the dry case. Apparently, precipitation reduces the total condensate on the windward side of the barrier so that the descending flow warms adiabatically sooner than in the non-precipitating case.

The instant rainout case removes all condensed water, leaving no water available for evaporation on the lee slope. The results for this case exhibit a stronger wave response, which is closer to the dry case.

Smith and Lin (1982) examined the influence of asymmetric heating profiles across a mountain barrier induced by precipitation. Without precipitation, air following a streamline experiences heating as it rises over the windward slope and evaporative cooling as it descends along the lee slope. Smith and Lin imposed various heating functions at different locations across the mountain barrier. The amplitude of the heating function was inferred from observed surface precipitation rates. Figure 11.18 illustrates the simulated flow field for (a) the case with heating above the windward slope and cooling above the lee slope, (b) the precipitating case with heating only over the windward slope, and (c) the precipitating case with heating well upwind of the mountain crest. Smith and Lin claim that case (c) corresponds to the observation that the region of maximum condensation rate often occurs well upstream of the mountain.

These examples show that precipitation-induced asymmetric heating can make significant changes to the flow fields. The magnitudes of the imposed heating rates correspond to small precipitation rates. Precipitation rates of 5-10 mm $h^{-1}$ would produce a disturbance so large that the assumptions of linear theory would be violated.

Some other consequences of clouds and precipitation on stable orographic flows have been suggested. Hill (1978) inferred, from special serial rawinsonde ascents upwind of a mountain barrier, that precipitation can induce a rotor-like blocked flow. A conceptual model of the hypothesized process is shown in Fig. 11.19. During Stage 1 and just prior to frontal passage, a thick orographic cloud is depicted and precipitation is confined to the mountains. A few hours after frontal passage (Stage 2), the cloud bases are lower and precipitation is heaviest. At this stage, a downdraft begins at low levels on the windward side of the barrier. Hill speculates that water loading is the major factor in initiating the downdraft. The fact that the precipitation-forced circulation occurs when the lapse rate is at least neutral, if not unstable, for wet processes, suggests that evaporation of precipitation helps to drive the circulation as well. Stage 3 is characterized by a deepening of the downdraft circulation and lighter precipitation. Finally, in Stage 4—the postfrontal stage—the lapse rate stabilizes, cloud bases rise, and the low-level circulation terminates.

The flow in Stage 3 is similar to the flow when blocking takes place, as dry stable air rises over a mountain barrier. The flow reversal and downdraft coincide with the frontal passage when the air becomes increasingly stabilized, suggesting that blocking may be a factor. Without further study, it is not obvious whether the flow is first established by a dry blocking process and then accentuated by precipitation loading the evaporation, or if the latter processes initiate the circulation. We shall examine blocking processes in subsequent sections.

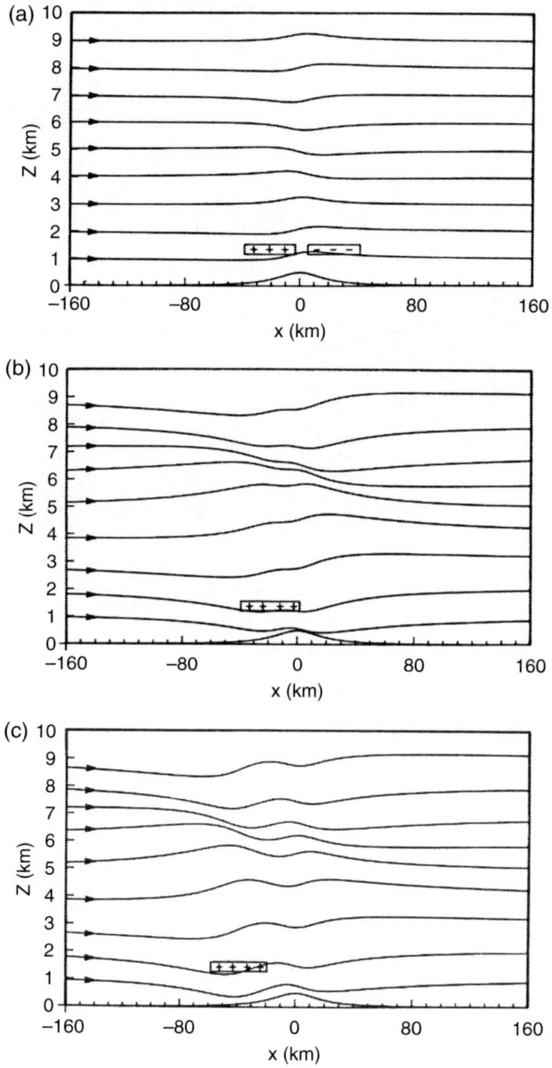


**FIGURE 11.18 (a) Hydrostatic flow with combined thermal and orographic forcing.** Heating occurs over the windward slope while cooling is specified over the leeward slope. This flow is calculated with a heating rate $Q = 1107$ W m kg$^{-1}$, a wind speed $\overline{U} = 10$ m s$^{-1}$, a Brunt-Väisälä frequency $N = 0.01$ s$^{-1}$ and a mountain with half-width of 20 km and maximum height of 0.5 km. (b) Same as (a), except heating is applied only over windward slopes. (c) Same as (b), but with the isolated heating centered farther upstream ($c$ = 40 km) to correspond to the observation that the region of maximum condensation rate often occurs well upstream of the mountain. *(From Smith and Lin (1982))*

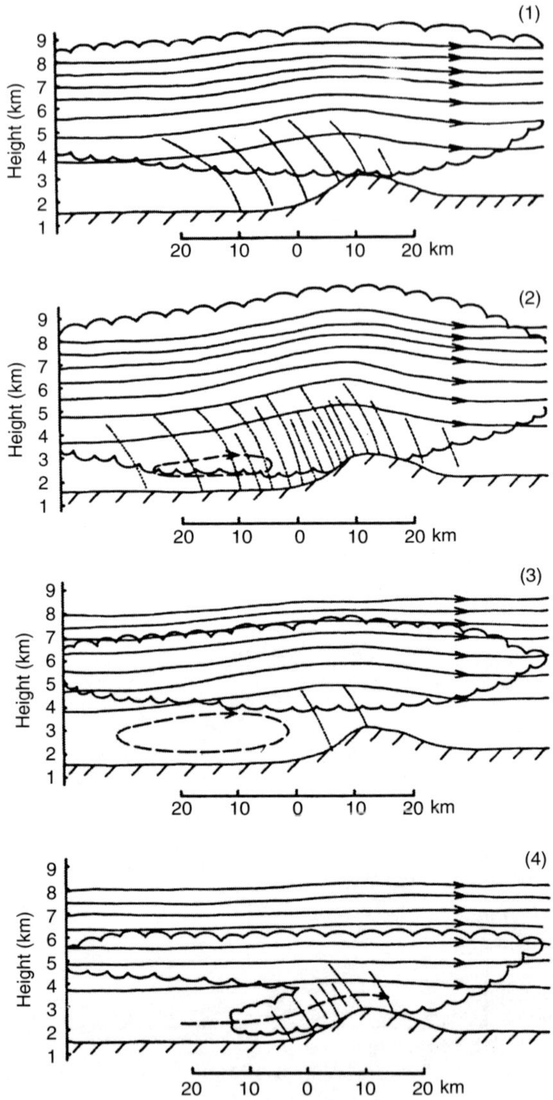


**FIGURE 11.19 Conceptual model of the development and dissipation of a precipitation-forced circulation during the course of a winter orographic storm.** Airflow is indicated by streamlines with direction arrows also shown; cloud outlines are depicted by wavy outlines and precipitation is depicted by short dashed curves. Four stages are (1) just prior to or at time of frontal passage, (2) period of heavy precipitation and development of low-level precipitation forced circulation, (3) time of strong precipitation forced circulation and sharply reduced precipitation, and (4) return of low-level orographic flow; increased precipitation, but with drier conditions aloft compared to stage 2. *(From Hill (1978))*

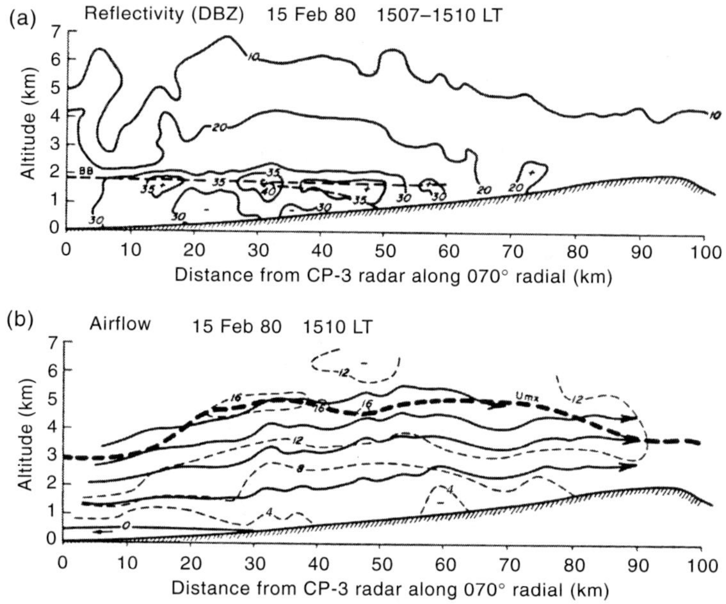


**FIGURE 11.20 Cross sections of (a) reflectivity (dBZ), and (b) airflow along 070° radial at 1507-1510 LST.** The dashed line in (a) is the bright band. The wide dashed line in (b) is the altitude of the maximum wind ($U_{\mathrm{mx}}$); the solid lines are streamlines and the thin dashed lines are isotachs in m s$^{-1}$. *(From Marwitz (1983))*

Marwitz (1983) argued that melting can significantly alter the flow field over the Sierra Nevada barrier. A common characteristic of flow over the Sierra Nevada in winter is that the 0 °C isotherm intersects the barrier roughly midway up the barrier slope (Fig. 11.20). Melting produces an isothermal layer; Atlas et al. (1969) estimated that 4 mm of melted snow produces a 700 m deep isothermal layer. A melting-induced isothermal layer can decouple the flow above the melting layer from the flow below it—evident in the wind field in Fig. 11.20b. Marwitz calculated that melting should contribute to a 3.3-mbar positive pressure perturbation on the barrier at levels sandwiched below the melting level and the barrier slope. This enhances the blocking effect of stable air being lifted over the barrier, and, as a result, winds below 2 km are slowed appreciably.

Marwitz et al. (1985) applied a two-dimensional version of the Anthes and Warner (1978) hydrostatic model to the simulation of the effects of melting on the flow over the Sierra Nevada. In one simulation, melting decreased the westerly component of flow upwind of the middle of the barrier by 7.5 m s$^{-1}$ and increased the southerly component of flow by 10 m s$^{-1}$. This southerly component of flow is referred to as the "barrier jet" (Parish, 1982) and is common along the western slopes of the Sierra Nevada. Similar barrier-parallel flows are observed along the Brooks Range in Alaska (Schwerdtfeger, 1974) and

Antarctica (Schwerdtfeger, 1975). The barrier-parallel jet is produced as nearly geostrophic flow impinges against a mountain barrier. As stably stratified air is lifted up the mountain slope, it becomes cooler than the undisturbed air at similar levels, resulting in a positive pressure perturbation along the windward slopes. Schwerdtfeger (1979) estimated pressure perturbations of the order of 4-8 mbar in the boundary layer away from the barrier, depending on initial winds and stratification. The resultant adverse-pressure gradient slows the wind component directed perpendicular to the barrier. This causes an imbalance in the previously geostrophically balanced flow, which turns the wind toward low pressure or, in the case of the Sierra barrier, into a southerly component of flow. If the flow persists for several hours or more, a new southerly, terrain-locked, steady flow is established between the local mountain blocking-induced pressure gradient and the Coriolis turning of the winds.

Marwitz et al. (1985) calculated that melting can result in a doubling in intensity of the barrier jet if a stable orographic storm is sufficiently persistent. They concluded that for melting to produce such dramatic and consistent effects, two conditions must be met. First, the component of flow normal to the barrier should be less than 10 m $s^{-1}$ within the melting layer. This is necessary to prevent the adiabatically cooled air from being driven over the barrier crest and not significantly affecting the flow on the windward side. Second, the barrier-normal component of flow must be 15-20 m $s^{-1}$ immediately above the 0 °C isotherm to produce an orographic ascent rate large enough to yield precipitation rates of 3-4 mm $h^{-1}$. The higher precipitation rates, in turn, increase the cooling by melting. Marwitz et al. noted that the effects of melting on the airflow are maximized when the maximum vertical shear of the horizontal wind is centered near 0 °C.

## 11.4. OROGENIC PRECIPITATION

We turn our attention to the processes involved in the formation of orogenic precipitation—precipitation caused by orography. The distribution of orogenic precipitation is influenced by forcing by both orography and microphysical responses. Simple orographic flow models can be useful predictors of orogenic precipitation (e.g. Elliott and Shaffer, 1962; Rhea, 1978). Such models are strictly forced by direct barrier lifting and are not at all affected by gravity waves. Their apparent success is a result of two factors. First, the heaviest precipitation comes from large orographic barriers. While trapped lee waves may form to the lee of the larger mountain barriers, the air flows through the lee wave clouds so quickly that there is not enough time for significant precipitation to form. Second, in a stably stratified air mass, the highest moisture-mixing ratios are generally confined to the lowest levels. As a result, liquid-water production is largest at the lower levels of the air mass lifted over the barrier. A model forced simply by continuity of air flowing over a barrier can capture the dominant lifting of the moisture-rich air capable of producing precipitation.

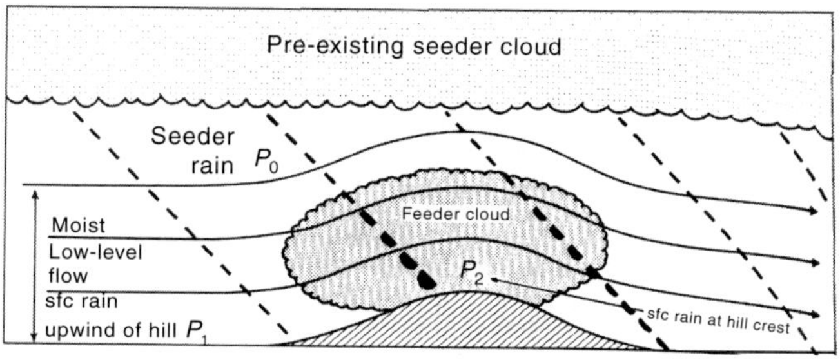


**FIGURE 11.21** **Conceptual model illustrating the orographic enhancement of rain.** *(From Browning's (1979), adaptation of Bergeron's (1965) figure)*

Whether or not the moisture condensed by lifting on the windward side of the barrier is realized as precipitation depends on air motions and precipitation processes occurring at higher levels and upstream of the mountains. That is, the actual precipitation efficiency of the cloud system, which may vary from 0% to 100%, is controlled to a large degree by the more detailed and complicated air motions and cloud microphysical processes, especially those at higher elevations upstream of the main mountain barrier.

## 11.4.1. Seeder-Feeder Process and Distribution of Precipitation

Relatively small hills, only 50 m or so above the general terrain height, can produce an increase in precipitation of the order of 25%-50% (Douglas and Glasspoole, 1947; Bergeron, 1949, 1960, 1967-1973; Holgate, 1973; Browning et al., 1975). These observations pertain to relatively warm low-level clouds in which the dominant precipitation process is collision and coalescence. Because the time scale for air flowing through such small clouds is so short, there is insufficient time for precipitation to be initiated by collision and coalescence in the low-level orographic cloud. Bergeron (1949, 1965) hypothesized that a significant fraction of the water produced in the orographic cloud may be washed out by precipitation settling from higher level clouds that form by large-scale ascent. Precipitation in the higher level clouds may originate by collision and coalescence or by ice-phase precipitation processes. A conceptual model of the process developed by Bergeron (1965) (Fig. 11.21) depicts a moist, saturated low-level flow impinging on a hill. The forced ascent of the flow over the hill causes the formation of an orographic "feeder" cloud. Above the feeder cloud is a "seeder" cloud, which forms as a result of large-scale ascent. Precipitation from the seeder cloud ($P_0$) is steady stratiform rainfall of light to moderate intensity. In the absence of the orographic cloud, it results in a surface rainfall ($P_1$). If the seeder cloud rainfall encounters the water-rich environment of the feeder cloud, the precipitation elements collect cloud droplets, thereby enhancing the rainfall to an amount $P_2$. The precipitation is enhanced by an amount $P_2 - P_0$, although Browning has noted that what is normally measured is $P_2 - P_1$. A number of investigators have developed quantitative models of the process (Hobbs et al., 1973; Storebo, 1976; Bader and Roach, 1977; Gocho, 1978; Carruthers and Choularton, 1983; Choularton and Perry, 1986). Factors important to the orographic enhancement of precipitation are (1) relative humidity and $\theta_e$ or $\theta_w$ of low-level air, (2) slope of the hill perpendicular to the wind direction, (3) strength of the wind component normal to the mountain, (4) depth of the feeder cloud, (5) precipitation rate from the feeder cloud, (6) rate of production of condensate in the feeder cloud, (7) cloud water content, and (8) rate of accretion or washout by the precipitation emanating from the seeder cloud. The relative humidity of the layer of air between the seeder and feeder clouds is also key in determining whether or not the seed particles survive descent through the layer.

The seeder-feeder mechanism is most effective in the warm sector of an extratropical cyclone. In this region, the warm, moist conveyor belt, often characterized by a low-level jet, results in nearly saturated air impinging on small hills and mountains. This is common in the winter in western Europe. While the magnitude of $\theta_w$ or $\theta_e$ seems to be a factor in determining the liquid-water production of the feeder cloud, Browning (1979) noted that the highest frequency of seeder-feeder rainfall in the British Isles occurs in December, when $\theta_w$ is relatively low. This only points out the overwhelming importance of the passage of cyclonic storms that reach a maximum frequency from November to February. The warm sector of a cyclonic storm is typified by low-level convergence which moistens the air mass impinging on the hills. The evaporation of stratiform rainfall from middle and high clouds also contributes to moistening the low-level air mass.

For saturated flow crossing a given orographic barrier, Bader and Roach's model calculations demonstrated that the rate of condensation is proportional to the strength of the low-level winds. In noting that the horizontal drift of precipitation at higher wind speeds is also significant, especially for smaller hills, Carruthers and Choularton (1983) identified two factors of importance to wind drift of precipitation:

1. The horizontal drift of a raindrop of radius $r$ falling at a speed $V_r$ through a cloud of depth $Z_c$ by a wind of speed $u_0$,

$$L_d \approx Z_c u_o / V_r. \tag{11.33}$$

2. The scale of horizontal variation of liquid-water content ($r_c$),

$$L_{r_c} \approx r_c/(dr_c/dx) \simeq L, \tag{11.34}$$

where $L$ is the half-width of the hill.

If $L_{r_c} \leq L_d$, such as over a short hill, wind drift will decrease the magnitude of maximum orographic enhancement because some of the seeder drops entering the feeder cloud summit would fall downwind in the region where both the feeder cloud droplets and the seeder drops will evaporate. For hills having $L < 10$ km, reduction of rainfall as a result of wind drift can be appreciable.

Bader and Roach and Carruthers and Choularton concluded that orographic enhancement of rainfall increased with the seeder cloud precipitation rate $P_0$. Browning (1979) noted that not only is $P_0$ of consequence, but so also is the size distribution of the seeder cloud precipitation elements. Rainfall rate is proportional to the mean drop volume, whereas the net washout rate is proportional to the net cross-sectional area of the droplets. For a given $P_0$, relatively smaller drops would collect feeder cloud drops more efficiently than large drops. If the seeder drops are too small, the collection efficiencies diminish and the smaller drops become more susceptible to wind drift. Thus, an optimum combination of wind speed, hill size, precipitation rate, and precipitation particle size determines the efficiency of washout.

Figure 11.21 illustrates a separation of the seeder and feeder cloud. This is not necessarily the case or important to the seeder-feeder concept. In some intense cyclonic storms, the atmosphere may be saturated through a great depth. In that case, the feeder cloud may be just a low-level maximum in liquid-water content on the windward side of the barrier.

In the modeling studies of the seeder-feeder process discussed thus far, the airflow is first specified or calculated; then a low-level feeder cloud is calculated on the basis of the given low-level airflow and moisture content of the air mass; finally, an upper-level seeder cloud precipitation rate is specified. Richard et al. (1987) simulated the seeder-feeder process in a two-dimensional hydrostatic cloud model. In their model, variations in moisture content of the low-level airflow, the flow speeds, and the resultant cloud microphysical processes could feed back into the airflow dynamics. Only the seeder-cloud precipitation rates and particle sizes were specified. As a result of the nonlinear interactions among the flow dynamics, thermodynamics, and microphysical processes, factors such as the cloud droplet spectrum, the size of seeding raindrops, and the seeder cloud precipitation rate exhibited only a small influence on the orographic enhancement of rainfall. However, orographic rainfall increased dramatically with low-level wind speed because the wind speed enhanced the maximum vertical velocity upstream of the barrier crest. Moisture also had a large effect because of its influence on mountain-wave dynamics. As noted previously, moisture reduces the Brunt-Väisälä frequency which, in turn, reduces the mountain-wave amplitude and, consequently, reduces orographic enhancement of precipitation. The modeling investigation by Richard et al. (1987) illustrates that models in which cloud microphysics and cloud dynamics are not allowed to interact yield an exaggerated importance to cloud microphysical processes.

Orographic enhancement of precipitation by the seeder-feeder concept increases slowly with the precipitation rate $P_0$ (Carruthers and Choularton, 1983). In the simple models such as Bader and Roach's, $P_0$ is a specified value based on estimates of stratiform precipitation rates well upstream of hills or mountains. Operationally, one could determine $P_0$ well upstream of the orographic features by radar, and then apply Bader and Roach's simple feeder model to estimate orogenic rainfall. This application presumes that orographic influences do not extend upward to the level of the seeder cloud. In the case of larger orographic barriers, blocking of the upstream flow can result in upward motion extending to the seeder cloud levels. Moreover, such orographically-induced vertical motion can trigger the release of potential instability which can create or enhance $P_0$. Elliott and Shaffer (1962) and Elliott and Hovind (1964) noted that heavy orographic rain is frequently associated with potential instability. Potential instability can also be released by large-scale vertical motion. Blocking of the low-level flow can induce differential thermal advection, causing convective instability and seeder cloud precipitation. Because hills and mountains are rarely isolated, upward motion induced by upstream hills can trigger seeder clouds (Rauber, 1981; Cotton et al., 1983, 1986).

#### 11.4.1.1. *The Seeder-Feeder Process in Ice-Phase Clouds*

The seeder-feeder process is also relevant to ice-phase precipitation processes. Choularton and Perry (1986) considered precipitation formation by ice-crystal vapor deposition growth and by riming of cloud droplets. They ignored aggregation processes. Riming growth of ice particles is similar to accretion of cloud droplets by seeder cloud precipitation elements in an all-water cloud. The major differences are due to the variety of cross-sectional areas, fall velocities, and collection efficiencies presented by the various forms of ice particles. Owing to the lower saturation vapor pressure with respect to ice compared to water, significant vapor-deposition growth of ice particles can occur in a water-saturated cloud having no significant liquid-water content. Depending on atmospheric pressure, the largest vapor-deposition growth rates occur between −14 and −16 °C. Ice-crystal habit also modulates the vapor-deposition growth rates. Dendritic crystal habits favored in a water-supersaturated environment at temperatures between −12 and −16 °C yield the highest vapor-deposition growth rates. A secondary maximum in vapor deposition growth occurs near −6 °C, where needle growth prevails. The secondary maximum occurs in spite of the fact that, in a water-saturated cloud, the supersaturation with respect to ice is relatively small. Choularton and Perry, however, ignored the effects of ice-particle habit on crystal growth rates. A seeder-feeder process can operate quite effectively in an ice-phase cloud even though the feeder cloud may contain very small amounts of liquid water. This is true if the feeder cloud has a top in the −12 to −16 °C temperature range, the range of dendrite crystal growth. Nonetheless, the maximum orographic enhancement of precipitation occurs when the feeder cloud has sufficient liquid water production to support riming growth. Choularton and Perry found that the orographic enhancement of precipitation by snowfall exceeded that of rainfall. In the case of a small hill, they calculated the maximum enhancement of precipitation of the ice-phase

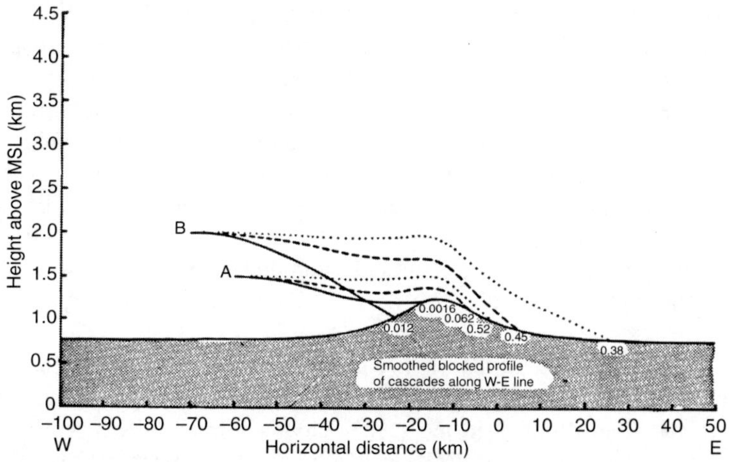


**FIGURE 11.22 Calculated trajectories for precipitation particles originating at A and B and growing by deposition riming over the Cascade Mountains in a westerly airstream with simulated blocking for the following specified concentrations (liter$^{-1}$) of ice particles: 1 (solid line), 25 (dashed line), 100 (dotted line).** The number at the end point of each trajectory is the total mass (milligrams) of precipitation that reaches the ground at that point originating in a volume of 1 liter at the starting point of the trajectory. *(From Hobbs et al. (1973))*

cloud to be 3.2 mm h$^{-1}$ while under similar conditions the enhancement for rainfall was 1 mm h$^{-1}$. The effects of wind drift are greater in an ice-phase system. As a result, increasing wind speed increases the maximum enhancement of snowfall over a long hill, whereas it has a reverse effect over a short hill.

Wind-drift effects and the trajectories of precipitation particles are complicated in ice clouds by the variety of crystal habits and particle forms. Variation in ice-crystal concentrations has an impact on particle trajectories; if concentrations are low, individual precipitation elements are larger and fall faster. They also grow by riming of cloud droplets at faster rates, which, in turn, increases their settling rates. Using the orographic flow model developed by Fraser et al. (1973), Hobbs et al. (1973) computed the trajectories of ice particles growing by vapor deposition and riming. Figure 11.22 illustrates the computed trajectories of precipitation elements starting at points A and B. For this case, the particles are carried farther downwind with increasing particle concentration, and the amount of precipitation reaching the ground increases. In some cases, ice particles settling in the dry sinking air on the lee side of a mountain may sublimate, resulting in a reduction of precipitation with increasing particle concentration. Aggregation did not substantially alter the particle trajectories relative to riming growth. However, if particle concentrations were high and if aggregation occurred over a thick layer, aggregation growth could be more rapid than growth by riming.

Rauber (1981) also determined air motions and liquid-water production using a linear, orographic flow model and calculated trajectories of ice particles.

Windward of the orographic barrier, he found that the particle trajectories can sometimes be nearly horizontal. That is, the increase in particle fall speed as the crystal grows to larger dimensions is largely offset by the increase in updraft speed as the air flows over the barrier. Rauber calculated that, in order for upper-level clouds to contribute to surface precipitation over the mountain slopes, the precipitation would have to settle out of those clouds as much as 100 km upwind of the mountain. This is because the precipitation falling from upper-level wintertime seeder clouds is generally in the form of pristine crystals with low terminal velocities. As in the clouds studied by Browning and colleagues over the British Isles, such distant seeder clouds would be formed by large-scale lifting processes.

As will be discussed more fully in a later section, the efficiency of the seeder-feeder process in mixed-phase clouds is also dependent upon the concentrations and size of cloud droplets in the feeder clouds. Enhanced concentrations of aerosol pollution serving as CCN can increase the concentrations of cloud droplets and reduce their size which can reduce the efficiency of riming by seeder cloud ice crystals settling into a lower level supercooled feeder cloud (Saleeby et al., 2009).

## 11.5. TURBULENCE EDDIES AND EMBEDDED CONVECTION IN OROGRAPHIC CLOUDS

There is a tendency to think of orographic clouds and precipitation as forming in smooth, slantwise ascent over terrain, producing steady, stratiform precipitation. This picture, however, is much too simplified. Orographic precipitation is often quite showery (e.g. Browning et al., 1975) which is thought to be due to either embedded cellular convection (Kirshbaum and Durran, 2004; Fuhrer and Schar, 2005) or shear-induced turbulence (Houze and Medina, 2005). Lifting of moist, stable air can lead to embedded cellular convection if the airmass is potentially unstable or $D\theta_e/DZ < 0.0$. Kirshbaum and Durran (2004) argued that the moist Brunt-Väisälä frequency $N_m^2$ is a better measure of the instability of moist flow impinging on a mountain slope than $D\theta_e/DZ$. They demonstrated with numerical simulations that a statically-stable cloud layer could form when $D\theta_e/DZ$ is less than 0 as long as $N_m^2$ is greater than 0.0. They showed that the presence of embedded cellular convection enhanced rainfall rates, precipitation efficiencies, and precipitation accumulations relative to pure stable flow.

Kirshbaum and Durran (2004) found in their simulations that, in addition to instability identified by $N_m^2 < 0.0$, cellular convection increased if the residence time of air parcels increased, or wider mountains were more likely to produce embedded convection than narrow ridges. Moreover, the greater depth of the unstable cap cloud favored the growth of embedded cells. Likewise vertical wind shear is important as stronger shear suppressed the formation of embedded cells, at least in two-dimensional simulations. Furthermore, Cosma et al. (2002) and Fuhrer and Schar (2005) found that small-scale topographic features and/or small amplitude upstream perturbations in the upstream flow favored the formation of embedded cellular convection.

Houze and Medina (2005) provided evidence that the presence of strong vertical windshear, where the Richardson number is less than 0.25, can induce turbulent cells that can enhance precipitation fall-out over the windward slopes of a mountain. They argue that without such embedded turbulent cells, precipitation particles would more likely settle further up or even on the downslope side of a mountain. They also note that the shear can be a direct result of topographic flow, as blocking of the low-level flow by terrain (see next section) can lead to enhancement of vertical windshear.

## 11.6. BLOCKING IMPACTS ON OROGRAPHIC PRECIPITATION

As stably stratified flow impinges on a mountain barrier, adiabatic cooling of the low-level lifted air results in the formation of a positive pressure perturbation along the windward slopes of the mountain. This produces a pressure gradient force which is directed upstream of the mountain barrier. The incoming airflow at low levels is thereby slowed down, and in some cases it actually reverses in direction. The flow is then said to be *blocked*.

The most important parameter for describing conditions when flow over a mountain will be blocked is the Froude number (Pierrehumbert and Wyman, 1985; Lin and Wang, 1996). The Froude number ($Fr$) represents the ratio of the square root of the kinetic energy of the horizontal flow impinging on a mountain barrier to the energy required to lift an air parcel from the base of a mountain to its top in a stably stratified environment.

$$Fr = U/Nh, \tag{11.35}$$

where $U$ is the speed of the incoming flow, $N$ is the Brunt-Väisälä frequency, and $h$ is the height of the mountain. Therefore blocking is more likely to occur when winds are weak or stability is large. Precise values for the onset of blocking are not available as the value of $Fr$ when blocking occurs varies with the details of orography (shape of the mountains) and whether it is more three dimensional or two-dimensional. Pierrehumbert and Wyman (1985) found that for two-dimensional flow about a Gaussian-shaped mountain blocking occurs for $Fr < 0.67$, and for a bell-shaped mountain blocking was initiated for $Fr < 0.57$. Lin and Wang (1996) found that the onset of blocking took place at $Fr < 0.9$, while Smith (1985) found a value of 0.985, both being larger than threshold values found by Baines and Hoinka (1985) and Pierrehumbert and Wyman (1985).

The question we address here is how does low-level blocking influence orographic precipitation intensity and location? Grossman and Durran (1984) conclude that blocking induces upward motion well upstream of the barrier, and results in precipitation more upstream. Marwitz (1980) argues that blocking

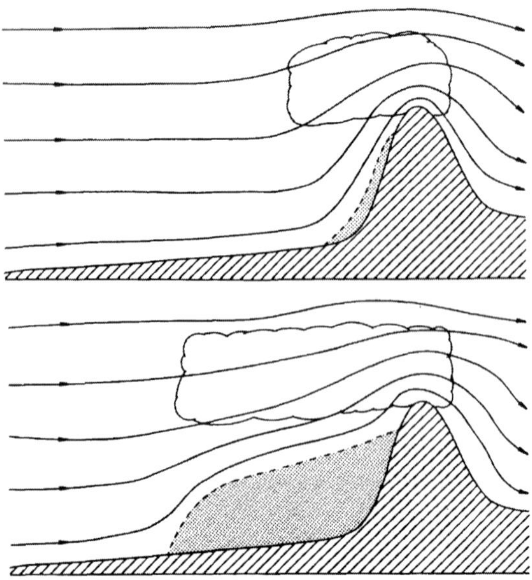

**FIGURE 11.23 Conceptual model for the working hypothesis that low-level decoupled flow (stippled area) acts as an extension of the mountain barrier for orographic lift purposes, which would then alter the location of condensate production and hence precipitation.** A small amount of decoupled low-level flow (a) allows parcel lift to occur near the barrier, while a large amount of low-level decoupled flow (b) forces parcel lift to occur upstream of the barrier. *(From Peterson et al. (1991))*

reduces the effective height of the barrier and thereby decreases precipitation. Using a combination of observational analysis and two-dimensional modeling studies, Peterson et al. (1991) concluded that low-level decoupling changes the effective shape of a mountain barrier and thereby induces ascent further upstream of the mountain. This results in condensate production further upstream than would occur by pure orographic lifting. They found that blocking thus increased the efficiency of precipitation formation by allowing ice crystals to grow over a longer trajectory. A conceptual model of how blocking can alter cloud structure is shown in Fig. 11.23.

## 11.7. OROGRAPHIC MODIFICATION OF EXTRATROPICAL CYCLONES AND PRECIPITATION

The heaviest precipitation events in mountainous terrain in the winter are associated with extratropical cyclones. In the case of low-amplitude hills or mountains or deep intense storm systems passing over larger mountains, the organization of vertical motion and precipitation associated with the passage of a cyclonic storm is similar to that in the model developed by the Norwegian school (Bjerknes, 1919; Bjerknes and Solberg, 1921). Many cyclones are modified significantly as they pass over the larger mountain barriers. In some cases, the storm is so strongly disturbed that certain characteristics of the Norwegian model, such as the warm front, are absent or not discernible in the sounding data (Hobbs, 1973). Many large mountain barriers are responsible for the high frequency of cyclogenesis located to the lee of the mountains (Petterssen, 1956; Radinovic and Lalic, 1959; Radinovic, 1965; Reitan, 1974; Chung et al., 1976). Here, we concentrate on the effects of blocking by large mountains on the evolution of cyclonic storms and precipitation. We also examine the development of upslope storms on the eastern slopes of the Rocky Mountains.

### 11.7.1. Differential Thermal Advection Caused by Orographic Blocking

When a baroclinic zone or cold front advances toward a mountain barrier, at low levels the cold air slows as it encounters a mountain-induced pressure gradient directed upstream of the barrier. That is, the low-level flow is said to be blocked. At higher levels, the adverse mountain-induced pressure gradient is reduced in magnitude or even reversed in direction. The cold air advances at a greater speed at higher levels than it does below the mountain crest. This mountain-induced differential thermal advection process results in cooler air being advected over lower-level warmer air, leading to the formation of deep convection or "embedded" stratocumulus-type convection. Smith (1982) estimates changes in the horizontal wind using Queney's (1947) model for flow over a bell-shaped mountain. He assumed that the motion of the front was primarily due to horizontal advection, rather than wavelike propagation. If vertical motion is neglected, the horizontal position of the front as a function of altitude $x_{f(z,t)}$ is altered by advection as follows:

$$dx_f/dt = u_f(x, z). \tag{11.36}$$

The advecting wind field ($u_f$) determined by Queney's (1947) model is

$$u_f(x, z) = u_0\left[1 + (hbl)\left|\frac{-b\sin lz + x\cos lz}{b^2 + x^2}\right|\right]. \tag{11.37}$$

The parameters $h$ and $b$ are the height and half-width of the mountain. Figure 11.24 illustrates the model's position of a cold front with an unperturbed slope of 1:50. Near the ground, the flow is slowed and the movement of the front is retarded. Because the flow is actually increased in speed aloft, the front moves more rapidly there. Approaching the mountain crest, the frontal surface becomes distorted and eventually turns over. Because the cold air has overridden the warmer low-level air, the atmosphere has been transformed to convectively unstable.

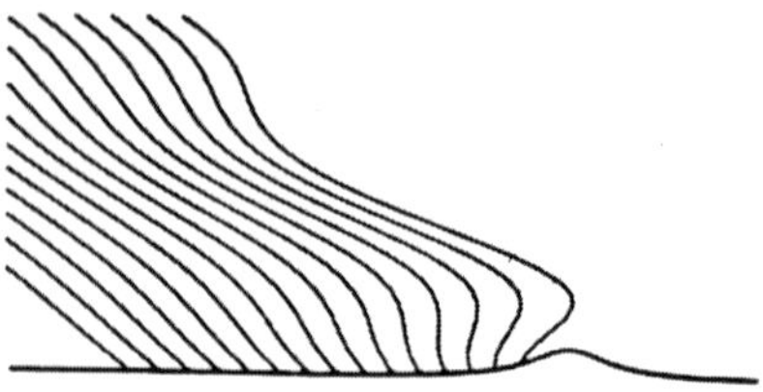

FIGURE 11.24 **A schematic depiction of the position of a cold front, at 2-h intervals, as it approaches and is influenced by a mountain range.** The distortion of the frontal surface is from slowing of the low-level flow by the mountain and the acceleration aloft. This differential advection causes the cold air behind the front to override the warm air, producing an unstable air column. The resulting small-scale convection enhances precipitation upstream of the mountain and on its windward slopes. This diagram is constructed for $u_0 = 10$ m s$^{-1}$, $N = 0.01$ s$^{-1}$, $b = 20$ km, $h = 800$ m, $x_0 = -100$ km, and $a = 1/50$. The vertical exaggeration is 12:1. *(From Smith (1982))*

Convective instability is common as a frontal system approaches a mountain barrier (Hobbs et al., 1975; Marwitz, 1980; Rauber et al., 1986; Reynolds and Dennis, 1986). In some cases, convective instability may be a result of local orographic lifting; in others it may be a property of the unperturbed thermodynamic structure of an extratropical cyclone as it passes over mountainous terrain. Heggli and Reynolds (1985) and Reynolds and Dennis (1986), for example, applied Browning and Monk's (1982) split-front model to the description of the evolution of precipitation and liquid water during frontal passages in that region. The split-front model adapted to mountainous terrain is shown schematically in Fig. 11.25. Browning (1985) noted that split fronts are common in the United Kingdom. Whether or not local orography or the perturbing influence of the British Isles affects their formation in that region has not been determined. Browning and Monk's criterion for a split front is that the upper- and lower-level fronts be separated by more than 100 km; many are separated by more than 500 km. D.W. Reynolds (personal communication) noted that cyclonic storms often exhibit split-frontal characteristics in satellite imagery several hundred kilometers off the west coast of the United States. It is unknown if the continental/orographic massif of the North American continent plays a role in generating the split-front characteristics by inducing blocking well west of the continent. Although the horizontal separation of upper-and lower-level fronts is often less over the Sierras, the progression of cloud and precipitation events is consistent with the observations reported by Heggli and Reynolds and Reynolds and Dennis. As the cold front encounters low-level blocking, low $\theta_w$ air aloft advances ahead of the surface cold front, creating the upper-level front identified as UU in Fig. 11.25a. Frequently, air ahead of the upper cold front is warm and moist because of the slantwise slow ascent of the warm conveyor belt of air. Precipitation in this region is typically steady and stratiform in character. Associated with the passage of the upper cold front, convective instability ensues and convective precipitation-generating cells prevail throughout much of the depth of the troposphere (Fig. 11.25b). The

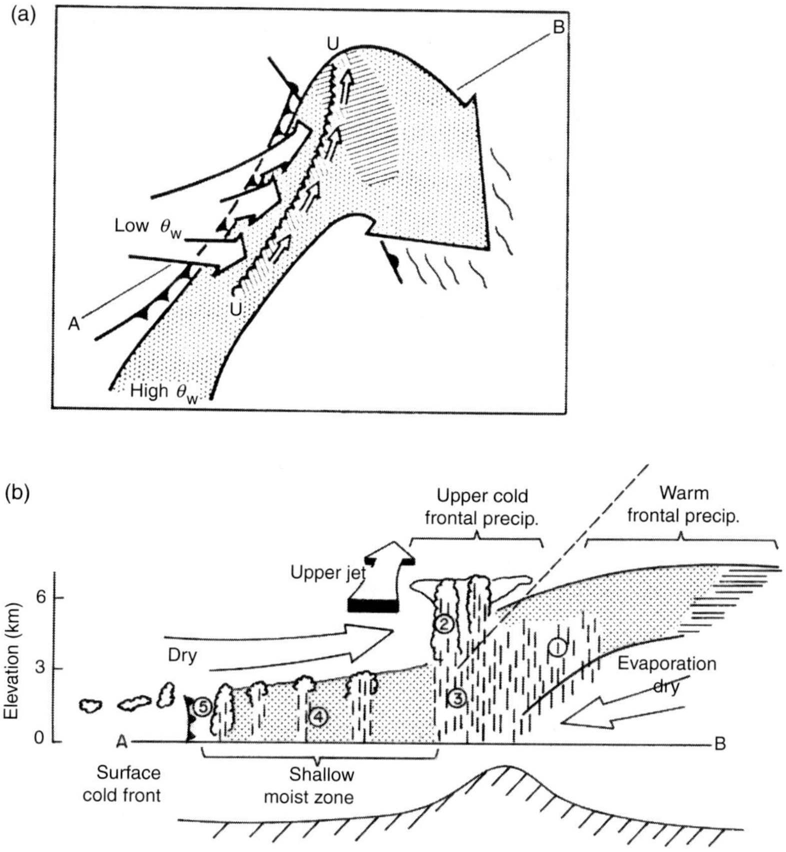


**FIGURE 11.25 Schematic portrayal of a split front with the warm conveyor belt undergoing forward-sloping ascent, but drawing attention to the split-front characteristic and the overall precipitation distribution: (a) plan view, (b) vertical section along AB in (a).** In (a) UU represents the upper cold front. The hatched shading along UU and ahead of the warm front represents precipitation associated with the upper cold front and warm front, respectively. Numbers in (b) represent precipitation type as follows: (1) warm-frontal precipitation; (2) convective precipitation-generating cells associated with the upper cold front; (3) precipitation for the upper cold-frontal convection descending through an area of warm advection; (4) shallow moist zone between the upper and surface cold fronts characterized by warm advection and scattered outbreaks of mainly light rain and drizzle; (5) shallow precipitation at the surface cold front itself. *(Adapted from Browning and Monk (1982), Browning (1985), and Reynolds and Dennis (1986). Reproduced with the permission of the Controller of Her Britannic Majesty's Stationery Office)*

seeder-feeder mechanism operates very efficiently in this region. A shallow moist zone prevails behind or west of the upper cold front. Orographic lifting of this air leads to the formation of shallow orographic clouds and stratocumulus clouds. Because of the dry air mass aloft, seeder clouds and the seeder-feeder process are absent.

In some cases, low-level orographic blocking cuts off the warm moist conveyor belt on the windward side of the barrier. Hobbs et al. (1975) found that a shallow tongue of high $\theta_W$ air became trapped along the windward slopes of the Cascade Mountains. As a result, cloudiness decreased and precipitation was delayed until the front passed over the higher terrain. Leeward of the barrier, orographic sinking motion combined with the trapping of the high $\theta_W$ air on the windward side led to a complete disappearance of any frontal precipitation by the time the front was 37 km east of the Cascade crest.

Not surprisingly, the mesoscale features of extratropical cyclones such as prefrontal and postfrontal rainbands can be altered to a greater or lesser extent, depending on the size of the orographic features, the airflow in the various sectors of the storm relative to the topography, and the type of rainband. Parsons and Hobbs (1983) described a cyclone in which the warm-sector and postfrontal rainbands appeared to be triggered by orographic lifting of potentially unstable air along the windward slopes of a mountain barrier. In contrast, preexisting, postfrontal rainbands completely dissipated to the lee of the Olympic Mountains, Washington. Parsons and Hobbs noted that smaller hills did not have any substantial impact on precipitation from warm-sector rainbands. Both narrow and wide cold-frontal rainbands appeared to be affected less by the larger mountain barriers. The orientation and magnitude of the precipitation associated with these bands were altered somewhat, but the bands remained intact as they passed over the large orographic features.

### 11.7.2. Lee Upslope Storms

Orogenic precipitation events that occur on the lee side of a barrier, with respect to the prevailing middle-to upper-level large-scale flow, are frequent on the eastern slopes of the Rocky Mountains extending from Alberta, Canada, to northern New Mexico in the United States. Lilly (1981) suggests that a similar phenomenon occurs on the north side of the Alps, along the east coast of southern Mexico, and in the so-called "backdoor" cold fronts and subsequent high-pressure ridges that develop along the southern Appalachians. In an extensive review of upslope precipitation events, Reinking and Boatman (1986) distinguished between anticyclonic systems, which are shallow, and cyclonic systems, which are deeper and produce the major precipitation events. As in most meteorological classification schemes, these two classifications represent the extremes of a continuum of events; nonetheless, we adhere to this classification.

#### *11.7.2.1. Anticyclonic Upslope Storms*

In the case of lee upslope events along the eastern slopes of the Rocky Mountains, cold Pacific and Arctic air masses are often not deep enough to cross the high mountain ridges of the Continental Divide. The barrier is hardly uniform in vertical extent, however; instead, the barrier is "leaky" allowing

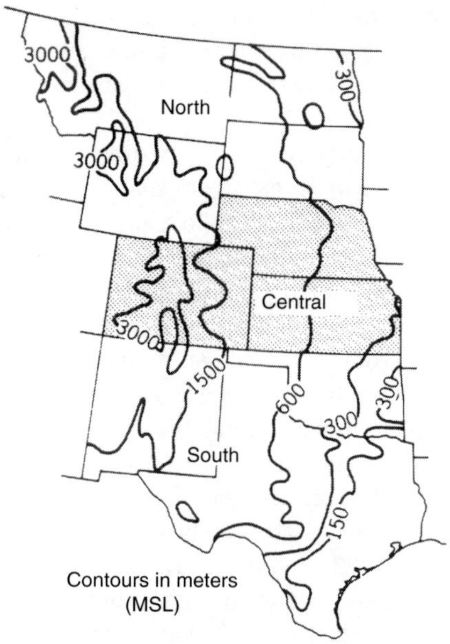


**FIGURE 11.26** **Elevation high-plains topography doubles with each contour.** *(From Reinking and Boatman (1986), after Whiteman (1973))*

cold air to spill over the lower-level ridges and drain southeastward. One of the more leaky portions of the Continental Divide is the region just north of the Colorado border in southern Wyoming (Fig. 11.26). Here, cold Pacific air masses can leak through the lower-lying terrain and, as they flow southeastward, the turning of the flow to the right of its direction of motion by the earth's rotation (Coriolis turning) causes the cold air to become trapped along the eastern slopes of the Rocky Mountains. A similar phenomenon occurs with the deeper Arctic air masses, except the topographic trapping of the air mass may not take place locally but occurs all along the north-south extent of the Rocky Mountain barrier from Alberta southward into New Mexico. The speed of southward movement of the topographically trapped cold air mass has been likened to a density current (Charba, 1974; Shapiro et al., 1985; Mass and Albright, 1986) and to a topographically trapped Kelvin wave, such as ascribed to coastal lows off southern Africa (Gill, 1977). The southward surge of the cold air mass occasionally resembles the behavior of the "Southerly Buster" that moves northward along the southeast coast of Australia (Baines, 1980; Colquhoun et al., 1985). The rapid southward progression of the cold air mass along the eastern slopes of the Rocky Mountains often gives rise to blizzard-like conditions.

Some evidence suggests that shallow cold fronts move faster during the night than during the day over the High Plains of the United States (Wiesmueller, 1982; Toth, 1987). A shallow cold front will be stalled during the day by low-lying ridges such as the Cheyenne Ridge extending along the Colorado-Wyoming border. As darkness ensues, the shallow cold air mass commences its southward movement again. Thus shallow upslope events occur more frequently during nighttime.

Toth (1987) showed by numerical experiment that, because heating is distributed over a deeper layer on the warm side of a cold front during the day, frontolysis occurs at the surface, leading to a slowing of the frontal movement. Furthermore, heating at the western higher terrain side of the cold air mass inhibits the development of horizontal pressure gradients favorable to northerly flow. Westerly winds over the higher terrain are heated during the day and flow downslope, mixing and eroding the cold air at the surface.

Figure 11.27 illustrates the southward progression of a topographically trapped cold front which moved along the eastern slopes of the Colorado Rockies. In anticyclonic events, the cloud mass forms as the air lifts upward along the eastern slopes of the barrier. It is shallow, being about 2 km in depth, and the moist air mass is absolutely stable. Using single-Doppler radar data, Lilly and Durran (1983) estimated average ascent rates of the cold air at 0.02-0.03 m $s^{-1}$ over the gently sloping plains and 0.05-0.1 m $s^{-1}$ over the foothills to the west. Lilly and Durran examined what happens to the shallow rising air mass as it reaches its maximum vertical extent. No easterly or southerly return flow is evident below the capping inversion. Lilly and Durran speculated that the rising cold air is entrained into the westerlies above the inversion. The cold air entrained through the inversion is then carried away to the east by the overriding westerlies. They provided some Doppler radar evidence that turbulent processes take place near the inversion, although details of the entrainment process were lacking. They argued that much of the entrainment took place over the foothills where ground clutter prevented adequate radar coverage. The foothills are a preferred location for convective instability induced by differential thermal advection. The resultant convection would efficiently vent the cold air mass and mix it with the overlying westerly airstream.

For the period of steady upslope analyzed, Lilly and Durran noted that the weak upslope must be balanced by downward heat flux, through turbulent entrainment, in a self-regulating way. If a greater flux of cold air was moved up the mountain than could be entrained away by the upper westerlies, an adverse pressure gradient would be established that would inhibit the upslope flow. Conversely, excessive entrainment would result in a shallowing of the inversion and allow more cold air to be drawn up the slope.

Precipitation from shallow anticyclonic upslope storms can range from none to light to moderate. Reinking and Boatman argued that the seeder-feeder process is important to controlling the occurrence and amounts of precipitation from these shallow storms. Both aircraft observations (Walsh, 1977) and radar data (Lilly and Durran, 1983) suggest that precipitation settling from the overlying westerly airstream into the water-rich, upslope feeder cloud can serve

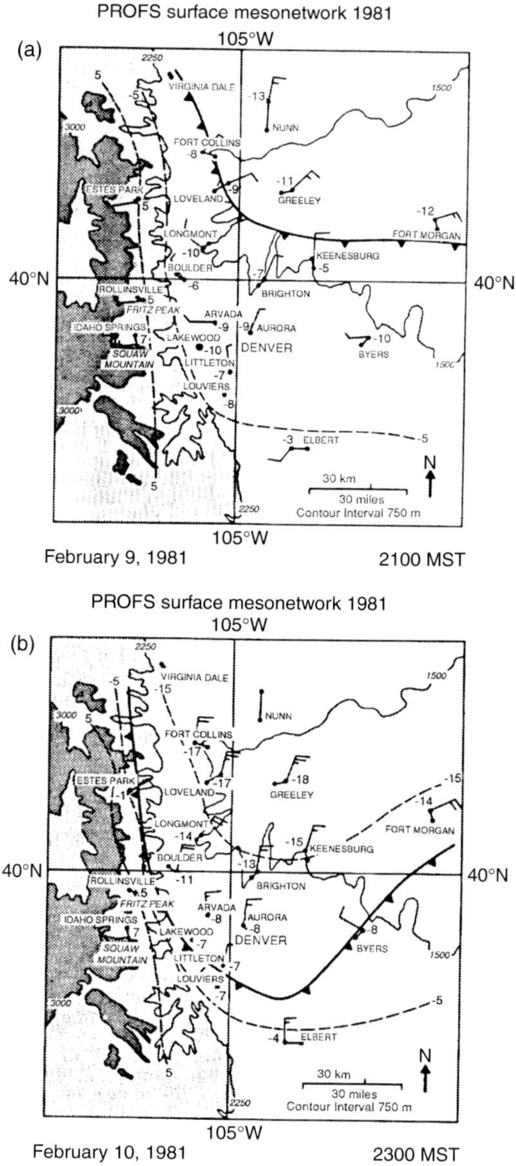


**FIGURE 11.27 Surface analysis at (a) 2100, (b) 2300, and (c) 0100 local time, 9-10 February (04 – 08Z), from PROFS surface net in north-central Colorado.** Altitude contours are in meters, temperatures in °C, reduced dry adiabatically to the altitude in Boulder, Colorado. A full wind barb is 10 kt. The frontal position is based mainly on wind speed. The station locations and the nearest towns are denoted by solid dots. *(From Lilly and Durran (1983))*

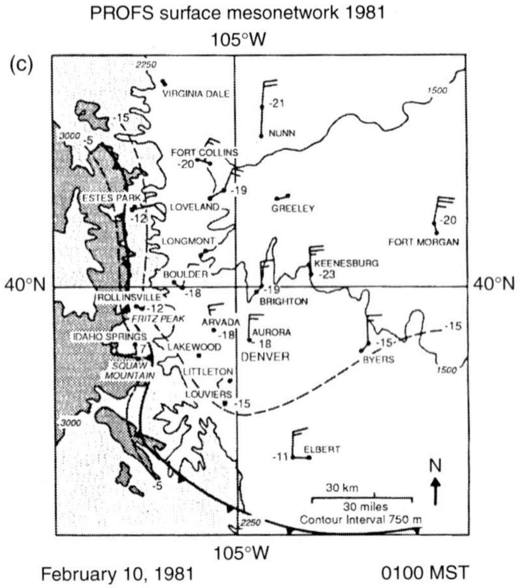


**FIGURE 11.27** (*continued*)

as seeder elements. The seeder-feeder process is fundamental to precipitation formation in the shallow anticyclonic upslope events in which cloud tops are likely to be relatively warm (i.e. warmer than $-10$ °C). For such warm-topped clouds in which liquid-water contents are relatively low (i.e. typically less than 0.2 g m$^{-3}$), primary and secondary production of ice crystals is quite low in the upslope feeder cloud. Ice crystals formed over the higher terrain at cold temperatures and advected eastward by the overlying westerlies can be effective seeder crystals.

Weickmann (1981) noted that small, heavily rimed graupel pellets are common in shallow upslope events with cloud tops in the $-6$ to $-10$ °C range. Referring to the laboratory experiments of Fukuta et al. (1982, 1984), Reinking and Boatman (1986) suggested that this temperature range is a region of suppressed vapor-deposition growth of ice crystals. The preferred ice-crystal shapes are rather isometrically shaped columnar forms of crystals. Fukuta et al.'s experiments suggest that the faster fall speeds of these crystals enhance the ice particles' ability to switch over to the heavily rimed graupel mode of growth. This more rapid switch-over to the graupel mode of growth compensates for some of the otherwise suppressed precipitation growth by vapor deposition.

### *11.7.2.2. Cyclonic Systems*

By far, the heaviest upslope precipitation events along the eastern slopes of the Colorado Rockies are associated with lee cyclogenesis that often commences in

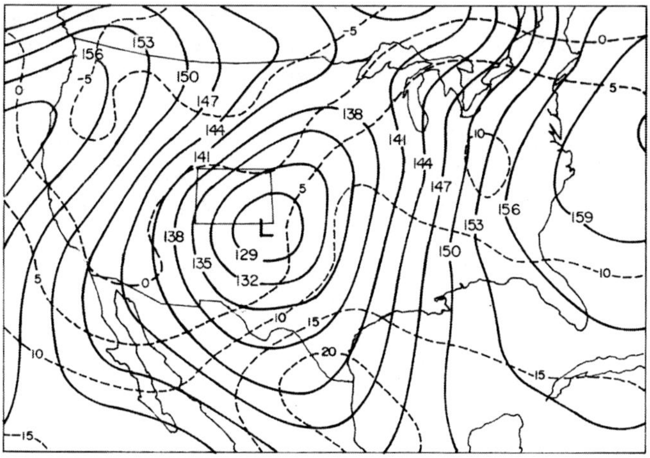


**FIGURE 11.28 The 850-mbar analysis for 0500 MST, 24 December 1982.** *(From Abbs and Pielke (1986))*

the Four Corners region (intersection of the states of Colorado, Utah, Arizona, and New Mexico). Figure 11.28 illustrates the 850-mbar analysis at the time a closed low had moved into southeastern Colorado. In the more intense storm systems, such as the 1982 Christmas Eve blizzard illustrated here, a closed low may extend upward to 500 or even 300 mbar. This results in vigorous northeasterly upslope flow through a large depth of the atmosphere. The circulation of the surface cyclone also draws polar air southward where it becomes topographically trapped along the eastern slopes of the Rocky Mountains. The advection of cold air up the eastern slopes of the Rocky Mountains can be particularly strong when the cold conveyor belt injects a jet of cold air against the eastward-facing slopes. Warm moist air drawn from the Gulf of Mexico is undercut by cold polar air, and the interaction of these two air masses can produce complex and heavy snowfall patterns. The distribution, intensity, and maximum amounts of snowfall depend upon the location of formation of the cyclone, its track and moisture supply, speed of movement, and local terrain effects.

Condensate production in these storms is a result of large-scale lifting by the cyclonic storm as well as terrain lifting. Auer and White (1982) extensively analyzed the Thanksgiving Day storm of 20-21 November 1979 along the Front Range of the Rocky Mountains. Reported snowfall amounts were 36 cm at Casper, Wyoming, 25 cm at Laramie, Wyoming, 66 cm in Cheyenne, Wyoming, 46 cm in Fort Collins, Colorado, and 43 cm in Denver, Colorado. To show local terrain influences, one author (WRC) measured 111 cm of snowfall at his house in the foothills at ~2400 m MSL, 24 km west of Fort Collins, Colorado. Vertical velocities forced by the large-scale dynamics of the storm were estimated at 0.02-0.06 m $s^{-1}$ in the northeast quadrant of the storm.

Low-level, terrain-induced vertical velocities due to small-amplitude ridges over the plains were estimated at 0.02 m $s^{-1}$ near Denver and 0.04-0.07 m $s^{-1}$ near Cheyenne. Higher terrain-induced vertical velocities were obviously present over the foothills.

While computing the large-scale divergence field and vertical motion for the 1979 Thanksgiving Day storm, Auer and White noted that the level of nondivergence, and hence the level of maximum vertical velocity, corresponded to the temperature range of $-13$ to $-17$ °C. This roughly corresponds to the dendritic mode of ice-crystal growth, in which the maximum rate of vapor depletion occurs. Moreover, the presence of dendritic crystals favors the operation of an aggregation process through a deep layer. Auer and White (1982) and Reinking and Boatman (1986) note that aggregates are the dominant precipitation form in the heaviest upslope precipitation events. Auer and White conclude that whenever the maximum vertical velocity of a storm system corresponds to roughly the $-13$ to $-17$ °C level, the maximum in water production rate coincides with the maximum in ice-crystal growth rate (other factors remaining equal). Auer and White (1982) provided climatological evidence that the maximum snowfall events across the United States, correspond to the level of maximum large-scale vertical motion in the temperature range $-13$ to $-17$ °C. They also provide evidence that the heavier orographic snowfalls are characterized by the occurrence of the maximum orographic lifting being coincident with the dendritic ice-crystal growth habit.

Although no major scientific studies have dealt with the role of convection in either shallow or deep upslope storm events, there is circumstantial evidence that embedded convection occasionally occurs. Reinking and Boatman noted that variations in precipitation rates suggest the presence of a banded storm structure similar to that frequently found in extratropical cyclones. Moreover, it is common to observe large heavily rimed graupel particles during the early stages of major upslope events. Such large graupel particles are generally favored in convective cells with high liquid water content. One would expect blocking induced differential thermal advection to be the most pronounced during the early stages of an upslope storm event. The instability so caused would favor the formation of embedded convective elements.

## 11.8. DISTRIBUTION OF SUPERCOOLED LIQUID-WATER IN OROGRAPHIC CLOUDS

The magnitude and spatial distribution of supercooled liquid-water in wintertime orographic clouds are important to a number of applications. The presence of supercooled water is a major factor in determining the amount and type of precipitation. The presence of supercooled water is also a hazard to aircraft operations. Thus an assessment of the amount and distribution of supercooled water is important to predicting aircraft icing conditions. The radiative properties of a cloud are also affected by the amount and spatial distribution of liquid-water. The acidity of precipitation is affected by the degree of ice particle riming versus vapor deposition growth (Borys et al., 1983). As will be discussed later, the impact of pollution aerosols on suppressing precipitation is strongly dependent upon the availability of supercooled liquid-water. Finally, the opportunity for precipitation enhancement by cloud seeding is dependent upon the amount and spatial distribution of supercooled liquid-water. It is this last application which has motivated a number of observational and diagnostic studies of the amount and distribution of supercooled liquid-water in the western United States.

The liquid-water content at any location in an orographic cloud is the integrated consequence of several liquid-water production processes (FP) and liquid-water depletion processes (FD). Liquid-water production is determined by the cooling rate of the air as it rises over a mountain. The cooling rate, in turn, is primarily due to adiabatic ascent. Net radiative cooling is also a factor, at least in weaker wind situations. The major liquid-water depletion processes are due to vapor deposition and riming growth of ice crystals, coalescence among cloud droplets, and removal as precipitation and by entrainment of dry air. The depletion rates are therefore a function of the concentration of ice particles, the temperature and habits of the ice particles, the cloud supersaturation and liquid-water content, concentrations of cloud condensation nuclei, and the characteristics of cloud-mixing processes.

Whenever FP is greater than FD, liquid-water will accumulate in the cloud to levels that can be detected by direct or remote-probing systems. If FP is less than or equal to FD, liquid-water will be depleted so rapidly that only solid precipitation elements will be detectable. As a simple example of the competition between FP and FD, consider the diagram in Fig. 11.29 (Chappell, 1970). Estimated water-production rates for airflow over the Climax, Colorado, mountain barrier are a function of the 500-mbar temperature in that region. Chappell used the 500-mbar temperature as a crude index of the cloud-top temperature in that region. As the cloud-top temperatures become colder, cloud bases do too, because the bases are limited by terrain height. Saturation mixing ratios diminish and thereby decrease FP. The long-dashed curve in Fig. 11.29 illustrates Chappell's estimate of FD, assuming that the major depletion process was vapor-deposition growth of ice particles. He also assumed that the concentration of ice crystals could be determined by the so-called Fletcher ice nuclei spectra (Chapter 4). This figure illustrates that, at temperatures warmer than −20 °C, FP is much greater than FD; thus, the opportunities for significant liquid-water production are large. At temperatures colder than −20 °C, the more numerous ice crystals deplete liquid-water at a rate faster than it is produced. Also shown in Fig. 11.29 are the observed surface precipitation rates. At temperatures colder than −20 °C, the observed precipitation is comparable in magnitude to the calculated FP. This simply shows that, at colder temperatures, FD is limited by FP and little excess liquid-water is produced. At temperatures warmer than −20 °C, the observed precipitation rate is considerably less

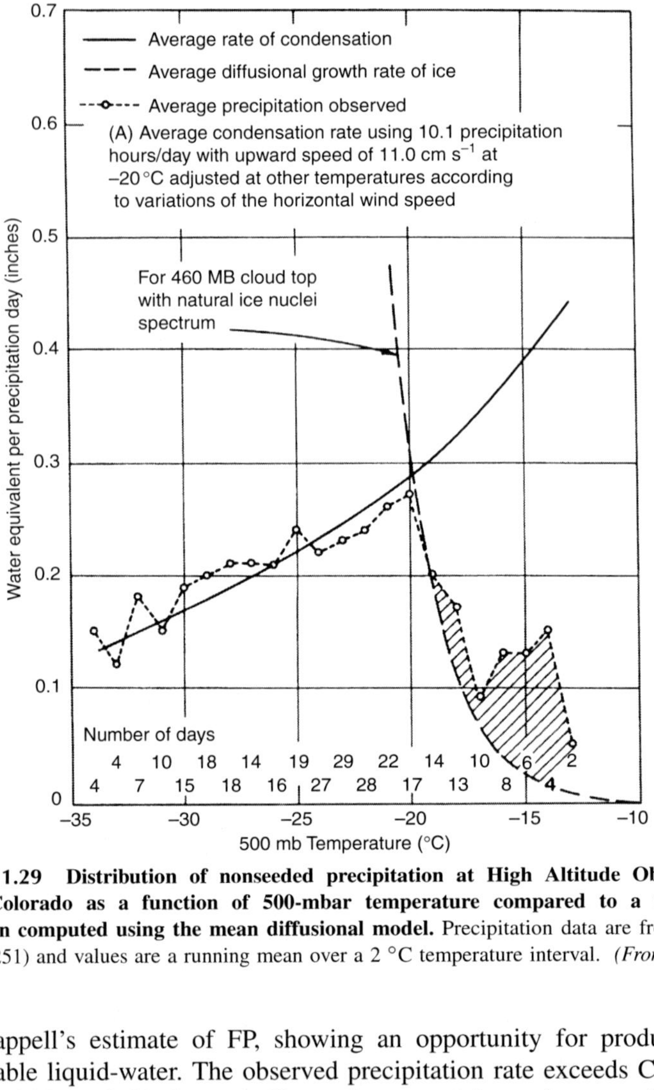


**FIGURE 11.29 Distribution of nonseeded precipitation at High Altitude Observatory, Climax, Colorado as a function of 500-mbar temperature compared to a theoretical distribution computed using the mean diffusional model.** Precipitation data are from Climax I sample (251) and values are a running mean over a 2 °C temperature interval. *(From Chappell (1970))*

than Chappell's estimate of FP, showing an opportunity for production of considerable liquid-water. The observed precipitation rate exceeds Chappell's estimate of FD probably because Chappell assumed that only vapor deposition contributed to FD. However, riming growth of ice crystals could have been significant at these warmer temperatures, where high liquid-water contents are likely. Also noted in Chapter 4, the Fletcher ice nuclei curve is frequently a poor estimator of observed ice-crystal concentrations at warmer temperatures because of the effects of ice multiplication.

## 11.8.1. Stable Orographic Clouds

The system that allows easiest interpretation of supercooled water is the "pure" orographic cloud system which occurs over mountainous regions when large-scale frontal activity is absent. It occurs when a strong cross-barrier pressure gradient drives strong winds containing moist air across the barrier. These cloud systems are well-simulated by linear orographic flow models. As found by Hobbs (1973) over the Cascade Mountains of Washington and by Rauber (1985) over the Park Range of northern Colorado, the supercooled liquid-water content is largest in regions of strong orographic forcing on the windward side of steep rises in topography. Rauber also noted that in some cases liquid-water varied uniformly through the depth of the orographic cloud layer without any discernible changes in the airflow properties of the cloud. The only detectable change in cloud structure was a variation in precipitation rate, probably a result of variations in ice-crystal concentrations. The causes of such concentration variations are not well known, but P.J. DeMott (personal communication) suggests that enhanced nucleation of ice particles occurs in localized regions of high supersaturation associated with upward vertical motions near cloud top.

### 11.8.2. Supercooled Liquid-Water in Orographically Modified Cyclonic Storms

Variations in the distribution of liquid-water in orographic clouds are frequently associated with variations in thermodynamic stratification and vertical motion as cyclonic storm systems pass over a mountain. Based on aircraft observations of the liquid-water and microphysical structure of orographic clouds over the San Juan Mountains of southern Colorado, Marwitz (1980) and Cooper and Marwitz (1980) identified four stages in the thermodynamic stratification of a passing storm system. The four stages shown at the top of Fig. 11.30 can roughly be identified with the passage of a cyclonic storm. Figure 11.31 illustrates a conceptual model of the liquid-water distribution for three of the four stages. During the "stable stage," which corresponds to the prefrontal stage of a cyclonic storm, moisture supplied by the rising warm conveyor belt contributes to a deep stable cloud layer with cold tops. The cloud system is nearly completely glaciated owing to the high ice-crystal concentrations associated with the cold cloud tops. Liquid-water occurs only in short-lived patches where gravity waves create transient regions of rapid ascent. Any liquid-water that is produced is small, less than 0.1 g $m^{-3}$. Substantial blocking of the low-level flow occurs during this period (Fig. 11.31a). As a cyclonic storm advances across the mountain barrier, the atmosphere transforms to near-neutral stability throughout much of the depth of the troposphere. Liquid-water contents rise to values of the order of 0.3 g $m^{-3}$ (Fig. 11.31b).

Following the passage of the 500-mbar trough and in some cases a discernible front, the atmosphere becomes convectively unstable at lower levels upwind of the mountains and above a surface zone of convergence. Cooper and Marwitz reported that liquid-water contents were highest during this stage of the storm system, approaching 1.0 g $m^{-3}$ in the upper cloud levels.

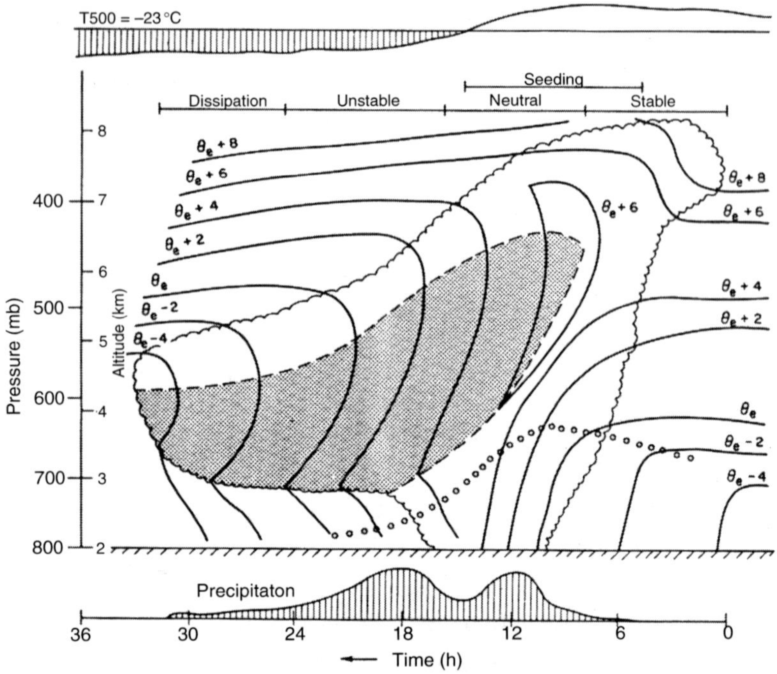


**FIGURE 11.30 Schematic height-time cross section for a typical storm over the San Juans.** The solid lines are lines of constant equivalent potential temperature ($\theta_\epsilon$), and the shaded area indicates convective instability. The typical sequence of 500-mbar temperature is shown at the top. The reference baseline is −23 °C. The level below which the wind directions are <160° is indicated by circles and is defined as blocked flow. The four stages of the storm and a typical seeding period are indicated. *(From Cooper and Marwitz (1980))*

Hobbs (1975) and Heggli et al. (1983) also found the highest liquid-water contents in postfrontal convective clouds observed over the Cascade Mountains and the California Sierras, respectively. Like Cooper and Marwitz, their observations were obtained primarily by aircraft penetrations. One deficiency of aircraft observations is that, for safety reasons, aircraft are limited to flying 600 to 1000 m above mountain tops. Also, for convective cells embedded in stratiform clouds, time continuity of the convective cells cannot be obtained. Remote-probing systems, such as microwave radiometers (Hogg et al., 1983) and polarization-diversity lidar (Sassen, 1984), provide an opportunity to sense the presence of liquid-water in regions previously inaccessible by aircraft. The data from these new observational systems are substantially altering our view of the distribution of liquid-water in orographic clouds.

Using a combination of aircraft and passive microwave radar observations, Rauber (1985), Rauber et al. (1986), and Rauber and Grant (1986) described the distribution of supercooled liquid-water observed in the Park Range of northern Colorado during passage of extratropical cyclones. In the region

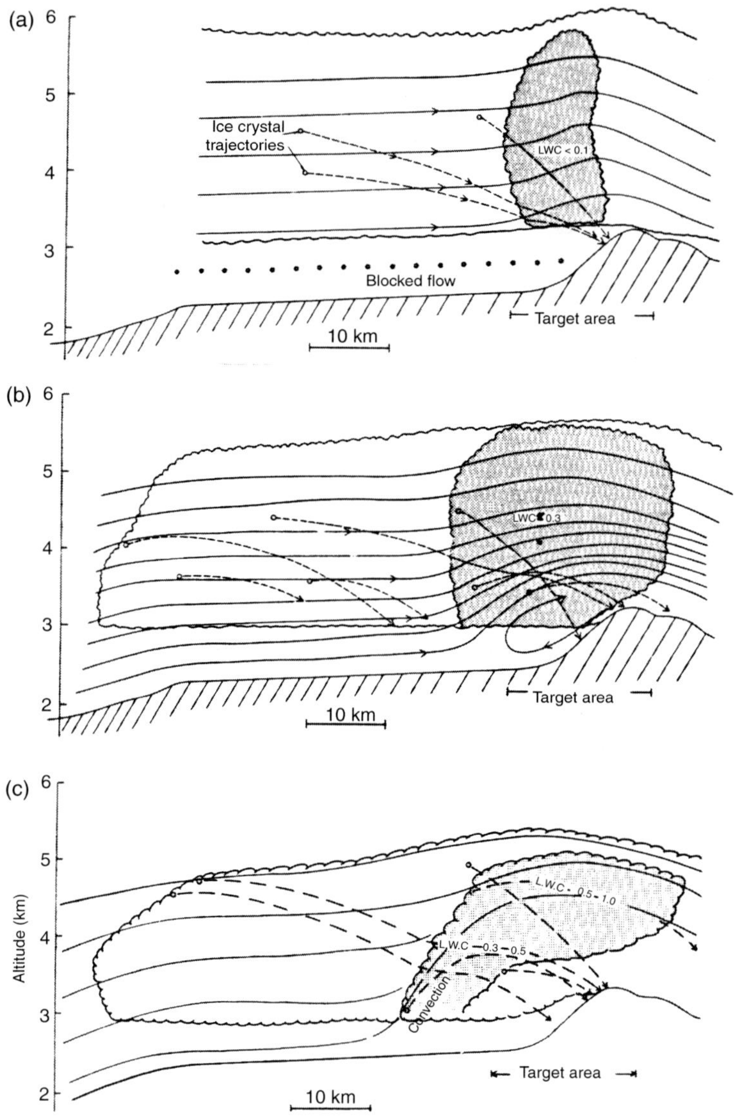


**FIGURE 11.31 Typical appearance of San Juan storms during the (a) stable, (b) neutral, and (c) unstable storm stages.** The airflow is from the left at 15 m $s^{-1}$. The shaded regions are regions where liquid-water is generally found, and some typical liquid-water contents (g $m^{-3}$) are indicated. Solid lines with arrows show air trajectories, and the irregular lines indicate cloud boundaries. Some typical ice crystal trajectories are also shown as dashed lines. The farthest downwind trajectory shows the fall of a hydrometeor with a terminal velocity of 1 m $s^{-1}$. *(From Cooper and Marwitz (1980))*

just ahead of and during frontal passage, clouds were characterized by wide areas of stratiform precipitation and clouds with frequently superimposed deeper clouds and heavier precipitation. Many of the heavier precipitating regions were convective, but some occurred in a stably stratified or neutral atmosphere. Liquid-water production in the more stratiform region of the prefrontal environment was strongly influenced by the orientation of the approaching front relative to the orientation of the mountain barrier. Over the Park Range, midlevel flow with a strong westerly component generated orographic lifting and was generally of higher moisture content. Convection, sometimes organized in bands was often observed in the prefrontal environment. Liquid-water was found in the convective clouds during their initial stages of development, but it was rapidly depleted by efficient precipitation processes. In general, liquid-water amounts reached a minimum within 15-20 min after the onset of convection.

The postfrontal clouds over the Park Range were also characterized by wide areas of stratiform precipitation. Orographic forcing was particularly strong because the westerly to northwesterly midlevel winds are normal to the barrier orientation. The highest liquid-water amounts were 10-15 km upwind of the barrier crest and reached a maximum over the windward slopes where the condensation rates were a maximum. Two regions of convective instability were observed in the postfrontal region; the first occurred just after frontal passage, and the second occurred during the passage of the 500-mbar trough as the system dissipated. The distribution of liquid-water in the postfrontal zone was complicated by the fact that superimposed on steady orographic production of liquid-water was transient convective activity. Again, the convection produced liquid-water early in its life cycle, but later produced substantial numbers of ice crystals which depleted liquid-water in the convective regions as well as in the lower-level feeder regions.

Figure 11.32 is a schematic of the observed distribution of supercooled liquid-water in (a) shallow orographic clouds, (b) a deep stratiform cloud system, and (c) a convective region embedded in a shallow stratiform cloud system. In all three systems, liquid-water was observed in the feeder zone directly upwind and over the barrier crest. A second region of liquid-water was inferred between cloud base and $-10$ °C. In this region, planar crystals falling into it did not grow rapidly by vapor-deposition growth. The more rapidly growing needles and columnar form of crystals were characteristically absent. The liquid-water depletion mechanisms (FD) were at a minimum, yet water production (FP) at these warmer temperatures was quite high. A third region near the cloud tops in Fig. 11.32 was also a region of frequent liquid-water. Rauber and Grant (1986) suggested that this zone of liquid-water results from an imbalance between liquid-water production by adiabatic cooling and depletion by vapor deposition. Large ice crystals, capable of extracting large quantities of vapor as they grow (and hence depleting liquid-water by the Bergeron-Feindeisen process), rapidly settled out of the cloud-top layer.

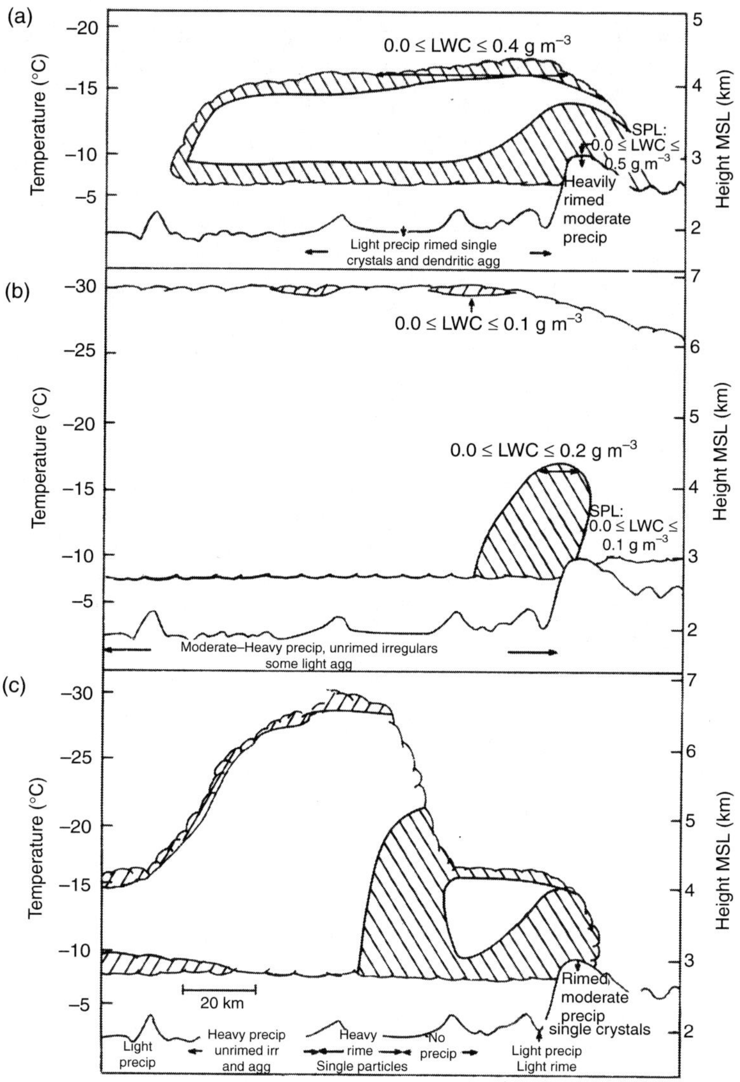


**FIGURE 11.32 Conceptual models of supercooled liquid-water distribution in Park Range cloud systems for (a) a shallow stratiform cloud system with a warm cloud-top temperature (CTT > −22 °C); (b) a deep stratiform system with a cold cloud-top temperature; and (c) a deep convective region embedded in a shallow stratiform system.** The characteristic precipitation type is listed for each case along the bottom of the figure. Magnitudes of liquid-water contents measured with aircraft or at SPL are shown on the figure. *(From Rauber and Grant (1986))*

Remaining were numerous small ice crystals incapable of extracting water substance at the rates being supplied by adiabatic ascent. Another possible contribution to the observed liquid-water near cloud top was that liquid-water production could be intensified by radiative cooling. Radiative cooling rates estimated by Chen and Cotton (1987) for stratocumulus clouds having relatively warm cloud-top temperatures can approach 200 °C $day^{-1}$ in a layer of 5-mbar thickness near cloud top. For wet adiabatic ascent, the cooling rate (CR) is,

$$CR = 665\ w \quad (°C\ day^{-1}), \tag{11.38}$$

where $w$ is the vertical motion in meters per second. Thus, an updraft of 0.3 m $s^{-1}$ is required to produce the same rate of cooling as cloud-top radiation. Cloud-top radiative cooling, therefore, can contribute to an imbalance between FP and FD. Preliminary numerical experiments by Mulvilhill over the Colorado Park Range suggest that cloud-top radiative cooling enhanced the development of a high-level blanket-type cloud with enhanced liquid-water relative to a case without radiative effects.

One may ask if a shallow layer of liquid-water near cloud top is of any consequence to precipitation formation, especially because it occurs at a level inaccessible to accretion growth and because the residence time of ice particles growing by vapor deposition is so short. Consider a cloud whose top is in the −12 to −16 °C temperature range. This temperature range is suitable for the growth of dendrites, provided that the cloud is supersaturated with respect to water. Rauber (1985) showed that, under suitable cloud-top temperature conditions, this cloud-top layer of liquid-water can favor the formation of dendrites. As the dendrites settle into lower levels, they rapidly deplete the available moisture and can readily grow by aggregation. This thin liquid-water layer can trigger more efficient precipitation processes, and if it extends well upwind of the main mountain barrier in a blanket-type cloud, ice crystals formed in the layer can serve as seeder crystals in the seeder-feeder process where the feeder cloud resides at low levels over the barrier crest.

### 11.8.3. Observations of Supercooled Liquid-Water over the Sierra Nevada

Aircraft observations of supercooled liquid-water in winter cloud systems over the Sierra Nevada (Lamb et al., 1976; Heggli et al., 1983) suggested that the largest amounts of supercooled liquid-water resided in convective towers. More recent observations, using passive microwave radiometers as well as observations of heavily rimed ice crystals from shallow orographic clouds, led Reynolds and Dennis (1986) to conclude that the shallow nonconvective clouds contained substantial amounts of supercooled liquid-water. Heggli (1986) noted that, about 75% of the time, saturated clouds possibly containing liquid-water existed below minimum aircraft flight altitudes and that the vertical distribution of liquid-water content was bimodal. A low-level maximum was observed at temperatures between −2 and −4 °C and a secondary maximum similar to that observed byRauber and Grant (1986) occurred between −10 and −12 °C. Heggli also noted that supercooled liquid-water occurred in multiple layers separated by 100 m or more of unsaturated air. The mechanisms responsible for such layering of the liquid-water zones have not been identified.

Heggli and Reynolds (1985) and Reynolds and Dennis (1986) interpreted the behavior of liquid-water variations during the passage of cyclonic storms over the Sierra Nevada in terms of Browning and Monk's (1982) split-front model (Fig. 11.25). Following the passage of the frontal rainband where the heaviest precipitation was observed, a shallow orographic cloud, often a stratocumulus cloud, prevails. The highest liquid-water contents are observed in this shallow cloud. Reynolds (1986) attributed this increase in supercooled liquid-water to the disappearance of ice crystals settling from higher clouds. Deep clouds prevailing during the passage of the elevated frontal band yield an environment favorable for the seeder-feeder process. High-level seeder crystals readily deplete liquid-water being generated by the lower-level, embedded feeder clouds. Following the passage of the frontal rainband, only shallow cloud layers prevail and the seeder-feeder process ends. Heggli and Rauber (1988) described the results of a comprehensive analysis of the occurrence of supercooled liquid-water in Sierra Nevada orographic clouds using a dual-channel microwave radiometer. The supercooled liquid-water content varied depending on the strength, evolution, and trajectory of motion of extratropical cyclonic storms. They examined the presence of supercooled liquid-water relative to two basic types of cyclonic storm, those having a predominant zonal flow component and those having a predominant meridional flow component. Storms with a zonal component of flow contained supercooled liquid-water in postfrontal shallow stratocumulus and cumulus clouds. Storms with a meridional component of flow exhibited supercooled liquid-water within prefrontal bands in the warm conveyor belt region. Little supercooled liquid-water was found in the narrow or broad cold-frontal bands, presumably because the seeder-feeder process operates efficiently in that region.

In summary, supercooled liquid-water in winter orographic clouds varies with the passage of cyclonic storm systems, the depth of the cloud system, the organization and type of cyclonic storm system, and proximity to mountain slopes. Liquid-water is also found near the tops of the cloud system and at temperatures between $-10$ and $-12$ °C, where liquid-water depletion processes are not efficient. There is also evidence of multiple layering of liquid-water in some orographic clouds.

## 11.9. EFFICIENCY OF OROGRAPHIC PRECIPITATION AND DIURNAL VARIABILITY

### 11.9.1. Precipitation Efficiency of Orographic Clouds

Knowledge of the precipitation efficiency of orographic clouds and the factors causing its variability is important to the development of simple physical models for diagnosing or predicting orographic precipitation and in evaluating the efficacy of complex models that treat explicitly cloud microphysical processes. The effectiveness of an orographic cloud in scavenging pollutants and contributing to acid deposition is also controlled largely by the precipitation efficiency of the cloud system. The opportunity for artificially enhancing precipitation from orographic clouds is determined by the natural precipitation efficiency of the cloud system. If the precipitation efficiency is low, then an opportunity exists to artificially enhance precipitation, perhaps by cloud-seeding techniques.

Browning et al. (1975) examined the efficiency of orographic clouds from low hills over the British Isles and determined the efficiency to be

$$E_1 E_2 = W/C, \tag{11.39}$$

where $E_1$ is the efficiency influenced by microphysical processes and $E_2$ is a factor that takes into account any subsaturation in the initial flow upwind of the hills. The depth $W$ of orographic water reaching the ground per unit time per unit width for a section across the mountain of length $x_0$ is given by

$$W = \int_{-x_0}^{x=0} R\mathrm{d}x, \tag{11.40}$$

where $R$ is the orographic component of rainfall. The parameter $C$ represents the total depth of water condensed over a layer of 3-km depth. Browning et al. estimated rainfall efficiencies ranging from 0.1 to 0.7 and interpreted their results in terms of the efficiency of the seeder-feeder process and whether or not the low-level airstream was saturated. Cases of high efficiency were characterized by efficient natural seeding and an initially saturated flow due to large-scale ascent.

Elliott and Hovind (1964) calculated precipitation efficiencies over the San Gabriel and Santa Ynez ranges in California. They estimated the condensation rate by applying the Myers (1962) hydraulic airflow model to those ranges. Precipitation efficiencies ranged from 17% to 25% for cases in which the environment was stably stratified and from 26% to 28% for unstable cases. The deeper vertical penetration of the convective elements compensated for water loss by entrainment, yielding greater condensate and precipitation production.

Chappell (1970) estimated condensate supply rates by assuming values of orographic vertical motion, and $\mathrm{d}r_s/\mathrm{d}t$ through the estimated mean cloud depth for Wolf Creek Pass and Climax, Colorado. Figure 11.29, showing the observed precipitation rates and inferred supply rates, suggests that precipitation efficiencies were near 100% at temperatures colder than −20 °C and were about 25% to 35% at the warmer temperatures.

Rather than use a model for estimated condensate production, Dirks (1973) employed aircraft data to measure directly precipitation efficiencies over a relatively isolated mountain in Wyoming. Aircraft soundings (Fig. 11.33) were used to calculate condensate production and precipitation rates. In all observed

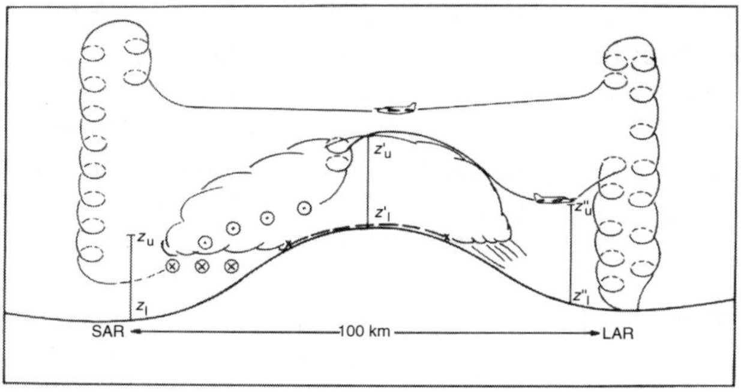


**FIGURE 11.33 Schematic of experimental design showing aircraft operations with respect to the orographic cloud.** Airflow is from left to right. The columns represent the cloud-producing layer upstream ($z_1 - z_u$), at maximum vertical lifting ($z_1' - z_u'$), and the downstream ($z_1'' - z_u''$) along a vertical stream surface. The path, $x - - - - x$, represents the region in which ground teams operate. The $x$ in a circle represents cloud base flights normal to the airflow and the dots in a circle represents cloud penetrations normal to the airflow. *(FromDirks (1973))*

cases, the soundings were slightly unstable. Precipitation efficiencies ranged from 25% to 80%. The lowest efficiencies were at very cold temperatures and strong winds, and the highest efficiencies were under moderately cold cloud top temperatures and moderate wind speeds.

Normally, precipitation measurements for precipitation efficiency calculations are based on surface-precipitation gauge measurements or aircraft-precipitation samples. Evidence suggests that a significant contributor to the water content of the snowpack arises from rime deposits on trees and other objects located on mountain peaks; this deposit is then shed to the snow surface. Hindman et al. (1982), for example, estimated that between 4% and 11% of the water content of snowpack near the peak of the Park Range, Colorado, was the result of rime deposits.

Smith et al. (2003) proposed an alternate strategy for estimating precipitation efficiencies. They propose using the drying ratio (DR),

$$\text{DR} = P/I = \text{area integrated total precipitation/impinging horizontal vapor flux.} \tag{11.41}$$

Kirshbaum and Smith (2009) proposed an alternate expression for DR in the form

$$\text{DR} = \text{CR} \times \text{PE}, \tag{11.42}$$

where CR is the condensation ratio ($C/I$) where $C$ is the area integrated condensation rate.

Because PE requires an estimate of vertical air velocity at cloud base, it is subject to rather large measurement errors. Since DR uses the ratio of precipitation to water vapour flux, it is easier to quantify than PE. An important advantage of DR is that it can be estimated using hydrogen and oxygen isotope analysis, which permits evaluation using stream flow or sapwood collected near a stream. The idea is that condensation preferentially removes the heavier isotopes so that the remaining vapor becomes progressively lighter such that

$$R/R_0 = \Theta^{\alpha-1} \tag{11.43}$$

where $R$ and $R_0$ are the ratios of heavy-to-light isotope concentrations in the final and initial state of vapor, $\Theta$ is the fraction of water vapor remaining, and $\alpha$ is the fractionation coefficient (see Smith and Evans, 2007). The drying ratio can thus be computed using

$$\mathrm{DR} = 1 - \Theta = 1 - (R_p/R_{p0})^{1(a-1)}, \tag{11.44}$$

where $R_P$ and $R_{P0}$ are the ratios of heavy to light isotope concentrations in the upwind and downwind precipitation. DR estimates using isotope analysis range from 43% for the combination of the coastal and Cascade mountain ranges in western Oregon and 43% for a case in the of flow against the Alps (Smith et al., 2003) and nearly 50% for the southern Andes (Smith and Evans, 2007).

Kirshbaum and Smith (2009) examined several factors that influence the drying ratio. They found that for flow over a two-dimensional ridge, DR decreases as surface temperature increases. They showed that this was due to the fact that (1) the normalized condensate rate decreases as the flow gets warmer owing to the nonlinear behavior of the Clausius-Clapeyron equation, (2) the ice-phase precipitation processes are more efficient than warm cloud collection processes, and (3) embedded convection which is likely at warmer temperatures mainly acts to vertically-redistribute moisture rather than to increase precipitation. As noted by Kirshbaum and Smith, mountain geometry is an important factor regulating DR. In two dimensions, wide mountains excite deep mountain-waves which produce deep wide areas of condensate while narrow mountains induce evanescent perturbations that generate less condensation. Moreover, narrower mountains are more susceptible to blow-over effects in which precipitation settles downwind into subsaturated regions where evaporative loss is high. They also showed that when a three dimensional ridge becomes shorter, larger amounts of windward moisture deflects around a mountain thus lowering DR.

### 11.9.2. Diurnal Variations in Wintertime Orographic Precipitation

Some limited observations suggest a significant diurnal variation in orographic precipitation. Grant et al. (1969) analyzed the diurnal variations in the percentage of snowfall amounts and the frequency of snowfall events for a number of mountain precipitation stations in Colorado. Figure 11.34 depicts the percentage of total snowfall and frequency for Climax, Colorado, a central Colorado mountain site. A pronounced early-morning maximum at 0300 local time is evident in the frequency and amount of precipitation measured. A minimum occurs shortly after sunrise, and a secondary maximum is evident in the afternoon and early evening. An early morning maximum was also found at several other stations in the central and northern Colorado mountains. The secondary maxima in the afternoon were generally less pronounced in the more northern Colorado stations. A different behavior in the diurnal variation of precipitation applies to the southern Colorado San Juan Mountains. Stations on the lee slopes of the San Juans exhibited an early morning maximum and a secondary maximum in the afternoon in frequency and amount. Stations upwind of the mountain crest and over the mountain peaks, however, did not show an early morning maximum in precipitation frequency. Instead, a pronounced afternoon maximum in precipitation frequency was evident (Fig. 11.35). Unfortunately, the relative frequency of daily amounts of snowfall was not measured at the higher-elevation stations. The differences in behavior between the northern and southern Colorado precipitation stations are related to the climatological differences in airflow between the two regions. Southern Colorado experiences a generally moister higher value $\theta_e$ airflow than does northern Colorado. As a result, afternoon heating is more likely to lead to convective destabilization over the windward side of the San Juans than in the more northern Colorado mountains.

The pronounced early-morning maximum in precipitation frequency and amounts, in the more northern Colorado stations are more difficult to explain; one can only speculate on possible mechanisms. It is likely that the early-morning maximum is the result of longwave radiation cooling at the earth's surface or at the top of a moist layer and/or orographic cloud layer. We noted earlier that orographic clouds often exhibit a large blanket stratiform cloud. One expects that, after sunset, when solar heating in the orographic cloud layer ceases, the longwave radiative cooling leads to a slow buildup of liquid-water in the upper levels of the cloud deck. Eventually, the liquid-water can build to levels where precipitation can commence or be enhanced. The actual presence of a blanket cloud deck may not be necessary for a radiative enhancement of liquid-water production. One auther (WRC) has frequently observed a clear afternoon sky at mountain locations slowly transform to a gradually thickening cloud deck after sunset which becomes deep enough to precipitate lightly in the early morning hours. The cloud deck then rapidly dissipates after sunrise. This suggests that radiative cooling at the top of a moist layer may trigger the onset of a stratiform cloud which, through cloud-top cooling, transforms to a precipitating cloud. The role of orographic lifting here is to lift the air to near water saturation so that only modest rates of radiative cooling will transform the layer into a saturated layer.

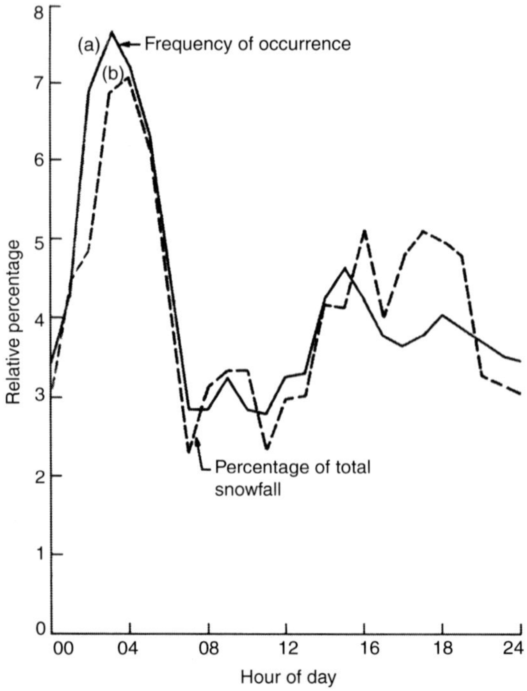


**FIGURE 11.34 Distribution of snowfall at Climax, Colorado, as a function of the hour of day, November through May for 1964 to 1967.** *(From Grant et al. (1969))*

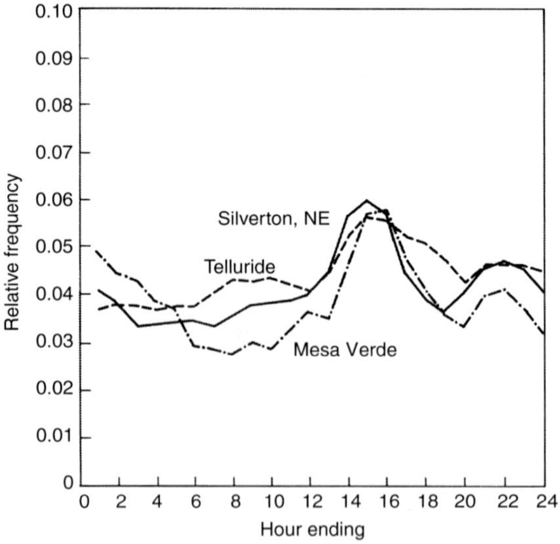


**FIGURE 11.35 Diurnal frequency of snowfall at Silverton, Telluride, and Mesa Verde, Colorado, November through April for 1948 to 1968.** *(From Grant et al. (1969))*

An even more complex dynamic linkage to radiative cooling may be considered. Tripoli (1986) found that enhanced cloud-top cooling at night can lead to local destabilization of the cloud top or moist layer and a reduction of the Brunt-Väisälä frequency (Chapter 9). As a result, gravity-wave energy dispersed from the lifting of air over the mountain barrier becomes trapped by the cloud-top unstable layer resulting from radiative cooling. This leads to higher amplitude gravity waves between the trapping cloud-top zone and the surface, and results in enhanced orographic lifting, thus favoring deeper, wetter clouds. Radiative effects, and perhaps a nocturnal maximum in precipitation, may be less in higher wind situations. In a simulation of a Sierra Nevada orographic cloud in a high-wind situation, M. Meyers (personal communication) found that radiative cooling had little influence on the simulated cloud microstructure.

Another likely important factor of nocturnal radiative cooling is that surface cooling will lead to enhancement of the positive pressure perturbation on the windward side of the barrier, thus strengthening orographic blocking. As suggested by Marwitz (1980), this will increase the effective orographic lifting of the mountain. In most cases, more localized lifting near the windward side of the barrier during the afternoon will be transformed by a larger cold pool on the windward side of the barrier into smoother, larger scale ascent commencing well upwind of the mountain. This favors the formation of a blanket-type cloud, which can respond to cloud-top radiative cooling. Moreover, initiation of cloud and precipitation farther upwind of a barrier favors operation of the seeder-feeder process near the barrier crest.

Observations of an early-morning maximum in orographic precipitation are not limited to the Colorado Mountains. Lee (1986) analyzed the climatology of precipitation events over the Sierra Nevada in the Sierra Cooperative Pilot Project (SCPP). A well-defined early morning maximum of heavy precipitation events was observed during December. No consistent pattern in heavy precipitation amounts was found in the other winter months, however. Light precipitation amounts exhibited a midday maximum, probably reflecting the greater tendency to develop convective instability in the warmer, moister air masses typical of the Sierra Mountains. Lee also determined the diurnal variations in precipitation echo types (Reynolds and Dennis, 1986). Surprisingly, major bands or well-organized radar-echo features, which exhibit elongated areas of reflectivity, occur about 35% of the time between 0300 to 0500 local time from January through March. The characteristic orographic cloud radar-echo type near mountain top occurs between 0430 and 1100 PST 25% of the time. Not surprisingly, the more convective echo types exhibit a midday maximum. Preference for an early morning maximum in the banded precipitation echo type is not well understood. It may be linked to changes in blocking of fronts or other factors noted earlier that favor the faster movement of fronts at night.

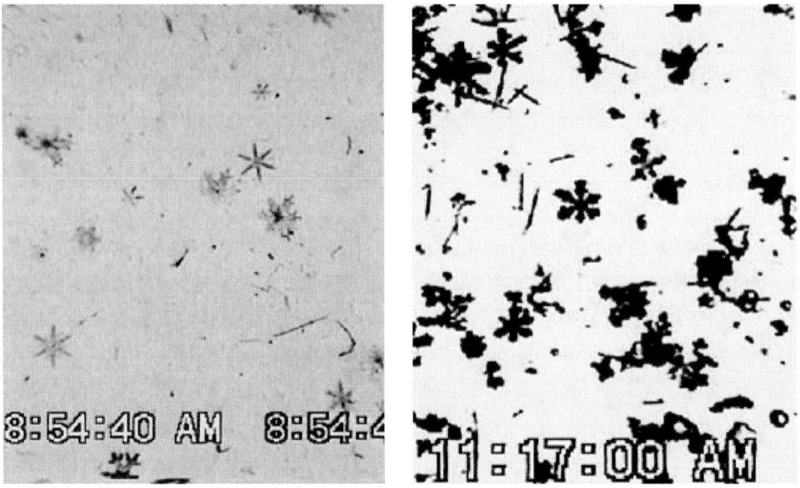


**FIGURE 11.36 Light riming of ice crystals in clouds affected by pollution (left) compared to heavier riming in non-polluted clouds (right) (Borys et al., 2003).**

## 11.10. AEROSOL INFLUENCES ON OROGRAPHIC PRECIPITATION

There is observational evidence (Borys et al., 2000, 2003) that pollution can delay precipitation in winter orographic clouds in the Rocky Mountains. They show that pollution increases the concentration of CCN and therefore cloud drops, leading to the formation of smaller cloud drops and less efficient riming. Reduced riming results in smaller, more pristine ice crystals (see Fig. 11.36), with smaller fall velocities, and less snowfall.

Saleeby et al. (2009) further examined the Borys hypothesis by performing three dimensional simulations of the influence of varying concentrations of CCN on simulated wintertime orographic clouds and precipitation over the Park Range of Colorado. Consistent with the observations by Borys et al. (2000, 2003) they found that higher CCN concentrations lead to the formation of smaller, more numerous droplets and reduced riming. They also found that with higher CCN concentrations the smaller droplets evaporate more readily when ice crystals grow at the expense of cloud droplets (the Bergeron-Findeisen process) depleting liquid-water contents and further decreasing riming growth. They found that reduced riming lowered snow water equivalent precipitation amounts on the windward side of the mountain barrier and increased it on the lee slopes. Overall total precipitation was reduced only a small amount but in the case of the Park Range a downstream shift moved water from the Pacific watershed to the Atlantic watershed. They also showed that this affect was only important for relatively wet storms where riming is important. Low supercooled liquid-water content storms are less influenced by aerosol pollution.

Muhlbauer and Lohmann (2008) performed simulations of pollution impacts on precipitation over Switzerland. Their simulations were for non-freezing clouds. They likewise found a reduction of precipitation on the windward slopes and increase on the lee slopes. They found that for unblocked flow the loss of upslope precipitation was not compensated by downslope precipitation gains. Near compensation for upslope precipitation loss on the lee slopes was found for blocked flow, however.

Further observational support for the hypothesis that aerosol pollution reduces orographic precipitation can be found in the studies by Givati and Rosenfeld (2004) who defined a precipitation enhancement factor, $R0$, which is the ratio between precipitation over mountains with respect to the upwind lowland precipitation amount. Examining about 100 years of precipitation records, they showed that downwind(in a climatological sense) of pollution sources in California and Israel, the windward slopes of mountains and mountain tops exhibited precipitation losses of 20% and 7%, respectively. Farther downwind on the lee side of mountains, the amount of precipitation was found to increase by 14%. However, they hypothesized that the integrated rainfall amount over the whole mountain range was reduced by the progressively increased pollution over the years. Subsequent studies show similar decreasing trends in R0 over a few western States in the US (Rosenfeld and Givati, 2006; Griffith et al., 2005) and for upslope events over the east slopes of the Colorado Rockies (Jirak and Cotton, 2006). Alpert et al. (2008) noted that one should not interpret $R0$ too quantitatively as it is very sensitive to the selection of precipitation measurement sites, and a decrease in the orographic ratio can also be achieved by an increase in the denominator, or if both the numerator and denominator increase by the same amount. Griffith et al. (2005), for example, found $R0$ decreased for the Salt Lake City area but as a result of an increase of upwind precipitation. This is not consistent with the basic hypothesis.

Lynn et al. (2007) presented results of modeling studies designed to interpret the study of Givati and Rosenfeld (2004) and Rosenfeld and Givati (2006). Using the Weather and Research Forecast model (WRF) coupled to their bin microphysics model they performed a series of two-dimensional simulations of a Sierra mountain (California) case study, in which they varied CCN input, horizontal wind velocity, and relative humidity, to examine the effect of these factors on the nature and distribution of condensate and precipitation relative to the mountain slope. In the control case, a five-fold increase in CCN (at 1% supersaturation) resulted in greater amounts of suspended ice condensate, and significant downwind displacement of this condensate, to the lee of the mountain. The associated decrease in surface precipitation was 27% (consistent with the analysis of precipitation trends in the Sierra Mountains by (Givati and Rosenfeld, 2004), although there is no observational evidence to support the assumed five-fold increase in CCN. Interestingly, the downwind shift in the surface precipitation associated with the highest topographical feature was small (10 km). Sensitivity studies showed that under high relative humidity the surface precipitation amounts were a factor of about 10 higher than for the control (drier) conditions, and the differences in precipitation between clean and polluted orographic clouds were negligible. A decrease in horizontal wind speed to 0.75 of that in the control case resulted in a very significant reduction in precipitation and a stronger difference between clean and polluted conditions. This study illustrates that dynamical/thermodynamical factors, such as horizontal wind speed and relative humidity,dominate over the impact of even a very large, five-fold increase in CCN.

In summary, there is considerable evidence suggesting that aerosol pollution can lead to orographic precipitation reductions, or a downwind shift of precipitation amounts. The impacts vary with storm systems, with the wetter storms being the most susceptible to aerosol pollution impacts. What has not been determined, however, is how significant the pollution impacts are on total water resources for water sheds potentially impacted by pollution.

## REFERENCES

Abbs, D. J., and Pielke, R. A. (1986). Numerical simulation of orographic effects on NE Colorado snowstorms. Meteorol. Atmos. Phys. 37, 1–10.

Alpert, P., Halfon, N., and Levin, Z. (2008). Does air pollution really suppress precipitation in Israel? J. Appl. Meteor. Clim. 47, 933–943.

Anthes, R. A., and Warner, T. T. (1978). Development of hydrodynamic models suitable for air pollution and other mesometeorological studies. Mon. Weather Rev. 196, 1045–1078.

Atlas, D., Tetehira, R. C., Srivastava, R. C., Marker, W., and Carbone, R. (1969). Precipitation induced mesoscale wind perturbations in the melting layer. Q. J. R. Meteorol. Soc. 95, 544–560.

Auer Jr., A. H., and White, J. M. (1982). The combined role of kinematics, thermodynamics and cloud physics associated with heavy snowfall episodes. J. Meteorol. Soc. Jpn 60, 500–507.

Bader, M. J., and Roach, W. T. (1977). Orographic rainfall in warm sectors of depressions. Q. J. R. Meteorol. Soc. 103, 269–280.

Baines, P. B. (1980). The dynamics of the southerly buster. Aust. Meteorol. Mag. 28, 175–200.

Baines, P. G., and Hoinka, (1985). Stratified flow over two-dimensional topography in fluid of infinite depth: A laboratory simulation. J. Atmos. Sci. 42, 1614–1630.

Barcilon, A., Jusem, J. C., and Drazin, P. G. (1979). On the two-dimensional hydrostatic flow of a stream of moist air over a mountain ridge. Geophys. Astrophys. Fluid Dyn. 13, 125–140.

Bergeron, T. (1949). The problem of artificial control of rainfall on the globe. II. The coastal orographic maxima of precipitation in autumn and winter. Tellus 1, 15–32.

Bergeron, T. (1960). Operation and results of "Project Pluvius". In "Physics of Precipitation." In Geophys. Monogr., vol. 5. pp. 152–157. Am. Geophys. Union, Washington, DC.

Bergeron, T. (1965). On the low-level redistribution of atmospheric water caused by orography. In "Suppl., Proc. Int. Conf. Cloud Phys. Tokyo." pp. 96–100.

Bergeron, T. (1967-1973). Mesometeorological Studies of Precipitation, vols. I–V. Meteorol. Inst., Uppsala.

Bjerknes, J. (1919). On the structure of moving cyclones. Geofys. Publ. 1, 1–8.

Bjerknes, J., and Solberg, H. (1921). Meteorological conditions for the formation of rain. Geofys. Publ. 2, 1–60.

Booker, J. R., and Bretherton, F. P. (1967). The critical layer for internal gravity waves in shear flow. Q. J. R. Meteorol. Soc. 27, 513–539.

Borys, R. D., DeMott, P. J., Hindman, E. E., and Feng, D. (1983). The significance of snow crystal and mountain-surface riming to the removal of atmospheric trace constituents from cold clouds. In "Precipitation Scavenging, Dry Deposition, and Resuspension, vol. 1, Precipitation Scavenging" (H. R. Pruppacher, R. G. Semonin, W. G. N. Stinn and coords, Eds.), pp. 181–189. Elsevier, New York.

Borys, R. D., Lowenthal, D. H., and Mitchell, D. L. (2000). The relationships among cloud microphysics, chemistry, and precipitation rate in cold mountain clouds. Atmos. Environ. 34, 2593–2602.

Borys, R. D., Lowenthal, D. H., Cohn, S. A., and Brown, W. O. J. (2003). Mountaintop and radar measurements in anthropogenic aerosol effects on snow growth and snowfall rate. Geophys. Res. Ltrs. 30 (10), 1538, doi:10.1029/2002GL016855.

Brinkmann, W. A. R. (1974). Strong downslope winds at Boulder, Colorado. Mon. Weather Rev. 102, 592–602.

Browning, K. A. (1979). Structure, mechanism and prediction of orographically enhanced rain in Britian. In "Global Atmos. Res. Programme Ser., vol. 23." pp. 88–114. World Meteorol. Organ.

Browning, K. A. (1985). Conceptual models of precipitation systems. Meteorol. Mag. 114, 293–318.

Browning, K. A., and Monk, G. A. (1982). A simple model for the synoptic analysis of cold fronts. Q. J. R. Meteorol. Soc. 108, 435–452.

Browning, K. A., Pardoe, C. W., and Hill, F. F. (1975). The nature of orographic rain at wintertime cold fronts. Q. J. R. Meteorol. Soc. 101, 333–352.

Carruthers, D. J., and Choularton, W. T. (1983). A model of the feeder-seeder mechanism of orographic rain including stratification and wind-drift effects. Q. J. R. Meteorol. Soc. 109, 575–588.

Chappell, C. F. (1970). Modification of cold orographic clouds. Ph.D. Thesis, Atmos. Sci. Pap. No. 173, Dep. Atmos. Sci., Colorado State University.

Charba, J. (1974). Application of gravity current model to analysis of squall line gust front. Mon. Weather Rev. 102, 140–156.

Chen, C., and Cotton, W. R. (1987). The physics of the marine stratocumulus-capped mixed layer. J. Atmos. Sci. 44, 2951–2977.

Choularton, T. W., and Perry, S. J. (1986). A model of the orographic enhancement of snowfall by the seeder-feeder mechanism. Q. J. R. Meteorol. Soc. 112, 335–345.

Chung, Y. S., Hage, K. D., and Reinelt, E. R. (1976). On lee cyclogenesis and airflow in the Canadian Rocky Mountains and the east Asian mountains. Mon. Weather Rev. 104, 879–891.

Clark, T. L., and Farley, R. D. (1984). Severe downslope windstorm calculations in two and three spatial dimensions using anelastic interactive grid nesting: A possible mechanism for gustiness. J. Atmos. Sci. 41, 329–350.

Clark, T. L., and Gall, R. (1982). Three-dimensional numerical model simulations of airflow over mountainous-terrain: A comparison with observations. Mon. Weather Rev. 110, 766–791.

Clark, T. L., and Peltier, W. R. (1977). On the evolution and stability of finite-amplitude mountain waves. J. Atmos. Sci. 34, 1715–1730.

Clark, T. L., and Peltier, W. R. (1984). Critical level reflection and the resonant growth of nonlinear mountain waves. J. Atmos. Sci. 41, 3122–3134.

Colquhoun, J. R., Shepherd, D. J., Coulman, C. E., Smith, R. K., and McInnes, K. (1985). The southerly buster of South Eastern Australia: An orographically forced cold front. Mon. Weather Rev. 113, 2090–2107.

Colson, DeVer (1954). Meteorological problems in forecasting mountain lee waves. Bull. Am. Meteorol. Soc. 35, 363–371.

Cooper, W. A., and Marwitz, J. D. (1980). Winter storms over the San Juan Mountains. Part III: Seeding potential. J. Appl. Meteorol. 19, 942–949.

Corby, G. A., and Sawyer, J. S. (1958). The air flow over a ridge—The effects of the upper boundary and high-level conditions. Q. J. R. Meteorol. Soc. 84, 25–37.

Cosma, S., Richard, E., and Miniscloux, F. (2002). The role of small-scale orographic features in the spatial distributions of precipitation. Q. J. R. Meteorol. Soc. 128, 75–92.

Cotton, W. R., George, R. L., Wetzel, P. J., and McAnelly, R. L. (1983). A long-lived mesoscale convective complex. Part I: The mountain-generated component. Mon. Weather Rev. 111, 1893–1918.

Cotton, W. R., Tripoli, G. J., Rauber, R. M., and Mulvihill, E. A. (1986). Numerical simulation of the effects of varying ice crystal nucleation rates and aggregation processes on orographic snowfall. J. Climate Appl. Meteorol. 25, 1658–1680.

Crapper, G. D. (1962). Waves in the lee of a mountain with elliptical contours. Philos. Trans. R. Soc. London, Ser. A 254, 601–624.

Danielsen, E. F., and Bleck, R (1970). Tropospheric and stratospheric ducting of stationary mountain lee waves. J. Atmos. Sci. 27, 758–772.

Deaven, D. G. (1976). A solution for boundary problems in isentropic coordinate models. J. Atmos. Sci. 33, 1702–1713.

Dirks, R. A. (1973). The precipitation efficiency of orographic clouds. J. Rech. Atmos. 7, 177–184.

Douglas, C. K. M., and Glasspoole, J. (1947). Meteorological conditions in heavy orographic rainfall. Q. J. R. Meteorol. Soc. 73, 11–38.

Durran, D. R. (1986). Another look at downslope windstorms. Part I: The development of analogs to supercritical flow in an infinitely deep, continuously stratified fluid. J. Atmos. Sci. 43, 2527–2543.

Durran, D. R., and Klemp, J. B. (1982a). On the effects of moisture on the Brunt-Väisälä frequency. J. Atmos. Sci. 39, 2152–2158.

Durran, D. R., and Klemp, J. B. (1982b). The effects of moisture on trapped mountain lee waves. J. Atmos. Sci. 39, 2490–2506.

Durran, D. R., and Klemp, J. B. (1983). A compressible model for the simulation of moist mountain waves. Mon. Weather Rev. 111, 2341–2361.

Eliassen, A. (1968). On meso-scale mountain waves on the rotating earth. Geophys. Norv. 27, 1–15.

Eliassen, A., and Palm, E. (1960). On the transfer of energy in stationary mountain waves. Geophys. Norv. 22, 1–23.

Eliassen, A., and Rekustad, J.-E. (1971). A numerical study of meso-scale mountain waves. Geophys. Norv. 28, 1–13.

Elliott, R. D., and Hovind, E. L. (1964). On convection bands within Pacific Coast storms and their relation to storm structure. J. Appl. Meteorol. 3, 143–154.

Elliott, R. D., and Shaffer, R. W. (1962). The development of quantitative relationships between orographic precipitation and airmass parameters for use in forecasting and cloud seeding evaluation. J. Appl. Meteorol. 1, 218–228.

Foldvik, A., and Wurtele, M. G. (1967). The computation of the transient gravity wave. Geophys. J. R. Astron. Soc. 13, 167–185.

Fraser, A. B., Easter, R. C., and Hobbs, P. V. (1973). A theoretical study of the flow of air and fallout of solid precipitation over mountainous terrain: Part I. Air flow model. J. Atmos. Sci. 30, 801–812.

Fuhrer, O., and Schar, C. (2005). Embedded cellular convection in moist flow past topography. J. Atmos. Sci. 62, 2810–2828.

Fukuta, N., Kowa, M., and Gong, N.-H. (1982). Determination of ice crystal growth parameters in a new supercooled cloud tunnel. In "Conf. Cloud Phys. Chicago, Ill." pp. 325–328. Am. Meteorol. Soc., Boston, Massachusetts. Preprints.

Fukuta, N., Gong, H.-H., and Wang, A.-S. (1984). A microphysical origin of graupel and hail. In "Proc., Int. Conf Cloud Phys., 9th., Tallinn, USSR." pp. 257–260.

Givati, A., and Rosenfeld, D. (2004). Quantifying precipitation suppression due to air pollution. J. Appl. Meteorol. 43, 1038–1056.

Gill, A. E. (1977). Coastally trapped waves in the atmosphere. Q. J. R. Meteorol. Soc. 103, 431–440.

Gjevik, B., and Marthinsen, T. (1977). Three-dimensional lee-wave pattern. Q. J. R. Meteorol. Soc. 104, 947–957.

Gocho, Y. (1978). Numerical experiment of orographic heavy rainfall due to a stratiform cloud. J. Meteorol. Soc. Jpn 56, 405–422.

Grant, L. O., Chappell, C. F., Crow, L. W., Mielke Jr., P. W., Rasmussen, J. L., Shobe, W. E., Stockwell, H., and Wykstra, R. A. (1969). An operational adaptation program of weather modification for the Colorado River Basin. Interim Rep. to Bur. Reclam., July 1968-June 1969, Contract No. 14-06-0-6467.

Griffith, D. A., Solak, M. E., and Yorty, D. P. (2005). Is air pollution impacting winter orographic precipitation in Utah? Weather modification association. J. Wea. Modif. 37, 17–20.

Grossman, R. L., and Durran, D. R. (1984). Interaction of low-level flow with the western Ghat Mountains and offshore convection in the summer monsoon. Mon. Weather Rev. 112, 652–672.

Heggli, M. F. (1986). A ground based approach used tq determine cloud seeding opportunity. In "Proc. Conf. Weather Modif. 10th, Arlington, Va." pp. 64–67. Am. Meteorol. Soc., Boston, Massachusetts.

Heggli, M. F., and Rauber, R. M. (1988). The characteristics and evolution of supercooled water in wintertime storms over the Sierra Nevada: A summary of radiometric measurements taken during the Sierra Cooperative Pilot Project. J. Appl. Meteorol. 27, 989–1015.

Heggli, M. F., and Reynolds, D. W. (1985). Radiometric observations of supercooled liquid water within a split front over the Sierra Nevada. J. Climate Appl. Meteorol. 24, 1258–1261.

Heggli, M. F., Vardiman, L., Stewart, R. E., and Huggins, A. (1983). Supercooled liquid water and ice crystal distributions within Sierra Nevada winter storms. J. Climate Appl. Meteorol. 22, 1875–1886.

Hill, G. E. (1978). Observations of precipitation-forced circulations in winter orographic storms. J. Atmos. Sci. 35, 1463–1472.

Hindman, E. E., Borys, R. D., and DeMott, P. J. (1982). Hydrometeorological significance of rime ice deposits on trees in the Colorado Rockies. In "Int. Symp. Hydrpmeteorol." pp. 95–99. Am. Water Resour. Assoc. Preprint.

Hobbs, P. V. (1973). Anomalously high ice particle concentrations in clouds. In "Invited Rev. Pap., Int. Conf. Nucleation, 8th, Leningrad".

Hobbs, P. V. (1975). The nature of winter clouds and precipitation in the Cascade Mountains and their modification by artificial seeding. J. Appl. Meteorol. 14, 783–858.

Hobbs, P. V., Easter, R. C., and Fraser, A. B. (1973). A theoretical study of the flow of air and fallout of solid precipitation over mountainous terrain: Part II. Microphysics. J. Atmos. Sci. 30, 813–823.

Hobbs, P. V., Houze Jr., R. A., and Matejka, T. J. (1975). The dynamical and microphysical structure of an occluded frontal system and its modification by orography. J. Atmos. Sci. 32, 1542–1562.

Hogg, D. C., Guiraud, F. O., Snider, J. B., Decker, M. T., and Westwater, E. R. (1983). A steerable dual-channel microwave radiometer for measurement of water vapor and liquid in the troposphere. J. Appl. Meteorol. 22, 789–806.

Holgate, H. T. D. (1973). Rainfall forecasting for river authorities. Meteorol. Mag. 102, 33–38.

Holmboe, J., and Klieforth, H. (1957). Investigation of mountain lee waves and the air flow over the Sierra Nevada. Final Rep. to Geophys. Res. Dir., Air Force Cambridge Res. Cent. Contract No. AF 19(604)-728, March.

Houze Mr., R. A., and Medina, S. (2005). Turbulence as mechanism for orographic precipitation enhancement. J. Atmos. Sci. 62, 3599–3623.

Hovermale, J. B. (1965). A non-linear treatment of the problem of airflow over mountains. Ph.D. Thesis, Dep. Meteorol. Pennsylvania State University.

Jirak, I. L., and Cotton, W. R. (2006). Effect of air pollution on precipitation along the front range of the Rocky mountains. J. Appl. Meteorol. 45, 236–245.

Kelvin, Lord (1886). On stationary waves in flowing water. Philos. Mag. 5, 353–357, 445–552, 517–530.

Kirshbaum, D. J., and Durran, D. R. (2004). Factors governing of three-dimensional convective storm dynamics. J. Atmos. Sci. 61, 682–698.

Kirshbaum, D. J., and Smith, R. B. (2009). Temperature and moist-stability effects on midlatitude orographic precipitation and air-mass transformation. Q. J. R. Meteorol. Soc.

Klemp, J. B., and Lilly, D. K. (1975). The dynamics of wave-induced downslope winds. J. Atmos. Sci. 32, 320–339.

Klemp, J. B., and Lilly, D. K. (1978). Numerical simulation of hydrostatic mountain waves. J. Atmos. Sci. 35, 78–107.

Klemp, J. B., and Wilhelmson, R. B. (1978a). The simulation of theee-dimensional convective storm dynamics. J. Atmos. Sci. 35, 1070–1096.

Klemp, J. B., and Wilhelmson, R. B. (1978b). Simulations of right- and left-moving storms produced through storm splitting. J. Atmos. Sci. 35, 1097–1110.

Krishnamurti, T. N. (1964). The finite amplitude mountain wave problem with entropy as a vertical coordinate. Mon. Weather Rev. 92, 147–160.

Kuettner, J. (1958). Moazagotl und Foehnwell. Contrib. Atmos. Phys. 25, 79–114.

Lalas, D. P., and Einaudi, F. (1974). On the correct use of the wet adiabatic lapse rate in stability criteria of a saturated atmosphere. J. Appl. Meteorol. 13, 318–324.

Lamb, D., Nielsen, K. W., Klieforth, H. E., and Hallett, J. (1976). Measurements of liquid water content in winter cloud systems over the Sierra Nevada. J. Appl. Meteorol. 15, 763–775.

Lee, T. F. (1986). Seasonal and interannual trends of Sierra Nevada Clouds and precipitation. J. Climate Appl. Meteorol. 26, 1270–1276.

Lilly, D. K. (1981). Doppler radar observations of upslope snowstorms. In "Proc. Conf Radar Meteorol., 20th." pp. 638–645. Am. Meteorol. Soc., Boston, Massachusetts.

Lilly, D. K., and Durran, D. R. (1983). Stably stratified moist airflow over mountainous terrain. In "Proc. Sino-Am. Workshop Mountain Meteorol., 1st, 1982, Beijing" (E. R. Reiter, Z. Baozhen and Q. Yongfu, Eds.), pp. 569–608. Science Press and Am. Meteorol. Soc, Beijing, Boston, Massachusetts.

Lilly, D. K., and Zipser, E. J. (1972). The Front Range windstorm of 11 January 1972. A meteorological narrative. Weatherwise 25, 56–63.

Lin, Y.-L., and Wang, T.-A. (1996). Flow regimes and transient dynamics of two-dimensional stratified flow over an isolated mountain ridge. J. Atmos. Sci. 53, 139–158.

Long, R. R. (1953a). Some aspects of the flow of stratified fluids. I. A Theoreticalinvestigation. Tellus 5, 42–58.

Long, R. R. (1953b). A laboratory model resembling the "Bishop-Wave"phenomenon. Bull. Am. Meteorol. Soc. 34, 205–211.

Long, R. R. (1954). Some aspects of the flow of stratified fluids. II. Experiments with a two-fluid system. Tellus 6, 97–115.

Long, R. R. (1955). Some aspects of the flow of stratified fluids. III. Continuous density gradients. Tellus 7, 341–357.

Long, R. R. (1972). Finite amplitude disturbances in the flow of inviscid rotating and stratified fluids over obstacles. Annu. Rev. Fluid Mech. 4, 69–92.

Lynn, B., Khain, A., Rosenfeld, D., and Woodley, W. L. (2007). Effects of aerosols on precipitation from orographic clouds. J. Geophys. Res. 112, DIO225, doi:10.1029/2006JD007537.

Lyra, G. (1943). Theorie der stationaren Leewellenstromung in freier Atmosphare. Z. Angew. Math. Mech. 23, 1–28.

Mahrer, Y., and Pielke, R. A. (1975). A numerical study of the air flow over mountains using the two-dimensional version of the University of Virginia mesoscale model. J. Atmos. Sci. 32, 2144–2155.

Marwitz, J. D. (1980). Winter storms over the San Juan Mountains. Part I: Dynamical processes. J. Appl. Meteorol. 19, 913–926.

Marwitz, J. D. (1983). The kinematics of orographic airflow during Sierra storms. J. Atmos. Sci. 40, 1218–1227.

Marwitz, J., Waight, K., Martner, B., and Gordon, G. (1985). Cloud physics studies in SCPP during 1984-85. Rep. to Div. Atmos. Resour. Res., Bur. Reclam., US Dep. Inter. Contract No. 2-07-81-V0256, September.

Mass, C. F., and Albright, M. D. (1986). Coastal southerlies and alongshore surges of the West Coast of North America: Evidence of mesoscale topographically trapped response to synoptic forcing. Mon. Weather Rev. 115, 1707–1738.

Miles, J. W., and Huppert, H. E. (1969). Lee waves in a stratified flow. Part 4. Perturbation approximations. J. Fluid Mech. 35, 497–525.

Muhlbauer, A., and Lohmann, U. (2008). Sensitivity studies of the role of aerosols in warm-phase orographic precipitation in different dynamic flow regimes. J. Atmos. Sci. 65, 2522–2542.

Myers, V. A. (1962). Airflow of the windward side of a large ridge. J. Geophys. Res. 67, 4267–4291.

Nickerson, E. C. (1979). On the numerical simulation of airflow and clouds over mountainous terrain. Contrib. Atmos. Phys. 52, 161–175.

Nickerson, E. C., Richard, E., Rosset, R., and Smith, D. R. (1986). The numerical simulation of clouds, rain, and airflow over the Vosges and Black Forest Mountains: A meso-$\beta$ model with parameterized microphysics. Mon. Weather Rev. 114, 398–414.

Palm, E. (1955). Multiple-layer mountain wave models with constant static stability and shear. Sci. Rep. No. 3, Contract No. AF 19(604)–728, Air Force Cambridge Res. Cent., Cambridge, Massachusetts.

Palm, E. (1958). Two-dimensional and three-dimensional mountain waves. Geophys. Norv. 20.

Paluch, I. R. (1979). The entrainment mechanism in Colorado cumuli. J. Atmos. Sci. 36, 2467–2478.

Parish, T. (1982). Barrier winds along the Sierra Nevada Mountains. J. Appl. Meteorol. 21, 925–930.

Parsons, D. B., and Hobbs, P. V. (1983). The mesoscale and microscale structure and organization of clouds and precipitation in midlatitude cyclones. IX: Some effects of orography on rainbands. J. Atmos. Sci. 40, 1930–1949.

Peltier, W. R., and Clark, T. L. (1979). The evolution and stability of finite-amplitude mountain waves. Part II: Surface wave drag and severe downslope windstorms. J. Atmos. Sci. 36, 1498–1529.

Peltier, W. R., and Clark, T. L. (1980). Reply. J. Atmos. Sci. 37, 2122–2125.

Peterson, T. C., Grant, L. O., Cotton, W. R., and Rogers, D. C. (1991). The effectof decoupled low-level flow on winter orographic clouds and precipitation in theYampa River Valley. J. Appl. Meteorol. 30, 368–386.

Petterssen, S. (1956). "Weather Analysis and Forecasting." 2nd Ed., McGraw-Hill, New York.

Pierrehumbert, R. T., and Wyman, B. (1985). Upstream effects of mesoscale mountains. J. Atmos. Sci. 42, 977–1003.

Queney, P. (1947). Theory of perturbations in stratified currents with applications to air flow over mountain barriers, Misc. Rep. No. 23. University of Chicago Press, Chicago, Illinois.

Queney, P. (1948). The problem of air flow over mountains: A summary of theoretical studies. Bull. Am. Meteorol. Soc. 29, 16–26.

Radinovic, D. (1965). Cyclonic activity in Yugoslavia and surrounding areas. Arch. Meteorol. Geophys. Bioklimatol. A14, 391–408.

Radinovic, D., and Lalic, D. (1959). Ciklonska aktivnost a Zapadnom Sredozemlliju. Rasprave Stud.-Mem. 7, 1–57.

Rayleigh, R. J. (1883). The form of standing waves on the surface of running water. Proc. London Math. Soc. 15, 69–78.
Rauber, R. M. (1981). Microphysical processes in two stably stratified orographic cloud systems. M.S. Thesis, Atmos. Sci. Pap. No. 337, Dep. Atmos. Sci., Colorado State University.
Rauber, R. M. (1985). Physical structure of northern Colorado river basin cloud systems. Ph.D. Thesis, Atmos. Sci. Pap. No. 390, Dep. Atmos. Sci., Colorado State University.
Rauber, R. M., and Grant, L. O. (1986). The characteristics and distribution of cloud water over the mountains of northern Colorado during wintertime storms. Part II: Spatial distribution and microphysical characteristics. J. Climate Appl. Meteorol. 25, 489–504.
Rauber, R. M., Feng, D., Grant, L. O., and Snider, J. B. (1986). The characteristics and distribution of cloud water over the mountains of northern Colorado during wintertime storms. Part I: Temporal variations. J. Climate Appl. Meteorol. 25, 468–480.
Reinking, R. F., and Boatman, J. F. (1986). Upslope precipitation events. In "Mesoscale Meteorology and Forecasting" (P. S. Ray, Ed.), pp. 437–471. Am. Meteorol. Soc., Boston, Massachusetts.
Reitan, C. H. (1974). Frequencies of cyclones and cyclogenesis for North America, 1951-1970. Mon. Weather Rev. 102, 861–868.
Reynolds, D. W. (1986). A randomized exploratory seeding experiment on widespread shallow orographic clouds: Forecasting suitable cloud conditions. In "Proc. Conf. Weather Modif., 10th, Arlington, Va." pp. 7–12. Am. Meteorol. Soc., Boston, Massachusetts.
Reynolds, D. W., and Dennis, A. S. (1986). A review of the Sierra Cooperative Pilot Project. Bull. Am. Meteorol. Soc. 67, 513–523.
Rhea, J. O. (1978). Orographic precipitation model for hydrometeorological use. Atmos. Sci. Paper No. 287, Dept. of Atmospheric Science, Colorado State University.
Richard, E., Chaumerliac, N., Mahfouf, J. F., and Nickerson, E. C. (1987). Numerical simulation of orographic enhancement of rain with a mesoscale model. J. Climate Appl. Meteorol. 26, 661–669.
Rosenfeld, D., and Givati, A. (2006). Evidence of orographic precipitation suppression by air pollution-induced aerosols in the western United States. J. Appl. Meteorol. Climate 455, 893–911.
Saleeby, S. M., Cotton, W. R., Lowenthal, D., Borys, R. D., and Wetzel, M. A. (2009). Influence of cloud condensation nuclei on orographic snowfall. J. Appl. Meteorol. Climate 48, 903–922.
Sassen, K. (1984). Deep orographic' cloud structure and composition derived from comprehen sive remote sensing measurements. J. Climate Appl. Meteorol. 23, 568–583.
Sawyer, J. S. (1960). Numerical calculation of the displacements of a stratified airstream crossing a ridge of small height. Q. J. R. Meteorol. Soc. 86, 326–345.

Sawyer, J. S. (1962). Gravity waves in the atmosphere as a three-dimensional problem. Q. J. R. Meteorol. Soc. 88, 412–425.

Schwerdtfeger, W. (1974). Mountain barrier effect on the flow of stable air north of the Brooks Range. In "Proc. Alaskan Sci. Conf, 24th, Geophys. Inst., Univ. Alaska, Fairbanks." pp. 204–208.

Schwerdtfeger, W. (1975). The effect of the Antarctic Peninsula on the temperature regime of the Weddell Sea. Mon. Weather Rev. 103, 45–51.

Schwerdtfeger, W. (1979). Meteorological aspects of the drift of ice from the Weddell Sea toward the mid-latitude westerlies. J. Geophys. Res. 84, 6321–6327.

Scorer, R. S. (1949). Theory of waves in the lee of mountains. Q. J. R. Meteorol. Soc. 75, 41–56.

Scorer, R. S. (1953). Theory of airflow over mountains: II-The flow over a ridge. Q. J. R. Meteorol. Soc. 79, 70–83.

Scorer, R. S. (1954). Theory of airflow over mountains: III-Airstream characteristics. Q. J. R. Meteorol. Soc. 80, 417–428.

Scorer, R. S., and Klieforth, H. (1959). Theory of mountain waves of large amplitude. Q. J. R. Meteorol. Soc. 85, 131–143.

Scorer, R. S., and Wilkinson, M. (1956). Waves in the lee on an isolated hill. Q. J. R. Meteorol. Soc. 82, 419–427.

Shapiro, M. A., Hampel, T., Rotzoll, D., and Mosher, F. (1985). The frontal hydraulic head: A micro-a-scale (−1 km) triggering mechanism for mesoconvective weather systems. Mon. Weather Rev. 113, 1166–1183.

Smith, R. B. (1977). The steepening of hydrostatic mountain waves. J. Atmos. Sci. 34, 1634–1654.

Smith, R. B. (1979). The influence of mountains on the atmosphere. Adv. Geophys. 21, 87–230.

Smith, R. B. (1982). A differential advection model of orographic rain. Mon. Weather Rev. 110, 306–309.

Smith, R. B. (1985). On severe downslope winds. J. Atmos. Sci. 42, 2597–2603.

Smith, R. B., and Lin, Y.-H. (1982). The addition of heat to a stratified airstream with application to the dynamics of orographic rain. Q. J. R. Meteorol. Soc. 108, 353–378.

Smith, R. B., and Evans, J. P. (2007). Orographic precipitation and water fractionalization over the southern Andes. J. Hydromet 8, 3–18.

Smith, R. B., and Evans, J. P. (2007). Orographic precipitation and water vapor fractionation over the Southern Andes. J. Hydromet. 8, 3–19.

Smith, R. B., Jiang, Q., Fearon, M., Tabary, P., Dorninger, M., Doyle, J., and Benoit, R. (2003). Orographic precipitation and airmass transformation: An alpine example. Q. J. R. Meteorol. Soc. 129B, 433–454.

Sommerfeld, A. (1912). Die greensche funktion der schwingungsgleichung. Jahresber. Dtsch. Math.- Ver. 21, 309–353.

Sommerfeld, A. (1948). Vorlesungen uber theoretische Physik, 2nd Rev. Ed., vol. VI. Akad. Verlagsges, Leipzig.

Starr, J. R., and Browning, K. A. (1972). Observations of lee waves by high-power radar. Q. J. R. Meteorol. Soc. 98, 73–85.
Storebo, P. B. (1976). Small scale topographical influences on precipitation. Tellus 28, 45–59.
Toth, J. J. (1987). Interaction of shallow cold surges with topography on scales of 100-1000 km. Ph.D. Thesis, Dep. Atmos. Sci., Colorado State University.
Tripoli, G. J. (1986). A numerical investigation of an orogenic mesoscale convective system. Ph.D. Thesis, Atmos. Sci. Pap. No. 401, Dep. Atmos. Sci., Colorado State University.
Tripoli, G. J., and Cotton, W. R. (1982). The Colorado State University three-dimensional cloud/ mesoscale model-1982. Part I: General theoretical framework and sensitivity experiments. J. Rech. Atmos. 16, 185–220.
Walsh, P. A. (1977). Cloud droplet measurements in wintertime clouds. M.S. Thesis, Dep. Atmos. Sci., University of Wyoming.
Weickmann, H. (1981). Mechanism of shallow winter-type stratiform cloud systems. NOAA Environ. Res. Lab. V.S. Gov. Print. Off. 1982-576-001/1220.
Whiteman, C. D. (1973). Some climatological characteristics of seed able upslope cloud systems in the high plains. NOAA Tech. Rep. ERL 268-APCL-27, V.S.D.C., Boulder, Colorado. US Gov. Print. Off., 1972-784214/1150 Region No. 8, Washington, DC.
Wiesmueller, J. L. (1982). The effect of diurnal heating on the movement of cold fronts through Eastern Colorado. NWS Tech. Memo., NWS-CR-66.
Wurtele, M. (1953). On lee wave in the interface separating two barotropic layers. Final Rep., Sierra Wave Proj., Contract No. AF 19(122)-263, Air Force Cambridge Res. Cent., Cambridge, Massachusetts.

Part III

# Clouds, Storms and Global Climate

Chapter 12

# Clouds, Storms, and Global Climate

## 12.1. INTRODUCTION

Throughout this book we have examined clouds and storms from both a detailed observational view and from the perspective of models that resolve cloud physics and cloud dynamics explicitly. We have also examined the impacts of clouds and storms on a locale level including rainfall and severe weather. In this chapter we step back and view clouds and storms from a global perspective and consider their impacts on the global radiation budget, on the energetics of the tropical atmosphere, their impacts on the global hydrological cycle, and how clouds and storms transport pollutants out of the boundary layer and into the upper troposphere. We also examine how aerosol pollution interacts with clouds and storms to potentially alter the climate and how clouds respond to a varying climate. We end by considering how clouds are represented in general circulation models.

## 12.2. CLOUDS AND THE GLOBAL RADIATION BUDGET

The moderate climate of planet earth is largely a consequence of the hydrological cycle and the associated presence of clouds. High clouds can be viewed as greenhouse warming agents in that they reduce outgoing longwave radiation flux. The longwave cloud radiative forcing (LWCRF) or longwave radiative flux at the top of the atmosphere (TOA) by high clouds largely depends on their cloud top temperature. Shortwave cloud radiative forcing (SWCRF) or outgoing shortwave radiative flux at TOA is usually negative, since clouds reflect solar radiation. The amount of reflected sunlight depends on the liquid-water paths and ice water paths of the clouds as well as particle sizes and liquid or ice phases. The sum of LWCRF and SWCRF at TOA is the net cloud radiative forcing (NCRF). High thin clouds tend to have a positive NCRF as their albedos are generally low such that SWCRF does not offset LWCRF. Thick tropical cirrus clouds, on the other hand, particularly the remnants of thick stratiform-anvil clouds of MCSs, exhibit a negative NCRF owing to the high albedo of the optically thicker clouds.

Low clouds such as boundary layer stratocumuli and trade-wind cumuli contribute to a NCRF that is negative since they exhibit high albedo and owing to their warm cloud tops, they emit longwave radiation similar to the surface of the earth. NCRF for tropical deep convective clouds is nearly zero as LWCRF and SWCRF nearly cancel each other (Bretherton and Hartmann, 2009). Likewise, middle-level stratus clouds tend to exhibit near zero NCRF, again owing to the canceling affects of LWCRF and SWCRF. The net globally-averaged NCRF is negative largely due to the high coverage and albedo of marine boundary layer stratocumuli and trade-wind cumuli.

Because clouds play such an important role in regulating the radiative budget of the planet, there is considerable interest and debate regarding how clouds might change with a warming planet. Will they reinforce, say, greenhouse gas warming, or provide a negative feedback? Ramanathan and Collins (1991) argue that increases in upper level optically thick anvil clouds, as sea surface temperatures (SST's) rise, will increase planetary albedo or produce a negative NCRF, and limit further rise in SST's, and thus serve as a "thermostatic" to the climate system. Lindzen (1990) and Lindzen et al. (2001) argue and provide evidence that in a warming climate, convective clouds will increase in coverage and intensity in the tropics which will result in enhanced compensating subsidence and thus warming and drying of the upper troposphere. The warming and drying in the upper troposphere will permit larger amounts of longwave radiation to escape to space, yielding a negative NCRF. This is referred to as an "infrared iris" effect. Neither cloud resolving modeling (Tompkins and Craig, 1999) nor satellite-based observations over higher SST regions (Hartmann and Michelsen, 2002; Rapp et al., 2005) support the iris hypothesis. It must be recognized that large scale circulations have a major control over cloud properties. Any major changes in circulations like the Hadley cell, Walker circulations, or shift in middle latitude storm tracks associated with changes in global climate, will have a major impact on cloud properties and, as a consequence, NCRF (e.g. Seager et al., 2007; Vecchi and Soden, 2007). Likewise changes in areal coverage and strength of subtropical high pressure regions (possibly in response to alterations in Hadley cells) will have a major impact on marine stratocumulus coverage and optical thickness and hence NCRF. Clearly the impact of clouds on the radiative properties of the climate system, puts a major burden on general circulation models (GCMs) to correctly represent the detailed macrostructure and microstructure of clouds, their interactions with their immediate environments, and the need to represent regional circulations that have a major control on cloud properties correctly.

## 12.3. HOT TOWERS AND TROPICAL CIRCULATIONS

Since the early studies of tropical convection by Riehl and Malkus (1958) over 50 years ago, it is generally believed that latent heating in deep convective clouds fuels the equatorial, upward branch of the tropical Hadley cell. They calculated that 1500 to 2500 cumulonimbus "hot towers", having nearly undiluted updrafts with speeds of 2 to 4 m $s^{-1}$. would supply enough latent heat release to drive the Hadley circulation. Riehl and Simpson (1979) repeated the energy budget calculations and again concluded that undiluted hot towers were essential for driving the Hadley cell. Fierro et al. (2009) performed three dimensional simulations and observational analysis of a tropical squall line. They performed parcel trajectory analysis using both Doppler radar observed and simulated storm velocities and interpreted the results relative to the hot tower hypothesis. The original Riehl and Malkus (Riehl and Malkus, 1958; Riehl and Simpson, 1979) hypothesis was based on the assumption that much of the transport of boundary layer air into the upper troposphere was achieved by undiluted protected cores within the stronger/deeper cumulonimbi. As we noted in Chapter 8, there is little evidence supporting the prevalence of undilute protected cores ascending from the boundary layer to the tropopause in the tropical marine atmosphere. Fierro et al. 's analysis supported Zipser's (2003) conclusion that the hot tower hypothesis should be modified to include the prevalence of heavily entrained cumulus towers, particularly below about 5 km altitude. Fierro et al. concluded that ice-phase related latent heating invigorated rising parcels giving rise to a secondary maximum in updrafts at 10-11 km levels. Thus the latent heating associated with the ice phase offset the effects of dilution by entrainment in the lower troposphere. They suggested that in the future the concept of hot towers should be redefined as "any deep convective tower rooted on the boundary layer and topping the upper troposphere".

The Riehl and Malkus (1958) studies recognized that much of the convection in the tropics is organized into clusters of clouds which we now call MCSs. We have seen in Chapters 8 and 9, as well as in studies by Tao et al. (2002) and Johnson et al. (2007), that the heating profiles differ appreciably between isolated upright ordinary convection or isolated cumulonimbi, and MCSs. Cotton et al. (1995) estimated that the global annual contribution of MCSs to total precipitation is roughly a factor of 5 greater than that from ordinary cumulonimbi (hot towers). Because latent heating scales with precipitation, this suggests that heating associated with MCSs dominates that from ordinary upright convection. Modeling and analysis studies by Mapes and Houze (1995), Donner et al. (2001) and Schumacher et al. (2004) all suggest that the elevated heating associated with the stratiform-anvil circulations of MCSs has a major impact on tropical circulations. Donner et al. (2001) concluded that heating profiles associated with the mesoscale circulations of MCSs produced stronger Hadley and Walker circulations, warmer upper tropospheric Tropics, and moister Tropics than that produced by ordinary upright convection(hot towers) in simulations with the GFDL GCM. As noted by Bellon and Sobel (2010) the character of the Hadley cell and the simulated intertropical convergence zone (ITCZ) depends strongly on how convection is parameterized. It is therefore important to represent the relative contributions of cumulus congestus, isolated cumulonimbi, versus MCSs, versus TCs, correctly in global simulations of the Earth's climate.

## 12.4. CLOUDS AND THE GLOBAL HYDROLOGICAL CYCLE

We have seen that clouds associated with extratropical cyclones, cumulonimbi, and MCSs are major contributors to rainfall and therefore to regional hydrology. We now ask how clouds and cloud systems may respond to a changing global climate, particularly a warming climate in response to warming induced by anthropogenically-produced greenhouse gases, and how those responses may alter the global hydrological cycle? This is not a simple question to answer as it requires greater skill in observing the earth's components to the hydrological cycle than is presently possible, and it places major demands on global climate models which can only represent the major cloud contributions to global hydrology through rather crude parameterizations.

Stephens and Ellis (2008) analyzed simulated model output data from coupled atmosphere/ocean models used in the IPCC Fourth Assessment Report to diagnose the simulated response of the global hydrological cycle to a 1% increase in $CO_2$ per year until its level doubled. As found in previous modeling studies (e.g. Trenberth et al., 2007), atmospheric water vapor increases in response to the applied heating at roughly $7\%\ K^{-1}$. This is largely a response to increases in SST's and following the Clausius-Clapeyron relation for saturated air (see Chapter 2) the water vapor content of the atmosphere must increase. Many studies have shown that changes in column-mean water vapor content follow the expected behavior of the Clausius-Clapeyron equation (Stephens, 1990; Wentz and Schabel, 2000; Trenberth et al., 2005). Of course there are large regional variations in column-integrated water vapor in response to global warming, largely due to changes in global circulations. For example, stronger winds over the southern oceans poleward of 40 degrees south in response to global warming, lead to regionally greatly enhanced water vapor fluxes.

One would expect that precipitation should increase at the same globally-averaged rate (e.g. Wentz et al., 2007). But Stephens and Ellis (2008) show that global precipitation increases at only roughly $1\%\ K^{-1}$. Figure 12.1 illustrates how the model-predicted changes in precipitation rate are much less than would be expected by the increase in column-integrated water vapor content. That is, the globally-averaged change in precipitation efficiency is defined as

$$\epsilon = \frac{W}{P}\frac{\Delta P}{\Delta W}, \tag{12.1}$$

where $W$ and $P$ are global mean values of column water vapor and precipitation, respectively, and $\delta P$ and $\delta W$ are the increased precipitation and column water vapor related to global warming. They show that $\epsilon$ is much less than one. This is because increases in global precipitation tend to track increases in radiative heat loss (see Figure 12.2). Thus as water vapor increases, the atmosphere cannot

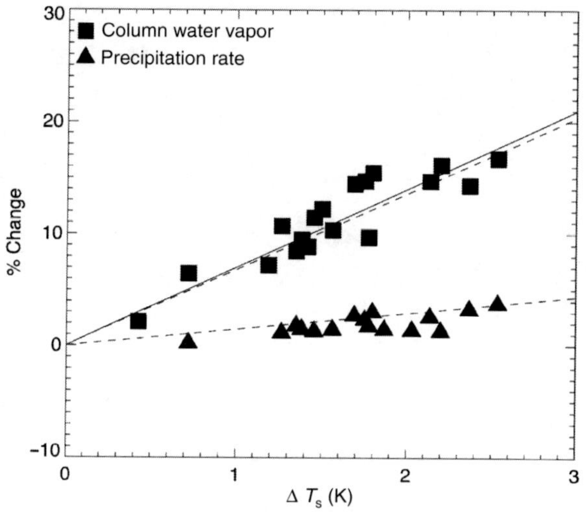


**FIGURE 12.1** **The relative changes in column water vapor amount and precipitation rate, expressed as percentage changes, as functions of global temperature change derived from the AR4 models.** The change in column water vapor derived assuming the CC relationship corresponds to an increase of 7.4% $K^{-1}$. The sensitivity of global precipitation rate changes to temperature changes is approximately 2.3% $K^{-1}$. The discrepancy between these two sensitivities indicates that the ratio of precipitation sensitivity to water vapor sensitivity in these models must be much less than unity. Note: Not all models had both column water vapor and precipitation data. *(From Stephens and Ellis (2008).)*

admit radiation at a large enough rate to permit precipitation increases at the same rate as water vapor increases. The change in the increase of clear sky emission associated with increases in column-integrated water vapor dominates the change in the energy balance of the atmosphere and $\epsilon$. Nonetheless, a negative cloud radiative feedback occurs through reductions in cloud amount in the middle troposphere. The reduction of middle troposphere cloud amount exposes the warmer atmosphere below to high clouds resulting in a net warming of the atmospheric column by clouds. This leads to a negative feedback on precipitation. Unfortunately, as will be discussed later, we are not confident that GCMs can represent changes in cloudiness in response to global warming, realistically. This could account for the fact that observationally-based studies (Gu et al., 2007; Allan and Soden, 2007; Zhang et al., 2007; Wentz et al., 2007) infer $\epsilon$ values closer to unity. Stephens and Ellis (2008) however point out the limitations of these observational studies. Thus we return to our introductory comments that such analysis requires greater skill in observing the earth's components to the hydrological cycle than is presently possible and it places major demands on global climate models which can only represent the major cloud contributions to global hydrology through rather crude parameterizations.

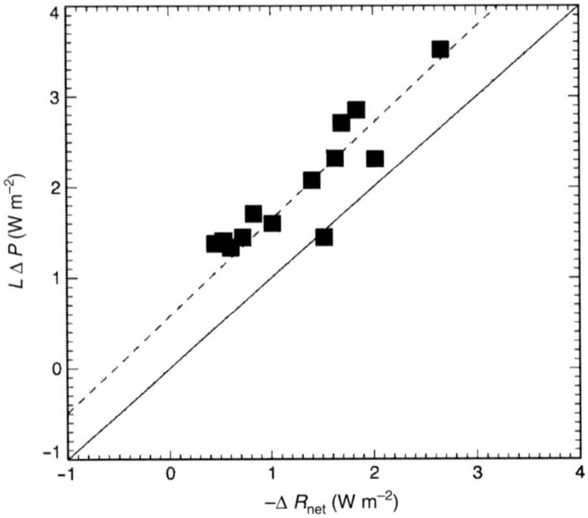


**FIGURE 12.2** **The relationship between changes in latent heating ($L\Delta P$) vs. changes in atmospheric column cooling ($\Delta R_{net}$) for the AR4 models.** The dotted line represents the linear relationship between the two quantities, and the offset between that line and the solid line representing a one-to-one correspondence reflects the contribution of sensible heating to the energy balance. *(From Stephens and Ellis (2008).)*

## 12.5. CLOUD VENTING

"Cloud venting" refers to the process of transporting gaseous matter and aerosols from the lower troposphere into the middle and upper troposphere (Ching, 1982). In this section we review global estimates of cloud venting. For the most part cloud venting studies have been restricted to a few observational cases and the use of parameterized or cloud resolving models to make estimates of venting by convective clouds. The only global estimates that we are aware of were made by Cotton et al. (1995). They used the archived results of cloud resolving model simulations using RAMS to make estimates of cloud venting rates for various cloud types. They found that the characteristic extratropical cyclone exhibited the highest boundary layer mass flux of all the cloud systems considered. In terms of annual boundary layer mass flux, extratropical cyclones still contribute the most. Owing to their great numbers, MCSs and ordinary thunderstorms rank second and third, respectively, to the total boundary layer air being vented. This is followed by TCs and MCCs which, while they vent large amounts of boundary layer air per storm event, are much fewer in numbers. The estimated total annual boundary layer mass flux of $4.95 \times 10^{19}$ kg by all these cloud systems represents a venting of the entire boundary layer about 90 times per year.

## 12.6. AEROSOL POLLUTION IMPACTS ON GLOBAL CLIMATE

As we have seen in earlier chapters, aerosol pollution can impact clouds in a number of ways, and if the impact on clouds occurs over a large enough area, aerosol pollution can impact global climate. Aerosols can impact climate directly by absorbing and reflecting solar radiation. They can also impact climate indirectly by increasing the albedo of clouds, the so-called Twomey effect, and also by altering the precipitation process which can potentially impact cloud lifetimes and optical thicknesses. The latter is sometimes referred to as the second indirect effect or Albrecht effect. While GCMs treat these processes as if they are two independent processes, they are, in fact, intimately coupled. We have seen that changes in precipitation rates associated with pollution aerosols can result in very nonlinear responses including invigoration of entrainment processes, changes in boundary layer stability, and cloud and storm propagation changes by alteration of low-level cold-pools. Virtually none of these nonlinear responses are represented in global climate models.

Rough estimates of the global mean magnitude of the Twomey (or cloud albedo) effect since pre-industrial times lies between $-0.5$ and $-1.9$ W $m^{-2}$ from different climate models, whereas the cloud lifetime effect is estimated to be between $-0.3$ and $-1.4$ W $m^{-2}$. Kristjánsson (2002) and Williams et al. (2001) concluded that the Twomey effect at the top-of-the atmosphere is four times as important as the cloud lifetime effect whereas Ghan et al. (2001) and Quaas et al. (2004) simulated a cloud lifetime effect that is larger than the Twomey effect.

A few GCMs have been used to examine the impacts of aerosol acting as ice nuclei (IN). Lohmann (2002) found that increases in contact IN result in more frequent glaciation of clouds and increase the amount of precipitation via the ice phase. This effect can offset, in part at least, the solar indirect aerosol effect (Twomey effect) on water clouds and oppose the suppression of drizzle by enhanced CCN concentrations.

Several GCMs have simulated changes in the general circulation, which then affects precipitation over large areas (Rotstayn et al., 2000; Williams et al., 2001; Rotstayn and Lohmann, 2002). These models were coupled to an ocean mixed-layer model so that enhanced cloud albedo produced lower ocean surface temperatures in the northern hemisphere. In addition, suppressed rainfall resulted in more extensive cloud cover that also contributed to cooler ocean surfaces. The models responded by shifting the Intertropical Convergence Zone (ITCZ) southward, which enhanced precipitation in tropical regions of the southern hemisphere (Rotstayn et al., 2000) and drying in the Sahel zone in Africa (Rotstayn and Lohmann, 2002). The latter response is consistent with the observed reduction in rainfall in the Sahel zone during the 20th century (Giannini et al., 2003). Williams et al. (2001) also found a similar response to both the direct and indirect effects of pollutant aerosols; in addition, they reported a reduction in precipitation associated with the Indian monsoon during June, July, and August. The cooling in their model also resulted in expanded sea-ice coverage in the Arctic Ocean in summer. This was in response to the southward displacement of storm tracks associated with the shift of the ITCZ southward. Thus, the greatest impacts of enhanced aerosol concentrations were over the north Polar regions and secondarily around 40 °N. However, one should not interpret the results of these simulations as being quantitative forecasts of the effects of aerosols on patterns and amounts of regional precipitation. As noted previously, there are many uncertaintiesin the distribution and concentrations of aerosols in the past, and even in the present. In addition, there are many simplifications in the models that limit their ability to realistically simulate the indirect effects of aerosols. However, these model simulations demonstrate the potential effects of direct and indirect aerosol forcing on clouds and precipitation in regions well beyond those directly influenced by changes in radiation produced by the aerosol.

We have seen that greenhouse warming, as a result of enhanced $CO_2$ concentrations, is only significant when greater amounts of water vapor are evaporated into the air principally over the oceans but also over land. Recent GCM simulations of greenhouse warming and direct and indirect aerosol effects (Liepert et al., 2004) suggest that the indirect and direct cooling effects of aerosols reduce surface latent and sensible heat transfer and, as a consequence, act to lower water vapor amounts in the troposphere, and thereby substantially weaken the impacts of greenhouse gas warming. This is important since most investigators compare top of the atmosphere radiative differences for greenhouse gas warming and aerosol direct and indirect effects separately. Since greenhouse warming depends on enhancement of the water vapor content of the atmosphere, and aerosol direct and indirect cooling reduces it, the potential influence of aerosols on climate could be far more significant than previously thought.

## 12.7. REPRESENTING CLOUDS IN GENERAL CIRCULATION MODELS

### 12.7.1. Introduction

Representing clouds in general circulation models (GCMs) is a major challenge as they typically have a limited number of vertical levels (typically 12-15) which provides insufficient resolution to represent cirrus, middle tropospheric stratus, and boundary layer clouds well. Moreover, their horizontal resolution is generally about 100 km which means all forms of convective clouds must be parameterized in some way. In the first edition to this book we devoted an entire chapter to discussing large scale cloud diagnostics and cumulus parameterization. Since that time there has been a growing awareness that cumulus parameterization schemes do not represent the behavior of many tropical weather systems properly, such as equatorially trapped waves which in GCMs propagate more like dry waves rather than slower convectively-coupled waves. Moreover, GCMs do a rather poor job of representing intraseasonal variability associated with monsoons which have strong deep convective forcing and convectively-coupled wave systems like the Madden-Julian oscillation.

### 12.7.2. Representing Boundary Layer Clouds

GCMs have been shown to be highly sensitive to the representation or parameterization of boundary layer clouds, such as marine stratocumulus clouds and trade-wind cumuli (Bony and Dufresne, 2005). Large scale controls on boundary layer clouds include subsidence, inversion strength, and boundary layer humidity. These are features that a GCM can crudely represent although inversion strength is a feature not represented well owing to their limited vertical resolution. As described in Chapter 6 the approaches to modeling boundary layer clouds includes one-dimensional layer-averaged or mixed-layer models, entity-type or plume models, higher ordered closure models, and large-eddy simulation models. LES models require grid spacings of 50 m or finer and thus are too costly to represent boundary layer clouds in GCMs.

Most GCMs use mixed-layer models to represent stratocumuli and some combination of plume or mass flux models and mixed layer models to simulate both stratocumuli and trade-wind cumuli. These models are limited in their application to drizzling boundary layer clouds and their impacts on boundary layer stability and cloud organization and coverage. They are also limited in their ability to represent aerosol influences on the cloudy boundary layer except for the *pure* Twomey hypothesis. As noted previously, the distinction between the first and second indirect aerosol effect is quite artificial and once cloud droplet concentrations are modified by varying aerosol amounts, feedbacks through entrainment and drizzle impacts on boundary layer stability, and cloud organization (coverage) have major impacts on cloud properties. These nonlinear dynamic consequences of aerosol variability are not represented well, if at all, in GCMs.

Only a few GCMs use higher-order closure models to represent boundary layer clouds including stratocumuli and trade-wind cumuli. These models require higher vertical resolution and higher temporal resolution than mixed layer models and thus are computationally quite expensive. They do offer the potential for making a smooth transition from solid stratus to cumulus regimes without adjusting parameters, although they are unable to provide predictions of changes in cloud organization such as open-cell versus closed cell organization. Using a PDF approach, subgrid quantities, such as vertical velocity and LWP, are determined from prescribed basis functions in which various moments of the basis functions are predicted in the models (Pincus and Klein, 2000; Golaz et al., 2002a,b; Larson et al., 2005). The prediction of PDFs of vertical velocity also provides information for use in the activation of CCN to form cloud droplets.

In spite of their importance to global climate, GCMs still do not represent boundary layer clouds well, especially in a drizzling boundary layer.

### 12.7.3. Representing Middle and High Clouds

GCMs represent layer clouds in the middle and upper troposphere through explicitly-resolved vertical motions and large-scale moistening. Unfortunately these cloud systems are often only a few hundred meters in depth, while the limited vertical resolution of GCMs smear these features over depths of several kilometers. Moreover, these clouds often exhibit embedded convection or turbulence which must be represented by some form of turbulence closure model either mixed layer models or higher-order closure models. Unfortunately these models have not benefited from the large observational data base that is available for boundary layer clouds. Much of the effort in representing middle and high clouds in GCMs has focused on representing the cloud microphysics of mixed-phase and ice-phase clouds. This is often done without representing the cloud-scale updrafts in those clouds which is important in determining the nucleation of cloud particles, the sedimentation rates of hydrometeors, and the coverage of clouds. Most of the GCMs represent warm cloud microphysics following a Kessler-type of approach to bulk microphysics parameterization such as overviewed in Chapter 4. For ice-phase clouds, a 1970s era cloud microphysics parameterization is often followed in which, at a certain temperature, the cloud is immediately converted into an ice saturated cloud where non-precipitating "cloud ice" is represented. Thus the precipitation shafts or fall streaks that are characteristic of higher-level clouds is not represented. Moreover, the affects of varying aerosol in the middle and upper troposphere and their impacts of cloud radiative and hydrological properties are not represented in current generation GCMs.

### 12.7.4. Representing deep convective clouds

While deep convective clouds may not play a major role in controlling net cloud radiative forcing, they still are major contributors in driving tropical circulations including the Walker and Hadley circulations. In the first edition of this book we devoted an entire chapter to discussing convective parameterization schemes, including their detailed formulations and the fundamental theories and the concepts they are based on. While there certainly have been major advances in the formulation of convective parameterization schemes in the intervening 20 years, as noted by Grabowski and Petch (2009), it has become recognized that GCMs using convective parameterization schemes have major shortcomings including:

- They typically do not represent the transition from ordinary upright convection to MCSs properly, if at all.
- They do a poor job of simulating intraseasonal variability including monsoonal circulations and the Madden-Julian oscillation.
- They misrepresent the phase of the diurnal cycle of warm season precipitation over the continents.

- They misrepresent the frequency and intensity of convective precipitation.
- Changes in cloud-scale processes are immediately felt on the global scale whereas in the real world responses occur on mesoscales and regional scales with only the residual imbalances on these scales felt by the global scales.

The latter point may be more a result of GCM resolution than convective parameterization schemes.

What are the options that can overcome these deficiencies of convective parameterization schemes? One approach is to use a cloud resolving model (CRM) embedded in GCM grid points, or what has been called "super parameterization schemes" (Grabowski and Smolarkiewicz, 1999; Grabowski, 2001; Randall et al., 2003; Wyant et al., 2006), or the multiscale modeling framework (MMF) by Khairoutdinov et al. (2007). We should note that the use of the term CRM is an abuse of the concept as these embedded models typically have 4-5 km grid spacings and are two-dimensional. As noted in earlier chapters, a true cloud-resolving model should have resolution of about 100 m and be three-dimensional. We will use the term cloud-representing-models as CRM. According to Grabowski and Petch (2009) this approach is about 2 to 3 orders of magnitude more computationally demanding than use of convective parameterization schemes. It is still not practical for use in longer term climate simulations. One problem with this approach is that there is not a natural continuum of cloud responses from upright convection to the mesoscale to global scales so that the coupling between the convective scales and global scales is artificial (see Khairoutdinov et al., 2007). MCSs that form in a given grid cell in the GCM cannot propagate into neighboring cells and, moreover, the deck is reshuffled at the end of each time step such that at the next time step the represented convection must undergo evolution from scratch from upright convection to MCSs. In other words, there is no grid point memory of previous convective organization.

A second approach is to use a nonhydrostatic GCM with high enough resolution to represent deep convection explicitly. We will refer to this as a GCRM. These are normally referred to as cloud-resolving GCMs but, since they use grid spacings of roughly 7 km and occasionally 3 km (see Collins and Satoh, 2009), we will use the term cloud-representing GCMs in the spirit of the discussion in the previous section. The Earth Simulator (called NICAM) developed at Japan's Frontier Research Center (Miura et al., 2007) is an example of a GCRM. NICAM has been able to reproduce a Madden-Julian Oscillation event (Miura et al., 2007) and is purported to produce global climatologies close to those observed (Iga et al., 2007). It is the experience of the lead author (Cotton) with running RAMS in realtime mesoscale forecasts over Colorado with 3 km grid spacing for about 10 years, that during the convective season, convection is delayed until CAPE is large enough to support resolved convection and the subsequent simulated storms are more vigorous than observed (too little entrainment) and produce too much precipitation. If CAPE does not build up, then no convection forms and an under-prediction of precipitation results.

NCAM exhibits a similar over-prediction bias in precipitation in the tropics. Furthermore, according to Grabowski and Petch (2009) the use of a GCRM in NICAM is roughly 6 orders of magnitude more computationally demanding than a conventional GCM.

Clearly we still have a long way to go in representing clouds properly in climate models used for decadal and centuries-long simulations.

## REFERENCES

Allan, R. P., and Soden, B. J. (2007). Large discrepancy between observed and simulated precipitation trends in the ascending and descending branches of the tropical circulation. Geophys. Res. Lett. 34, L18705, doi:10.1029/2007GL031460.

Bellon, G., and Sobel, A. H. (2010). Multiple equilibria of the Hadley circulation in an intermediate-complexity axisymmetric model. J. Climate 23, 1760–1778.

Bony, S., and Dufresne, J.-L. (2005). Marine boundary layer clouds at the heart of tropical cloud feedback uncertainties in climate models. Geophys. Res. Lett. 32, L20806, doi:10.1029/2005GL023851.

Bretherton, C. S., and Hartmann, D. L. (2009). Large-scale controls on cloudiness. In "Clouds in the Perturbed Climate System: Their Relationship to Energy Blance, Atmospheric Dynamics, and Precipitation" (J. Heintzenberg and R. J. Charlson, Eds.), pp. 217–234. The MIT Press, Cambridge, MA.

Ching, J. (1982). The role of convective clouds in venting ozone from the mixed layer. In "3rd Joint Conf. Applications of Air Pollution Meteorology, 12-15 Jan 1982", San Antonio, TX, Amer. Meteor. Soc. Preprints.

Collins, W. D., and Satoh, M. (2009). Simulating global clouds: Past, present and future. In "Clouds in the Perturbed Climate System: Their Relationship to Energy Blance, Atmospheric Dynamics, and Precipitation" (J. Heintzenberg and R. J. Charlson, Eds.), pp. 469–486. The MIT Press, Cambridge, MA.

Cotton, W. R., Alexander, G. D., Hertenstein, R., Walko, R. L., McAnelly, R. L., and Nicholls, M. (1995). Cloud Venting–Areview and some new global annual estimates. Earth Sci. Rev. 39, 169–206.

Donner, L. J., Seman, C. J., Hemler, R. S., and Fan, S. (2001). A cumulus parameterization including mass fluxes, convective vertical velocities, and mesoscale effects: Thermodynamic and hydrological aspects in a general circulation model. J. Climate 14, 3444–3463.

Fierro, Alexandre O., Simpson, Joanne, LeMone, Margaret A., Straka, Jerry M., and Smull, Bradley F. (2009). On how hot towers fuel the Hadley cell: An observational and modeling study of line-organized convection in the equatorial trough from TOGA COARE. J. Atmos. Sci. 66, 2730–2746.

Ghan, S. J., Easter, R. C., Chapman, E. G., Abdul-Razzak, H., Zhang, Y., Leung, L. R., Laulainen, N., Saylor, R. D., and Zaveri, R. A. (2001). A physically based estimate of forcing by anthropogenic sulfate aerosol. J. Geophys. Res. 106, 5279–5293.

Giannini, A., Saravanan, R., and Chang, P. (2003). Oceanic forcing of Sahel rainfall on interannual to interdecadal time scales. Science 302, 1027–1030, doi:10.1126/science.1089357.

Golaz, J.-C., Larson, V. E., and Cotton, W. R. (2002a). A PDF-based model for boundary layer. Part I: Method and model description. J. Atmos. Sci. 59, 3540–3551.

Golaz, J.-C., Larson, V. E., and Cotton, W. R. (2002b). A PDF-based model for boundary layer clouds. Part II: Model results. J. Atmos. Sci. 59, 3552–3571.

Grabowski, W. W., and Smolarkiewicz, P. K. (1999). CRCP: A cloud resolving convection parameterization for modeling the tropical convecting atmosphere. Physica D 133, 171–178.

Grabowski, W. W., and Petch, J. C. (2009). Deep convective clouds. In "Clouds in the Perturbed Climate System: Their Relationship to Energy Blance, Atmospheric Dynamics, and Precipitation" (J. Heintzenberg and R. J. Charlson, Eds.), pp. 197–215. The MIT Press, Cambridge, MA.

Grabowski, W. (2001). Coupling cloud processes with the large-scale dynamics using the cloud-resolving convective parameterization (CRCP). J. Atmos. Sci. 58, 978–997.

Gu, G., Adler, R. F., Huffman, G. J., and Curtis, S. (2007). Tropical rainfall variability on interannual-to-interdecadal and longer time scales derived from the GPCP monthly product. J. Climate 20, 4033–4046.

Hartmann, D. L., and Michelsen, M. L. (2002). No evidence for iris. Bull. Amer. Meteor. Soc. 83, 249–254.

Iga, S., Tomita, H., Tsushima, Y., and Satoh, M. (2007). Climatology of a nonhydrostatic global model with explicit cloud processes. Geophys. Res. Lett. 34, L22814.

Johnson, D. E., Tao, W.-K., Simpson, J., and Sui, C.-H. (2007). A study of the response of deep tropical clouds to large-scale thermodynamic forecings. Part I: Modeling strategies and simulations of TOGA COARE convective systems. J. Atmos. Sci. 59, 3492–3518.

Khairoutdinov, M., DeMott, C., and Randall, D. (2007). Evaluation of the simulated interannual and subseasonal variability in an AMIP-style simulation using the CSU multi-sclae modeling framework. J. Climate 21, 413.

Kristjánsson, J. E. (2002). Studies of the aerosol indirect effect from sulfate and black carbon aerosols. J. Geophys. Res. 107, doi:10.1029/2001JD000887.

Larson, V. E., Golaz, J.-C., Jiang, H., and Cotton, W. R. (2005). Supplying local microphysics parameterizations with information about subgrid variability: Latin hypercube sampling. J. Atmos. Sci. 62, 4010–4026.

Liepert, B. G., Feichter, J., Lohmann, U., and Roeckner, E. (2004). Can aerosols spin down the watercycle in a warmer and moister world? Geophys. Res. Lett. 31, doi:10.1029/2003GL019060.

Lindzen, R. S. (1990). Some coolness concerning global warming. Bull. Amer. Meteor. Soc. 71, 288–299.

Lindzen, R. S., Chou, M.-D., and Hou, A. Y. (2001). Does the earth have an adaptive infrared iris? Bull. Amer. Meteor. Soc. 82, 417–432.

Lohmann, U. (2002). A glaciation indirect aerosol effect caused by soot aerosols. Geophys. Res. Lett. 29, doi:10.1029/2001GL014357.

Mapes, B. E., and Houze Jr., R. A. (1995). Diabatic divergence profiles in western Pacific mesoscale convective systems. J. Atmos. Sci. 52, 1807–1828.

Miura, H., Satoh, M., Nasuno, T., Noda, A. T., and Oouchi, K. (2007). A Madden-Julian Oscillation event simulation by a global cloud-resolving model. Science 318, 1763–1765.

Pincus, R., and Klein, S. A. (2000). Unresolved spatial variability and microphysical process rates in large-scale models. J. Geophys. Res. 105, 27059–27065.

Quaas, J., Boucher, O., and Bréon, F.-M. (2004). Aerosol indirect effects in POLDER satellite data and the Laboratoire de Météorologie Dynamique-Zoom (LMDZ) general circulation model. J. Geophys. Res. 10, doi:10.1029/2003JD004317.

Ramanathan, V., and Collins, W. (1991). Thermodynamic regulation of ocean warming by cirrus clouds deduced from observationsl of the 1987 El Niño. Nature 351, 27–32, doi:10.1038/351027a0.

Randall, D., Khairoutdinov, M., Arakawa, A., and Grabowski, W. (2003). Breaking the cloud parameterization deadlock. Bull. Amer. Meteor. Soc. 84, 1547–1564, doi:10.1175/BAMS-84-11-1547.

Rapp, A. D., Kummerow, C., Berg, W., and Griffith, B. (2005). An evaluation of the proposed mechanism of the adaptive infrared Iris hypothesis using TRIMM VIRS and PR measurements. J. Climate 18, 4185–4194.

Riehl, H., and Malkus, J. S. (1958). On the heat balance in the equatorial trough zone. Geophysica 6, 503–538.

Riehl, H., and Simpson, J. S. (1979). On the heat balance in the equatorial trough zone, revisited. Contrib. Atmos. Phys. 52, 287–305.

Rotstayn, L. D., and Lohmann, U. (2002). Tropical rainfall trends and the indirect aerosol effect. J. Clim. 15, 2103–2116.

Rotstayn, L. D., Ryan, B. F., and Penner, J. E. (2000). Precipitation changes in a GCM resulting from the indirect effects of anthropogenic aerosols. Geophys. Res. Lett. 27, 3045–3048.

Schumacher, C., Houze Jr., R. A., and Kraucunas, I. (2004). The tropical dynamical response to latent heating estimates derived from the TRMM precipitation radar. J. Atmos. Sci. 61, 1341–1358.

Seager, R., and coauthors, (2007). Model projections of an imminent transition to a more arid climate in southwestern North America. Science 316, 1181–1184, doi:10.1126/science.1139601.

Stephens, G. L. (1990). On the relationship between water vapor over the oceans and sea surface temperature. J. Climate 3, 634–645.

Stephens, G. L., and Ellis, T. D. (2008). Controls of global-mean precipitation increases in global warming GCM experiments. J. Climate 21, 6141–6155.

Tao, W.-K., Lang, S., Olson, W. S., Meneghini, R., Yang, S., Simpson, J., Kummerow, C., Smith, E., and Halverson, J. (2002). Retrieved vertical profiles of latent heat release using TRMM rainfall products for February 1988. J. Appl. Meteor. 40, 957–982.

Tompkins, A. M., and Craig, G. C. (1999). Sensitivity of tropical convection to sea surface temperature in the absence of large-scale flow. J. Climate 12, 462–476.

Trenberth, K. E., Fasullo, J., and Smith, L. (2005). Trends and variability in column-integrated atmospheric water vapor. Climate Dyn. 24, 741–758, doi:10.1007/s00382-005-0017-4.

Trenberth, K. E., and coauthors, (2007). Observations: Surface and atmospheric climate change. In "Climate Change 2007: The Physical Science Basis" (S. Solomon et al., Eds.), pp. 235–336. Cambridge University Press.

Vecchi, G. A., and Soden, B. J. (2007). Global warming and the weakening of the tropical circulation. J. Climate 20, 4316–4340.

Wentz, F. J., and Schabel, M. (2000). Precise climate monitoring using complementary satellite data sets. Nature 403, 414–416.

Wentz, F. J., Ricciardulli, L., Hilburn, K., and Mears, C. (2007). How much more rain will global warming bring? Science 317, 233–235, doi:10.1126/science.1140746.

Williams, K. D., Jones, A., Roberts, D. L., Senior, C. A., and Woodage, M. J. (2001). The response of the climate system to the indirect effects of anthropogenic sulfate aerosol. Clim. Dyn. 17, 845–856.

Wyant, M. C., Khairoutdinov, M., and Bretherton, C. S. (2006). Climate sensitivity and cloud response of a GCM with a superparameterization. Geophy. Res. Lett. 33, L06714.

Zhang, X., Zwiers, F. W., Hegerl, G. C., Lambert, F. H., Gillett, N. P., Solomon, S., Stott, P. A., and Nozawa, T. (2007). Detection of human influence on twentieth-century precipitation trends. Nature 448, 461–465, doi:10.1038/nature06025.

Zipser, E. J. (2003). Some views on "Hot Towers" after 50 years of tropical field programs and two years of TRMM data. Meteor. Monogr. 29, 49.

# International Geophysics Series

Edited by

RENATA DMOWSKA
School of Engineering and Applied Sciences
Harvard University
Cambridge, Massachusetts

DENNIS HARTMANN
Department of Atmospheric Sciences
University of Washington
Seattle, Washington

H. THOMAS ROSSBY
Graduate School of Oceanography
University of Rhode Island
Narragansett, Rhode Island

Volume 1 BENO GUTENBERG. Physics of the Earth's Interior. 1959[1]

Volume 2 JOSEPH W. CHAMBERLAIN. Physics of the Aurora and Airglow. 1961[1]

Volume 3 S. K. RUNCORN (ed.). Continental Drift. 1962[1]

Volume 4 C. E. JUNGE. Air Chemistry and Radioactivity. 1963[1]

Volume 5 ROBERT G. FLEAGLE AND JOOST A. BUSINGER. An Introduction to Atmospheric Physics. 1963[1]

Volume 6 L. DEFOUR AND R. DEFAY. Thermodynamics of Clouds. 1963[1]

Volume 7 H. U. ROLL. Physics of the Marine Atmosphere. 1965[1]

Volume 8 RICHARD A. CRAIG. The Upper Atmosphere: Meteorology and Physics. 1965[1]

Volume 9 WILLIS L. WEBB. Structure of the Stratosphere and Mesosphere. 1966[1]

Volume 10 MICHELE CAPUTO. The Gravity Field of the Earth from Classical and Modern Methods. 1967[1]

Volume 11 S. MATSUSHITA AND WALLACE H. CAMPBELL (eds.). Physics of the Geomagnetic Phenomena (In two volumes). 1967[1]

Volume 12 K. YA KONDRATYEV. Radiation in the Atmosphere. 1969[1]

Volume 13 E. PALMAN AND C. W. NEWTON. Atmospheric Circulation Systems: Their Structure and Physical Interpretation. 1969[1]

Volume 14 HENRY RISHBETH AND OWEN K. GARRIOTT. Introduction to Ionospheric Physics. 1969[1]

Volume 15 C. S. RAMAGE. Monsoon Meteorology. 1971[1]
Volume 16 JAMES R. HOLTON. An Introduction to Dynamic Meteorology. 1972[1]
Volume 17 K. C. YEH AND OWEN K. GARRIOTT. Theory of Ionospheric Waves. 1972[1]
Volume 18 M. I. BUDYKO. Climate and Life. 1974[1]
Volume 19 MELVIN E. STERN. Ocean Circulation Physics. 1975
Volume 20 J. A. JACOBS. The Earth's Core. 1975[1]
Volume 21 DAVID H. MILLER. Water at the Surface of the Earth: An Introduction to Ecosystem Hydrodynamics. 1977
Volume 22 JOSEPH W. CHAMBERLAIN. Theory of Planetary Atmospheres: An Introduction to their Physics and Chemistry. 1978[1]
Volume 23 JAMES R. HOLTON. An Introduction to Dynamic Meteorology, 2nd Ed. 1979[1]
Volume 24 ARNETT S. DENNIS. Weather Modification by Cloud Seeding. 1980[1]
Volume 25 ROBERT G. FLEAGLE AND JOOST A. BUSINGER. An Introduction to Atmospheric Physics, 2nd Ed. 1980[1]
Volume 26 KUO-NAN LIOU. An Introduction to Atmospheric Radiation. 1980[1]
Volume 27 DAVID H. MILLER. Energy at the Surface of the Earth: An Introduction to the Energetics of Ecosystems. 1981
Volume 28 HELMUT G. LANDSBERG. The Urban Climate. 1991
Volume 29 M. I. BUDKYO. The Earth's Climate: Past and Future. 1982[1]
Volume 30 ADRIAN E. GILL. Atmosphere-Ocean Dynamics. 1982
Volume 31 PAOLO LANZANO. Deformations of an Elastic Earth. 1982[1]
Volume 32 RONALD T. MERRILL AND MICHAEL W. McELHINNY. The Earth's Magnetic Field: Its History, Origin, and Planetary Perspective. 1983[1]
Volume 33 JOHN S. LEWIS AND RONALD G. PRINN. Planets and Their Atmospheres: Origin and Evolution. 1983
Volume 34 ROLF MEISSNER. The Continental Crust: A Geophysical Approach. 1986
Volume 35 M. U. SAGITOV, B. BODKI, V. S. NAZARENKO, AND KH. G. TADZHIDINOV. Lunar Gravimetry. 1986
Volume 36 JOSEPH W. CHAMBERLAIN AND DONALD M. HUNTEN. Theory of Planetary Atmospheres, 2nd Ed. 1987
Volume 37 J. A. JACOBS. The Earth's Core, 2nd Ed. 1987[1]
Volume 38 J. R. APEL. Principles of Ocean Physics. 1987
Volume 39 MARTIN A. UMAN. The Lightning Discharge. 1987[1]
Volume 40 DAVID G. ANDREWS, JAMES R. HOLTON AND CONWAY B. LEOVY. Middle Atmosphere Dynamics. 1987

Volume 41 PETER WARNECK. Chemistry of the National Atmosphere. 1988[1]

Volume 42 S. PAL ARYA. Introduction to Micrometeorology. 1988[1]

Volume 43 MICHAEL C. KELLEY. The Earth's Ionosphere. 1989[1]

Volume 44 WILLIAM R. COTTON AND RIHCARD A. ANTHES. Storm and Cloud Dynamics. 1989

Volume 45 WILLIAM MENKE. Geophysical Data Analysis: Discrete Inverse Theory, Revised Edition. 1989

Volume 46 S. GEORGE PHILANDER. El Niño, La Niña, and the Southern Oscillation. 1990

Volume 47 ROBERT A. BROWN. Fluid Mechanics of the Atmosphere. 1991

Volume 48 JAMES R. HOLTON. An Introduction to Dynamic Meteorology, 3rd Ed. 1992

Volume 49 ALEXANDER A. KAUFMAN. Geophysical Field Theory and Method.
Part A: Gravitational, Electric, and Magnetic Fields. 1992[1]
Part B: Electromagnetic Fields I. 1994[1]
Part C: Electromagnetic Fields II. 1994[1]

Volume 50 SAMUEL S. BUTCHER, GORDON H. ORIANS, ROBERT J. CARLSON, AND GORDON V. WOLFE. Global Biogeochemical Cycles. 1992

Volume 51 BRIAN EVANS AND TENG-FONG WONG. Fault Mechanics and Transport Properties of Rocks. 1992

Volume 52 ROBERT E. HUFFMAN. Atmospheric Ultraviolet Remote Sensing. 1992

Volume 53 ROBERT E. HOUZE, JR. Cloud Dynamics. 1993

Volume 54 PETER V. HOBBS. Aerosol-Cloud-Climate Interactions. 1993

Volume 55 S. J. GIBOWICZ AND A. KIJKO. An Introduction to Mining Seismology. 1993

Volume 56 DENNIS L. HARTMANN. Global Physical Climatology. 1994

Volume 57 MICHAEL P. RYAN. Magmatic Systems. 1994

Volume 58 THORNE LAY AND TERRY C. WALLACE. Modern Global Seismology. 1995

Volume 59 DANIEL S. WILKS. Statistical Methods in the Atmospheric Sciences. 1995

Volume 60 FREDERIK NEBEKER. Calculating the Weather. 1995

Volume 61 MURRY L. SALBY. Fundamentals of Atmospheric Physics. 1996

Volume 62 JAMES P. McCALPIN. Paleoseismology. 1996

Volume 63 RONALD T. MERRILL, MICHAEL W. McELHINNY, AND PHILLIP L. McFADDEN. The Magnetic Field of the Earth: Paleomagnetism, the Core, and the Deep Mantle. 1996

Volume 64 NEIL D. OPDYKE AND JAMES E. T. CHANNELL. Magnetic Stratigraphy. 1996

Volume 65 JUDITH A. CURRY AND PETER J. WEBSTER. Thermodynamics of Atmospheres and Oceans. 1988

Volume 66 LAKSHMI H. KANTHA AND CAROL ANNE CLAYSON. Numerical Models of Oceans and Oceanic Processes. 2000

Volume 67 LAKSHMI H. KANTHA AND CAROL ANNE CLAYSON. Small Scale Processes in Geophysical Fluid Flows. 2000

Volume 68 RAYMOND S. BRADLEY. Paleoclimatology, 2nd Ed. 1999

Volume 69 LEE-LUENG FU AND ANNY CAZANAVE. Satellite Altimetry and Earth Sciences: A Handbook of Techniques and Applications. 2000

Volume 70 DAVID A. RANDALL. General Circulation Model Development: Past, Present, and Future. 2000

Volume 71 PETER WARNECK. Chemistry of the Natural Atmosphere, 2nd Ed. 2000

Volume 72 MICHAEL C. JACOBSON, ROBERT J. CHARLSON, HENNING RODHE, AND GORDON H. ORIANS. Earth System Science: From Biogeochemical Cycles to Global Change. 2000

Volume 73 MICHAEL W. McELHINNY AND PHILLIP L. McFADDEN. Paleomagnetism: Continents and Oceans. 2000

Volume 74 ANDREW E. DESSLER. The Chemistry and Physics of Stratospheric Ozone. 2000

Volume 75 BRUCE DOUGLAS, MICHAEL KEARNY AND STEPHEN LEATHERMAN. Sea Level Rise: History and Consequences. 2000

Volume 76 ROMAN TEISSEYRE AND EUGENIUSZ MAJEWSKI. Earthquake Thermo-dynamics and Phase Transformation in the Interior. 2001

Volume 77 GEROLD SIEDLER, JOHN CHURCH, AND JOHN GOULD. Ocean Circulation and Climate: Observing and Modelling the Global Ocean. 2001

Volume 78 ROGER A. PIELKE, SR. Mesoscale Meteorological Modeling, 2nd Ed. 2001

Volume 79 S. PAL ARYA. Introduction to Micrometeorology. 2001

Volume 80 BARRY SALTZMAN. Dynamical Paleoclimatology: Generalized Theory of Global Climate Change. 2002

Volume 81A WILLIAM H. K. LEE, HIROO KANAMORI, PAUL JENNINGS, AND CARL KISSLINGER. International Handbook of Earthquake and Engineering Seismology, Part A. 2002

Volume 81B WILLIAM H. K. LEE, HIROO KANAMORI, PAUL JENNINGS, AND CARL KISSLINGER. International Handbook of Earthquake and Engineering Seismology, Part B. 2003

Volume 82 GORDON G. SHEPPERD. Spectral Imaging of the Atmosphere. 2002

Volume 83 ROBERT P. PEARCE. Meteorology at the Millennium. 2001
Volume 84 KUO-NAN LIOU. An Introductionto Atmospheric Radiation, 2nd Ed. 2002
Volume 85 CARMEN J. NAPPO. An Introduction to Atmospheric Gravity Waves. 2002
Volume 86 MICHAEL E. EVANS AND FRIEDRICH HELLER. Environmental Magnetism: Principles and Applications of Enviromagnetics. 2003
Volume 87 JOHN S. LEWIS. Physics and Chemistry of the Solar System, 2nd Ed. 2003
Volume 88 JAMES R. HOLTON. An Introduction to Dynamic Meteorology, 4th Ed. 2004
Volume 89 YVES GUÉGUEN AND MAURICE BOUTÉCA. Mechanics of Fluid Saturated Rocks. 2004
Volume 90 RICHARD C. ASTER, BRIAN BORCHERS, AND CLIFFORD THURBER. Parameter Estimation and Inverse Problems. 2005
Volume 91 DANIEL S. WILKS. Statistical Methods in the Atmospheric Sciences, 2nd Ed. 2006
Volume 92 JOHN M. WALLACE AND PETER V. HOBBS. Atmospheric Science: An Introductory Survey, 2nd Ed. 2006
Volume 93 JOHN MARSHALL AND RAYMOND ALAN PLUMB. Atmosphere, Ocean and Climate Dynamics: An Introductory Text. 2008
Volume 94 EIICHI FUKUYAMA. Fault-Zone Properties and Earthquake Rupture Dynamics. 2009
Volume 95 JAMES MCCALPIN. Paleoseismology, 2nd Ed. 2009.
Volume 96 MICHAEL C. KELLEY. The Earth's Ionosphere: Plasma Physics and Electrodynamics, 2nd Ed. 2009.
Volume 97 ALFRED R. OSBORNE. Nonlinear Ocean Waves and the Inverse Scattering Transform. 2010.
Volume 98 BENOIT CUSHMAN-ROISIN AND JEAN-MARIE BECKERS. Introduction to Geophysical Fluid Dynamics, Physical and Numerical Aspects. In Press.

[1] Out of Print

# Index

## A

## B

## C

## D

## E

## F

## G

## H

## M

## N

P

## Q

## R

## S

## U



## V

## W

## Z

Made in United States
North Haven, CT
04 November 2026

21764584R00450